Advanced Foundation Engineering

Geotechnical Engineering Series

Geotechnical Engineering Series

- **Textbook of Soil Mechanics and Foundation Engineering**

- **Advanced Foundation Engineering**

Advanced Foundation Engineering

Geotechnical Engineering Series

VNS Murthy

Consulting Geotechnical Engineer
Bangalore, India

CBS

CBS Publishers & Distributors Pvt Ltd

New Delhi • Bengaluru • Chennai • Kochi • Kolkata • Lucknow • Mumbai
Hyderabad • Jharkhand • Nagpur • Patna • Pune • Uttarakhand

Advanced Foundation Engineering

Geotechnical Engineering Series

ISBN-13: 978-81-239-1506-7
ISBN-10: 81-239-1506-3

Copyright © Author

First Edition: 2007
Reprint: 2009, 2010, 2011, 2014, 2016, 2022

Published by Satish Kumar Jain and produced by Varun Jain for

CBS Publishers & Distributors Pvt Ltd
4819/XI Prahlad Street, 24 Ansari Road, Daryaganj, New Delhi 110 002, India
Ph: 011-23289259, 23266861, 23266867   Website: www.cbspd.com
Fax: 011-23243014   e-mail: delhi@cbspd.com;
                                          cbspubs@airtelmail.in.

Corporate Office: 204 FIE, Industrial Area, Patparganj, Delhi 110 092, India
Ph: 011-4934 4934   Fax: 011-4934 4935   e-mail: publishing@cbspd.com;
                                            publicity@cbspd.com

Branches

- **Bengaluru:** Seema House 2975, 17th Cross, KR Road, Banasankari 2nd Stage, Bengaluru 560 070, Karnataka, India
  Ph: +91-80-26771678/79   Fax: +91-80-26771680   e-mail: bangalore@cbspd.com
- **Chennai:** 7, Subbaraya Street, Shenoy Nagar, Chennai 600 030, Tamil Nadu, India
  Ph: +91-44-26680620, 26681266   Fax: +91-44-42032115   e-mail: chennai@cbspd.com
- **Kochi:** 42/1325, 1326, Power House Road, Opp KSEB, Power House, Ernakulum Kochi 682 018, Kerala, India
  Ph: +91-484-4059061-65,67   Fax: +91-484-4059065   e-mail: kochi@cbspd.com
- **Kolkata:** 147, Hind Ceramics Compound, 1st Floor, Nilgunj Road, Belghoria, Kolkata-700056, West Bengal, India
  Ph: +91-9096713055/7798394118, 9836841399   e-mail: kolkata@cbspd.com
- **Lucknow:** Basement, Khushnuma Complex, 7 Meerabai Marg (Behind Jawahar Bhawan),Lucknow-226001, UP, India
  Ph: +0522-4000032   e-mail: tiwari.lucknow@cbspd.com
- **Mumbai:** PWD Shed, Gala no 25/26, Ramchandra Bhatt Marg, Next to JJ Hospital Gate no. 2, Opp. Union Bank of India,
  Noorbaug, Mumbai-400009, Maharashtra, India
  Ph: 022-66661880/89   e-mail: mumbai@cbspd.com

Representatives

• Hyderabad	0-9885175004	• Jharkhand	0-9811541605	• Nagpur	0-9421945513
• Patna	0-9334159340	• Pune	0-9623451994	• Uttarakhand	0-9716462459

Printed at India Binding House, Noida, UP (India)

Dedicated

to

the Cause of Students

Foreword

*A*fter his first book *Textbook of Soil Mechanics and Foundation Engineering*, Dr Murthy takes the readers deeper into the realms of foundation engineering in this second book *Advanced Foundation Engineering*. As the author himself states, the objective is to provide to the students, teachers and practising engineers a comprehensive review of all the relevant theories in the field of foundation engineering. The author has amply met this objective.

As Terzaghi pointed out more than 50 years ago, the state of maturity in foundation engineering is the semi-empirical stage after the initial empirical and scientific stages. To appreciate the complexities and the need for semi-empiricism and to be a successful practitioner of the art of foundation engineering, one needs to have a thorough theoretical background. Dr Murthy's book serves this purpose admirably.

The book has 18 chapters. The first three chapters cover the basics of geotechnical properties and soil exploration. In the next four chapters, bearing capacity and settlement aspects of shallow foundations are dealt with. Different aspects of deep foundations — piles, piers, and caissons, are treated in Chapters 8 to 12. The elaborate treatment of deep foundations is perhaps the speciality of this book. Chapters 13 to 18 are somewhat disconnected but still cover useful topics of foundations in expansive soils, cellular cofferdams, machine foundations, reinforced earth and ground anchors, soil improvement and braced cuts. Overall, Dr Murthy has succeeded in laying a good foundation for the challenging subject of foundation engineering.

In each chapter, a large number of example problems have been worked out to help students grasp the concepts. A number of problems have also been set for solution by the students, which when completed will enable them to understand the subject. The famous Chinese philosopher Confucius (fifth century BC) has said: "I hear, I forgot; I see, I remember; I do, I understand."

With pleasure, I recommend the book to all interested in geotechnical engineering.

K.S. Subba Rao
Emeritus Professor

Preface

*A*dvanced *F*oundation *Engineering* is the second book in the series of Geotechnical Engineering. The first book in the series is *A Textbook of Soil Mechanics and Foundation Engineering* written to satisfy the requirements of undergraduate students studying geotechnical engineering as a subject. The present book goes deeper into the various aspects of Foundation Engineering. To make the book self-sufficient in all respects, certain portions of the first book have been repeated.

The objective of writing this book is to bring all the relevant advanced theories on foundation engineering in a book form at one place. Many books are available on foundation engineering in advanced countries but the author has not yet come across a book based purely on the subject matter. Most of the books available are written just to satisfy certain categories of students or professional's needs. It must be understood well that foundations have to be designed to simulate field conditions. With all the efforts put in by numerous reputed research workers and designers in the field, the author has still to come across one single theory that satisfies the field conditions. If that is the case, what could be the solution? As things stand, the only way is to try a few more theories appropriate to the field conditions. It is, therefore, essential that the consultants must be conversant with the theories that are available in this field and students also must know the available approaches. It is the opinion of the author that the present book meets most of the requirements.

Research work is another field of importance. The author has presented in detail his work on laterally loaded pile foundations to solve many of the problems confronted in this field. The approach is direct and simple as compared to the complicated methods proposed by many leading and well-known advocates in this field. It is the author's ardent opinion that there is always a simple solution to a complicated problem.

Finally, the author wishes to convey through this book that he would be extremely happy if this book serves the purpose for which it is intended.

V.N.S. Murthy

Acknowledgements

I sincerely thank Mr S.K. Jain, Managing Director, and Mr Y.N. Arjuna, Publishing Director, CBS Publishers & Distributors, for the splendid work they have done in publishing my book. I also thank their supporting staff members.

I thank Prof. K.S. Subba Rao for his excellent Foreword to this book. He is an internationally known geotechnical engineer and his opinion on this book has a lot of weight. The review of this book has been done by an upcoming well-known geotechnical engineer Prof. T.G. Seetharam who is the Professor in the Department of Civil Engineering, Indian Institute of Science, Bangalore. His analysis of the subject matter is highly appreciated. I thank him very much.

V.N.S. Murthy

Department of Civil Engineering
Indian Institute of Science
Bangalore - 560 012, India

Review

This book is one of the best reference books for the undergraduate students, postgraduate students and working professionals in the area of foundation engineering. It presents both theoretical and practical knowledge of foundation engineering. Each topic has been developed in logical progression, exhaustive and up-to-date and it will be very useful to students, teachers and practitioners. Prof. Murthy has brought out a comprehensive review of all the relevant theories required for the practice of foundation engineering in his book. The book has 18 chapters, in which the author has comprehensively covered the topics of study for undergraduate curriculum on "foundation engineering" and also to some extent to the postgraduate curriculum. Prof. Murthy has developed a practical and pragmatic approach to the foundation design and suggested construction features at some locations keeping in view the safety and economics of the proposed methodology.

Geotechnical properties and soil exploration have been covered in Chapters 2 and 3 respectively. Four chapters are devoted to shallow foundations and five chapters to deep foundations. Prof. Murthy has brought out the principles of soil mechanics, field and laboratory testing to highlight the importance of these topics in foundation engineering design. Prof. Murthy's own research work on vertical and batter piles has been brought out eloquently for advanced reading in Chapter 9. The other chapters presented are machine foundations, drilled pier foundations, caisson foundations, cofferdams, foundations on expansive and collapsible soils, braced cuts and drainage including soil improvement. He has covered extensively drilled pier foundations, caisson foundations and cellular cofferdams and this book is unique in this respect. A number of chosen problems have been solved to illustrate the concepts in most of the chapters. Also, relevant questions and problems are given at the end of some of the chapters for the benefit of students. Though the book is not designed for any one particular level of students, it is very useful for all levels of students due to its clarity of presentation and list of problems solved. I strongly recommend this excellent book *Advanced Foundation Engineering* to the students, teachers, and practitioners. I feel this book will also serve as a valuable reference for students who take competitive examinations like Graduate Aptitude Test for Engineers (GATE), UPSC examinations, and other national selection entrance tests/examinations.

Prof. T. G. Sitharam PhD (Canada), FIGS, FIE

Professor, Department of Civil Engineering,
Indian Institute of Science, Bangalore 560012
email: sitharam@civil.iisc.ernt.in

Publisher's Note

The publisher is thankful to Prof T.G. Sitharam who has reviewed this book very objectively and given his comments on the coverage and presentation of the text. A brief biosketch of Prof Sitharam is given here.

Prof T.G. SITHARAM PhD, FIGS, FIE is currently Professor, Department of Civil Engineering, Indian Institute of Science, Bangalore. He obtained his BE (Civil Engg) from Mysore University in 1984, Masters from IISc, Bangalore, in 1986, and PhD from University of Waterloo, Canada, in 1991. He was a postdoctoral researcher at University of Texas at Austin, USA, until 1994.

Prof Sitharam is an emerging national and international leader in the area of soil dynamics and earthquake geotechnical engineering. His research work on evaluation of dynamic properties of soils, liquefaction behaviour of Indian soils, regional seismicity, site response and seismic microzonation of Bangalore city, are significant contributions to the country. He was responsible for the indigenous development of the state-of-the-art cyclic triaxial testing, piezo vibro cone system with a large calibration chamber, shake table facility and development of laminar box. He has successfully completed several sponsored research projects.

He has published more than 70 papers in international/national journals and presented 140 papers at international/national conferences. He has delivered lectures invited / keynote lectures and also chaired/cochaired technical sessions in several international/ national conferences.

He is an active member of the task committee on "microzonation of Bangalore city" set up by Seismology Division of Department of Science and Technology (DST). Recognizing his contribution to this area, he has been inducted as Member, Programme Advisory and Monitoring Committee (PAMC), for the nationally coordinated programme on Seismicity by DST, Government of India.

Prof Sitharam is Associate Editor, *ASCE Journal of Materials in Civil Engineering*, USA, and also Member, Committee on Soils and Rock Instrumentation (AFS 20), Transportation Research Board of the National Academies, Division of National Research Council (NRC), USA, for the period 2007-09. He is also a member of TC 29 Laboratory Stress–Strain Strength Testing of Geomaterials, International Society of Soil Mechanics and Geotechnical Engineering (ISSMGE), for the year 2001-09. Prof Sitharam has guided 8 PhD students, 3 MSc (Engg) students and several ME project students. Currently he has 7 doctoral students working under him.

Prof Sitharam has written two textbooks: *Applied Elasticity* and *Soil Mechanics and Foundation Engineering*, and also guest-edited a volume on "Geotechnics and Earthquake Hazards" for *Current Science*. He is an excellent consultant and has carried out more than 50 projects related to specialized geotechnical investigations (measurement of dynamic properties and vibration isolation), slope stability in rocks and soils, underground spaces in rocks/soils and design of earth dams and tailing ponds for ash and redmud including ground improvement.

Contents

Foreword by Prof K.S. Subba Rao   *vii*
Preface   *ix*
Review by Prof T.G. Sitharam   *xi*

Chapter 1
Introduction   **1**

1.1	Foundation engineering defined	1
1.2	The subject matter	1
1.3	Requirements for foundation design	2
1.4	The objective	2

Chapter 2
Geotechnical Properties of Soil   **3**

2.1	Introduction	3
2.2	Soil weight–volume relationship	3
2.3	Index properties of soils	4
2.4	Sieve sizes	4
2.5	Grain size distribution curves	5
2.6	Relative density and consistency	6
2.7	Identification and classification of soil	8
2.8	Hydraulic properties of soil	9
2.9	Stress distribution in soils	10
2.10	Consolidation and settlement	12
2.11	Shear strength	15
2.12	Stress paths	16
2.13	Lateral pressures by theory of elasticity for surcharge loads on the surface of backfill	24

Chapter 3
Soil Exploration   **29**

3.1	Introduction	29
3.2	Boring of holes	30
3.3	Sampling in soil	36
3.4	Rock core sampling	41

3.5	Standard penetration test	42
3.6	Corrections to observed SPT values in cohesionless soils	46
3.7	SPT values related to relative density of cohesionless soils	51
3.8	SPT values related to consistency of clay soil	51
3.9	Static cone penetration test (CPT)	53
3.10	Pressuremeter	67
3.11	The flat dilatometer test	80
3.12	Field vane shear test (VST)	83
3.13	Field-plate load test (PLT)	83
3.14	Ground water conditions	83
3.15	Geophysical exploration	88
3.16	Planning of soil exploration	94
3.17	Execution of soil exploration programme	95
3.18	Report	98
3.19	Problems	98

Chapter 4
Shallow Foundation 1:
Depth of Foundation and Other Considerations   **103**

4.1	Shallow and deep foundations	103
4.2	Requirements for a stable foundation	104
4.3	Foundation location and depth	106
4.4	Minimum depth for shallow foundation	106
4.5	Selection of type of foundation	109

Chapter 5
Shallow Foundation 2:
Ultimate Bearing Capacity   **111**

5.1	Introduction	111
5.2	The ultimate bearing capacity of soil defined	111
5.3	Some of the terms defined	112
5.4	Types of failure in soil	113
5.5	An overview of bearing capacity theories	115
5.6	Terzaghi's bearing capacity theory	116
5.7	Skempton's bearing capacity factor N_c	122
5.8	Effect of water table on bearing capacity	123
5.9	The general bearing capacity equation	132
5.10	Effect of soil compressibility on bearing capacity of soil	138
5.11	Bearing capacity of foundations subjected to eccentric loads	144
5.12	Ultimate bearing capacity of footings based on SPT values (N)	147
5.13	The CPT method of determining ultimate bearing capacity	147
5.14	Ultimate bearing capacity of footings resting on stratified deposits of soil	150
5.15A	Meyerhof's method of computing ultimate bearing capacity of foundations on slopes	157
5.15B	Bearing capacity of foundations on top of a slope	160

5.16 The pressuremeter method of determining ultimate
bearing capacity 163
5.17 Foundations on rock 172
5.18 Case history of failure of the transcona grain elevator 174
5.19 Problems 177

Chapter 6
Shallow Foundation 3:
Safe Bearing Pressure and Settlement Calculation 185

6.1 Introduction 185
6.2 Field plate load tests 186
6.3 Effect of size of footings on settlement 193
6.4 Design charts from SPT values for footings on sand 194
6.5 Empirical equations based on SPT values for footings on
cohesionless soils 198
6.6 Safe bearing pressure from empirical equations based
on CPT values for footings on cohesionless soil 199
6.7 Foundation settlement 200
6.8 Evaluation of modulus of elasticity 202
6.9 Methods of computing settlements 204
6.10 Elastic settlement beneath the corner of a uniformly loaded
flexible area based on the theory of elasticity 204
6.11 Janbu, Bjerrum and Kjaernsli's method of determining elastic
settlement under undrained conditions 206
6.12 Schmertmann's method of calculating settlement in granular
soils by using CPT values 207
6.13 Pressuremeter method of estimating settlement in cohesionless
and cohesive soils 214
6.14 Estimation of consolidation settlement by using
oedometer test data 219
6.15 Skempton-Bjerrum method of calculating consolidation
settlement (1957) 220
6.16 Consolidation settlement by Lambe's
stress path method 224
6.17 Problems 228

Chapter 7
Shallow Foundation 4:
Combined Footings and Mat Foundation 233

7.1 Introduction 233
7.2 Safe bearing pressures for mat foundations on sand and clay 234
7.3 Eccentric loading 235
7.4 The coefficient of subgrade reaction 236
7.5 Proportioning of cantilever footing 238
7.6 Design of combined footings by rigid method (conventional method) 239
7.7 Design of mat foundation by rigid method 241

7.8	Design of combined footings by elastic line method	241
7.9	Design of mat foundations by elastic plate method	242
7.10	Floating foundation	243
7.11	Problems	250

Chapter 8
Deep Foundation 1:
Vertical Load Bearing Capacity of Single Vertical Pile   251

8.1	Introduction	251
8.2	Classification of piles	251
8.3	Types of piles according to the method of installation	252
8.4	Uses of piles	254
8.5	Selection of pile	256
8.6	Installation of piles	256
8.7	Load transfer mechanism	259
8.8	Methods of determining ultimate load bearing capacity of a single vertical pile	263
8.9	General theory for ultimate bearing capacity	263
8.10	Ultimate bearing capacity in cohesionless soils	265
8.11	Critical depth	266
8.12	Tomlinson's solution for Q_b in sand	267
8.13	Meyerhof's method of determining Q_b for piles in sand	269
8.14	Vesic's method of determining Q_b	270
8.15	Janbu's method of determining Q_b	273
8.16	Coyle and Castello's method of estimating Q_b in sand	273
8.17	The ultimate skin resistance of a single pile in cohesionless soil	274
8.18	Skin resistance Q_f by Coyle and Castello method (1981)	275
8.19	Static bearing capacity of piles in clay soil	277
8.20	Bearing capacity of piles in granular soils based on SPT value	280
8.21	Bearing capacity of piles based on static cone penetration tests (CPT)	295
8.22	Bearing capacity of a single pile by load test	306
8.23	Pile bearing capacity from dynamic pile driving formulas	309
8.24	Bearing capacity of piles founded on a rocky bed	313
8.25	Uplift resistance of piles	314
8.26	Problems	316

Chapter 9
Deep Foundation 2 :
Behaviour of Single Vertical and Batter Piles
Subjected to Lateral Loads   319

9.1	Introduction	319
9.2	Winkler's hypothesis	321
9.3	The differential equation	321

Part A
Vertical Piles Subjected To Lateral Loads · · **325**

9.4	Solution for laterally loaded single piles	325
9.5	Modulus of subgrade reaction	326
9.6	Closed-form solution for pile of infinite length	329
9.7	Finite difference method of solving the differential equation for a laterally loaded long pile (Glesser, 1953)	332
9.8	Non-dimensional method of analysis of vertical piles subjected to lateral loads	339
9.9	Broms method for the analysis of laterally loaded piles (1964a, 1964b)	351
9.10	Lateral deflections at working loads in saturated cohesive soils (Broms, 1964a)	351
9.11	Ultimate lateral resistance of piles in saturated cohesive soils (Broms, 1964a)	354
9.12	Lateral deflections at working loads in cohesionless soils (Broms, 1964b)	356
9.13	Ultimate lateral resistance of piles in cohesionless soils (Broms, 1964b)	360
9.14	A direct method for solving the non-linear behaviour of laterally loaded flexible pile problems	367
9.15	Case studies for laterally loaded vertical piles in sand	374
9.16	Case studies for laterally loaded vertical piles in clay	378
9.17	p-y curves for the solution of laterally loaded piles	383
9.18	Solution for the laterally loaded piles by the use of p-y curves	397
9.19	Pressuremeter method to solve laterally loaded pile problems	409
9.20	Poulos method of elastic analysis for laterally loaded single piles	414

Part B
Batter Piles in Cohesionless Soils · · **423**

9.21	Mechanism of failure of batter piles under lateral loads in cohesionless soils	423
9.22	Statement of the problem of batter piles subjected to lateral loads	426
9.23	Model tests on instrumented batter piles in cohesionless soil (Murthy, 1965)	426
9.24	Variation of soil modulus along batter piles	427
9.25	Non-dimensional solutions for laterally loaded batter piles in sand (Murthy, 1965)	428
9.26	Relative stiffness factor for batter piles in sand (Murthy, 1965)	431
9.27	Ultimate lateral bearing capacity of batter piles in sand (Murthy, 1965)	432
9.28	Lateral resistance of batter piles as a ratio to that of vertical pile in sand (Murthy, 1965)	434
9.29	Coefficients of passive earth pressure for batter piles in sand	435
9.30	Behaviour of laterally loaded batter piles in sand (Murthy, 1965)	440
9.31	Problems	454

Chapter 10
Deep Foundation 3:
Pile Groups Subjected to Vertical and Lateral Loads 457

10.1	Introduction	457
10.2	Number and spacing of piles in a group	457
10.3	Pile group efficiency	459
10.4	Efficiency of pile groups in sand	460
10.5	Pile group efficiency equation	461
10.6	Vertical bearing capacity of pile groups embedded in sands and gravels	461
10.7	Bored pile groups in sand and gravel	462
10.8	Pile groups in cohesive soils	462
10.9	Settlement of piles and pile groups in sands and gravels	463
10.10	Settlement of pile groups in cohesive soils	465
10.11	Allowable loads on groups of piles	466
10.12	Negative friction on piles	466
10.13	Analysis of pile foundations comprising vertical and batter piles and subjected to vertical and lateral loads	469
10.14	Pile groups subjected to eccentric vertical loads	479
10.15	Anchor piles	481
10.16	Uplift capacity of a pile group	482
10.17	Examples	483
10.18	Problems	488

Chapter 11
Deep Foundations 4:
Drilled Pier Foundations 491

11.1	Introduction	491
11.2	Types of drilled piers	491
11.3	Advantages and disadvantages of drilled pier foundations	492
11.4	Methods of construction	493
11.5	Design considerations	498
11.6	Vertical load transfer mechanism	500
11.7	Vertical bearing capacity of drilled piers	503
11.8	The general bearing capacity equation for the base resistance q_b $(= q_{max})$	505
11.9	Bearing capacity equations for the base in cohesive soil	505
11.10	Bearing capacity equation for the base in granular soil	506
11.11	Bearing capacity equations for the base in cohesive IGM or rock (O'Neill and Reese, 1999)	508
11.12	The ultimate skin resistance of cohesive and intermediate materials	510
11.13	Ultimate skin resistance in cohesionless soil and gravelly sands (O'Neill and Reese, 1999)	513
11.14	Ultimate side and total resistance in rock (O'Neill and Reese, 1999)	514
11.15	Estimation of settlements of drilled piers at working loads	514
11.16	Uplift capacity of drilled piers	526
11.17	Lateral bearing capacity of drilled piers	527

11.18   Case study of a drilled pier subjected to lateral loads   534
11.19   Problems   535

Chapter 12
Deep Foundation 5:
Caisson Foundations   **539**

12.1   Introduction   539
12.2   Types of wells or caissons   539
12.3   Stability analysis of well foundations   541
12.4   Limit equilibrium method of determining the grip
       Length of wells in cohesionless soils   542
12.5   Grip lengths of wells in cohesive soils   545
12.6   Determination of scour depth in cohesionless soils   546
12.7   Thickness of steining of wells   548
12.8   Examples   550

Chapter 13
Foundations On Collapsible And Expansive Soils   **555**

13.1   General considerations   555

Part A
Collapsible Soils   **556**

13.2   General observations   556
13.3   Collapse potential and settlement   558
13.4   Computation of collapse settlement   559
13.5   Foundation design   563
13.6   Treatment methods for collapsible soils   564

Part B
Expansive Soils   **564**

13.7   Distribution of expansive soils   564
13.8   General characteristics of swelling soils   565
13.9   Clay mineralogy and mechanism of swelling   566
13.10   Definition of some parameters   567
13.11   Evaluation of the swelling potential of expansive soils by single
       index method (Chen, 1988)   568
13.12   Classification of swelling soils by indirect measurement   568
13.13   Swelling pressure by direct measurement   575
13.14   Effect of initial moisture content and initial dry density on
       swelling pressure   576
13.15   Estimating the magnitude of swelling   577
13.16   Design of foundations in swelling soils   579
13.17   Drilled pier foundations in expansive, soils   580
13.18   Elimination of swelling   589
13.19   Problems   590

Chapter 14
Cellular Cofferdams   593

14.1	Introduction	593
14.2	Cellular cofferdams	594
14.3	Components of cellular cofferdams	595
14.4	Dimensions of cellular cofferdam	596
14.5	Stability of cellular cofferdams	596
14.6	Examples	603
14.7	Questions and problems	607

Chapter 15
Machine Foundations Subjected to Dynamic Loads   609

15.1	Introduction	609
15.2	Basic theories of vibration	610
15.3	Simple harmonic motion	610
15.4	Free vibration of a mass-and-spring system without damping	612
15.5	Free vibrations with viscous damping	615
15.6	Forced vibrations of mass-and-spring system without damping	619
15.7	Forced vibrations of mass-and-spring system with viscous damping	621
15.8	Forced frequency dependent exciting force with viscous damping	625
15.9	Properties of response curves	626
15.10	Machine foundations subjected to steady state vibrations	628
15.11	Vibration analysis of rigid circular footings by elastic half-space analog method	632
15.12	Elastic-soil-spring method of vibration analysis of foundations (Barkan, 1962)	646
15.13	Vibration analysis of foundations subjected to simultaneous vertical, sliding and rocking oscillations by elastic soil-spring method (Barkan, 1962)	654
15.14	Machine foundations subjected to impact loads (Barkan, 1962)	657
15.15	Design criteria for machine foundations	664
15.16	Screening vibrations	666
15.17	Examples	668

Chapter 16
Geotextiles Reinforced Earth and Ground Anchors   683

16.1	Geotextiles	683
16.2	Reinforced earth and general considerations	685
16.3	Backfill and reinforcing materials	688
16.4	Construction details	691
16.5	Design consideration for a reinforced earth wall	692
16.6	Design method	693
16.7	External stability	698
16.8	Examples of measured lateral earth pressures	711

16.9   Ground anchors   712
16.10   Problems   717

Chapter 17
Soil Improvement   **721**

17.1   Introduction   721
17.2   Mechanical compaction   722
17.3   Laboratory tests on compaction   722
17.4   Effect of compaction on engineering behaviour   728
17.5   Field compaction and control   731
17.6   Compaction for deeper layers of soil   740
17.7   Preloading   741
17.8   Sand compaction piles and stone columns   747
17.9   Soil stabilisation by the use of admixtures   748
17.10   Soil stabilisation by injection of suitable groups   749
17.11   Soil stabilisation by electrical and thermal methods   750
17.12   Problems   751

Chapter 18
Braced-Cuts and Drainage   **753**

Part A
Braced-cuts   **753**

18.1   General considerations   753
18.2   Lateral earth pressure distribution on braced-cuts   754
18.3   Stability of braced-cuts in saturated clay   758
18.4   Bjerrum and Eide (1956) method of analysis   760
18.5   Piping failures in sand cuts   764
18.6   Problems   765

Part B
Drainage   **766**

18.7   Introduction   766
18.8   Ditches and sumps   766
18.9   Well points   767
18.10   Deep-well pumps   769
18.11   Sand drains   769

Appendix A
SI Units in Geotechnical Engineering   **771**

References   **777**

Index   **793**

Advanced Foundation Engineering

Geotechnical Engineering Series

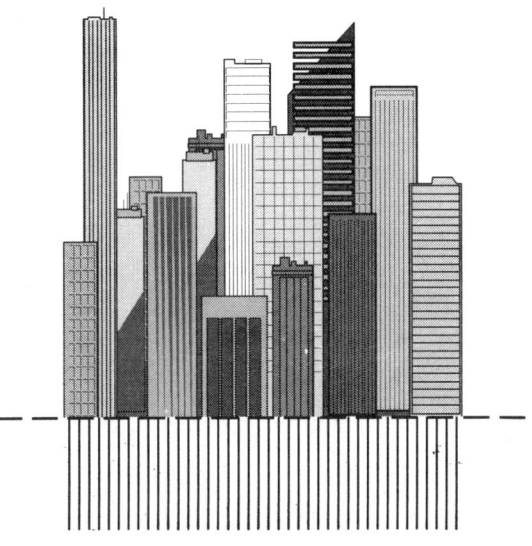

Introduction

1

1.1 FOUNDATION ENGINEERING DEFINED

Foundation Engineering is a subject built on the basic principles of Soil Mechanics, Soil Hydraulics and Structural Mechanics. All these three together may be considered as the pillars of Foundation Engineering. A wrong application of the principles of any one of the three subjects may lead to a faulty design of the foundation.

Theories have been developed for the design of foundations to suit ideal soil conditions. However, such conditions rarely exist in nature since soils found in natural conditions are mostly heterogeneous in character. Theories may have to be modified or adjusted to suit field conditions.

A foundation is a part of a superstructure. The stresses and strains that are brought to the foundation from the superstructure would lead to interaction between the foundation structural element and the soil surrounding it. It is this interaction which is very difficult to evaluate as this is quite a complex phenomenon. The theories that have been developed for ideal conditions do not take into account all the variables that would lead to the interaction between the soil and the foundation element. The presence of water table would make the interaction problem all the more difficult to solve. It is therefore essential that a *design engineer* should have a thorough knowledge of the theories he wants to use for the design of foundations and also its limitations. A knowledge of the theories and its limitations by themselves would not lead to the design of a safe and sound foundation if the environmental conditions, the strength and settlement characteristics of the soil are not properly known in advance. An *ideal design engineer,* therefore, is the one who has a thorough knowledge of the theories and the field conditions and also who can modify or adjust the design to suit the field conditions. This requires therefore a **practical and pragmatic approach** to the problem of design and construction while keeping in view the *safety and economics of the* project.

1.2 THE SUBJECT MATTER

The subject matter pertaining to the field of foundation engineering has been dealt with in a logical manner. A brief review of geotechnical properties of soil is given in Chapter 2. Soil Exploration finds an important place in this book as the design of a foundation will be meaningless if the strength and settlement characteristics of the soil *in-situ* are not properly understood. This is possible only if the Soil Exploration is properly planned and executed by **competent geotechnical consultants.**

Shallow and deep foundations have been dealtwith in detail. Four chapters have been devoted to shallow foundations and five chapters to deep foundations. All the relevant basic theories have been discussed in detail for the benefit of students and teachers of technical colleges. The limitations of the theories have also been mentioned for the benefit of practising engineers. The author's own research work on vertical and batter piles have been included in Chapter 9.

Foundation soil improvements, the use of geotextiles, reinforced earth and ground anchors have also been discussed briefly.

The other matters discussed are—machine foundations subject to vibratory loads, drilled pier foundations, foundations on collapsible and expansive soils, caisson foundations, cofferdams, and braced cuts and drainage.

1.3   REQUIREMENTS FOR FOUNDATION DESIGN

When once a site is selected for a particular project, the job of the foundation engineer is to design the foundations for the structures. The following information is needed for this purpose.

1. A lay out plan of the project.
2. A plan of load-bearing elements such as columns, walls, caissons, etc. with the estimated dead and live loads.
3. The strength and settlement characteristics of the subsoil.
4. The hydraulic conditions of the site.

The first two of the informations have to be provided by structural engineers. A detailed soil exploration provides the informations pertaining to the last two. Based on the above data, the depth and type of foundations have to be decided. Foundation engineer will be then be in a position to design a foundation. The foundations so designed should satisfy all the requirements of safety.

1.4   THE OBJECTIVE

The objective of the author is to provide for the students, teachers and practising engineers a comprehensive review of all the relevant theories in the field of foundation engineering in one book which is at present not available in any book published so far (2005). The book is, therefore, not designed for any one particular level of students and as such is useful for all levels of students.

Geotechnical Properties of Soil ①

2.1 INTRODUCTION

This chapter deals very briefly with the principal engineering properties of soil, and its strength and settlement characteristics. Readers may refer to Soil Mechanics and Foundation Engineering by the same author for a detailed discussion on the subject.

2.2 SOIL WEIGHT–VOLUME RELATIONSHIP

The soil weight–volume relationships that are of practical interest are the following,

$$\text{Void ratio,} \qquad e = \frac{V_v}{V_s} \tag{2.1}$$

$$\text{Porosity,} \qquad n = \frac{V_v}{V} \times 100 \text{ percent} \tag{2.2}$$

$$\text{The degree of saturation,} \quad S = \frac{V_w}{V_v} \times 100 \text{ percent} \tag{2.3}$$

where, V = total volume of the mass,
  V_v = volume of the voids,
  V_s = volume of the solid particles,
  V_w = volume of water.

$$\text{Water content,} \qquad w = \frac{W_w}{W_s} \times 100 \text{ percent} \tag{2.4}$$

where, W_w = weight of water in a soil mass of volume V,
  W_s = weight of solid particles in the same volume.

$$\text{Total unit weight,} \qquad \gamma_t = \frac{W}{V} = \frac{G\gamma_o(1+w)}{(1+e)} \tag{2.5}$$

where, W = total weight of a soil mass of volume V.

Dry unit weight $\qquad \gamma_d = \dfrac{W_s}{V}$ $\qquad\qquad\qquad\qquad$ (2.6)

Saturated unit weight,

$$\gamma_{sat} = \frac{W}{V} = \frac{\gamma_o\,(G+e)}{1+e} \qquad\qquad (2.7)$$

Specific gravity of the solids,

$$G = \frac{\gamma_s}{\gamma_o} = \frac{W_s}{V_s\,\gamma_o} \qquad\qquad (2.8)$$

where, $\quad \gamma_s$ = unit weight of solids = $\dfrac{W_s}{V_s}$ $\qquad\qquad\qquad$ (2.9)

γ_o = unit weight of water at 4° C.

Submerged unit weight $= \gamma_{sat} - \gamma_w = \gamma_b = \dfrac{\gamma_w\,(G-e)}{1+e}$ $\qquad\qquad$ (2.10)

where, $\quad \gamma_o = \gamma_w$ the unit weight of water for all practical purposes.

Relative density $D_r = \dfrac{e_{max} - e}{e_{max} - e_{min}} \times 100$ percent $\qquad\qquad$ (2.11)

$$= \frac{\gamma_{dM}}{\gamma_d}\ \frac{\gamma_d - \gamma_{dm}}{\gamma_{dM} - \gamma_{dm}} \times 100 \text{ percent} \qquad\qquad (2.12)$$

where, e_{max} = void ratio of soil in the loosest state having unit weight γ_{dm},

$\quad e_{min}$ = void ratio of soil in the densest state having unit weight γ_{dM},

$\quad e$ = void ratio of the soil in the field having unit weight γ_d.

The specific gravity of a mixture of soil varies from 2.50 to 2.70. For sandy soils $G = 2.65$ and clayey soils $G = 2.70$ may be assumed.

2.3 INDEX PROPERTIES OF SOILS

The various properties of soil which would be considered as index properties are,

1. The specific gravity.
2. The size and shape of particles.
3. The relative density or consistency of soil.

2.4 SIEVE SIZES

The sieve and mechanical analysis gives an idea of the size and shape of particles. The sieve analysis is normally carried out on cohesionless soils and wet mechanical analysis (normally hydrometer analysis) on clay soils. Table 2.1 shows the standard ASTM (1961) and IS Sieves (1962).

Tabel 2.1 ASTM (1961) and the IS (1962) Sieves

ASTM		IS	
Designation	*Aperture (mm)*	*Designation*	*Aperture (mm)*
2 in	50.80	50.00 mm	50.00
1½ in	38.10	40.00	40.00
¾ in	19.00	20.00	20.00
$^3/_8$ in	9.51	10.00	10.00
4	4.76	4.75	4.75
7	2.83	2.80	2.80
10	2.00	2.00	2.00
14	1.41	1.40	1.40
16	1.19	1.18	1.18
18	1.00	1.00	1.00
30	0.595	600.00 µ	0.60
35	0.500	500.00	0.50
40	0.420	425.00	0.425
45	0.354	355.00	0.355
60	0.250	250.00	0.250
70	0.210	212.00	0.212
80	0.177	180.00	0.180
100	0.149	150.00	0.150
120	0.125	125.00 µ	0.125
170	0.088	90.00	0.090
200	0.074	75.00	0.075
325	0.044	45.00 µ	0.045

2.5 GRAIN SIZE DISTRIBUTION CURVES

The shapes of the grain size distribution curves indicate the nature of the soil tested. On the basis of the shapes, the soil may be classified as

1. Uniformly graded or poorly graded,
2. Well graded,
3. Gap graded.

The uniformity coefficient (C_u), which is a ratio of D_{60} to D_{10}, gives an idea of the grading of the soil as shown in Table 2.2.

Table 2.2 Soil grading according to $C_u (= D_{60}/D_{10})$

C_u	*Type of soil*
< 5	Uniform size particles
5–10	Medium graded soil
> 15	Well graded soil

2.6 RELATIVE DENSITY AND CONSISTENCY

Relative Density

The cohesionless soils are classified according to relative density D_r as in Table 2.3.

Table 2.3 Classification of sandy soils

Type of sand	Relative density D_r%
Loose	0–33
Medium	33–66
Dense	66–100

Consistency Based on Plasticity Index

The consistency is a term used to indicate the degree of firmness of cohesive soils. The consistency of natural cohesive soil deposits is expressed qualitatively by such terms as very soft, stiff, very stiff and hard. The consistency of a soil can be expressed in terms of Atterberg limits and unconfined compressive strengths of soil. The Atterberg limits are liquid, plastic and shrinkage limits. The range of water content between the liquid and plastic limits which is an important measure of plastic behaviour, is called as the *plasticity index, I_p*, i.e. $I_p = w_l - w_p$,

where, w_l = liquid limit, w_p = plastic limit.

According to the range of plasticity index, the soil is classified as per Table 2.4.

Table 2.4 Soil classification according to plasticity index

Plasticity Index, I_p	Plasticity
0	Non-plastic
< 7	Low plastic
7–17	Medium plastic
> 17	Highly plastic

A liquid limit greater than 100 is uncommon for inorganic clays of non-volcanic origin. However, for clays containing considerable amount of organic matter and clays of volcanic region, the liquid limit may considerably exceed 100. Bentonite, a material consisting of chemically disintegrated volcanic ash, has a liquid limit ranging from 400 to 600.

Degree of Shrinkage

The shrinkge limit w_s indicates whether a soil is a swelling type or not. The soil having a shrinkage limit around 10 is of highly swelling type. The *degree of shrinkage* is expressed as

$$S_r = \frac{V_o - V_d}{V_o} \times 100 \tag{2.13}$$

where, V_o = original volume, .

V_d = final volume at the shrinkage limit.

The type of soil according to the degree of shrinkage may be classified as per Table 2.5.

Table 2.5 Soil classification according to S_r

$S_r \%$	Quality of soil
< 5	Good
5–10	Medium good
10–15	Poor
> 15	Very poor

The soils that belong to the montmorillonite group, such as black cotton soil in India, shrink more than the soils of the Kaolinite and illite groups. These soils are of expansive type also when in contact with water.

Activity of Soil

Skempton classifies the various swelling characteristics of soils based on a number called as *active number,* A_c. The activity of a clay soil may be expressed as

$$A_c = \frac{\text{Plasticity index, } I_p}{\text{Percent finer than 2 micron}}$$

Table 2.6 gives the type of soil according to the value of A_c.

The clay soil which has an activity value greater than 1.4 can be considered as belonging to the swelling type.

Table 2.6 Soil classification based on activity number

A_c	Soil type
< 0.75	Inactive
0.75–1.4	Normal
> 1.4	Active

Consistency Based on Unconfined Compressive Strength

The method that is recommended is to express consistency qualitatively on the basis of unconfined compressive strength q_u as per Table 2.7.

Table 2.7 Consistency based on q_u

Consistency	q_u, kPa
Very soft	< 25
Soft	25–50
Medium	50–100
Stiff	100–200
Very stiff	200–400
Hard	> 400

Classification of Soil Based on Sensitivity

The degree of disturbance of undisturbed clay sample due to remoulding can be expressed as

$$\text{Sensitivity,} \qquad S_t = \frac{q_u, \text{undisturbed}}{q_u{}', \text{remoulded}}$$

When q_u' is very low as compared to q_u, the clay is highly sensitive. On the basis of the values of S_t, clays can be classified as in Table 2.8.

Table 2.8 Classification of soil on the basis of S_t
(After Skempton and Northey)

S_t	Nature of clay
1	Insensitive
1–2	Low sensitive clays
2–4	Medium sensitive clays
4–8	Sensitive clays
8–16	Extra sensitive clays
> 16	Quick clays

2.7 IDENTIFICATION AND CLASSIFICATION OF SOIL

The coarse grained soils can be identified primarily on the basis of grain size since these soils are non-plastic. The classification of the soil on the basis of IS: 1498–1970 is as per Table 2.9.

Table 2.9 Classification of soil as per grain size IS: 1498–1970

Grain size (mm)	Soil type
> 300	Boulder
80–300	Cobble
4.75–80	Gravel
0.075–4.75	Sand
2–4.75	Coarse sand
0.475–2.0	Medium sand
0.075–2.0	Fine sand
< 0.075	Finer fractions silt and clay

The identification of clay soil in the field is based on dry strength test, shaking test, plasticity test and dispersion test. The characteristics of clay soil is based only on the plasticity of the soil. The Unified Soil Classification System is the one that is used for classifying coarse grained and fine grained soils.

The classification systems do not take into account the properties of intact materials as found in nature. Since the foundation materials of most of the engineering structures are undisturbed, the properties of intact materials only, determine the soil behaviour during and after construction. The classification of a soil according to any of the accepted systems does not in itself enable detailed studies of soils to be dispensed with altogether.

2.8 HYDRAULIC PROPERTIES OF SOIL

Inter-granular and Porewater Pressures

The pressure transmitted through grain to grain at the contact points through a soil mass is termed as *inter-granular* or *effective pressure*. If the pores of a soil mass are filled with water, and if a pressure is induced the porewater tries to separate the grains, then this pressure is termed as *porewater pressure* or *neutral stress*.

When in a soil mass, if the effective pressure reduces to zero due to the increase in the porewater pressure the soil will be in a state of *quicksand condition*. This phenomenon is also known as *boiling*.

Capillary Phenomenon

If the lower part of a mass of dry soil comes in contact with water, the water rises in the voids to a certain height above the free water surface. The upward flow into the voids of the soil is attributed to the *surface tension* in the water. The height to which water rises against the force of gravity is called as *capillary rise*. The water held in the pores of the soil above the free water surface is retained in a state of reduced pressure, and this pressure is called as *capillary* or *soil moisture suction pressure*.

Permeability of Soil

Methods of Determining coefficients of Permeability of soils.

The normal laboratory methods are:

1. Constant head permeability test.
2. Falling head permeability test.

The field methods are:

1. Pumping tests.
2. Bore hole tests.

The constant head permeability test is normally used for determining the coefficients of permeability in cohesionless soils, whereas the falling head method is used for cohesive soils.

A rough method that is normally used for cohesionless soils for computing k is the equation

$$k = CD_{10}^2 \qquad\qquad (2.14)$$

where C a factor taken as equal to 100 and D_{10} is the effective grain size in cm at 10 percent finer. Pumping test is normally used in major projects and bore hole tests in smaller projects for determining k.

The coefficients of permeability of coarse grained soils vary from a minimum of 10^{-4} cm/sec (clean sand and gravel mixtures) to 10^2 cm/sec (clean gravels).

For very fine sand k is around 10^{-5} cm/sec, for silt 10^{-6} cm/sec, and for clay soils less than 10^{-7} cm/sec.

Seepage Flow

The computation of seepage loss under or through a dam, the uplift pressures caused by the water on the base of a concrete dam and the effect of seepage on the stability of earth slopes can be studied by constructing *flow nets*.

Piping failures caused by *heave* can be expected to occur on the downstream side of a hydraulic structure when the uplift forces of seepage exceed the downward forces due to the submerged weight of the soil.

2.9 STRESS DISTRIBUTION IN SOILS

Introduction

Estimation of vertical stress at any point in a soil mass due to external vertical loadings are of great significance in the prediction of settlements of buildings, bridges, etc. Equations have been developed to compute stresses at any point in a soil mass on the basis of theory of elasticity. The equation that is quite popular is the Boussinesq's equation which has been extended from point loads to distributed loads. The basic equation of Boussinesq is

$$\sigma_z = \frac{Q}{z^2} \frac{3}{2\pi} \frac{1}{\left[1 + (r/z)^2\right]^{5/2}} \qquad (2.15)$$

where, Q = point load on the surface,
z = depth of point where the vertical stress σ_z is required,
r = radial distance of the point from the axis of symmetry.

The Stress Under the Corner of a Rectangular Foundation

The stress below the corner of a footing may be computed by making use of the graph in Fig. 2.1. The vertical stress σ_z at a depth z below the corner of a footing may be expressed as

$$\sigma_z = qI \qquad (2.16)$$

where, q = stress per unit area
I = influence value for any known values of m and n
$m = b/z$, where b = width of footing
$n = l/z$, where l = length of footing

The stress below any point O, either within the loaded area or outside may be calculated with reference to Fig. 2.2.

When the point O is inside.

$$\sigma_z = q \, (I_1 + I_2 + I_3 + I_4)$$

where, I_1, I_2, I_3 and I_4 are the influence values for rectangles 1, 2, 3, and 4 respectively.

When the point O is outside

$$\sigma_z = q \, (I_1 - I_2 - I_3 + I_4)$$

where, I_1, I_2, I_3 and I_4 refer to rectangles OB_1CD_1, OB_1BD_2, OD_1DA_1 and OA_1AD_2 respectively.

Pressure Isobars

Pressure isobars of square, rectangular and circular footings may sometimes conveniently be used for determining vertical pressure, σ_z at any depth z below the base of the footings. The depth z from the ground surface and the distance r (or x) from the centre of the footing are expressed as a function of the width of the footings B (or $B//2 = b$) or radius R_o. Figures 2.3 and 2.4 gives the pressure isobars for square and long footings, and circular footings respectively.

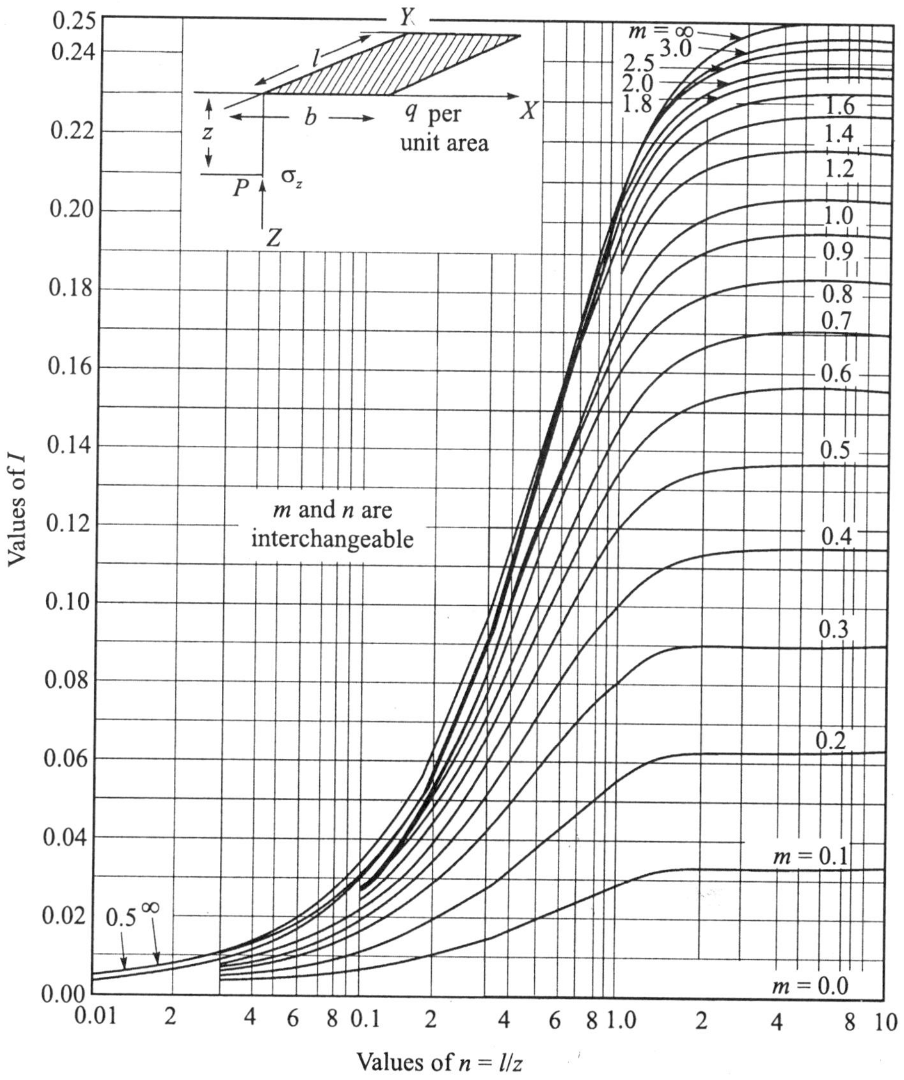

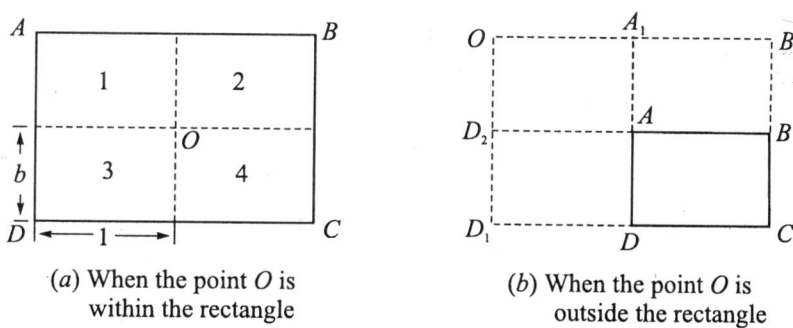

Fig. 2.1 Graph for determining influence value I for vertical normal stress σ_z at point P located beneath one corner of a uniformely loaded rectangular area (after Fadum, 1941)

(a) When the point O is within the rectangle

(b) When the point O is outside the rectangle

Fig. 2.2 Computation of vertical stress below a point

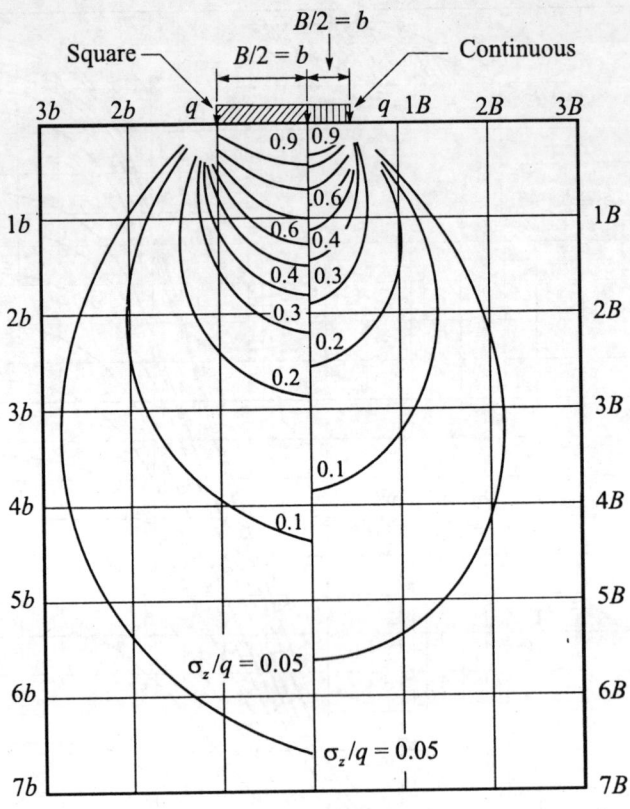

Fig. 2.3 Pressure isobars based on Boussinesq's equation for square and continuous footings

Principal Stresses σ_1 and σ_3 at any Point in a Soil Mass Due to External Uniform Loading q Per Unit Area on a Circular Footing

Figure 2.5 gives charts for computing the principal stresses σ_1 and σ_3 as ratios of the uniformly distributed external load q on a circular area. Along the vertical line $\sigma_1 = \sigma_z$ and $\sigma_3 = \sigma_h = K_o \sigma_z$ where K_o = coefficient of earth pressure for the at rest condition. These charts help to compute the principal stresses (excluding the geostatic stresses) at any point in the soil mass due to a uniformly distributed vertical load q on a circular footing.

2.10 CONSOLIDATION AND SETTLEMENT

Introduction

When a saturated clay–water system is subjected to an external pressure, the pressure applied is initially taken by the water in the pores resulting thereby an excess porewater pressure. With the advance of time, a portion of the applied pressure is transferred to the soil skeleton, which in turn, causes a reduction in the porewater pressure. This process, involving a gradual compression occurring simultaneously with a flow of water out of the mass and with the gradual transfer of the applied pressure from the porewater to mineral skeleton is called *consolidation*. The settlement of a structure founded on soil is due to the consolidation of the underlying clay strata. The total compression of a clay strata under excess effective pressure may be considered as the sum of

1. Immediate compression,
2. Primary compression,
3. Secondary compression.

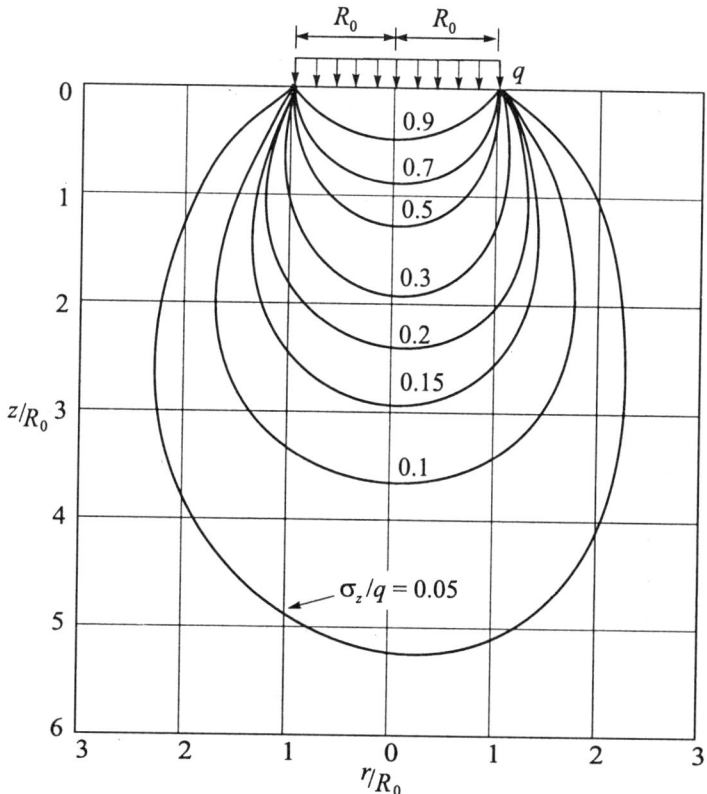

Fig. 2.4 Pressure isobars based on Boussinesq's equation for uniformly loaded circular footings

The *immediate settlement* occurs in clay soils without any change in the water content. The *primary compression* is due to the expulsion of water. The *secondary compression* starts after the primary compression ceases (except in organic soils where the primary and secondary compressions may overlap). At the present time (2005), there is no satisfactory way of computing secondary compression.

Computation of Consolidation Settlement

The equation that is normally used for computing consolidation settlement S_c is

$$S_c = \frac{C_c}{1 + e_o} H \log_{10} \frac{p_o + \Delta p}{p_o} \qquad (2.17)$$

where, C_c = compression index to be determined in the laboratory or calculated from some equations,

e_o = *in situ* void ratio,

H = thickness of clay strata,

p_o = effective overburden pressure at the centre of clay strata,

Δp = increase in pressure at the middle of clay strata due to external loading from elastic theory (Section 2.9).

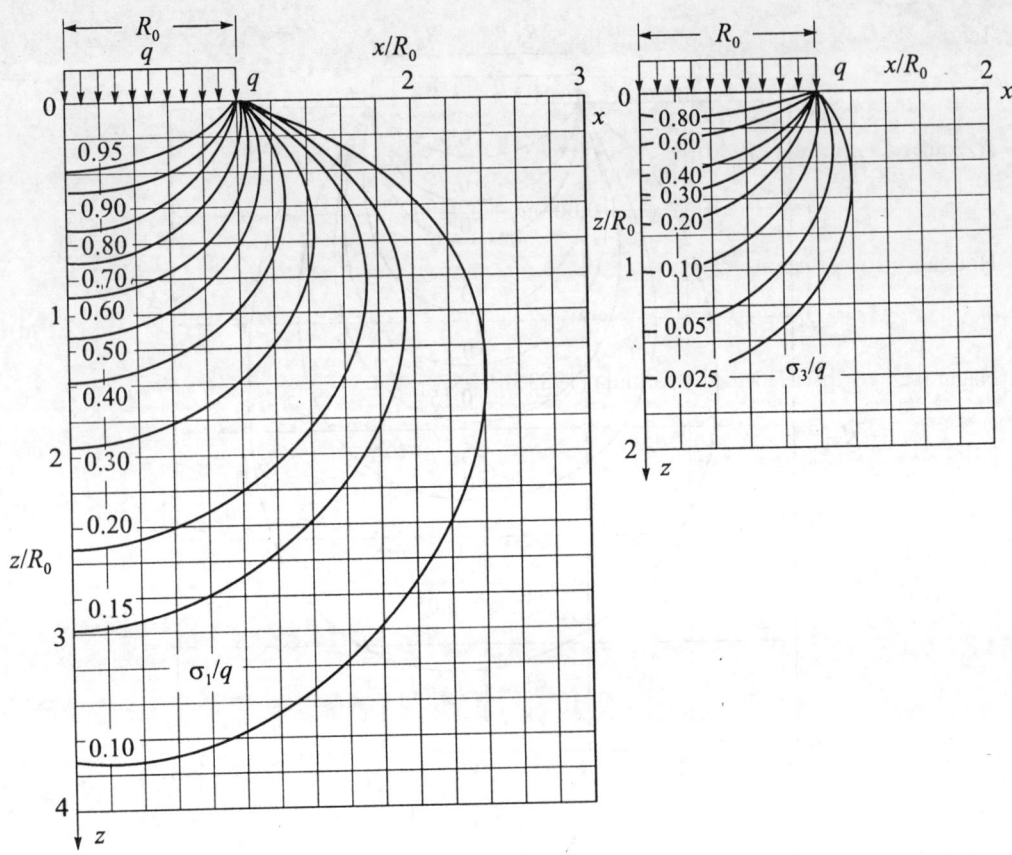

Fig. 2.5 Principal stresses σ_1 and σ_3 under uniform load on circular area

If the thickness of the clay strata is too large, the strata may be divided with smaller layers of thickness less than 3 m. The net change in pressure Δp at the middle of each layer will have to be determined. The compression index C_c may have to be found out for each of the layers if required. The equation for the total consolidation settlement is

$$S_c = \sum H_i \frac{C_c}{1+e_o} \log_{10} \frac{p_o + \Delta p}{p_o} \qquad (2.18a)$$

Equation (2.17) may also be expressed in a different way as

$$S_c = \Sigma H_i \,(m_v \,\Delta p) \qquad (2.18b)$$

where, m_v = coefficient of volume compressibility,

$\qquad$ = $1/E_c$, where E_c is called as the compression modulus.

The value of C_c may vary from 4.5 for peat to less than 0.03 for hard clay. The value of compression modulus E_c may vary from about 0.1 MPa for peat to more than 15 MPa for hard clay.

Empirical Relationships for Computing C_c

Skempton's (1944), formula for remoulded clay

$$C_c = 0.007 \,(w_l - 10) \qquad (2.19a)$$

Terzaghi and Peck formula for normally consolidated clays

$$C_c = 0.009 (w_l - 10) \tag{2.19b}$$

Azzouz *et al* formula (1976)

$$C_c = 0.37 (e_o + 0.003 \, w_l + 0.0004 \, w_n - 0.34) \tag{2.19c}$$

Hough's (1957) formula

$$C_c = 0.3 (e_o - 0.27) \tag{2.19d}$$

Nagaraj and Srinivasa Murthy formula (1983)

$$C_c = 0.2343 e_l \tag{2.19e}$$

$$C_c = 0.39 e_o \tag{2.19f}$$

where, w_l = liquid limit,

w_n = natural moisture content,

e_o = initial void ratio,

e_l = void ratio at liquid limit.

2.11   SHEAR STRENGTH

Coulomb Equation

The fundamental shear strength equation proposed by Coulomb is

$$s = c + \sigma \tan \phi \tag{2.20}$$

where, c = cohesion,

σ = total normal pressure on the failure plane,

ϕ = angle of shearing resistance.

Equation (2.20) may be expressed in terms of effective stresses as

$$s = c' + (\sigma - u) \tan \phi' = c' + \sigma' \tan \phi' \tag{2.21}$$

where c' = apparent cohesion in terms of effective stresses,

σ = total normal pressure on the failure plane,

u = porewater pressure,

ϕ' = angle of shearing resistance in terms of effective stresses.

Types of Laboratory Tests

The laboratory tests for determining shear strength parameters of soils may be on

1. undistorted samples for cohesive soils,
2. disturbed samples on cohesionless soils.

In the case of cohesive soils, the soil may be fully saturated or partially saturated. For cohesionless soils, it does not make much difference if the soil is fully saturated or fully dry (but it should not be partially saturated).

The various type of tests normally used are

1. Undrained or quick tests,
2. Consolidated undrained or consolidated quick tests,
3. Drained or slow tests.

The drainage condition of a sample is generally the deciding factor in choosing a particular type of test in the laboratory. The purpose of carrying out a particular test is to simulate the field conditions as for as possible. Because of high permeability of sand, consolidation occurs relatively rapidly and is usually completed during the application of the load. Tests on sand are therefore generally carried out under drained conditions.

2.12   STRESS PATHS

Definition of Stress Path

Stress path is a path that depicts graphically the state of stress in the test specimen or in a soil mass at any stage of loading from the equilibrium state to the failure state. Stress path gives a better insight into soil behaviour at any stage or successive stages of loading. There are a number of ways by which the locus of stresses or the stress paths can be depicted graphically. Only two methods are discussed briefly in this book. The methods are

1. Lambe's *p-q* diagram (1964, 1967).
2. Rendulic's diagram (1937).

Lambe's *p-q* Diagram

It is often necessary to depict the changes in stresses at a point in a soil mass or in a test specimen under different stages of loading. One way of doing it is to draw a series of stress circles as shown in Fig. 2.6 (b). However, a diagram with many circles can become quite confusing, especially if the results of several tests are plotted on the same diagram. An alternate method for plotting the state of stress is to plot only a series of stress points *a, b, c* and *d*, as shown in Fig. 2.6 (b), and connect these points with a line or curve. Such a line or curve is called a *stress path*. The various stress circles designated as 1, 2, and 3 in Fig. 2.6 (b) are obtained for the different stages of loading of a specimen shown in Fig. 2.6 (a). The lateral pressure σ_3 is assumed as constant for the duration of the test. The different stages of loading marked as 1, 2 and 3 on the stress–strain curve in Fig. 2.6 (a) are shown as the corresponding stress circles in Fig. 2.6 (b). The points *b, c,* and *d* give the maximum shear stress for the corresponding stages of loading. The stress circle 3 is the Mohr circle of failure. The point *a* on the abscissa represent the initial condition of the specimen before the application of the deviator load on the specimen. The points *a, b, c* and *d* [Fig. 2.6 (b)] may be plotted as a *p-q* diagram as shown in Fig. 2.6 (c). The coordinates of the points may be found out from the equations.

$$p = \frac{\sigma_1 + \sigma_3}{2} \tag{2.22a}$$

$$q = \frac{\sigma_1 - \sigma_3}{2} \tag{2.22b}$$

The line connecting the points *a, b, c* and *d* in Fig. 2.6 (c) is a *stress path* and the diagram as such is called as *p-q diagram*. The stress paths may be plotted by using either total or effective stresses. If effective stresses are used, we have

$$\frac{\sigma_1' - \sigma_3'}{2} = \frac{\sigma_1 - \sigma_3}{2} = q \text{ or } q' \tag{2.23a}$$

$$\frac{\sigma_1' + \sigma_3'}{2} = \frac{\sigma_1 + \sigma_3}{2} - u = p - u$$

or

$$p' = p - u \qquad\qquad (2.23b)$$

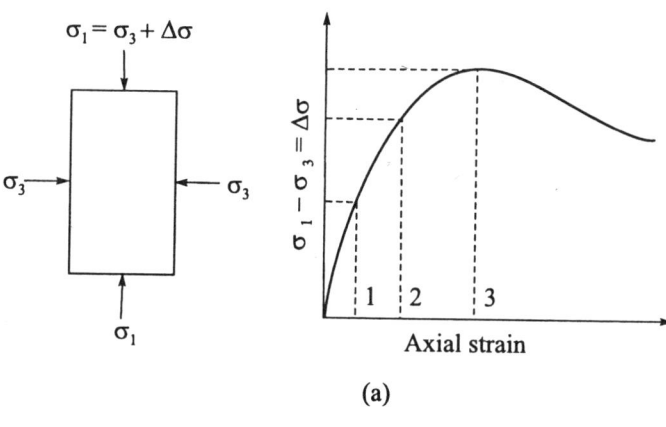

(a)

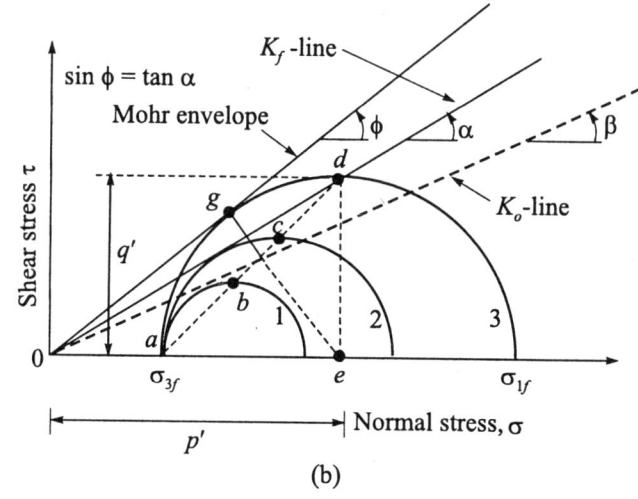

(b)

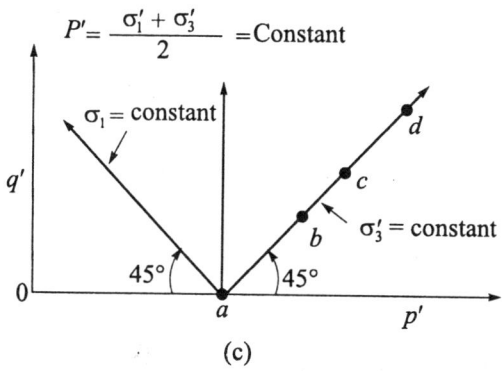

(c)

Fig. 2.6 Lambe's stress path for normally consolidated clay soil: (a) Stress–strain curve, (b) concept of strees path, (c) stress paths

The use of effective stress coordinates simply shifts the p-q plot along the p-axis by the magnitude of the pore-pressure u. It should be noted that the p, q, coordinates represent the centre and radius of the stress circles respectively. The stress path for a specimen where $\sigma_1 = \sigma_3$ at the initial stage of the test, and σ_1 is increased while keeping σ_3 constant, is a 45°-line as shown in Fig. 2.6 (c). The stress paths for the other variations of σ_1 and σ_3 are also shown in the figure.

Now in Fig. 2.6 (b), the line joining the origin of coordinates '0' to point d is called as K_f-line which makes an angle α with the horizontal axis. This line is applicable both for granular soils and normally consolidated clay soils.

From Fig. 2.6 (b), we have

$$\tan \alpha = \frac{de}{0e} = \frac{\left(\sigma_{1f} - \sigma_{3f}\right)/2}{\left(\sigma_{1f} + \sigma_{3f}\right)/2}$$

$$\sin \phi = \frac{eg}{0e} = \frac{\left(\sigma_{1f} - \sigma_{3f}\right)/2}{\left(\sigma_{1f} + \sigma_{3f}\right)/2}$$

Therefore, $\tan \alpha = \sin \phi$ (2.24)

p-q Diagram for One-Dimensional Consolidation Test

Equations for K_0: Consider a case where a soil specimen is subjected to one-dimensional consolidation as shown in an oedometer type of loading test [Fig. 2.7 (a)]. We may write for this case as

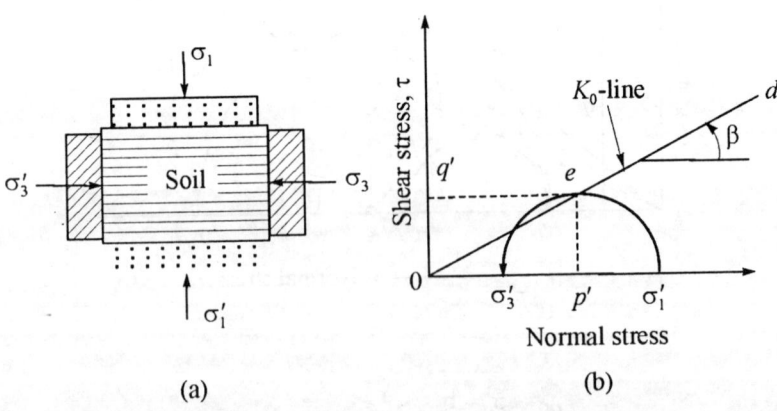

(a) (b)

Fig. 2.7 Determination of K_0-line: (a) Consolidation test, (b) K_0-line

$$\sigma_3' = K_0 \, \sigma_1'$$ (2.25)

where K_0 is the at-rest earth pressure coefficient which can be expressed as (Jaky 1944),

$$K_0 = 1 - \sin \phi$$ (2.26a)

for both granular soils and normally consolidated (NC) clays. For normally consolidated clays, Brooker and Ireland (1965) proposed

$$K_0 = 0.95 - \sin \phi$$ (2.26b)

Alpan (1967), expresses a relationship between K_0 and I_p (plasticity index) for normally consolidated soils as

$$K_0 = 0.19 + 0.233 \log I_p \qquad (2.26c)$$

Since K_0 varies with over-consolidation ratio, K_0 for over-consolidated (OC) clays has been expressed as

$$K_0 (OC) = K_0 (NC) R_{oc}^n \qquad (2.26d)$$

where R_{oc} = OCR; the value of the exponent n for cohesive soils has been expressed as

$$n = \frac{0.54}{10^{I_p / 281}} \qquad (2.26e)$$

Wroth (1975), analysed a number of soils reported in literatures by others and proposes for K_0 as

$$K_0 (OC) = R_{oc} \times K_0 (NC) - \frac{\mu'}{1 - \mu'} (R_{oc} - 1) \qquad (2.26f)$$

where μ' = Poisson's ratio in terms of effective stresses. The equation suggested by Bowles (1986) for μ' is

$$\mu' = 0.23 + 0.003 I_p \qquad (2.26g)$$

There are, therefore, many equations that have been suggested by different investigators at different times. The user must apply his own engineering judgement to decide what value to use for K_0.

Equations for *p* and *q*

For the soil *in-situ* condition,

$$\sigma_1' = \sigma_v' \text{ and } \sigma_3' = \sigma_h' = K_0 \sigma_v'$$

Therefore,

$$q' = \frac{\sigma_1' - \sigma_3'}{2} = \frac{\sigma_v' - \sigma_h'}{2} = \frac{\sigma_v' (1 - K_0)}{2} \qquad (2.27a)$$

$$p' = \frac{\sigma_1' + \sigma_3'}{2} = \frac{\sigma_v' + \sigma_h'}{2} = \frac{\sigma_v' (1 + K_0)}{2} \qquad (2.27b)$$

Hence,

$$\tan \beta = \frac{q'}{p'} = \frac{1 - K_0}{1 + K_0} \qquad (2.27c)$$

where p' and q' are the effective stress coordinates of point e on the Mohr circle in Fig. 2.7 (b), and the line Od is called as the K_0-line which makes an angle β with the horizontal axis. For the purpose of comparison, the K_0-line is also drawn in Fig. 2.6 (b). It is obvious from this figure that the K_f-line falls below Mohr envelope, and the K_0-line below the K_f-line. The latter is obvious since K_0 is an equilibrium *in-situ* stress state and K_f is a failure state.

Stress path

The stress path for consolidation test as shown in Fig. 2.8 (a) may be explained as follows:

1. Let point A represent the state of effective stress at the end of primary consolidation for a particular load increment in a consolidation test. Let the effective vertical pressure at point A

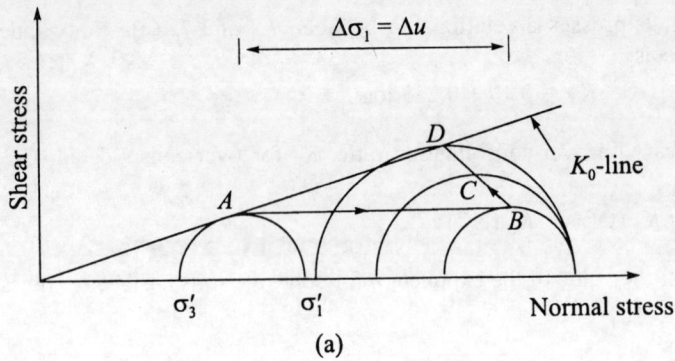

(a)

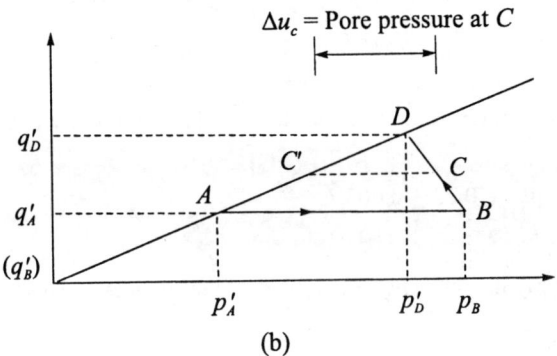

(b)

Fig. 2.8 Stress path for consolidation test: (a) Stress circles, (b) stress path

be σ_1'. The corresponding lateral pressure may be expresses as $\sigma_3' = K_0 \sigma_1'$. Therefore for point A, p, q coordinates may be expressed as

$$p_A' = \frac{\sigma_1'(1 + K_0)}{2}$$

$$q_A' = \frac{\sigma_1'(1 - K_0)}{2}$$

2. Let $\Delta\sigma_1$ be the next load increment applied instantaneously. Since drainage occurs only after sometime, there will be no change in volume and correspondingly no change in shearing stress including the maximum value q during this period. Thus

$$q_A' = \frac{\sigma_1'(1 - K_0)}{2}$$

remains unchanged but causing the diameter of the circle at point A [Fig. 2.8 (a)] displaced to position B by an amount equal to the change in vertical pressure. For the vertical increment $\Delta\sigma_1$, the pore-pressure developed is equal to

$$\Delta\sigma_1 = \Delta u$$

for saturated clay soils. The p, q coordinates of point B are

$$P_B = \frac{\sigma_1'\left(1 + K_0\right)}{2} + \Delta\sigma_1$$

$$q_B = q_A' = \frac{\sigma_1'\left(1 - K_0\right)}{2}$$

3. As drainage takes place, the pore-pressure decreases and effective stress increases as indicated by circles C and D after some elapsed times. Circle D is the effective stress state when primary consolidation under the load increment $\Delta\sigma_1$ is complete. Now in Fig. 2.8 (b), the path $ABCD$ represents the total stress path (TSP) for the increment of load $\Delta\sigma_1$ from the initial position A and line AD represents the effective stress path (ESP). The horizontal distance between AD and BCD represent the residual excess pore pressure at some instant of time, for example, at point C on BD, the excess pore pressure is represented by the distance CC'. The p-q coordinates of point D are

$$p_D' = \frac{\sigma_1'\left(1 + K_0\right)}{2} + \frac{\Delta\sigma_1'\left(1 + K_0\right)}{2}$$

$$q_D' = \frac{\sigma_1'\left(1 - K_0\right)}{2} - \frac{\Delta\sigma_1'\left(1 - K_0\right)}{2}$$

Stress-Path for Normally Consolidated Clay Under Consolidated Undrained Condition

Consider a clay specimen [Fig. 2.9 (a)] consolidated under an all-round pressure $\sigma_3 = \sigma_3'$ in a triaxial test apparatus represented as point a in Fig. 2.9 (b). When deviator stress $\Delta\sigma_1$ is applied on the

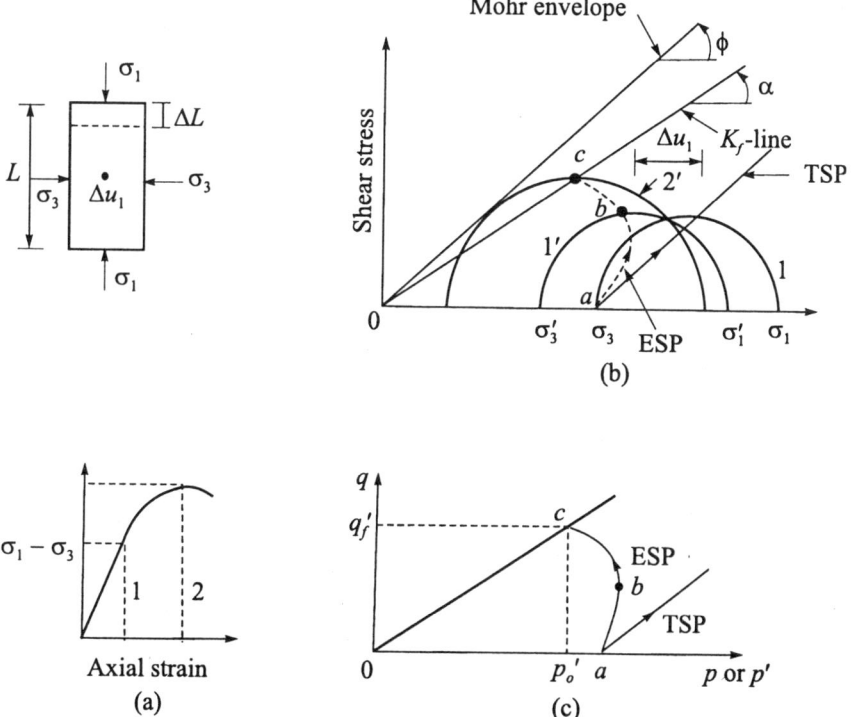

Fig. 2.9 Stress path for consolidated undrained triaxial test for normally consolidated clay .

sample under no drainage condition, there will be an increase in the pore water pressure equal to Δu_1. The major and minor effective principal stresses for this stage of loading may be written as follows:

Consider the stress–strain curve given in Fig. 2.9 (a) for the sample tested. The first stage of loading is marked as 1 on the stress–strain curve and the corresponding total stress circle also as 1 in Fig. 2.9 (b). Let the pore water pressure measured at this stage be Δu_1. The effective principal stresses for this stage may be written as

$$\sigma_1' = \sigma_1 - \Delta u_1 \; ; \; \sigma_3' = \sigma_3 - \Delta u_1$$

The effective stress circle marked as 1′ in Fig. 2.9 (b) is shifted to the left by an amount equal to Δu_1. In the same way the effective stress circle 2′ for the peak strength can be obtained. Let points b and c represent the maximum shear stresses on the stress circles 1′ and 2′ respectively. Now the efective stress path (ESP) is represented by the curve abc which is the path for the undrained condition with constant water content. The total stress path (TSP) and the effective stress path (ESP) are given in Fig. 2.9 (c) as a p-q diagram.

Characteristics of Stress Paths Under Undrained Conditions

Some of the characteristics of effective stress paths for normally consolidated clays are

1. The effective stress paths also indicate the contours of constant water content.
2. The stress paths are geometrically similar as shown in Fig. 2.10.
3. If points are marked on each of the stress paths representing equal axial strains, and if the points of equal strains are joined, the contours of equal strains are more or less straight lines passing through the origin of coordinates Fig. 2.10.

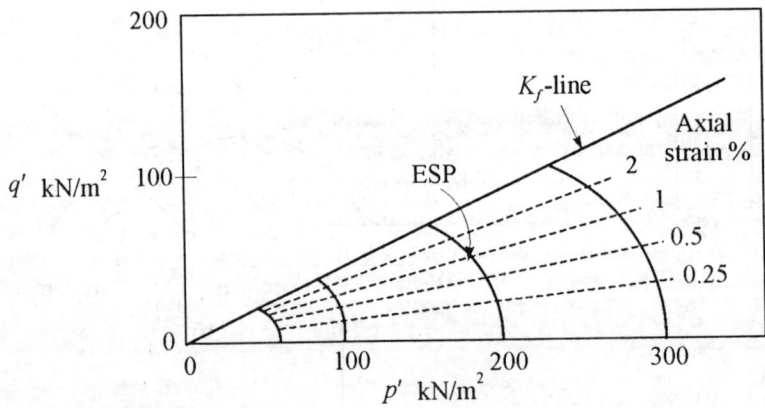

Fig. 2.10 Typical effective stress paths of a normally consolidated clay under undrained condition

Stress-Paths for Over-Consolidated Clays

K_f-**line :** The K_f-line for over consolidated clays on a p-q diagram is shown in Fig. 2.11. From Fig. 2.11, the intercept a of K_f-line on the vertical axis can be obtained as follows:

From triangle $0'cd$,

$$\frac{a}{cd} = \frac{0'0}{0'c} ; \text{ Therefore } a = \frac{0'0 \times cd}{0'c}$$

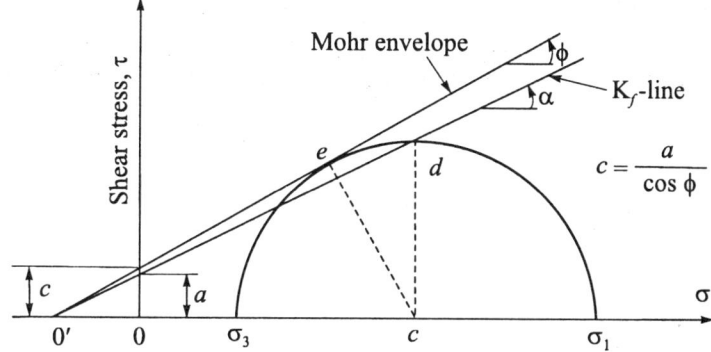

Fig. 2.11 K_f-line of over-consolidated clay

From triangle $0'ce$,

$$\sin \phi = \frac{ce}{0'c} = \frac{cd}{0'c}$$

Therefore, $0'c = \dfrac{cd}{\sin \phi}$. The equation for a may now be written as

$$a = \frac{0'0 \times cd \sin \phi}{cd} = 0'0 \sin \phi$$

Since $0'0 = c \cot \phi$, we have

$$a = c \cot \phi \sin \phi = c \cos \phi \qquad\qquad (2.28)$$

It is obvious from Fig. 2.11

$$\tan \alpha = \sin \phi$$

Stress path

The stress paths for over-consolidated clays for drained and consolidated undrained tests are given in Fig. 2.12 on the p-q diagram.

The stress path *ad in* Fig. 2.12 (a) for drained test is not only a TSP but also an ESP. For over-consolidated clays, the shape of ESP for the consolidated undrained test [Fig. 2.12 (b)] is different from that for normally consolidated clays [Fig. 2.9 (c)].

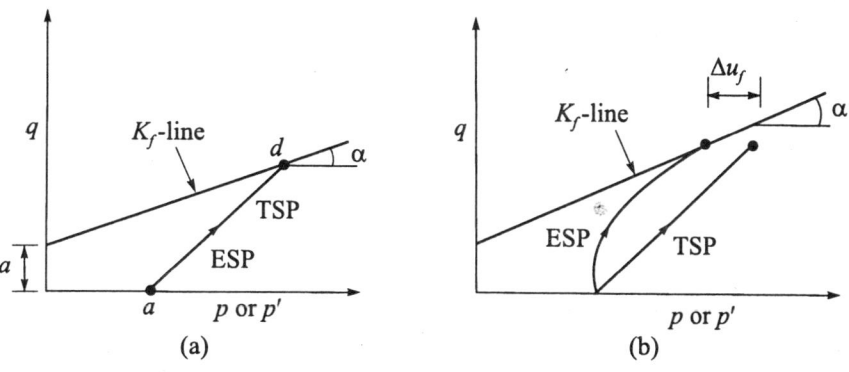

Fig. 2.12 Stress paths for over-consolidated clays: (a) Drained test, (b) consolidated undrained test

Stress Paths for Compression and Extension Tests

The following types of triaxial tests can be carried out in a triaxial test apparatus.

1. Increase of vertical pressure by keeping the lateral pressure constant (standard compression test).
2. Decrease of lateral pressure by keeping the vertical pressure constant (compression test with decreasing lateral pressure).
3. Decrease of vertical pressure by holding lateral pressure constant (Extension test – decreasing vertical pressure).
4. Increase of lateral pressure by holding vertical pressure constant (Extension test with constant vertical pressure).

The stress paths for the above four types of triaxial tests on p-q diagram are shown in Fig. 2.13. The initial stress condition is represented by point A on the p-axis for all the types of tests.

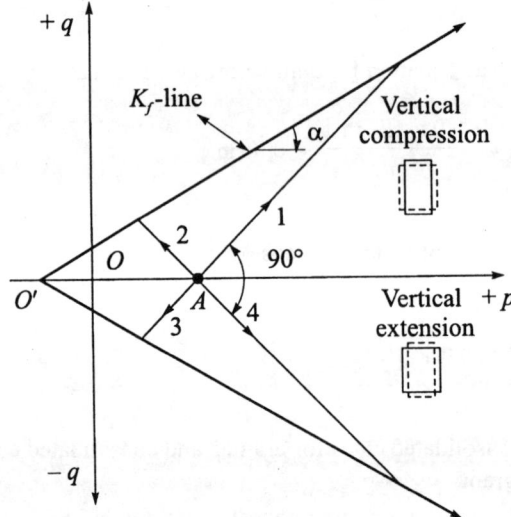

Fig. 2.13 Stress paths for triaxial compression and extension tests

2.13 LATERAL PRESSURES BY THEORY OF ELASTICITY FOR SURCHARGE LOADS ON THE SURFACE OF BACKFILL

The surcharges on the surface of a backfill parallel to the retaining wall may be any one of the following:

1. A concentrated load.
2. A line load.
3. A strip load.

Lateral Pressure at a Point in a Semiinfinite Mass due to a Concentrated Load on the Surface

Tests by Spangler (1938), and others indicate that lateral pressures on the surface of rigid walls can be computed for various types of surcharges by using modified forms of the theory of elasticity equations. Lateral pressure on an element in a semi-infinite mass at depth z from the surface may be calculated by Boussinesq theory for a concentrated load Q acting at a point on the surface. The equation may be expressed as

$$p_h = \frac{Q}{2\pi z^2} \left[3 \sin^2\beta \cos^3\beta - \frac{(1 - 2\mu) \cos^2\beta}{1 + \cos\beta} \right] \tag{2.29}$$

Figure 2.14 (a) gives the notations used in Eq. (2.29)

If we write $r = x$ in Fig. 2.14, and redefine the terms as

$$x = mH \text{ and } z = nH$$

where H = height of rigid wall and take possion's ratio $\mu = 0.5$, we may write Eq. (2.29) as

$$p_h = \frac{3Q}{2\pi H^2} \frac{m^2 n}{\left(m^2 + n^2\right)^{5/2}} \tag{2.30}$$

Equation (2.30) is strictly applicable for computing lateral pressures at a point in a semi-infinite mass. However, this equation has to be modified if a rigid wall intervenes and breaks the continuity of soil mass. The modified forms are given below for various types of surcharge loads.

Lateral Pressure on a Rigid Wall Due to a Concentrated Load on the Surface

Let Q be the point load acting on the surface as shown in Fig. 2.14. The various equations are

(a) For $m > 0.4$

$$p_h = \frac{1.77 Q}{H^2} \frac{n^2}{\left(m^2 + n^2\right)^3} \tag{2.31a}$$

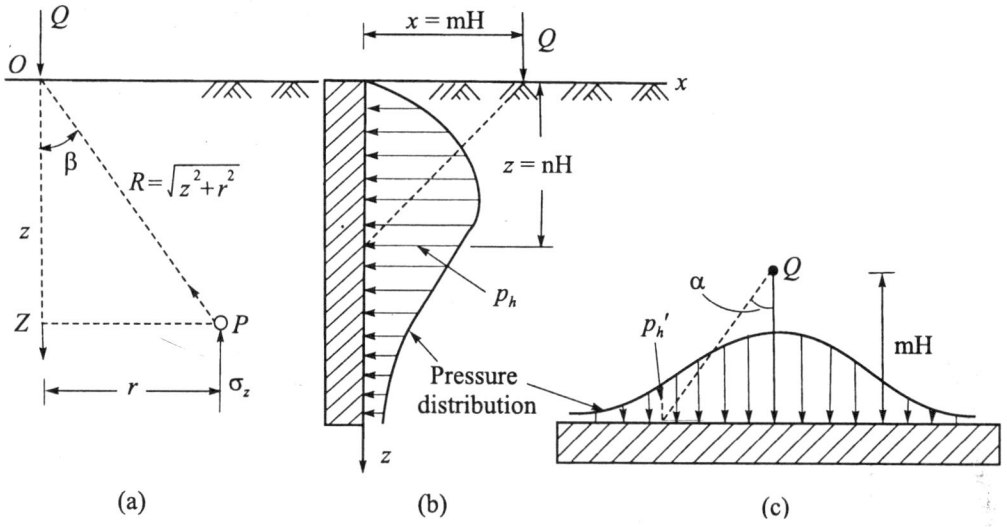

(a)   (b)   (c)

Fig. 2.14 Lateral pressure against a rigid wall due to a point load: (a) Vertical pressure within an earth mass, (b) vertical section, (c) horizontal section

(b) For $m \leq 0.4$

$$p_h = \frac{0.28Q}{H^2} \frac{n^2}{\left(0.16 + n^2\right)^3} \qquad (2.31b)$$

(c) Lateral pressure at points along the wall on each side of a perpendicular from the concentrated load Q to the wall [Fig. 2.14 (c)]

$$p'_h = p_h \cos^2 (1.1\alpha) \qquad (2.31c)$$

Lateral Pressure on a Rigid Wall Due to Line Load

A concrete block wall, conduit laid on the surface, or wide strip loads may be considered as series of parallel line loads as shown in Fig. 2.15. The modified equations for computing p_h are as follows.

(a) For $m > 0.4$

$$p_h = \frac{4}{\pi} \frac{q}{H} \left[\frac{m^2 n}{\left(m^2 + n^2\right)^2}\right] \qquad (2.32a)$$

(b) For $n \leq 0.4$

$$p_h = \frac{q}{H} \left[\frac{0.203 n}{\left(0.16 + n^2\right)^2}\right] \qquad (2.32b)$$

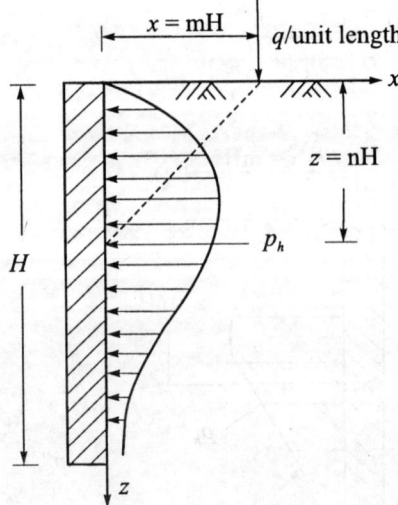

Fig. 2.15 Lateral pressure against a rigid wall due to a line load

Lateral Pressure on a Rigid Wall Due to Strip Load

A strip load is a load intensity with a finite width, such as a highway, railway line or earth embankment which is parallel to the retaining structure. The application of load is as given in Fig. 2.16.

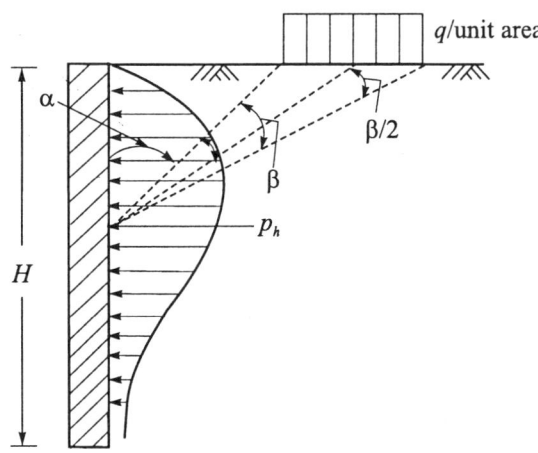

Fig. 2.16 Lateral pressure against a rigid wall due a to a strip load

The equation for computing p_h is

$$p_h = \frac{2q}{\pi} \, (\beta - \sin \beta \cos 2\alpha) \tag{2.33}$$

where β is in radians.

Soil Exploration

3

3.1 INTRODUCTION

The stability of the foundation of a building, a bridge, an embankment or any other structure built on soil depends on the strength and compressibility characteristics of the subsoil. The field and laboratory investigations required to obtain the essential information on the subsoil is called *Soil Exploration or Soil Investigation.* Soil exploration happens to be one of the most important parts of Foundation Engineering and at the same time the most neglected part of it. Terzaghi in 1951 (Bjerrum *et al* 1960) had rightly remarked, that *Building foundations have always been treated as step children.* His remarks are relevant even today. The success or failure of a foundation depends essentially on the reliability of the various soil parameters obtained from the field investigation and laboratory testing, and used as an input into the design of foundations. *Sophisticated theories alone will not give a safe and sound design.*

Soil exploration is a *must* in the present age for the design of foundations of any project. The extent of the exploration depends upon the magnitude and importance of the project. Projects such as buildings, power plants, fertilizer plants, bridges, etc. are localized in areal extent. The area occupied by such projects may vary from a few square meters to many square kilometers. Transmission lines, railway lines, roads and other such projects extend along a narrow path. The length of such projects may be several kilometers. Each project has to be treated as per its requirements. The principle of soil exploration remains the same for all the projects but the programme and methodology may vary from project to project.

The elements of soil exploration depend mostly on the importance and magnitude of the project, but generally should provide the following:

1. Information to determine the type of foundation required such as a shallow or deep foundation.
2. Necessary information with regards to the strength and compressibility characteristics of the subsoil to allow the Design Consultant to make recommendations on the safe bearing pressure or pile load capacity.

Soil exploration involves broadly the following:

1. Planning of a programme for soil exploration.
2. Collection of disturbed and undisturbed soil or rock samples from the holes drilled in the field. The number and depths of holes depend upon the project.

3. Conducting all the necessary *in-situ* tests for obtaining the strength and compressibility characteristics of the soil or rock directly or indirectly.
4. Study of ground-water conditions and collection of water samples for chemical analysis.
5. Geophysical exploration, if required.
6. Conducting all the necessary tests on the samples of soil/rock and water collected.
7. Preparation of drawings, charts, etc.
8. Analysis of the data collected.
9. Preparation of report.

3.2 BORING OF HOLES

Auger Method

Hand operated augers

Auger boring is the simplest of the methods. Hand operated or power driven augers may be used. Two types of hand operated augers are in use as shown in Fig. 3.1.

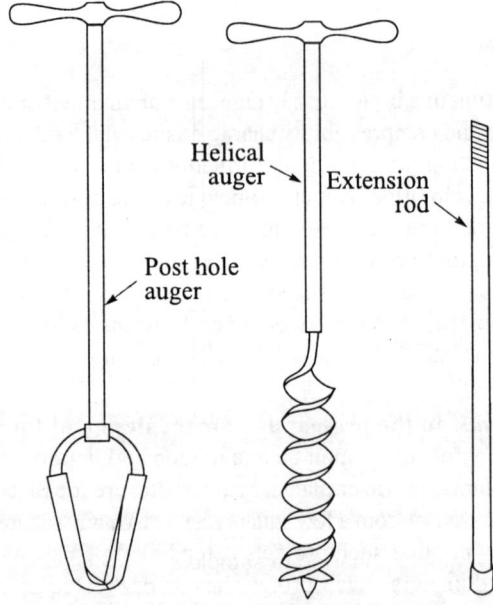

Fig. 3.1 Hand augers

The depths of the holes are normally limited to a maximum of 10 m by this method. These augers are generally suitable for all types of soil above the water table but suitable only in clayey soil below the water table (except for the limitations given below). A string of drill rods is used for advancing the boring. The diameters of the holes normally vary from 10 to 20 cm. Hand operated augers are not suitable in very stiff to hard clay nor in granular soils below the water table. Hand augering is not practicable in denses and nor in sand mixed with gravel even if the strata lies above the water table.

Power Driven Augers

In many countries the use of power driven continuous flight augers is the most popular method of soil exploration for boring holes. The flights act as a screw conveyor to bring the soil to the surface. This method may be used in all types of soil including sandy soils below the water table but is not suitable if the soil is mixed with gravel, cobbles, etc. The central stem of the auger flight may be hollow or solid. A hollow stem is sometimes preferred since standard penetration tests or sampling may be done through the stem without lifting the auger from its position in the hole. Besides, the flight of augers serves the purpose of casing the hole. The hollow stem can be plugged while advancing the bore and the plug can be removed while taking samples or conducting standard penetration tests (to be described) as shown in Fig. 3.2. The drilling rig can be mounted on a truck or a tractor. Holes may be drilled by this method rapidly to depths of 60 m or more.

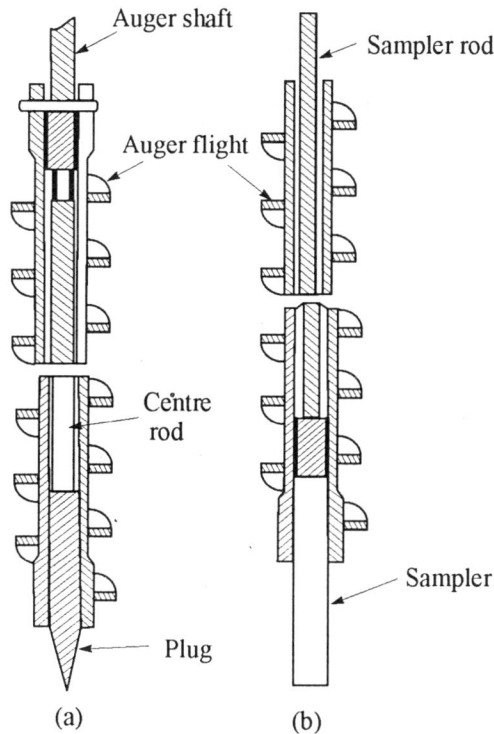

Fig. 3.2 Hollow-stem auger: (a) Plugged while advancing the auger, and (b) plug removed and sampler inserted to sample soil below auger

Shell and Auger Method

Shell and auger method of drilling holes is a popular method of boring in India. Shell, which is also called as a *sand bailer,* is nothing but a heavy duty pipe with a hard cutting edge and a flat valve which opens only inside as shown in Fig. 3.3. The length of the shell vary from 1 to 3 m or more depending on the weight required for cutting the soil in the hole. The weight may range from 30 to 60 kg or more. Sinker bars are sometimes added to increase the weight of the bailer. Sinker bars are nothing but solid rods fixed on the top of the bailer. The outside diameter of the bailer is less than the inside diameter of the casing pipe by at least 25 mm for easy operation.

Boring is always started first with augering. When further boring by the use of auger is not possible, the shell is used for advancing the bore. Boring by shell consists of raising it above the bottom of the hole and allow it to fall freely. The impact of the drop cuts the soil and pushes it into the tube. This process is continued till the shell is practically filled with the soil. The shell is then withdrawn from the hole and emptied. The lifting and lowering of the shell may be done either manually or by a power-driven winch.

Boring by shell is very useful even in dense sandy deposits or stiff to hard clay soils. Even sandy soil mixed with gravel can be bored by this method.

Wash Boring

Wash boring is commonly used for boring holes. Soil exploration below the ground water table is usually very difficult to perform by means of pits or auger-holes. Wash boring in such cases is a very convenient method provided the soil is either sand, silt or clay. The method is not suitable if the soil is mixed with gravel or boulders.

Figure 3.4 shows the assembly for a wash boring. To start with, the hole is advanced a short depth by auger and then a casing pipe is pushed to prevent the sides from caving in. The hole is then continued by the use of a chopping bit fixed at the end of a string of hollow drill rods. A stream of water under pressure is forced through the rod and the bit into the hole, which loosens the soil as the water flows-up around the pipe. The loosened soil in suspension in water is discharged into a tub. The soil in suspension settles down in the tub and clean water flows into a sump which is reused for circulation. The motive power for a wash boring is either mechanical or man power. The bit which is hollow is screwed to a string of hollow drill rods supported on a tripod by a rope or steel cable passing over a pulley and operated by a winch fixed on one of the legs of the tripod.

The purpose of wash boring is to drill holes only and not to make use of the disturbed washed materials for analysis. Whenever, an undisturbed sample is required at a particular depth, the boring is stopped, and the chopping bit is replaced by a sampler. The sampler is pushed into the soil at the bottom of the hole and the sample is withdrawn.

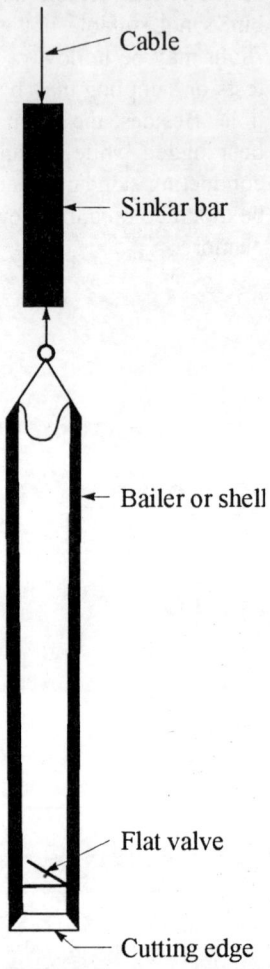

Fig. 3.3 Shell with sinker bar

(labels: Cable, Sinkar bar, Bailer or shell, Flat valve, Cutting edge)

Rotary Drilling

In the rotary drilling method a cutter bit or a core barrel with a coring bit attached to the end of a string of drill rods is rotated by a power rig. The rotation of the cutting bit shears or chips the material penetrated and the material is washed out of the hole by a stream of water just as in the case of a wash boring. Rotary drilling is used primarily for penetrating the overburden between the levels\ of which samples are required. Coring bits, on the other hand, cut an annular hole around an intact core which enters the barrel and is retrieved. Thus, the core barrel is used primarily in rocky strata to get rock samples.

As the rods with the attached bit or barrel are rotated, a downward pressure is applied to the drill string to obtain penetration, and drilling fluid under pressure is introduced into the bottom of the hole through the hollow drill rods and the passages in the bit or barrel. The drilling fluid serves the dual function of cooling the bit as it enters the hole and removing the cuttings from the bottom of the hole as it returns to the surface in the annular space between the drill rods and the walls of the hole. In an uncased hole, the drilling fluid also serves to support the walls of the hole. When boring in soil, the drilling bit is removed and replaced by a sampler when sampling is required, but in rocky strata

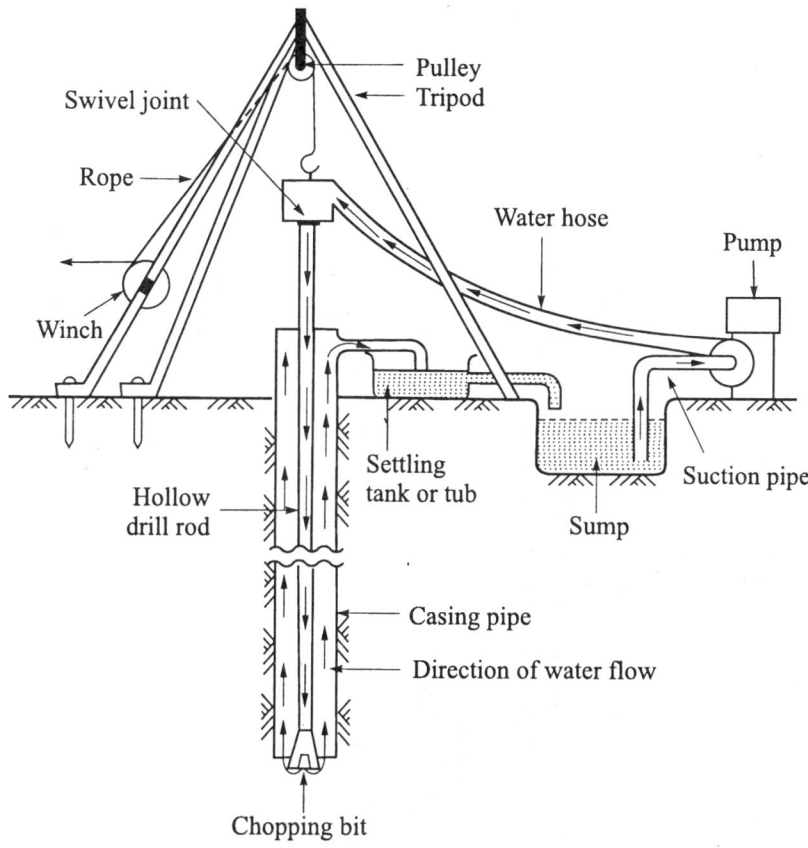

Fig. 3.4 Wash boring

the coring bit is used to obtain continuous rock samples. The rotary drilling rig of the type given in Fig. 3.5 can also be used for wash boring and auger boring.

Coring Bits

Three basic categories of bits are in use. They are diamond, carbide insert, and saw tooth. Diamond coring bits may be of the surface set or diamond impregnated type. Diamond coring bits are the most versatile of all the coring bits since they produce high quality cores in rock materials ranging from soft to extremely hard. Carbide insert bits use tungsten carbide in lieu of diamonds. Bits of this type are used to core soft to medium hard rock. They are less expensive than diamond bits but the rate of drilling is slower than with diamond bits. In saw-tooth bits, the cutting edge comprises a series of teeth. The teeth are faced and tipped with a hard metal alloy such as tungsten carbide to provide wear resistance and thereby increase the life of the bit. These bits are less expensive but normally used to core overburden soil and very soft rocks only. The coring bits are shown in Fig. 3.6.

Calyx or Shot Core Drilling

In the case of calyx or shot core drilling, cutting action is provided by a slotted bit of mild steel and by very hard steel shot, which is fed into the drill hole with the wash water and reaches the bit via the annular space between the core and the wall of the barrel. Slots cut into the bit at its lower end facilitate movement of the shot to the bottom and outside of the bit.

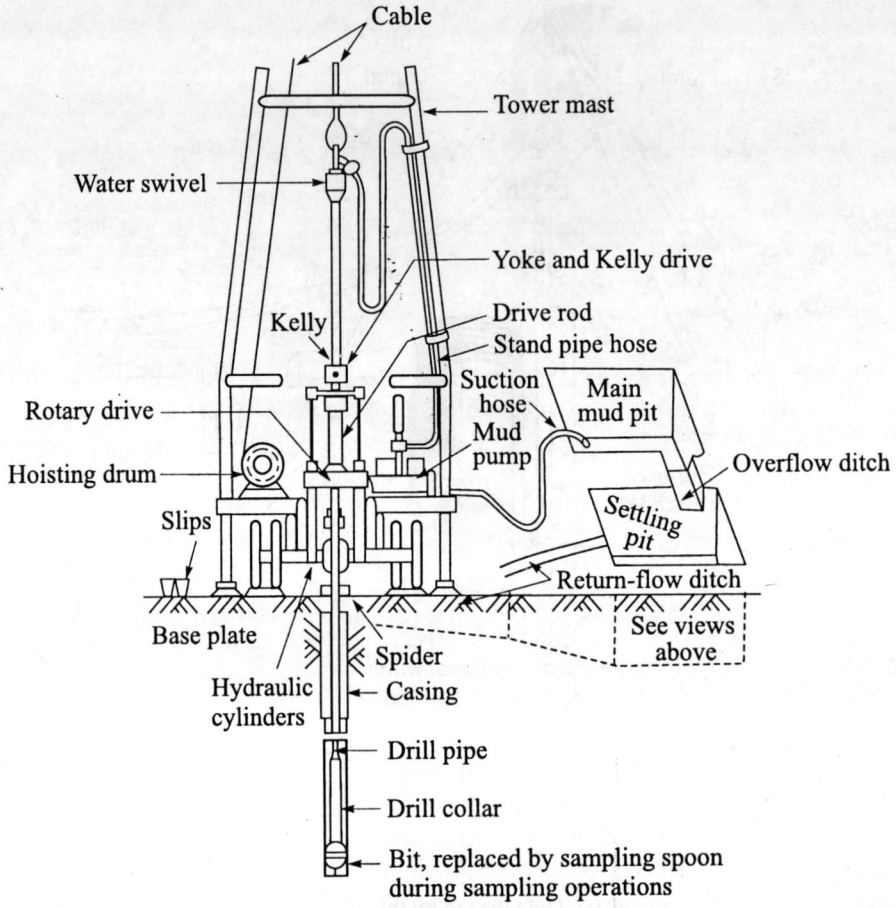

Fig. 3.5 Rotary drilling rig (After Hvorslev, 1949)

In operation, as the barrel is rotated, the shot which becomes wedged beneath and around the slotted bit is crushed into abrasive particles. These particles, some of which become embedded in the mild steel bit, provide the cutting action required. The cuttings of the rock are removed by the circulating water as in the case of wash boring. The diameters of the holes drilled by calyx may go up to about a metre or more and this method is largely used by the tubewell borers. However, this method is also suitable in soil exploration works in overburden stiff to hard soil, and in soft to medium hard rocks.

Percussion Drilling

Percussion drilling is possibly the only method of drilling in river sandy deposits mixed with hard boulders of the quartzitic type. Rotary drilling does not work in such deposits. *Percussion drilling* is also known as *cable-tool drilling*. In this method a heavy drilling bit is alternatively raised and dropped in such a manner that it powders the underlying material and forms it to a slurry to the consistency of sand or silt. When boring is required above the water table, water may have to be added periodically. When the accumulation of the slurry material interferes with the drilling, the drilling tools are removed from the hole, and the soil is cleaned out by means of bailers and sand pumps.

The hole may be cased if necessary to prevent the sides from collapsing during drilling. Percussion drilling is normally used for drilling tubewells, but generally not favoured in soil exploration where

Fig. 3.6 Coring bits: (a) Diamond with conventional waterways, (b) diamond with bottom discharge waterways, (c) carbide insert blade type, (d) carbide insert, pyramid type, and (e) saw tooth

undisturbed samples are required. The weight of the drilling bit may range up to 50 kN depending upon the size of the hole and the type of the strata met with. The shape of the bit used also depends to a large extent on the nature of the materials to be penetrated. The cutting edge of the bit is made of high carbon steel.

Changes in the nature of the material penetrated by this method are noted by observations similar to those used in wash boring. These include rate of progress, behaviour of the drilling tools, colour of the slurry, and the character of the cuttings.

Equipments Used for Making Borings

The machinery used to advance the hole and take samples, commonly referred to as a _drill rig,_ in general consists of:

1. A power-driven motor to operate a hammer to drive casing and to operate a winch to raise and lower the drilling and sampling equipments, provides rotary motion where required to turn augers or coring equipments and provide downward pressure to push samplers into the ground.
2. A water pump or air compressor which provides water or air under pressure for the removal of cuttings from the drill hole and for the cooling of rotary bits.
3. A winch to raise and lower drilling tools and casing pipes.
4. A tripod or a four legged derric equipped with a sheave for use in conjunction with the winch in raising and lowering drill tools.

The type and size of the rigs depend upon the size of the project, depth of hole, type of soil or rock met with, etc. In many cases hand operated winches fixed on one of the legs of a tripod is used. Mechanical drill rigs may be mounted on the back of a truck, or on a skid which may be dragged along the ground. For boring beneath the surface of water in lakes, rivers or oceans where the depths of water is shallow, land-type rigs mounted on barges, floating platforms supported by pontoons or oil drums or of platforms supported by piles or spuds may be used. For deep water borings, the equipments have to be fixed on the decks of ships which are to be berthed over the area to be explored.

Stabilisation of Bore Holes

Two problems that are normally met with in the drilling of holes are,

1. Caving of the sides of the hole,
2. Heaving of the bottom of the hole.

Caving of the sides of the hole is due to the release of the stresses caused by the removal of the overburden material from the hole. Holes drilled above the water table may not cave in even in sandy soil due to the presence of apparent cohesion present between the particles, but caving in sandy soils cannot be avoided below water table. But holes drilled in cohesive soils may remain stable even up to considerable depth below water table. Caving in fine sandy and silty soils below water table can be minimised to a certain extent by the use of *drilling mud.* A drilling mud is nothing but bentonite (which is a pure form of clay) mixed in water. The percentage of bentonite required in water has to be decided by trial and error according to the soil conditions met with. Even with the use of drilling mud, the sides may cave in during the operation of the drilling tools. The raising of a drilling tool in water creates a suction below the tool which drags the sides of the hole. If drilling mud is to be successful in stabilising the sides, the tools used for drilling must be so designed as not to create suction during the operation. Drilling mud is not to be used in cases where it is suspected that it would contaminate undisturbed samples.

Stabilising of the sides of the holes is best done by the use of casing pipes. Casing pipes may be standard or extra-duty black steel pipe or seamless steel pipe. Extra-heavy duty pipes are required where deep borings are required and where difficulty is anticipated in driving. Pipes of lengths ranging from 1 to 3 m are used for casing. Pipes may be joined either by external couplings or provided flush joints. Flush jointed casing are easy to drive or push into the hole. Casing pipes are either driven or pushed into the hole as the hole advances.

Heaving of the bottom of the hole is due to the differential head created between the water table levels outside and inside the hole which causes the water to flow from the surrounding region to the hole. Heaving of the bottom can be prevented by keeping the water level in the bore hole sufficiently above the water level outside the hole to counteract the inflow.

3.3 SAMPLING IN SOIL

Introduction

Soils met in nature are heterogeneous in character with a mixture of sand, silt and clay in different proportions. But in water deposits of soil, there are distinct layers of sand, silt and clay of varying thicknesses and alternating with depth. However, we can bring all the deposits of soil under two distinct groups for the purpose of study, namely, coarse-grained and fine grained soils. Soils with particles of size coarser than 0.075 mm are brought under the category of coarse grained and those finer than 0.075 mm under fine grained soils. Sandy soil falls in the group of coarse grained, and silt and clay soils in the fine grained group. A satisfactory design of foundation depends upon the accuracy with which the various soil parameters required for the design are obtained. The accuracy of the soil parameters depend upon the accuracy with which the representative soil samples are obtained from the field. We require mainly two types of samples. They are,

1. Disturbed samples,
2. Undisturbed samples.

Disturbed samples

Disturbed samples are representative samples which contain all the constituents in their proper proportions, but the structure of soil is not the same as in the *in-situ* condition. These soils are sufficient for identification and classification. The various laboratory tests that can be conducted on such soil samples are:

1. Mechanical Analysis.
2. Atterberg limits.
3. Specific gravity.
4. Chemical analysis.

Undisturbed samples

Undisturbed samples are those that represent the *in-situ* condition of the soil in all respects, such as structural arrangement of the particles, water content, density and stress conditions. Undisturbed samples are required for carrying out one or more of the following tests in a laboratory:

1. Shear strength.
2. Consolidation.
3. *In-situ* density and water content.
4. Permeability.

As it has already been explained elsewhere, it is not possible to obtain undisturbed samples of the pure form of coarse grained soils. We are here concerned with the undisturbed samples of fine grained soils or coarse grained soils mixed with cohesive soils. Experience indicates that it is practically impossible to get a truly undisturbed sample from bore hole sampling. The conditions that contribute for the disturbance of samples and unreliable test results are:

1. Distortion of the samples during pushing/driving of sampling tubes into the natural strata.
2. Relief of *in-situ* pressure leading to surface cracks when samples are extracted from sampling tubes for laboratory tests. This is particularly applicable to overconsolidated clay soils.
3. Disturbance caused to the samples during extraction from sampling tubes.
4. Disturbance to the samples during handling and transporting from the site to the laboratory.
5. Evaporation of moisture from the sample due to improper sealing.
6. Carelessness during sampling and testing.

Disturbed Samples

Auger samples may be used to identify soil strata and for field classification tests, but not useful for laboratory tests. The cuttings or choppings from wash borings are of little value except for indicating changes in stratification to the boring supervisor. The material brought up with the drilling mud is contaminated and usually unsuitable even for identification.

For proper identification and classification of a soil, representative samples are required at frequent intervals along the bore hole. Representative samples can usually be obtained by driving or pushing into the strata in a bore hole an open-ended sampling spoon or otherwise called a split spoon sampler which is used for conducting standard penetration tests (Fig. 3.7). It is made up of a driving shoe and a barrel. The barrel is split longitudinally into two halves with a coupling at the upper end for connection to the drill rods. The dimensions of the split spoon is given in the figure. In a test the sampler is driven into the soil a measured distance. After a sample is taken, the cutting shoe and the coupling are unscrewed and the two halves of the barrel separated to expose the material. Experience indicates that samples recovered by this device are likely to be highly

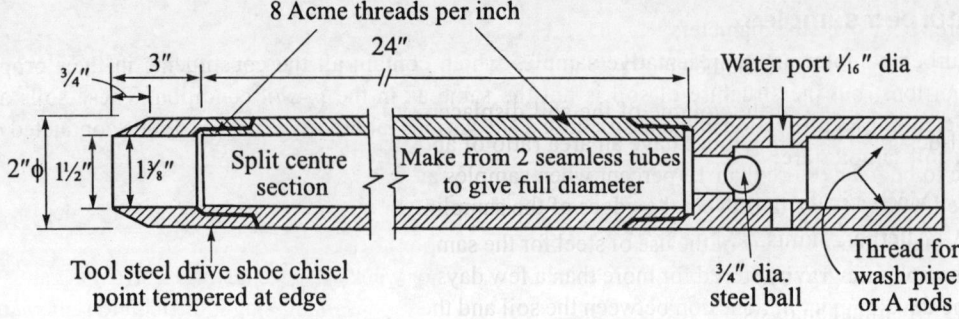

Fig. 3.7 Split barrel sampler for standard penetration test

disturbed and as such can only be used as disturbed samples. The samples so obtained are stored in plastic jars or polythene bags, referenced and sent to the laboratory for testing. If spoon samples are to be transported to the laboratory without examination in the field, the barrel often cored out to hold a cylindrical thin-walled tube known as a *liner.* After a sample has been obtained, the liner and the sample it contains are removed from the spoon and the ends are sealed with caps or with metal discs and wax. Samples of cohesionless soils below water table cannot be retained in conventional sampling spoons without the addition of a *spring core catcher.*

Undisturbed Samples

Many types of samplers are in use for extracting the so-called undisturbed samples. Only two types of samplers are described here. They are,

1. Open drive sampler,
2. Piston sampler.

Open Drive Sampler

The wall thickness of the open drive sampler used for sampling may be thin or thick according to the soil conditions met in the field. The samplers are made of seamless steel pipes. Thin-walled tube sampler is sometimes called as shelby tube sampler (Fig. 3.8). which consists of a thin wall metal tube connected to a sampler head. The sampler head contains a ball check valve and ports which permit the easy escape of water or air from the sample tube as the sample enters it. The thin wall tube, which is normally formed from 1/16 to 1/8 inch metal, is drawn in at the lower end and is reamed so that the inside diameter of the cutting edge is 0.5 to 1.5 percent less than that of the inside diameter of the tube. The exact percentage is governed by the size and wall thickness of the tube. The wall thickness is governed by the *area ratio,* (A_r), which is defined as

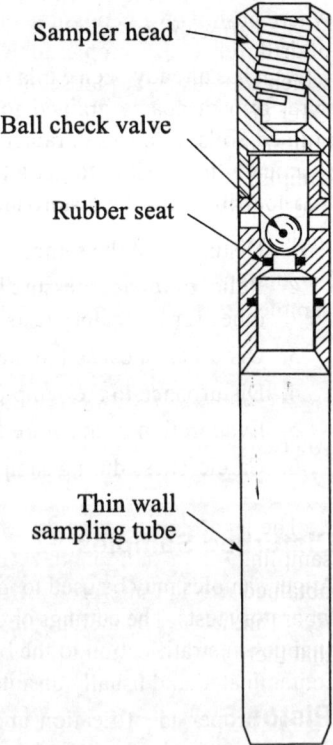

Fig. 3.8 Thin wall shelby tube sampler

$$A_r = \frac{d_o^2 - d_i^2}{d_i^2} \times 100 \text{ percent} \tag{3.1}$$

where, d_i = inside diameter,

d_o = outside diameter.

A_r is a measure of the volume of the soil displacement to the volume of the collected sample. Well-designed sampling tubes have an area ratio of about 10 percent. However, the area ratio may have to be much more than 10 percent when samples are to be taken in very stiff to hard clay soils mixed with kankars to prevent the edges of the sampling tubes getting distorted during sampling.

A major disadvantage of the use of steel for the sampler is the danger of corrosion in cases where the samples have to be stored for more than a few days or weeks prior to testing. Corrosion can lead to the development of adhesion between the soil and the tube wall, making it difficult to remove the sample from the tube without causing disturbance. Also, in certain soils chemical changes, which significantly alter the engineering properties of the soil, may occur. Steel tubes should therefore be coated with lacquer prior to use. The diameters of the sampling tubes may range from 50 to 100 mm and the lengths may range from 450 to 600 mm or more.

Sample extraction

The thin-wall tube sampler is primarily used for sampling in soft to medium stiff cohesive soils. The wall thickness has to be increased if sampling is to be done in very stiff to hard strata of soil. For best results it is better to push the sampler statically into the strata. Samplers are driven into the strata where pushing is not possible or practicable. The procedure of sampling involves attaching a string of drill rods to the sampler tube adopter and lowering the sampler to rest on the bottom of bore hole which was cleaned of loose materials in advance. The sampler is then pushed or driven into the soil. Over driving or pushing should be avoided. After the sampler is pushed to the required depth, the soil at the bottom of the sampler is sheared off by giving a twist to the drill rod at the top. The sampling tube is taken out of the bore hole and the tube is separated from the sampler head. The top and bottom of the sample is either sealed with molten wax or capped to prevent evaporation of moisture. The sampling tubes are suitably referenced for later identification.

Another term that is sometimes used for estimating the degree of disturbance of soil or rock samples is the *recovery ratio*, (R_r) expressed as

$$R_r = \frac{L_a}{L_t} \tag{3.2}$$

where, L_a = actual length of the recovered sample,

L_t = total length of sampling tube driven below the bottom of the bore hole.

The length of the tube driven (L_t) into the soil has to be measured during sampling. Soon after sampling, the actual length L_a of the sample in the tube should be measured. The type of sample obtained may be interpreted from the ratio, R_r. R_r = 1 indicates that the sample is not disturbed; R_r < 1 indicates that the sample is compressed due to friction of the sides of the tube; R_r > 1 indicates that there is expansion of the sample within the tube.

Piston Sampler (After Osterberg)

To improve the quality of samples and to increase the recovery of soft or slightly cohesive soils, a *piston sampler* is normally used. Such a sampler consists of a thin walled tube fitted with a piston that closes the end of the sampling tube until the apparatus is lowered to the bottom of the bore hole [Fig. 3.9 (a)]. The sampling tube is pushed into the soil hydraulically by keeping the piston stationary [Fig. 3.9 (b)]. The presence of the piston prevents the soft soils from squeezing rapidly into the tube and thus eliminates most of the distortion of the sample. The piston also helps to increase the length of sample that can be recovered by creating a slight vacuum that tends to retain the sample if the top of the column of soil begins to separate from the piston. During the withdrawal of the sampler, the

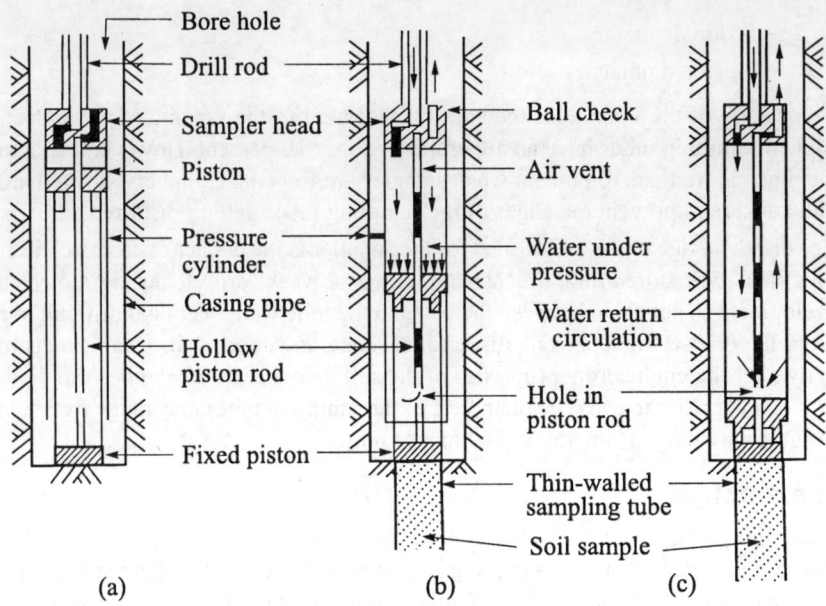

Fig. 3.9 Osterberg piston sampler: (a) Sampler is set in drilled hole, (b) sample tube is pushed hydraulically into the soil, (c) pressure is released through hole in piston rod

piston also prevents water pressure from acting on the top of the sample and thus increases the chances of recovery. The design of piston samplers has been refined to the extent that it is sometimes possible to take undisturbed samples of sand from below the water table. However, piston sampling is relatively a costly procedure and may be adopted only where its use is justified.

General Remarks on Undisturbed Samples in Soil

It is practically impossible to get undisturbed samples of cohesionless soils for strength testing purpose. If the soil lies above water table, it can be extracted in open drive sampling tubes because of the apparent cohesion that exists between particles under partially saturated state. The primary use of such undisturbed samples is to obtain unit weight and water content only. Fairly, undisturbed cohesionless samples below water table can be obtained by the use of piston sampler. Some attempts have been made to recover cohesionless undisturbed samples by freezing a zone around the sample (but not the sample as such) and then extracting the sample along with the freezed portion. Such procedures are rarely adopted. Any attempt to transfer a cohesionless sample from a tube to a testing machine for strength determination is not likely to meet with much success. A sample rebuilt in the laboratory to the *in-situ* unit weight is lacking in both the cementation and structural arrangement of particles.

Since it is almost impossible to get undisturbed samples from cohesionless deposits, density, strength and compressibility estimates are usually made from penetration tests or *in-situ* measurements. Permeability may be estimated or obtained from field tests.

Cohesive soils

The limitations of obtaining truly undisturbed soil samples from cohesive soil deposits have already been discussed. Disturbance to the samples is normally very high if the clay strata in the field happens to be very soft to soft consistency. In such soils, the shear strength in the field can be determined by the field vane tests. Whereas, there is no clearcut procedure for determining the shear strength and compressibility characteristics of very stiff to hard clays. Each consultant has to develop his own procedure for such soils according to local conditions. One of the attempts that is normally made is to relate the strength and compressibility characteristics of clay soil to penetration or pressuremeter test results.

Samples from Test Pits

Excavated test pits are one of the oldest methods of soil exploration since they permit a detailed visual examination of the subsurface material *in-situ* condition. Disturbed and practically undisturbed samples can be obtained from the pits. The depths of pits depend upon the soil condition and the position of the water table. Normally, the depths are limited to a maximum of 3 m or up to the water table level whichever is earlier. Since the undisturbed samples obtained above the water table are not fully saturated, the test results obtained from such samples have to be analysed with caution.

3.4 ROCK CORE SAMPLING

Rock coring is the process in which a sampler consisting of a tube (core barrel) with a cutting bit at its lower end cuts an annular hole in a rock mass, thereby creating a cylinder or core of rock which is recovered in the core barrel. Rock cores are normally obtained by rotary drilling.

The primary purpose of core drilling is to obtain intact samples. The behaviour of a rock mass is affected more by the presence of fractures in the rock. The size and spacing of fractures, the degree of weathering of fractures, and the presence of soil within the fractures are critical items. Figure 3.10

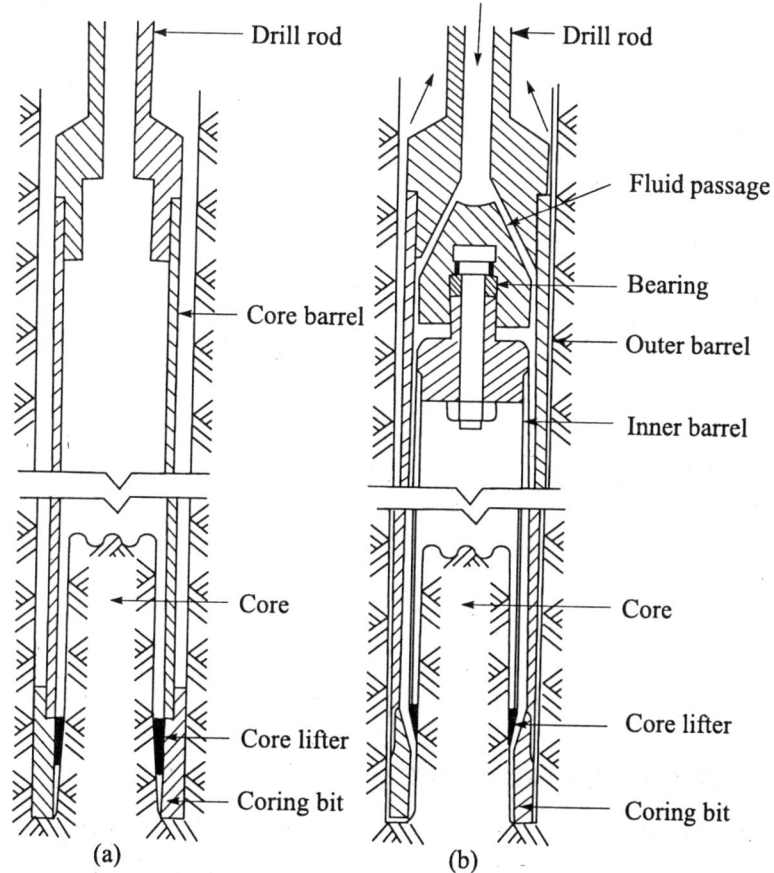

Fig. 3.10 Schematic diagram of core barrels: (a) Single tube, (b) double tube

gives a schematic diagram of core barrels with coring bits at the bottom. As discussed earlier, the cutting element may consist of diamonds, tungsten carbide inserts or chilled shot. The core barrel may consist of a single tube or a double tube. Samples taken in a single tube barrel are likely to experience considerable disturbance due to torsion, swelling and contamination with the drilling fluid, but these disadvantages are not there if the coring apparatus to be in hard, intact, rocky strata. However, if double tube barrel is used, the core is protected by the circulating fluid. Most core barrels are capable of retaining cores up to a length of 2 m. Single barrel is used in Calyx drilling. Standard rock cores range from about 1¼ inches to nearly 6 inches in diameter. The more common sizes are given in Table 3.1.

Table 3.1 Standard sizes of core barrels, drill rods, and compatible casing
(Peck *et al* 1974)

Core barrel			Drill rod		Casing		
Symbol	Hole dia (in)	Core dia (in)	Symbol	Outside dia (in)	Symbol	Outside dia (in)	Inside dia (in)
EWX, EWM	1½	$1^3/_{16}$	E	$1^5/_{16}$	–	–	–
AWX, AWM	$1^{15}/_{16}$	$1^3/_{16}$	A	$1^5/_8$	EX	$1^{13}/_{16}$	$1^1/_2$
BWX, BWM	$2^3/_8$	$1^5/_8$	B	$1^7/_8$	AX	$2^1/_4$	$1^{29}/_{32}$
NWX, NWM	3	$2^1/_8$	N	$2^3/_8$	BX	$2^7/_8$	$2^3/_8$
$2^3/_4 \times 3^7/_8$	$3^7/_8$	$2^{11}/_{16}$	–	–	NX	$3^1/_2$	3

Note: Symbol X indicates single barrel, M indicates double barrel.

The *recovery ratio* (R_r), defined as the percentage ratio between the length of the core recovered and the length of the core drilled on a given run, is related to the quality of rock encountered in boring, but it is also influenced by the drilling technique and the type and size of core barrel used. Generally, the use of a double tube barrel results in higher recovery ratios than can be obtained with single tube barrels. A better estimate of *in-situ* rock quality is obtained by a modified core recovery ratio known as the *Rock quality designation* (RQD) which is expressed as

$$RQD = \frac{\bar{L}_a}{L_t} \qquad (3.3)$$

where $\bar{L}_a$ = total length of intact hard and sound pieces of core of length greater than 100 mm arranged in its proper position,

L_t = total length of drilling.

Breaks obviously caused by drilling are ignored. The diameter of the core should preferably not less than $2^1/_8$ inches. Table 3.2 gives the rock quality description as related to RQD.

3.5 STANDARD PENETRATION TEST

Description of Test

The most commonly used *in-situ* test in a bore hole is the Standard Penetration Test (SPT). The test is made by making use of a split spoon sampler shown in Fig. 3.7. The method of carrying out this test is as follows:

Table 3.2 Relation of RQD and *in-situ* rock quality
(Peck *et al* 1974)

RQD	Rock quality
90–100	Excellent
75–90	Good
50–75	Fair
25–50	Poor
0–25	Very poor

1. The split spoon sampler, connected to a string of drill rods, is lowered into the bottom of bore hole cleaned of all loose materials in advance.

2. The sampler is driven into the soil strata up to a maximum depth of 450 mm by making use of 63.5 kg weight (called also as a monkey) falling freely from a height of 760 mm on to an anvil fixed on the top of drill rod. The weight is guided to fall along a guide rod. The weight is raised and allowed to fall by means of a manila rope, one end tied to the monkey and the other end passing over a pulley on to a hand operated winch or a motor driven cathead.

3. The number of blows required to penetrate each of the successive 150 mm depths is counted to produce a total penetration of 450 mm.

4. To avoid seating errors, the blows required for the first 150 mm of penetration are not taken into account; those required to increase the penetration from 150 mm to 450 mm constitute the *N*-value.

As per some codes of practice if the *N*-value exceeds 100, it is termed as refusal, and the test is stopped even if the total penetration falls short of the last 300 mm depth of penetration. Standardisation of refusal at 100 blows allows all the drilling organisations to standardise costs so that higher blows if required may be eliminated to prevent the excessive wear and tear of the equipment. The SPT is conducted normally at 1 to 2 m intervals. The intervals may be increased at greater depths if necessary.

The Validity of SPT

The validity of the SPT has been the subject of study and research by many authors since this test was introduced by Terzaghi in the year 1927. The basic conclusion is that the test results are difficult to reproduce. Some of the factors that affect the reproducibility are

1. Variation in the height of free fall of drop weight during the test.
2. Interference in the free fall of drop weight by the guide rod which can be out of plumb during the test.
3. Diameter and condition of the drum of the hand operated winch or cathead (rusty, clean, etc.).
4. The number of turns of rope around cathead or drum of the hand operated winch.
5. The actual condition of the manila rope used for the test, whether new or old, etc.
6. Use of badly damaged drive shoe,
7. Improper seating of the sampler on the bottom of the hole.
8. Effect of isolated stones met during driving.
9. Effect of overburden pressure.
10. Carelessness in conducting the test.
11. Length and diameter of drill rod.

Possibly there may be many more factors which affect the reproducibility and which cannot be accounted for. The number of turns around the drum affect the frictional resistance offered for the free fall of the weight. It appears that a nominal two turns of rope around the drum is the optimum and is widely used. A new rope reduces the frictional resistance whereas an old one increases the resistance.

Of late some research centres and organisations have developed automatic free fall hammer which eliminates the first five irritants mentioned above. Studies conducted by Gibbs and Holtz (1957) and Brown (1977), indicate that stiffness and weight of the drill rod does not affect the N value. Studies of others indicate that if SPT is conducted at depths greater than about 60 m, the N value obtained will be greater than the actual value. This discrepancy between the actual and the apparent has been attributed to the weight and flexibility of drill rods.

Standardisation of Test

If an automatic free fall hammer is not used which is normally the case with most of the organisations, then the first five factors mentioned earlier affect the driving energy. The energy imparted to the sampler depends upon the velocity of the drop weight at the impact level which can be compared to the theoretical velocity or energy. For a standardised fall of height of 760 mm for the standard penetration test, a large discrepancy in the impact velocity exists between one or two and three turns of the rope around the cathead.

The theoretical kinetic energy, E_t, of a falling body is given by

$$E_t = \frac{1}{2} m v_t^2 = \frac{1}{2} \frac{W}{g} v_t^2 \qquad (3.4)$$

$$v_t = \sqrt{2gh} \qquad (3.5)$$

where, m = mass of the falling weight,
$\quad\quad\quad v_t$ = the velocity at impact,
$\quad\quad\quad g$ = acceleration due to gravity,
$\quad\quad\quad h$ = height of fall.

If E_m is the measured energy and v_m actual velocity at impact, the ratio of the measured kinetic energy to the theoretical may be defined as the energy ratio (R_e) and is given by

$$R_e = \frac{E_m}{E_t} = \frac{v_m^2}{v_t^2} \qquad (3.6)$$

The studies conducted by Kovacs *et al* (1982) indicate that for a fall of 760 mm, the energy delivered to the sampler decreases with the increase in the number of turns of rope around the cathead. The energy ratio, R_e for 1, 2 and 3 turns are given in Table 3.3.

Table 3.3 Energy ratio for a given number of turns of rope around cathead

No. of turns	Impact velocity in cm/sec	R_e %
1	318	67
2	312	65
3	254	43

Generally speaking, the use of one or two turns of rope around the cathead results in about 66 percent of the kinetic energy delivered as compared to about 40 percent for three turns. It is clear from the study of Kovacs (1977), the number of turns around cathead should be limited to two only. It should also be remembered here that these are only approximate values and the other factors mentioned earlier also affect the energy transfer. By merely changing some of the SPT conditions, wide variations in delivered energy could occur.

It may be noted here that large values of energy ratios, R_e, decreases the blow count N nearly linearly that is, if $N = 32$ for $R_e = 50\%$, the blow count for $R_e = 80\%$ is

$$N = \frac{50}{80} \times 32 = 20$$

or in other words, we may write a general equation

$$R_{e1} \times N_1 = R_{e2} \times N_2 = \text{constant} \tag{3.7}$$

or

$$N_1 = \frac{R_{e2}}{R_{e1}} \times N_2 \tag{3.8}$$

If Re_2 and N_2 are the known energy ratio and the blow count respectively of drill rig 2, the blow count N_1 for the drill rig 1, can be found out if its energy ratio R_{e1}, is known. Equation (3.8) helps to standardise the energy imparted to a sampler. Kovac *et al* (1977) proposed that the impact velocity or energy standard be a criteria for the standard penetration test. The selected velocity or energy should be that which least disturbs the existing correlations between N values and engineering properties, performance of foundations, etc.

Standard penetration tests is a very popular *in-situ* test all over the world. There are variations from country to country in the equipments used and the methodology adopted for the test. This has lead to a wide scatter of energy ratios R_e and the resulting blow counts N, therefore, vary for the same type of strata. It is the opinion of some authors that the drill system dependent on R_e be referenced to a standard energy ratio R_{es}. In this way a drill rig with, say $R_e = 50$ percent would, on adjustment to the standard R_{es}, compute the same N count as from a drill rig with $R_e = 70$ percent. The standard energy ratios suggested by different authors are as given in Table 3.4 (Ref. Bowles, 1988).

Table 3.4 Energy ratio R_{es}

Author	Average R_{es} %
Schemertmann	30–55
Seed *et al* (1985)	60
Skempton (1986)	60
Rigs (1986)	70–80

Bowles (1988) propose to use $R_{es} = 55$ percent as he found this was close to the energy ratio R_e obtained in North American practice, since most of the existing correlations between N value and engineering properties have emanated from America. $R_{es} = 55$ percent seems to be reasonable. If R_e in other countries is different from R_{es}, the blow counts N obtained with energy ratio R_e should be converted to blow count for the standard energy ratio adopted.

Example 3.1

Standard penetration test was carried out at a place in India by using manually operated winch for conducting the test. If the N value at a particular depth is 30, what is the blow count for the standard energy ratio $R_{es} = 70\%$.

Here we have to know R_e for the Indian conditions. Since no data is available, R_e is assumed as equal to 50%.

Solution

Since,
$$\overline{N}_s \times R_{es} = N \times R_e$$

we have,
$$\overline{N}_s = \frac{R_e}{R_{es}} \times N = \frac{50}{70} \times 30 = 21.4 \text{ or } 21$$

In deposits containing gravels, cobbles, etc. the results of SPT are quite unrealiable. In sensitive clays the SPT may lead to a gross mis-conception of the consistency. If the water level in a bore hole is allowed to drop below the ground water level, as may easily occur for instance when the drill rods are removed rapidly, an upward hydraulic gradient is created in the sand beneath the drill hole. Consequently, the sand may become quick and its relative density may be greatly reduced. The N value will accordingly be much lower than in the actual case under the undisturbed condition. Care is required to see that the water level in the bore hole is always maintained at or slightly above the normal water table level.

3.6 CORRECTIONS TO OBSERVED SPT VALUES IN COHESIONLESS SOILS

Two types of corrections are normally applied to the observed N values in cohesionless soil. They are,

1. Correction due to dilatancy.
2. Correction due to overburden pressure.

Correction Due to Dilatancy

In saturated fine or silty dense or very dense sand deposits, the N value observed may be greater than the actual value because of the tendency of such materials to dilate during shear under undrained conditions. Terzaghi and peck recommended that if the observed N value is greater than 15, it should be corrected for dilatation effect as

$$N' = 15 + 1/2 \, (N_o - 15) \tag{3.9a}$$

where, N_o = observed SPT value
N' = corrected value for dilatation effect

Bazara (1967), gives the following equation for correcting N

$$N' = 0.6 \, N_o \tag{3.9b}$$

where N_o is greater than 15.

The correction is based on the assumption that critical void ratio occurs at approximately $N = 15$, and in fine grained cohesionless soils, the coefficient of permeability is so low that the excess pore-water pressure developed by the driving impedes the penetration of the split spoon sampler, thus increasing the N value. Since sufficient experimental evidence is not forthcoming to confirm this correction, many designers do not use this correction. However, the corrected (or uncorrected) N value for dilatancy has to be corrected for overburden pressure effect.

Correction Due to Overburden Pressure

The effect of overburden pressure on the N value in cohesionless soils has been recognised for long. Gibbs and Holtz (1957), proposed an equation for correcting N value based on tests carried out in a laboratory. The equation as developed by them (in a modified form) is

$$N = \frac{4\,N'}{0.01\,p_o' + 0.8} = C_N\,N' \tag{3.10}$$

where, N' = SPT value corrected for dilatancy,

$\quad\quad p_o'$ = the effective overburden pressure in kPa,

$\quad\quad C_N$ = correction factor.

In Eq. (3.10) the maximum value of C_N assumed is 2.

Equation by Bazara

Bazara (1967), made an analysis of a large number of SPT values taken in bore holes by others. He proposed the following equations for the effect of overburden pressure.

For $\quad\quad\quad\quad\quad\quad p_o' \le 75$ kPa,

$$N = \frac{4\,N'}{0.04\,p_o' + 1} = C_N\,N' \tag{3.11a}$$

For $\quad\quad\quad\quad\quad\quad p_o' > 75$ kPa,

$$N = \frac{4\,N'}{0.01\,p_o' + 3.25} = C_N\,N' \tag{3.11b}$$

With regards to the use of Eq. (3.11), the following points may be kept in view:

1. For $p_o' \le 75$ kPa, $N > N'$.
2. For $p_o' > 75$ kPa, $N < N'$.
3. When N' indicates a relative density $D_r < 0.5$, no correction is required. C_N may be taken as equal to unity up to $D_r = 0.5$.
4. C_N is limited to 2.

Equation by Peck *et al*

Peck *et al* (1974), proposed the following equation for correcting the N value which is given here in a modified form as

$$N = 0.77\,N'\,\log_{10}\frac{2000}{p_o'} = C_N\,N' \tag{3.12}$$

where, p_o' is in kPa. This equation has the following characteristics:

1. $N \to \infty$ as $p_o' \to 0$.
2. $N \approx N'$ at $p_o' = 100$ kPa.
3. $N < N'$ for $p_o' > 100$ kPa.
4. Equation (3.12) is valied for $p_o' \ge 25$ kPa.

Equation (3.12) modified by the author

To overcome the ambiguity that the Eq. (3.12) is not valid for $p_o' \leq 25$ kPa, the author proposes the following modified form:

For $\qquad\qquad p_o' \leq 25$ kPa,

$$N = \frac{4N'}{2 + 0.034\, p_o'} = C_N N' \qquad\qquad (3.13a)$$

For $\qquad\qquad p_o' > 25$ kPa,

$$N = \frac{4N'}{2.5 + 0.015\, p_o'} = C_N N' \qquad\qquad (3.13b)$$

The maximum value of C_N is 2 when $p_o' = 0$.

Equation by Liao and Whitman (1986)

Liao and Whitman have proposed the following equation

$$N = N' \sqrt{\frac{\overline{p}_o}{p_o'}} = C_N N' \qquad\qquad (3.14)$$

where, $\overline{p}_o$ = standard overburden pressure = 95.75 kPa.

C_N values may be computed by making use of Eqs (3.10) to (3.14) for a comparative study for various values of p_o'.

It will be clear that Eq. (3.10) gives values of C_N greater than 2 up to an overburden pressure $p_o' \approx 125$ kPa which is not the same with the other equations. The present practice is not to use this equation for the correction of N for overburden effect. Some engineers use Eq. (3.11) by limiting the value of C_N to 2. Equation (3.12) is presented in the form of a curve in Fig. 3.11.

Corrections to the Observed SPT Value (USA Practice)

Three types of corrections are normally applied to the observed N values. They are:

1. Hammer efficiency correction.
2. Drillrod, sampler and borehole corrections.
3. Correction due to overburden pressure.

1. Hammer efficiency correction, E_h

Different types of hammers are in use for driving the drill rods. Two types are normally used in USA. They are (Bowles, 1996).

1. Donut with two turns of manila rope on the cathead with a hammer efficiency $E_h = 0.45$.
2. Safety with two turns of manila rope on the cathead with a hammer efficiency as follows:
   Rope-pulley or cathead = 0.7 to 0.8;
   Trip or automatic hammer = 0.8 to 1.0.

2. Drill rod, Sampler and Borehole corrections

Correction factors are used for correcting the effects of length of drill rods, use of split spoon sampler with or without liner, and size of bore holes. The various correction factors are (Bowles, 1996).

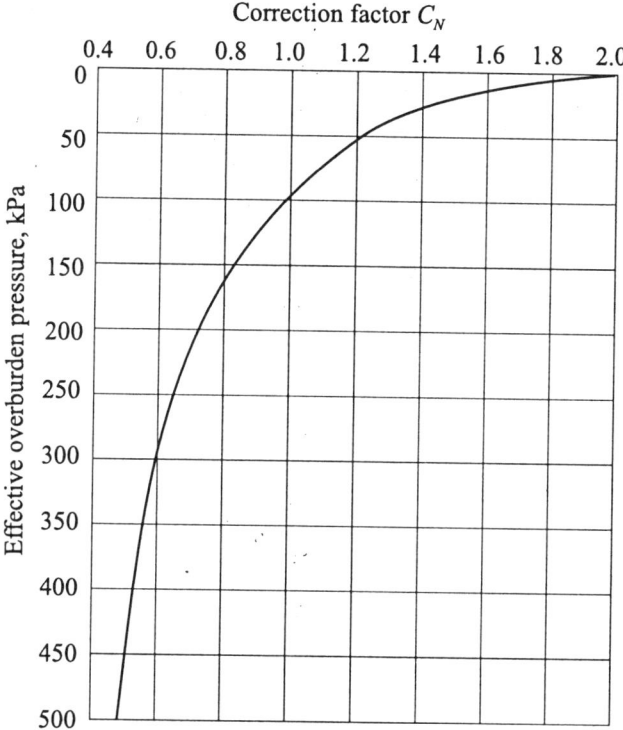

Fig. 3.11 Chart for correction of *N*-values in sand for influence of overburden pressure (After Peck *et al*)

(a) Drill rod length correction factor C_d

Length (m)	Correction factor (C_d)
> 10 m	1.0
4 – 10 m	0.85 – 0.95
< 4.0 m	0.75

(b) Sampler correction factor, C_s
Without linear $C_s = 1.00$
With liner,
Dense sand, clay = 0.80
Loose sand = 0.90

(c) Bore hole diameter correction factor, C_b

Bore hole diameter	Correction factor, C_b
60 – 120 mm	1.0
150 mm	1.05
200 mm	1.15

3. Correction factor for overburden pressure in granular soils, C_N

The C_N as per Liao and Whitman (1986) is

$$C_N = \left(\frac{95.76}{p_o'}\right)^{1/2} \tag{3.15a}$$

where, p_o' = effective overburden pressure in kN/m^2.

There are a number of empirical relations proposed for C_N. However, the most commonly used relationship is the one given by Eq. (3.15a).

N_{cor} may be expressed as

$$N_{cor} = C_N N_o E_h C_d C_s C_b \tag{3.15b}$$

N_{cor} is related to the standard energy ratio used by the designer. N_{cor} may be expressed as N_{70} or N_{60} according to the designer's choice.

In Eq. (3.15b) $C_N N_o$ is the corrected value for overburden pressure only. The value of C_N as per Eq. (3.15a) is applicable for granular soils only, whereas $C_N = 1$ for cohesive soils for all depths.

Note: In this text N or N_{cor} is used for the corrected value.

Example 3.2

The observed standard penetration test value in a deposit of fully submerged sand was 45 at a depth of 6.5 m. The average effective unit weight of the soil is 9.69 kN/m^3. The other data given are: (*a*) hammer efficiency = 0.8, (*b*) drill rod length correction factor = 0.9, and (*c*) borehole correction factor = 1.05. Determine the corrected SPT value for standard energy: (*a*) R_{es} = 60 percent, and (*b*) R_{es} = 70%.

Solution

Per Eq. (3.15b), the equation for N_{60} may be written as

$$N_{60} = C_N N_o E_h C_d C_s C_b$$

where N_o = observed SPT value

C_N = overburden correction

Per Eq. (3.15a), we have

$$C_N = \left(\frac{95.76}{p_o'}\right)^{1/2}$$

where p_o' = effective overburden pressure

$$= 6.5 \times 9.69 = 63 \text{ kN/m}^2$$

Substituting for p_o',

$$C_N = \left(\frac{95.76}{60}\right)^{1/2} = 1.233$$

Substituting the known values, the corrected N_{60} is

$$N_{60} = 1.233 \times 45 \times 0.8 \times 0.9 \times 1.05 = 42$$

For 70 percent standard energy

$$N_{70} = 42 \times \frac{0.6}{0.7} = 36$$

3.7 SPT VALUES RELATED TO RELATIVE DENSITY OF COHESIONLESS SOILS

Although the SPT is not considered as a refined and completely reliable method of investigation, the N_{cor} values give useful information with regard to consistency of cohesive soils and relative density of cohesionless soils. The correlation between N_{cor} values and relative density of granular soils suggested by Peck *et al*, (1974) is given in Table 3.5 (a).

Before using Table 3.5 (a) the observed N value has to be corrected for standard energy and overburden pressure. The correlations given in Table 3.5 (a) are just a guide and may vary according to the fineness of the sand.

Table 3.5 (a) N_{cor} and ϕ related to relative density

N_{cor}	*Compactness*	*Relative density, D_r (%)*	$\phi°$
0–4	Very loose	0–15	< 28
4–10	Loose	15–35	28–30
10–30	Medium	35–65	30–36
30–50	Dense	65–85	36–41
> 50	Very dense	> 85	> 41

Meyerhof (1956), suggested the following approximate equations for computing the angle of friction ϕ from the known value of D_r.

For granular soil with fine sand and more than 5 percent silt,

$$\phi° = 25 + 0.15D_r \tag{3.16a}$$

For granular soils with fine sand and less than 5 percent silt,

$$\phi° = 30 + 0.15D_r \tag{3.16b}$$

where D_r is expressed in percent.

3.8 SPT VALUES RELATED TO CONSISTENCY OF CLAY SOIL

Peck *et al*, (1974) have given for saturated cohesive soils, correlations between N_{cor} value and consistency. This correlation is quite useful but has to be used according to the soil conditions met in the field. Table 3.5 (b) gives the correlations.

The N_{cor} value to be used in Table 3.5 (b) is the blow count corrected for standard energy ratio R_{es}. The present practice is to relate q_u with N_{cor} as follows,

$$q_u = \bar{k} N_{cor} \text{ kPa} \tag{3.16c}$$

where q_u is the unconfined compressive strength.

or

$$\bar{k} = \frac{q_u}{N_{cor}} \text{ kPa} \tag{3.16d}$$

where, $\bar{k}$ is the proportionality factor. A value of $\bar{k} = 12$ has been recommended by Bowles (1996).

Table 3.5 (b) Relation between N_{cor} and q_u

Consistency	N_{cor}	$q_u' kPa$
Very soft	0–2	< 25
Soft	2–4	25–50
Medium	4–8	50–100
Stiff	8–15	100–200
Very stiff	15–30	200–400
Hard	> 30	> 400

Example 3.3

For the corrected N values in Ex. 3.2, determine the (a) relative density, and (b) the angle of friction. Assume the percent of fines in the deposit is less than 5%.

Solution

Per Table 3.5 (a) the relative density and ϕ are

For $\qquad N_{60} = 42, D_r = 77\%, \phi = 39°$
For $\qquad N_{70} = 36, D_r = 71\%, \phi = 37.5°$

Per Eq. (3.16b)

For $\qquad D_r = 77\%, \phi = 30 + 0.15 \times 77 = 41.5°$
For $\qquad D_r = 71\%, \phi = 30 + 0.15 \times 71 = 40.7°$

Example 3.4

For the corrected values of N given in Ex. 3.3, determine the unconfined compressive strength q_u in a clay deposit.

Solution

(a) From Table 3.5 (b)

For $\qquad N_{60} = 42$
For $\qquad N_{70} = 36$ $\qquad$ $q_u > 400$ kPa – The soil is of a hard consistency.

(b) Per Eq. (3.16c) $\quad q_u = \overline{k} N_{cor}$, where $\overline{k} = 12$ (Bowles, 1996)

For $\qquad N_{60} = 42, q_u = 12 \times 42 = 504$ kPa
For $\qquad N_{70} = 36, q_u = 12 \times 36 = 432$ kPa

Example 3.5

Refer to Ex. 3.2. Determine the corrected SPT value for $R_{es} = 100$ percent, and the corresponding values of D_r and ϕ. Assume the percent of fine sand in the deposit is less than 5%.

Solution

From Ex. 3.2, $\qquad N_{60} = 42$

Hence $\qquad N_{100} = 2 \times \dfrac{0.6}{1.0} \approx 25$

From Table 3.3, $\qquad \phi = 34.5°$ and $D_r = 57.5\%$
From Eq. (3.16b) for

$\qquad D_r = 57.5\%, \phi = 30 + 0.15 \times 57.5 = 38.6°$

3.9 STATIC CONE PENETRATION TEST (CPT)

The static cone penetration test normally called the *Dutch cone penetration test* (CPT). It has gained acceptance rapidly in many countries. The method was introduced nearly 50 years ago. One of the greatest values of the CPT consists of its function as a scale model pile test. Empirical correlations established over many years permit the calculation of pile bearing capacity directly from the CPT results without the use of conventional soil parameters.

The CPT has proved valuable for soil profiling as the soil type can be identified from the combined measurement of end resistance of cone and side friction on a jacket. The test lends itself to the derivation of normal soil properties such as density, friction angle and cohesion. Various theories have been developed for foundation design.

The popularity of the CPT can be attributed to the following three important factors:

1. General introduction of the electric penetrometer providing more precise measurements, and improvements in the equipment allowing deeper penetration.
2. The need for the penetrometer testing *in-situ* technique in offshore foundation investigations in view of the difficulties in achieving adequate sample quality in marine environment.
3. The addition of other simultaneous measurements to the standard friction penetrometer such as pore pressure and soil temperature.

The Penetrometer

There are a variety of shapes and sizes of penetrometers being used. The one that is standard in most countries is the cone with an apex angle of 60° and a base area of 10 cm². The sleeve (jacket) has become a standard item on the penetrometer for most applications. On the 10 cm² cone penetrometer the friction sleeve should have an area of 150 cm² as per standard practice. The ratio of side friction and bearing resistance, the *friction ratio,* enables identification of the soil type (Schmertmann 1975) and provides useful information in particular when no bore hole data are available. Even when borings are made, the friction ratio supplies a check on the accuracy of the boring logs.

Two types of penetrometers are used which are based on the method used for measuring cone resistance and friction. They are,

1. The mechanical type,
2. The electrical type.

Mechanical type

The Begemann Friction Cone Mechanical type penetrometer is shown in Fig. 3.12. It consists of a 60° cone with a base diameter of 35.6 mm (sectional area 10 cm²). A sounding rod is screwed to the base. Additional rods of one metre length each are used. These rods are screwed or attached together to bear against each other. The sounding rods move inside mantle tubes. The inside diameter of the mantle tube is just sufficient for the sounding rods to move freely whereas the outside diameter is equal to or less than the base diameter of the cone. All the dimensions in Fig. 3.12 are in mm.

Jacking system

The rigs used for pushing down the penetrometer consist basically of a hydraulic system. The thrust capacity for cone testing on land varies from 20 to 30 kN for hand operated rigs and 100 to 200 kN for mechanically operated rigs as shown in Fig. 3.13. Bourden gauges are provided in the driving mechanism for measuring the pressures exerted by the cone and the friction jacket either individually or collectively during the operation. The rigs may be operated either on the ground or mounted on heavy duty trucks. In either case, the rig should take the necessary upthrust. For ground based rigs screw anchors are provided to take up the reaction thrust.

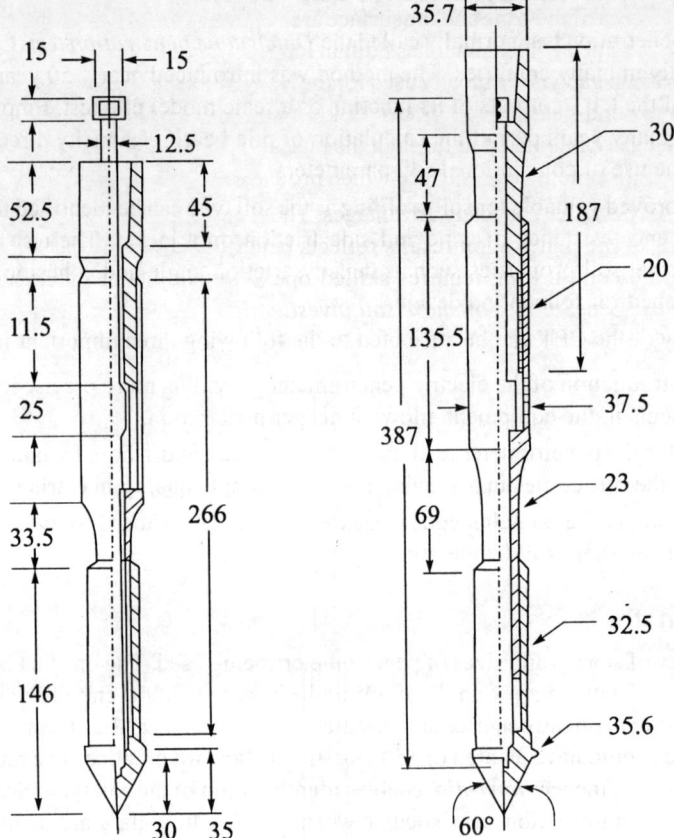

Fig. 3.12 Begemann friction-cone mechanical type penetrometer (All dimensions are in mm)

Operation of Penetrometer

The sequence of operation of the penetrometer shown in Fig. 3.14 is explained as follows:

Position 1: The cone and friction jacket assembly in a collapsed position.

Position 2: The cone is pushed down by the inner sounding rods to a depth a until a collar engages the cone. The pressure gauge records the total force Q_c to the cone. Normally, $a = 40$ mm.

Position 3: The sounding rod is pushed further to a depth b. This pushes the friction jacket and the cone assembly together; the force is Q_t. Normally, $b = 40$ mm.

Position 4: The outside mantle tube is pushed down a distance $a + b$ which brings the cone assembly and the friction jacket to position 1. The total movement = $a + b = 80$ mm.

The process of operation illustrated above is continued until the proposed depth is reached. The cone is pushed at a standard rate of 20 mm per second. The mechanical penetrometer has its advantage as it is simple to operate and the cost of maintenance is low. The quality of the work depends on the skill of the operator. The depth of CPT is measured by recording the length of the sounding rods that have been pushed into the ground.

The electric penetrometer

The electric penetrometer is an improvement over the mechanical one. Mechanical penetrometers operate incrementally whereas the electric penetrometer is advanced continuously.

Figure 3.15 shows an electric-static penetrometer with the friction sleeve just above the base of the cone. The sectional area of the cone and the surface area of the friction jacket remain the same as those of a mechanical type. The penetrometer has built in load cells that record separately the cone bearing and side friction. Strain gauges are mostly used for the load cells. The load cells have a normal capacity of 50 to 100 kN for end bearing and 7.5 to 15 kN for side friction, depending on the soils to be penetrated. An electric cable inserted through the push rods (mantle tube) connect the penetrometer with the recording equipment at the surface which produces graphs of resistance *versus* depth.

The electric penetrometer has many advantages. The repeatability of the cone test is very good. A continuous record of the penetration results reflects better the nature of the soil layers penetrated. However, electronic cone testing requires skilled operators and better maintenance. The electric penetrometer is indispensable for *offshore soil investigation.*

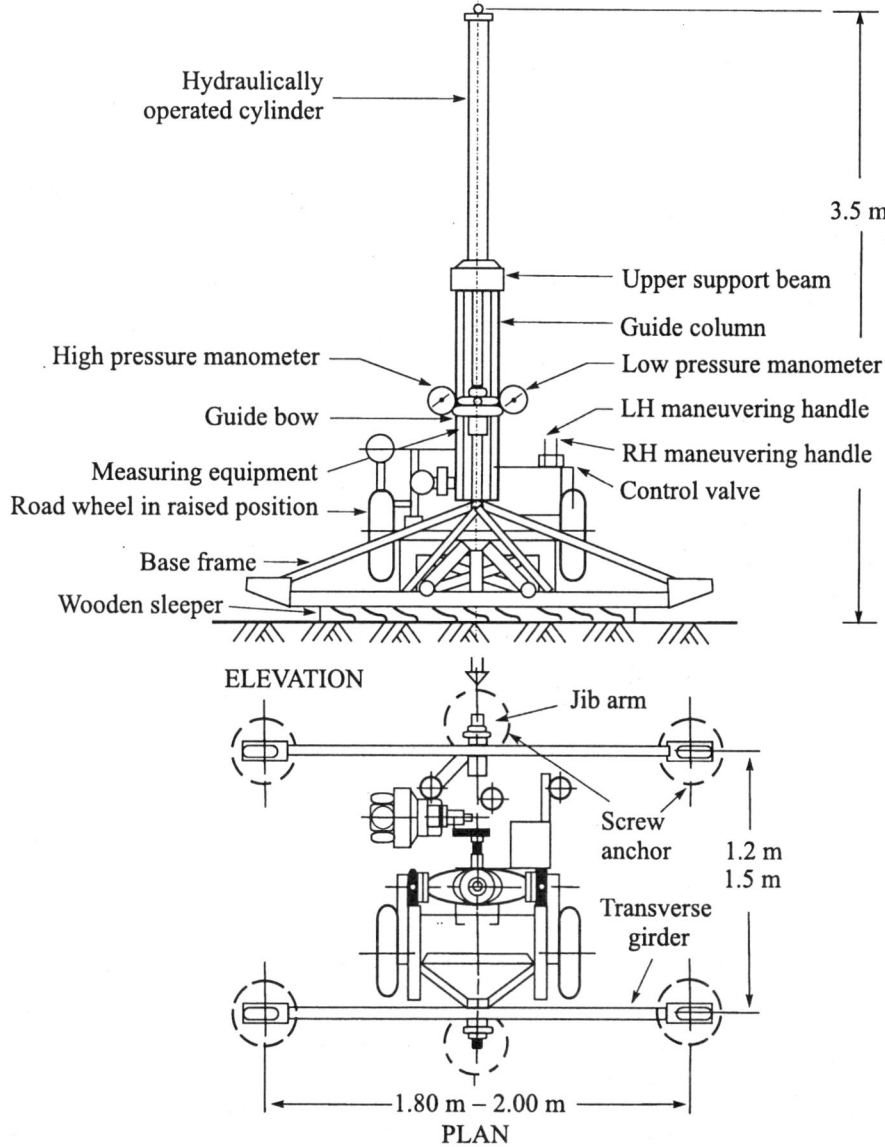

Fig. 3.13 Static cone penetration testing equipment

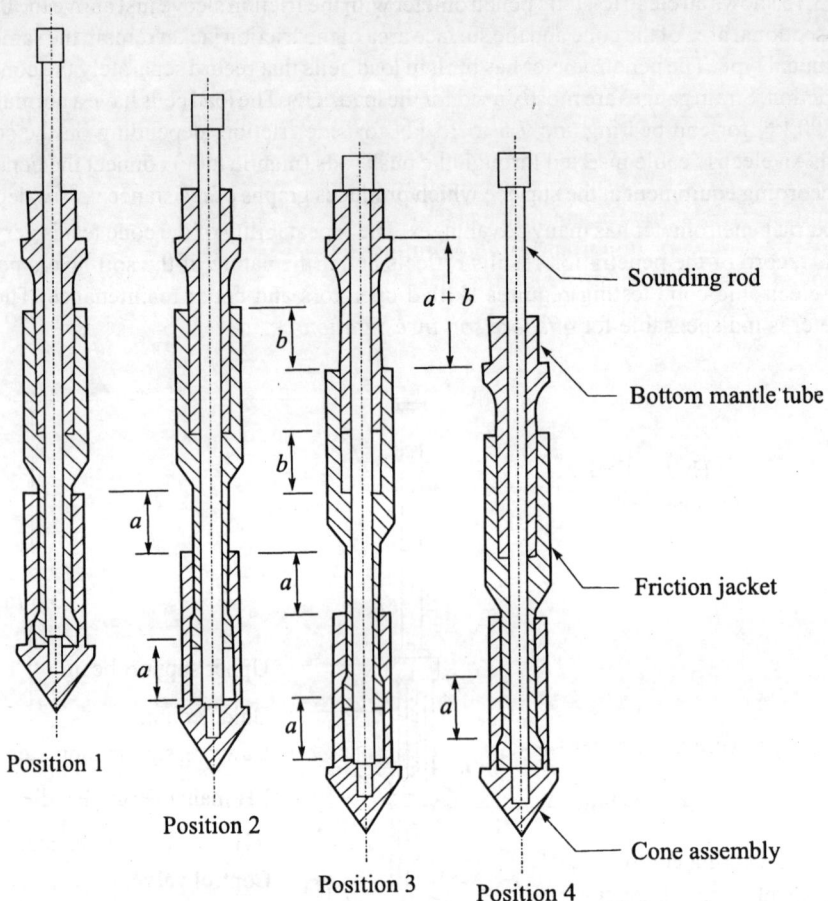

Fig. 3.14 Four positions of the sounding apparatus with friction jacket

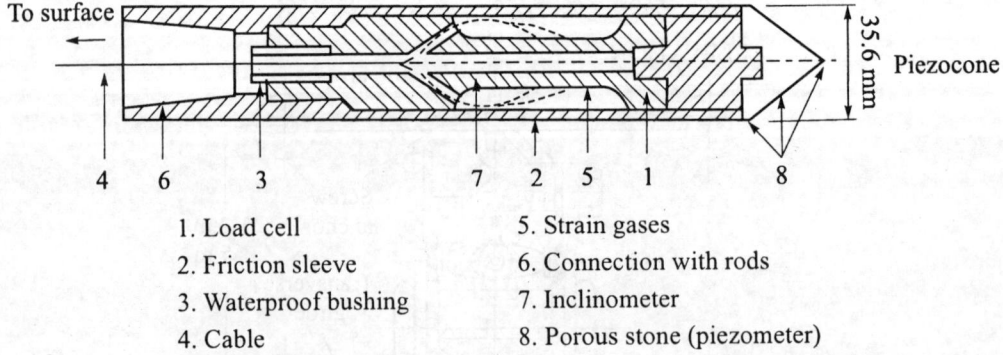

1. Load cell
2. Friction sleeve
3. Waterproof bushing
4. Cable

5. Strain gases
6. Connection with rods
7. Inclinometer
8. Porous stone (piezometer)

Fig. 3.15 An-electric-static cone penetrometer

Operation of penetrometer

The electric penetrometer is pushed continuously at a standard rate of 20 mm per second. A continuous record of the bearing resistance q_c and frictional resistance f_s against depth is produced in the form of a graph at the surface in the recording unit.

Piezocone

A piezometer element included in the cone penetrometer is called a *piezocone* (Fig. 3.16). There is now a growing use of the piezocone for measuring pore pressures at the tips of the cone. The porous element is mounted normally midway along the cone tip allowing pore water to enter the tip. An electric pressure transducer measures the pore pressure during the operation of the CPT. The pore pressure record provides a much more sensitive means to detect thin soil layers. This could be very important in determining consolidation rates in a clay within the sand seams.

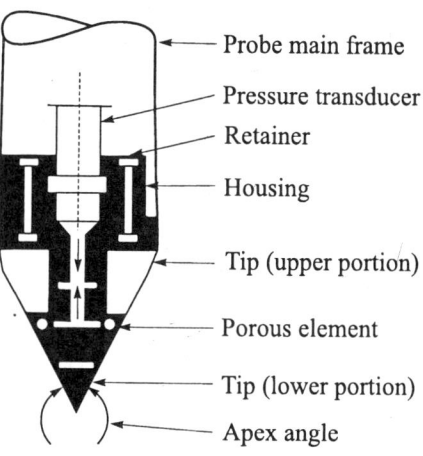

Probe main frame

Pressure transducer

Retainer

Housing

Tip (upper portion)

Porous element

Tip (lower portion)

Apex angle

Temperature Cone

The temperature of a soil is required at certain localities to provide information about environmental changes. The temperature gradient with depth may offer possibilities to calculate the heat conductivity of the soil. Measurement of the temperature during CPT is possible by incorporating a temperature sensor in the electric penetrometer. Temperature measurements have been made in permafrost, under blast furnaces, beneath undercooled tanks, along marine pipe lines, etc.

Fig. 3.16 Details of 60°/10 cm² piezocone

Effect of Rate of Penetration

Several studies have been made to determine the effect of the rate of penetration on cone bearing and side friction. Although the values tend to decrease for slower rates, the general conclusion is that the influence is insignificant for speeds between 10 and 30 mm per second. The standard rate of penetration has been generally accepted as 20 mm per second.

Cone Resistance q_c and Local Side Friction f_c

Cone penetration resistance q_c is obtained by dividing the total force Q_c acting on the cone by the base area A_c of the cone.

$$q_c = \frac{Q_c}{A_c}$$ (3.17a)

In the same way, the local side friction f_c is

$$f_c = \frac{Q_f}{A_f}$$ (3.17b)

where, $Q_f = Q_t - Q_c =$ force required to push the friction jacket,

$Q_t =$ the total force required to push the cone and friction jacket together in the case of a mechanical penetrometer,

$A_f =$ surface area of the friction jacket.

Friction Ratio, R_f

Friction ratio, R_f is expressed as

$$R_f = \frac{f_c}{q_c}$$ (3.17c)

where f_c and q_c are measured at the same depth. R_f is expressed as a percentage. Friction ratio is an important parameter for classifying soil (Fig. 3.19).

Relationship between q_o, Relative Density D_r and Friction Angle ϕ for Sand

Research carried out by many indicates that a unique relationship between cone resistance, relative density and friction angle valid for all sands does not exist. Robertson and Campanella (1983a) have provided a set of curves (Fig. 3.17) which may be used to estimate D_r based on q_c and effective overburden pressure. These curves are supposed to be applicable for normally consolidated clean sand. Figure 3.18 gives the relationship between q_c and ϕ (Robertson and Campanella, 1983b).

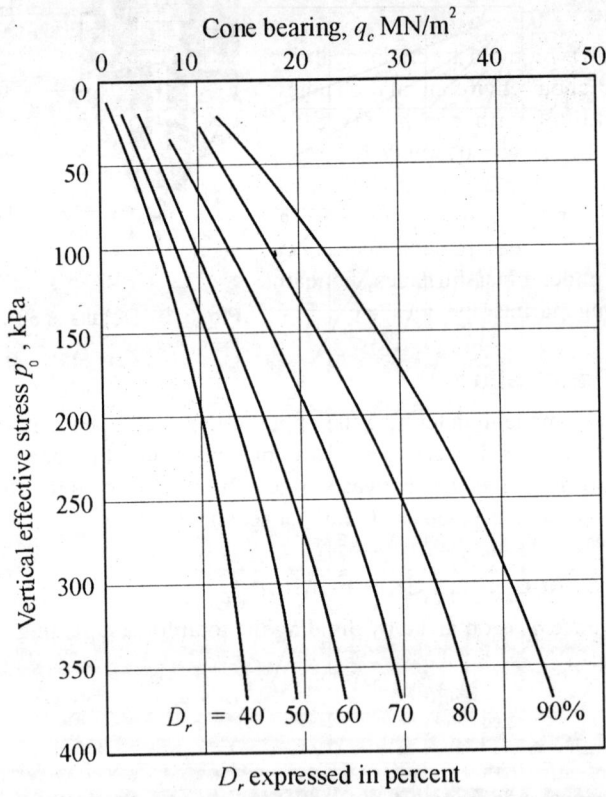

Fig. 3.17 Relationship between relative density D_r and penetration resistance q_c for uncemented quartz sands (Robertson and Campanella, 1983a)

Relationship between q_c and Undrained Shear Strength, c_u of Clay

The cone penetration resistance q_c and c_u may be related as

$$q_c = N_k c_u + p_o \quad \text{or} \quad c_u = \frac{q_c - p_o}{N_k} \tag{3.18}$$

where, N_k = cone factor,

$p_o = \gamma z$ = overburden pressure.

Lune and Kelven (1981), investigated the value of the cone factor N_k for both normally consolidated and overconsolidated clays. The values of N_k as obtained are given below:

Type of clay	Cone factor
Normally consolidated	11 to 19
Overconsolidated	
at shallow depths	15 to 20
at deep depths	12 to 18

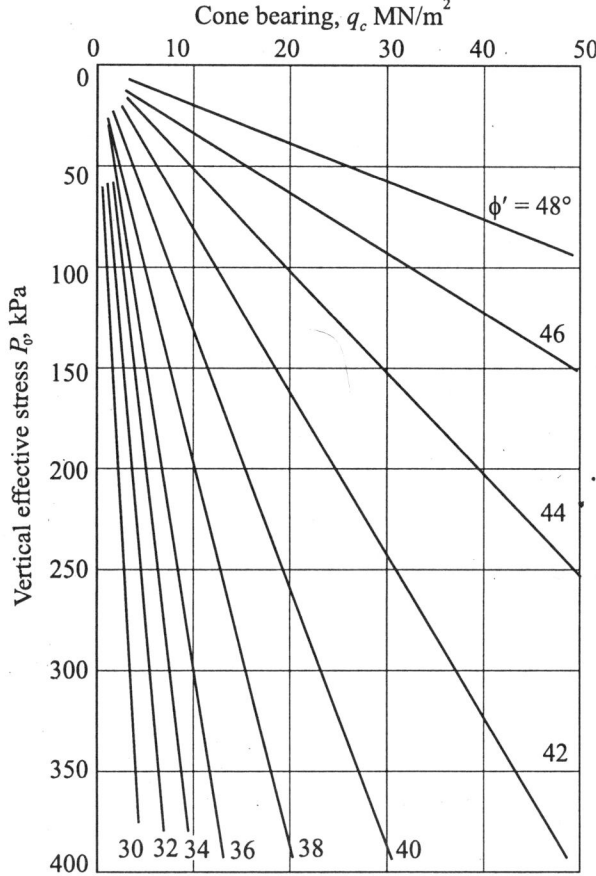

Fig. 3.18 Relationship between cone point resistance q_c and angle of internal friction ϕ for uncemented quartz sands (Robertson and Campanella, 1983b)

Possibly a value of 20 for N_k for both types of clays may be satisfactory. Sanglerat (1972), recommends the same value for all cases where an overburden correction is of negligible value.

Soil Classification

One of the basic uses of CPT is to identify and classify soils. A CPT-Soil Behaviour Type Prediction System has been developed by Douglas and Olsen (1981) using an electric-friction cone penetrometer. The classification is based on the *friction ratio f_c/q_c*. The ratio f_c/q_c varies greatly depending on whether it applies to clays or sands. Their findings have been confirmed by hundreds of tests.

For clay soils, it has been found that the friction ratio decreases with increasing liquidity index I_l. Therefore, the friction ratio is an indicator of the soil type penetrated. It permits approximate identification of soil type though no samples are recovered.

Douglas (1984), presented a simplified classification chart shown in Fig. 3.19. His chart uses cone resistance normalized (q_{cn}) for overburden pressure using the equation

$$q_{cn} = q_c (1 - 1.25 \log p'_o) \qquad (3.19)$$

where, p'_o = effective overburden pressure in tsf, and q_c = cone resistance in tsf,

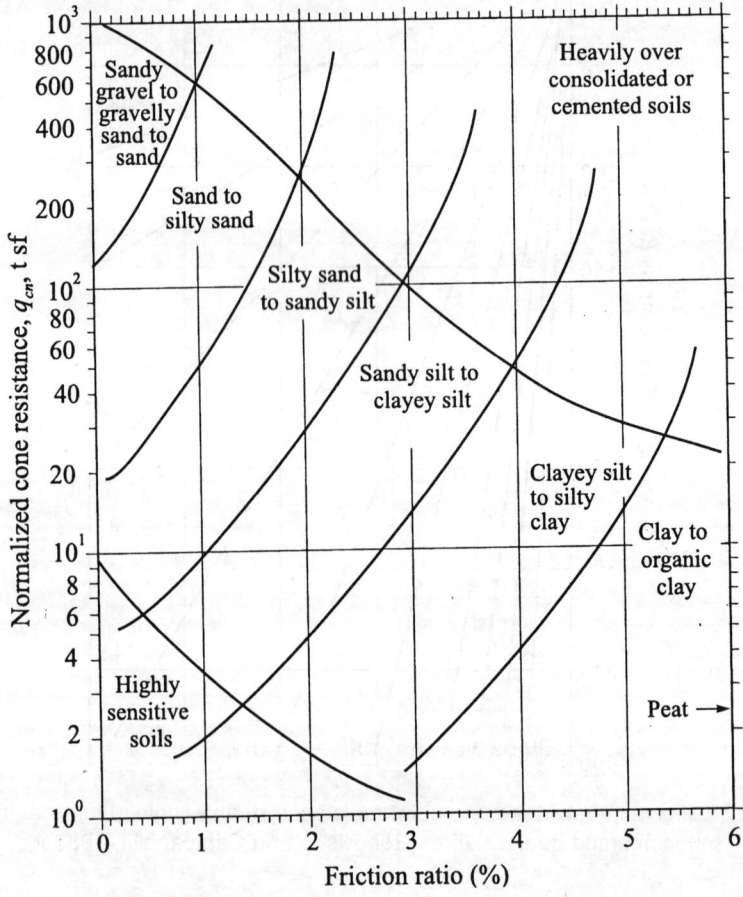

Fig. 3.19 A simplified classification chart (Douglas, 1984)

In conclusion, CPT data provide a repeatable index of the aggregate behavior of *in-situ* soil. The CPT classification method provides a better picture of overall subsurface conditions than is available with most other methods of exploration.

A typical sounding log is given in Fig. 3.20.

The friction ratio R_f varies greatly with the type of soil. The variation of R_f for the various types of soils is generally of the order given in Table 3.6.

Table 3.6 Soil classification based on friction ratio R_f (Sanglerat, 1972)

R_f %	Type of soil
0–0.5	Loose gravel fill
0.5–2.0	Sands or gravels
2–5	Clay sand mixtures and silts
> 5	Clays, peats, etc.

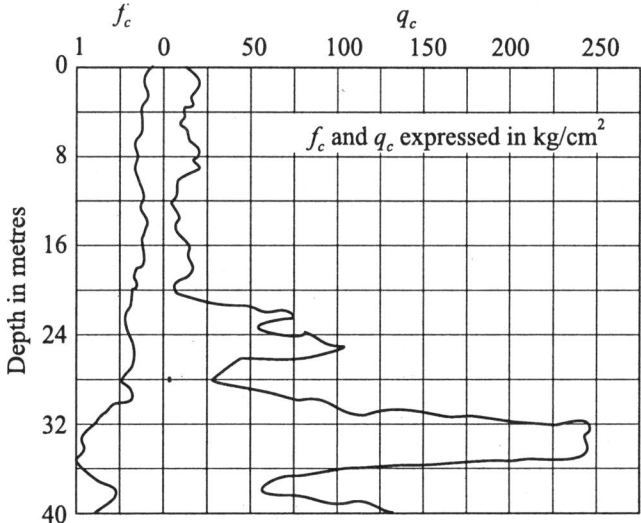

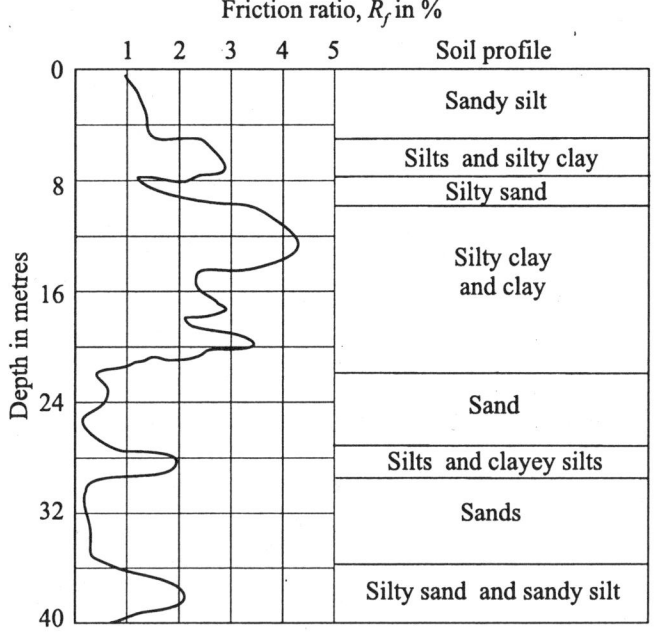

Fig. 3.20 A typical sounding log

Correlation between SPT and CPT

Meyerhof (1965), presented comparative data between SPT and CPT. For fine or silty medium loose to medium dense sands, he presents the correlation as

$$q_c = 0.4\,N\,\text{MN}/\text{m}^2 \tag{3.20}$$

His findings are as given in Table 3.7.

The lowest values of the angle of internal friction given in Table 3.7 are conservative estimates for uniform, clean sand and they should be reduced by at least 5° for clayey sand. These values, as well as the upper values of the angles of internal friction which apply to well graded sand, may be increased by 5° for gravelly sand.

Table 3.7 Approximate relationship between relative density of fine sand, the SPT, the static cone resistance and the angle of internal fraction (Meyerhof, 1965)

State of sand	D_r	N_{cor}	q_c (MPa)	$\phi°$
Very loose	< 0.2	< 4	< 2.0	< 30
Loose	0.2–0.4	4–10	2–4	30–35
Medium dense	0.4–0.6	10–30	4–12	35–40
Dense	0.6–0.8	30–50	12–20	40–45
Very dense	0.8–1.0	> 50	> 20	45

Figure 3.21 shows a correlation presented by Robertson and Campanella (1983) between the ratio of q_c/N and mean grain size, D_{50}. It can be seen from the figure that the ratio varies from 1 at $D_{50} = 0.001$ mm to a maximum value of 8 at $D_{50} = 1.0$ mm. The soil type also varies from clay to sand.

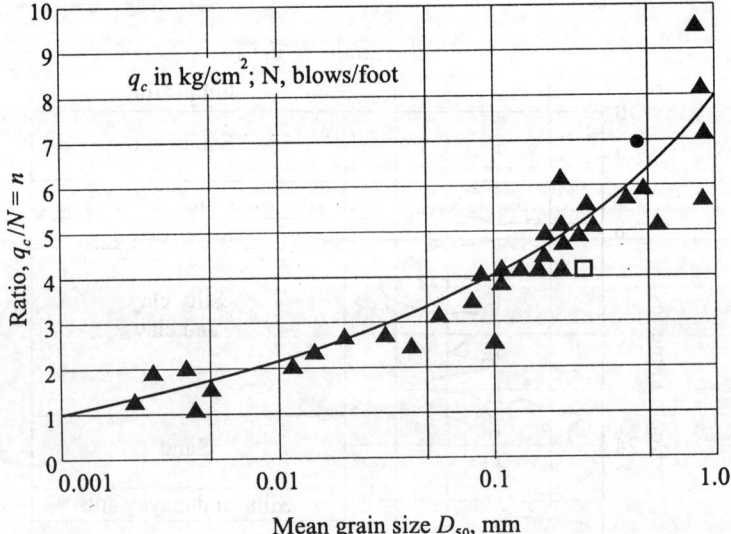

Fig. 3.21 Relationship between q_c/N and mean grain size D_{50} (mm) (Robertson and Campanella, 1983a)

It is clear from the above discussions that the value of $n\,(= q_c'/N)$ is not a constant for any particular soil. Designers must use their own judgement while selecting a value for n for a particular type of soil.

Example 3.6

If a deposit at a site happens to be a saturated overconsolidated clay with a value of $q_c = 8.8$ MN/m², determine the unconfined compressive strength of clay given $p_0 = 127$ kN/m²

Solution

Per Eq. (3.18)

$$c_u = \frac{q_c - p_o}{N_k} \text{ or } q_u = \frac{2(q_c - p_o)}{N_k}$$

Assume $N_k = 20$. Substituting the known values and simplifying

$$q_u = \frac{2(8800 - 127)}{20} = 867 \text{ kN/m}^2$$

If we neglect the overburden pressure p_0

$$q_u = \frac{2 \times 8800}{20} = 880 \text{ kN/m}^2$$

It is clear that the value of q_u is little affected by neglecting the overburden pressure in Eq. (3.18).

Example 3.7

Static cone penetration tests were carried out at a site by using an electric-friction cone penetrometer. The following data were obtained at a depth of 12.5 m.

Cone resistance $q_c = 19.152$ MN/m² (200 tsf)

Friction ratio $R_f = \dfrac{f_c}{q_c} = 1.3$

Classify the soil as per Fig. 3.19. Assume γ (effective) = 16.5 kN/m³.

Solution

The values of $q_c = 19.152 \times 10^3$ kN/m² and $R_f = 1.3$. From Eq. (3.19)

$$q_{cn} = 200 \times \left[1 - 1.25 \log \left(\frac{16.5 \times 12.5}{100} \right) \right] = 121 \text{ tsf}$$

The soil is sand to silty sand (Fig. 3.19) for $q_{cn} = 121$ tsf and $R_f = 1.3$.

Example 3.8

Static cone penetration test at a site at depth of 30 ft revealed the following

Cone resistance $q_c = 125$ tsf
Friction ratio $R_f = 1.3\%$

The average effective unit weight of the soil is 115 psf. Classify the soil per Fig. 3.19.

Solution

The effective overburden pressure

$$p_0' = 30 \times 115 = 3450 \text{ lb/ft}^2 = 1.725 \text{ tsf}$$

From Eq. (3.19)

$$q_{cn} = 125 \, (1 - 1.25 \log 1.725) = 88 \text{ tsf}$$
$$R_f = 1.3\%$$

From Fig. 3.19 the soil is classified as sand to silty sand for $q_{cn} = 88$ tsf and $R_f = 1.3\%$

Example 3.9

The static cone penetration resistance at a site at 10 m depth is 2.5 MN/m^2. The friction ratio obtained from the test is 4.25%. If the unit weight of the soil is 18.5 kN/m^3, what type of soil exists at the site.

Solution

$$q_c = 2.5 \times 1000 \text{ kN/m}^2 = 2500 \text{ kN/m}^2 = 26.1 \text{ tsf}$$

$$p_0' = 10 \times 18.5 = 185 \text{ kN/m}^2 = 1.93 \text{ tsf}$$

$$q_{cn} = 26.1 \ (1 - 1.25 \log 1.93) = 16.8 \text{ tsf}$$

$$R_f = 4.25\%$$

From Fig. 3.19, the soil is classified as clayey silt to silty clay to clay.

Offshore Investigation

A majority of offshore soil exploration consists of conventional drilling and percussion type soil sampling methods. The inadequacy of bore hole samples for an accurate determination of soil stratigraphy and soil properties have been noticed by many investigators in the past. The use of electronic CPT for the last nearly 15 years have been found to give quite accurately the soil stratigraphy, soil type, strength, density and compressibility. As a further development, special cones are used which incorporates in its body devices for measuring porewater pressure and soil temperature also.

Two methods are in use for sea bed subsoil investigation by the use of electronic CPT. They are,

1. By the use of sea floor supported CPT rig (called as Seacalf),
2. By the use of downhole penetrometer.

Seafloor supported CPT rig

The CPT starts from the seabed supported rig which was lowered in advance to the required position by a ship berthed directly overhead on the water surface. The penetrometer is pushed at a constant rate through a string of drill rods and remotely controlled electronically from the deck of the ship. The test results are recorded at the control unit in the form a continuous graph. The reaction capacity of the jacking system on the sea bed varies from 200 to 300 kN. The exploration of sea bed subsoil can be carried out in water depths of over 500 m. The depth of CPT below sea bed depends on the soil condition and the reaction provided by the weight of the rig. A typical CPT results graph along with the rig is given in Fig. 3.22.

Downhole penetrometer

The depth of subsoil exploration below sea bed by the seafloor supported rig depends upon the anchoring capacity of the rig which is normally limited to a maximum of 300 kN. When exploration is required beyond this depth, downhole penetrometer method is followed that can perform CPT from the bottom of a bore hole.

Down hole penetration test, or sometimes also called as *Wire Line CPT*, is continued below the depth achieved by the seafloor rigs. First hole is drilled up to this depth by a drilling bit attached to the end of a string of hollow drill rods operated from the deck of a ship berthed vertically above. The electronic penetrometer is lowered to the bottom of the hole through the hollow drillrods. The penetrometer is pushed down at a constant rate by a hydraulic jacking system which forms also a part

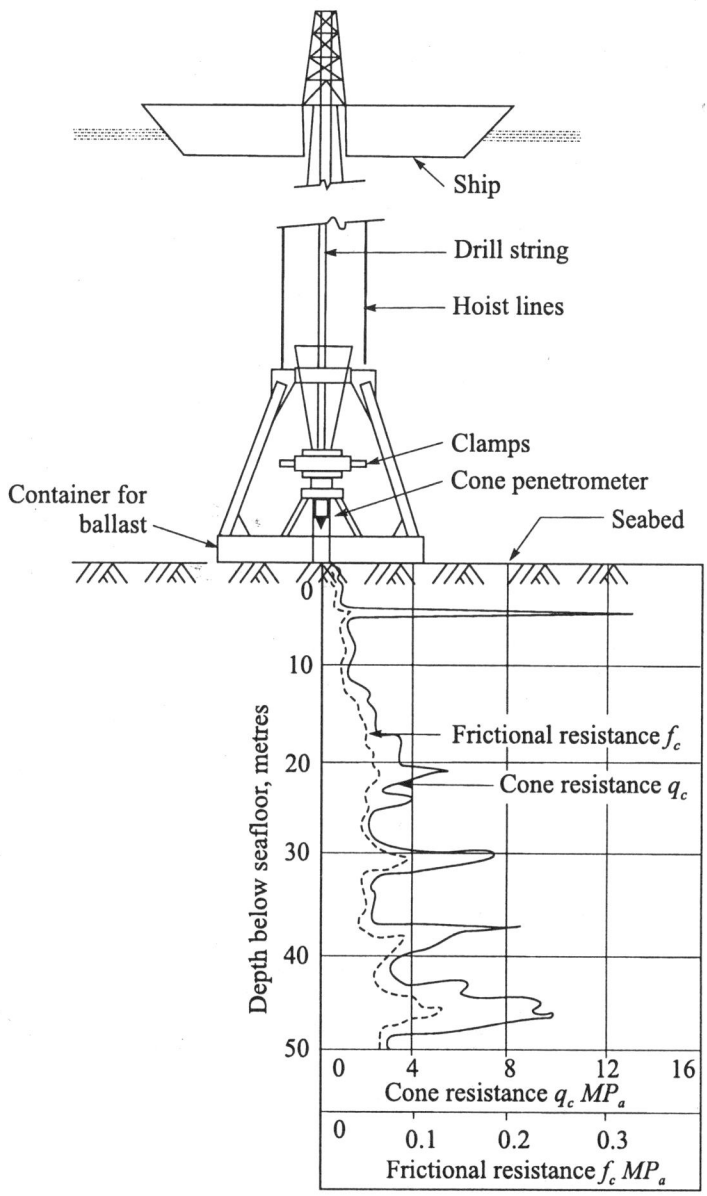

Fig. 3.22 Sea-bed-supported CPT rig with a typical graph of test results

of the drilling system within the bore hole. The weight of the drill string acts as a reaction force [Fig. 3.23 (a)]. The test results are recorded in the control unit on the deck of the strip. Figure 3.23 (a) shows the wire line cone penetrometer and drill string anchor at the bottom of the bore hole.

Sampling below seabed

Sampling under the seabed in boreholes are done using the same equipment as used for wire line CPT. The only difference is that a sampler takes the place of penetrometer. The sampler is pushed into the soil at a constant rate and then retrieved. Figure 3.23 (b) shows the push sampler with the wire line equipment.

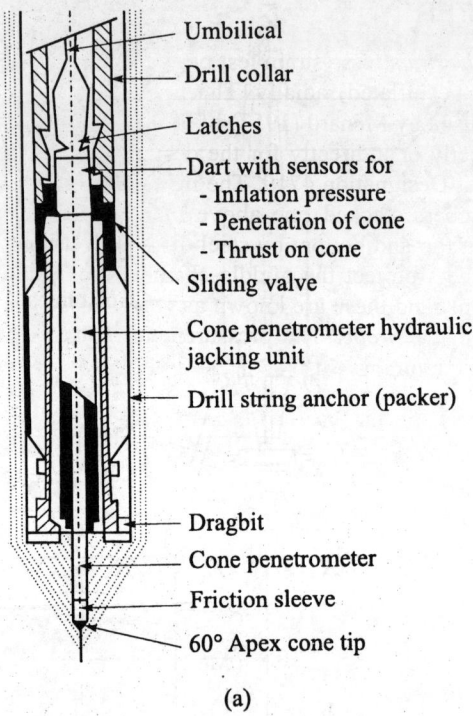

Umbilical

Drill collar

Latches

Dart with sensors for
- Inflation pressure
- Penetration of cone
- Thrust on cone

Sliding valve

Cone penetrometer hydraulic jacking unit

Drill string anchor (packer)

Dragbit

Cone penetrometer

Friction sleeve

60° Apex cone tip

(a)

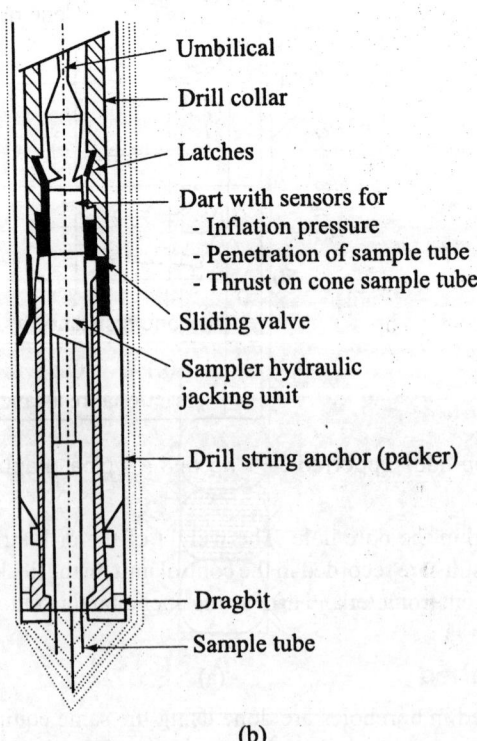

Umbilical

Drill collar

Latches

Dart with sensors for
- Inflation pressure
- Penetration of sample tube
- Thrust on cone sample tube

Sliding valve

Sampler hydraulic jacking unit

Drill string anchor (packer)

Dragbit

Sample tube

(b)

Fig. 3.23 Downhole penetrometer and sampling unit: (a) Wire line push CPT equipment and drill string anchor, (b) wire line push sampling equipment and drill string anchor

3.10 PRESSUREMETER

A pressuremeter test is an *in-situ* stress–strain test performed on the walls of a bore hole using a cylindrical probe that can be inflated radially. The pressuremeter, which was first conceived, designed, constructed and used by Menard (1957) of France, has been in use since 1957. The test results are used either directly or indirectly for the design of foundations. The Menard test has been adopted as ASTM Test Designation 4719. The instrument as conceived by Menard consists of three independent chambers stacked one above the other (Fig. 3.24) with inflatable user membranes held together at top and bottom by steel discs with a rigid hollow tube at the centre. The top and bottom chambers protect the middle chamber from the end effects caused by the finite length of the apparatus, and these are known as *guard cells*. The middle chamber with the end cells together is called the *Probe*. The pressuremeter consists of three parts, namely, the probe, the control unit and the tubing.

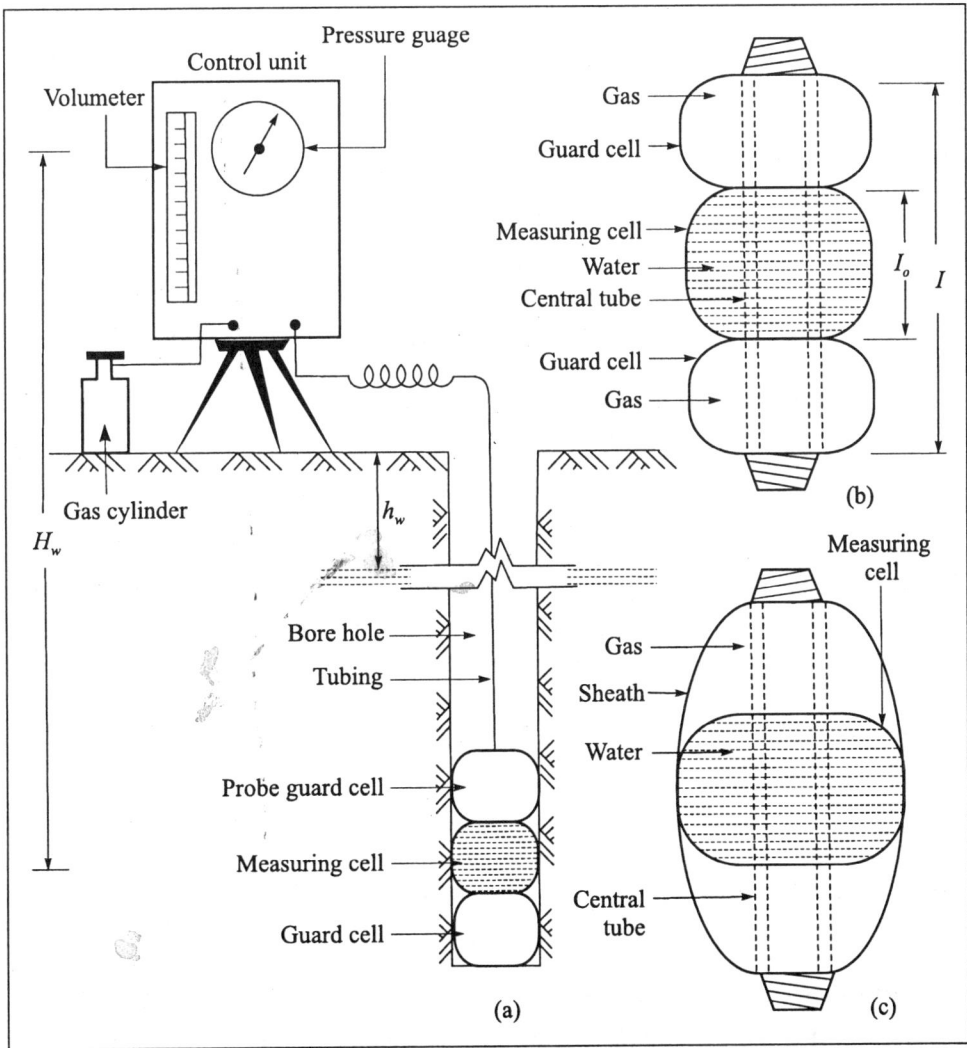

Fig. 3.24 Components of Menard pressuremeter: (a) Basic principles of the pressuremetre, (b) pressuremetre with three independent cells, (c) pressuremetre with independent measuring cell

Ever since Menard used his first pressuremeter, many changes have taken place in the design and use of the probe. Schematic sketches of two types of pressuremeters are shown in Fig. 3.24 (b) and (c). The pressuremeter (b) has three independent cells. Water is used for inflating the measuring cell, whereas gas (air, carbon dioxide or nitrogen) is used for the guard cells. The pressuremeter (c) has a separate measuring cell which is self contained, and located inside a sheath which runs the whole length of the probe. The voids created at either end of the measuring cell form the guard cells. Many organisations have broughtout a pressuremeter with monocellular probe with no guard cells.

The practice since 1957 has been to conduct tests in pre-bored holes. It has been found that the results obtained from pre-bored hole tests do not represent truly the soil behaviour under the *in-situ* conditions. Some of the important factors that affect the results are:

1. Diameter of bore hole is either too large or too small as compared to the diameter of the probe under deflated condition.
2. Disturbance caused to the sides of the bore hole during drilling is considerable.
3. There will be yield of the soil due to the release in the *in-situ* pressure in the bore hole.

To overcome the above deficiencies to a certain extent Self-Boring Pressuremeters have been introduced, Fig. 3.25. The probe used in this pressuremeter is the same type as that used in pre-bored hole except that the probe is equipped with an extension of a thin walled tube with a cutting edge and a grinder inside the tube.

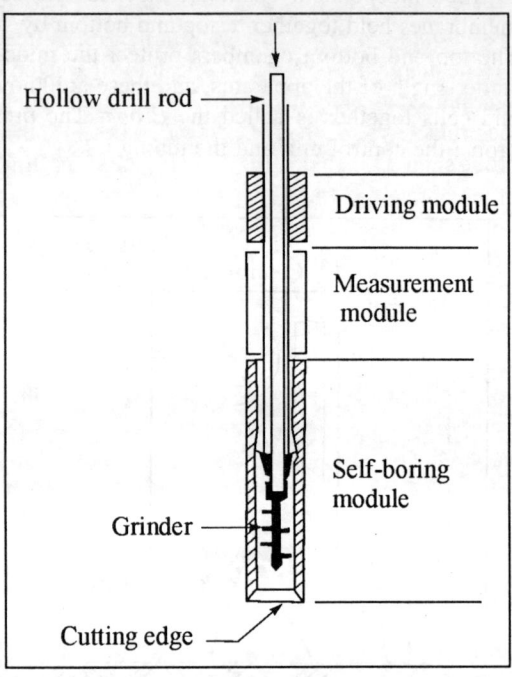

Fig. 3.25 Principles of self-boring pressuremeter

In this case the probe is forced into the ground at a constant rate. The core of soil which penetrates the tube is immediately powdered by a grinder (usually a system of rotating blades). The cuttings which are produced are transported up through the centre of the apparatus in suspension in a fluid which is injected at the level of the grinder.

The rubber membranes are sometimes protected from damage by gluing on it with thin stainless steel strips.

Pressuremeter test which is an *in-situ* test gives valuable information for the design of foundations. This test sets up a stress field in the ground much like the one that will be induced by the foundation. Future progress in estimating soil properties will come from refinements in *in-situ* testing rather than from the laboratory. The discussions in this section is confined to Menard pressuremeter only.

Basic Features of Menard Pressuremeter

The pressuremeter consists of three parts as shown schematically in Fig. 3.24. They are,

1. The probe,
2. The control unit,
3. The tubing.

The probe

A section of a probe is shown schematically in Fig. 3.24 (b). The core of the probe is a hollow metal tube. When the probe is in position in the bore hole, water is used to inflate the measuring cell and gas to inflate the guard cells. The inflated guard cells effectively seal off the bore hole and prevent the measuring cell membrane from expanding into the void of hole. The rubber membrane is sufficiently flexible to ensure that a uniform pressure is applied to the walls of the hole, and the presence of guard cells means that a longer length of bore hole is pressurised than would be the case with just the measuring cell alone. Thus plane strain conditions are created in the soil around the cavity.

The diameter of the probe, D_b, under depleted condition varies widely from approximately 25 to 125 mm. The combined length of measuring and guard cells shall be at least 7 times the diameter of the probe. The length of the measuring cell is normally about half the total length. The design of the probe shall be such that the drilling fluid may flow freely past the probe without disturbing the sides of the bore hole.

The control unit

A typical control unit used for the tests with a gas cylinder and probe is shown schematically. The whole control unit is fixed on a panel which can be mounted on a tripod as shown in the figure. The unit contains a volumeter, and its graduated scale. Plastic tubing of the required length is used for the volumeter which is connected in parallel with a reservoir. The scale has graduations for every 5 cm³ of water in the reservoir and can be read to the nearest cm³. The zero is part-way down the scale. Valves are provided at the bottom of the reservoir to control the flow of water from volumeter to the probe. The capacity of the reservoir to hold water may be of the order of 1000 cm³. The water pressure is measured by two gauges. The range of pressure of these gauges vary according to the purpose for which they are used. For soil one of the pressure gauges may have a range of 0–2500 kPa and for rocks 0–10,000 kPa. Gas is provided in a separate bottle under pressure and the pressure of gas in the bottle is measured by the gauge. The monitoring of the pressure difference between the measuring and guard cells is done by means of the differential pressure gauge which is located between the water and gas circuits. The tubing leading to the probe is attached to the control unit by means of a connector.

The tubing

The tubing is required between the control unit and the probe to allow the flow of water and gas from one to the other. Both water and gas may flow in separate tubings or in coaxial tubings. In the coaxial tubings, water is carried to the measuring cell through the inner tubing, while the outer annulus conducts gas (or water) to the guard cells. The tubing is normally made of semi-rigid polymide which is transparent. The coaxial tubing reduces the possibility of the water circuit expanding under pressure reducing thereby the volume correction to a negligible amount. In the case of co-axial tubing, the diameters of the tubings may be of the following order.

Inner tubing	: Inside diameter	=	4.5 mm
	Outer diameter	=	6.0 mm
Outside tubing :	Inside diameter	=	8 mm
	Outside diameter	=	13.5 mm

The Pressuremeter Test

The pressuremeter test involves the following:

1. Drilling of a hole.
2. Lowering the probe into the hole and clamping it at the desired elevation.
3. Conducting the test.

Drilling and positioning of probe

Menard pressure test is carried out in a hole drilled in advance. The drilling of hole is done by the use of a suitable drilling rig which disturbs the soil the least. Drilling mud may be used if required for stabilising the sides of the hole. The diameter of the bore hole, D_h, in which the test is to be conducted shall satisfy the condition.

$$1.03 \, D_p < D_h < 1.20 \, D_p \qquad (3.21)$$

where, D_p is the diameter of the probe under deflated condition.

Typical sizes of probe and bore hole are given in Table 3.8.

Table 3.8 Typical sizes of probe and bore hole for pressuremeter test

Hole dia Designation	Probe dia (mm)	l_0 cm	l cm	Bore hole dia	
				Nominal mm	Max mm
AX	44	36	66	46	52
BX	58	21	42	60	66
NX	70	25	50	72	84

Note: l_0 = length of measuring cell; l = length of probe. See Fig. 3.24 (b).

The probe is lowered down the hole soon after boring to the desired elevation and held in position by a clamping device. Pressuremeter tests are usually carried out at 1 m intervals in all the bore holes.

Conducting the test

With the probe in position in the bore hole, the test is started by opening the valves in the control unit for admitting water and gas (or water) to the measuring cell and the guard cells respectively. The pressure in the guard cells is normally kept equal to the pressure in the measuring cell. The pressures to the soil through the measuring cell is applied by any one of the following methods:

1. Equal pressure increment method.
2. Equal volume increment method.

If pressure is applied by the first method, each equal increment of pressure is held constant for a fixed length of time, usually one minute. Volume readings are made after one minute elapsed time. Normally, ten equal increments of pressure are applied for soil to reach the limit pressure, p_l.

If pressure is applied by the second method, the volume of the probe shall be increased in increments equal to 5 percent of the nominal volume of the probe (in the deflated condition) and held constant for 30 seconds. Pressure readings are taken after 30 seconds of elapsed time.

Steps in both the methods are continued until the maximum probe volume to be used in the test is reached. The test may last at each position from 10 to 15 minutes. This means that the test is essentially an undrained test in clay soils and drained test in a freely draining material.

Typical test result

First a typical curve based on the observed readings in the field may be plotted. The plot is made of the volume of the water read at the volumeter in the control unit, v, as abscissa for each increment of pressure, p, as ordinate. The curve is a result of the test conducted on the basis of equal increments of pressure and each pressure held constant for a period of one minute. This curve is a raw curve which requires some corrections. The pressuremeter has, therefore, to be calibrated before it is used in the design.

Calibration of Pressuremeter and Hydrostatic Pressure Correction, p_w

A pressuremeter has to be calibrated for

1. Pressure loss, p_c.
2. Volume loss, v_c.

Pressure loss, p_c

The pressure loss, p_c, occurs due to the rigidity of the probe membranes. As the probe is inflated, a certain amount of pressure is necessary to overcome the resistance of the rubber membrane. The pressure readings obtained on the control unit during field tests includes pressures required to expand the membranes which must be deducted to obtain the actual pressure exerted on the soil. The probe should therefore be calibrated to determine the losses at different inflated pressures. This is done by inflating the probe under atmospheric pressure outside the hole. The pressure p required to expand the probe and the corresponding volume of water, v, injected into the probe are measured. A plot of pressure p *versus* v gives a calibration curve for the pressure loss as shown in Fig. 3.26 (a).

Volume loss, v_c

Volume loss, v_c, occurs due to the expansion of the tubing system and the compressibility of any part of the testing equipment including the probe and the liquid. Calibration for volume loss is done by keeping the probe vertically in close-fitting in a thick steel tube so that the probe may not expand during the application of water pressure to the walls of the tubing. Initially, the tubing and the probe is filled with water. Pressure is then applied to the water in the system in increments and the volumeter readings are taken. Figure 3.26 (a) gives a plot of pressure *versus* volume reading curve. The volume correction is the volume loss, v_c, obtained from the curve for any stage of pressure.

Hydrostatic pressure correction, p_w

As can be seen in Fig. 3.24 (a), H_w is the difference in head between the centre of the measuring cell in the bore hole and the pressure gauge in the control unit. This difference in height is the height of column of water in the tubing which exerts an additional pressure on the soil. This pressure is not recorded by the gauge in the control unit. Therefore, the hydrostatic pressure correction, p_w, is

$$p_w = \gamma_w H_w \tag{3.22a}$$

where, γ_w is the unit weight of water.

Corrected Plot of Pressure–Volume Curve

The raw field curve has to be corrected for

1. Pressure loss, p_c.
2. Volume loss, v_c.
3. Pressure loss, p_w, due to differential head.

The corrected pressure, p, and volume, v, for any point on the curve may be obtained as per the equations given below:

$$p = p_r + p_w - p_c \tag{3.22b}$$

where, p_r is the actual pressure reading of the gauge in the control unit.

$$v = v_r - v_c \tag{3.22c}$$

where, v_r is the actual volume reading of the volumeter in the control unit.

A typical corrected plot of the pressure-volume curve is given in Fig. 3.26 (b). The characteristic parts of this curve are three in number. They are:

1. The initial part of the curve OA. This curve is a result of pushing the yielded wall of the hole back to the original position. At point A, the at rest condition is supposed to have been restored. The expansion of the cavity is considered only from point A. v_0 is the volume of water required to be injected over and above the volume V_c of the probe under the deflated condition. If V_0 is the total volume of cavity at point A, we can write

$$V_0 = V_c + v_0 \tag{3.22d}$$

where, v_0 is the abscissa of the point A. The horizontal pressure at point A is represented as p_{om}.

2. The second part of the curve is AB. This is supposed to be a straight line portion of the curve and may represent the elastic range. Since AB gives an impression of an elastic range, it is called the *pseudo-elastic* phase of the test. The point A is considered to be the start of the pressuremeter test in most theories. The point B marks the end of the straight line portion of the curve. The coordinates of the point B are p_f and v_f, where p_f is known as the *creep pressure*.

3. The curve BC marks the final phase. The plastic phase is supposed to start from point B, and the curve becomes eventually asymptotic at point C at a large deformation of the cavity. The *limit pressure*, p_l, is usually defined as the pressure that is required to double the initial volume of the cavity. It occurs at a volume such that

$$v_l - v_0 = V_0 = V_c + v_0 \tag{3.22e}$$

or
$$v_l = V_c + 2v_0 \tag{3.22f}$$

v_0 is normally limited to about 300 cm^3 for probes used in AX and BX holes. The initial volume of these probes is of the order of 535 cm^3. This means that $(V_c + 2v_0)$ is of the order of 1135 cm^3. These values may vary according to the design of the pressuremeter.

It is, therefore, necessary that the reservoir capacity in the control unit should be of the order of 1135 cm^3. In case the reservoir capacity is limited and p_l is not reached within its limit, the test, therefore, has to be stopped at that level. In such a case, the limit value, p_l, has to be extrapolated.

In some of the pressuremeters, provision is made to measure the increase in the probe radius at every increment of pressure applied. This system replaces the measurement of change in volume of the probe. Gas is used to inflate the measuring cell in such pressuremeters. Measurement of change of volume by the change in the probe radius is also shown in Fig. 3.26 (b).

At Rest Horizontal Pressure

The at rest total horizontal pressure, p_{oh}, at any depth, z, under the *in-situ* condition before drilling a hole may be expressed as

$$p_{oh} = (\gamma z - u) K_0 + u \tag{3.22g}$$

where, u = pore pressure at depth z,

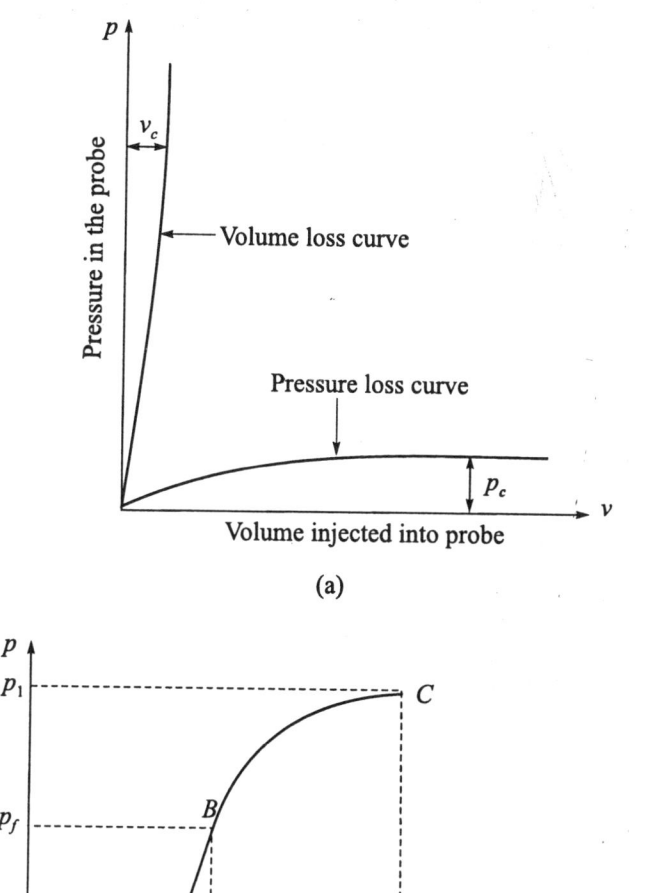

Fig. 3.26 Typical corrected pressuremeter curve: (a) Calibration for volume and pressure losses, (b) corrected pressuremetre curve

γ = gross unit weight of the soil.

K_0 = coefficient of earth pressure for the at rest condition.

The values of γ and K_0 are generally assumed taking into account the type and condition of the soil. The porewater pressure under the hydrostatic condition is

$$u = \gamma_w \, (z - h_w) \tag{3.22h}$$

where, γ_w = unit weight of water,

h_w = depth of water table from ground surface.

As per Fig. 3.26 (b), p_{om} is the pressure which corresponds to the volume v_0 at the start of the straight line portion of the curve. Since it has been found that it is very difficult to determine accurately p_{om}, p_{oh} may not be equal to p_{om}. As such, p_{om} bears no relation to what it is to be the true earth pressure at rest. In Eq. (3.22g), K_0 has to be assumed and its accuracy is doubtful. In such circumstances it is not possible to calculate p_{oh} also. However, p_{om} can be used for calculating the pressuremeter modulus E_m. The experience of many investigator is that a self-boring pressuremeter gives more reliable values for p_{oh}.

The Pressuremeter Modulus E_m

Since the curve between points A and B in Fig. 3.26 (b) is approximately a straight line, the soil in this region may be assumed to behave a more or less elastic material. The equation for the radial expansion of a cylindrical cavity in an infinite elastic medium is (Lame, 1852),

$$G = V \frac{\Delta p}{\Delta V} \tag{3.23a}$$

where, G = the shear modulus,

V = the volume of the cavity,

p = pressure in the cavity.

For the pressuremeter test, we may write $\Delta V = \Delta v$, and as such Eq. 3.23 (a) becomes,

$$G = V \frac{\Delta p}{\Delta v} \tag{3.23b}$$

Between points A and B in Fig. 3.26 (b), the slope of the curve is constant, whereas the volume of the cavity changes from v_0 at A to v_f at B. The value of G, therefore, depends on the location on the line AB. By convention, the volume at mid point of AB is chosen for computing G. If V_m is the volume at mid-point, we may write,

$$V = V_m = V_c + \frac{v_0 + v_f}{2} \tag{3.24a}$$

where, V_c is the volume of the deflated portion of the measuring cell at zero volume reading on the Volumeter in the control unit.

Since, Menard first proposed this procedure, the shear modulus G is now called as G_m (here the subscript m stand for Menard), Therefore,

$$G_m = V_m \frac{\Delta p}{\Delta v} \tag{3.24b}$$

The shear modulus G_m may now be converted to pressuremeter deformation modulus E_p by making use of the well-known relationship

$$G_m = \frac{E_p}{2 \, (1 + \mu)} \tag{3.24c}$$

where, μ is the Poisson's ratio.

Now substituting for G_m and transposing, we have

$$E_p = 2 (1 + \mu) V_m \frac{\Delta p}{\Delta v} \qquad (3.24d)$$

Suitable value for μ may be assumed in the above equation depending on the type of soil. For saturated clay soils μ is taken as equal to 0.5 and for freely draining soils, the value is less. Since it has been found out that G_m is not very much affected by a little variation in μ, Menard proposed a constant value of 0.33 for μ. As such the resulting deformation modulus is called as *Menards Modulus* E_m. The equation for E_m reduces to

$$E_m = 2.66 \, V_m \frac{\Delta p}{\Delta v} \qquad (3.25)$$

As per Fig. 3.26 (b), we have

$$\left. \begin{array}{l} \Delta p = p_f - p_{om} \\ \Delta v = v_f - v_0 \\ V_m = V_c + \dfrac{v_f + v_0}{2} \end{array} \right\} \qquad (3.26)$$

Substituting these values in Eq. (3.25), we have

$$E_m = 2.66 \left(V_c + \frac{v_f + v_0}{2} \right) \left(\frac{p_f - p_{om}}{v_f - v_0} \right) \qquad (3.27)$$

It is to be noted here that if a break in the straight line AB (Fig. 3.26b) portion of the pressuremeter curve is observed, calculations shall include a pressuremeter modulus for each straight line section of the pressuremeter test curve.

E_m in Terms of the Increase in the Radius of the Cavity

For the tests where the measuring cell radius R at every stage of the test is measured, the pressuremeter modulus, E_m, may be calculated as per the equation

$$E_m = 2.66 \left(R_c + \frac{r_f + r_0}{2} \right) \left(\frac{p_f - p_{om}}{r_f - r_0} \right) \qquad (3.28)$$

where, R_c = radius of the measuring cell at the deflated condition corresponding to the start of the test [Fig. 3.26 (b)],

r_0 and r_f = the increase in the radii corresponding to the positions A and B respectively.

Relationship between Menard Modulu E_m and the Young's Modulus E

Menard (1975), stated that the pressuremeter modulus, E_m, cannot be compared directly with a compression modulus such as the Young's modulus for the following reasons:

1. The stress path followed in the soil around the pressuremeter probe is different from that in a compression test or under a plate or footing.

2. Since elastic theory indicates that the increase in compressive stress in a radial direction equals the increase in tensile stress, it is likely that the compression modulus is different from the tension modulus. As such, Menard feels that E_m probably lies between the two.

3. The pressuremeter modulus E_m is not a measure of what Menard calls the modulus of *microdeformation E*, which is defined as the modulus of the soil skeleton when it is subjected to small strains.

Because of the afforesaid reasons, Menard proposes the relationship between E_m and E (the Young's modulus) for soils as

$$E = \frac{E_m}{\alpha} \tag{3.29}$$

Table 3.9 (a) gives the values of α which is called as a rheological factor for various types of soils which depends on the ratios $E_m / \bar{p}_l$ where $\bar{p}_l = p_l - p_{oh}$.

Table 3.9 (a) Rheological factor α for various soils

Type of soil	Peat $E_m/\bar{p}_l$	Peat α	Clay $E_m/\bar{p}_l$	Clay α	Silt $E_m/\bar{p}_l$	Silt α	Sand $E_m/\bar{p}_l$	Sand α	Sand and gravel $E_m/\bar{p}_l$	Sand and gravel α
Over-consolidated			> 16	1	> 14	0.67	> 12	0.5	> 10	0.33
Normally consolidated		1	9–16	0.67	8–14	0.50	7–12	0.33	6–10	0.25
Weathered and remoulded			7–9	0.50		0.25		0.33		0.25

Rock extremely fractured	Other	Slightly fractured or extremely weathered
$\alpha = 1/3$	$\alpha = 1.2$	$\alpha = 0.67$

Relationship between E_m and p_l

Menard (1975) has suggested some typical values for E_m and p_l which are given in Table 3.9 (b). According to him the ratio E_m/p_l is a characteristic of the type of soil under examination. For overconsolidated clay soils, the ratio E_m/p_l may range from 12 to 30, whereas for sands, gravel silty sands under water, the ratio are lower and are in the range of 5 to 8.

Table 3.9 (b) Typical values for E_m and p_l

Type of soil	E_m MPa	p_l MPa
Mud, peat	0.2 – 1.5	0.02 – 0.15
Soft clay	0.5 – 3	0.05 – 0.3
Medium clay	3 – 8	0.3 – 0.8
Stiff clay	8 – 40	0.6 – 2.0
Loose silty sand	0.5 – 2	0.1 – 0.5
Silt	2 – 10	0.2 – 1.5
Sand and gravel	8 – 40	1.2 – 5
Lime stone	80 – 20,000	3 –> 10

Undrained Shear Strength c_u from the Net Limit Pressure $\bar{p}_l$

The method of using the net limit pressure $\bar{p}_l$ to compute c_u has received a lot of attention. Based on an ideal elastic-plastic assumptions, the following three theoretical solutions were proposed.

1. Bishop, Hill and Mott (1945)

$$\bar{p}_l = c_u \left[1 + \log_e \frac{E}{2 c_u (1 + \mu)} \right] \tag{3.30a}$$

2. Hill (1950)

$$\bar{p}_l = c_u \left[1 + \log_e \frac{E}{c_u (5 - 4\mu)} \right] \tag{3.30b}$$

3. Salencon (1966)

$$\bar{p}_l = c_u \left[1 + \log_e \frac{E}{4 c_u (1 - \mu^2)} \right] \tag{3.30c}$$

where, $\bar{p}_l = p_l - p_{oh}$

p_{oh} = total horizontal earth pressure for at-rest condition

E = theoretical modulus of deformation

μ = Poisson's ratio

If $\mu = 0.5$ is assumed for undrained condition, all the three formulae get reduced to

$$\bar{p}_l = c_u \left(1 + \log_e \frac{E}{3 c_u} \right) \tag{3.31}$$

$$c_u = \frac{\bar{p}_l}{1 + \log_e (E / 3 c_u)} = \frac{\bar{p}_l}{\beta} \tag{3.32a}$$

Experimental investigations have indicated that the value of β lies between 6.5 and 12. An average value of 9 has been suggested. The final equation for c_u may be written as

$$c_u = \frac{\bar{p}_l}{9} \tag{3.32b}$$

Amar and Jézéquel (1972) have suggested another equation of the form

$$c_u = \frac{\bar{p}_l}{10} + 25 \text{ kPa} \tag{3.32c}$$

where, both $\bar{p}_l$ and c_u are in kPa.

Drained Shear Strength

As per Baguelin *et al* (1978), there is at present no theoretical way to convert pressuremeter test results to effective strength parameters c' and ϕ' for freely draining soils. However, an empirical relationship between $\bar{p}_l$ and ϕ' has been suggested by Muller (1970) in the following form

$$\bar{p}_l = b \times 2^n \tag{3.33a}$$

where, $\qquad n = \dfrac{\phi' - 24}{4}$ $\qquad\qquad\qquad\qquad\qquad\qquad\qquad$ (3.33b)

$\qquad$ $n = 1.8$ for homogeneous wet soil,

$\qquad$ $= 3.5$ for dry, heterogeneous soil,

$\qquad$ $= 2.5$ an average value.

Sufficient data is not available to confirm the above suggestion.

Relationship between MPT and CPT

The relationship between $\bar{q}_c$ and $\bar{p}_l$ can be expressed in the form of a ratio $\bar{q}_c/\bar{p}_l$ where, $\bar{q}_c = p_l - p_{oh}$ and $\bar{p}_l = p_l - p_{oh}$. Baguelin *et al* (1978), give the values of $\bar{q}_c/\bar{p}_l$ for various types of soils as shown in Table 3.10.

Table 3.10 Values of $\bar{q}_c/\bar{p}_l$

Type of soil	$\bar{q}_c/\bar{p}_l$
Very soft to soft clay	Close to 1 or between 2.5 and 3.5
Firm to very stiff clay	2.5 to 3.5
Very stiff to hard clay	3 to 4
Very loose to loose sand and compressible silt	1 to 1.5 or 3 to 4
Compact silt	3 to 5
Sand and gravel	5 to 12

Relationship between $\bar{p}_l$ and SPT *N* Value for Granular Soils

Baguelin *et al* (1978), provide certain information about the relationship between $\bar{p}_l$ and *N* values for granular soils under different densities as shown in Table 3.11.

Table 3.11 Relationship between $\bar{p}_l$ and *N* for granular soils

Density of soil	$\bar{p}_l$ kPa	N value
Very loose	0 – 200	0 – 4
Loose	200 – 500	4 – 10
Medium dense	500 – 1500	10 – 30
Dense	1500 – 2500	30 – 50
Very dense	> 2500	> 50

Limitations of Determining Soil Conditions from Pressuremeter Test Results

The pressuremeter test is a very short duration test in the field and therefore these tests by itself cannot give any direct information concerning the type of soil in which the test is carried out. This means that the test results alone cannot be used to classify the soil. It cannot even indicate whether the soil is pervious or not. As per the knowledge available at present, it is not possible to identify the soil purely on the basis of the values of p_f, p_l, and E_m. If, however, the soil can be classified by some other method as clay, sand or whatever, then the condition of the soil can be determined from the pressuremeter test results. However, pressuremeter test results can be applied directly for the design of foundations.

Example 3.10

A pressuremeter test was carried out at a site at a depth of 7 m below the ground surface. The water table level was at a depth of 1.5 m. The average unit weight of saturated soil is 17.3 kN/m³. The corrected pressuremeter curve is given in Fig. Ex. 3.10 and the depleted volume of the probe is $V_c = 535$ cm³. Determine the following.

(a) The coefficient of earth pressure for the at-rest condition.

(b) The Menard pressuremeter modulus E_m.

(c) The undrained shear strength c_u. Assume that $p_{oh} = p_{om}$ in this case.

Solution

From Fig. Ex. 3.10, $p_{oh} = p_{om} = 105$ kPa

The effective overburden pressure is

$$p_0' = 17.3 \times 7 - 5.5 \times 9.81 = 67.2 \text{ kPa}$$

The effective horizontal pressure is

$$p_{oh}' = 105 - 5.5 \times 9.81 = 51.0 \text{ kPa}$$

(a) From Eq. (3.22g)

$$K_0 = \frac{p_{oh}'}{p_o'} = \frac{51.0}{67.2} = 0.76$$

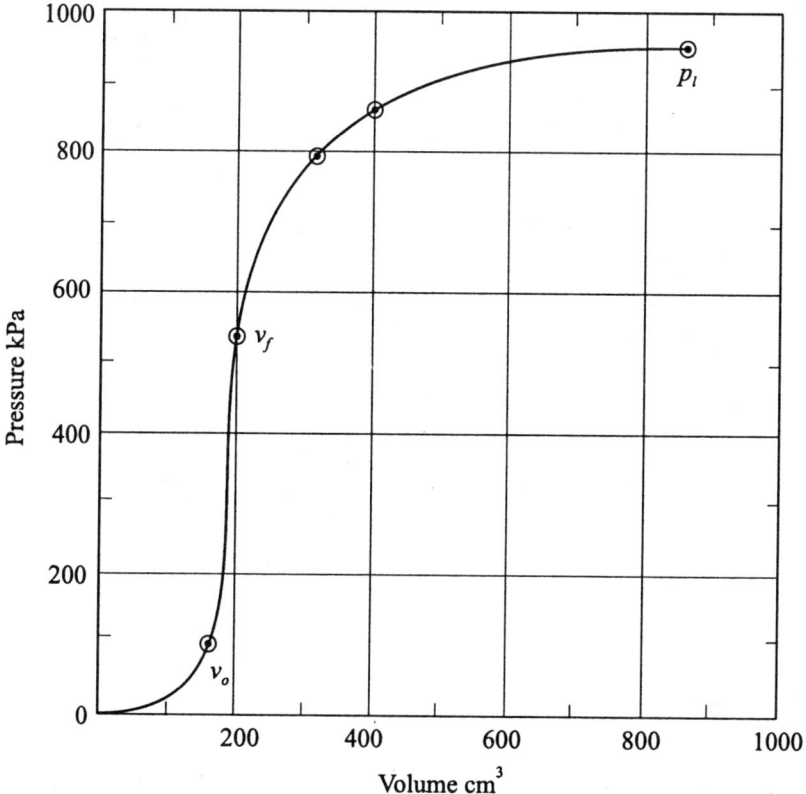

Fig. Ex. 3.10

(b) From Eq. (3.25)

$$E_m = 2.66 \, V_m \frac{\Delta p}{\Delta v}$$

From Fig. Ex. 3.10

$$v_f = 200 \text{ cm}^3 \qquad p_f = 503 \text{ kPa}$$
$$v_o = 160 \text{ cm}^3 \qquad p_{om} = 105 \text{ kPa}$$

From Eq. (3.24a)

$$V_m = 535 + \frac{200 + 160}{2} = 715 \text{ cm}^3$$

$$\frac{\Delta p}{\Delta v} = \frac{530 - 105}{200 - 160} = 10.625$$

Now $\qquad E_m = 2.66 \times 715 \times 10.625 = 20,208 \text{ kPa}$

(c) From Eq. (3.32b)

$$c_u = \frac{\overline{p}_l}{9} \quad \frac{p_l - p_{oh}}{9}$$

From Fig. Ex. 3.10

$$\overline{p}_l = 950 - 105 = 845 \text{ kPa}$$

Therefore $\qquad c_u = \frac{845}{9} = 94 \text{ kPa}$

From Eq. (3.22c)

$$c_u = \frac{\overline{p}_l}{10} + 25 = \frac{845}{10} + 25 = 109.5 \text{ kPa}$$

3.11 THE FLAT DILATOMETER TEST

The *flat dilatometer* is an *in-situ* testing device developed in Italy by Marchetti (1980). It is a penetration device that includes a lateral expansion arrangement after penetration. The test, therefore, combines many of the features contained in the cone penetration test and the pressuremeter test. This test has been extensively used for reliable, economical and rapid *in-situ* determination of geotechnical parameters. The flat plate dilatometer (Fig. 3.27) consists of a stainless steel blade with a flat circular expandable membrane of 60 mm diameter on one side of the stainless steel plate, a short distance above the sharpened tip. The size of the plate is 220 mm long, 95 mm wide and 14 mm thick. When at rest the external surface of the circular membrane is flush with the surrounding flat surface of the blade.

The probe is pushed to the required depth by making use of a rig used for a static cone penetrometer. The probe is connected to a control box at ground level through a string of drill rods, electric wires for power supply and nylon tubing for the supply of nitrogen gas. Beneath the membrane is a measuring device which turns a buzzer off in the control box. The method of conducting the DMT is as follows:

1. The probe is positioned at the required level. Nitrogen gas is pumped into the probe. When the membrane is just flush with the side of the surface, a pressure reading is taken which is called the *lift-off* pressure. Approximate zero corrections are made. This pressure is called p_1.

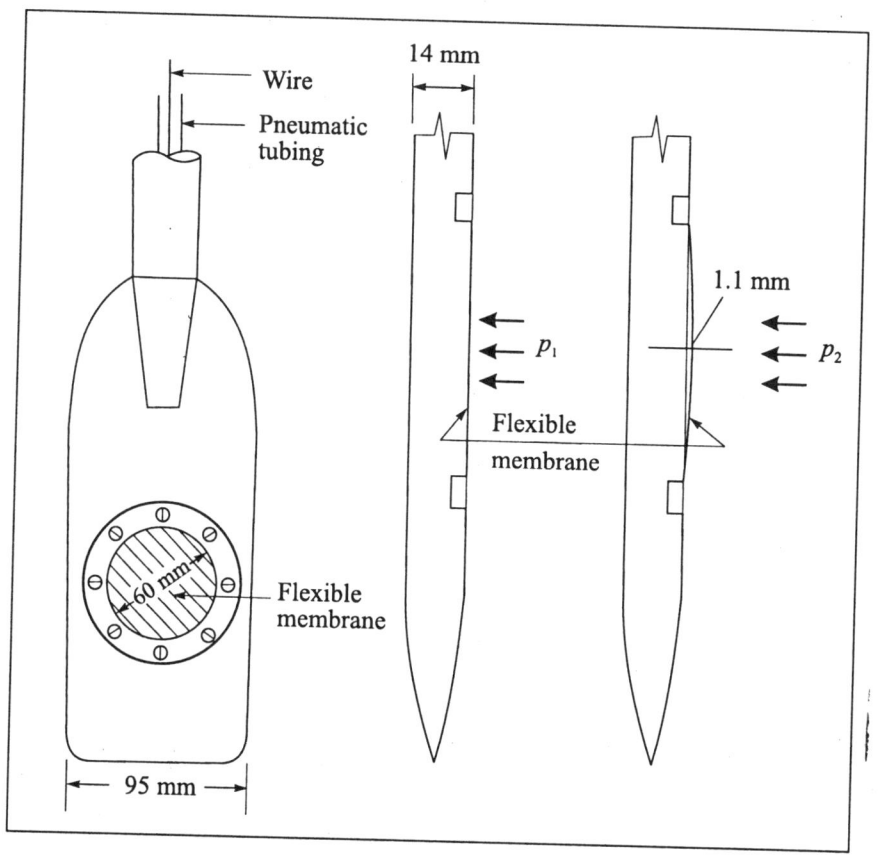

Fig. 3.27 Flat plate dilatometer (after Marchetti, 1980)

2. The probe pressure is increased until the membrane expands by an amount $\Delta l = 1.1$ mm. The corrected pressure is p_2.

3. The next step is to decrease the pressure until the membrane returns to the lift off position. This corrected reading is p_3. This pressure is related to excess pore water pressure (Schmertmann, 1986).

The details of the calculation lead to the following equations.

1. Material index,
$$I_D = \frac{p_z - p_1}{p_2 - u} \tag{3.24a}$$

2. The lateral stress index,
$$K_D = \frac{p_1 - u}{p_o'} \tag{3.24b}$$

3. The dilatometer modulus,
$$E_D = 34.7 \, (p_2 - p_1) \text{ kN/m}^2 \tag{3.24c}$$

where, p_o' = effective overburden pressure = $\gamma'z$

u = pore water pressure equal to static water level pressure

γ' = effective unit weight of soil

z = depth of probe level from ground surface

The lateral stress index K_D is related to K_0 (the coefficient of earth pressure for the at-rest condition) and to *OCR* (overconsolidation ratio).

Marchetti (1980) has correlated several soil properties as follows

$$E_s = (1 - \mu^2) E_D \tag{3.25}$$

$$K_0 = \left(\frac{K_D}{1.5}\right)^{0.47} - 0.6 \tag{3.26}$$

$$OCR = (0.5 \, K_D)^{1.6} \tag{3.27a}$$

$$\left(\frac{c_u}{p_o{}'}\right)_{oc} = \left(\frac{c_u}{p_o{}'}\right)_{nc} \times (0.5 \, K_D)^{1.25} \tag{3.27b}$$

where E_s is the modulus of elasticity.

The soil classification as developed by Schmertmann (1986) is given in Fig. 3.28. I_D is related with E_D in the development of the profile.

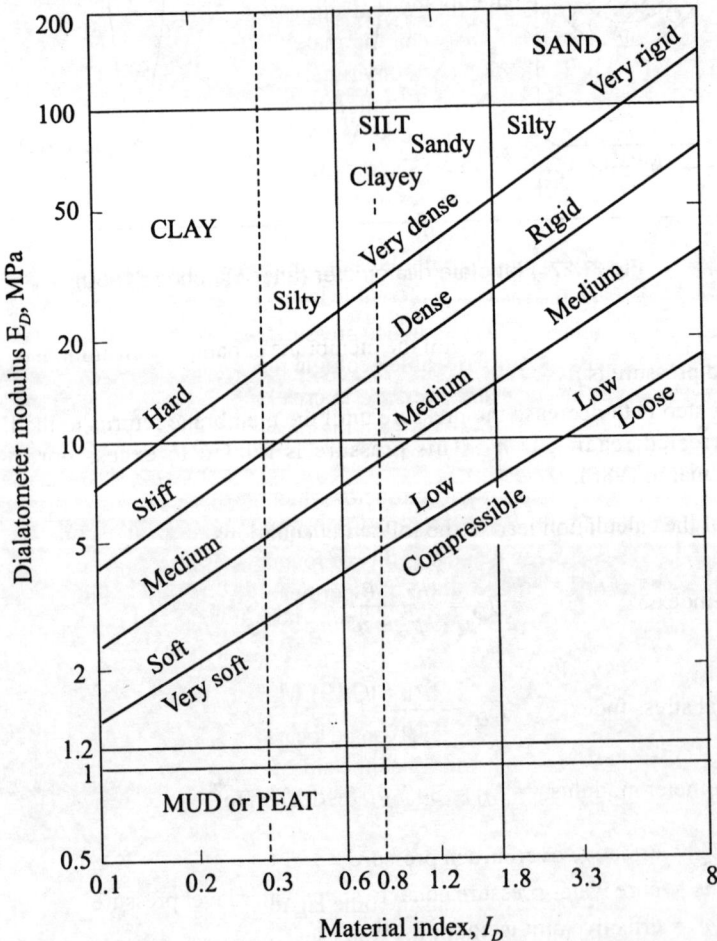

Fig. 3.28 Soil profile based on dilatometer test (after Schmertmann, 1986)

3.12 FIELD VANE SHEAR TEST (VST)

The vane shear test is one of the *in-situ* tests used for obtaining the undrained shear strength of soft sensitive clays. It is in deep beds of such material that the vane test is most valuable for the simple reason that there is at present no other method known by which the shear strength of these clays can be measured. The details of the VST have been explained in the book *Soil Mechanics and Foundation Engineering* by the same author.

3.13 FIELD PLATE LOAD TEST (PLT)

The field plate test is the oldest of the methods for determining either the bearing capacity or settlement of footings. The details of PLT are discussed under Shallow Foundations in Chapter 6.

3.14 GROUND WATER CONDITIONS

Introduction

Ground water conditions play an important part in the stability of foundations. If the water table lies very close to the base of footings, the bearing capacity and settlement characteristics of the soil would be affected. The level of the water table fluctuates with season. During the end of monsoons, the water table level will be closer to the ground surface as compared to the period just before the monsoons. The difference in levels between the maximum and the minimum may fluctuate from year to year. In many big projects, it is sometimes very essential to know these fluctuations. Piezometers are therefore required to be installed in such areas for measuring the level of water table for one or more years. In some cases clients may demand the depth of water table during the period of site investigation. The depth can be measured fairly accurately during boring operation. Normally, during boring, the water table drops down in the bore hole and attains equilibrium condition after a period of time. In a fairly draining material such as sand and gravel, the water level returns to its original position in a matter of a few minutes or hours, whereas, in soils of low permeability it may take several days. In such cases, the water table level has to be located by some reliable method.

In some cases, the ground water flows under pressure through a pervious layer of soil confined from its top and bottom between impermeable geologic formations. If the water flows from a higher elevation to a lower level, an *artesian pressure* is created and such a ground water is termed as *artesian water.* It is very essential to investigate the possibility of existance of artesian water in a project area.

Permeability of soils is another important factor which needs to be known in many of the major projects. Selection of pumps for pumping out water from excavated trenches or pits depends on the permeability of soils. The settlement and stability of foundations also depend on the permeability of soils. The different method for determining the permeability of soils have been discussed in the book of *Soil Mechanics and Foundation Engineering* by the same author.

Water Table Location by Hvorselev (1949) Method

As per the Hvorselev method, water table level can be located in a bore hole used for soil investigation. The bore hole should have a casing to stabilise the sides. The method normally used, is the Rising Water Level Method for determining the water table locations.

Rising water level method

This method most commonly referred to as the time lag method consists of bailing the water out of the casing and observing the rate of rise of the water level in the casing at intervals of time until the rise in water level becomes negligible. The rate is observed by measuring the elapsed time and the depth of the water surface below the top of the casing. The intervals at which the readings are required will vary some what with the permeability of the soil. In no case should the total elapsed time for the readings be less than 5 minutes. In freely draining materials such as sands, gravels, etc. the interval of time between successive readings may not exceed 1 to 2 hours, but in soils of low permeability such as fine sand silts and clays, the intervals may rise from 12 to 24 hours, and it may take a few days to determine the stabilised water table level.

Let the time be t_0 when the water table level was at depth H_o below the normal water table level (Fig. 3.29). Let the successive rise in water levels be h_1, h_2, h_3, etc. at times t_1, t_2, t_3 respectively wherein the difference in time $(t_1 - t_0), (t_2 - t_1), (t_3 - t_2)$, etc. is kept constant.

Now, from Fig. 3.29

$$H_0 - H_1 = h_1$$

$$H_1 - H_2 = h_2$$

$$H_2 - H_3 = h_3$$

Let $\qquad (t_1 - t_0) = (t_2 - t_1) = (t_3 - t_2)$, etc. $= \Delta t$

The depths H_0, H_2, H_3 of the water level in the casing from the normal water table level can be computed as follows.

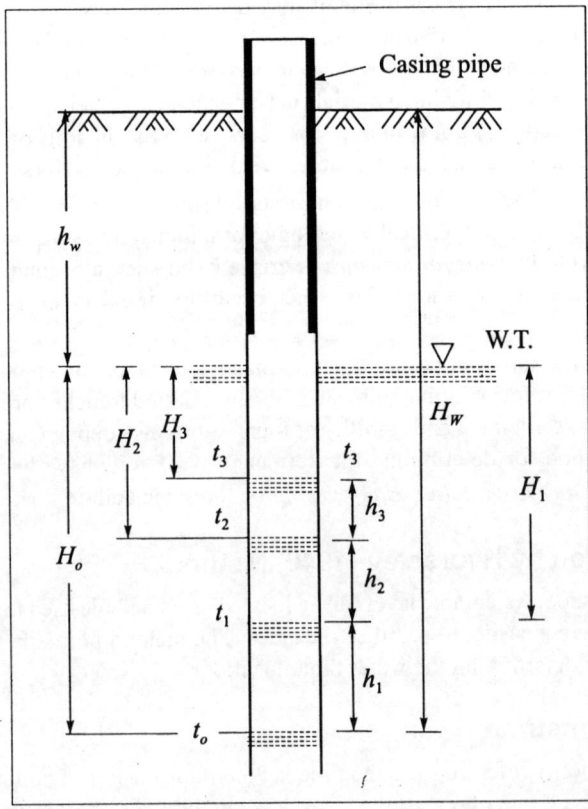

Fig. 3.29 Water table level location by rising water level method

$$H_0 = \frac{h_1^2}{h_1 - h_2}$$

$$H_2 = \frac{h_2^2}{h_1 - h_2} \Bigg\} , \text{etc.} \tag{3.28}$$

$$H_3 = \frac{h_3^2}{h_2 - h_3}$$

Let the corresponding depths of water table level below the ground surface be h_{w1}, h_{w2}, h_{w3}, etc. Now, we have

$$h_{w1} = H_w - H_o$$

$$h_{w2} = H_w - (h_1 + h_2) - H_2$$

$$h_{w3} = H_w - (h_1 + h_2 + h_3) - H_3$$

where, H_w is the depth of water level in the casing from the ground surface at the start of the test. Normally, $h_{w1} = h_{w2} = h_{w3} = h_w$; if not an average value gives h_w.

Example 3.11

Establish the location of ground water in a clayey strata. Water in bore hole was bailed out to a depth of 10.67 m below ground surface, and the rise of water was recorded at 24 hour intervals as follows Fig. 3.29.

$$h_1 = 64.0 \text{ cm}, h_2 = 57.9 \text{ cm and } h_3 = 51.8 \text{ cm.}$$

Solution

$$H_0 = \frac{0.64^2}{0.64 - 0.579} = 6.714 \text{ m,}$$

$$H_2 = \frac{0.579^2}{0.64 - 0.579} = 5.496 \text{ m,}$$

$$H_3 = \frac{0.518^2}{0.579 - 0.518} = 4.399 \text{ m,}$$

$$h_{w1} = 10.67 - 6.714 = 3.953 \text{ m,}$$

$$h_{w2} = 10.67 - (0.64 + 0.579 + 5.496) = 3.955,$$

$$h_{w3} = 10.67 - (0.64 + 0.579 + 0.518 + 4.399) = 4.534.$$

Average, $$h_w = \frac{3.953 + 3.955 + 4.534}{3} = 4.147 \text{ m.}$$

Ground Water Level Observation

When measurement of water table levels has to be made over a long period of time, the best way of doing this is to instal a series of stand pipes or piezometers over the area in bore holes and observe the water levels. A simple stand pipe [Fig. 3.30 (a)] consisting of a PVC tubing with a slotted end and surrounded by granular filter or plastic fabrics is satisfactory for granular soils or permeable rocks. In silts or clays more sensitive equipment is required. The hydraulic piezometer [Fig. 3.30 (b)] consists of a porous element connected by twin small-bore plastic tubing to a remote reading station where pressures are measured by a mercury monometer or a Bourden gauge.

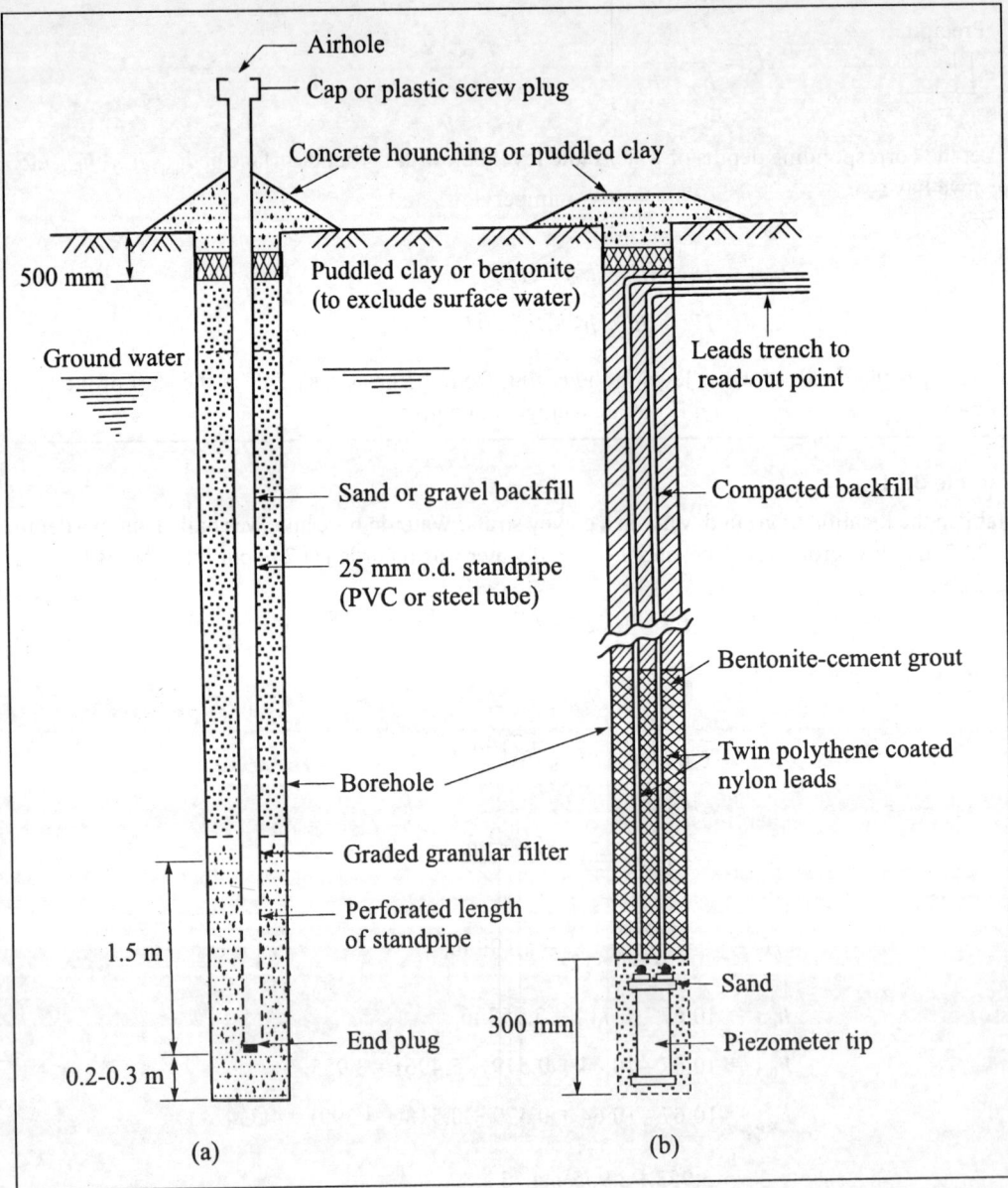

Fig. 3.30 Types of piezometer for ground water level observations: (a) Standpipe, and (b) hydraulic piezometer

Artesian Ground Water Flow

Figure 3.31 (a) is a schematic sketch which shows the presence of artesian ground water flow and artesian pressure. If a well is made in the soil puncturing the top impermeable stratum of soil, the water in the well rises to a height h above the bottom of the well. If p is the pressure acting at the bottom of the well, the water rises to a height h, which is equal to

$$h = \frac{p}{\gamma_w}$$

(3.29)

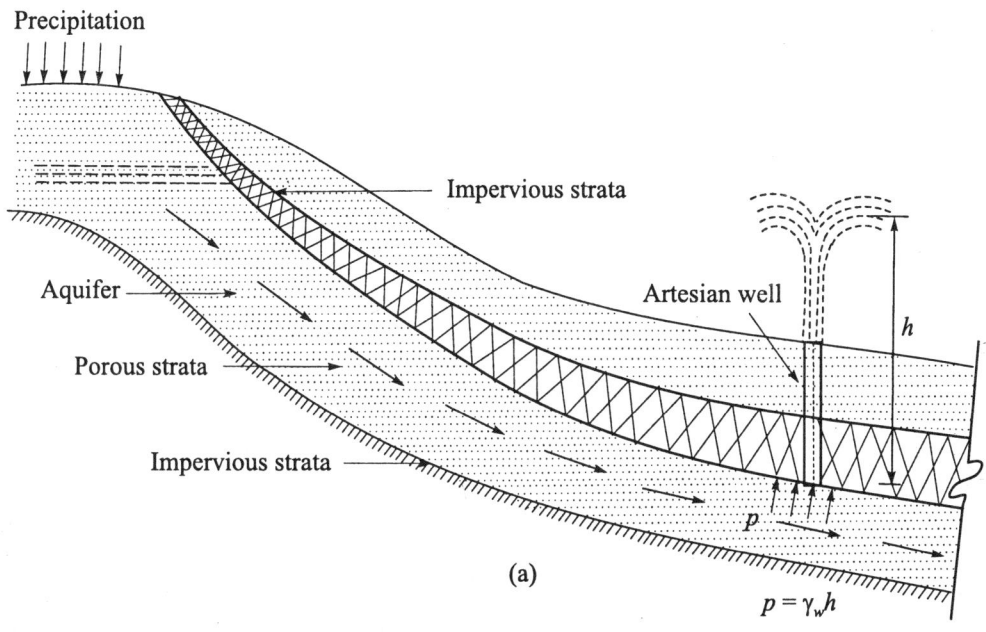

(a)

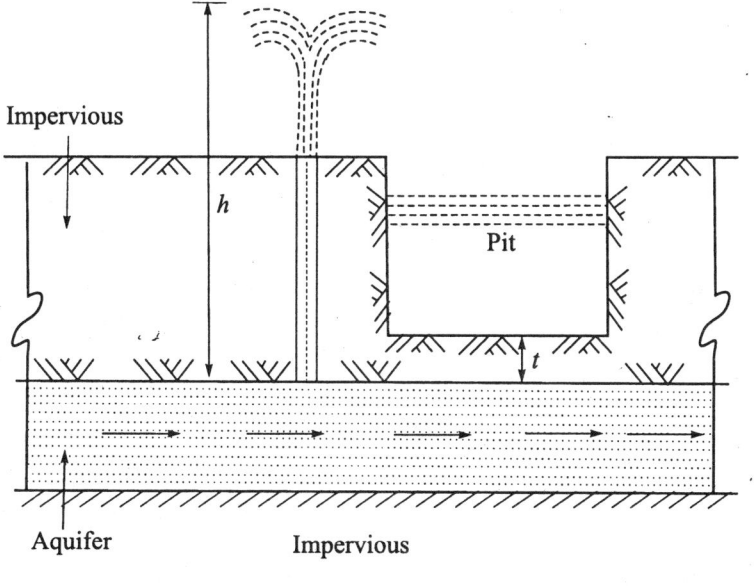

(b)

Fig. 3.31 Artesian ground water flow: (a) Artesian well, (b) effect of artesian water on the stability of the bottom of an excavation

Such a well is called as *artesian well*. Ground water flow under artesian pressure is quite common in valleys and in areas close to hilly tracts. In such a locality, ground water flows from higher elevations to lower elevations thereby, give rise to *artesian water* under high pressures. The artesian water endangers the stability of foundations if the bottom of the foundations lie close to the artesian acquifer.

Figure 3.31 (b), indicates the possible destruction of a foundation pit excavated in a soil below which artesian condition prevails. If the impermeable soil strata below the bottom of the trench (or foundation) is of insufficient thickness, *t*, the *artesian water pressure* may break through the bottom of the pit inundate it, or even destroy it totally.

3.15   GEOPHYSICAL EXPLORATION

The stratification of soils and rocks can be determined by geophysical methods of exploration which measure changes in certain physical characteristics of these materials, for example, the magnetism, density, electrical resistivity, elasticity or a combination of these properties. However, the utility of these methods in the field of foundation engineering is very limited since the methods do not quantify the characteristics of the various substrata. Vital information on ground water conditions is usually lacking. Geophysical methods at best provide some missing information between widely spaced bore holes but they cannot replace bore holes. Two methods of exploration which are sometimes useful are discussed briefly in this section. They are

1. Seismic refraction method,
2. Electrical resistivity method.

Seismic Refraction Method

The seismic refraction method is based on the fact that seismic waves have different velocities in different types of soils (or rock). The waves refract when they cross boundaries between different types of soils. If artificial impulses are produced either by detonation of explosives or mechanical blows with a heavy hammer at the ground surface or at shallow depth within a hole, these shocks generate three types of waves. In general, only compression waves (longitudinal waves) are observed. These waves are classified as either direct, reflected or refracted. Direct waves travel in approximately straight lines from the source of the impulse to the surface. Reflected or refracted waves undergo a change in direction when they encounter a boundary separating media of different seismic velocities. The *seismic refraction method* is more suited to shallow exploration for civil engineering purposes.

The method starts by inducing impact or shock waves into the soil at a particular location. The shock waves are picked up by geophones. In Fig. 3.32 (a), point A is the source of seismic impulse. The points D_1 through D_8 represent the locations of the geophones or detectors which are installed in a straight line. The spacings of the geophones are dependent on the amount of detail required and the depth of the strata being investigated. In general, the spacing must be such that the distance from D_1 to D_8 is three to four times the depth to be investigated. The geophones are connected by cable to a central recording device. A series of detonations or impacts are produced and the arrival time of the first wave at each geophone position is recorded in turn. When the distance between source and geophone is short, the arrival time will be that of a direct wave. When the distance exceeds a certain value (depending on the thickness of the stratum), the refracted wave will be the first to be detected by the geophone. This is because the refracted wave, although longer than that of the direct wave, passes through a stratum of higher seismic velocity.

A typical plot of test results for a three layer system is given in Fig. 3.32 (a), with the arrival time plotted against the distance source and geophone. As in the figure, if the source-geophone spacing is more than the distance d_1, which is the distance from the source to point B, the direct wave reaches the geophone in advance of the refracted wave and the time-distance relationship is represented by

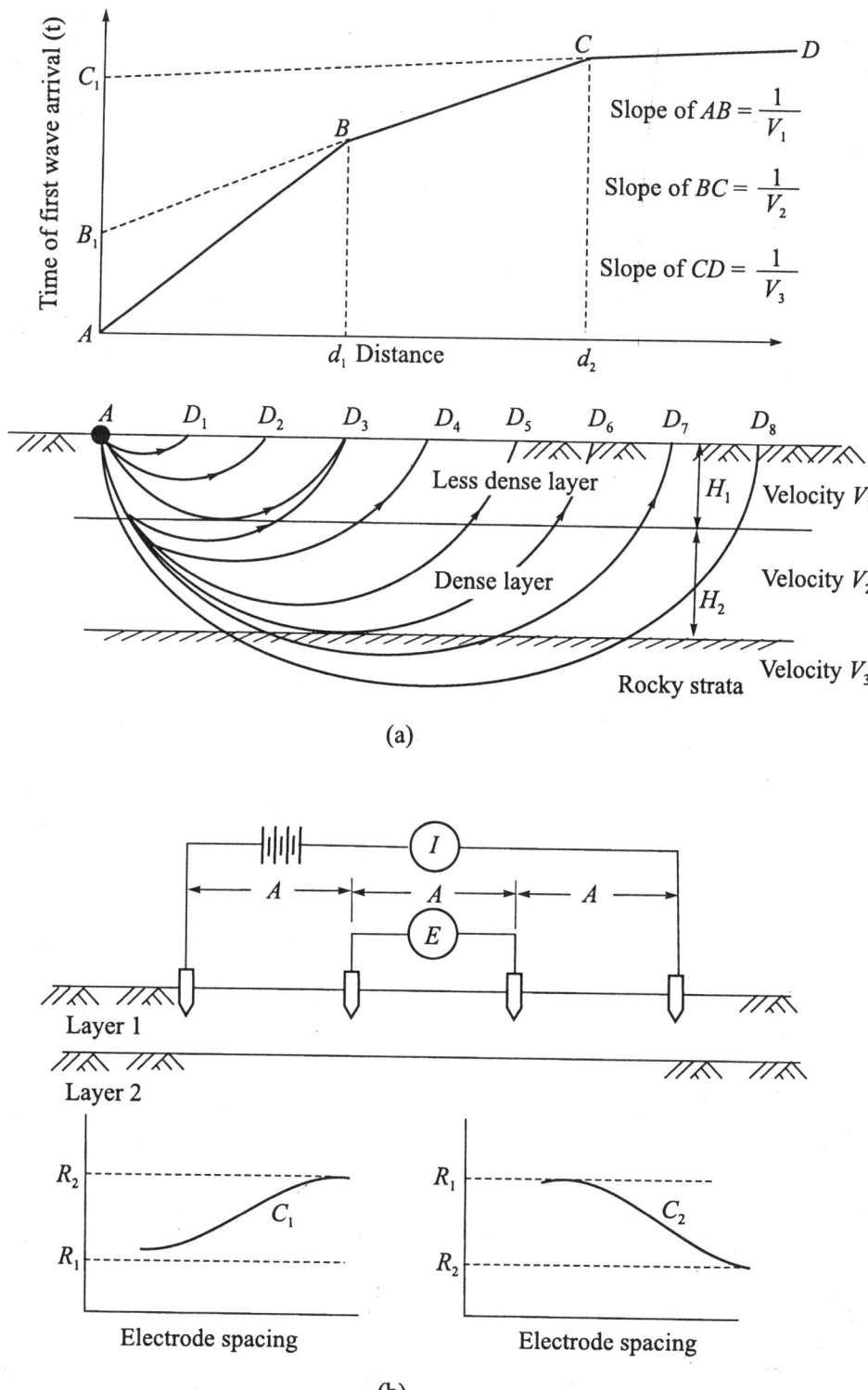

Fig. 3.32 Geophysical methods of exploration: (a) Schematic representation of refraction method, (b) schematic representation of electrical resistivity method

a straight line AB through the origin represented by A. If on the other hand, the source geophone distance is greater than d_1, the refracted waves arrive in advance of the direct waves and the time-distance relationship is represented by another straight line BC which will have a slope different from that of AB. The slopes of the lines AB and BC are represented by $1/V_1$, and $1/V_2$ respectively, where V_1 and V_2 are the velocities of the upper and lower strata respectively. Similarly, the slope of the third line CD is represented by $1/V_3$ in the third strata.

The general types of soil or rocks can be determined from a knowledge of these velocities. The depth H_1 of the top strata (provided the thickness of the stratum is constant) can be estimated from the formula

$$H_1 = \frac{d_1}{2} \sqrt{\frac{V_2 - V_1}{V_2 + V_1}} \tag{3.30a}$$

The thickness of the second layer (H_2) is obtained from

$$H_2 = 0.85 \, H_1 + \frac{d_2}{2} \sqrt{\frac{V_3 - V_2}{V_3 + V_2}} \tag{3.30b}$$

The procedure is continued if there are more than three layers.

If the thickness of any stratum is not constant, average thickness is taken.

The following equations may be used for determining the depths H_1 and H_2 in a three layer strata:

$$H_1 = \frac{t_1 V_1}{2 \cos \alpha} \tag{3.31}$$

$$H_2 = \frac{t_2 V_2}{2 \cos \beta} \tag{3.32}$$

where $t_1 = AB_1$, Fig. 3.32 (a); the point B_1 is obtained on the vertical passing through A by extending the straight line CB,

$t_2 = (AC_1 - AB_1)$; AC_1 is the intercept on the vertical through A obtained by extending the straight line DC,

$\alpha = \sin^{-1}(V_1/V_2)$,

$\beta = \sin^{-1}(V_2/V_3)$ (3.33)

α and β are the angles of refraction of the first and second stratum interfaces respectively.

The formulae used to estimate the depths from seismic refraction survey data are based on the following assumptions:

1. Each stratum is homogeneous and isotropic.
2. The boundaries between strata are either horizontal or inclined planes.
3. Each stratum is of sufficient thickness to reflect a change in velocity on a time-distance plot.
4. The velocity of wave propagation for each succeeding stratum increases with depth.

Table 3.12 gives typical seismic velocities in various materials. Detailed investigation procedures for refraction studies are presented by Jakosky (1950).

Table 3.12 Range of seismic velocities in soils near the surface or at shallow depths
(after Peck *et al*, 1974)

Material	Velocity	
	ft/sec	*m/sec*
1. Dry silt, sand, loose gravel, loam, loose rock talus, and moist fine-grained top soil	600 – 2500	180 – 760
2. Compact till, indurated clays, compact clayey gravel, cemented sand and sand clay	2500 – 7500	760 – 2300
3. Rock, weathered, fractured or partly decomposed	2000 – 10000	600 – 3000
4. Shale, sound	2500 – 11000	760 – 3350
5. Sandstone, sound	5000 – 14000	1500 – 4300
6. Limestone, chalk, sound	6000 – 20000	1800 – 6000
7. Igneous rock, sound	12000 – 20000	3650 – 6000
8. Metamorphic rock, sound	10000 – 16000	3000 – 4900

Electrical Resistivity Method

The method depends on differences in the electrical resistance of different soil (and rock) types. The flow of current through a soil is mainly due to electrolytic action and therefore depends on the concentration of dissolved salts in the pores. The mineral particles of soil are poor conductors of current. The resistivity of soil, therefore, decreases as both water content and concentration of salts increase. A dense clean sand above the water table, for example, would exhibit a high resistivity due to its low degree of saturation and virtual absence of dissolved salts. A saturated clay of high void ratio, on the other hand, would exhibit a low resistivity due to the relative abundance of pore water and the free ions in that water.

There are several methods by which the field resistivity measurements are made. The most popular of the methods is the Wenner method.

Wenner method

The Wenner arrangement consists of four equally spaced electrodes driven approximately 20 cm into the ground as shown in Fig. 3.32 (b). In this method a dc current of known magnitude is passed between the two outer (current) electrodes, thereby producing within the soil an electric field, whose pattern is determined by the resistivities of the soils present within the field and the boundary conditions. The potential drop E for the surface current flow lines is measured by means of the inner electrodes. The apparent resistivity, R, is given by the equation

$$R = \frac{2\pi \, AE}{I} \tag{3.34}$$

It is customary to express A in centimeters, E in volts, I in amperes, and R ohm-cm. The apparent resistivity represents a weighted average of true resistivity to a depth A in a large volume of soil, the soil close to the surface being more heavily weighted than the soil at greater depths. The presence of a stratum of low resistivity forces the current to flow closer to the surface resulting in a higher voltage drop and hence a higher value of apparent resistivity. The opposite is true if a stratum of low resistivity lies below a stratum of high resistivity.

The method know as *sounding* is used when the variation of resistivity with depth is required. This enables rough estimates to be made of the types and depths of strata. A series of readings are taken, the (equal) spacing of the electrodes being increased for each successive reading. However, the centre of the four electrodes remains at a fixed point. As the spacing is increased, the apparent resistivity is influenced by a greater depth of soil. If the resistivity increases with the increasing electrode spacings, it can be concluded that an underlying stratum of higher resistivity is beginning to influence the readings. If increased separation produces decreasing resistivity, on the other hand, a lower resistivity is beginning to influence the readings.

Apparent resistivity is plotted against spacing, preferably, on log paper. Characteristic curves for a two layer structure are shown in Fig. 3.32 (b). For curve C_1, the resistivity of layer 1 is lower than that of 2; for curve C_2, layer 1 has a higher resistivity than that of layer 2. The curves become asymptotic to lines representing the true resistance R_1, and R_2 of the respective layers. Approximate layer thickness can be obtained by comparing the observed curves of resistivity *versus* electrode spacing with a set of standard curves.

The procedure known as *profiling* is used in the investigation of lateral variation of soil types. A series of readings is taken, the four electrodes being moved laterally as a unit for each successive reading; the electrode spacing remains constant for each reading of the series. Apparent resistivity is plotted against the centre position of the four electrodes, to natural scale; such a plot can be used to locate the position of a soil of high or low resistivity. Contours of resistivity can be plotted over a given area.

The electrical method of exploration has been found to be not as reliable as the seismic method as the apparent resistivity of a particular soil or rock can vary over a wide range of values.

Representative values of resistivity are given in Table 3.13.

Table 3.13 Representative values of resistivity. The values are expressed in units of 10^3 ohm-cm (after Peck *et al*, 1974)

Material	Resistivity ohm-cm
Clay and saturated silt	0 – 10
Sandy clay and wet silty sand	10 – 25
Clayey sand and saturated sand	25 – 50
Sand	50 – 100
Gravel	150 – 500
Weathered rock	100 – 200
Sound rock	150 – 4000

Example 3.12

A seismic survey was carried out for a large project to determine the nature of the substrata. The results of the survey are given in Fig. Ex. 3.12 in the form of a graph. Determine the depths of the strata.

Solution

Two methods may be used

1. Use of Eq. (3.30)
2. Use of Eqs (3.31) and (3.32)

First we have to determine the velocities in each stratum (Fig. Ex. 3.12).

$$V_1 = \frac{\text{Distance}}{\overline{AB}} = \frac{2.188}{12.75 \times 10^{-3}} = 172 \text{ m/sec}$$

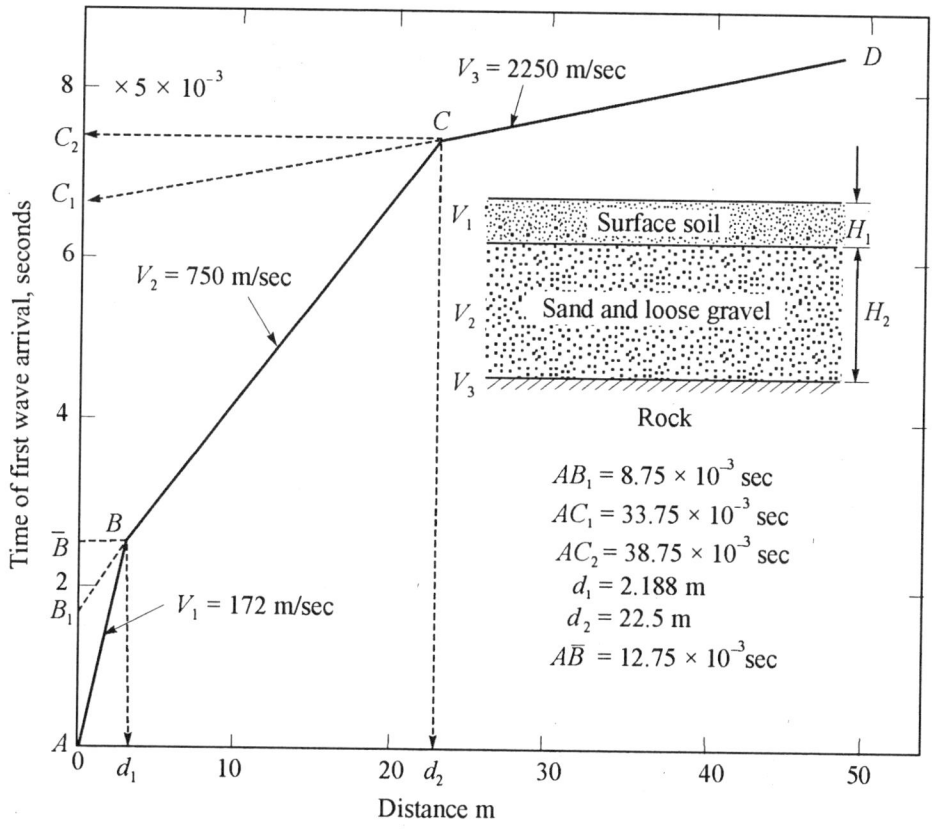

Fig. Ex. 3.12

$$V_2 = \frac{d_2}{AC_2 - AB_1} = \frac{22.5}{(7.75 - 1.75)\, 5 \times 10^{-3}} = 750 \text{ m/sec}$$

In the same way, the velocity in the third stata can be determined. The velocity obtained is $V_3 = 2250$ m/sec.

Method 1

From Eq. (3.30a), the thickness H_1 of the top layer is

$$H_1 = \frac{d_1}{2}\sqrt{\frac{V_2 - V_1}{V_2 + V_1}} = \frac{2.188}{2}\sqrt{\frac{750 - 172}{1000}} = 0.83 \text{ m}$$

From Eq. (3.30b) the thickness H_2 is

$$H_2 = 0.85\, H_1 + \frac{d_2}{2}\sqrt{\frac{V_3 - V_2}{V_s + V_2}}$$

$$H_2 = 0.85 \times 0.83 + \frac{22.5}{2}\sqrt{\frac{2250 - 750}{3000}}$$

$$= 0.71 + 7.955 = 8.67 \text{ m}$$

Method 2

From Eq. (3.31)

$$H_1 = \frac{t_1 V_1}{2 \cos \alpha}$$

$$t_1 = AB_1 = 1.75 \times 5 \times 10^{-3} \text{ sec (Fig. Ex. 3.12)}$$

$$\alpha = \sin^{-1} \frac{V_1}{V_2} = \sin^{-1} \frac{172}{750} = 13.26°$$

$$\cos \alpha = 0.9733$$

$$H_1 = \frac{12.75 \times 10^{-3} \times 172}{2 \times 0.9737} = 1.13 \text{ m}$$

From Eq. (3.32)

$$H_2 = \frac{t_2 V_2}{2 \cos \beta}$$

$$t_2 = 5 \times 5 \times 10^{-3} \text{ sec}$$

$$\beta = \sin^{-1} \frac{750}{2250} = 19.47°; \cos \beta = 0.9428$$

$$H_2 = \frac{5 \times 5 \times 10^{-3} \times 750}{2 \times 0.9428} = 9.94 \text{ m}$$

3.16 PLANNING OF SOIL EXPLORATION

The planner has to consider the following points before making a programme:

1. Type, size and importance of the project.
2. Whether the site investigation is preliminary or detailed.

In the case of large projects, a preliminary investigation is normally required for the purpose of

1. Selecting a site and making a feasibility study of the project.
2. Making tentative designs and estimates of the cost of the project.

Preliminary site investigation needs only a few bore holes distributed suitably over the area for taking samples. The data obtained from the field and laboratory tests must be adequate to provide a fairly good idea of the strength characteristics of the subsoil for making preliminary drawings and design. In case a particular site is found unsuitable on the basis of the study, an alternate site may have to be chosen.

Once a site is chosen, a detailed soil investigation is undertaken. The planning of a soil investigation includes the following steps:

1. A detailed study of the geographical condition of the area which include
    (a) Collection of all the available information about the site, including the collection of existing topographical and geological maps.
    (b) General topographical features of the site.
    (c) Collection of the available hydraulic conditions, such as water table fluctuations, flooding of the site, etc.
    (d) Access to the site.

2. Preparation of a layout plan of the project.
3. Preparation of a borehole layout plan which includes the depths and the number of bore holes suitably distributed over the area.
4. Marking on the layout plan any additional types of soil investigation.
5. Preparation of specifications and guidelines for the field execution of the various elements of soil investigation.
6. Preparation of specifications and guide lines for laboratory testing of the samples collected, presentation of field and laboratory test results, writing of report, etc.

The planner can make an intelligent, practical and pragmatic plan if he is conversant with the various elements of soil investigation.

Depths and Number of Bore Holes

Depths of bore holes

The depth up to which bore holes should be driven is governed by the depth of soil affected by the foundation bearing pressures. The standard practice is to take the borings to a depth (called the significant depth) at which the excess vertical stress caused by a fully loaded foundation is of the order of 20 percent or less of the net imposed vertical stress at the foundation base level. The depth the borehole as per this practice works out to about 1.5 times the least width of the foundation from the base level of the foundation as shown in Fig. 3.33 (a). Where strip or pad footings are closely spaced which results in the overlapping of the stressed zones, the whole loaded area becomes in effect a raft foundation with correspondingly deep borings as shown in Fig. 3.33 (b) and (c). In the case of pile or pier foundations the subsoil should be explored to the depths required to cover the soil lying even below the tips of piles (or pile groups) and piers which are affected by the loads transmitted to the deeper layers, Fig. 3.33 (d). In case rock is encountered at shallow depths, foundations may have to rest on rocky strata. The boring should also explore the strength characteristics of rocky strata in such cases.

Number of bore holes

An adequate number of bore holes is needed to

1. Provide a reasonably accurate determination of the contours of the proposed bearing stratum.
2. Locate any soft pockets in the supporting soil which would adversely affect the safety and performance of the proposed design.

The number of bore holes which need to be driven on any particular site is a difficult problem which is closely linked with the relative cost of the investigation and the project for which it is undertaken. When the soil is homogeneous over the whole area, the number of bore holes could be limited, but if the soil condition is erratic, limiting the number would be counter productive.

3.17 EXECUTION OF SOIL EXPLORATION PROGRAMME

The three limbs of a soil exploration are:

1. Planning,
2. Execution,
3. Report writing.

All three limbs are equally important for a satisfactory solution of the problem. However, the execution of the soil exploration programme acts as a bridge between planning and report writing,

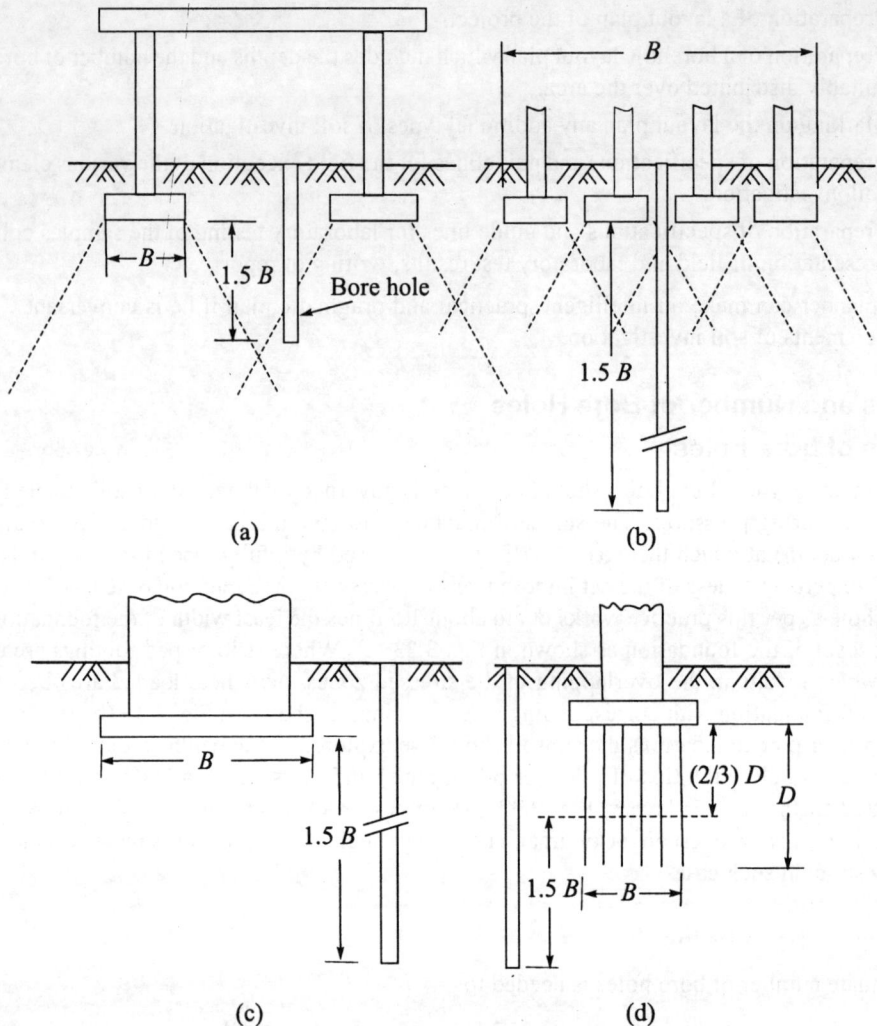

Fig. 3.33 Depth of bore holes: (a) Footings placed far apart, (b) footings placed at close intervals, (c) raft foundation, (d) pile foundation

and as such occupies an important place. No amount of planning would help report writing, if the field and laboratory works are not executed with diligence and care. It is essential that the execution part should always be entrusted to well-qualified, reliable and resourceful geotechnical consultants, who will also be responsible for report writing.

Deployment of Personnel and Equipment

The geotechnical consultant should have well-qualified and experienced engineers and supervisors, who complete the work per the requirements. The firm should have the capacity to deploy an adequate number of rigs and personnel for satisfactory completion of the job on time.

Boring Logs

A detailed record of boring operations and other tests carried out in the field is an essential part of the field work. The bore hole log is made during the boring operation. The soil is classified based on the visual examination of the disturbed samples collected. A typical example of a bore hole log is given in Fig. 3.34. The log should include the difficulties faced during boring operations including the occurrence of sand boils, and the presence of artesian water conditions if any, etc.

BORE-HOLE LOG

Job No.	Date: 06-04-84
Project: Farakka STPP	BH No.: 1
	GL: 64.3 m
Location: WB	WTL: 63.0 m
Boring Method: Shell and Auger	Supervisor: X
Dia. of BH 15 cm	

Soil type		Level m	Depth m	SPT				Sample type	Remarks
				15 cm	15 cm	15 cm	N		
Yellowish stiff clay			1.0	4	6	8	14	D U	
		62.3							
Greyish sandy silt med. dense			3.3	7	10	16	26	D W	
		59.8							
Greyish silty sand dense			5.0	14	16	21	37	D	
			7.5	15	18	23	41	D U	
		56.3							
Blackish very stiff clay			9.0	9	10	14	24	D	
		53.3	11.0						

D = disturbed sample,   U = undisturbed sample,
W = water sample,   N = SPT value.

Fig. 3.34 A typical bore-hole log

In-situ Tests

The field work may also involve one or more of the *in-situ* tests discussed earlier. The record should give the details of the tests conducted with exceptional clarity.

Laboratory Testing

A preliminary examination of the nature and type of soil brought to the laboratory is very essential before deciding upon the type and number of laboratory tests. Normally, the SPT samples are used for this purpose. First the SPT samples should be arranged borewise and depthwise. Each of the samples should be examined visually. A chart should be made giving the bore hole numbers and the types of tests to be conducted on each sample depthwise. An experienced geotechnical engineer can do this job with diligence and care.

Once the types of tests are decided, the laboratory assistant should carry out the tests with all the care required for each of the tests. The test results should next be tabulated on a suitable format borewise and the soil is classified according to standard practice. The geotechnical consultant should examine each of the tests before being tabulated. Unreliable test results should be discarded.

Graphs and Charts

All the necessary graphs and charts are to be made based on the field and laboratory test results. The charts and graphs should present a clear insight into the subsoil conditions over the whole area. The charts made should help the geotechnical consultant to make a decision about the type of foundation, the strength and compressibility characteristics of the subsoil, etc.

3.18 REPORT

A report is the final document of the whole exercise of soil exploration. A report should be comprehensive, clear and to the point. Many can write reports, but only a very few can produce a good report. A report writer should be knowledgable, practical and pragmatic. No theory, books or codes of practice provide all the materials required to produce a good report. It is the experience of a number of years of dedicated service in the field which helps a geotechnical consultant make report writing an art. A good report should normally comprise the following:

1. A general description of the nature of the project and its importance.
2. A general description of the topographical features and hydraulic conditions of the site.
3. A brief description of the various field and laboratory tests carried out.
4. Analysis and discussion of the test results.
5. Recommendations.
6. Calculations for determining safe bearing pressures, pile loads, etc.
7. Tables containing borelogs, and other field and laboratory test results.
8. Drawings which include an index plan, a site-plan, test results plotted in the form of charts and graphs, soil profiles, etc.

3.19 PROBLEMS

3.1 Compute the area ratio of a sampling tube given the outside diameter = 100 mm and inside diameter = 94 mm. In what types of soil can this tube be used for sampling ?

3.2 A standard penetration test was carried out at a site. The soil profile is given in Fig. Prob. 3.2 with the penetration values. The average soil data are given for each layer. Compute the corrected values of N and plot showing the

   (a) variation of observed values with depth.

   (b) variation of corrected values with depth for standard energy 60%.

   Assume: $E_h = 0.7$, $C_d = 0.9$, $C_s = 0.85$ and $C_b = 1.05$

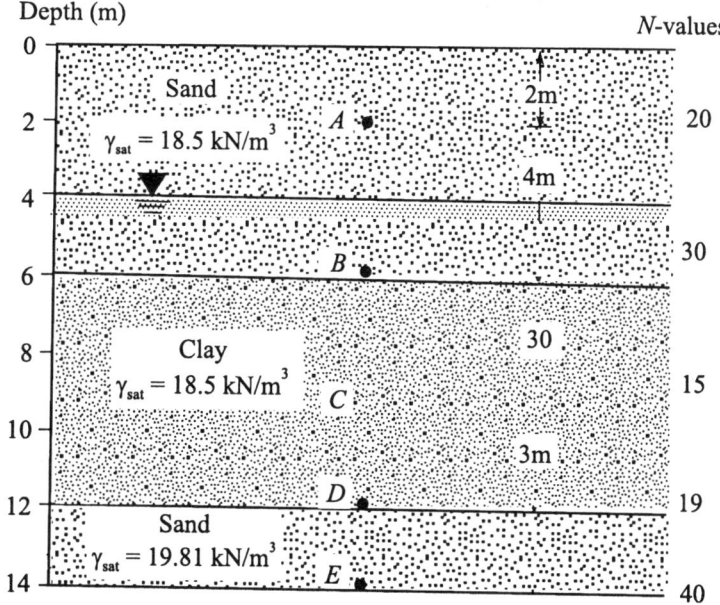

Fig. Prob. 3.2

3.3 For the soil profile given in Fig. Prob. 3.2, compute the corrected values of N for standard energy 70%.

3.4 For the soil profile given in Fig. Prob 3.2, estimate the average angle of friction for the sand layers based on the following:

(a) Table 3.5 (a).

(b) Equation (3.16b) by assuming the profile contains less than 5% fines [D_r may be taken from Table 3.5 (a)].

Estimate the values of ϕ and D_r for 60 percent standard energy.

Assume: $N_{cor} = N_{60}$.

3.5 For the corrected values of N_{60} given in Prob. 3.2, determine the unconfined compressive strengths of clay at points C and D in Fig. Prob. 3.2 by making use of Table 3.5 (b) and Eq. (3.16c). What is the consistency of the clay ?

3.6 A static cone penetration test was carried out at a site using an electric-friction cone penetrometer. Figure Prob. 3.6 gives the soil profile and values of q_c obtained at various depths.

(a) Plot the variation of q_c with depth.

(b) Determine the relative density of the sand at the points marked in the figure by using Fig. 3.17.

(c) Determine the angle of internal friction of the sand at the points marked by using Fig. 3.18.

3.7 For the soil profile given in Fig. Prob. 3.6, determine the unconfined compressive strength of the clay at the points marked in the figure using Eq. (3.18).

3.8 A static cone penetration test carried out at a site at a depth of 50 ft gave the following results:

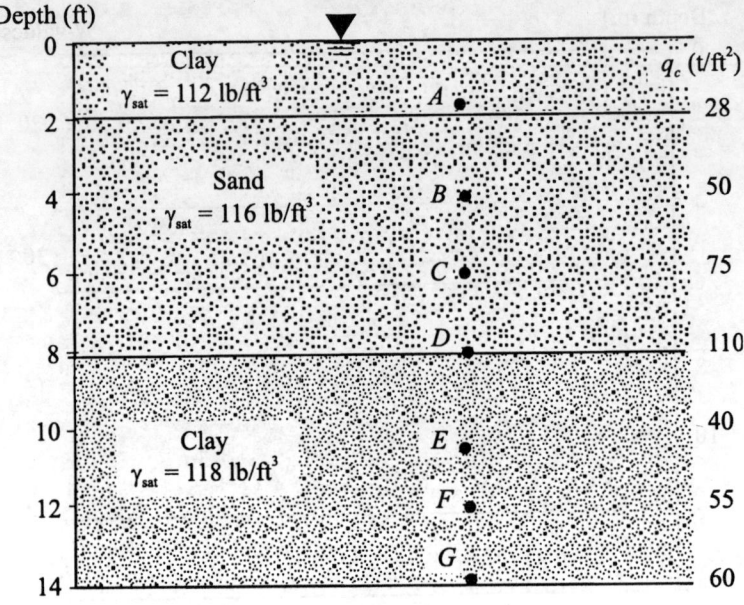

Fig. Prob. 3.6

(a) cone resistance $q_c = 250 \, t/ft^2$.

(b) average effective unit weight of the soil $= 115 \, lb/ft^3$.

Classify the soil for friction ratios of 0.9 and 2.5 percent.

3.9 A static cone penetration test was carried out at a site using an electric-friction cone penetrometer. Classify the soil for the following data obtained from the site

$q_c \ (MN/m^2)$	Friction ratio $R_f\%$
25	5
6.5	0.50
12.0	0.25
1.0	5.25

Assume in all the above cases that the effective overburden pressure is 50 kN/m².

3.10 Determine the relative density and the friction angle if the corrected SPT value N_{60} at a site is 30 from Eq. (3.20) and Table 3.7. What are the values of D_r and ϕ for N_{70}?

3.11 Figure Prob. 3.11 gives a corrected pressuremeter curve. The values of p_{om}, p_f and p_l and the corresponding volumes are marked on the curve. The test was conducted at a depth of 5 m below the ground surface. The average unit weight of the soil is 18.5 kN/m³. Determine the following:

(a) The coefficient of earth pressure for the at-rest condition.

(b) The Menard pressuremeter modulus.

(c) The undrained shear strength c_u.

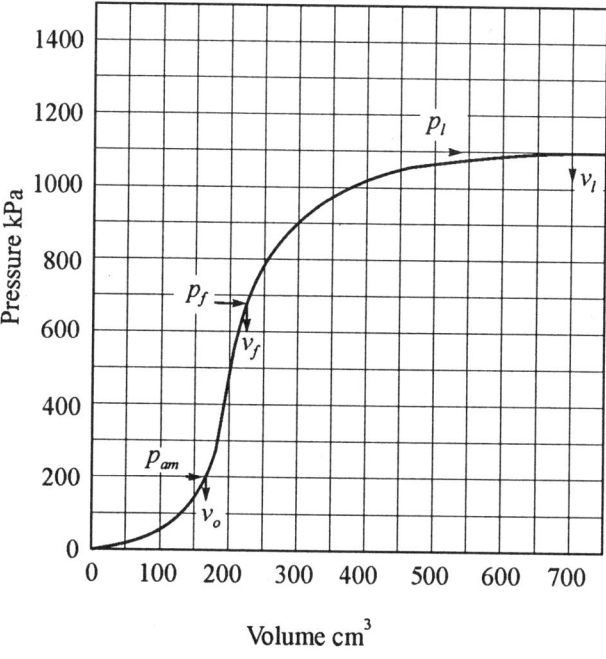

$p_{om} = 200$ kPa, $v_o = 180$ cm^3; $p_f = 660$ kPa; $v_f = 220$ cm^3;

$p_l = 1100$ kPa; $v_l = 700$ cm^3

Fig. Prob. 3.11

3.12 A seismic refraction survey of an area gave the following data:
   (*i*) Distance from impact point to geophone in m   15   30   60   80   100
   (*ii*) Time of first wave arrival in sec   0.025   0.25   0.10   0.11   0.12
      (*a*) Plot the time travel *versus* distance and determine velocities of the top and underlying layer of soil.
      (*b*) Determine the thickness of the top layer.
      (*c*) Using the seismic velocities evaluate the probable earth materials in the two layers.

Shallow Foundation 1
Depth of Foundation and Other Considerations

4

4.1 SHALLOW AND DEEP FOUNDATIONS

Introduction

Foundation is that part of a structure which serves exclusively to transmit loads from the structure on to the sub-soil. All structures, bridges, towers, etc. are built on soil. If the structure of soil lying close to the ground surface possesses adequate power to take on the loads from the superstructure, the foundations of such structures may lie at shallow depths. However, if the upper strata are too weak to take the load, the loads have to be transmitted to deeper depths by means of piles, piers, etc. Foundations can, therefore, be studied broadly two headings. They are:

1. Shallow Foundations.
2. Deep Foundations.

Shallow Foundation

For the purpose of study, shallow foundations are considered as those that are placed at a depth D_f, not exceeding the width, B, of the foundation. From the point of view of design, the shallow foundations are classified into four types. They are:

1. Spread footings or pad foundations.
2. Strap footings.
3. Combined footings.
4. Raft or mat foundation.

A *spread footing* is that in which the base of a column or wall is enlarged. The footing of a column is also called as *pad foundation*. A pad foundation may consist of a simple circular, square or rectangular slab of uniform thickness, or they may be stepped or launched to distribute the load from a heavy column. A wall footing is also called as a continuous footing or a strip footing.

If the footing supports more than one column, it is called as a strap or combined footing. A row of column foundation connected together by a beam is called as a continuous footing. Wide strip footings or foundations are necessary where the bearing capacity of the soil is low enough to necessitate a strip so wide that transverse bending occurs in the projecting portion of the foundation beam, and reinforcements are required to prevent cracking.

Raft or mat foundations are normally required on soils of low bearing capacity or where the structural columns or other loaded areas are so close in both the directions that individual footing foundations would nearly touch each other. A normal practice is to use a raft foundation where the sum of the areas covered by the conventional individual spread footings is more than about 50 percent of the loaded area of the structure. Raft foundations are useful in reducing differential settlements on variable soil or where there is wide variation in loading between adjacent columns or other applied loads. Raft foundations are commonly used beneath multistoried buildings, storage tanks, silo clusters, chimneys, etc. It is common to use mat foundation and to provide the floor slab for the basement. Mat foundations may be supported by piles in situations such as high ground water to control buoyancy or where the base soil is susceptible to large settlements.

In between column footings and raft foundations, comes combined footings where several footings (two or more) are joined to form a small mat. A combined footing may have either rectangular or trapezoidal shape or a series of pads connected by narrow rigid beams called straps. Such footings are called as strap footings.

The various types of shallow foundations are given in Fig. 4.1.

Deep Foundations

Deep foundations are normally defined as those that have depth width (D_f/B) ratio greater than 2. Very deep foundations have D_f/B ratio greater than 4. Piles, piers and caissons fall in this category. They are used to transmit loads to deeper layers of soil. Piles and drilled piers are used both on land and under water for supporting structures, whereas caissons are normally used for bridges and sometimes for multistoried buildings also. The terms foundation pier and caisson are interchangeably used by engineers to denote a cylindrical foundation with or without steel reinforcement. Piers are constructed with or without enlarged bottom which is concreted in place after excavation or drilling. Piers, which are sometimes called as drilled piers, are nothing but large diameter piles, bored and *cast-in-situ*. Whereas, a caisson is a large monolith which is built above ground and sunk in stages to the required founding level as a single unit. In some countries, the caisson type of foundation is also called as *well foundation*.

4.2 REQUIREMENTS FOR A STABLE FOUNDATION

A foundation is an integral part of the superstructure. The stability of a structure depends upon the stability of supporting soil. Whether a foundation is shallow or deep, the following basic requirements must be satisfied.

1. The foundation structure must be properly located with respect to any future influence which could adversely affect its performance.
2. The foundation (including the earth beneath) must be stable or safe from failure.
3. The foundation must not settle or deflect sufficiently to damage the structure or impair its usefulness.

These requirements should ordinarily be considered in the order given above. The first is rather nebulous; it involves many different factors, some of which cannot be evaluated analytically but which must be determined by engineering judgement. The second is specific. It is analogous to the requirement that a beam in the superstructure must be safe against breaking under its working load. The third requirement is both specific and nebulous. It is analogous to the requirement that a beam in the superstructure should not deflect enough to be objectionable; but how much is objectionable cannot always be defined accurately. These three requirements are independent of one another, and each must be satisfied; that is, if only two of the three have been met, the foundation is inadequate.

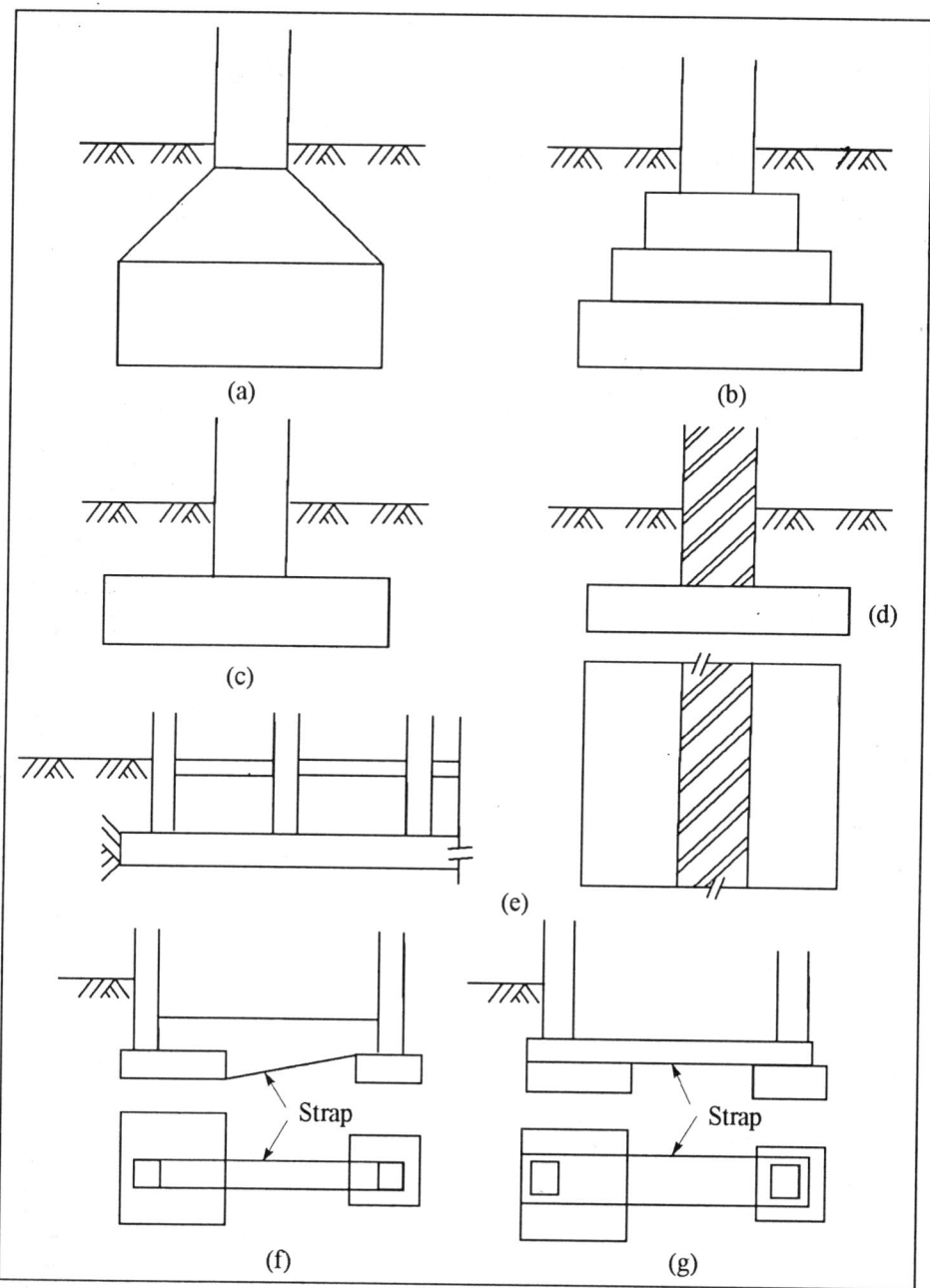

Fig. 4.1 Types of shallow foundations: (a) RC foundation, (b) stepped RC foundation, (c) plain reinforced concrete foundation, (d) plain concrete wall foundation, (e) mat or raft foundiation, and (f, g) strap foundation

The first requirement is considered in this Chapter. Requirements 2 and 3 relate to stress and strain, that is, possible overloading; and undue deformation of the soil. The depth of embedment required to prevent such overload or undue deformation is determined by analysis. These matters are discussed in Chapters 5 and 6. Foundations may rest in a rocky strata also. This matter is discussed in Chapter 5. Chapter 7 deals with combined footings and raft foundations.

4.3 FOUNDATION LOCATION AND DEPTH

The location and depth of foundation of a structure play an important point in the overall stability of the foundation. The location of the foundation in an area should not affect either its future expansion or its foundation should not be affected by the constructions in the adjoining areas. The depth of foundation depends upon the type of soil in the area, size of structure, the magnitudes of loads, and the environmental conditions.

In this chapter, the factors that affect depths for a shallow foundation only are considered. The depths for deep foundations are considered under the relevant chapters.

4.4 MINIMUM DEPTH FOR SHALLOW FOUNDATION

The foundations that govern the minimum depth for shallow foundations are:

1. Local erosion of soil due to flowing water.
2. Underground defects such as root holes, cavities, mine shafts, etc.
3. Unconsolidated filledup soil.
4. Adjacent structures, property lines, excavations and future construction operations.
5. Ground water level.
6. Depth of frost action.
7. Depth of volume change due to the presence of expansive soils.
8. Dessication due to the heat from boilers, furnaces, etc.
9. Dessication due to drawing of water by the roots of trees.

1. Local Erosion

In areas where there is heavy rainfall, there is every possibility of the top soil getting eroded due to the flowing water. The erosion will be particularly severe, if the top soil is loose, and further if the structure lies on a sloping ground. Local experience should indicate the possible depths of erosion in such cases and the foundations should be located below such depths.

2. Underground Defects

Presence of root holes, cavities of burrowing animals, buried old vaults, and mining shafts, etc. should be investigated before deciding the depths for foundations. The collapse of the roof of a buried cave or an old mine can be a serious problem in some regions. Man-made discontinuities, including old wells, sewers, cables are frequently encountered in cities and established industrial sites. No footing be located above such a discontinuity or on its backfill unless both are known to be capable of carrying the imposed load.

3. Unconsolidated Filledup Soil

Filledup soils are quite common in low lying areas. Soils washed down by flowing rain water from higher elevations and deposited at low level areas remain unconsolidated for a number of years. Man made fillings of old ponds, abandoned quarries, old dried up nallas, etc. pose serious problems for foundations as such soils normally remain unconsolidated. The strength characteristics of all the filledup soils with respect to depth should be investigated before deciding the type and depth of foundation.

4. Presence of Adjacent Structure, Property Lines, etc.

The proximity of existing structures and the possibility of future construction are important factors in the location and depth of foundations. When the foundation of an existing structure is close to the external wall of a proposed building, the problems that arise are:

(*a*) The type of foundation to be adopted for the proposed building.

(*b*) The depth at which the foundation is to be located.

(*c*) The construction method to be adopted.

The construction of a new foundation close to the existing one can cause damage to it due to vibration (if piles are driven), shocks of blasting, undermining by excavation or the lowering of water table. The deeper the new foundation and nearer it is to the old, the greater the damage is likely to be. The load imposed on the soil by a new structure can cause settlement of the existing one because the stresses spread through the soil in both the horizontal and vertical directions. If it is possible, the new foundation should be kept sufficiently away from the old, and at a depth greater than the old. An empirical rule for the minimum spacing of footings to avoid interference between the old and the new is given in Fig. 4.2. By rule of thumb, the minimum horizontal spacing the old and new footing should be equal to the width, *B*, of the wider one. Further a line drawn at a 45° angle (30° for soft soil) with the horizontal should not intersect the base of the lower one as shown in Fig. 4.2. While the use of this rule will help minimize damage to adjacent foundations, a better answer to the foundation location and depth in such a case is an analysis of bearing capacity and settlement and a study of the proposed construction procedure.

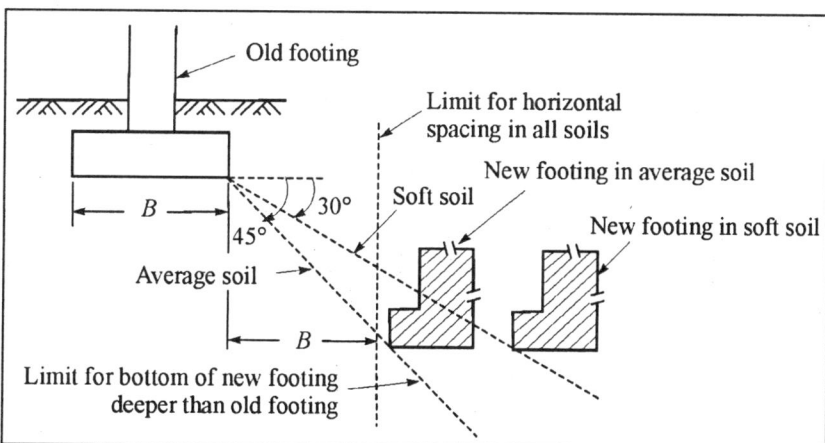

Fig. 4.2 New foundation adjacent to an existing foundation

Foundation depth must be selected with future nearby excavation in mind. This is particularly true close to the property lines, where only limited legal control over the construction operation on the adjoining site may be possible. The effect of the transfer of stresses from the footings placed at higher levels to footings at lower levels is shown in Fig. 4.3.

5. Ground Water Level

The level of ground water is a factor in foundation in three ways as follows:

(*a*) The construction below ground water level often presents difficulties. In cohesionless sands and silts, for example, upward flow of water into a footing excavation can create a quick condition, and construction is impossible without predrainage.

(*b*) The presence of ground water close to the foundation level can reduce the ability of some soils to carry high foundation pressures.

(*c*) When the ground water level is above the lowest floor, for example, basement floor, water proofing and resistance against hydrostatic uplift become serious considerations.

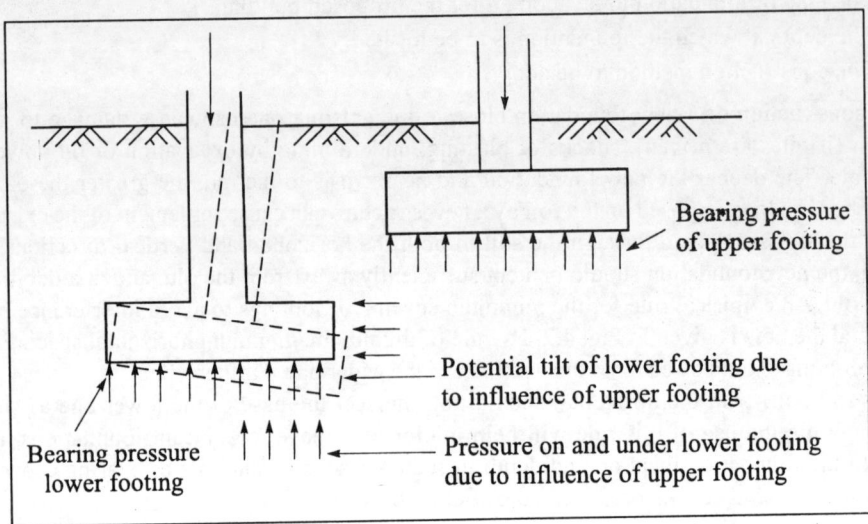

Fig. 4.3 Effect of transfer of stresses from one foundation to the other lower down

Ordinarily, spread foundations are placed above the highest ground water level unless the additional expense of greater depth is fully justified.

6. Frost Action

In regions where the air temperature falls below 32° F (0° C) far more than a few days, the ground freezes and heave of soil may occur. Foundations placed within the zone of heave are slowly lifted during cold weather and suddenly dropped when the frozen mass thaws. The zone of frost heave may extend to depths 2 to 3 m beneath the ground surface in areas of extreme cold. To be free of frost heave under average conditions, the base of foundation should be placed at a depth equal to about *three-fourths* of the maximum frost penetration. In highly susceptible soils, such as saturated silty sands and silts, the full depth of penetration is normally used; while in gravels and dry sands, even less than three-fourths of the maximum depth of penetration is often adequate.

Normally, clay soils are insensitive to frost heave as their permeability is low thereby limiting the amount of moisture which can be drawn up into the soil to form ice lenses which cause the heave.

The question sometimes arises whether the requirement for embedment below the depth of frost penetration applies to pile caps, grade beams, etc. If the loads coming on pile caps or grade beams from the superstructure are more than the frost heave pressure exerted on the bottom of caps and beams, there will not be any danger to the structures. Otherwise, there should be free surface underside the pile caps and grade beams to preclude any potential heave action.

The heave pressure is the pressure required to prevent expansion of soil during the formation of ice. This pressure, as an extreme case, is estimated to be about 2 to 3 MPa. This is a large force but for high capacity piles in the 2000 kN range or more, the column load (dead load only) is usually sufficient to offset the heave pressure. Where such is the case, some saving in the cost can be effected by reducing the embedment of the pile caps, raising the soffit into the zone of frost penetration.

A serious frost penetration problem may occur beneath cold storage, refrigirator, and similar spaces within a building. The depth of freezing in such cases can be quite large, often making it impractical to carry the foundations to adequate depth to get below frost level. In cases of this type, the only sure protection is the installation of warm air ducts or pipes with a circulating heated fluid to prevent loss of heat from the underlying soil.

7. Volume Change in Expansive Soils

The effect of change of volume in expansive soils on foundations is discussed in detail in Chapter 13.

8. Dessication Due to the Heat from Boilers, Furnaces, Sun, etc.

In humid regions where soils are ordinarily moist, severe dessication may cause susceptible soils to shrink and bring about severe settlement of structures. Accelerated dessication accompanied by rapid and irregular settlement can be caused by many local conditions such as the heat transferred to soil from boilers, ovens, furnaces, etc. that are inadequately insulated from ground. Dessication can happen due to sun's heat which is severe in fine silts and clay soils. Foundations should be kept below the depths of dessication due to sun's heat as these soils become very soft during rainy seasons.

9. Dessication Due to the Roots of Existing Trees

In many instances, presence of large trees and even some shrubs, close to the buildings has resulted in soil dessication as the roots of these vegetation are capable of removing sufficient amount of moisture from soils, and hence cause settlement of foundations placed above or adjacent to their major root system. It is therefore essential to see that such vegetation is not grown close to buildings.

4.5 SELECTION OF TYPE OF FOUNDATION

Factors to be Considered

The selection of type of foundation for a given site depends an many factors. The most important factors are:

1. The function of the structure and the loads it must carry.
2. The subsurface condition of the soil.
3. The cost of the superstructure.

All the factors mentioned above are interrelated. If the structure is of an important type and carries very heavy loads, the type of foundation must be such that it gives stability under all adverse conditions. Possibly in such cases, cost might not be a major factor for consideration. The type of foundation and cost generally depend on the type of soil met at the site.

In selecting the type of foundation, the design load plays an important part which again depends on the subsoil conditions. The various loads that are likely to be considered are:

(*i*) Dead loads.

(*ii*) Live loads.

(*iii*) Wind and earthquake forces.

(*iv*) Lateral pressures exerted by the foundation earth on the embedded structural elements.

(*v*) Impact equivalents relating to moving and dynamic loads.

In addition to the above loads, it might be necessary, under special circumstances, to consider the following loads based on the subsoil conditions.

(*i*) Lateral or uplift forces on the foundation elements due to high water table level.

(*ii*) Swelling pressures on the foundations in expansive soils.

(*iii*) Heave pressures on the foundations in areas subjected frost heave.

(*iv*) Negative frictional drag on piles where pile foundations are used in highly compressible soils.

Dead loads include the weight of the structure and all materials permanently attached to it, such as floor finish, exterior walls, permanent and fixed service equipments, such as plumbing stacks and risers, etc. If the weight of the earth is directly supported by the elements of the structure, it should be considered as dead load.

Live loads include all equivalent vertical loads that are not a permanent part of a structure but are expected to superimpose on the structure during a part or all of its useful life. Vertical loads due to wind and snow are also included. Human occupancy, furnitures, warehouse goods, mechanical equipments, etc. are some of the other major live loads. The magnitude of live loads to be used in the design of buildings, industrial structures, etc. are usually stipulated in local building codes.

Railway and highway bridges as well as other structures subjected to traffic loads are designed as per the local codes in practice. Industrial floors, subjected to a special type of truck traffic must be designed to suit each specific truck loading. Reaction from industrial cranes sometimes constitute a large portion of the live load.

The live loads due to human occupancy including furniture and appliances are often reduced for the design of large girders, columns and foundations. The amount of reduction varies with the floor area and the number of floors. Local building codes give the permissible reduction factors.

Wind load acts on all exposed surfaces of structure. The magnitude of design pressure is usually stipulated in local building codes.

Earthquake motion may result in lateral forces. Every structure lying in earthquake zones must be designed to resist the lateral forces generated by earthquakes. Earth pressure is a lateral force acting permanently against a certain part of the substructure below ground surface. It should be treated as a basic load similar to dead load.

Water pressure may act laterally against basement walls and vertically on base slabs. Considering the substructure as a whole, the lateral hydrostatic pressures are always balanced; but the hydrostatic uplift or buoyancy force must be counteracted by the dead load of the structure. If the dead load is insufficient, some provision must be made to anchor the structure.

Swelling and heave pressures are also very important in the design of foundations. The former occurs in expansive soils and the latter in cold regions.

Negative friction on piles poses a serious problem if piles are constructed in recently filled up soils or in marine regions. The negative frictional force in the design of pile foundations is an important factor which will be considered in the chapter under pile foundations subjected to vertical loads.

Steps for the Selection of the Type of Foundation

In choosing the type of foundation, the design engineer must perform five successive steps.

1. Obtain the required information concerning the nature of the superstructure and the loads to be transmitted to the foundation.

2. Obtain from soil investigation the subsurface soil conditions.

3. Explore the possibility of constructing any one of the types of foundation under the existing conditions by taking into account: (*i*) the bearing power of the soil to carry to required load, and (*ii*) the adverse effect on the structure due to differential settlements. Eliminate in this way, the unsuitable types.

4. Once one or two types of foundation are selected on the basis of preliminary studies, make more detailed studies. These studies may require more accurate determination of loads subsurface conditions and footing sizes. It may also be necessary to make more refined estimates of settlement to predict the behaviour of the structure.

5. Estimate the cost of each of the promising type of foundation, and choose the type that represents the most acceptable compromise between performance and cost.

Shallow Foundation 2
Ultimate Bearing Capacity

5

5.1 INTRODUCTION

It is the customary practice to regard a foundation as shallow if the depth of the foundation is less than or equal to the width of the foundation. The different types of footings that we normally come across are given in Fig. 4.1. A foundation is an integral part of a structure. The stability of a structure depends upon the stability of the supporting soil. Two important factors that are to be considered are:

1. The foundation must be stable against shear failure of the supporting soil.
2. The foundation must not settle beyond a tolerable limit to avoid damage to the structure.

The other factors that require consideration are the location and depth of the foundation. In deciding the location and depth, one has to consider the erosions due to flowing water, underground defects such as root holes, cavities, unconsolidated fills, ground water level, presence of expansive soils, etc.

In selecting a type of foundation, one has to consider the functions of the structure and the load it has to carry, the subsurface condition of the soil, and the cost of the superstructure.

Design loads also play an important part in the selection of the type of foundation. The various loads that are likely to be considered are: (*i*) dead loads, (*ii*) live loads, (*iii*) wind and earthquake forces, (*iv*) lateral pressures exerted by the foundation earth on the embedded structural elements, and (*v*) the effects of dynamic loads.

In addition to the above loads, the loads that are due to the subsoil conditions are also required to be considered. They are: (*i*) lateral or uplift forces on the foundation elements due to high water table, (*ii*) swelling pressures on the foundations in expansive soils, (*iii*) heave pressures on foundations in areas subjected to frost heave, and (*iv*) negative frictional drag on piles where pile foundations are used in highly compressible soils. The steps that are required to be considered in the selection of the type of foundation are discussed in Chapter 4.

5.2 THE ULTIMATE BEARING CAPACITY OF SOIL DEFINED

Consider the simplest case of a shallow foundation subjected to a central vertical load. The footing is founded at a depth D_f below the ground surface Fig. 5.1 (a). If the settlement, S, of the footing is recorded against the applied load, Q, load-settlement curves, similar in shape to a stress–strain curve, may be obtained as shown in Fig. 5.1 (b).

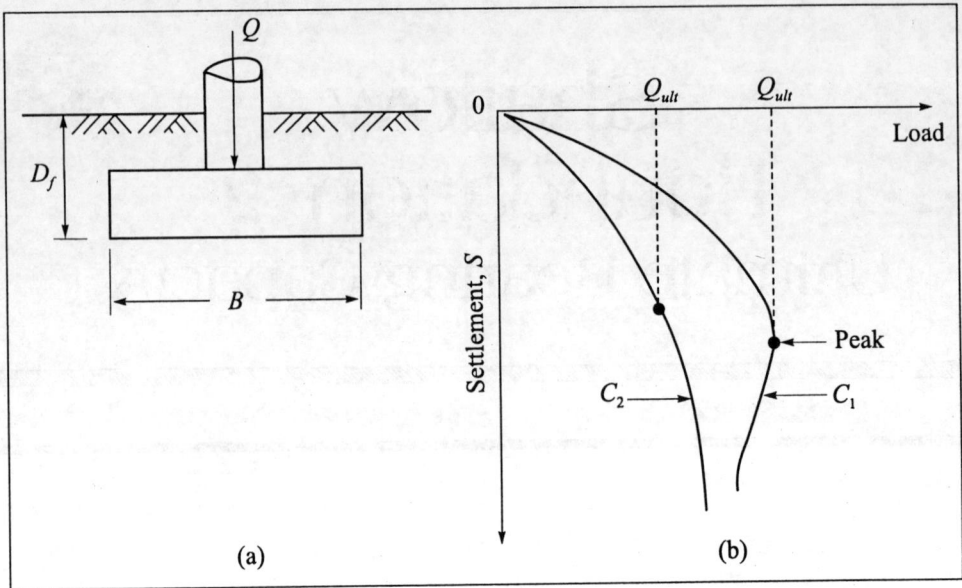

Fig. 5.1 Typical load-settlement curves: (a) Footing, (b) load-settlement curves

The shape of the curve depends generally on the size and shape of the footing, the composition of the supporting soil, and the character, rate, and frequency of loading. Normally, a curve will indicate the ultimate load Q_u that the foundation can support. If the foundation soil is a dense sand or a very stiff clay, the curve passes fairly abruptly to a peak value and then drops down as shown by curve C_1 in Fig. 5.1 (b). The peak load Q_u is quite pronounced in this case. On the other hand, if the soil is loose sand or soft clay, the settlement curve continues to descend on a slope as shown by curve C_2 which shows that the compression of soil is continuously taking place without giving a definite value for Q_u. On such a curve, Q_u may be taken at a point beyond which there is a constant rate of penetration.

5.3   SOME OF THE TERMS DEFINED

It will be useful to define, at this stage, some of the terms relating to bearing capacity of foundations (refer to Fig. 5.2).

(a) Total Overburden Pressure q_o

q_o is the intensity of total overburden pressure due to the weight of both soil and water at the base level of the foundation.

$$q_o = \gamma D_{w1} + \gamma_{sat} \bar{D}_w \qquad (5.1)$$

(b) Effective Overburden Pressure q'_o

q'_o is the effective overburden pressure at the base level of the foundation.

$$q'_o = \gamma D_{w1} + \gamma_b \bar{D}_w \qquad (5.2)$$

when $\bar{D}_w = 0,\ q'_o = \gamma D_{w1} = \gamma D_f$

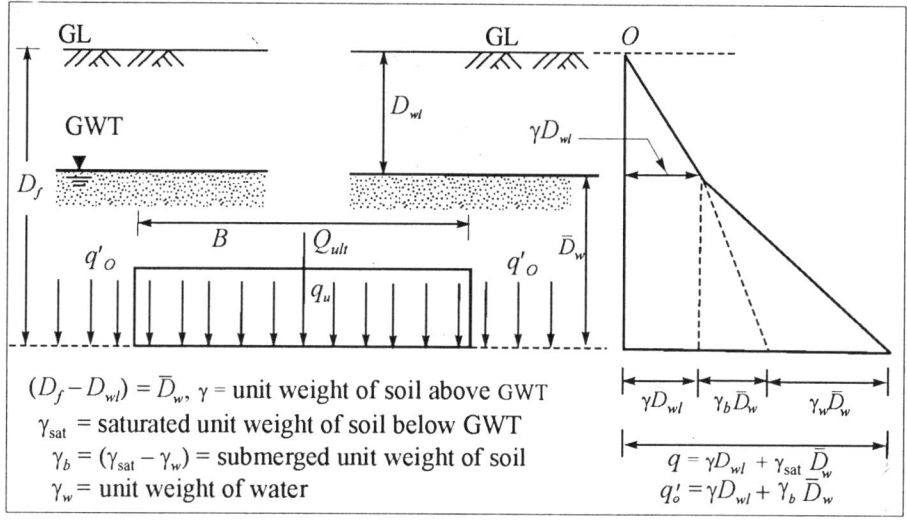

Fig. 5.2 Total and effective overburden pressures

(c) The Ultimate Bearing Capacity of Soil, q_u

q_u is the maximum bearing capacity of soil at which the soil fails by shear.

(d) The Net Ultimate Bearing Capacity, q_{nu}

q_{nu} is the bearing capacity in excess of the effective overburden pressure q'_o, expressed as

$$q_{nu} = q_u - q'_o \qquad (5.3)$$

(e) Gross Allowable Bearing Pressure, q_a

q_a is expressed as

$$q_a = \frac{q_u}{F_s} \qquad (5.4)$$

where F_s = factor of safety.

(f) Net Allowable Bearing Pressure, q_{na}

q_{na} is expressed as

$$q_{na} = \frac{q_u - \gamma D_f}{F_s} = \frac{q_{nu}}{F_s} \qquad (5.5)$$

(g) Safe Bearing Pressure, q_s

q_s is defined as the net safe bearing pressure which produces a settlement of the foundation which does not exceed a permissible limit.

Note: In the design of foundations, one has to use the least of the two values of q_{na} and q_s.

5.4 ˙TYPES OF FAILURE IN SOIL

Experimental investigations have indicated that foundations on dense sand with relative density greater than 70 percent fail suddenly with pronounced peak resistance when the settlement reaches about 7 percent of the foundation width. The failure is accompanied by the appearance of failure surfaces and by considerable bulging of a sheared mass of sand as shown in Fig. 5.3 (a). This type of failure is designated as general shear failure by Terzaghi (1943). Foundations on sand of relative density lying between 35 and 70 percent do not show a sudden failure. As the settlement exceeds about 8 percent of the foundation width, bulging of sand starts at the surface. At settlements of about 15 percent of foundation width, a visible boundary of sheared zones at the surface appears. However, the peak of base resistance may never be reached. This type of failure is termed local shear failure, Fig. 5.3 (b), by Terzaghi (1943).

Foundations on relatively loose sand with relative density less than 35 percent penetrate into the soil without any bulging of the sand surface. The base resistance gradually increases as settlement progresses. The rate of settlement, however, increases and reaches a maximum at a settlement of about 15 to 20 percent of the foundation width. Sudden jerks or shears can be observed as soon as the settlement reaches about 6 to 8 percent of the foundation width. The failure surface, which is vertical or slightly inclined and follows the perimeter of the base, never reaches the sand surface. This type of failure is designated as punching shear failure by Vesic (1963) as shown in Fig. 5.3 (c).

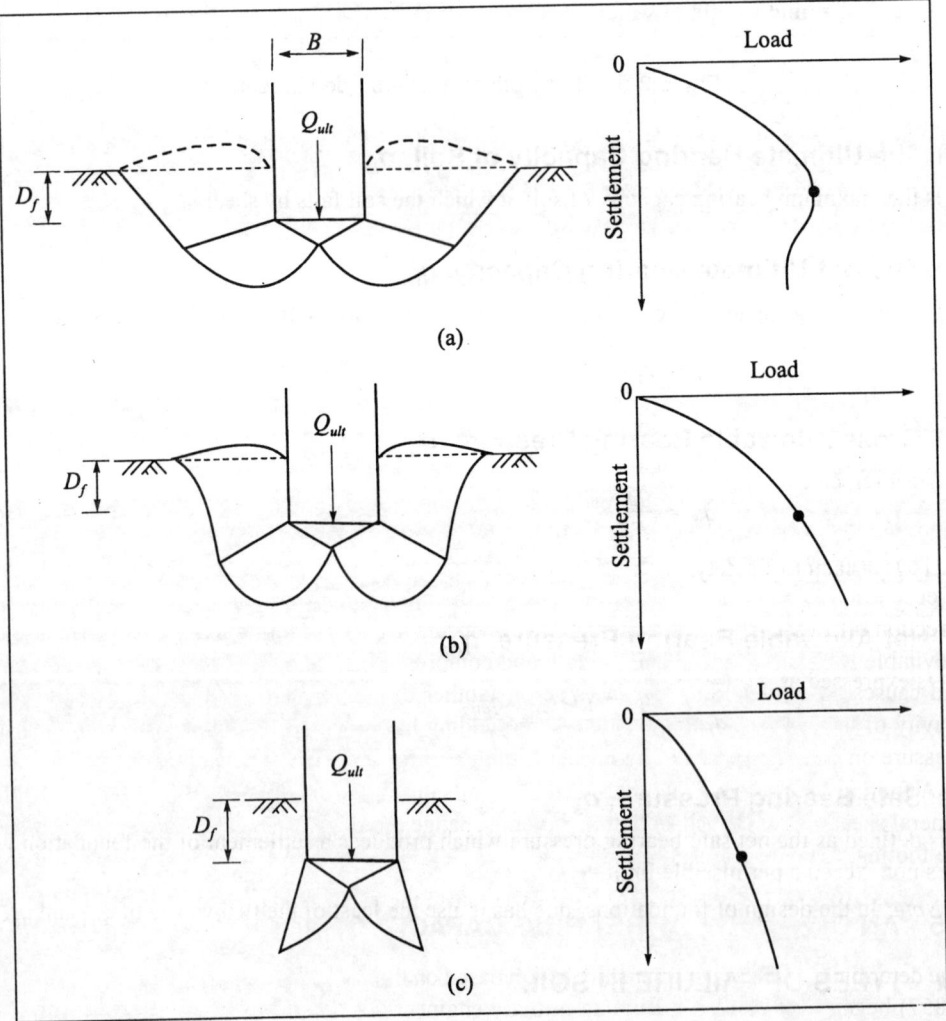

Fig. 5.3 Modes of bearing capacity failure (Vesic, 1963): (a) General shear, (b) local shear failure, (c) punching shear failure

The three types of failure described above were observed by Vesic (1963) during tests on model footings. It may be noted here that as the relative depth/width ratio increases, the limiting relative densities at which failure types change increase. The approximate limits of types of failure to be affected as relative depth D_f/B, and relative density of sand, D_r, vary are shown in Fig. 5.4 (Vesic, 1963). The same figure shows that there is a critical relative depth below which only punching shear failure occurs for circular foundations, this critical depth, D_f/B, is around 4 and for long rectangular foundations around 8.

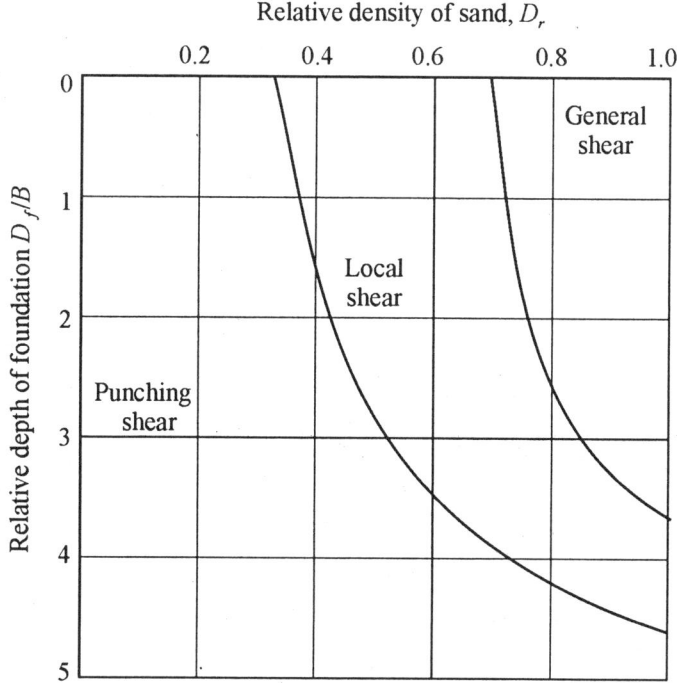

Fig. 5.4 Modes of failure of model footings in sand (after Vesic, 1963)

The surfaces of failures as observed by Vesic are for concentric vertical loads. Any small amount of eccentricity in the load application changes the modes of failure and the foundation tilts in the direction of eccentricity. Tilting nearly always occurs in cases of foundation failures because of the inevitable variation in the shear strength and compressibility of the soil from one point to another and causes greater yielding on one side or another of the foundation. This throws the center of gravity of the load towards the side where yielding has occurred, thus increasing the intensity of pressure on this side followed by further tilting.

A footing founded on precompressed clays or saturated normally consolidated clays will fail in general shear, if it is loaded so that no volume change can take place and fails by punching shear if the footing is founded on soft clays.

5.5 AN OVERVIEW OF BEARING CAPACITY THEORIES

The determination of bearing capacity of soil based on the classical earth pressure theory of Rankine (1857) began with Pauker, a Russian military engineer (1889), and was modified by Bell (1915). Pauker's theory was applicable only for sandy soils but the theory of Bell took into account cohesion also. Neither theory took into account the width of the foundation. Subsequent developments led to the modification of Bell's theory to include width of footing also.

The methods of calculating the ultimate bearing capacity of shallow strip footings by plastic theory developed considerably over the years since Terzaghi (1943) first proposed a method by taking into account the weight of soil by the principle of superposition. Terzaghi extended the theory of Prandtl (1921). Prandtl developed an equation based on his study of the penetration of a long hard metal punch into softer materials for computing the ultimate bearing capacity. He assumed the material was weightless possessing only cohesion and friction. Taylor (1948) extended the equation of Prandtl by taking into account the surcharge effect of the overburden soil at the foundation level.

No exact analytical solution for computing bearing capacity of footings is available at present because the basic system of equations describing the yield problems is nonlinear. On account of these reasons, Terzaghi (1943) first proposed a semiempirical equation for computing the ultimate bearing capacity of strip footings by taking into account cohesion, friction and weight of soil, and replacing the overburden pressure with an equivalent surcharge load at the base level of the foundation. This method was for the general shear failure condition and the principal of superposition was adopted. His work was an extension of the work of Prandtl (1921). The final form of the equation proposed by Terzaghi is the same as the one given by Prandtl.

Subsequent to the work by Terzaghi, many investigators became interested in this problem and presented their own solutions. However, the form of the equation presented by all these investigators remained the same as that of Terzaghi, but their methods of determining the bearing capacity factors were different.

Of importance in determining the bearing capacity of strip footings is the assumption of plane strain inherent in the solutions of strip footings. The angle of internal friction as determined under an axially symmetric triaxial compression stress state, ϕ_t, is known to be several degrees less than that determined under plane strain conditions under low confining pressures. Thus, the bearing capacity of a strip footing calculated by the generally accepted formulas, using ϕ_t, is usually less than the actual bearing capacity as determined by the plane strain footing tests which leads to a conclusion that the bearing capacity formulas are conservative.

The ultimate bearing capacity, or the allowable soil pressure, can be calculated either from bearing capacity theories or from some of the *in-situ* tests. Each theory has its own good and bad points. Some of the theories are of academic interest only. However, it is the purpose of the author to present here only such theories which are of basic interest to students in particular and professional engineers in general. The application of field tests for determining bearing capacity are also presented which are of particular importance to professional engineers since present practice is to rely more on field tests for determining the bearing capacity or allowable bearing pressure of soil.

Some of the methods that are discussed in this chapter are:

1. Terzaghi's bearing capacity theory.
2. The general bearing capacity equation.
3. Pressuremeter.
4. Field tests.

5.6 TERZAGHI'S BEARING CAPACITY THEORY

Terzaghi (1943) used the same form of equation as proposed by Prandtl (1921) and extended his theory to take into account the weight of soil and the effect of soil above the base of the foundation on the bearing capacity of soil. Terzaghi made the following assumptions for developing an equation for determining q_u for a c-ϕ soil.

(1) The soil is semiinfinite, homogeneous and isotropic, (2) the problem is two-dimensional, (3) the base of the footing is rough, (4) the failure is by general shear, (5) the load is vertical and symmetrical, (6) the ground surface is horizontal, (7) the overburden pressure at foundation level is

equivalent to a surcharge load $q_0' = \gamma D_f$, where γ is the effective unit weight of soil, and D_f, the depth of foundation less than the width B of the foundation, (8) the principle of superposition is valid, and (9) Coulomb's law is strictly valid, that is, $\sigma = c + \sigma \tan \phi$.

Mechanism of Failure

The shapes of the failure surfaces under ultimate loading conditions are given in Fig. 5.5. The zones of plastic equilibrium represented in this figure by the area *gedcf* may be subdivided into

1. Zone I of elastic equilibrium.
2. Zones II of radial shear state.
3. Zones III of Rankine passive state.

When load q_u per unit area acting on the base of the footing of width B with a rough base is transmitted into the soil, the tendency of the soil located within zone I is to spread but this is counteracted by friction and adhesion between the soil and the base of the footing. Due to the existence of this resistance against lateral spreading, the soil located immediately beneath the base remains permanently in a state of elastic equilibrium, and the soil located within this Central Zone I behaves as if it were a part of the footing and sinks with the footing under the superimposed load. The depth of this wedge shaped body of soil *abc* remains practically unchanged, yet the footing sinks. This process is only conceivable if the soil located just below point *c* moves vertically downwards. This type of movement requires that the surface of sliding *cd* (Fig. 5.5) through point *c* should start from a vertical tangent. The boundary *bc* of the zone of radial shear *bcd* (Zone II) is also the surface of sliding. As per the theory of plasticity, the potential surfaces of sliding in an ideal plastic material intersect each other in every point of the zone of plastic equilibrium at an angle ($90° - \phi$). Therefore, the boundary *bc* must rise at an angle ϕ to the horizontal provided the friction and adhesion between the soil and the base of the footing suffice to prevent a sliding motion at the base.

The sinking of Zone I creates two zones of plastic equilibrium, II and III, on either side of the footing. Zone II is the radial shear zone whose remote boundaries *bd* and *af* meet the horizontal surface at angles ($45° - \phi/2$), whereas Zone III is a passive Rankine zone. The boundaries *de* and *fg* of these zones are straight lines and they meet the surface at angles of ($45° - \phi/2$). The curved parts *cd* and *cf* in Zone II are parts of logarithmic spirals whose centres are located at *b* and *a* respectively.

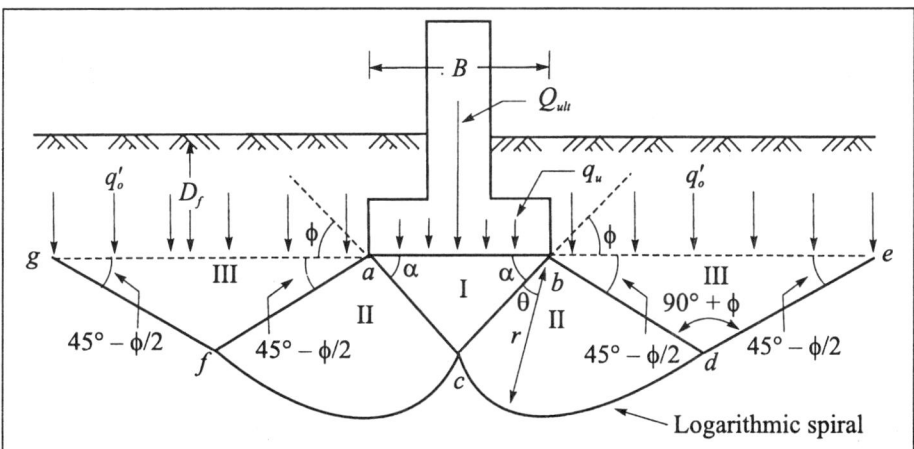

Fig. 5.5 General shear failure surface as assumed by Terzaghi for a strip footing

Ultimate Bearing Capacity of Soil

Strip footings

Terzaghi developed his bearing capacity equation for strip footings by analysing the forces acting on the wedge *abc* in Fig. 5.5. The equation developed for the ultimate bearing capacity q_u is

$$q_u = \frac{Q_{ult}}{B} = cN_c + \gamma D_f N_q + \frac{1}{2}\gamma BN_\gamma \qquad (5.6)$$

where Q_{ult} = ultimate load per unit length of footing, c = unit cohesion, γ the effective unit weight of soil, B = width of footing, D_f = depth of foundation, N_c, N_q and N_γ are the bearing capacity factors. They are functions of the angle of friction, ϕ.

The bearing capacity factors are expressed by the following equations

$$\left.\begin{array}{l} N_c = \left(N_q - 1\right)\cot\phi \\[2mm] N_q = \dfrac{a_\theta^2}{2\cos^2\left(45° + \phi/2\right)} \\[2mm] \text{where } a_\theta = e^{\eta\tan\phi}, \ \eta = \left(0.75\pi - \phi/2\right) \\[2mm] N_\gamma = \dfrac{1}{2}\tan\phi\left(\dfrac{K_{p\gamma}}{\cos^2\phi} - 1\right) \end{array}\right\} \qquad (5.7)$$

where $K_{p\gamma}$ = passive earth pressure coefficient.

Table 5.1 gives the values of N_c, N_q and N_γ for various values of ϕ and Fig. 5.6 gives the same in a graphical form.

Table 5.1 Bearing capacity factors of Terzaghi

$\phi°$	N_c	N_q	N_γ
0	5.7	1.0	0.0
5	7.3	1.6	0.14
10	9.6	2.7	1.2
15	12.9	4.4	1.8
20	17.7	7.4	5.0
25	25.1	12.7	9.7
30	37.2	22.5	19.7
35	57.8	41.4	42.4
40	95.7	81.3	100.4
45	172.3	173.3	360.0
50	347.5	415.1	1072.8

Equations for Square, Circular, and Rectangular Foundations

Terzaghi's bearing capacity Eq. (5.6) has been modified for other types of foundations by introducing the shape factors. The equations are

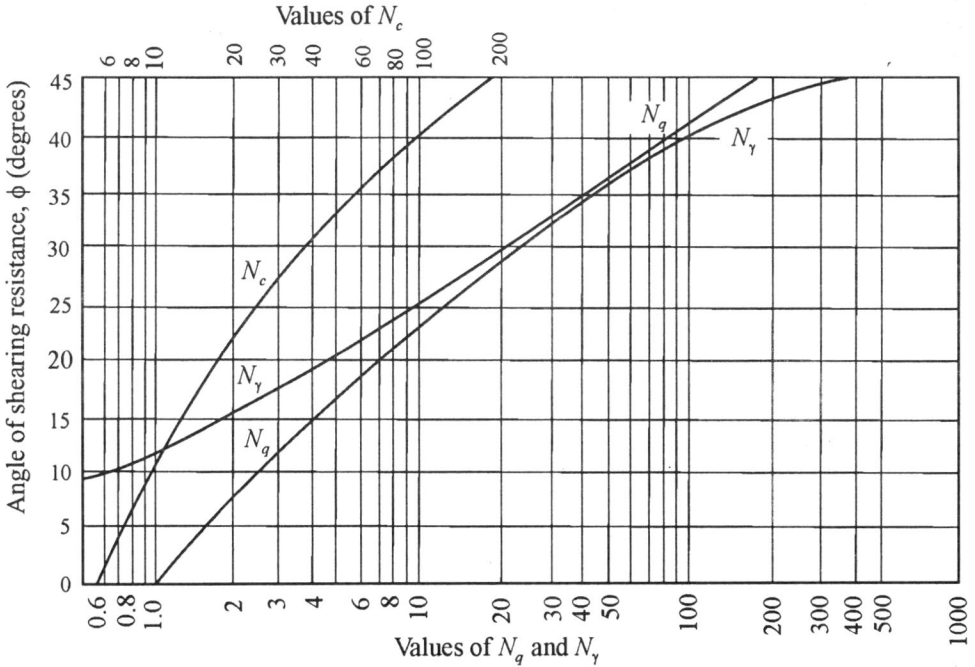

Fig. 5.6 Terzaghi's bearing capacity factors for general shear failure

Square foundations

$$q_u = 1.3c\,N_c + \gamma\,D_f\,N_q + 0.4\gamma\,BN_\gamma \qquad (5.8)$$

Circular foundations

$$q_u = 1.3c\,N_c + \gamma\,D_f\,N_q + 0.3\gamma\,BN_\gamma \qquad (5.9)$$

Rectangular foundations

$$q_u = cN_c\left(1 + 0.3 \times \frac{B}{L}\right) + \gamma\,D_f\,N_q + \frac{1}{2}\,\gamma\,BN_\gamma\left(1 - 0.2 \times \frac{B}{L}\right) \qquad (5.10)$$

where B = width or diameter, L = length of footing.

Ultimate Bearing Capacity for Local Shear Failure

The reasons as to why a soil fails under local shear have been explained under Section 5.4. When a soil fails by local shear, the actual shear parameters c and ϕ are to be reduced as per Terzaghi (1943). The lower limiting values of c and ϕ are

$$\bar{c} = 0.67c$$

and $\qquad \tan \bar{\phi} = 0.67 \tan \phi \text{ or } \bar{\phi} = \tan^{-1}(0.67 \tan \phi) \qquad (5.11)$

The equations for the lower found values for the various types of footings are given below.

Strip foundation

$$q_u = 0.67c\,\bar{N}_c + \gamma\,D_f\bar{N}_q + \frac{1}{2}\,\gamma\,B\bar{N}_\gamma \qquad (5.12)$$

Square Foundation

$$q_u = 0.867c\,\bar{N}_c + \gamma\,D_f\bar{N}_q + 0.4\gamma\,B\bar{N}_\gamma \qquad (5.13)$$

Circular Foundation

$$q_u = 0.867c\,\bar{N}_c + \gamma\,D_f\bar{N}_q + 0.3\gamma\,B\bar{N}_\gamma \qquad (5.14)$$

Rectangular Foundation

$$q_u = 0.67c\left(1 + 0.3\times\frac{B}{L}\right)\bar{N}_c + \gamma\,D_f\bar{N}_q + \frac{1}{2}\,\gamma\,B\bar{N}_\gamma\left(1 - 0.2\times\frac{B}{L}\right) \quad (5.15)$$

where $\bar{N}_c$, $\bar{N}_q$ and $\bar{N}_\gamma$ are the reduced bearing capacity factors for local shear failure. These factors may be obtained either from Table 5.1 or Fig. 5.7 by making use of the friction angle $\bar{\phi}$.

Ultimate Bearing Capacity q_u in Purely Cohesionless and Cohesive Soils Under General Shear Failure

Equations for the various types of footings for $(c - \phi)$ soil under general shear failure have been given earlier. The same equations can be modified to give equations for cohesionless soil (for $c = 0$) and cohesive soils (for $\phi = 0$) as follows.

It may be noted here that for $c = 0$, the value of $N_c = 0$, and for $\phi = 0$, the value of $N_c = 5.7$ for a strip footing and $N_q = 1$.

(a) Strip footing

For $\qquad c = 0, \qquad q_u = \gamma\,D_f N_q + \frac{1}{2}\,\gamma\,BN_\gamma \qquad (5.16)$

For $\qquad \phi = 0, \qquad q_u = 5.7c + \gamma\,D_f$

(b) Square footing

For $\qquad c = 0, \qquad q_u = \gamma\,D_f N_q + 0.4\gamma\,BN_\gamma \qquad (5.17)$

For $\qquad \phi = 0, \qquad q_u = 7.4c + \gamma\,D_f$

(c) Circular Footing

For $\qquad c = 0, \qquad q_u = \gamma\,D_f N_q + 0.3\gamma\,BN_\gamma \qquad (5.18)$

For $\qquad \phi = 0, \qquad q_u = 7.4c + \gamma\,D_f$

(d) Rectangular footing

For $\qquad c = 0, \qquad q_u = \gamma\,D_f N_q + \frac{1}{2}\,\gamma\,BN_\gamma\left(1 - 0.2\times\frac{B}{L}\right) \qquad (5.19a)$

For $\qquad \phi = 0, \qquad q_u = 5.7c \left(1 + 0.3 \times \dfrac{B}{L}\right) + \gamma\, D_f$ $\qquad\qquad$ (5.19b)

Similar types of equations as presented for general shear failure can be developed for local shear failure also.

Transition from Local to General Shear Failure in Sand

As already explained, local shear failure normally occurs in loose and general shear failure occurs in dense sand. There is a transition from local to general shear failure as the state of sand changes from loose to dense condition. There is no bearing capacity equation to account for this transition from loose to dense state. Peck *et al*, (1974) have given curves for N_γ and N_q which automatically incorporate allowance for the mixed state of local and general shear failures as shown in Fig. 5.7.

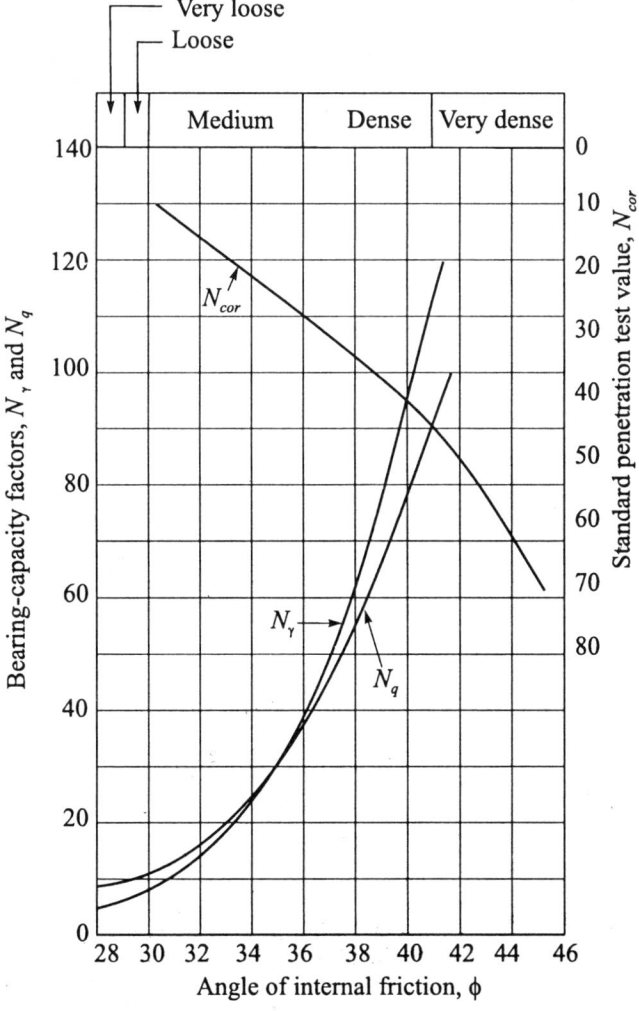

Fig. 5.7 Terzaghi's bearing capacity factors which take care of mixed state of local and general shear failures in sand (Peck *et al*, 1974)

The curves for N_q and N_γ are developed on the following assumptions.

1. Purely local shear failure occurs when $\phi \leq 28°$.
2. Purely general shear failure occurs when $\phi \geq 38°$.
3. Smooth transition curves for values of ϕ between 28° and 38° represent the mixed state of local and general shear failures.

N_q and N_γ for values of $\phi \geq 38°$ are as given in Table 5.1. Values of $\overline{N}_q$ and $\overline{N}_\gamma$ for $\phi \leq 28°$ may be obtained from Table 5.1 by making use of the relationship $\overline{\phi} = \tan^{-1}(2/3)\tan\phi$.

In the case of purely cohesive soil local shear failure may be assumed to occur in soft to medium stiff clay with an unconfined compressive strength $q_u \leq 100$ kPa.

Figure 5.7 also gives the relationship between SPT value N_{cor} and the angle of internal friction ϕ by means of a curve. This curve is useful to obtain the value of ϕ when the SPT value is known.

Net Ultimate Bearing Capacity and Safety Factor

The net ultimate bearing capacity q_{nu} is defined as the pressure at the base level of the foundation in excess of the effective overburden pressure $q'_o = \gamma D_f$ as defined in Eq. (5.3). The net q_{nu} for a strip footing is

$$q_{nu} = (q_u - \gamma D_f) = cN_c + \gamma D_f(N_q - 1) + \frac{1}{2}\gamma BN_\gamma \tag{5.20}$$

Similar expressions can be written for square, circular, and rectangular foundations and also for local shear failure conditions.

Allowable bearing pressure

Per Eq. (5.4), the gross allowable bearing pressure is

$$q_a = \frac{q_u}{F_s} \tag{5.21a}$$

In the same way the net allowable bearing pressure q_{na} is

$$q_{na} = \frac{q_u - \gamma D_f}{F_s} = \frac{q_{nu}}{F_s} \tag{5.21b}$$

where F_s = factor of safety which is normally assumed as equal to 3.

5.7 SKEMPTON'S BEARING CAPACITY FACTOR N_C

For saturated clay soils, Skempton (1951), proposed the following equation for a strip foundation

$$q_u = cN_c + \gamma D_f \tag{5.22a}$$

or $\qquad q_{nu} = q_u - \gamma D_f = cN_c \tag{5.22b}$

$$q_{na} = \frac{q_{nu}}{F_s} = \frac{cN_c}{F_s} \tag{5.22c}$$

The N_c values for strip and square (or circular) foundations as a function of the D_f/B ratio are given in Fig. 5.8. The equation for rectangular foundation may be written as follows

$$(N_c)_R = \left(0.84 + 0.16 \times \frac{B}{L}\right)(N_c)_S \qquad\qquad (5.22d)$$

where $(N_c)_R = N_c$ for rectangular foundation, $(N_c)_S = N_c$ for square foundation.

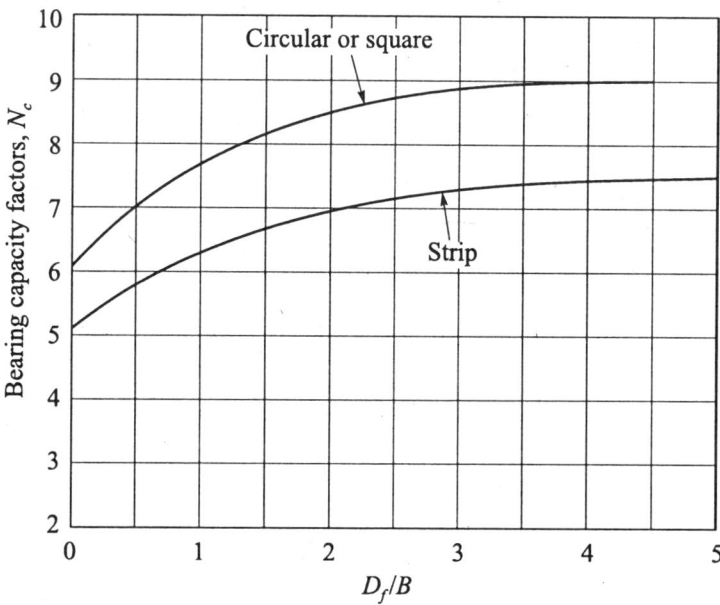

Fig. 5.8 Skempton's bearing capacity factor N_c for clay soils

The lower and upper limiting values of N_c for strip and square foundations may be written as follows:

Type of foundation	Ratio D_f/B	Value of N_c
Strip	0	5.14
	≥ 4	7.5
Square	0	6.2
	≥ 4	9.0

5.8 EFFECT OF WATER TABLE ON BEARING CAPACITY

The theoretical equations developed for computing the ultimate bearing capacity q_u of soil are based on the assumption that the water table lies at a depth below the base of the foundation equal to or greater than the width B of the foundation or otherwise the depth of the water table from ground surface is equal to or greater than $(D_f + B)$. In case the water table lies at any intermediate depth less than the depth $(D_f + B)$, the bearing capacity equations are affected due to the presence of the water table.

Two cases may be considered here,

Case 1: When the water table lies above the base of the foundation.

Case 2: When the water table lies within depth B below the base of the foundation.

We will consider the two methods for determining the effect of the water table on bearing capacity as given below.

Method 1

For any position of the water table within the depth $(D_f + B)$, we may write Eq. (5.6) as

$$q_u = cN_c + \gamma D_f N_q R_{w1} + \frac{1}{2} \gamma BN_\gamma R_{w2} \qquad (5.23)$$

where R_{w1} = reduction factor for water table above the base level of the foundation,

$\quad\quad R_{w2}$ = reduction factor for water table below the base level of the foundation,

$\quad\quad \gamma = \gamma_{sat}$ for all practical purposes in both the second and third terms of Eq. (5.23).

Case 1: When the water table lies above the base level of the foundation or when $D_{w1}/D_f \leq 1$ [Fig. 5.9 (a)] the equation for R_{w1} may be written as

$$R_{w1} = \frac{1}{2}\left(1 + \frac{D_{w1}}{D_f}\right) \qquad (5.24a)$$

For $D_{w1}/D_f = 0$, we have $R_{w1} = 0.5$, and for $D_{w1}/D_f = 1.0$, we have $R_{w1} = 1.0$.

Case 2: When the water table lies below the base level or when $D_{w2}/B \leq 1$ [Fig. 5.9 (b)] the equation for R_{w2} is

$$Rw_2 = \frac{1}{2}\left(1 + \frac{D_{w2}}{B}\right) \qquad (5.24b)$$

For $D_{w2}/B = 0$, we have $R_{w2} = 0.5$, and for $D_{w2}/B = 1.0$, we have $R_{w2} = 1.0$.

Figure 5.9 shows in a graphical form the relations D_{w1}/D_f vs. R_{w1} and D_{w2}/B vs. R_{w2}.

Equations (5.24a) and (5.24b) are based on the assumption that the submerged unit weight of soil is equal to half of the saturated unit weight and the soil above the water table remains saturated.

Method 2: Equivalent effective unit weight method

Equation (5.6) for the strip footing may be expressed as

$$q_u = cN_c + \gamma_{e1} D_f N_q + \frac{1}{2} \gamma_{e2} BN_\gamma \qquad (5.25)$$

where γ_{e1} = weighted effective unit weight of soil lying above the base level of the foundation,

$\quad\quad \gamma_{e2}$ = weighted effective unit weight of soil lying within the depth B below the base level of the foundation,

$\quad\quad \gamma_m$ = moist or saturated unit weight of soil lying above WT (case 1 or case 2)

$\quad\quad \gamma_{sat}$ = saturated unit weight of soil below the WT (case 1 or case 2)

$\quad\quad \gamma_b$ = submerged unit weight of soil = $\gamma_{sat} - \gamma_w$

Case 1: An equation for γ_{e1} may be written as

$$\gamma_{e1} = \gamma_b + \frac{D_{w1}}{D_f}(\gamma_m - \gamma_b) \qquad (5.26a)$$

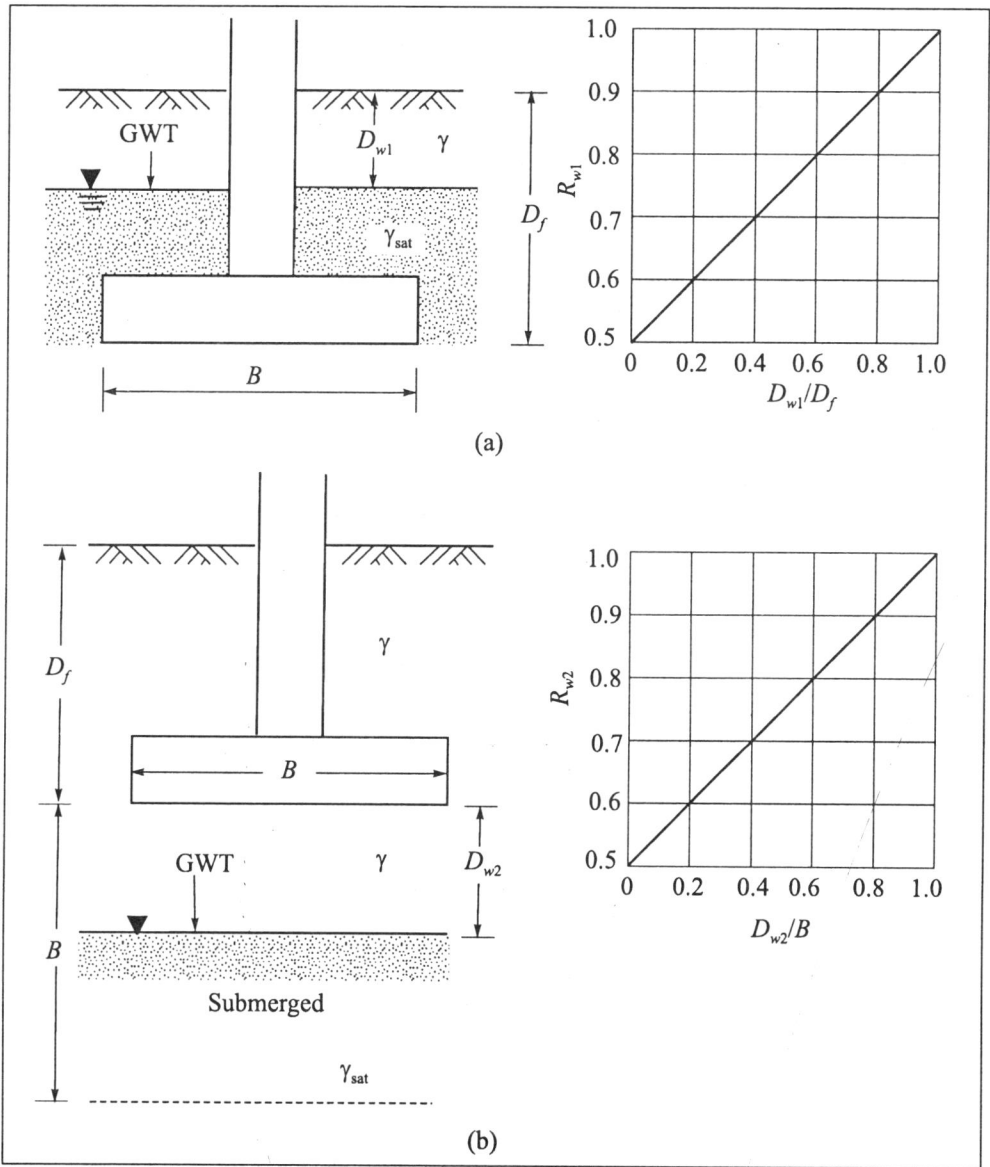

Fig. 5.9 Effect of WT on bearing capacity: (a) Water table above base level of foundation, (b) water table below base level of foundation

Case 2:

$$\gamma_{e2} = \gamma_b$$

$$\gamma_{e1} = \gamma_m$$

$$\gamma_{e2} = \gamma_b + \frac{D_{w2}}{B}(\gamma_m - \gamma_b) \tag{5.26b}$$

Example 5.1

A strip footing of width 3 m is founded at a depth of 2 m below the ground surface in a $(c - \phi)$ soil having a cohesion $c = 30$ kN/m² and angle of shearing resistance $\phi = 35°$. The water table is at a depth of 5 m below ground level. The moist weight of soil above the water table is 17.25 kN/m³. Determine (*a*) the ultimate bearing capacity of the soil, (*b*) the net bearing capacity, and (*c*) the net allowable bearing pressure and the load/m for a factor of safety of 3. Use the general shear failure theory of Terzaghi.

Solution

For $\phi = 35°$, $N_c = 57.8$, $N_q = 41.4$, and $N_\gamma = 42.4$

From Eq. (5.6),

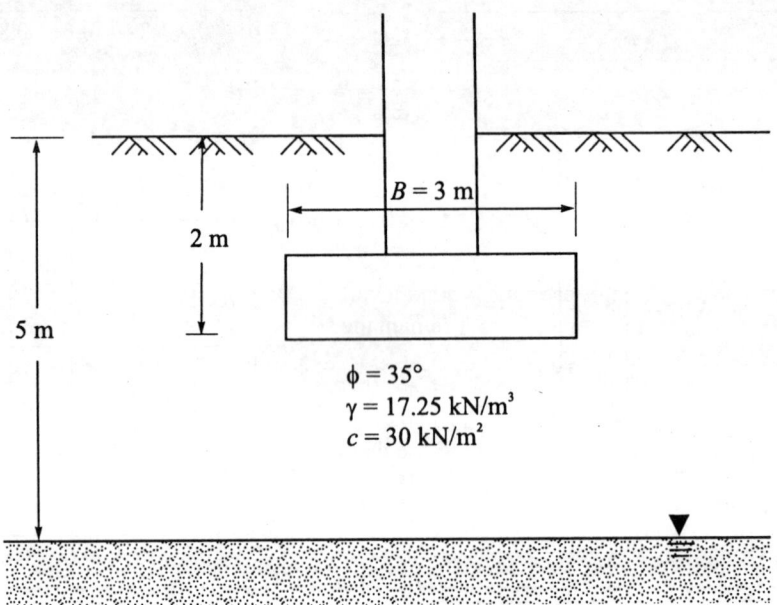

Fig. Ex. 5.1

$$q_u = cN_c + \gamma D_f N_q + \frac{1}{2} \gamma BN_\gamma$$

$$= 30 \times 57.8 + 17.25 \times 2 \times 41.4 + \frac{1}{2} \times 17.25 \times 3 \times 42.4 = 4259 \text{ kN/m}^2$$

$$q_{nu} = q_u - \gamma D_f = 4259 - 17.25 \times 2 \approx 4225 \text{ kN/m}^2$$

$$q_{na} = \frac{q_{nu}}{F_s} = \frac{4225}{3} \approx 1408 \text{ kN/m}^2$$

$$Q_a = q_{na} B = 1408 \times 3 = 4225 \text{ kN/m}$$

Example 5.2

If the soil in Ex. 5.1 fails by local shear failure, determine the net safe bearing pressure. All the other data given in Ex. 5.1 remain the same.

Solution

For local shear failure:

$$\bar{\phi} = \tan^{-1} 0.67 \tan 35° = 25°$$

$$\bar{c} = 0.67c = 0.67 \times 30 = 20 \text{ kN/m}^2$$

From Table 5.1, for $\bar{\phi} = 25°$, $\bar{N}_c = 25.1$, $\bar{N}_q = 12.7$, $\bar{N}_\gamma = 9.7$

Now from Eq. (5.12)

$$q_u = 20 \times 25.1 + 17.25 \times 2 \times 12.7 + \frac{1}{2} \times 17.25 \times 3 \times 9.7 = 1191 \text{ kN/m}^2$$

$$q_{nu} = 1191 - 17.25 \times 2 = 1156.5 \text{ kN/m}^2$$

$$q_{na} = \frac{1156.50}{3} = 385.5 \text{ kN/m}^2$$

$$Q_a = 385.5 \times 3 = 1156.5 \text{ kN/m}$$

Example 5.3

If the water table in Ex. 5.1 rises to the ground level, determine the net safe bearing pressure of the footing. All the other data given in Ex. 5.1 remain the same. Assume the saturated unit weight of the soil $\gamma_{sat} = 18.5 \text{ kN/m}^3$.

Solution

When the WT is at ground level we have to use the submerged unit weight of the soil.

Therefore $\gamma_b = \gamma_{sat} - \gamma_w = 18.5 - 9.81 = 8.69 \text{ kN/m}^3$

The net ultimate bearing capacity is

$$q_{nu} = 30 \times 57.8 + 8.69 \times 2 (41.4 - 1) + \frac{1}{2} \times 48.69 \times 3 \times 42.4 \approx 2992 \text{ kN/m}^2$$

$$q_{na} = \frac{2992}{3} = 997.33 \text{ kN/m}^2$$

$$Q_a = 997.33 \times 3 = 2992 \text{ kN/m}$$

Example 5.4

If the water table in Ex. 5.1 occupies any of the positions: (*a*) 1.25 m below ground level or (*b*) 1.25 m below the base level of the foundation, what will be the net safe bearing pressure?

Assume $\gamma_{sat} = 18.5 \text{ kN/m}^3$, γ (above WT) = 17.5 kN/m^3. All the other data remain the same as given in Ex. 5.1.

Solution

Method 1

By making use of reduction factors R_{w1} and R_{w2} and using Eqs (5.20) and (5.23), we may write

$$q_{nu} = cN_c + \gamma D_f (N_q - 1) R_{w1} + \frac{1}{2} \gamma BN_\gamma R_{w2}$$

Given: $N_q = 41.4$, $N_\gamma = 42.4$ and $N_c = 57.8$

Case 1: When the WT is 1.25 m below the GL

From Eq. (5.24), we get $R_{w1} = 0.813$ for $D_{w1}/D_f = 0.625$, $R_{w2} = 0.5$ for $D_{w2}/B = 0$.

By substituting the known values in the equation for q_{nu}, we have

$$q_{na} = 30 \times 57.8 + 18.5 \times 2 \times 40.4 \times 0.813 + \frac{1}{2} \times 18.5 \times 3 \times 42.4 \times 0.5 = 3538 \text{ kN/m}^2$$

$$q_{na} = \frac{3538}{3} = 1179 \text{ kN/m}^2$$

Case 2: When the WT is 1.25 m below the base of the foundation

$R_{w1} = 1.0$ for $D_{w1}/D_f = 1$, $R_{w2} = 0.71$ for $D_{w2}/B = 0.42$.

Now the net bearing capacity is

$$q_{nu} = 30 \times 57.8 + 18.5 \times 2 \times 40.4 \times 1 + \frac{1}{2} \times 18.5 \times 3 \times 42.4 \times 0.71 = 4064 \text{ kN/m}^2$$

$$q_{na} = \frac{4064}{3} = 1355 \text{ kN/m}^2$$

Method 2

Using the equivalent effective unit weight method.

Submerged unit weight $\gamma_b = 18.5 - 9.81 = 8.69 \text{ kN/m}^3$.

Per Eq. (5.25), the net ultimate bearing capacity is

$$q_{nu} = cN_c + \gamma_{e1} D_f(N_q - 1) + \frac{1}{2}\gamma_{e2} BN_\gamma$$

Case 1: When $D_{w1} = 1.25$ m (Fig. Ex. 5.4)

From Eq. (5.26a)

$$\gamma_{e1} = \gamma_b + \frac{D_{w1}}{D_f}(\gamma_m - \gamma_b)$$

where $\gamma_m = \gamma_{sat} = 18.5 \text{ kN/m}^3$

$$\gamma_{e1} = 8.69 + \frac{1.25}{2}(18.5 - 8.69) = 14.82 \text{ kN/m}^3$$

$$\gamma_{e2} = \gamma_b = 8.69 \text{ kN/m}^3$$

$$q_{nu} = 30 \times 57.8 + 14.82 \times 2 \times 40.4 + \frac{1}{2} \times 8.69 \times 3 \times 42.4 = 3484 \text{ kN/m}^2$$

$$q_{na} = \frac{3484}{3} = 1161 \text{ kN/m}^2$$

Case 2: When $D_{w2} = 1.25$ m (Fig. Ex. 5.4)

From Eq. 5.26 (b)

$$\gamma_{e1} = \gamma_m = 18.5 \text{ kN/m}^3$$

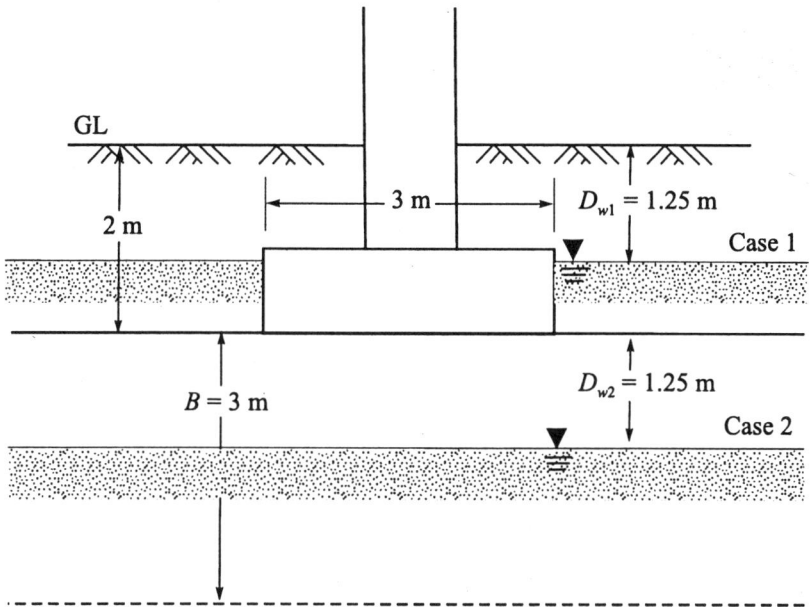

Fig. Ex. 5.4 Effect of WT on bearing capacity

$$\gamma_{e2} = 8.69 + \frac{1.25}{3}(18.5 - 8.69) = 12.78\ \text{kN/m}^3$$

$$q_{nu} = 30 \times 57.8 + 18.5 \times 2 \times 40.4 + \frac{1}{2} \times 12.78 \times 3 \times 42.4 = 4042\ \text{kN/m}^2$$

$$q_{na} = \frac{4042}{3} = 1347\ \text{kN/m}^2$$

Example 5.5

A square footing fails by general shear in a cohesionless soil under an ultimate load of $Q_{ult} = 1687.5$ kips. The footing is placed at a depth of 6.5 ft below ground level. Given: $\phi = 35°$, and $\gamma = 110$ lb/ft^3, determine the size of the footing if the water table is at a great depth (Fig. Ex. 5.5).

Solution

For a square footing Eq. (5.17) for $c = 0$, we have

$$q_u = \gamma D_f N_q + 0.4\gamma\, BN_\gamma$$

For $\phi = 35°$, $N_q = 41.4$, and $N_\gamma = 42.4$ from Table 5.1.

$$q_u = \frac{Q_u}{B^2} = \frac{1687.5 \times 10^3}{B^2}$$

By substituting known values, we have

$$\frac{1687.5 \times 10^3}{B^2} = 110 \times 6.5 \times 41.4 + 0.4 \times 110 \times 42.4B$$

$$= (29.601 + 1.866B)\,10^3$$

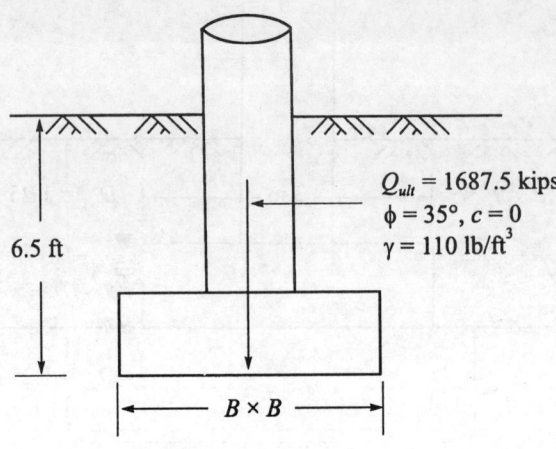

Q_{ult} = 1687.5 kips
ϕ = 35°, c = 0
γ = 110 lb/ft^3

6.5 ft

$B \times B$

Fig. Ex. 5.5

Simplifying and transposing, we have

$$B^3 + 15.863B^2 - 904.34 = 0$$

Solving this equation yields, $B = 6.4$ ft.

Example 5.6

A rectangular footing of size 10 × 20 ft is founded at a depth of 6 ft below the ground surface in a homogeneous cohesionless soil having an angle of shearing resistance $\phi = 35°$. The water table is at a great depth. The unit weight of soil $\gamma = 114$ lb/ft^3. Determine: (1) the net ultimate bearing capacity, (2) the net allowable bearing pressure for $F_s = 3$, and (3) the allowable load Q_a the footing can carry. Use Terzaghi's theory (Refer to Fig. Ex. 5.6).

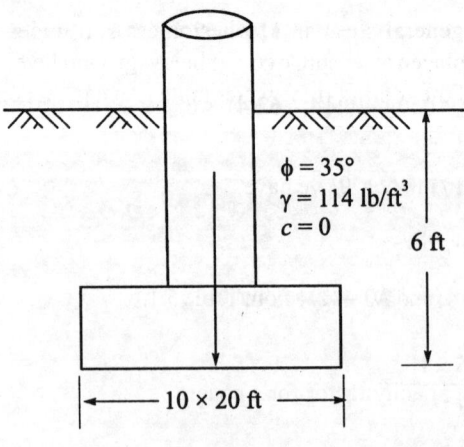

ϕ = 35°
γ = 114 lb/ft^3
c = 0

6 ft

10 × 20 ft

Fig. Ex. 5.6

Solution

Using Eqs (5.19) and (5.20) for $c = 0$, the net ultimate bearing capacity for a rectangular footing is expressed as

$$q_{nu} = \gamma D_f (N_q - 1) + \frac{1}{2} \gamma BN_\gamma \left(1 - 0.2 \frac{B}{L}\right)$$

From Table 5.1,

$$N_q = 41.4,\ N_\gamma = 42.4 \text{ for } \phi = 35°$$

By substituting the known values,

$$q_{nu} = 114 \times 6\,(41.4 - 1) + \frac{1}{2} \times 114 \times 10 \times 42.4 \left(1 - 0.2 \times \frac{10}{20}\right) = 49385 \text{ lb/ft}^2$$

$$q_{na} = \frac{49385}{3} = 16462 \text{ lb/ft}^2$$

$$Q_a = (B \times L)\, q_{na} = 10 \times 20 \times 16462 \approx 3292 \times 10^3 \text{ lb} = 3292 \text{ kips}$$

Example 5.7

A rectangular footing of size 10 × 20 ft is founded at a depth of 6 ft below the ground level in a cohesive soil ($\phi = 0$) which fails by general shear. Given: $\gamma_{sat} = 114 \text{ lb/ft}^3$, $c = 945 \text{ lb/ft}^2$. The water table is close to the ground surface. Determine q_u, q_{nu} and q_{na} by (a) Terzaghi's method, and (b) Skempton's method (use $F_s = 3$).

Solution

(a) Terzaghi's method

Use Eq. (5.19)

For $\phi = 0°$, $N_c = 5.7$, $N_q = 1$

$$q_u = cN_c \left(1 + 0.3 \times \frac{B}{L}\right) + \gamma_b\, D_f$$

Substituting the known values,

$$q_u = 945 \times 5.7 \left(1 + 0.3 \times \frac{10}{20}\right) + (114 - 62.4) \times 6 = 6504 \text{ lb/ft}^2$$

$$q_{nu} = (q_u - \gamma_b\, D_f) = 6504 - (114 - 62.4) \times 6 = 6195 \text{ lb/ft}^2$$

$$q_{na} = \frac{q_{na}}{F_s} = \frac{6195}{3} = 2065 \text{ lb/ft}^2$$

(b) Skempton's method

From Eqs (5.22a) and (5.22b), we may write

$$q_u = cN_{cr} + \gamma D_f$$

where N_{cr} = bearing capacity factor for a square foundation.

$$N_{cr} = \left(0.84 + 0.16 \times \frac{B}{L}\right) \times N_{cs}$$

where N_{cs} = bearing capacity factor for a square foundation.

From Fig. 12.9, $N_{cs} \doteq 7.2$ for $D_f/B = 0.60$.

Therefore

$$N_{cr} = \left(0.84 + 0.16 \times \frac{10}{20}\right) \times 7.2 = 6.62$$

Now $q_u = 945 \times 6.62 + 114 \times 6 = 6940 \text{ lb/ft}^2$

$q_{nu} = (q_u - \gamma D_f) = 6940 - 114 \times 6 = 6256 \text{ lb/ft}^2$

$$q_{na} = \frac{q_{nu}}{F_s} = \frac{6256}{3} = 2085 \text{ lb/ft}^2$$

Note: Terzaghi's and Skempton's values are in close agreement for cohesive soils.

Example 5.8

If the soil in Ex. 5.6 is cohesionless ($c = 0$), and fails in local shear, determine: (*i*) The ultimate bearing capacity, (*ii*) the net bearing capacity, and (*iii*) the net allowable bearing pressure. All the other data remain the same.

Solution

From Eqs (5.15) and (5.20), the net bearing capacity for local shear failure for $c = 0$ is

$$q_{nu} = (q_u - \gamma D_f) = \gamma D_f\, (\bar{N}_q - 1) + \frac{1}{2}\gamma B \bar{N}_\gamma \left(1 - 0.2 \times \frac{B}{L}\right)$$

where $\quad \bar{\phi} = \tan^{-1} 0.67 \tan 35° \approx 25°$, $\bar{N}q = 12.7$, and $\bar{N}_\gamma = 9.7$ for $\bar{\phi} = 25°$ from Table 5.1.

By substituting known values, we have

$$q_{nu} = 114 \times 6\,(12.7 - 1) + \frac{1}{2} \times 114 \times 10 \times 9.7 \left(1 - 0.2 \times \frac{10}{20}\right) = 12979 \text{ lb/ft}^2$$

$$q_{na} = \frac{12979}{3} = 4326 \text{ lb/ft}^2$$

5.9 THE GENERAL BEARING CAPACITY EQUATION

The bearing capacity Eq. (5.6), developed by Terzaghi is for a strip footing under general shear failure. Equation (5.6) has been modified for other types of foundations such as square, circular and rectangular by introducing shape factors. Meyerhof (1963), presented a general bearing capacity equation which takes into account the shape and the inclination of load. The general form of equation suggested by Meyerhof for bearing capacity is

$$q_u = c N_c s_c d_c i_c + q_o' N_q s_q d_q i_q + \frac{1}{2}\gamma B N_\gamma s_\gamma d_\gamma i_\gamma \qquad (5.27)$$

where
c = unit cohesion,
q_o' = effective overburden pressure at the base level of the foundation = $\bar{\gamma} D_f$,
$\bar{\gamma}$ = effective unit weight above the base level of foundation,
γ = effective unit weight of soil below the foundation base,
D_f = depth of foundation,

s_c, s_q, s_γ = shape factors,

d_c, d_q, d_γ = depth factor,

i_c, i_q, i_γ = load inclination factors,

B = width of foundation,

N_c, N_q, N_γ = bearing capacity factors.

Hansen (1970) extended the work of Meyerhof by including in Eq. (5.27) two additional factors to take care of base tilt and foundations on slopes. Vesic (1973, 1974) used the same form of equation suggested by Hansen. All three investigators use the equations proposed by Prandtl (1921) for computing the values of N_c and N_q wherein the foundation base is assumed as smooth with the angle $\alpha = 45° + \phi/2$ (Fig. 5.5). However, the equations used by them for computing the values of N_γ are different. The equations for N_c, N_q and N_γ are:

$$N_q = e^{\pi \tan \phi} N_\phi,$$

$$N_c = (N_q - 1) \cot \phi,$$

$$N_\gamma = (N_q - 1) \tan (1.4\ \phi) \quad \text{(Meyerhof)}$$

$$N_\gamma = 1.5\ (N_q - 1) \tan \phi \quad \text{(Hansen)}$$

$$N_\gamma = 2\ (N_q + 1) \tan \phi \quad \text{(Vesic)}$$

Table 5.2 gives the values of the bearing capacity factors. Equations for shape, depth and inclination factors are given in Table 5.3. The tilt of the base and the foundations on slopes are not considered here.

Table 5.2 The values of N_c, N_q, and Meyerhof (M), Hansen (H), and Vesic (V) N_γ factors

ϕ	N_c	N_q	N_γ (H)	N_γ (M)	N_γ (V)
0	5.14	1.0	0.0	0.0	0.0
5	6.49	1.6	0.1	0.1	0.4
10	8.34	2.5	0.4	0.4	1.2
15	10.97	3.9	1.2	1.1	2.6
20	14.83	6.4	2.9	2.9	5.4
25	20.71	10.7	6.8	6.8	10.9
26	22.25	11.8	7.9	8.0	12.5
28	25.79	14.7	10.9	11.2	16.7
30	30.13	18.4	15.1	15.7	22.4
32	35.47	23.2	20.8	22.0	30.2
34	42.14	29.4	28.7	31.1	41.0
36	50.55	37.7	40.0	44.4	56.2
38	61.31	48.9	56.1	64.0	77.9
40	72.25	64.1	79.4	93.6	109.4
45	133.73	134.7	200.5	262.3	271.3
50	266.50	318.5	567.4	871.7	762.84

In Table 5.3 following terms are defined with regard to the inclination factors.

Q_h = horizontal component of the inclined load,

Q_u = vertical component of the inclined load,

c_a = unit adhesion on the base of the footing,

A_f = effective contact area of the footing.

Table 5.3 Shape, depth and load inclination factors of Meyerhof, Hansen and Vesic

Factors	Meyerhof	Hansen	Vesic
s_c	$1 + 0.2\, N_\phi \left(\dfrac{B}{L}\right)$	$1 + \dfrac{N_q}{N_c} \left(\dfrac{B}{L}\right)$	
s_q	$1 + 0.1\, N_\phi \left(\dfrac{B}{L}\right)$ for $\phi > 10°$	$1 + \dfrac{B}{L}\tan\phi$	
s_γ	$s_\gamma = s_q$ for $\phi > 10°$ $s_\gamma = s_q = 1$ for $\phi = 0$	$1 - 0.4\,\dfrac{B}{L}$	The shape and depth factors of Vesic are the same as those of Hansen
d_c	$1 + 0.2\,\sqrt{N_\phi}\left(\dfrac{D_f}{B}\right)$	$1 + 0.4\left(\dfrac{D_f}{B}\right)$	
d_q	$1 + 0.1\,\sqrt{N_\phi}\left(\dfrac{D_f}{B}\right)$ for $\phi > 10°$	$1 + 2\tan\phi\,(1 - \sin\phi)^2 \left(\dfrac{D_f}{B}\right)$	
d_γ	$d_\gamma = d_q$ for $\phi > 10°$ $d_\gamma = d_q = 1$ for $\phi = 0$	1 for all ϕ *Note:* Vesic's s and d factors = Hansen's s and d factors	
i_c	$\left(1 - \dfrac{\alpha°}{90}\right)^2$ for any ϕ	$i_q - \dfrac{1 - i_q}{N_q - 1}$ for $\phi > 0$ $0.5\left(1 - \dfrac{Q_h}{A_f c_a}\right)^{\frac{1}{2}}$ for $\phi = 0$	Same as Hansen for $\phi > 0$ $1 - \dfrac{mQ_h}{A_f c_a N_c}$
i_q	$i_q = i_c$ for any ϕ	$\left(1 - \dfrac{0.5 Q_h}{Q_u + A_f c_a \cot\phi}\right)^5$	$\left(1 - \dfrac{Q_h}{Q_u + A_f c_a \cot\phi}\right)^m$
i_γ	$\left(1 - \dfrac{\alpha°}{\phi°}\right)^2$ for $\phi > 0$ $i_\gamma = 0$ for $\phi = 0$	$\left(1 - \dfrac{0.7 Q_h}{Q_u + A_f c_a \cot\phi}\right)^5$	$\left(1 - \dfrac{Q_h}{Q_u + A_f c_a \cot\phi}\right)^{m+1}$

$$m = m_B = \frac{2 + B/L}{1 + B/L} \text{ with } Q_h \text{ parallel to } B$$

$$m = m_L = \frac{2 + B/L}{1 + B/L} \text{ with } Q_h \text{ parallel to } L$$

The general bearing capacity Eq. (5.27) has not taken into account the effect of the water table position on the bearing capacity. Hence, Eq. (5.27) has to be modified according to the position of water level in the same way as explained in Section 5.7.

Validity of the Bearing Capacity Equations

There is currently no method of obtaining the ultimate bearing capacity of a foundation other than as an estimate (Bowles, 1996). There has been little experimental verification of any of the methods except by using model footings. Up to a depth of $D_f \approx B$ the Meyerhof q_u is not greatly different from the Terzaghi value (Bowles, 1996). The Terzaghi equations, being the first proposed, have been quite popular with designers. Both the Meyerhof and Hansen methods are widely used. The Vesic method has not been much used. It is a good practice to use at least two methods and compare the computed values of q_u. If the two values do not compare well, use a third method.

Example 5.9

Refer to Ex. 5.1. Compute using the Meyerhof equation (a) the ultimate bearing capacity of the soil, (b) the net bearing capacity, and (c) the net allowable bearing pressure. All the other data remain the same.

Solution

Use Eq. (5.27). For $i = 1$ the equation for net bearing capacity is

$$q_{nu} = q_u - \gamma D_f = cN_c s_c d_c + \gamma D_f (N_q - 1) s_q d_q + \frac{1}{2} \gamma B N_\gamma s_\gamma d_\gamma$$

From Table 5.3

$$s_c = 1 + 0.2N_\phi \left(\frac{B}{L} \right) = 1 \text{ for strip footing}$$

$$s_q = 1 + 0.1N_q \left(\frac{B}{L} \right) = 1 \text{ for strip footing}$$

$$s_\gamma = s_q = 1$$

$$d_c = 1 + 0.2 \sqrt{N_\phi} \left(\frac{D_f}{B} \right) = 1 + 0.2 \tan \left(45° + \frac{35}{2} \right)\left(\frac{2}{3} \right) = 1.257$$

$$d_q = 1 + 0.1 \sqrt{N_\phi} \left(\frac{D_f}{B} \right) = 1 + 0.1 \tan \left(45° + \frac{35}{2} \right)\left(\frac{2}{3} \right) = 1.129$$

$$d_\gamma = d_q = 1.129$$

From Ex. 5.1, $c = 30 \text{ kN/m}^2$, $\gamma = 17.25 \text{ kN/m}^3$, $D_f = 2 \text{ m}$, $B = 3 \text{ m}$.

From Table 5.2 for $\phi = 35°$, we have $N_c = 46.35$, $N_q = 33.55$, $N_\gamma = 37.75$. Now substituting the known values, we have

$$q_{nu} = 30 \times 46.35 \times 1 \times 1.257 + 17.25 \times 2 \times (33.55 - 1) \times 1 \times 1.129$$

$$+ \frac{1}{2} \times 17.25 \times 3 \times 37.75 \times 1 \times 1.129$$

$$= 1,748 + 1,268 + 1,103 = 4,119 \text{ kN/m}^2$$

$$q_{na} = \frac{4,119}{3} = 1,373 \text{ kN/m}^2$$

There is very close agreement between Terzaghi's and Meyerhof's methods.

Example 5.10

Refer to Ex. 5.6. Compute by Meyerhof's method the net ultimate bearing capacity and the net allowable bearing pressure for $F_s = 3$. All the other data remain the same.

Solution

From Ex. 5.6, we have $B = 10$ ft, $L = 20$ ft, $D_f = 6$ ft, and $\gamma = 114$ lb/ft^3. From Eq. (5.27) for $c = 0$ and $i = 1$, we have

$$q_{nu} = q_u - \gamma D_f = \gamma D_f (N_q - 1) s_q d_q + \frac{1}{2} \gamma B N_\gamma s_\gamma d_\gamma$$

From Table 5.3

$$s_q = 1 + 0.1 N_\phi \left(\frac{B}{L} \right) = 1 + 0.1 \tan^2 \left(45° + \frac{35}{2} \right) \left(\frac{10}{20} \right) = 1.185$$

$$s_\gamma = s_q = 1.185$$

$$d_q = 1 + 0.1 \sqrt{N_\phi} \left(\frac{D_f}{B} \right) = 1 + 0.1 \tan \left(45° + \frac{35}{2} \right) \left(\frac{6}{20} \right) = 1.115$$

$$d_\gamma = d_q = 1.115$$

From Table 5.2 for $\phi = 35°$, we have $N_q = 33.55$, $N\gamma = 37.75$. By substituting the known values, we have

$$q_{nu} = 114 \times 6 \, (33.55 - 1) \times 1.185 \times 1.115 + \frac{1}{2} \times 114 \times 10 \times 37.75 \times 1.185 \times 1.115$$

$$= 29,417 + 28,431 = 57,848 \text{ lb/ft}^2$$

$$q_{na} = \frac{57,848}{3} = 19,283 \text{ lb/ft}^2$$

By Terzaghi's method $q_{na} = 16,462$ lb/ft^2.

Meyerhof's method gives a higher value for q_{na} by about 17%.

Example 5.11

Refer to Ex. 5.1. Compute by Hansen's method: (*a*) Net ultimate bearing capacity, and (*b*) the net safe bearing pressure. All the other data remain the same.

Given for a strip footing

$$B = 3 \text{ m}, D_f = 2 \text{ m}, c = 30 \text{ kN/m}^2 \text{ and } \gamma = 17.25 \text{ kN/m}^3, F_s = 3.$$

From Eq. (5.27) for $i = 1$, we have

$$q_{nu} = q_u - \gamma D_f = c N_c s_c d_c + \gamma D_f (N_q - 1) s_q d_q + \frac{1}{2} \gamma B N_\gamma s_\gamma d_\gamma$$

From Table 5.2 for Hansen's method, we have for $\phi = 35°$

$N_c = 46.35$, $N_q = 33.55$, and $N_\gamma = 34.35$.

From Table 5.3, we have

$$s_c = 1 + \frac{N_q}{N_c}\left(\frac{B}{L}\right) = 1 \text{ for a strip footing}$$

$$s_q = 1 + \frac{B}{L}\tan\phi = 1 \text{ for a strip footing}$$

$$s_\gamma = 1 - 0.4\frac{B}{L} = 1 \text{ for a strip footing}$$

$$d_c = 1 + 0.4\left(\frac{D_f}{B}\right) = 1 + 0.4 \times \frac{2}{3} = 1.267$$

$$d_q = 1 + 2\tan\phi\,(1 - \sin\phi)^2\left(\frac{D_f}{B}\right)$$

$$= 1 + 2\tan 35°\,(1 - \sin 35°)^2 \times \frac{2}{3} = 1 + 2 \times 0.7\,(1 - 0.574)^2 \times \frac{2}{3} = 1.17$$

$$d_\gamma = 1$$

Substituting the known values, we have

$$q_{nu} = 30 \times 46.35 \times 1 \times 1.267 + 17.25 \times 2 \times (33.55 - 1) \times 1 \times 1.17 +$$

$$\frac{1}{2} \times 17.25 \times 3 \times 34.35 \times 1 \times 1$$

$$= 1{,}762 + 1{,}314 + 889 = 3{,}965 \text{ kN/m}^2$$

$$q_{na} = \frac{3{,}965}{3} = 1{,}322 \text{ kN/m}^2$$

The values of q_{na} by Terzaghi, Meyerhof and Hansen methods are

Example	Author	q_{na} kN/m^2
5.1	Terzaghi	1,408
5.9	Meyerhof	1,373
5.11	Hansen	1,322

Terzaghi's method is higher than Meyerhof's by 2.5% and Meyerhof's higher than Hansen's by 3.9%. The difference between the methods is not significant. Any of the three methods can be used.

Example 5.12

Refer to Ex. 5.6. Compute the net safe bearing pressure by Hansen's method. All the other data remain the same.

Solution

Given: Size 10×20 ft, $D_f = 6$ ft, $c = 0$, $\phi = 35°$, $\gamma = 114$ lb/ft^3, $F_s = 3$.

For $\phi = 35°$, we have from Table 5.2, $N_q = 33.55$ and $N_\gamma = 34.35$

From Table 5.3, we have

$$s_q = 1 + \frac{B}{L} \tan \phi = 1 + \frac{10}{20} \times \tan 35° = 13.5$$

$$s_\gamma = 1 - 0.4 \frac{B}{L} = 1 - 0.4 \times \frac{10}{20} = 0.80$$

$$d_q = 1 + 2 \tan 35° (1 - \sin 35°)^2 \times \frac{6}{10} = 1.153$$

$$d_\gamma = 1$$

Substituting the known values, we have

$$q_{nu} = q_u - \gamma D_f = \gamma D_f (N_q - 1) s_q d_q + \frac{1}{2} \gamma B N_\gamma s_\gamma d_\gamma$$

$$= 114 \times 6 (33.55 - 1) \times 1.35 \times 1.153 + \frac{1}{2} \times 114 \times 10 \times 34.55 \times 0.8 \times 1$$

$$= 34,655 + 15,664 = 50,319 \text{ lb/ft}^2$$

$$q_{na} = \frac{50,319}{3} = 16,773 \text{ lb/ft}^2$$

The values of q_{na} by other methods are

Example	Author	q_{na} kN/m^2
5.6	Terzaghi	16,462
5.10	Meyerhof	19,283
5.12	Hansen	16,773

It can be seen from the above, the values of Terzaghi and Hansen are very close to each other, whereas the Meyerhof value is higher than that of Terzaghi by 17 percent.

5.10 EFFECT OF SOIL COMPRESSIBILITY ON BEARING CAPACITY OF SOIL

Terzaghi (1943), developed Eq. (5.6) based on the assumption that the soil is incompressible. To take into account the compressibility of soil, he proposed reduced strength characteristics $\bar{c}$ and $\bar{\phi}$ defined by Eq. (5.11). As per Vesic (1973) a flat reduction of ϕ in the cas of local and punching shear failures is too conservative and ignores the existence of scale effects. It has been conclusively established that the ultimate bearing capacity q_u of soil does not increase in proportion to the increase in the size of the footing as shown in Fig. 5.10 or otherwise the bearing capacity factor N_γ decrease with the increase in the size of the footing as shown in Fig. 5.11.

To take into account the influence of soil compressibility and the related scale effects, Vesic (1973) proposed a modification of Eq. (5.27) by introducing compressibility factors as follows.

$$q_u = c N_c d_c s_c C_s + q'_o N_q d_q s_q C_q + \frac{1}{2} \gamma B N_\gamma d_\gamma s_\gamma C_\gamma \tag{5.28}$$

where, C_c, C_q and C_γ are the soil compressibility factors. The other symbols remain the same as before.

For the evaluation of the relative compressibility of a soil mass under loaded conditions, Vesic introduced a term called *rigidity index* I_r, which is defined as

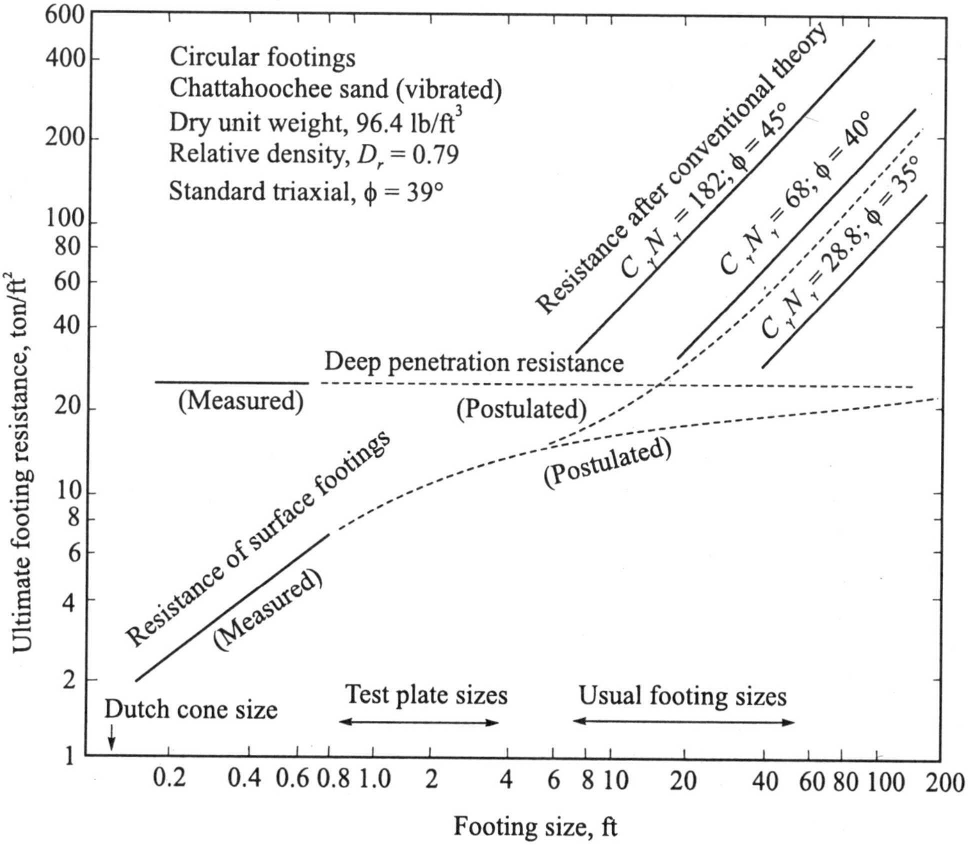

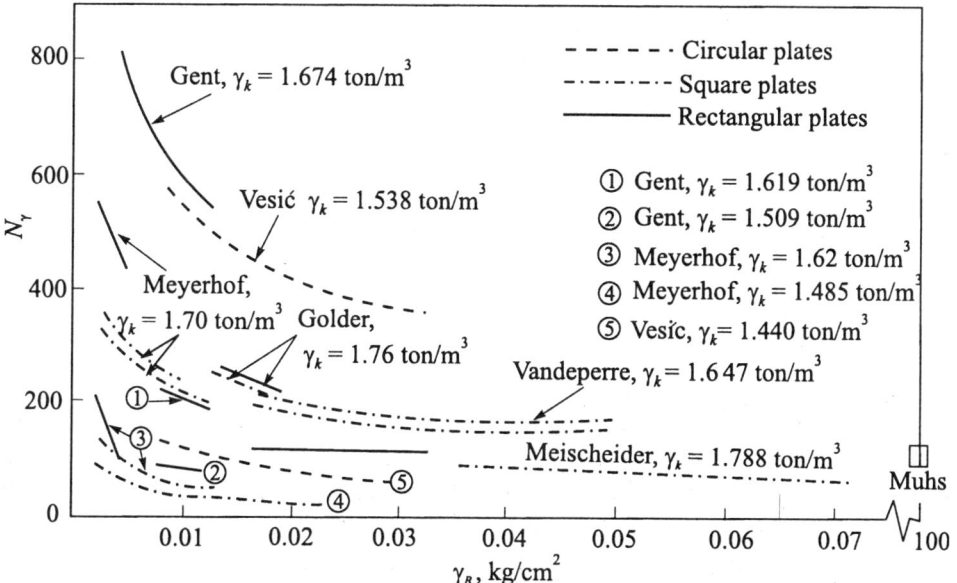

Fig. 5.10 Variation of ultimate resistance of footings with size (after Vesic, 1969)

Fig. 5.11 Effect of size on bearing capacity of surface footings in sand (after De Beer, 1965)

$$I_r = \frac{G}{c + \bar{q} \tan \phi} \tag{5.29}$$

where, G = shear modulus of soil = $\dfrac{E_s}{2(1+\mu)}$

E_s = modulus of elasticity

$\bar{q}$ = effective overburden pressure at a depth equal to $(D_f + B/2)$

μ = Poisson's ratio

c, ϕ = shear strength parameters

Equation (5.29) was developed on the basis of the theory of expansion of cavities in an infinite solid with the assumed ideal elastic properties behaviour of soil. To take care of the volumetric strain Δ in the plastic zone, Vesic (1965), suggested that the value of I_r, given by Eq. (5.29), be reduced by the following equation.

$$I_{rr} = F_r I_r \tag{5.30}$$

where F_r = reduction factor = $\dfrac{1}{1 + I_r \Delta}$

It is known that I_r varies with the stress level and the character of loading. A high value of I_{rr}, for example over 250, implies a relatively incompressible soil mass, whereas a low value of say 10 implies a relatively compressible soil mass.

Based on the theory of expansion of cavities, Vesic developed the following equation for the compressibility factors.

$$C_q = \exp\left[\left(-4.4 + 0.6\frac{B}{L}\right) \tan \phi + \left(\frac{3.07 \sin \phi \log 2I_r}{1 + \sin \phi}\right)\right] \tag{5.31}$$

For $\phi > 0$, one can determine from the theorem of correspondence

$$C_c = C_q - \frac{1 - C_q}{N_q \tan \phi} \tag{5.32}$$

For $\phi = 0$, we have

$$C_c = 0.32 + 0.12\frac{B}{L} + 0.6 \log I_r \tag{5.33}$$

For all practical purposes, Vesic suggests

$$C_q = C_\gamma \tag{5.34}$$

Equations (5.30) through Eq. (5.34) are valid as long as the values of the compressibility factors are less than unity. Figure 5.12 shows graphically the relationship between $C_q\ (=C_\gamma)$ and ϕ for two extreme cases of $L/B > 5$ (strip footing) and $B/L = 1$ (square) for different values of I_r (Vesic 1970). Vesic gives another expression called the *ctitical rigidity index* $(I_r)c_r$ expressed as

$$(I_r)_{cr} = \frac{1}{2} \exp\left[\left(3.3 - 0.45\frac{B}{L}\right) \cot \left(45 - \phi/2\right)\right] \tag{5.35}$$

$$I_r = \frac{G}{C + \overline{q}\tan\phi}$$

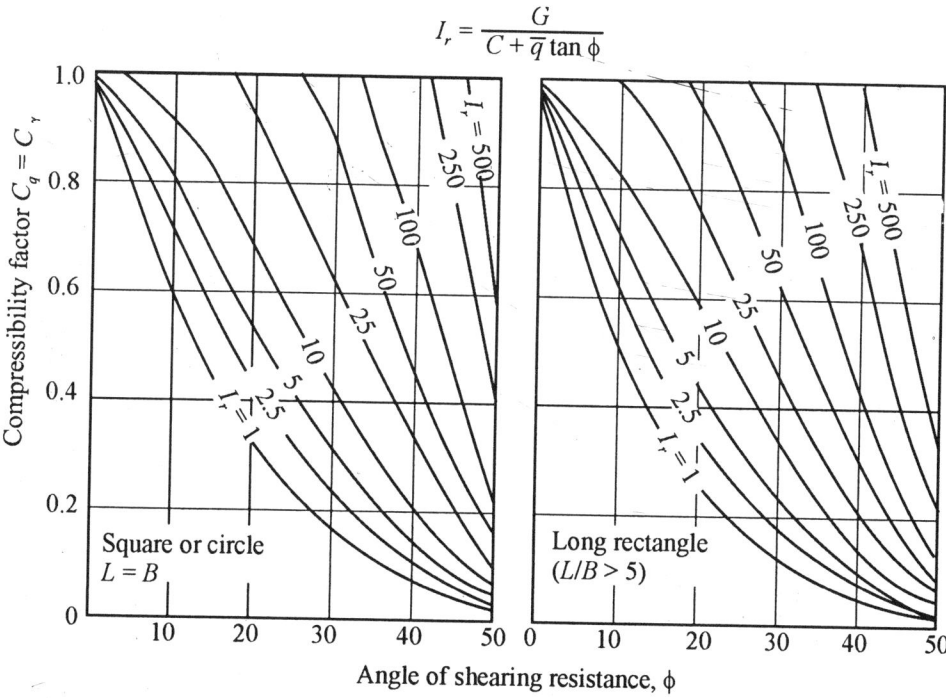

Fig. 5.12 Theoretical compressibility factors (after Vesic, 1970)

The magnitude of $(I_r)_{cr}$ for any angle of ϕ and any foundation shape reduces the bearing capacity because of compressibility effects. Numerical values of $(I_r)_{cr}$ for two extreme cases of $B/L = 0$ and $B/L = 1$ are given in Table 5.4 for various values of ϕ.

Table 5.4 Values of critical rigidity index

Angle of shearing resistance	Critical rigidity index for	
ϕ	Strip foundation $B/L = 0$	Square foundation $B/L = 1$
0	13	8
5	18	11
10	25	15
15	37	20
20	55	30
25	89	44
30	152	70
35	283	120
40	592	225
45	1442	486
50	4330	1258

Application of I_r (or I_{rr}) and $(I_r)_{crit}$

1. If I_r (or I_{rr}) $\geq (I_r)_{crit}$, assume the soil is incompressible and $C_c = C_q = C_\gamma = 1$ in Eq. (5.28).

2. If I_r (or I_{rr}) $< (I_r)_{crit}$, assume the soil is compressible. In such a case the compressibility factors C_c, C_q and C_γ are to be determined and used in Eq. (5.28).

The concept and analysis developed above by Vesic (1973) are based on a limited number of small scale model tests and need verification in field conditions.

Example 5.13

A square footing of size 12 × 12 ft is placed at a depth of 6 ft in a deep stratum of medium dense sand. The following soil parameters are available:

$\gamma = 100$ lb/ft^3, $c = 0$, $\phi = 35°$, $E_s = 100$ t/ft^2, Poissons' ratio $\mu = 0.25$.

Estimate the ultimate bearing capacity by taking into account the compressibility of the soil (Fig. Ex. 5.13).

Solution

Rigidity $I_r = \dfrac{E_s}{2(1+\mu)\,\bar{q}\,\tan\phi}$ for $c = 0$ from Eq. (5.29)

$$\bar{q} = \gamma(D_f + B/2) = 100\left(6 + \frac{12}{2}\right) = 1{,}200 \text{ lb/ft}^2 = 0.6 \text{ ton/ft}^2$$

Neglecting the volume change in the plastic zone

$$I_r = \frac{100}{2(1+0.25)\,0.6\tan 35°} = 95$$

From Table 5.4, $(I_r)_{crit} = 120$ for $\phi = 35°$

Since $I_r < (I_r)_{crit}$, the soil is compressible.

From Fig. 5.12, $C_q\,(= C_\gamma) = 0.90$ (approx) for square footing for $\phi = 35°$ and $I_r = 95$.

From Table 5.2, $N_q = 33.55$ and $N_\gamma = 48.6$ (Vesic's value)

Eq. (5.28) may now be written as

$$q_u = q_o' N_q d_q s_q C_q + \frac{1}{2}\gamma BN_\gamma d_\gamma s_\gamma C_\gamma$$

From Table 5.3

$$s_q = 1 + \frac{B}{L}\tan\phi = 1 + \tan 35° = 1.7 \text{ for } B = L$$

$$s_\gamma = 1 - 0.4 = 0.6 \text{ for } B = L$$

$$d_q = 1 + 2\tan 35°\,(1 - \sin 35°)^2 \times \frac{6}{12} = 1.127$$

$$d_\gamma = 1$$

$$q_o' = 100 \times 6 = 600 \text{ lb/ft}^2$$

Substituting

$$q_u = 600 \times 33.55 \times 1.127 \times 1.7 \times 0.90 + \frac{1}{2} \times 100 \times 12 \times 48.6 \times 1.0 \times 0.6 \times 0.90$$

$$= 34,710 + 15,746 = 50,456 \text{ lb/ft}^2$$

If the compressibility factors are not taken into account (That is, $C_q = C_\gamma = 1$) the ultimate bearing capacity q_u is

$$q_u = 38,567 + 17,496 = 56,063 \text{ lb/ft}^2$$

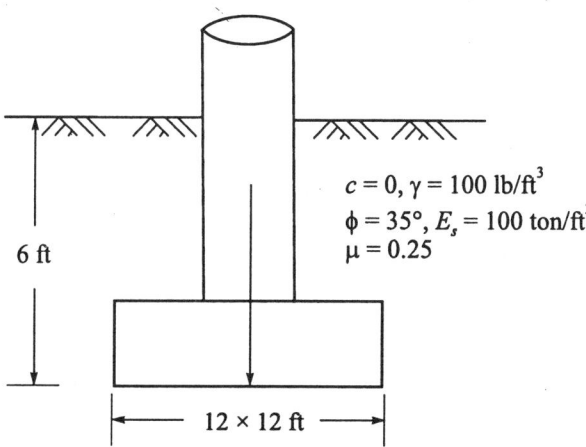

$c = 0$, $\gamma = 100$ lb/ft^3
$\phi = 35°$, $E_s = 100$ ton/ft^2
$\mu = 0.25$

6 ft

12 × 12 ft

Fig. Ex. 5.13

Example 5.14

Estimate the ultimate bearing capacity of a square footing of size 12 × 12 ft founded at a depth of 6 ft in a deep stratum of saturated clay of soft to medium consistency. The undrained shear strength of the clay is 400 lb/ft^2 ($= 0.2$ t/ft^2). The modulus of elasticity $E_s = 15$ ton/ft^2 under undrained conditions. Assume $\mu = 0.5$ and $\gamma = 100$ lb/ft^3.

Solution

Rigidity $I_r = \dfrac{E_s}{2(1 + \mu)c_u} = \dfrac{15}{2(1 + 0.5)\,0.2} = 25$

From Table 5.4, $(I_r)_{\text{crit}} = 8$ for $\phi = 0$

Since $I_r > (I_r)_{\text{crit}}$, the soil is supposed to be incompressible. Use Eq. (5.28) for computing q_u by putting $C_c = C_q = 1$ for $\phi = 0$

$$q_u = cN_c\,s_c\,d_c + q_o'\,N_q\,s_q\,d_q$$

From Table 5.2 for $\phi = 0$, $N_c = 5.14$, and $N_q = 1$

From Table 5.3

$$s_c = 1 + \frac{N_q}{N_c}\frac{B}{L} = 1 + \frac{1}{5.14} \approx 1.2$$

$$d_c = 1 + 0.4\,\frac{D_f}{B} = 1 + 0.4\,\frac{6}{12} = 1.2$$

$$s_q = d_q = 1$$

Substituting and simplifying, we have

$$q_u = 400 \times 5.14 \times 1.2 \times 1.2 + 100 \times 6 \times (1)(1)(1)$$
$$= 2,960 + 600 = 3,560 \, \text{lb}/\text{ft}^2 = 1.78 \, \text{ton}/\text{ft}^2$$

5.11 BEARING CAPACITY OF FOUNDATIONS SUBJECTED TO ECCENTRIC LOADS

Foundations Subjected to Eccentric Vertical Loads

If a foundation is subjected to lateral loads and moments in addition to vertical loads, eccentricity in loading results. The point of application of the resultant of all the loads would lie outside the geometric centre of the foundation, resulting in eccentricity in loading. The eccentricity e is measured from the centre of the foundation to the point of application normal to the axis of the foundation. The maximum eccentricity normally allowed is $B/6$ where B is the width of the foundation. The basic problem is to determine the effect of the eccentricity on the ultimate bearing capacity of the foundation. When a foundation is subjected to an eccentric vertical load, as shown in Fig. 5.13 (a), it tilts towards the side of the eccentricity and the contact pressure increases on the side of tilt and decreases on the opposite side. When the vertical load Q_{ult} reaches the ultimate load, there will be a failure of the supporting soil on the side of eccentricity. As a consequence, settlement of the footing will be associated with tilting of the base towards the side of eccentricity. If the eccentricity is very small, the load required to produce this type of failure is almost equal to the load required for producing a symmetrical general shear failure. Failure occurs due to intense radial shear on one side of the plane of symmetry, while the deformations in the zone of radial shear on the other side are still insignificant. For this reason the failure is always associated with a heave on that side towards which the footing tilts.

Research and observations of Meyerhof (1953, 1963) indicate that effective footing dimensions obtained (Fig. 5.13) as

$$L' = L - 2e_x, \quad B' = B - 2e_y \tag{5.36a}$$

should be used in bearing capacity analysis to obtain an effective footing area defined as

$$A' = B'L' \tag{5.36b}$$

The ultimate load bearing capacity of a footing subjected to eccentric loads may be expressed as

$$Q'_{ult} = q_u A' \tag{5.36c}$$

where q_u = ultimate bearing capacity of the footing with the load acting at the centre of the footing.

Determination of Maximum and Minimum Base Pressures Under Eccentric Loadings

The methods of determining the effective area of a footing subjected to eccentric loadings have been discussed earlier. It is now necessary to know the maximum and minimum base pressures under the same loadings. Consider the plan of a rectangular footing given in Fig. 5.14 subjected to eccentric loadings.

Let the coordinate axes XX and YY pass through the centre O of the footing. If a vertical load passes through O, the footing is symmetrically loaded. If the vertical load passes through O_x on the X-axis, the footing is eccentrically loaded with one way eccentricity. The distance of O_x from O, designated as e_x, is called the eccentricity in the X-direction. If the load passes through O_y on the Y-axis, the eccentricity is e_y in the Y-direction. If on the other hand the load passes through O_{xy}, the eccentricity is called *two-way eccentricity* or *double eccentricity*.

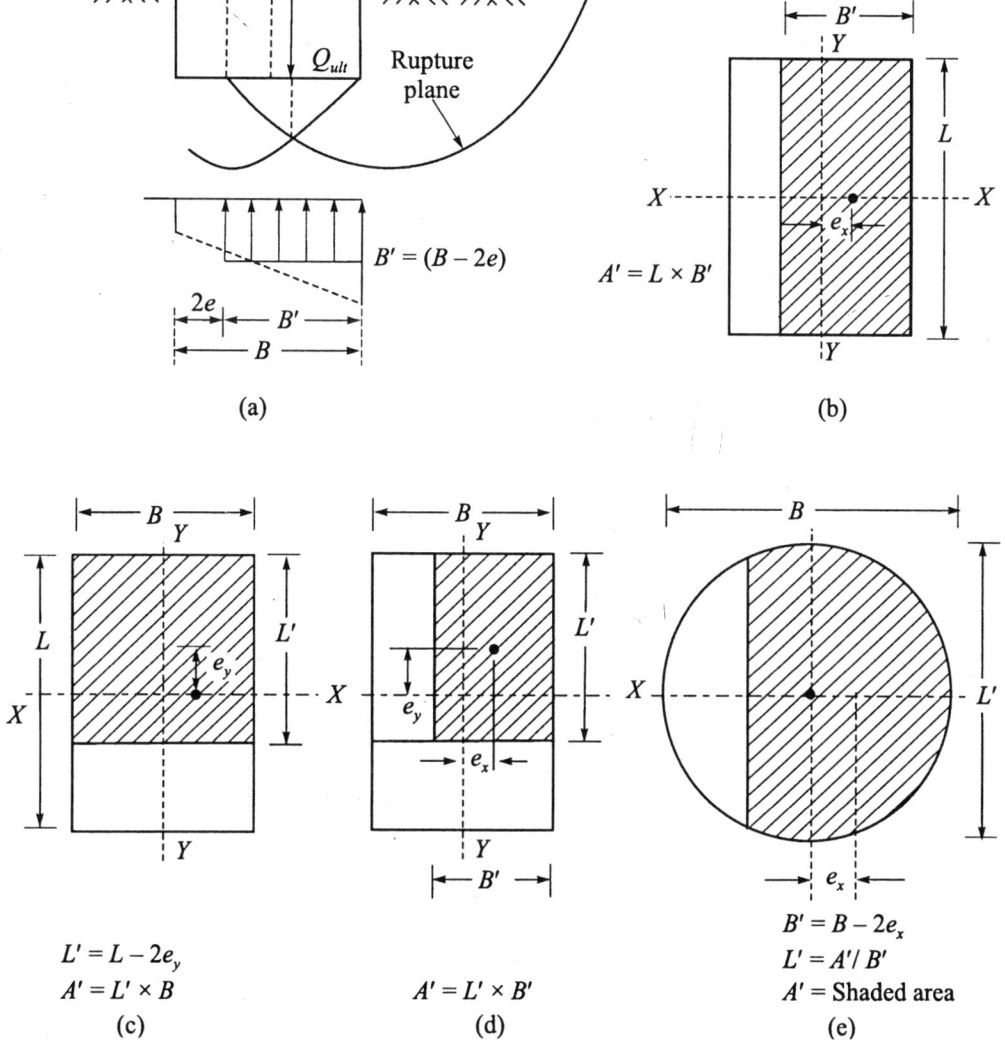

Fig. 5.13 Eccentrically loaded footing (Meyerhof, 1953)

When a footing is eccentrically loaded, the soil experiences a maximum or a minimum pressure at one of the corners or edges of the footing. For the load passing through O_{xy} (Fig. 5.14), the points C and D at the corners of the footing experience the maximum and minimum pressures respectively.

The general equation for pressure may be written as

$$q = \frac{Q}{A} \pm \frac{Qe_x}{I_y} x \pm \frac{Qe_y}{I_x} y \qquad (5.37)$$

or

$$q = \frac{Q}{A} \pm \frac{M_x}{I_y} x \pm \frac{M_y}{I_x} y \qquad (5.38)$$

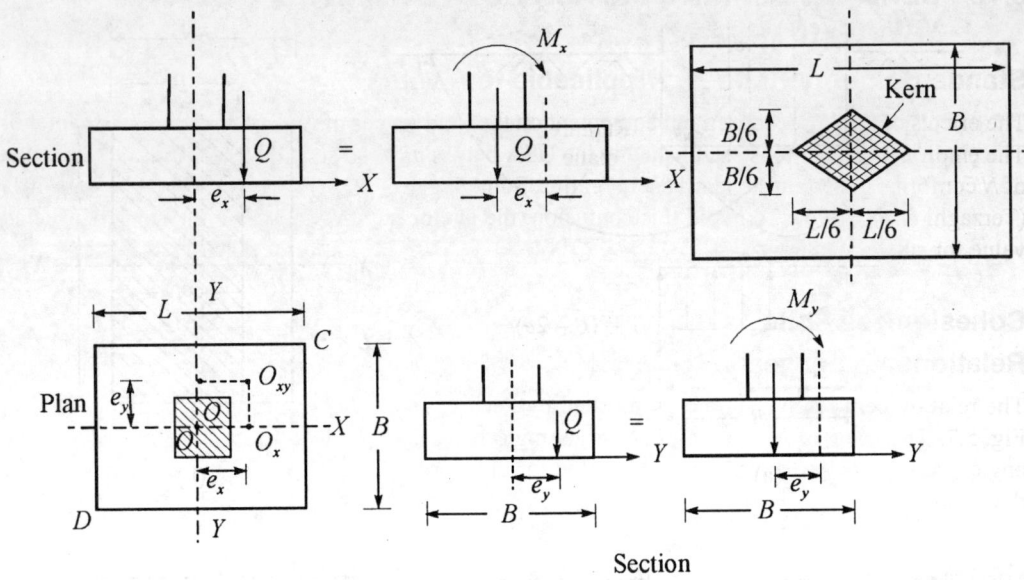

Fig. 5.14 Footing subjected to eccentric loadings

where q = contact pressure at a given point (x, y),
Q = total vertical load,
A = area of footing,
$Qe_x = M_x$ = moment about axis YY,
$Qe_y = M_y$ = moment about axis XX,
I_x, I_y = moment of inertia of the footing about XX and YY axes respectively.

q_{max} and q_{min} at points C and D respectively may be obtained by substituting in Eq. (5.37) or (5.38) for

$$I_x = \frac{LB^3}{12}, \quad I_y = \frac{BL^3}{12}, \quad x = \frac{L}{2}, \quad y = \frac{B}{2}$$

we have

$$q_{max} = \frac{Q}{A}\left(1 + 6\frac{e_x}{L} + 6\frac{e_y}{B}\right) \qquad (5.39a)$$

$$q_{min} = \frac{Q}{A}\left(1 - 6\frac{e_x}{L} - 6\frac{e_y}{B}\right) \qquad (5.39b)$$

Equation (5.39) may also be used for one way eccentricity by putting either $e_x = 0$, or $e_y = 0$.

When e_x or e_y exceed a certain limit, Eq. (5.39) gives a negative value of q which indicates tension between the soil and the bottom of the footing. Equations (5.39) are applicable only when the load is applied within a limited area which is known as the *Kern* as is shown shaded in Fig 5.14. so that the load may fall within the shaded area to avoid tension. The procedure for the determination of soil pressure when the load is applied outside the kern is laborious and as such not dealt with here. However, charts are available for ready calculations in references such as Teng (1969) and Highter and Anders (1985).

5.12 ULTIMATE BEARING CAPACITY OF FOOTINGS BASED ON SPT VALUES (N)

Standard Energy Ratio R_{es} Applicable to N Value

The effects of field procedures and equipment on the field values of N were discussed in Chapter 3. The empirical correlations established in the USA between N and soil properties indicate the value of N conforms to certain standard energy ratios, Some suggest 70% (Bowles, 1996) and others 60% (Terzaghi *et al*, 1996). To avoid this confusion, the author uses N_{cor} in this book as the corrected value for standard energy.

Cohesionless Soils
Relationship between N_{cor} and ϕ

The relation between N_{cor} and ϕ established by Peck *et al*, (1974) is given in a graphical form in Fig. 5.7. The value of N_{cor} to be used for getting ϕ is the corrected value for standard energy. The angle ϕ obtained by this method can be used for obtaining the bearing capacity factors, and hence the ultimate bearing capacity of soil.

Cohesive Soils
Relationship between N_{cor} and q_u (Unconfined compressive strength)

Relationships have been developed between N_{cor} and q_u (the undrained compressive strength) for the $\phi = 0$ condition. This relationship gives the value of c_u for any known value of N_{cor}. The relationship may be expressed as Eq. (3.16c),

$$q_u = 2c_u = \overline{k}N_{cor} \text{ (kPa)} \tag{5.40}$$

where the value of the coefficient $\overline{k}$ may vary from a minimum of 12 to a maximum of 25. A low value of 13 yields q_u given in Table 3.5 (b).

Once q_u is determined, the net ultimate bearing capacity and the net allowable bearing pressure can be found following Skempton's approach.

5.13 THE CPT METHOD OF DETERMINING ULTIMATE BEARING CAPACITY

Cohesionless Soils
Relationship between q_c, D_r and ϕ

Relationship between the static cone penetration resistance q_c and ϕ have been developed by Robertson and Campanella (1983b), [Fig. 3.18 (b)]. The value of ϕ can therefore be determined with the known value of q_c. With the known value of ϕ, bearing capacity factors can be determined and hence the ultimate bearing capacity. Experience indicates that the use of q_c for obtaining ϕ is more reliable than the use of N.

Bearing Capacity of Soil

As per Schmertmann (1978), the bearing capacity factors N_q and N_γ for use in the Terzaghi bearing capacity equation can be determined by the use of the equation

$$N_q = N_\gamma = 1.25q_c \tag{5.41}$$

where q_c = cone point resistance in kg/cm^2 (or tsf) averaged over a depth equal to the width below the foundation.

Undrained Shear Strength

The undrained shear strength c_u under $\phi = 0$ condition may be related to the static cone point resistance q_c as Eq. (3.18)

$$q_c = N_k c_u + p_o$$

or
$$c_u = \frac{q_c - p_o}{N_k} = \frac{\overline{q}_c}{N_k} \qquad (5.42)$$

where N_k = cone factor, may be taken as equal to 20 (Sanglerat, 1972) both for normally consolidated and preconsolidated clays.

  p_o = total overburden pressure.

When once c_u is known, the values of q_{nu} and q_{na} can be evaluated as per the methods explained in earlier sections.

Example 5.15

A water tank foundation has a footing of size 6×6 m, founded at a depth of 3 m below ground level in a medium dense sand stratum of great depth. The corrected average SPT value obtained from the site investigation is 20. The foundation is subjected to a vertical load at an eccentricity of $B/10$ along one of the axes. Figure Ex. 5.15 gives the soil profile with the remaining data. Estimate the ultimate load, Q_{ult}, by Meyerhof's method.

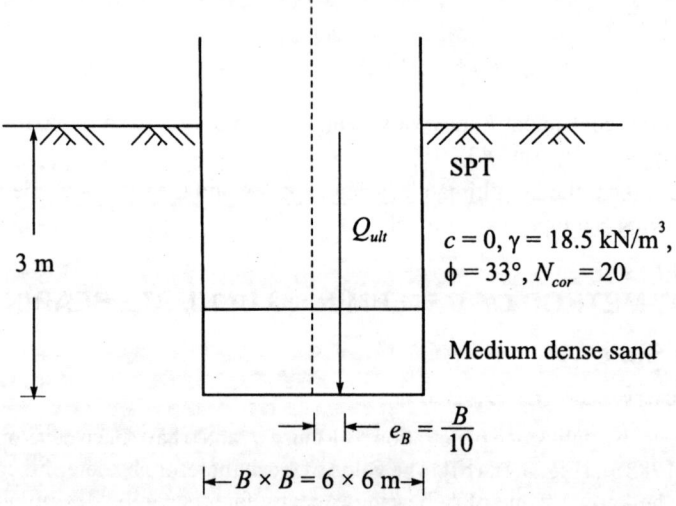

Fig. Ex. 5.15

Solution

From Fig. 5.7, $\phi = 33°$ for $N_{cor} = 20$.

$$B' = B - 2e = 6 - 2\,(0.6) = 4.8 \text{ m}$$

$$L' = L = B = 6 \text{ m}$$

For $c = 0$ and $i = 1$, Eq. (5.28) reduces to

$$q'_u = \gamma \, D_f N_q \, s_q \, d_q + \frac{1}{2}\gamma \, B' N_\gamma \, s_\gamma \, d_\gamma$$

From Table 5.2 for $\phi = 33°$, we have

$N_q = 26.3$, $N_\gamma = 26.55$ (Meyerhof)

From Table 5.3 (Meyerhof)

$$s_q = 1 + 0.1 N_\phi \left(\frac{B}{L}\right) = 1 + 0.1 \tan^2 \left(45° + \frac{33}{2}\right)(1) = 1.34$$

$$s_\gamma = s_q = 1.34 \quad \text{for } \phi > 10°$$

$$d_q = 1 + 0.1 \sqrt{N_\phi} \left(\frac{D_f}{B'}\right) = 1 + 0.1 \times 1.84 \left(\frac{3}{4.8}\right) = 1.115$$

Substituting $d_\gamma = d_q = 1.115$ for $\phi > 10°$

$$q'_u = 18.5 \times 3 \times 26.3 \times 1.34 \times 1.115 + \frac{1}{2} \times 18.5 \times 4.8 \times 26.55 \times 1.34 \times 1.115$$

$$= 2,181 + 1,761 = 3,942 \text{ kN/m}^2$$

$$Q'_{ult} = B \times B' \times q'_u = 6 \times 4.8 \times 3,942 = 113,530 \text{ kN} \approx 114 \text{ MN}$$

Example 5.16

Figure Ex. 5.16, gives the plan of a footing subjected to eccentric load with two way eccentricity. The footing is founded at a depth 3 m below the ground surface. Given $e_x = 0.60$ m and $e_y = 0.75$ m, determine Q_{ult}. The soil properties are: $c = 0$, $N_{cor} = 20$, $\gamma = 18.5$ kN/m³. The soil is medium dense sand. Use N_γ (Meyerhof) from Table 5.2 and Hansen's shape and depth factors from Table 5.3.

Solution

Figure Ex. 5.16 shows the two-way eccentricity. The effective lengths and breadths of the foundation from Eq. 5.36 (a) is

$$B' = B - 2e_y = 6 - 2 \times 0.75 = 4.5 \text{ m.}$$

$$L' = L - 2e_x = 6 - 2 \times 0.6 = 4.8 \text{ m.}$$

Effective area, $A' = L' \times B' = 4.5 \times 4.8 = 21.6 \text{ m}^2$

As in Example 5.15

$$q'_u = \gamma D_f N_q s_q d_q + \frac{1}{2} \gamma B' N_\gamma s_\gamma d_\gamma$$

For $\phi = 33°$, $N_q = 26.3$ and $N\gamma = 26.55$ (Meyerhof)

From Table 5.3 (Hansen)

$$s_q = 1 + \frac{B'}{L'} \tan 33° = 1 + \frac{4.5}{4.8} \times 0.65 = 1.61$$

$$s_\gamma = 1 - 0.4 \frac{B'}{L'} = 1 - 0.4 \times \frac{4.5}{4.8} = 0.63$$

$$d_q = 1 + 2 \tan 33° (1 - \sin 33°)^2 \times \frac{3}{4.5}$$

$$= 1 + 1.3 \times 0.21 \times 0.67 = 1.183$$

$$d_\gamma = 1$$

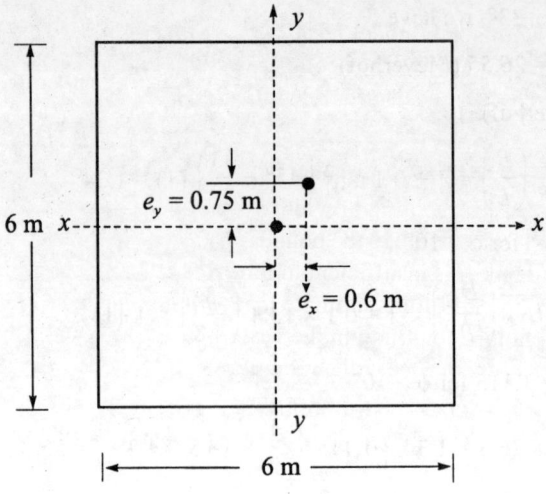

Fig. Ex. 5.16

Substituting

$$q'_u = 18.5 \times 3 \times 26.3 \times 1.61 \times 1.183 + \frac{1}{2} \times 18.5 \times 4.5 \times 26.55 \times 0.63 \times (1)$$

$$= 2,780 + 696 = 3,476 \text{ kN/m}^2$$

$$Q_{ult} = A'q'_u = 21.6 \times 3,476 = 75,082 \text{ kN}$$

5.14 ULTIMATE BEARING CAPACITY OF FOOTINGS RESTING ON STRATIFIED DEPOSITS OF SOIL

All the theoretical analysis dealt with so far is based on the assumption that the subsoil is isotropic and homogeneous to a considerable depth. In nature, soil is generally non-homogeneous with mixtures of sand, silt and clay in different proportions. In the analysis, an average profile of such soils is normally considered. However, if soils are found in distinct layers of different compositions and strength characteristics, the assumption of homogeneity to such soils is not strictly valid if the failure surface cuts across boundaries of such layers.

The present analysis is limited to a system of two distinct soil layers. For a footing located in the upper layer at a depth D_f below the ground level, the failure surfaces at ultimate load may either lie completely in the upper layer or may cross the boundary of the two layers. Further, we may come across the upper layer strong and the lower layer weak or vice versa. In either case, a general analysis for $(c - \phi)$ will be presented and will show the same analysis holds true if the soil layers are any one of the categories belonging to sand or clay.

The bearing capacity of a layered system was first analysed by Button (1953), who considered only saturated clay ($\phi = 0$). Later on Brown and Meyerhof (1969) showed that the analysis of Button leads to unsafe results. Vesic (1975) analyzed the test results of Brown and Meyerhof and others and gave his own solution to the problem.

Vesic considered both the types of soil in each layer, that is clay and $(c - \phi)$ soils. However, confirmations of the validity of the analysis of Vesic and others are not available. Meyerhof (1974) analysed the two layer system consisting of dense sand on soft clay and loose sand on stiff clay and supported his analysis with some model tests. Again Meyerhof and Hanna (1978) advanced the earlier analysis of Meyerhof (1974) to encompass $(c - \phi)$ soil and supported their analysis with model tests. The present section deals briefly with the analyses of Meyerhof (1974) and Meyerhof and Hanna (1978).

Case 1: A stronger layer overlying a weaker deposit

Figure 5.15 (a) shows a strip footing of width B resting at a depth D_f below ground surface in a strong soil layer (Layer 1). The depth to the boundary of the weak layer (Layer 2) below the base of the footing is H. If this depth H is insufficient to form a full failure plastic zone in layer 1 under the ultimate load conditions, a part of this ultimate load will be transferred to the boundary level _mn_. This load will induce a failure condition in the weaker layer (Layer 2). However, if the depth H is relatively large then the failure surface will be completely located in layer 1 as shown in Fig. 5.15 (b).

The ultimate bearing capacities of strip footings on the surfaces of homogeneous thick beds of layer 1 and layer 2 may be expressed as

Layer 1

$$q_1 = c_1 N_{c1} + \frac{1}{2}\gamma_1 BN_{\gamma 1} \tag{5.43}$$

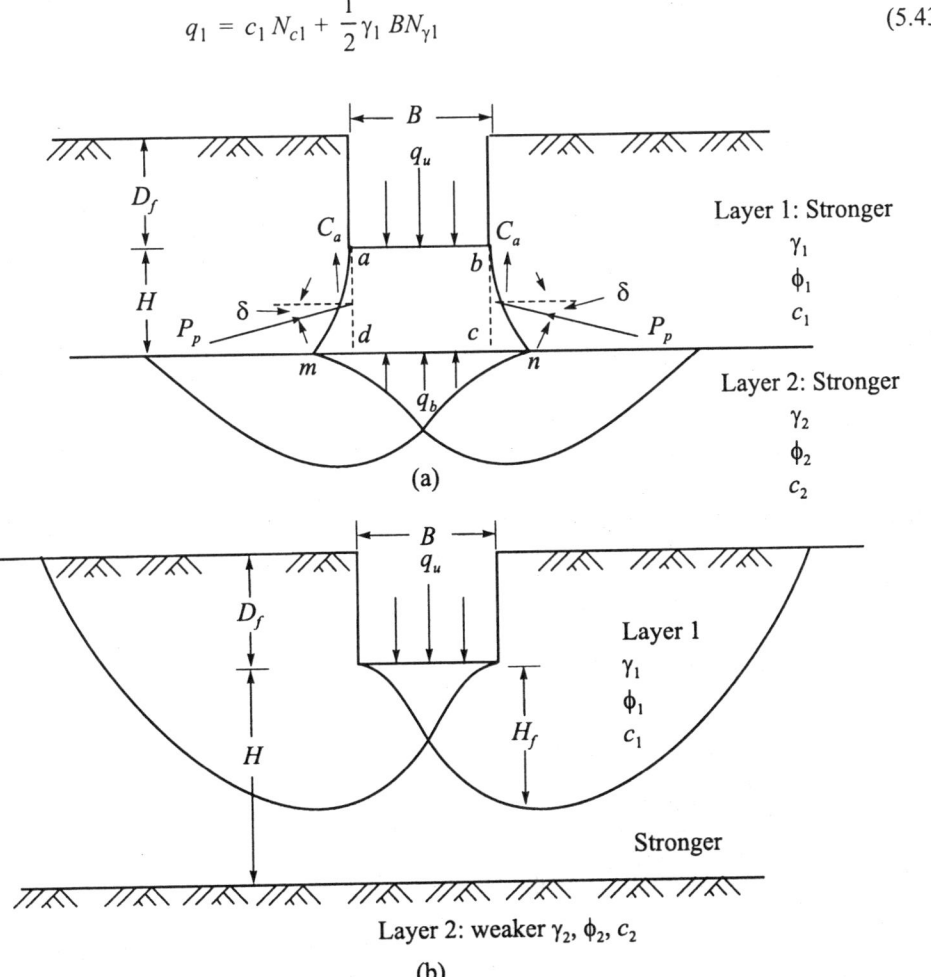

Fig. 5.15 Failure of soil below strip footing under vertical load on strong layer overlying weak deposit (after Meyerhof and Hanna, 1978)

Layer 2

$$q_2 = c_2 N_{c2} + \frac{1}{2} \gamma_2 BN_{\gamma 2} \tag{5.44}$$

where $N_{c1}, N_{\gamma 1}$ = bearing capacity factors for soil in layer 1 for friction angle ϕ_1.

$N_{c2}, N_{\gamma 2}$ = bearing capacity factors for soil in layer 2 for friction angle ϕ_2.

For the footing founded at a depth D_f, if the complete failure surface lies within the upper stronger Layer 1 [Fig. 5.15 (b)] an expression for ultimate bearing capacity of the upper layer may be written as

$$q_u = q_t = c_1 N_{c1} + q_o' N_{q1} + \frac{1}{2} \gamma_1 BN_{\gamma 1} \tag{5.45}$$

If q_1 is much greater that q_2 and if the depth H is insufficient to form a full failure plastic condition in layer 1, then the failure of the footing may be considered due to pushing of soil within the boundary ad and bc through the top layer into the weak layer. The resisting force of punching may be assume to develop on the faces ad and bc passing through the edges of the footing. The forces that act on these surfaces are (per unit length of footing),

Adhesive force, $C_a = c_a H$

Frictional force, $F_f = P_p \sin \delta$ $\tag{5.46}$

where c_a = unit cohesion,

P_p = passive earth pressure per unit length of footing, and

δ = inclination of P_p with the normal [Fig. 5.15 (a)].

The equation for the ultimate bearing capacity q_u for the two layer soil system may now be expressed as

$$q_u = q_b + \frac{2\left(C_a + P_p \sin \delta\right)}{B} - \gamma_1 H \tag{5.47}$$

where, q_b = ultimate bearing capacity of Layer 2

The equation for P_p may be written as

$$P_p = \frac{\gamma_1 H^2}{2 \cos \delta}\left(1 + \frac{2 D_f}{H}\right)K_p \tag{5.48}$$

Substituting for P_p and C_a, the equation for q_u may be written as

$$q_u = q_b + \frac{2 c_a H}{B} + \frac{\gamma_1 H^2}{B}\left(1 + \frac{2 D_f}{H}\right)K_p \tan \delta - \gamma_1 H \tag{5.49}$$

In practice, it is convenient to use a coefficient K_s of punching shearing resistance on the vertical plane through the footing edges, so that

$$K_s \tan \phi_1 = K_p \tan \delta \tag{5.50}$$

Substituting, the equation for q_u may be written as

$$q_u = q_b + \frac{2 c_a H}{B} + \frac{\gamma_1 H^2}{B}\left(1 + \frac{2 D_f}{H}\right)K_s \tan \phi_1 - \gamma_1 H \le q_t \tag{5.51}$$

Figure 5.16 gives the value of K_s for various values of ϕ_1 as a function of q_2/q_1. The variation of c_a/c_i with q_2/q_1 is shown in Fig. 5.17.

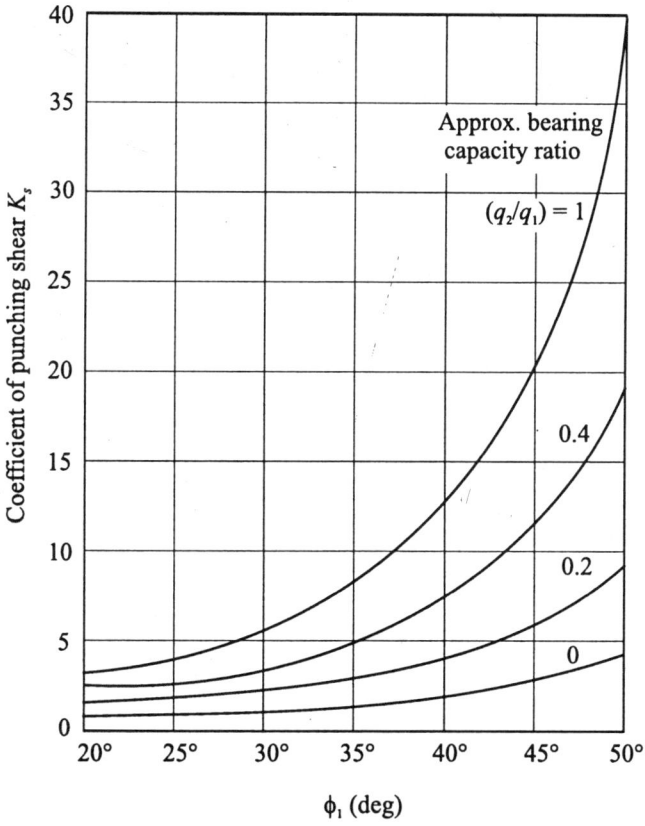

Fig. 5.16 Coefficients of punching shear resistance under vertical load (after Meyerhof and Hanna, 1978)

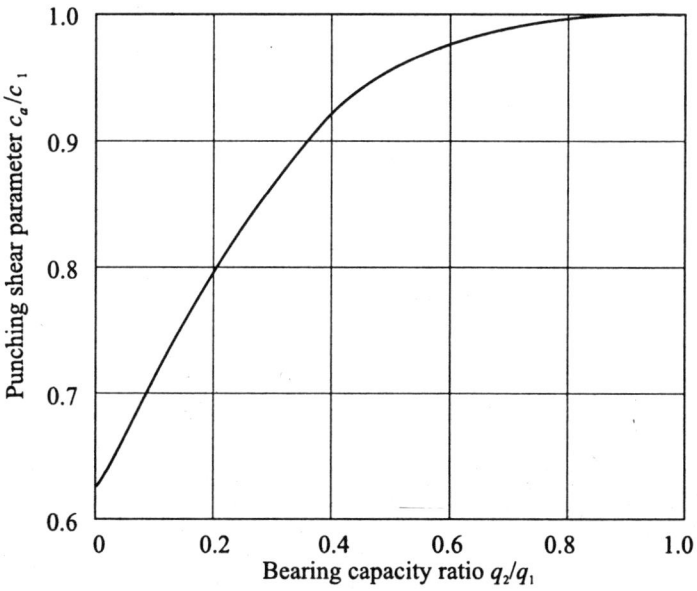

Fig. 5.17 Plot of c_a/c_1 *versus* q_2/q_1 (after Meyerhof and Hanna, 1978)

Equation (5.45) for q_t and q_b in Eq. (5.51) are for strip footings. These equations with shape factors may be written as

$$q_t = c_1 N c_1 s_{c1} + \gamma_1 D_f N_{q1} s_{q1} + \frac{1}{2} \gamma_1 B N_{\gamma 1} s_{\gamma 1} \qquad (5.52)$$

$$q_b = c_2 N_{c2} s_{c2} + \gamma_1 (D_f + H) N_{q2} s_{q2} + \frac{1}{2} \gamma_2 B N_{\gamma 2} s_{\gamma 2} \qquad (5.53)$$

where s_c, s_q and s_γ are the shape factors for the corresponding layers with subscripts 1 and 2 representing layers 1 and 2 respectively.

Equation (5.51) can be extended to rectangular foundations by including the corresponding shape factors.

The equation for a rectangular footing may be written as:

$$q_u = q_b + \left(\frac{2c_a H}{B}\right)\left(1 + \frac{B}{L}\right) + \frac{\gamma_1 H^2}{B}\left(1 + \frac{2D_f}{H}\right)\left(1 + \frac{B}{L}\right) K_s \qquad (5.54)$$

Case 2: Top layer dense sand and bottom layer saturated soft clay ($\phi_2 = 0$)
The value of q_b for the bottom layer from Eq. (5.53) may be expressed as

$$q_b = c_2 N_{c2} s_{c2} + \gamma_1 (D_f + H) \qquad (5.55)$$

From Table (5.3), $sc_2 = (1 + 0.2 \, B/L)$ (Meyerhof, 1963) and $N_c = 5.14$ for $\phi = 0$. Therefore

$$q_b = \left(1 + 0.2\frac{B}{L}\right) 5.14 c_2 + \gamma_1 (D_f + H) \qquad (5.56)$$

For $c_1 = 0$, q_t from Eq. (5.52) is

$$q_t = \gamma_1 D_f N_{q1} s_{q1} + \frac{1}{2} \gamma_1 B N_{\gamma 1} s_{\gamma 1} \qquad (5.57)$$

We may now write an expression for q_u from Eq. (5.24) as

$$q_u = \left(1 + 0.2\frac{B}{L}\right) 5.14 c_2 + \frac{\gamma_1 H^2}{B}\left(1 + \frac{2D_f}{H}\right)\left(1 + \frac{B}{L}\right) K_s \tan\phi_1 +$$

$$\gamma_1 D_f \le \gamma_1 D_f N_{q1} s_{q1} + \frac{1}{2} \gamma_1 B N_{\gamma 1} s_{\gamma 1} \qquad (5.58)$$

The ratio of q_2/q_1 may be expressed by

$$\frac{q_2}{q_1} = \frac{c_2 N_{c2}}{0.5 \, \gamma_1 B N_{\gamma 1}} = \frac{5.14 \, c_2}{0.5 \gamma_1 B N_{\gamma_1}} \qquad (5.59)$$

The value of K_s may be found from Fig. 5.16.

Case 3: When layer 1 is dense sand and layer 2 is loose sand ($c_1 = c_2 = 0$)
Proceeding in the same way as explained earlier the expression for q_u for a rectangular footing may be expressed as

$$q_u = \left[\gamma_1\left(D_f + H\right) N_{q2} s_{q2} + \frac{1}{2} \gamma_2 B N_{\gamma 2} s_{\gamma 2}\right] +$$

$$\frac{\gamma_1 H^2}{B}\left(1 + \frac{B}{L}\right)\left(1 + \frac{2D_f}{H}\right) K_s \tan\phi_1 - \gamma_1 H \le q_t \qquad (5.60)$$

where

$$q_t = \gamma_1 D_f N_{q1} s_{c1} + \frac{1}{2} \gamma_1 B N_{\gamma 1} s_{\gamma 1} \tag{5.61}$$

$$\frac{q_2}{q_1} = \frac{\gamma_2 N_{\gamma 2}}{\gamma_1 N_{\gamma 1}} \tag{5.62}$$

Case 4: Layer 1 is stiff saturated clay ($\phi_1 = 0$) and layer 2 is saturated soft clay ($\phi_2 = 0$)

The ultimate bearing capacity of the layered system can be given as

$$q_u = \left(1 + 0.2\frac{B}{L}\right) 5.14 c_2 + \left(1 + \frac{B}{L}\right) \frac{2 c_a H}{B} + \gamma_1 D_f \le q_t \tag{5.63}$$

$$q_t = \left(1 + 0.2\frac{B}{L}\right) 5.14 c_1 + \gamma_1 D_f \tag{5.64}$$

$$\frac{q_2}{q_1} = \frac{5.14 c_2}{5.14 c_1} = \frac{c_2}{c_1} \tag{5.65}$$

Example 5.17

A rectangular footing of size 3 × 2 m is founded at a depth of 1.5 m in a clay stratum of very stiff consistency. A clay layer of medium consistency is located at a depth of 1.5 m (= H) below the bottom of the footing (Fig. Ex. 5.17). The soil parameters of the two clay layers are as follows:

Top clay layer: $c_1 = 175 \text{ kN/m}^2$

$\gamma_1 = 17.5 \text{ kN/m}^3$

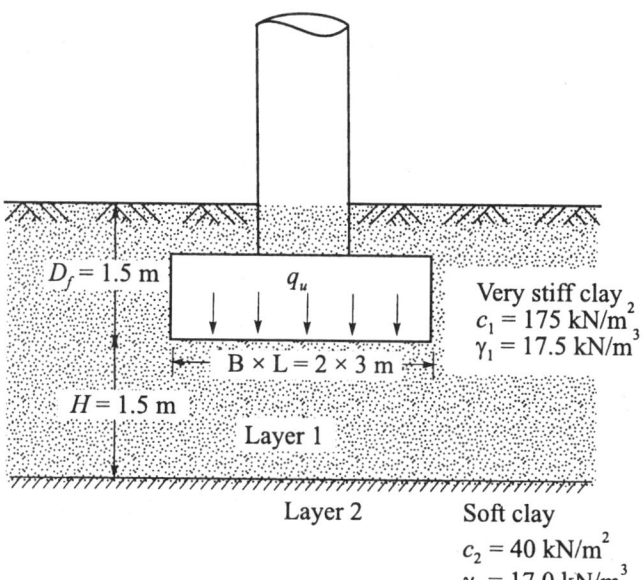

Fig. Ex. 5.17

Bottom layer: $c_2 = 40 \text{ kN/m}^2$

$\gamma_2 = 17.0 \text{ kN/m}^3$

Estimate the ultimate bearing capacity and the allowable bearing pressure on the footing with a factor of safety of 3.

Solution

The solution to this problem comes under Case 4 in Section 5.14. We have to consider here Eqs (5.63), (5.64) and (5.65).

The data given are:

$B = 2$ m, $L = 3$ m, $H = 1.5$ m [Fig. 5.15 (a)], $D_f = 1.5$ m, $\gamma_1 = 17.5 \text{ kN/m}^3$.

From Fig. 5.17, for $q_2/q_1 = c_2/c_1 = 40/175 = 0.23$, the value of $c_a/c_1 = 0.83$ or $c_a = 0.83$ $c_1 = 0.83 \times 175 = 145.25 \text{ kN/m}^2$.

From Eq. (5.63)

$$q_u = \left(1 + 0.2\frac{B}{L}\right) 5.14 c_2 + \left(1 + \frac{B}{L}\right) \frac{2 c_a H}{B} + \gamma_1 D_f \le q_t$$

Substituting the known values

$$q_u = \left(1 + 0.2 \times \frac{2}{3}\right) 5.14 \times 40 + \left(1 + \frac{2}{3}\right) \left(\frac{2 \times 145.25 \times 1.5}{2}\right) + 17.5 \times 1.5$$

$$= 233 + 364 + 26 = 623 \text{ kN/m}^2$$

From Eq. (5.64)

$$q_t = \left(1 + 0.2\frac{B}{L}\right) 5.14 c_1 + \gamma_1 D_f$$

$$= \left(1 + 0.2 \times \frac{2}{3}\right) 5.14 \times 175 + 17.5 \times 1.5$$

$$= 1{,}020 + 26 = 1{,}046 \text{ kN/m}^2$$

It is clear from the above that $q_u < q_t$ and as such q_u is the ultimate bearing capacity to be considered. Therefore

$$q_a = \frac{q_u}{F_s} = \frac{623}{3} = 208 \text{ kN/m}^2$$

Example 5.18

Determine the ultimate bearing capacity of the footing given in Example 5.17 in dense sand with the bottom layer being a clay of medium consistency. The parameters of the top layer are:

$$\gamma_1 = 18.5 \text{ kN/m}^3, \phi_1 = 39°$$

All the other data given in Ex. 5.17 remain the same. Use Meyerhof's bearing capacity and shape factors.

Solution

Case 2 of Section 5.14 is required to be considered here.

Given: Top layer: $\gamma_1 = 18.5$ kN/m^3, $\phi_1 = 39°$, Bottom layer: $\gamma_2 = 17.0$ kN/m^3, $c_2 = 40$ kN/m^2.

From Table 5.2

$N_{\gamma 1}(M) = 78.8$ for $\phi_1 = 39°$.

From Eq. (5.59)

$$\frac{q_2}{q_1} = \frac{5.14\, c_2}{0.5\, \gamma_1\, B N_{\gamma 1}} = \frac{5.14 \times 40}{0.5 \times 18.5 \times 2 \times 78.8} = 0.141$$

From Fig. 5.16

$K_s = 2.9$ for $\phi = 39°$

Now from Eq. (5.58), we have

$$q_u = \left(1 + 0.2\,\frac{B}{L}\right) 5.14 c_2 + \frac{\gamma_1 H^2}{B}\left(1 + \frac{2 D_f}{H}\right)\left(1 + \frac{B}{L}\right) K_s \tan \phi_1 + \gamma_1 D_f$$

$$= \left(1 + 0.2 \times \frac{2}{3}\right) 5.14 \times 40 + \frac{18.5 \times (1.5)^2}{2}\left(1 + \frac{2 \times 1.5}{1.5}\right)\left(1 + \frac{2}{3}\right) 2.9 \tan 39° + 18.5 \times 1.5$$

$$= 233 + 245 + 28 = 506 \text{ kN/m}^2$$

$q_u = 506$ kN/m^2

From Eq. (5.58), the limiting value q_t is

$$q_t = \gamma_1 D_f N_{q1} s_{q1} + \frac{1}{2}\gamma_1 BN_{\gamma 1} s_{\gamma 1}$$

where $\gamma_1 = 18.5$ kN/m^3, $D_f = 1.5$ m, $B = 2$ m.

From Table 5.2,

$N_{\gamma 1} = 78.8$ and $N_{q1} = 56.5$

From Table 5.3,

$$s_{q1} = 1 + 0.1\, N_\phi \left(\frac{B}{L}\right) = 1 + 0.1 \times \tan^2\left(45° + \frac{39}{2}\right) \times \frac{2}{3} = 1.29 = s_{\gamma 1}$$

Now $q_t = 18.5 \times 1.5 \times 56.5 \times 1.29 + \frac{1}{2} \times 18.5 \times 2 \times 78.8 \times 1.29 = 3903$ kN/m$^2 > q_u$

Hence $q_u = 506$ kN/m^2

5.15A   MEYERHOF'S METHOD OF COMPUTING ULTIMATE BEARING CAPACITY OF FOUNDATIONS ON SLOPES

Introduction

There are occasions where structure are required to be built on slopes or near the edges of slopes. Since full formations of shear zones under ultimate loading conditions are not possible on the sides close to the slopes or edges, the supporting capacity of soil on that side get considerably reduced. Meyerhof (1957) has extended his theories to include the effect of slopes on the stability of foundations.

Strip Foundations on Slopes

Figure 5.18 (a) gives a section of a foundation on a slope with the shapes of failure surfaces.

The zones of plastic flow is less than that of a similar foundation on a level ground. The region above the failure surface of a shallow strip foundation is assumed to be divided into a central elastic zone *abc*, a radial shear zone *bcd* and a mixed shear zone *bden*. The stresses in the zones of plastic equilibrium can be found out as for a horizontal ground surface (Meyerhof, 1951), by replacing the weight of the soil wedge *ben* by the equivalent stresses, p_0 and s_0 normal and tangential respectively, to the plane *be* inclined at an angle α to the horizontal. The form of equation for computing the bearing capacity given by Meyerhof is

$$q_u = cN_c + p_0 N_q + \frac{1}{2}\gamma B N_\gamma \tag{5.66}$$

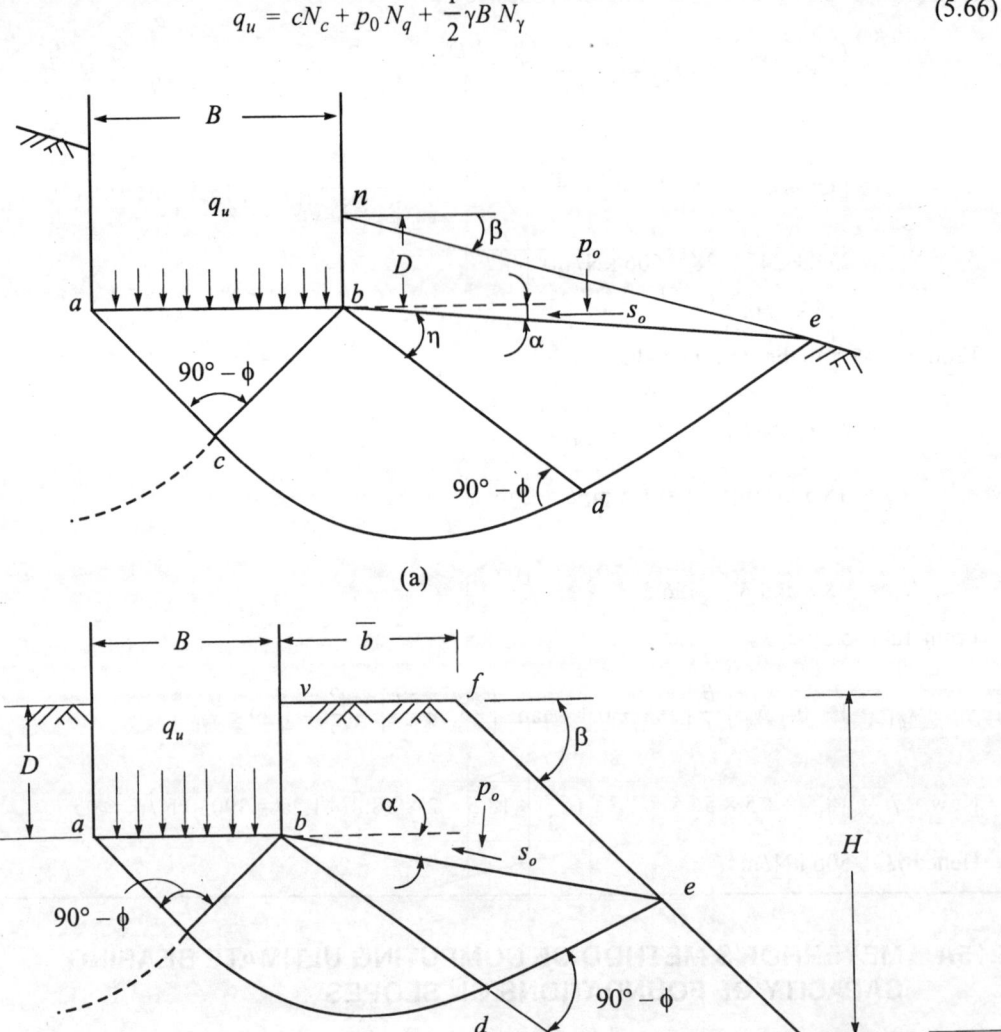

(a)

(b)

Fig. 5.18 Bearing capacity of strip footings on slopes (Meyerhof, 1957): (a) Strip foundation on slopes, (b) streep footings on the top of aslope

The same equation may also be expressed as Eq. (5.67)

$$q_u = cN_{cq} + \frac{1}{2}\gamma B\,N_{\gamma q} \tag{5.67}$$

The bearing capacity factors N_{cq} and $N_{\gamma q}$ for foundations on slope may be obtained from Figs 5.19 and 5.20 (Meyerhof, 1957), for purely cohesive soils ($\phi = 0$) and cohesionless soils ($c = 0$), respectively. It can be seen from the figures that the N factors decrease with greater inclination of slope. For inclination of slopes used in practice ($\beta \le 30°$), the decrease in bearing capacity is small in the case of clays and but is considerable for sand and gravel slopes.

The effect of flowing water on the bearing capacity may require to be analysed if the foundation is submerged with flowing water. If the foundation soil is cohesive with a small or no angle of shearing resistance, the stability of the foundation will have to be analysed by taking into account the stability of the slope. Slopes of purely cohesive soil of great depth may fail either by toe or base

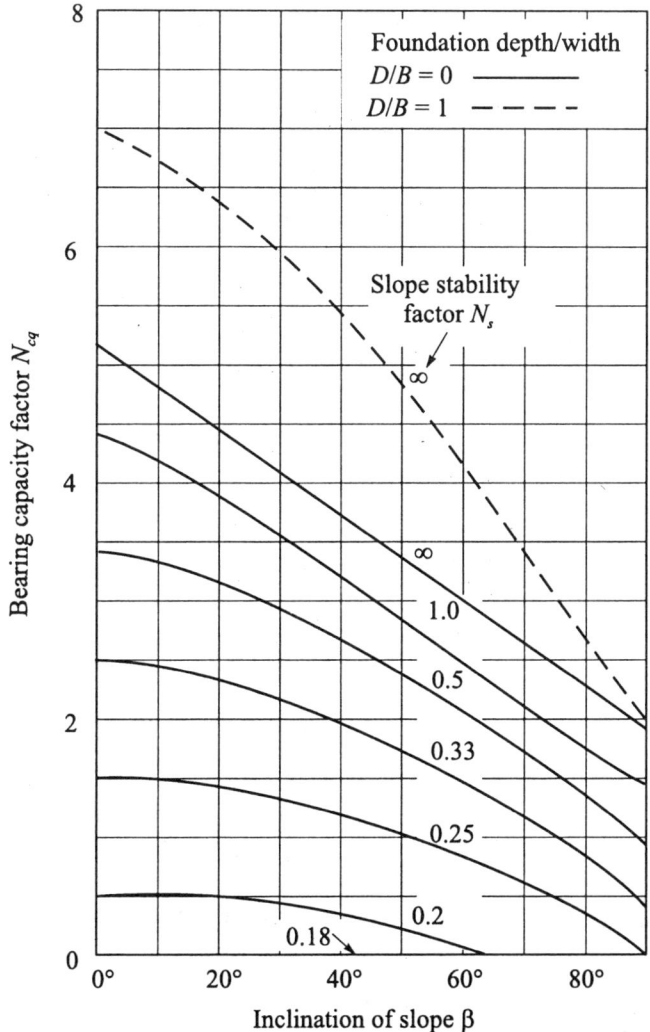

Fig. 5.19 Bearing capacity factors for strip foundation on face of slope of purely Cohesive material (Meyerhof, 1957)

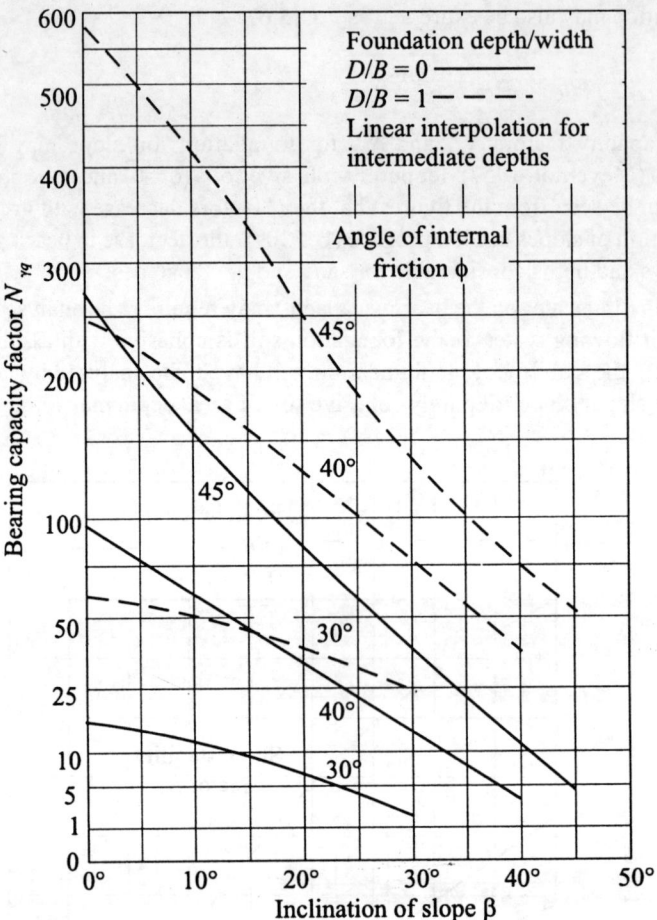

Fig. 5.20 Bearing capacity factors on face of slope of cohesionless material (Meyerhof, 1957)

failures. The upperlimit of the bearing capacity of a foundation in a purely cohesive soil may be estimated from the expression

$$q_u = cN_{cq} + \gamma D_f \tag{5.68}$$

where the factor N_{cq} is given in the upper part of Fig. 5.19.

5.15B BEARING CAPACITY OF FOUNDATIONS ON TOP OF A SLOPE

Figure 5.21 shows a section of a foundation with the failure surfaces under ultimate loading condition. The stability of the foundation depends on the distance $\bar{b}$ of the top edge of the slope from the face of the foundation.

The form of ultimate bearing capacity equation for a strip footing may be expressed as (Meyerhof, 1957)

$$q_u = cN_{cq} + \frac{1}{2}\gamma BN_{\gamma q} \tag{5.69}$$

The upper limit of the bearing capacity of a foundation in a purely cohesive soil may be estimated from

$$q_u = cN_{cq} + \gamma D_f \tag{5.70}$$

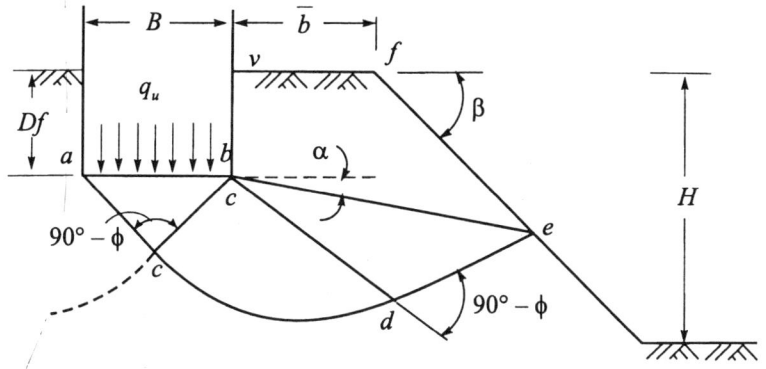

Fig. 5.21 Bearing capacity of a strip footing on top of a slope (Meyerhof, 1957)

The resultant bearing capacity factors N_{cq} and $N_{\gamma q}$ depend on the distance $\bar{b}$, β, ϕ and the D_f /B ratio. These bearing capacity factors are given in Figs 5.22 and 5.23 for strip foundation in purely cohesive and cohesionless soils respectively. It can be seen from the figures that the bearing capacity

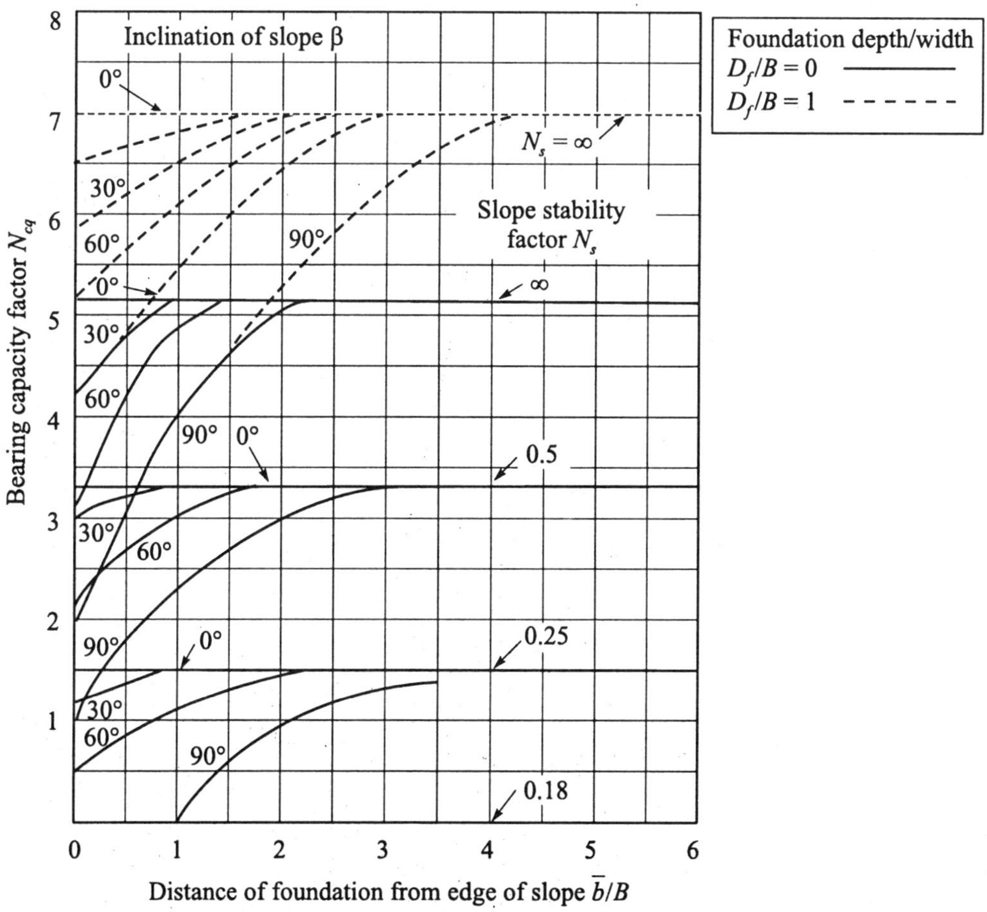

Fig. 5.22 Bearing capacity factors for strip foundation on top of slope of purely cohesive material (Meyerhof, 1957)

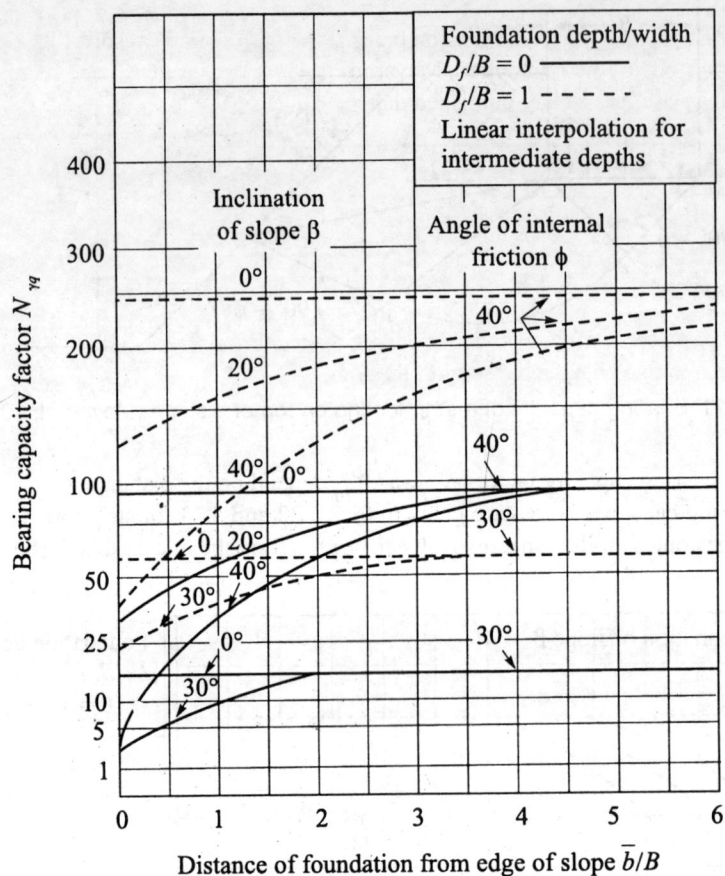

Fig. 5.23 Bearing capacity factors for strip foundation on top of slope of cohesionless material (Meyerhof, 1957)

factors increase with an increase of the distance $\bar{b}$. Beyond a distance of about 2 to 6 times the foundation width B, the bearing capacity is independent of the inclination of the slope, and becomes the same as that of a foundation on an extensive horizontal surface.

For a surcharge over the entire horizontal top surface of a slope, a solution of the slope stability has been obtained on the basis of dimensionless parameters called the stability number N_s, expressed as

$$N_s = \frac{c}{\gamma H} \tag{5.71}$$

The bearing capacity of a foundation on purely cohesive soil of great depth can be represented by Eq. (5.70), where the N_{cq} factor depends on $\bar{b}$ as well as β, and the stability number N_s. This bearing capacity factor, which is given in the lower parts of Fig. 5.22, decrease considerably with greater height and to a smaller extent with the inclination of the slope. For a given height and slope angle, the bearing capacity factor increases with an increase in $\bar{b}$, and beyond a distance of about 2 to 4 times the height of the slope, the bearing capacity is independent of the slope angle. Figure 5.22 shows that the bearing capacity of foundations on top of a slope is governed by foundation failure for small slope height (N_s approaching infinity) and by overall slope failure for greater heights.

The influence of ground water and tension cracks (in purely cohesive soils) shall also be taken into account in the study of the foundation. Meyerhof (1957) has not supported his theory with any practical examples of failure as any published data was not available for this purpose.

Example 5.19

A strip footing is to be constructed on the top of a slope as per Fig. 5.21. The following data are available:

$B = 3$ m, $D_f = 1.5$ m, $\bar{b} = 2$ m, $H = 8$ m, $\beta = 30°$, $\gamma = 18.5$ kN/m³, $\phi = 0$ and $c = 75$ kN/m²,

Determine the ultimate bearing capacity of the footing.

Solution

Per Eq. (5.70) q_u for $\phi = 0$ is

$$q_u = cN_{cq} + \gamma D_f$$

From Eq. (5.71)

$$N_s = \frac{c}{\gamma H} = \frac{75}{18.5 \times 8} = 0.51$$

and $\quad \dfrac{\bar{b}}{B} = \dfrac{2}{3} = 0.67$

From Fig. 5.22,

$$N_{cq} = 3.4 \text{ for } N_s = 0.51, \bar{b}/B = 0.67, \text{ and } \beta = 30°$$

Therefore

$$q_u = 75 \times 3.4 + 18.5 \times 1.5 = 255 + 28 = 283 \text{ kN/m}^2.$$

5.16   THE PRESSUREMETER METHOD OF DETERMINING ULTIMATE BEARING CAPACITY

Concept of the Method and Development of Equation

Pressuremeter is very much used in the European countries for determining ultimate bearing capacity and settlement of all types of foundations. The method of conducting the PMT in pre-bored holes has been explained in Chapter 3. The tests are normally conducted at 1 m intervals. The concept of the method is explained with reference to Fig. 5.24.

Consider a circular footing loaded at the top with a load $\bar{Q}_u$. This load is taken partly by the skin resistance and partly by the base resistance. Let the base resistance be q_u, per unit area under ultimate loading condition. The penetration of a circular footing is associated with the expansion of a spherical cavity as the soil particles move in the radial direction due to the penetration of the footing. Whereas, the pressuremeter test in a pre-bored hole is associated with the development of a cylindrical cavity with the ultimate net limit pressure $\bar{p}_l$. The ratio between the ultimate bearing capacity of a circular footing q_u, and the net limit pressure, $\bar{p}_l$ may be expressed as

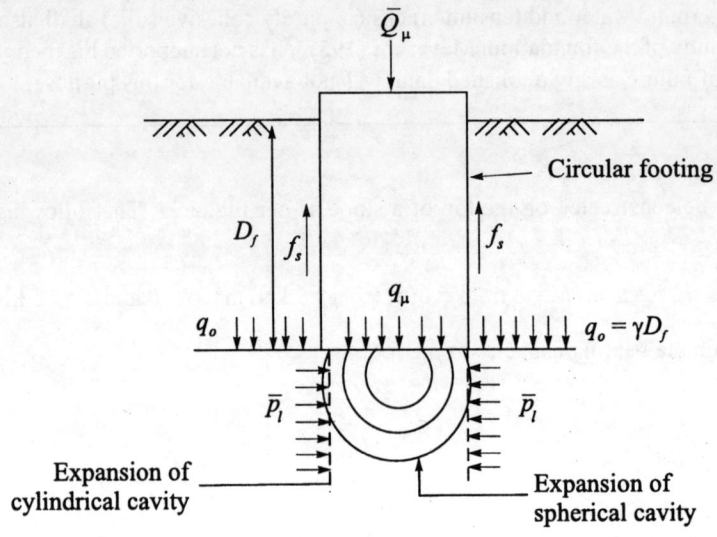

Fig. 5.24 The concept of bearing capacity by pressuremeter method

$$\frac{q_u - q_o}{p_l - p_{oh}} = \frac{q_{nu}}{\overline{p}_l} = k \qquad (5.72a)$$

or
$$q_{nu} = k\overline{p}_l \qquad (5.72b)$$

where, k is called as a bearing capacity factor, and

q_o = total overburden pressure adjacent to the foundation,

p_{oh} = the horizontal earth pressure in terms of total stress at the depth at which the pressuremeter test was carried out,

q_u = the gross ultimate bearing capacity of the foundation,

q_{nu} = the net ultimate bearing capacity,

p_l = the limit pressure from PMT,

$\overline{p}_l$ = the net limit pressure from PMT.

Based mainly on experimental evidence and experience, Baguelin *et al* (1978) have found that the factor k has no unique value, but varies from about 0.8 to 9 depending primarily upon,

1. the type of soil,

2. the relative depth of embedment,

3. the shape of foundation, and

4. the way in which the foundation is installed.

Adhesion or friction along the sides of a foundation such as a pier or pile is proportional to the net limit pressure, but the physical properties of the surface in contact with the soil are also important, as is the method with which the foundation is installed. For example, a concrete pile driven into a sand deposit would have a different capacity than would a steel pile driven into the same soil. In addition, if the concrete pile were cast *in-situ* rather than being driven, it would have a different capacity.

In Eq. (5.72), the limit pressure p_l and p_{oh} are measured with the pressuremeter whereas q_o is computed from the unit weight of soil and depth of foundation.

Values of *k* for Different Types of Soils

Since the pressuremeter bearing capacity factor k is influenced by many factors, such as type of soil, D_f/B ratio, shape of the foundation, etc. Baguelin *et al* (1978) have presented a set of curves for determining k for different types of soils for various D_f/B ratios as shown Figs 5.25 to 5.28. The curves are prepared for *cast-in-situ* shallow foundations. The values of k can be obtained from these curves for round, square and strip foundations. For a rectangular foundation, the equation for k is

$$k_r = k_1 + (k_2 - k_1) \times \frac{B}{2} \tag{5.73}$$

where $\quad k_r = k$ for rectangular foundation,

$\qquad k_1 = k$ for strip foundation,

$\qquad k_2 = k$ for square foundation,

$\qquad B = $ Width,

$\qquad L = $ Length.

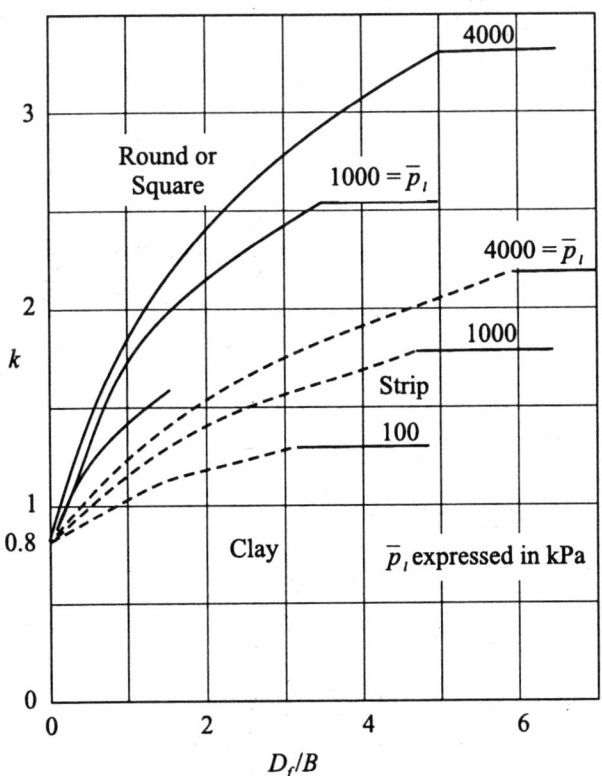

Fig. 5.25 *k* values for clay (Baguelin *et al*, 1978)

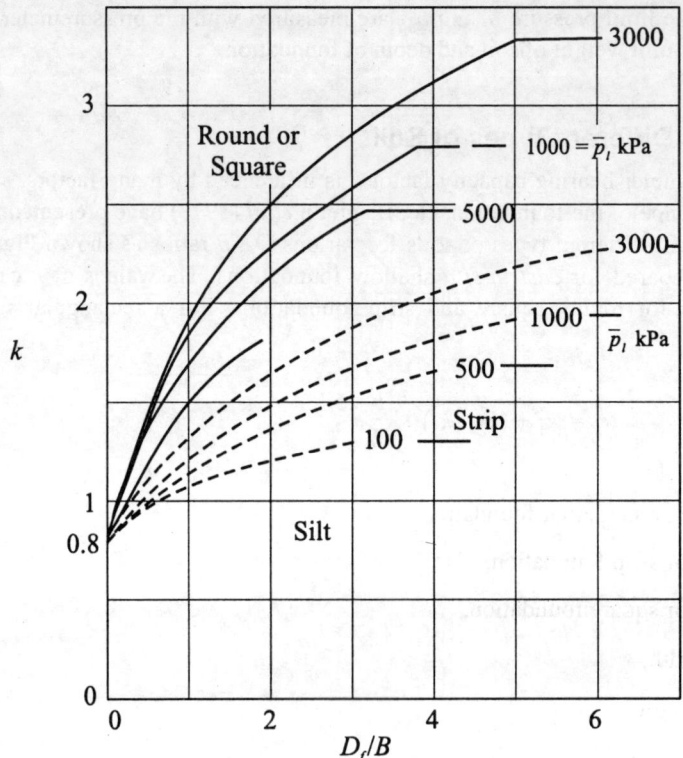

Fig. 5.26 k values for silt (Baguelin *et al*, 1978)

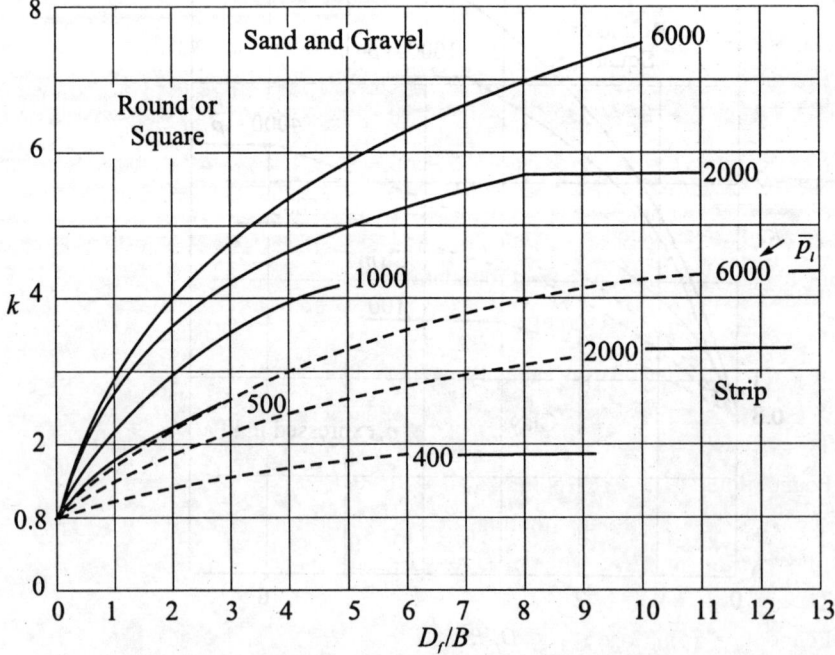

Fig. 5.27 k values for sand and gravel (Baguelin *et al*, 1978)

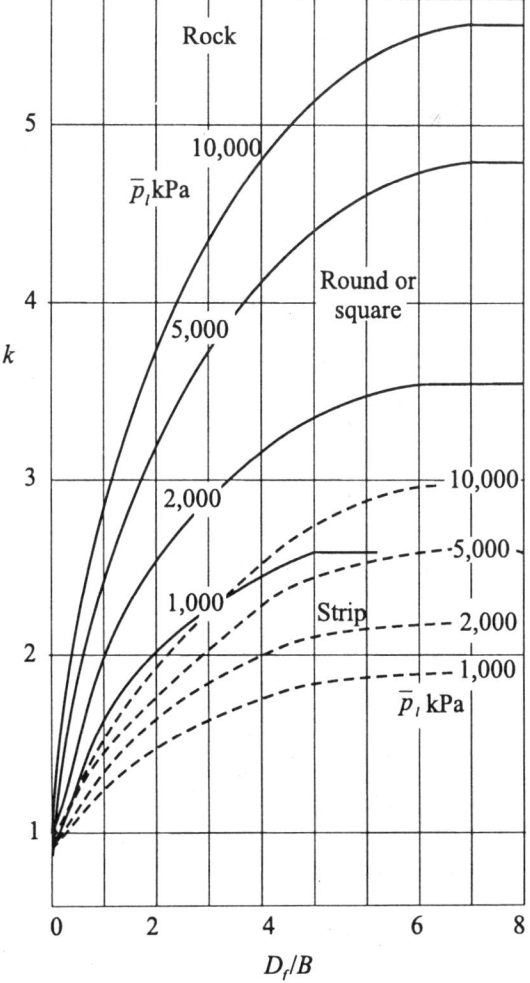

Fig. 5.28 *k* values for rock (Baguelin *et al*, 1978)

Example 5.20

The following data are given for a footing foundation,

$$D_f = 1 \text{ m}, B = 2 \text{ m}, \bar{p}_l = 700 \text{ kPa}.$$

Determine the net ultimate bearing capacity and net allowable bearing pressure for the following cases.

Case 1: Type of soil: Clay.

       Foundations   (*i*) Strip footing,

                    (*ii*) Square footing,

                  (*iii*) Rectangular footing $L = 8$ m.

Case 2: Type of soil: Sand.

       Foundation as in Case 1.

       Use a factor of safety of 3.

Solution

Case 1:   (*i*) For $B = 2$ m, $D_f = 1$ m, $D_f/B = 0.5$, we have $k = 1$ from Fig. 5.25.

$$q_{nu} = k\bar{p}_l = 1 \times 700 = 700 \text{ kPa},$$

$$q_{na} = \frac{700}{3} = 233 \text{ kPa}.$$

(*ii*) $D_f/B = 0.5$, $k = 1.5$

$$q_{nu} = 1.5 \times 700 = 1,050 \text{ kPa},$$

$$q_{na} = \frac{1,050}{3} = 350 \text{ kPa}.$$

(*iii*) $D_f/B = 0.5$, $B/L = 0.25$.

$$k_r = 1 + (1.5 - 1) \times 0.25 = 1.125,$$

$$q_{nu} = 1.125 \times 700 = 787 \text{ kPa},$$

$$q_{na} = \frac{787}{3} = 262 \text{ kPa}.$$

Case 2:   (*i*) For $D_f/B = 0.5$, $k = 1.1$ from Fig. 5.27.

$$q_{nu} = 1.1 \times 700 = 770 \text{ kPa},$$

$$q_{na} = \frac{770}{3} = 257 \text{ kPa}.$$

(*ii*) For $D_f/B = 0.5$, $k = 1.5$ from Fig. 5.27.

$$q_{nu} = 1.5 \times 700 = 1,050 \text{ kPa},$$

$$q_{na} = \frac{1,070}{3} = 350 \text{ kPa}.$$

(*iii*) For $B/L = 0.25$, $k_r = 1.125$.

$$q_{nu} = 1.125 \times 700 = 787 \text{ kPa},$$

$$q_{na} = \frac{787}{3} = 262 \text{ kPa}.$$

Determination of Equivalent Net Limit Pressure $\bar{p}_{le}$ in Nonhomogeneous Soil

Soil met in nature is not homogeneous. The composition of soil within the depth of influence may change from sand to clay, and then silt, etc. due to natural heterogeneity and variation in stress. If these variations are not too large, a value of $\bar{p}_l$ can be chosen which, for engineering purposes can be considered to be representative of the soil and the rules for homogeneous ground used to calculate foundation requirements. The problem is to determine this equivalent net limit pressure, $\bar{p}_{le}$. It may be calculated as follows.

The zone of influence is assumed to extend from a point, $1.5B$ above the base to $1.5B$ below the base as shown in Fig. 5.29. The $\bar{p}_{le}$ is taken as the geometric mean of $\bar{p}_l$ values within this zone of depth $3B$ provided they are measured at regular intervals. Thus,

$$\bar{p}_{le} = [(\bar{p}_l)_1 \times (\bar{p}_l)_2 \times \cdots (\bar{p}_l)_n]^{1/n} \tag{5.73a}$$

The zones of influence for both shallow and deep foundations with the positions of $\bar{p}_l$ are shown in Fig. 5.29. In the case of shallow foundations, Fig. 5.29 (a), the values of p_l to be considered are up to point 5, since they fall within the $3B$ zone.

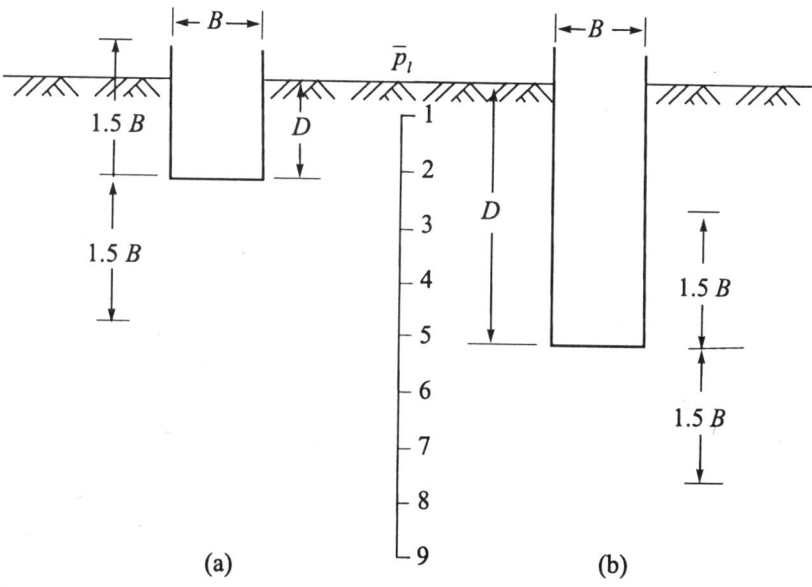

Fig. 5.29 Zones of influence of foundation: (a) Shallow foundation, (b) deep foundation

Therefore,

$$\bar{P}_{le} = (\bar{P}_{l1} \times \bar{P}_{l2} \times \bar{P}_{l3} \times \bar{P}_{l4} \times \bar{P}_{l5})^{1/5} \tag{5.73b}$$

In the case of deep foundations, Fig. 5.29 (b), $\bar{p}_l$ from point 3 to 8 are to be considered. Therefore,

$$\bar{P}_{le} = (\bar{P}_{l3} \times \bar{P}_{l4} \times \bar{P}_{l5} \times \bar{P}_{l6} \times \bar{P}_{l7} \times \bar{P}_{l8})^{1/6} \tag{5.73c}$$

This way of finding the equivalent, homogeneous $\bar{p}_{le}$ value must be used in cases where the individual $\bar{p}_l$ values do not differ more than 40 percent from the homogeneous $\bar{p}_{le}$ value. In a layered system where $\bar{p}_l$ in each layer is considerably different from the adjacent layers, $\bar{p}_{le}$ of each layer should be computed.

Determination of Equivalent Depth D_e in a Layered Soil System
Two layer soil system

Figure 5.30 represents a condition which is encountered frequently where there is layer (layer 2) of poor soil (loose sand or soft clay) lying over a layer of hard strata, such as dense sand or stiff clay, etc. (layer 1). In such a case, if the base of the foundation lies in the hard strata, up to a depth, D_1, the base capacity of a pile or a caisson can be calculated from the $\bar{p}_l$ values of the harder strata without neglecting the effect of the softer strata above. The procedure is to convert the softer strata to an equivalent thickness, Z_1 of good material. The thickness Z_1 of good soil which would be equivalent to Z_2 the thickness of the soft soil, can be calculated by using the expression,

$$Z_1 = Z_2 \times \frac{(\bar{p}_l)_2}{(\bar{p}_l)_1} \tag{5.74a}$$

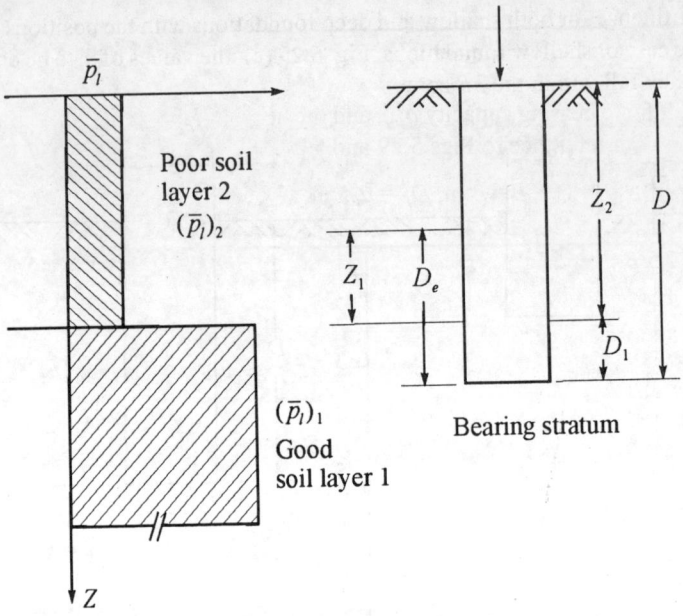

Fig. 5.30 Foundation on a two-layer soil

where, Z_1 = the equivalent thickness of goodsoil,

$(\bar{p}_l)_1$ = the net limit pressure of the good soil in layer 1,

$(\bar{p}_l)_2$ = the net limit pressure of the poor soil in layer 2.

The effective depth of embedment D_e, becomes

$$D_e = D_1 + Z_1 \qquad (5.74b)$$

The value of $\bar{p}_l$ within the depth D_e can be computed as if the soil were homogeneous with a $\bar{p}_l = (\bar{p}_l)_1$.

General case

To find the equivalent depth of embedment D_e in a general case, Eq. (5.74a) is applied to each elementary layer ΔZ_i, corresponding to test number i. The summation is made over the actual depth of embedment.

i.e.
$$D_e = \sum_{i=1}^{n} \Delta Z_i \qquad (5.74c)$$

$$D_e = \frac{1}{p_{le}} \sum_{i=1}^{n} \Delta Z_i \, \bar{p}_i \qquad (5.74d)$$

Note that, in general, $\Delta Z_i = 1$ m, since pressuremeter tests are carried out at a depth interval of one metre. The values at the surface ΔZ_1, and near the base of the foundation ΔZ_n are different, but usually these differences are not significant and they can be neglected.

Example 5.21

Calculate the net ultimate bearing capacity q_{nu}, and the net allowable bearing pressure q_{na} ($F_s = 3$), given the following with reference to Figs 5.29 and 5.30.

$$B = 3 \text{ m}, D_f = 2.5 \text{ m}, D_1 = 0.5 \text{ m}$$

Depth below GL (m)	$\bar{p}_l$ kPa	Depth below GL (m)	$\bar{p}_l$ kPa
0.5	350	6.5	700
1.5	500	7.5	725
2.5	600		
3.5	625		
4.5	600		
5.5	675		

The soil is assumed as clay and foundation square.

Solution

We have to consider $\bar{p}_l$ up to 6.5 m below ground level since the last value falls within the influence zone of $1.5B$ ($= 4.5$ m) below the base of the footing. First we have to calculate $\bar{p}_{le}$ from depth 0.5 m to 6.5 m below GL. Since the first value 350 kPa is very much different from the other values, the $\bar{p}_{le}$ for the whole depth may be calculated as follows.

Second layer, $(\bar{p}_{le})_2 = 350$ kPa.

First layer, $(\bar{p}_{le})_1 = (500 \times 600 \times 625 \times 600 \times 675 \times 700)^{1/6}$

$$= 615 \text{ kPa}.$$

Combined, $\bar{p}_{le} = (350 \times 615)^{1/2} = 464$ kPa.

Equivalent depth $D_e = 0.5 + \dfrac{2 \times 350}{464} = 0.5 + 1.5 = 2$ m.

From Fig. 5.25, for $D_e/B = 2/3 = 0.67$, and $\bar{p}_{le} = 464$ kPa, we have $k = 1.44$.

Therefore, $q_{nu} = 1.44 \times 464 = 668$ kPa,

$$q_{na} = \frac{q_{nu}}{3} = 223 \text{ kPa}.$$

Example 5.22

For the data given in Ex. 5.21, determine q_{nu} and q_{na} if the soil is sand.

Solution

For $D_e/B = 0.67$, $\bar{p}_{le} = 464$ kPa, we have $k = 1.6$ from Fig. 5.27. Therefore,

$$q_{nu} = 1.6 \times 464 = 742 \text{ kPa},$$

$$q_{na} = \frac{q_{nu}}{3} = 248 \text{ kPa}.$$

Example 5.23

Determine q_{nu} and q_{na} by the conventional method by assuming $\bar{p}_{le} = 464$ for the clay soil in Ex. 5.21.

Solution

As per Eq. (3.32c), the undrained cohesive strength c_u is

$$c_u = \frac{\bar{p}_{le}}{10} + 25 = \frac{464}{10} + 25 = 71 \text{ kPa}$$

For $D_e / B = 0.67$, from Fig. 5.8,

$$N_c = 7.5$$

$$q_{nu} = c_u N_c = 71 \times 7.5 = 533 \text{ kPa}$$

$$q_{na} = \frac{q_{nu}}{3} = 178 \text{ kPa}$$

Example 5.24

Use the results and data of Ex. 5.21. Determine the consistency of clay from Table 3.9 (b). From the relations given between consistency and SPT value N in Table 3.5 (b), determine q_{nu} and q_{na}.

Solution

From Ex. 5.21, $\bar{p}_{le} = 464$ kPa. From Table 3.9 (b), the consistency of clay is medium stiff ($\bar{p}_{le}$ ranges from 300 to 800 kPa).

The unconfined compressive strength q_u for medium stiff clay ranges from 50 to 100 kPa Table 3.5 (b). An average value $q_u = 75$ kPa may be taken.

Now,
$$q_{nu} = \frac{q_u}{2} \times N_c = \frac{75}{2} \times 7.5 = 281 \text{ kPa}$$

$$q_{na} = 94 \text{ kPa.}$$

Alternate method: The Spt value N lies between 4 and 8 for medium stiff clay from Table 3.5 (b). Now from Eq. (3.16c), the unconfined compressive strength q_u is

$$q_u = \bar{k} N \text{ kPa}$$

where the values of $\bar{k}$ varies from 13 to 25. Taking an average $\bar{k} = 19$, and an average $N = 6$, we have

$$q_u = 19 \times 6 = 114 \text{ kPa}$$

Therefore,
$$q_{nu} = \frac{q_u}{2} \times N_c = \frac{114}{2} \times 7.5 = 428 \text{ kPa}$$

$$q_{na} = \frac{q_{nu}}{3} = 143 \text{ kPa}$$

5.17 FOUNDATIONS ON ROCK

Rocks encountered in nature might be igneous, sedimentary or metamorphic. Granite and basalt belong to the first group. Granite is primarily composed of feldspar, quartz and mica possesses a massive structure. Basalt is a dark-coloured fine grained rock. Both basalt and granite under unweathered conditions serve as a very good foundation base. The most important rocks that belong to the second group are sandstones, limestones and shales. These rocks are easily weathered under hostile environmental conditions and as such, the assessment of bearing capacity of these types requires a little care. In the last group come gneiss, schist, slate and marble. Of these rocks gneiss serves as a good bearing material whereas schist and slate possess poor bearing quality.

All rocks weather under hostile environments. The ones that are close to the ground surface become weathered more than the deeper ones. If a rocky stratum is suspected close to the ground surface, the soundness or otherwise of these rocks must be investigated. The quality of the rocks is normally designated by RQD as explained in Chapter 3.

Joints are common in all rock masses. This word joint is used by geologists for any plane of structural weakness apart from faults within the mass. Within the sedimentary rock mass the joints are lateral in extent and form what are called *bedding planes*, and they are uniform throughout any one bed within igneous rock mass. Cooling joints are more closely spaced nearer the ground surface with wider spacings at deeper depths. Tension joints and tectonic joints might be expected to continue depthwise. Within metamorphic masses, open cleavage, open schistose and open gneissose planes can be of considerably further lateral extent than the bedding planes of the sedimentary masses.

Faults and fissures happen in rock masses due to internal disturbances. The joints with fissures and faults reduces the bearing strength of rocky strata.

Since most unweathered intact rocks are stronger and less compressible than concrete, the determination of bearing pressures on such materials may not be necessary. A confined rock possesses greater bearing strength than the rocks exposed at ground level.

Bearing Capacity of Rocks

Bearing capacities of rocks are often determined by crushing a core sample in a testing machine. Samples used for testing must be free from cracks and defects.

In the rock formation where bedding planes, joints and other planes of weakness exist, the practice that is normally followed is to classify the rock according to RQD (Rock Quality Designation). Table 3.2 gives the classification of the bearing capacity of rock according to RQD. Peck *et al*, (1974) have related the RQD to the allowable bearing pressure q_a as given in Table 5.5.

Table 5.5 Allowable bearing pressure q_a on jointed rock

RQD	q_a ton/ft^2	q_a MPa
100	300	29
90	200	19
75	120	12
50	65	6.25
25	30	3
0	10	0.96

The RQD for use in Table 5.5 should be the average within a depth below foundation level equal to the width of the foundation, provided the RQD is fairly uniform within that depth. If the upper part of the rock, within a depth of about $B/4$, is of lower quality, the value of this part should be used.

Another practice that is normally followed is to base the allowable pressure on the unconfined compressive strength, q_u, of the rock obtained in a laboratory on a rock sample. A factor of safety of 5 to 8 is used on q_u to obtain q_a. Past experience indicates that this method is satisfactory so long as the rocks *in-situ* do not possess extensive cracks and joints. In such cases a higher factor of safety may have to be adopted.

If rocks close to a foundation base are highly fissured/fractured, they can be treated by grouting which increases the bearing capacity of the material.

The bearing capacity of a homogeneous, and discontinuous rock mass cannot be less than the unconfined compressive strength of the rock mass around the footing and this can be taken as a lower bound for a rock mass with constant angle of internal friction ϕ and unconfined compressive strength q_{ur}. Goodman (1980), suggests the following equation for determining the ultimate bearing capacity q_u.

$$q_u = q_{ur} (N_\phi + 1) \tag{5.75}$$

where $N_\phi = \tan^2 (45° + \phi/2)$, q_{ur} = unconfined compressive strength of rock.

Recommendations by Building Codes

Where bedrock can be reached by excavation, the presumptive allowable bearing pressure is specified by Building Codes. Table 5.6 gives the recommendations of some buildings codes in the US.

Table 5.6 Presumptive allowable bearing pressures on rocks (MPa) as recommended by various building codes in USA (after Peck *et al*, 1974)

Rock type	Building codes			
	BOCA (1968)	National (1967)	Uniform (1964)	LOS Angeles (1959)
1. Massive crystalline bedrock, including granite diorite, gneiss, basalt, hard limestone and dolomite	10	10	$0.2 \, q_u^*$	1.0
2. Foliated rocks such as schist or slate in sound condition	4	4	$0.2 \, q_u$	0.4
3. Bedded limestone in sound condition sedimentary rocks including hard shales and sandstones	2.5	1.5	$0.2 \, q_u$	0.3
4. Soft or broken bedrock (excluding shale) and soft limestone	1.0	–	$0.2 \, q_u$	–
5. Soft shale	0.4	–	$0.2 \, q_u$	–

* q_u = unconfined compressive strength.

5.18 CASE HISTORY OF FAILURE OF THE TRANSCONA GRAIN ELEVATOR

One of the best known foundation failures occurred in October 1913 at North Transcona, Manitoba, and Canada. It was ascertained later on that the failure occurred when the foundation pressure at the base was about equal to the calculated ultimate bearing capacity of an underlaying layer of plastic clay (Peck and Byrant, 1953), and was essentially a shearing failure.

The construction of the silo started in 1911 and was completed in the autumn of 1913. The silo is 77 ft by 195 ft in plan and has a capacity of 10,00,000 bushels. It comprises 65 circular bins and 48 inter-bins. The foundation was a reinforced concrete raft 2 ft thick and founded at a depth of 12 ft below the ground surface. The weight of the silo was 20,000 tons, which was 42.5 percent of the total weight, when it was filled. Filling the silo with grain started in September 1913, and in October when the silo contained 8,75,000 bushels, and the pressure on the ground was 94 percent of the design pressure, a vertical settlement of 1 ft was noticed. The structure began to tilt to the west and within twenty-four hours was at an angle of 26.9° from the vertical, the west side being 24 ft below and the east side 5 ft above the original level (Szechy, 1961). The structure tilted as a monolith and there was no damage to the structure except for a few superficial cracks. Figure 5.31 shows a view of the tilted structure. The excellent quality of the reinforced concrete structure is shown by the fact that later it was underpinned and jacked up on new piers founded on rock. The level of the new foundation is 34 ft below the ground surface. Figure 5.32 shows the view of the silo after it was straightened in 1916.

Fig. 5.31 The tilted Transcona grain elevator
Courtesy: UMA Engineering Ltd., Manitoba, Canada.

During the period when the silo was designed and constructed, soil mechanics as a science had hardly begun. The behaviour of the foundation under imposed loads was not clearly understood. It was only during the year 1952 that soil investigation was carried out close to the silo and the soil properties were analyzed (Peck and Byrant, 1953). Figure 5.33 gives the soil classification and unconfined compressive strength of the soil with respect to depth. From the examination of undisturbed samples of the clay, it was determined that the average water content of successive layers of varved clay increased with their depth from 40 percent to about 60 percent. The average unconfined compressive strength of the upper stratum beneath the foundation was 1.13 tsf, that of the lower stratum was 0.65 tsf, and the weighted average was 0.93 tsf. The average liquid limit was found to be 105 percent; therefore, the plasticity index was 70 percent, which indicates that the clay was highly colloidal and plastic. The average unit weight of the soil was 120 lb/ft^3.

The contact pressure due to the load from the silo at the time of failure was estimated as equal to 3.06 tsf. The theoretical values of the ultimate bearing capacity by various methods are as follows:

Methods	q_u tsf
Terzaghi (Eq. 5.19b)	3.68
Meyerhof (Eq. 5.27)	3.30
Skempton (Eq. 5.22)	3.32

Fig. 5.32 The straightened Transcona grain elevator
Courtesy: UMA Engineering Ltd. Manitoba, Canada.

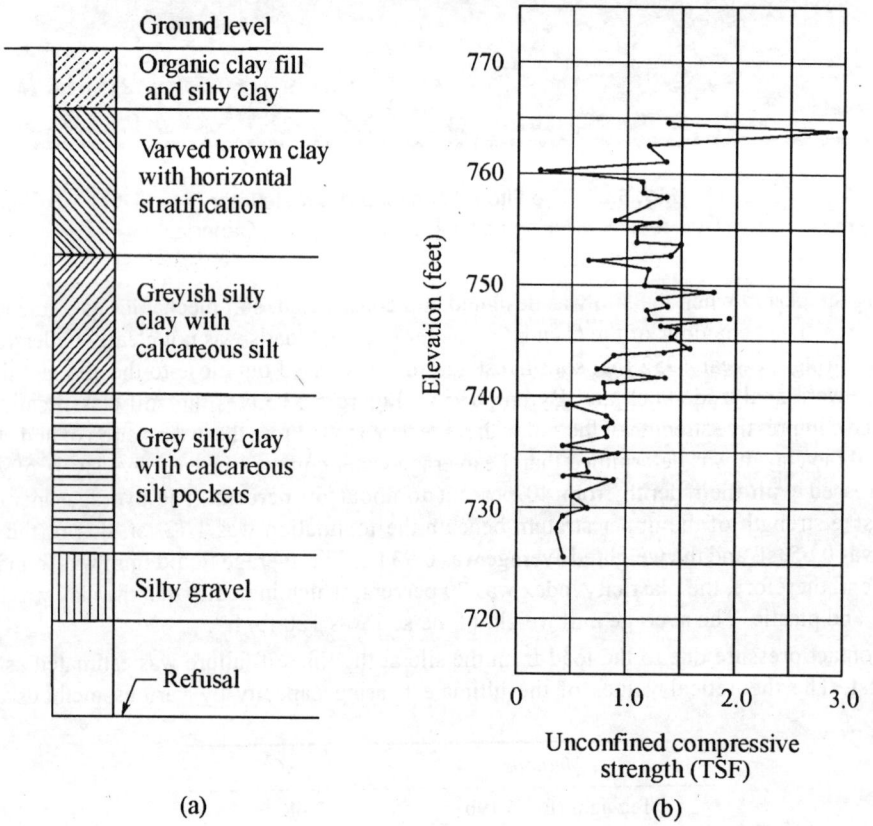

(a)

(b)

Fig. 5.33 Results of test boring at site of Transcona grain elevator: (a) Classification from test boring, (b) variation of unconfined compressive strength with depth (Peck and Byrant, 1953)

The above values compare reasonably well with the actual failure load 3.06 tsf. Perloff and Baron (1976), give details of failure of the Transcona grain elevator.

5.19 PROBLEMS

5.1 What will be the gross and net allowable bearing pressures of a sand having $\phi = 35°$ and an effective unit weight of 18 kN/m^3 under the following cases: (*a*) size of footing 1 × 1 m square, (*b*) circular footing of 1 m dia., and (*c*) 1 m wide strip footing.

The footing is placed at a depth of 1 m below the ground surface and the water table is at great depth. Use $F_s = 3$. Compute by Terzaghi's general shear failure theory.

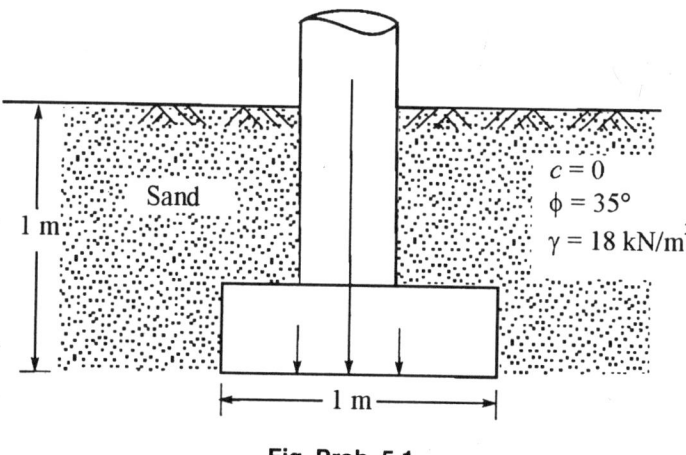

Fig. Prob. 5.1

5.2 A strip footing is founded at a depth of 1.5 m below the ground surface (Fig. Prob. 5.2). The water table is close to ground level and the soil is cohesionless. The footing is supposed to carry a net safe load of 400 kN/m^2 with $F_s = 3$. Given $\gamma_{sat} = 20.85$ kN/m^3 and $\phi = 35°$, find the required width of the footing, using Terzaghi's general shear failure criterion.

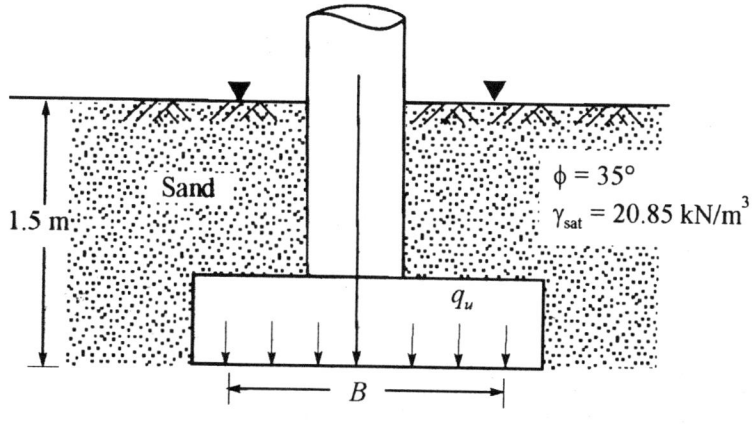

Fig. Prob. 5.2

5.3 At what depth should a footing of size 2 × 3 m be founded to provide a factor of safety of 3, if the soil is stiff clay having an unconfined compressive strength of 120 kN/m²? The unit weight of the soil is 18 kN/m³. The ultimate bearing capacity of the footing is 425 kN/m². Use Terzaghi's theory. The water table is close to the ground surface (Fig. Prob. 5.3).

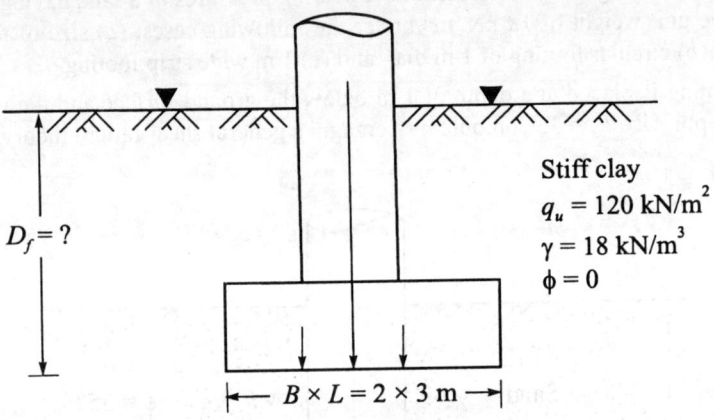

Stiff clay
$q_u = 120$ kN/m²
$\gamma = 18$ kN/m³
$\phi = 0$

$D_f = ?$

$B \times L = 2 \times 3$ m

Fig. Prob. 5.3

5.4 A rectangular footing is founded at a depth of 2 m below the ground surface in a $(c - \phi)$ soil having the following properties: porosity $n = 40\%$, $G = 2.67$, $c = 15$ kN/m², and $\phi = 30°$.

The water table is close to the ground surface. If the width of the footing is 3 m, what is the length required to carry a gross allowable bearing pressure $q_a = 455$ kN/m² with a factor of safety = 3? Use Terzaghi's theory of general shear failure (Fig. Prob. 5.4).

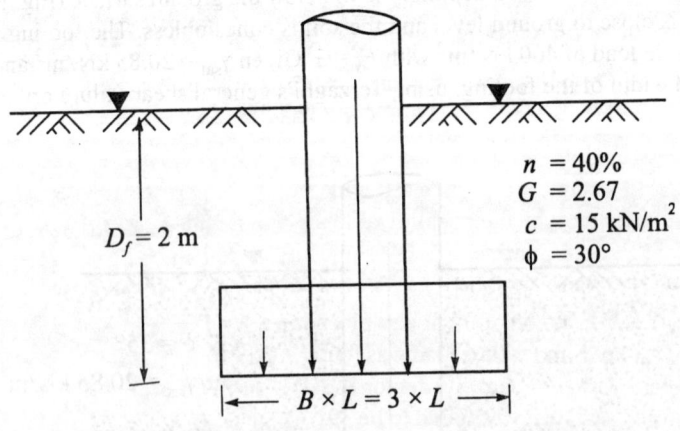

$n = 40\%$
$G = 2.67$
$c = 15$ kN/m²
$\phi = 30°$

$D_f = 2$ m

$B \times L = 3 \times L$

Fig. Prob. 5.4

5.5 A square footing located at a depth of 5 ft below the ground surface in a cohesionless soil carries a column load of 130 tons. The soil is submerged having an effective unit weight of 73 lb/ft³ and an angle of shearing resistance of 30°. Determine the size of the footing for $F_s = 3$ by Terzaghi's theory of general shear failure (Fig. Prob. 5.5).

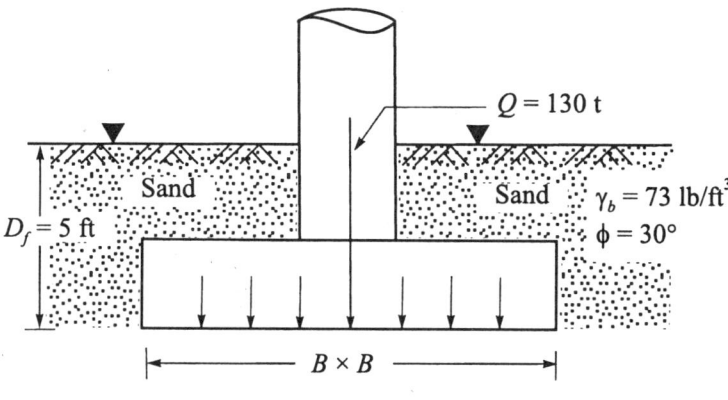

$Q = 130$ t

Sand

Sand $\gamma_b = 73$ lb/ft^3

$\phi = 30°$

$D_f = 5$ ft

$B \times B$

Fig. Prob. 5.5

5.6 A footing of 5 ft diameter carries a safe load (including its self weight) of 80 tons in cohesionless soil (Fig. Prob. 5.6). The soil has an angle of shearing resistance $\phi = 36°$ and an effective unit weight of 80 lb/ft^3. Determine the depth of the foundation for $F_s = 2.5$ by Terzaghi's general shear failure theory.

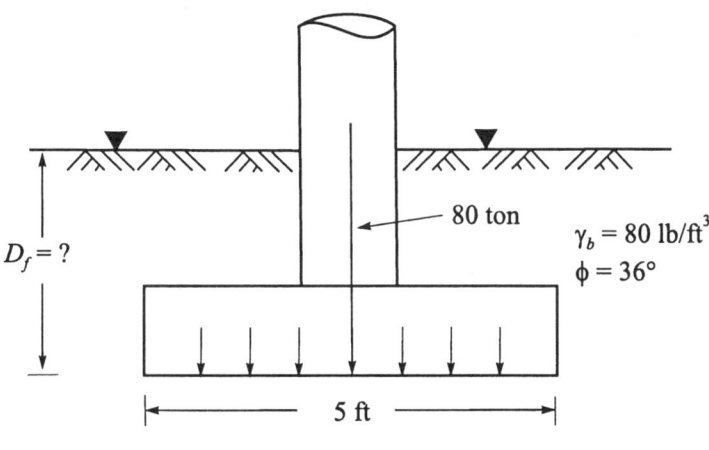

80 ton

$\gamma_b = 80$ lb/ft^3

$\phi = 36°$

$D_f = ?$

5 ft

Fig. Prob. 5.6

5.7 If the ultimate bearing capacity of a 4 ft wide strip footing resting on the surface of a sand is 5,250 lb/ft^2, what will be the net allowable pressure that a 10×10 ft square footing resting on the surface can carry with $F_s = 3$? Assume that soil is cohesionless. Use Terzaghi's theory of general shear failure.

5.8 A circular plate of diameter 1.05 m was placed on a sand surface of unit weight 16.5 kN/m^3 and loaded to failure. The failure load was found to give a pressure of 1,500 kN/m^2. Determine the value of the bearing capacity factor N_γ. The angle of shearing resistance of the sand measured in a triasial test was found to be 39°. Compare this value with the theoretical value of N_γ. Use Terzaghi's theory of general shear failure.

5.9 Find the net allowable bearing load per foot length of a long wall footing 6 ft wide founded on a stiff saturated clay at a depth of 4 ft. The unit weight of the clay is 110 lb/ft^3, and the shear strength is 2500 lb/ft^2. Assume the load is applied rapidly such that undrained conditions ($\phi = 0$) prevail. Use $F_s = 3$ and Skempton's method (Fig. Prob. 5.9).

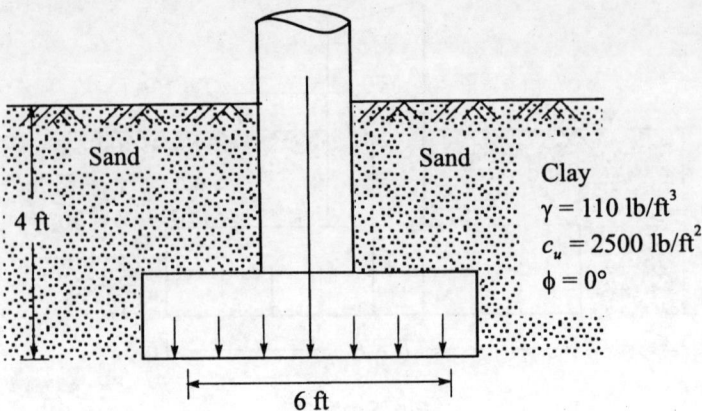

Fig. Prob. 5.9

5.10 The total column load of a footing near ground level is 5,000 kN. The subsoil is cohesionless soil with $\phi = 38°$ and $\gamma = 19.5 \text{ kN}/\text{m}^3$. The footing is to be located at a depth of 1.50 m below ground level. For a footing of size 3×3 m, determine the factor of safety by Terzaghi's general shear failure theory, if the water table is at a depth of 0.5 m below the base level of the foundation.

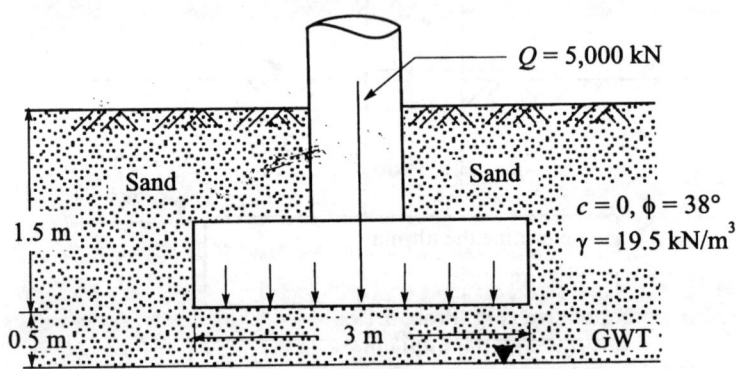

Fig. Prob. 5.10

5.11 What will be the factors of safety if the water table is met (*i*) at the base level of the foundation, and (*ii*) at the ground level in the case of the footing in Prob. 5.10, keeping all the other conditions the same? Assume that the saturated unit weight of soil $\gamma_{sat} = 19.5 \text{ kN}/\text{m}^3$, and the soil above the base of the foundation remains saturated even under (*i*) above.

5.12 If the factors of safety in Prob. 5.11 are other than 3, what should be the size of the footing for a minimum factor of safety of 3 under the worst condition ?

5.13 A footing of size 10×10 ft is founded at a depth of 5 ft in a medium stiff clay soil having an unconfined compressive strength of 2,000 lb/ft^2. Determine the net safe bearing capacity of the footing with the water table at ground level by Skempton's method. Assume $F_s = 3$.

5.14 If the average static cone penetration resistance, q_c, in Prob. 5.13 is 10 t/ft^2 determine q_{na} per Skempton's method. The other conditions remain the same as in Prob. 5.13. Ignore the effect of overburden pressure.

5.15 Refer to Prob. 5.10. Compute by Meyerhof theory: (*a*) The ultimate bearing capacity, (*b*) the net ultimate bearing capacity, and (*c*) the factor of safety for the load coming on the column. All the other data given in Prob. 5.10 remain the same.

5.16 Refer to Prob. 5.10. Compute by Hansen's method: (*a*) The ultimate bearing capacity, (*b*) the net ultimate bearing capacity, and (*c*) the factor of safety for the column load. All the other data remain the same. Comment on the results using the methods of Terzaghi, Meyerhof and Hansen.

5.17 A rectangular footing of size (Fig. Prob. 5.17) 12 × 24 ft is founded at a depth of 8 ft below the ground surface in a (*c* – ϕ) soil. The following data are available: (*a*) water table at a depth of 4 ft below ground level, (*b*) $c = 600$ lb/ft^2, $\phi = 30°$, and $\gamma = 118$ lb/ft^3. Determine the ultimate bearing capacity by Terzaghi and Meyerhof's methods.

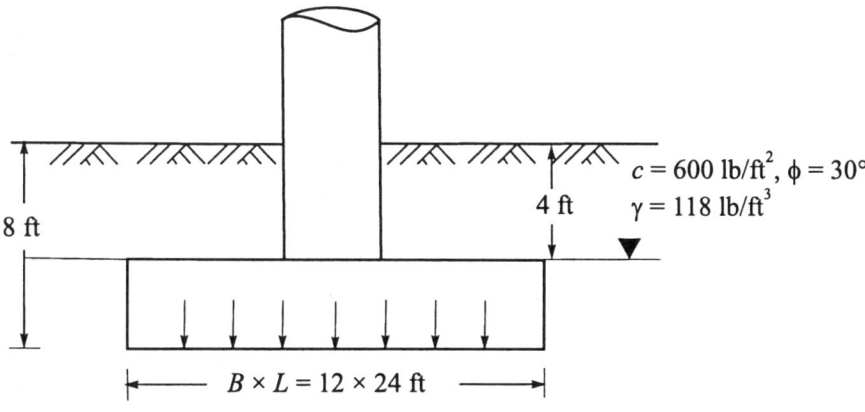

Fig. Prob. 5.17

5.18 Refer to Prob. 5.17 and determine the ultimate bearing capacity by Hansen's method. All the other data remain the same.

5.19 A rectangular footing of size (Fig. Prob. 5.19) 16 × 24 ft is founded at a depth of 8 ft in a deep stratum of (*c* – ϕ) soil with the following parameters:

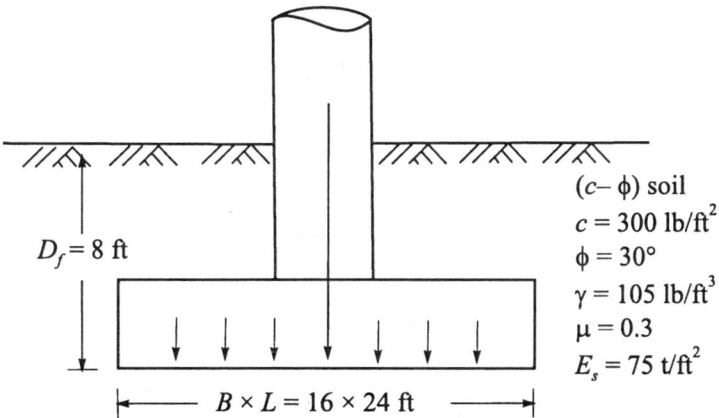

Fig. Prob. 5.19

$c = 300 \text{ lb/ft}^2$, $\phi = 30°$, $E_s = 75 \text{ t/ft}^2$, $\gamma = 105 \text{ lb/ft}^3$, $\mu = 0.3$.

Estimate the ultimate bearing capacity by: (*a*) Terzaghi's method, (*b*) Vesic's method by taking into account the compressibility factors.

5.20 A footing of size 10 × 15 ft (Fig. Prob. 5.20) is placed at a depth of 10 ft below the ground surface in a deep stratum of saturated clay of soft to medium consistency. The unconfined compressive strength of clay under undrained conditions is 600 lb/ft² and $\mu = 0.5$. Assume $\gamma = 95 \text{ lb/ft}^3$ and $E_s = 12 \text{ t/ft}^2$. Estimate the ultimate bearing capacity of the soil by the Terzaghi and Vesic methods by taking into account the compressibility factors.

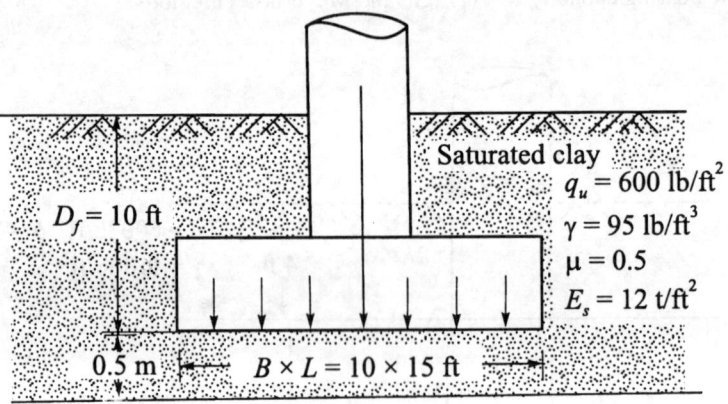

Fig. Prob. 5.20

5.21 Figure Prob. 5.21 gives a foundation subjected to an eccentric load in one direction with all the soil parameters. Determine the ultimate bearing capacity of the footing.

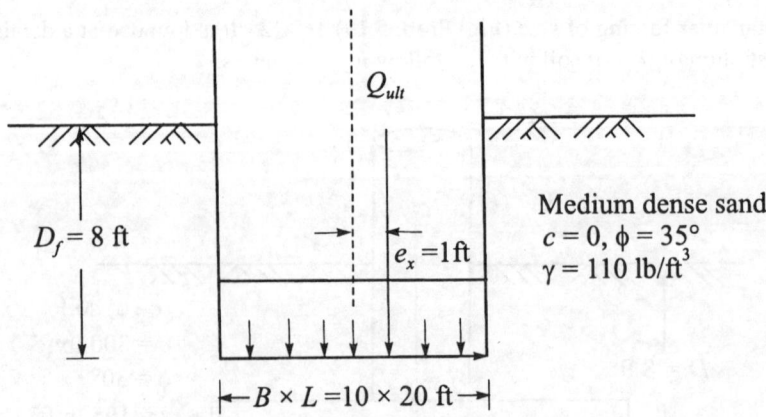

Fig. Prob. 5.21

5.22 Refer to Fig. Prob. 5.22. Determine the ultimate bearing capacity of the footing if $e_x = 3$ ft and $e_y = 4$ ft. What is the allowable load for $F_s = 3$?

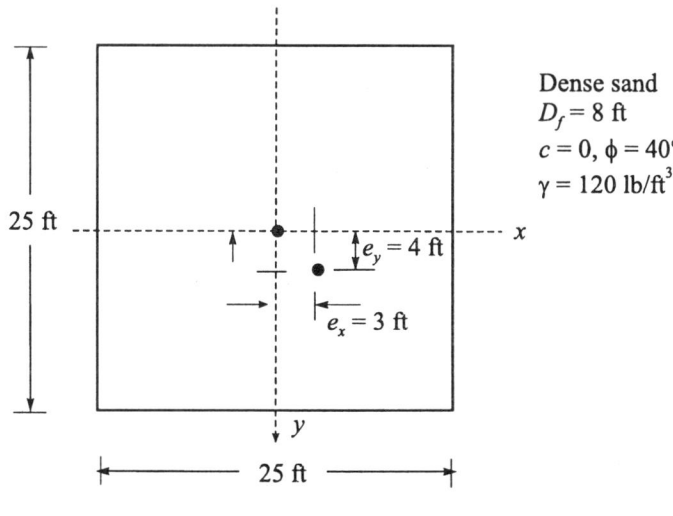

Fig. Prob. 5.22

5.23 Refer to Fig. Prob. 5.22. Compute the maximum and minimum contact pressures for a column load of $Q = 800$ tons.

5.24 A rectangular footing (Fig. Prob. 5.24) of size 6×8 m is founded at a depth of 3 m in a clay stratum of very stiff consistency overlying a softer clay stratum at a depth of 5 m from the ground surface. The soil parameters of the two layers of soil are:

Top layer: $c_1 = 200$ kN/m^2, $\gamma_1 = 18.5$ kN/m^3

Bottom layer: $c_2 = 35$ kN/m^2, $\gamma_2 = 16.5$ kN/m^3

Estimate the ultimate bearing capacity of the footing.

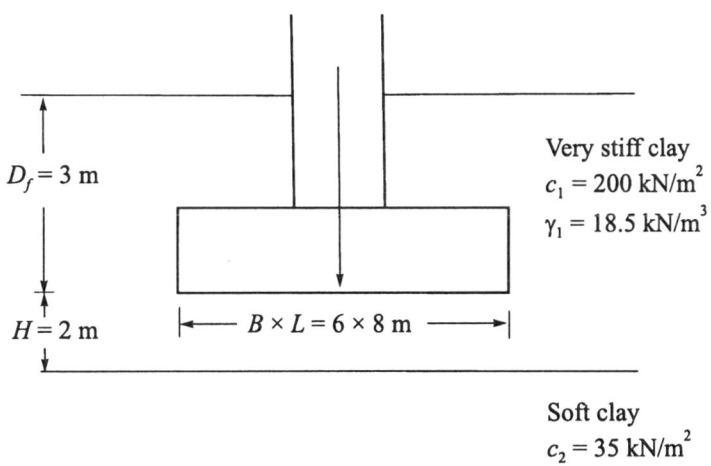

Fig. Prob. 5.24

5.25 If the top layer in Prob. 5.24 is dense sand, what is the ultimate bearing capacity of the footing ? The soil parameters of the top layer are:

$\gamma = 19.5 \text{ kN/m}^3, \phi 1 = 38°$

All the other data given in Prob. 5.24 remain the same.

5.26 A rectangular footing of size 3 × 8 m is founded on the top of a slope of cohesive soil as given in Fig. Prob. 5.26. Determine the ultimate bearing capacity of the footing.

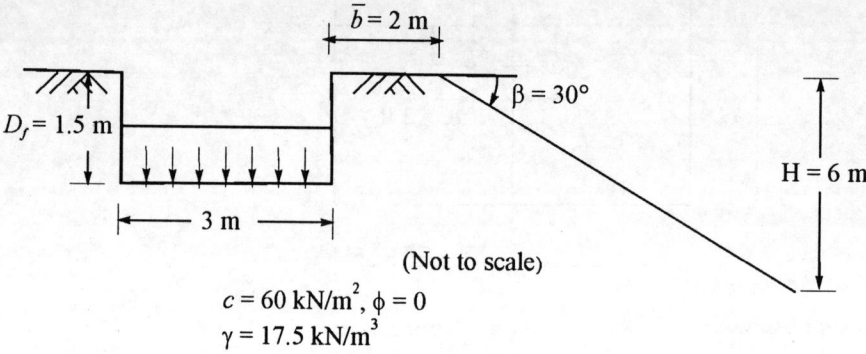

$c = 60 \text{ kN/m}^2, \phi = 0$

$\gamma = 17.5 \text{ kN/m}^3$

Fig. Prob. 5.26

Shallow Foundation 3

Safe Bearing Pressure and Settlement Calculation

6.1 INTRODUCTION

Allowable and Safe Bearing Pressures

The methods of calculating the ultimate bearing capacity of soil have been discussed at length in Chapter 5. The theories used in that chapter are based on shear failure criteria. They do not indicate the settlement that a footing may undergo under the ultimate loading conditions. From the known ultimate bearing capacity obtained from any one of the theories, the allowable bearing pressure can be obtained by applying a suitable factor of safety to the ultimate value.

When we design a foundation, we must see that the structure is safe on two counts. They are:

1. The supporting soil should be safe from shear failure due to the loads imposed on it by the superstructure.
2. The settlement of the foundation should be within permissible limits.

Hence, we have to deal with two types of bearing pressures. They are:

1. A pressure that is safe from shear failure criteria.
2. A pressure that is safe from settlement criteria.

For the sake of convenience, let us call the first the *allowable bearing pressure* and the second the *safe bearing pressure.*

In all our design, we use only the net bearing pressure and as such we call q_{na} the *net allowable bearing pressure* and q_s the net safe bearing pressure. In designing a foundation, we use the least of the two bearing pressures. In Chapter 5 we learnt that q_{na} is obtained by applying a suitable factor of safety (normally 3) to the net ultimate bearing capacity of soil. In this chapter we will learn how to obtain q_s. Even without knowing the values of q_{na} and q_s, it is possible to say from experience which of the two values should *be used in design* based upon the composition and density of soil and the size of the footing. The composition and density of the soil and the size of the footing decide the relative values of q_{na} and q_s.

The ultimate bearing capacity of footings on sand increases with an increase in the width, and in the same way the settlement of the footing increases with increases in the width. In other words for a given settlement S_1, the corresponding unit soil pressure decreases with an increase in the width of the footing. It is therefore, essential to consider that settlement will be the criterion for the design of footings in sand beyond a particular size. Experimental evidence indicates that for footings smaller than about 1.20 m, the allowable bearing pressure q_{na} is the criterion for the design of footings, whereas settlement is the criterion for footings greater than 1.2 m width.

The bearing capacity of footings on clay is independent of the size of the footings and as such the unit bearing pressure remains theoretically constant in a particular environment. However, the settlement of the footing increases with an increase in the size. It is essential to take into consideration both the shear failure and the settlement criteria together to decide the safe bearing pressure.

However, footings on stiff clay, hard clay, and other firm soils generally require no settlement analysis if the design provides a minimum factor of safety of 3 on the net ultimate bearing capacity of the soil. Soft clay, compressible silt, and other weak soils will settle even under moderate pressure, and therefore settlement analysis is necessary.

Effect of Settlement on the Structure

If the structure as a whole settles uniformly into the ground there will not be any detrimental effect on the structure as such. The only effect it can have is on the service lines, such as water and sanitary pipe connections, telephone and electric cables, etc. which can break if the settlement is considerable. Such uniform settlement is possible only if the subsoil is homogeneous and the load distribution is uniform. Buildings in Mexico City have undergone settlements as large as 2 m. However, the differential settlement if it exceeds the permissible limits will have a devastating effect on the structure.

According to experience, the differential settlement between parts of a structure may not exceed 75 percent of the normal absolute settlement. The various ways by which differential settlements may occur in a structure are shown in Fig. 6.1. Table 6.1 gives the absolute and permissible differential settlements for various types of structures.

Foundation settlements must be estimated with great care for buildings, bridges, towers, power plants and similar high cost structures. The settlements for structures such as fills, earthdams, levees, etc. can be estimated with a greater margin of error.

Approaches for Determining the Net Safe Bearing Pressure

Three approaches may be considered for determining the net safe bearing pressure of soil. They are:

1. Field plate load tests,
2. Charts,
3. Empirical equations.

6.2 FIELD PLATE LOAD TESTS

The plate load test is a semidirect method to estimate the allowable bearing pressure of soil to induce a given amount of settlement. Plates, round or square, varying in size, from 30 to 60 cm and thickness of about 2.5 cm are employed for the test.

The load on the plate is applied by making use of a hydraulic jack. The reaction of the jack load is taken by a cross beam or a steel truss anchored suitably at both the ends. The settlement of the plate is measured by a set of three dial gauges of sensitivity 0.02 mm placed 120° apart. The dial gauges are fixed to independent supports which remain undisturbed during the test.

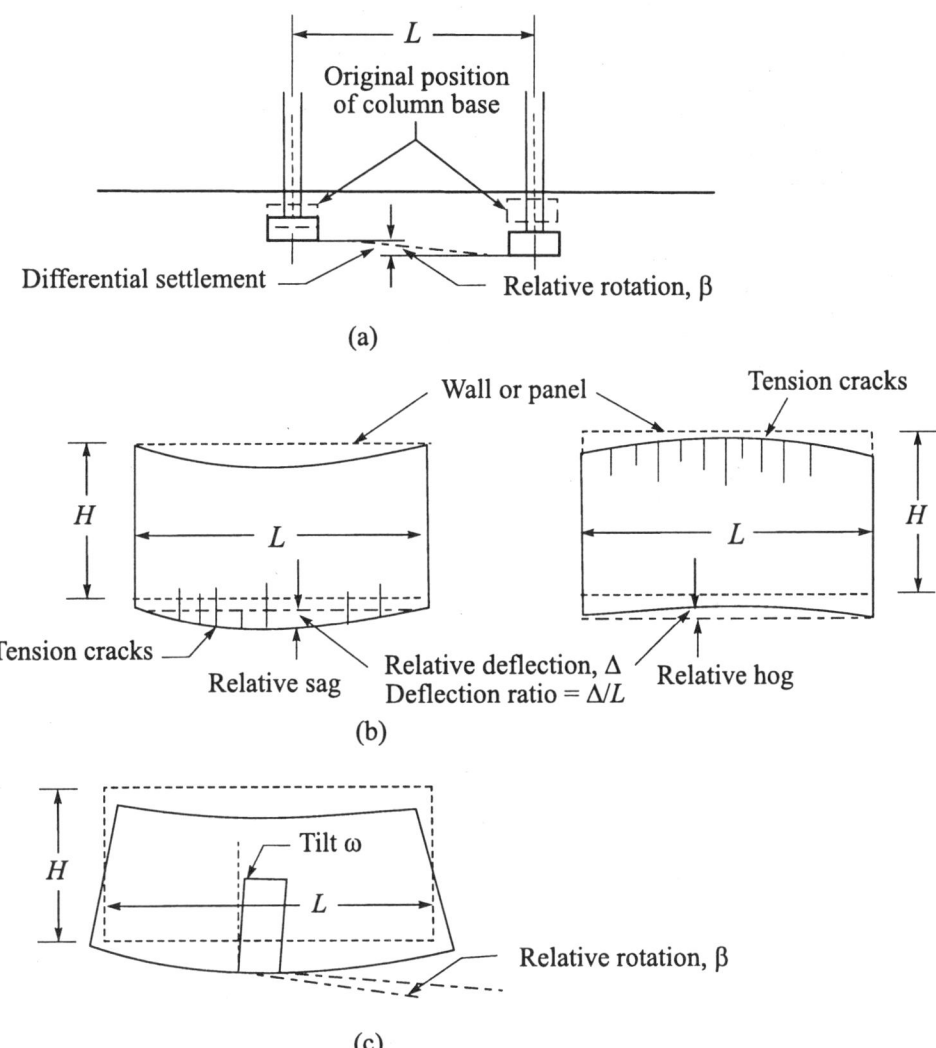

Fig. 6.1 Definitions of differential settlement for framed and load-bearing wall structures (after Burland and Wroth, 1974)

Table 6.1 (a) Maximum settlements and differential settlements of buildings in cm. (After McDonald and Skempton, 1955)

Sl. no.	Criterion	Isolated foundations	Raft
1.	Angular distortion	1/300	1/300
2.	Greatest differential settlements		
	Clays	4–5	4.5
	Sands	3–25	3.25
3.	Maximum settlements		
	Clays	7.5	10.0
	Sands	5.0	6.25

Table 6.1 (b) Permissible settlements (1955, USSR Building Code)

Sl. no.	Type of building	Average settlement (cm)
1.	Building with plain brickwalls on continuous and separate foundations with wall length L to wall height H	
	$L/H \geq 2.5$	7.5
	$L/H \leq 1.5$	10.0
2.	Framed building	10.0
3.	Solid reinforced concrete foundation of blast furnaces, water towers, etc.	30

Table 6.1 (c) Permissible differential settlement (USSR Building Code, 1955)

Sl. no.	Type of structure	Type of soil	
		Sand and hard clay	Plastic clay
1.	Steel and reinforced concrete structures	0.002 L	0.002 L
2.	Plain brick walls in multistorey buildings		
	for $L/H \leq 3$	0.0003 L	0.0004 L
	$L/H \geq 5$	0.0005 L	0.0007 L
3.	Water towers, silos, etc.	0.004 L	0.004 L
4.	Slope of crane way as well as track		
	for bridge crane track	0.003 L	0.003 L

where, L = distance between two columns or parts of structure that settle different amounts, H = height of wall.

Figure 6.2 (a) shows the arrangement for a plate load test. The method of performing the test is essentially as follows:

1. Excavate a pit of size not less than 4 to 5 times the size of the plate. The bottom of the pit should coincide with the level of the foundation.

2. If the water table is above the level of the foundation, pump out the water carefully and keep it at the level of the foundation.

3. A suitable size of plate is selected for the test. Normally, a plate of size 30 cm is used in sandy soils and a larger size in clay soils. The ground should be levelled and the plate should be seated over the ground.

4. A seating load of about 70 gm/cm^2 is first applied and released after sometime. A higher load is next placed on the plate and settlements are recorded by means of the dial gauges. Observations on every load increment shall be taken until the rate of settlement is less than 0.25 mm per hour. Load increments shall be approximately one-fifth of the estimated safe bearing capacity of the soil. The average of the settlements recorded by 2 or 3 dial gauges shall be taken as the settlement of the plate for each of the load increments.

5. The test should continue until a total settlement of 2.5 cm or the settlement at which the soil fails, whichever is earlier, is obtained. After the load is released, the elastic rebound of the soil should be recorded.

From the test results, a load-settlement curve should be plotted as shown in Fig. 6.2 (b). The allowable pressure on a prototype foundation for an assumed settlement may be found by making

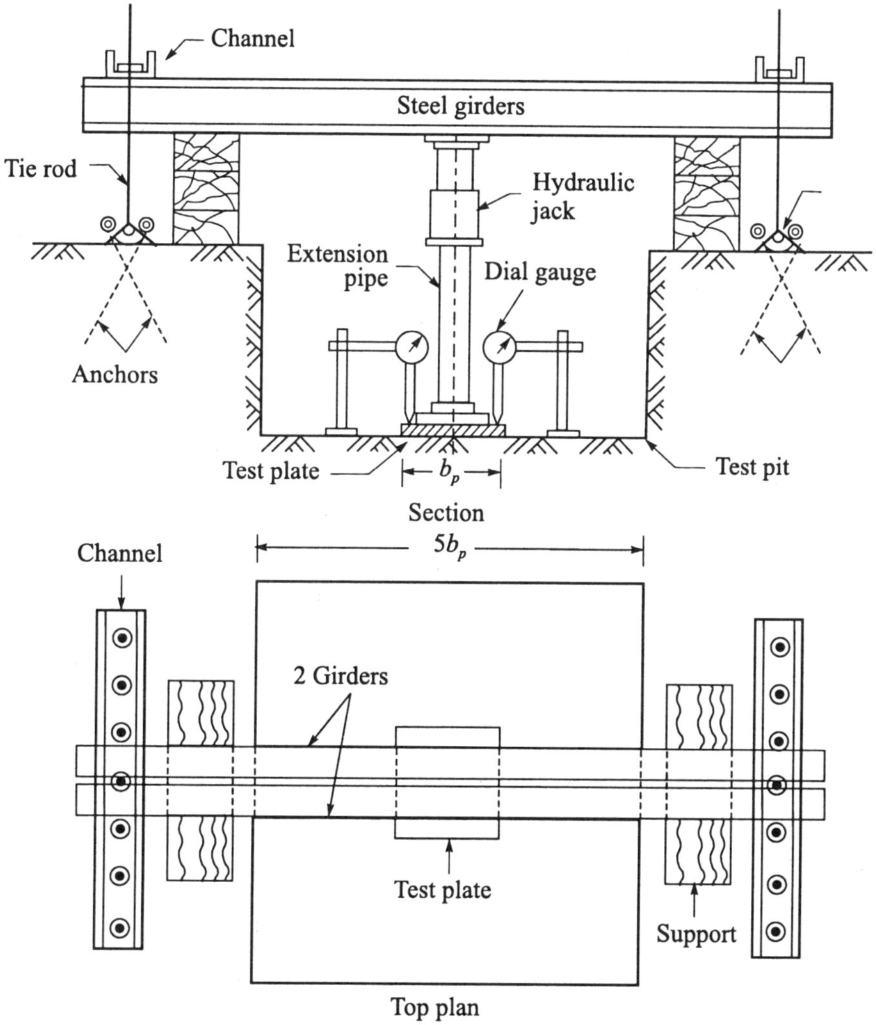

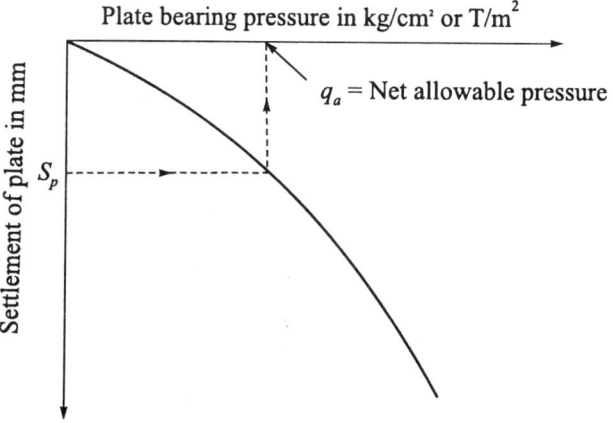

Fig. 6.2 (a) Plate load test arrangement

Fig. 6.2 (b) Load-settlement curve of a plate-load test

use of the following equations suggested by Terzaghi and Peck (1948) for square footings in granular soils.

$$S_f = S_p \left[\frac{B \left(b_p + 0.3 \right)}{b_p \left(B + 0.3 \right)} \right]^2 \qquad (6.1a)$$

in clay soils,

$$S_f = S_p \times \frac{B}{b_p} \qquad (6.1b)$$

where S_f = permissible settlement of foundation in mm,
   S_p = settlement of plate in mm,
   B = size of foundation in metres,
   b_p = size of plate in metres.

For a plate 1 ft square, Eq. (6.1a) may be expressed as

$$S_f = S_p \left(\frac{2B}{B+1} \right)^2 \qquad (6.2)$$

in which S_f and S_p are expressed in inches and B in feet.

The permissible settlement S_f for a prototype foundation should be known. Normally, a settlement of 2.5 cm is recommended. In Eqs (6.1a) or (6.2) the values S_f and b_p are known. The unknowns are S_p and B. The value of S_p for any assumed size B may be found from the equation. Using the plate load settlement curve Fig. 6.3, the value of the bearing pressure corresponding to the computed value of S_p is found. This bearing pressure is the safe bearing pressure for a given permissible settlement S_f. The principal shortcoming of this approach is the unreliability of the extrapolation of Eqs (6.1a) or (6.2).

Since a load test is of short duration, consolidation settlements cannot be predicted. The test gives the value of immediate settlement only. If the underlying soil is sandy in nature immediate settlement may be taken as the total settlement. If the soil is a clayey type, the immediate settlement is only a fraction of the total settlement. Load tests, therefore, do not have much significance in clayey soils to determine allowable pressure on the basis of a settlement criterion.

Plate load tests should be used with caution and the present practice is not to rely too much on this test. If the soil is not homogeneous to a great depth, plate load tests give very misleading results.

Assume, as shown in Fig. 6.2 (c), two layers of soil. The top layer is stiff clay whereas the bottom layer is soft clay. The load test conducted near the surface of the ground measures the characteristics of the stiff clay but does not indicate the nature of the soft clay soil which is below. The actual foundation of a building however has a bulb of pressure which extends to a great depth into the poor soil which is highly compressible. Here the soil tested by the plate load test gives results which are highly on the unsafe side.

A plate load test is not recommended in soils which are not homogeneous at least to a depth equal to 1½ to 2 times the width of the prototype foundation.

Plate load tests should not be relied on to determine the ultimate bearing capacity of sandy soils as the scale effect gives very misleading results. However, when the tests are carried on clay soils, the ultimate bearing capacity as determined by the test may be taken as equal to that of the foundation since the bearing capacity of clay is essentially independent of the footing size.

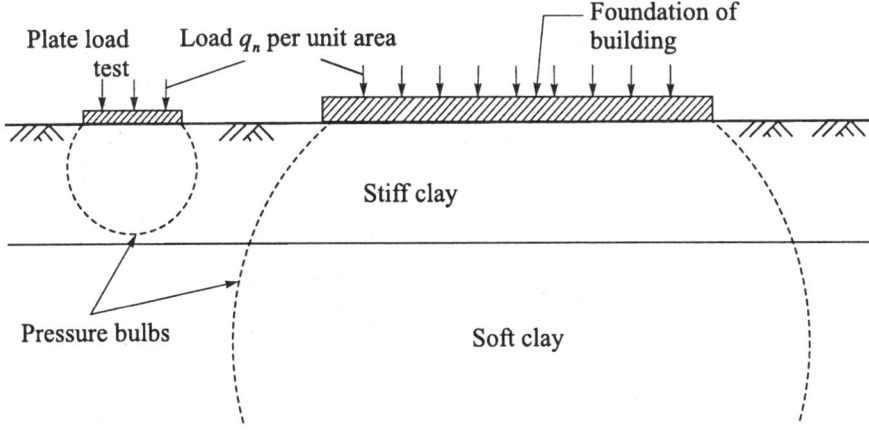

Fig. 6.2 (c) Plate load test on non-homogeneous soil

Housel's (1929), Method of Determining Safe Bearing Pressure from Settlement Consideration

The method suggested by Housel for determining the safe bearing pressure on settlement consideration is based on the following formula

$$Q = A_p m + P_p n \tag{6.3}$$

where Q = load applied on a given plate,
A_p = contact area of plate,
P_p = perimeter of plate,
m = a constant corresponding to the bearing pressure,
n = another constant corresponding to perimeter shear.

Objective

To determine the load Q_f and the size of a foundation for a permissible settlement S_f.

Housel suggests two plate load tests with plates of different sizes, say $B_1 \times B_1$ and $B_2 \times B_2$ for this purpose.

Procedure

1. Two plate load tests are to be conducted at the foundation level of the prototype as per the procedure explained earlier.
2. Draw the load-settlement curves for each of the plate load tests.
3. Select the permissible settlement S_f for the foundation.
4. Determine the loads Q_1 and Q_2 from each of the curves for the given permissible settlement S_f.

Now we may write the following equations

$$Q_1 = mA_{p1} + nP_{p1} \tag{6.4a}$$

for plate load test 1.

$$Q_2 = mA_{p2} + nP_{p2} \tag{6.4b}$$

for plate load test 2.

The unknown values of m and n can be found by solving the above Eqs (6.4a) and (6.5b). The equation for a prototype foundation may be written as

$$Q_f = mA_f + nP_f \tag{6.5}$$

where A_f = area of the foundation,

P_f = perimeter of the foundation.

When A_f and P_f are known, the size of the foundation can be determined.

Example 6.1

A plate load test using a plate of size 30 × 30 cm was carried out at the level of a prototype foundation. The soil at the site was cohesionless with the water table at great depth. The plate settled by 10 mm at a load intensity of 160 kN/m². Determine the settlement of a square footing of size 2 × 2 m under the same load intensity.

Solution

The settlement of the foundation S_f may be determined from Eq. (6.1a).

$$S_f = S_p \left[\frac{B\left(b_p + 0.3\right)}{b_p\left(B + 0.3\right)} \right]^2 = 10 \left[\frac{2\left(0.3 + 0.3\right)}{0.3\left(2 + 0.3\right)} \right]^2 = 30.24 \text{ mm}$$

Example 6.2

For Ex. 6.1 estimate the load intensity if the permissible settlement of the prototype foundation is limited to 40 mm.

Solution

In Ex. 6.1, a load intensity of 160 kN/m² induces a settlement of 30.24 mm. If we assume that the load-settlement is linear within a small range, we may write

$$\frac{q_2}{q_1} = \frac{S_{f2}}{S_{f1}} \text{ or } q_2 = q_1 \times \frac{S_{f2}}{S_{f1}}$$

where, q_1 = 160 kN/m²,

S_{f1} = 30.24 mm,

S_{f2} = 40 mm.

Substituting the known values

$$q_2 = 160 \times \frac{40}{30.24} = 211.64 \text{ kN/m}^2$$

Example 6.3

Two plate load tests were conducted at the level of a prototype foundation in cohesionless soil close to each other. The following data are given:

Size of plate	Load applied	Settlement recorded
0.3 × 0.3 m	30 kN	25 mm
0.6 × 0.6 m	90 kN	25 mm

If a footing is to carry a load of 1000 kN, determine the required size of the footing for the same settlement of 25 mm.

Solution

Use Eq. (6.3). For the two plate load tests we may write:

$$Q_1 = A_{p1}m + P_{p1}n$$

$$Q_2 = A_{p2}m + P_{p2}n$$

$PLT1$: $A_{p1} = 0.3 \times 0.3 = 0.09$ m^2; $\quad P_{p1} = 0.3 \times 4 = 1.2$ m; $\quad Q_1 = 30$ kN

$PLT2$: $A_{p2} = 0.6 \times 0.6 = 0.36$ m^2; $\quad P_{p2} = 0.6 \times 4 = 2.4$ m; $\quad Q_2 = 90$ kN

Now we have

$$30 = 0.09\, m + 1.2\, n$$

$$90 = 0.36\, m + 2.4\, n$$

On solving the equations, we have

$$m = 166.67, \text{ and } n = 12.5$$

For prototype foundation, we may write

$$Q_f = 166.67 A_f + 12.5 P_f$$

Assume the size of the footing as $B \times B$, we have

$$A_f = B^2, P_f = 4B \text{ and } Q_f = 1000 \text{ kN}$$

Substituting, we have

$$1000 = 166.67 B^2 + 50B$$

or $\quad\quad B^2 + 0.3B - 6 = 0$

The solution gives $B = 2.3$ m

The size of the footing = 2.3×2.3 m

6.3 EFFECT OF SIZE OF FOOTINGS ON SETTLEMENT

Figure 6.3 (a) gives typical load-settlement relationships for footings of different widths on the surface of a homogeneous sand deposit. It can be seen that the ultimate bearing capacities of the footings per unit area increase with the increase in the widths of the footings. However, for a given settlement S, such as 25 mm, the soil pressure is greater for a footing of intermediate width B_b than for a large footing with B_c. The pressures corresponding to the three widths intermediate, large and narrow, are indicated by points b, c and a respectively.

The same data is used to plot Fig. 6.3 (b) which shows the pressure per unit area corresponding to a given settlement S_1, as a function of the width of the footing. The soil pressure for settlement S_1 increases for increasing width of the footing, if the footings are relatively small, reaches a maximum at an intermediate width, and then decreases gradually with increasing width.

Although the relation shown in Fig. 6.3 (b) is generally valid for the behaviour of footings on sand, it is influenced by several factors including the relative density of sand, the depth at which the foundation is established, and the position of the water table. Furthermore, the shape of the curve suggests that for narrow footings small variations in the actual pressure, Fig. 6.3 (a), may lead to large variation in settlement and in some instances to settlements so large that the movement would be considered a bearing capacity failure. On the other hand, a small change in pressure on a wide footing has little influence on settlements as small as S_1, and besides, the value of q_1 corresponding to S_1 is far below that which produces a bearing capacity failure of the wide footing.

For all practical purposes, the actual curve given in Fig. 6.3 (b) can be replaced by an equivalent curve *omn* where *om* is the inclined part and *mn* the horizontal part. The horizontal portion of the curve indicates that the soil pressure corresponding to a settlement S_1 is independent of the size of the footing. The inclined portion *om* indicates the pressure increasing with width for the same given settlement S_1 up to the point *m* on the curve which is the limit for a bearing capacity failure. This means that the footings up to size B_1 in Fig. 6.3 (b) should be checked for bearing capacity failure also while selecting a safe bearing pressure by settlement consideration.

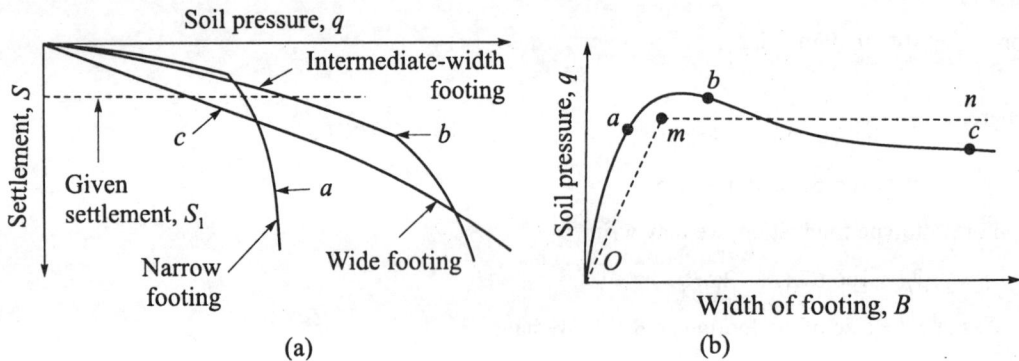

Fig. 6.3 Load-settlement curves for footings of different sizes (Peck *et al*, 1974)

The position of the broken lines *omn* differs for different sand densities or in other words for different SPT N values. The soil pressure that produces a given settlement S_1 on loose sand is obviously smaller than the soil pressure that produces the same settlement on a dense sand. Since N-value increases with density of sand, q_s therefore increases with an increase in the value of N.

6.4   DESIGN CHARTS FROM SPT VALUES FOR FOOTINGS ON SAND

The methods suggested by Terzaghi *et al*, (1996), for estimating settlements and bearing pressures of footings founded on sand from SPT values are based on the findings of Burland and Burbidge (1985). The SPT values used are corrected to a standard energy ratio. The usual symbol N_{cor} is used in all the cases as the corrected value.

Formulas for Settlement Calculations

The following formulas were developed for computing settlements for square footings.

For normally consolidated soils and gravels

$$S_c = B^{0.75}\,\frac{1.7\,q_s}{N_{cor}^{1.4}} \tag{6.6}$$

For preconsolidated sand and gravels

for $\qquad q_s > p_c \qquad S_c = B^{0.75}\,\frac{1.7}{N_{cor}^{1.4}}\,(q_s - 0.67p_c) \tag{6.7a}$

for $\qquad q_s < p_c \qquad S_c = \dfrac{1}{3} B^{0.75} \dfrac{1.7 q_s}{N_{cor}^{1.4}}$ $\qquad\qquad$ (6.7b)

If the footing is established at a depth below the ground surface, the removal of the soil above the base level makes the sand below the base preconsolidated by excavation. Recompression is assumed for bearing pressures up to preconstruction effective vertical stress q'_o at the base of the foundation. Thus, for sands normally consolidated with respect to the original ground surface and for values of q_s greater than q'_o, we have

for $\qquad q_s > q'_o \qquad S_c = B^{0.75} \dfrac{1.7}{N_{cor}^{1.4}} (q_s - 0.67 q'_o)$ $\qquad\qquad$ (6.8a)

for $\qquad q_s < p_c \qquad S_c = \dfrac{1}{3} B^{0.75} \dfrac{1.7 q_s}{N_{cor}^{1.4}} q_s$ $\qquad\qquad$ (6.8b)

where, $\quad S_c$ = settlement of footing, in mm, at the end of construction and application of permanent live load,

B = width of footing in m,

q_s = gross bearing pressure of footing = Q/A, in kN/m² based on settlement consideration,

Q = total load on the foundation in kN,

A = area of foundation in m²,

p_c = preconsolidation pressure in kN/m²,

q'_o = effective vertical pressure at base level,

N_{cor} = average corrected N value within the depth of influence Z_I below the base of the footing.

The depth of influence Z_I is obtained from

$$Z_I = B^{0.75} \qquad\qquad (6.9)$$

Figure 6.4 gives the variation of the depth of influence with depth based on Eq. (6.9) (after Burland and Burbidge, 1985).

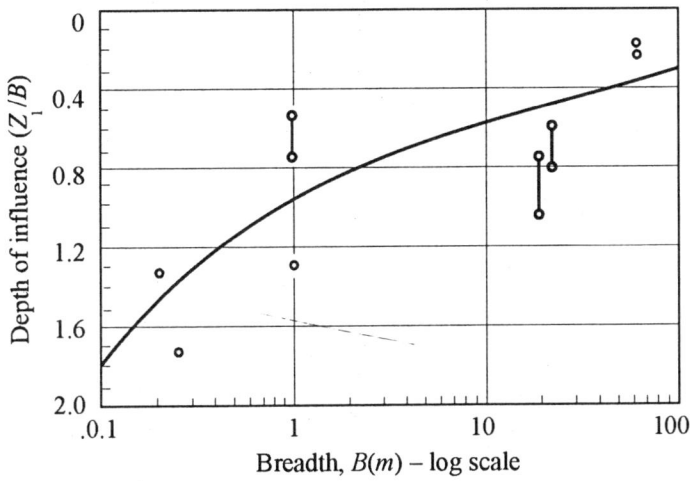

Fig. 6.4 Thickness of granular soil beneath foundation contributing to settlement, interpreted from settlement profiles (after Burland and Burbidge, 1985)

The settlement of a rectangular footing of size $B \times L$ may be obtained from

$$S_c (L/B > 1) = S_c \left(\frac{L}{B} = 1 \right) \left[\frac{1.25 (L/B)}{L/B + 0.25} \right]^2 \qquad (6.10)$$

when the ratio L/B is very high for a strip footing, we may write

$$\frac{S_c \,(\text{strip})}{S_c \,(\text{square})} = 1.56 \qquad (6.11)$$

It may be noted here that the ground water table at the site may lie above or within the depth of influence Z_I. Burland and Burbidge (1985), recommend no correction for the settlement calculation even if the water table lies within the depth of influence Z_I. On the other hand, if for any reason, the water table were to rise into or above the zone of influence Z_I after the penetration tests were conducted, the actual settlement could be as much as twice the value predicted without taking the water table into account.

Chart for Estimating Allowable Soil Pressure

Figure 6.5 gives a chart for estimating allowable bearing pressure q_s (on settlement consideration) corresponding to a settlement of 16 mm for different values of N (corrected). From Eq. (6.6), an expression for q_s may be written as (for normally consolidated sand)

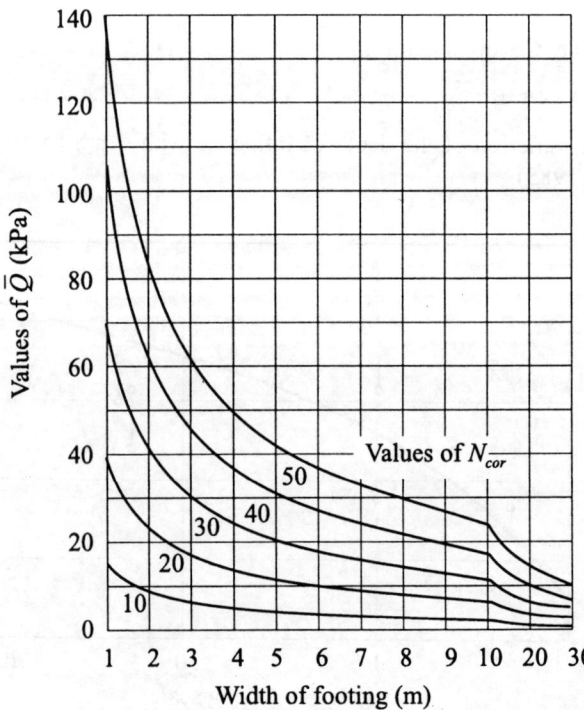

Fig. 6.5 Chart for estimating allowable soil pressure for footing on sand on the basis of results of standard penetration test (Terzaghi *et al*, 1996). This chart replace the chart proposed in 1974.

$$q_s = S_c \frac{N_{cor}^{1.4}}{1.7 B^{0.75}} = 16 \frac{N_{cor}^{1.4}}{1.7 B^{0.75}} = 16 \overline{Q} \qquad (6.12a)$$

where
$$\overline{Q} = \frac{N_{cor}^{1.4}}{1.7 B^{0.75}} \qquad (6.12b)$$

For sand having a preconsolidation pressure p_c; Eq. (6.7) may be written as

for $\qquad q_s > p_c \quad q_s = 16 \overline{Q} + 0.67 p_c \qquad (6.13a)$

for $\qquad q_s < p_c \quad q_s = 3 \times 16 \overline{Q} \qquad (6.13b)$

If the *s* and beneath the base of the footing is preconsolidated because excavation has removed a vertical effective stress q'_o Eq. (6.8) may be written as

for $\qquad q_s > q'_o \quad q_s = 16 \overline{Q} + 0.67 q'_o \qquad (6.14a)$

for $\qquad q_s < q'_o \quad q_s = 3 \times 16 \overline{Q} \qquad (6.14b)$

The chart in Fig. 6.5 gives the relationships between B and $\overline{Q}$. The value of q_s may be obtained from $\overline{Q}$ for any given width B. The $\overline{Q}$ to be used must conform to Eqs (6.12), (6.13) and (6.14).

The chart is constructed for square footings of width B. For rectangular footings, the value of q_s should be reduced in accordance with Eq. (6.10). The bearing pressures determined by this procedure correspond to a maximum settlement of 25 mm at the end of construction.

It may be noted here that the design chart Fig. 6.5 has been developed by taking the SPT values corrected for 60 percent of standard energy ratio.

Example 6.4

A square footing of size 4 × 4 m is founded at a depth of 2 m below the ground surface in loose to medium dense sand. The corrected standard penetration test value $N_{cor} = 11$. Compute the safe bearing pressure q_s by using the chart in Fig. 6.5. The water table is at the base level of the foundation.

Solution

From Fig. 6.5

$\qquad \overline{Q} = 5$ for $B = 4$ and $N_{cor} = 11$.

From Eq. (6.12a)

$\qquad q_s = 16 \overline{Q} = 16 \times 5 = 80$ kN/m^2

Example 6.5

Refer to Ex. 6.4. If the soil at the site is dense sand with $N_{cor} = 30$, estimate q_s for $B = 4$ m.

Solution

From Fig. 6.5

$\qquad \overline{Q} = 24$ for $B = 4$ m and $N_{cor} = 30$.

From Eq. (6.12a)

$\qquad q_s = 16 \overline{Q} = 16 \times 24 = 384$ kN/m^2

6.5 EMPIRICAL EQUATIONS BASED ON SPT VALUES FOR FOOTINGS ON COHESIONLESS SOILS

Footings on granular soils are sometimes proportioned using empirical relationships. Teng (1969), proposed an equation for a settlement of 25 mm based on the curves developed by Terzaghi and Peck (1948). The modified form of the equation is

$$q_s = 35 (N_{cor} - 3) \left(\frac{B + 0.3}{2B} \right)^2 R_{w2} F_d \, kN/m^2 \tag{6.15a}$$

where q_s = net allowable bearing pressure for a settlement of 25 mm in kN/m^2,

N_{cor} = corrected standard penetration value,

R_{w2} = water table correction factor (Refer Chapter 5),

F_d = depth factor = $(1 + D_f /B) \leq 2.0$,

B = width of footing in metres,

D_f = depth of foundation in metres.

Meyerhof (1956) proposed the following equations which are slightly different from that of Teng

$$q_s = 12 N_{cor} R_{w2} F_d \text{ for } B \leq 1.2 \text{ m} \tag{6.15b}$$

$$q_s = 8 N_{cor} \left(\frac{B + 0.3}{B} \right)^2 R_{w2} F_d \text{ for } B > 1.2 \text{ m} \tag{6.15c}$$

where $F_d = (1 + 0.33 \, D_f /B) \leq 1.33$.

Experimental results indicate that the equations presented by Teng and Meyerhof are too conservative. Bowles (1996) proposes an approximate increase of 50 percent over that of Meyerhof, which can also be applied to Teng's equations. Modified equations of Teng and Meyerhof are,

Teng's equation (modified),

$$q_s = 53 (N_{cor} - 3) \left(\frac{B + 0.3}{2} \right)^2 R_{w2} F_d \tag{6.16a}$$

Meyerhof's equation (modified),

$$q_s = 20 N_{cor} R_{w2} F_d \text{ for } B \leq 1.2 \text{ m} \tag{6.16b}$$

$$q_s = 12.5 N_{cor} \left(\frac{B + 0.3}{B} \right)^2 R_{w2} F_d \text{ for } B \geq 1.2 \text{ m} \tag{6.16c}$$

If the tolerable settlement is greater than 25 mm, the safe bearing pressure computed by the above equations can be increased linearly as,

$$q_s' = \frac{S'}{25} q_s \tag{6.16d}$$

where, q_s' = net safe bearing pressure for a settlement S' mm, q_s = net safe bearing pressure for a settlement of 25 mm.

6.6 SAFE BEARING PRESSURE FROM EMPIRICAL EQUATIONS BASED ON CPT VALUES FOR FOOTINGS ON COHESIONLESS SOIL

The static cone penetration test in which a standard cone of 10 cm² sectional area is pushed into the soil without the necessity of boring provides a much more accurate and detailed variation in the soil as discussed in Chapter 3. Meyerhof (1956) suggested a set of empirical equations based on the Terzaghi and Peck curves (1948). As these equations were also found to be conservative, modified forms with an increase of 50 percent over the original values are given below.

$$q_s = 3.6q_c\,R_{w2}\ \text{kPa for } B \le 1.2\ \text{m} \tag{6.17a}$$

$$q_s = 2.1q_c\left(1+\frac{1}{B}\right)^2 R_{w2}\ \text{kPa for } B > 1.2\ \text{m} \tag{6.17b}$$

An approximate formula for all widths

$$q_s = 2.7q_c\,R_{w2}\ \text{kPa} \tag{6.17c}$$

where q_c is the cone point resistance in kg/cm² and q_s in kPa.

The above equations have been developed for a settlement of 25 mm.

Meyerhof (1956) developed his equations based on the relationship $q_c = 4N_{cor}$ kg/cm² for penetration resistance in sand where N_{cor} is the corrected SPT value.

Example 6.6

Refer to Ex. 6.4 and compute q_s by modified (*a*) Teng's method, and (*b*) Meyerhof's method.

Solution

(*a*) Teng's equation (modified)—Eq. (6.16a)

$$q_s = 53\,(N_{cor}-3)\left(\frac{B+0.3}{2B}\right)^2 R_{w2}\,F_d\ \text{kN/m}^2$$

where $\qquad R_{w2} = \dfrac{1}{2}\left(1+\dfrac{D_{w2}}{B}\right) = 0.5$ since $D_{w2}=0$

$$F_d = \left(1+\frac{D_f}{B}\right) = \left(1+\frac{2}{4}\right) = 1.5 < 2$$

$$N_{cor} = 11,\ B = 4\ \text{m}.$$

By substituting

$$q_s = 53\,(11-3)\left(\frac{4.3}{8}\right)^2 \times 0.5 \times 1.5 = 92\ \text{kN/m}^2$$

(*b*) Meyerhof's equation (modified)—Eq. (6.16c)

$$q_s = 12.5N\left(\frac{B+0.3}{B}\right)^2 R_{w2}\,F_d$$

where $\quad R_{w2} = 0.5, F_d = \left(1 + 0.33 \times \dfrac{D_f}{B}\right) = \left(1 + 0.33 \times \dfrac{2}{4}\right) = 1.165 < 1.33$

By substituting

$$q_s = 12.5 \times 11 \left(\dfrac{4.3}{4}\right)^2 \times 0.5 \times 1.165 = 93 \text{ kN/m}^2$$

Note: Both the methods give the same result.

Example 6.7

A footing of size 3 × 3 m is to be constructed at a site at a depth of 1.5 m below the ground surface. The water table is at the base of the foundation. The average static cone penetration resistance obtained at the site is 20 kg/cm². The soil is cohesive. Determine the safe bearing pressure for a settlement of 40 mm.

Solution

Use Eq. (6.17b)

$$q_s = 2.1 q_c \left(1 + \dfrac{1}{B}\right)^2 R_{w2} \text{ kN/m}^2$$

where $\quad q_c = 20 \text{ kg/cm}^2, B = 3 \text{ m}, R_{w2} = 0.5.$

This equation is for 25 mm settlement. By substituting, we have

$$q_s = 2.1 \times 20 \left(1 + \dfrac{1}{3}\right)^2 \times 0.5 = 37.3 \text{ kN/m}^2$$

For 40 mm settlement, the value of q_s is

$$q_s = 37.3 \left(\dfrac{40}{25}\right) \approx 60 \text{ kN/m}^2$$

6.7 FOUNDATION SETTLEMENT

Components of Total Settlement

The total settlement of a foundation comprise three parts as follows

$$S = S_e + S_c + S_s \tag{6.18}$$

where $S = $ total settlement,
   $S_e = $ elastic or immediate settlement,
   $S_c = $ consolidation settlement,
   $S_s = $ secondary settlement,

Immediate settlement, S_e, is that part of the total settlement, S, which is supposed to take place during the application of loading. The consolidation settlement is that part which is due to the expulsion of pore water from the voids and is time-dependent settlement. Secondary settlement normally starts with the completion of the consolidation. It means, during the stage of this settlement, the pore water pressure is zero and the settlement is only due to the distortion of the soil skeleton.

Footing founded in cohesionless soils reach almost the final settlement, S, during the construction stage itself due to the high permeability of soil. The water in the voids is expelled simultaneously with the application of load and as such the immediate and consolidation settlements in such soils are rolled into one.

In cohesive soils under saturated conditions, there is no change in the water content during the stage of immediate settlement. The soil mass is deformed without any change in volume soon after the application of the load. This is die to the low permeability of the soil. With the advancement of time there will be gradual expulsion of water under the imposed excess load. The time required for the complete expulsion of water and to reach zero water pressure may be several years depending upon the permeability of the soil. Consolidation settlement may take many years to reach its final stage. Secondary settlement is supposed to take place after the completion of the consolidation settlement, though in some of the organic soils there will be overlapping of the two settlements to a certain extent.

Immediate settlements of cohesive soils and the total settlement of cohesionless soils may be estimated from elastic theory. The stresses and displacements depend on the stress–strain characteristics of the underlying soil. A realistic analysis is difficult because these characteristics are nonlinear. Results from the theory of elasticity are generally used in practice, it being assumed that the soil is homogeneous and isotropic and there is a linear relationship between stress and strain. A linear stress–strain relationship is approximately true when the stress levels are low relative to the failure values. The use of elastic theory clearly involves considerable simplification of the real soil.

Some of the results from elastic theory require knowledge of Young's modulus (E_s), here called the compression or deformation modulus, E_d, and Poisson's ratio μ, for the soil.

Seat of Settlement

Footings founded at a depth D_f below the surface settle under the imposed loads due to the compressibility characteristics of the subsoil. The depth through which the soil is compressed depends upon the distribution of effective vertical pressure p_o' of the overburden and the vertical induced stress Δ_p resulting from the net foundation pressure q_n as shown in Fig. 6.6.

In the case of deep compressible soils, the lowest level considered in the settlement analysis is the point where the vertical induced stress Δp is of the order of 0.1 to $0.2q_n$, where q_n is the net

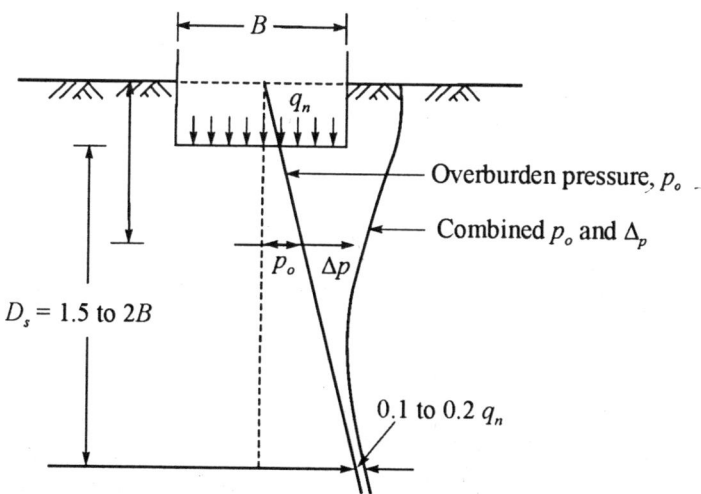

Fig. 6.6 Overburden pressure and vertical stress distribution

pressure on the foundation from the superstructure. This depth works out to about 1.5 to 2 times the width of the footing. The soil lying within this depth gets compressed due to the imposed foundation pressure and causes more than 80 percent of the settlement of the structure. This depth D_s is called as the *zone of significant stress*. If the thickness of this zone is more than 3 m, the steps to be followed in the settlement analysis are:

1. Divide the zone of significant stress into layers of thickness not exceeding 3 m.

2. Determine the effective overburden pressure p'_o at the centre of each layer.

3. Determine the increase in vertical stress Δp due to foundation pressure q_n at the centre of each layer along the centre line of the footing by the theory of elasticity.

4. Determine the average modulus of elasticity and other soil parameters for each of the layers.

6.8 EVALUATION OF MODULUS OF ELASTICITY

The most difficult part of a settlement analysis is the evaluation of the modulus of elasticity E_s, that would conform to the soil condition in the field. There are two methods by which E_s can be evaluated. They are:

1. Laboratory method.

2. Field method.

Laboratory Method

For settlement analysis the values of E_s at different depths below the foundation base are required. One way of determining E_s is to conduct triaxial tests on representative undisturbed samples extracted from the depths required. For cohesive soils, untrained triaxial tests and for cohesionless soils drained triaxial tests are required. Since, it is practically impossible to obtain undisturbed sample of cohesionless soils, the laboratory method of obtaining E_s can be ruled out. Even with regards to cohesive soils, there will be disturbance to the sample at different stages of handling it, and as such the values of E_s obtained from untrained triaxial tests do not represent the actual conditions and normally give very low values. A suggestion is to determine E_s over the range of stress relevant to the particular problem. Poulos *et al*, (1980), suggest that the undisturbed triaxial specimen be given a preliminary preconsolidation under K_0-conditions with an axial stress equal to the effective overburden pressure at the sampling depth. This procedure attempts to return the specimen to its original state of effective stress in the ground, assuming that the horizontal effective stress in the ground was the same as that produced by the laboratory K_0-condition. Simons and Som (1970) have shown that triaxial tests on London clay in which specimens were brought back to their original *in-situ* stresses gave elastic moduli which were much higher than those obtained from conventional untrained triaxial tests. This has been confirmed by Marsland (1971), who carried out 865 mm diameter plate loading tests in 900 mm diameter bored holes in London clay. Marsland found that the average moduli determined from the loading tests were between 1.8 to 4.8 times those obtained from undrained triaxial tests. A suggestion to obtain the more realistic value for E_s is,

1. Undisturbed samples obtained from the field must be reconsolidated under a stress system equal to that in the field (K_0-condition).

2. Samples must be reconsolidated isotropically to a stress equal to 1/2 to 2/3 of the *in-situ* vertical stress.

It may be noted here that reconsolidation of disturbed sensitive clays would lead to significant change in the water content and hence a stiffer structure which would lead to a very high E_s.

Because of the many difficulties faced in selecting a modulus value from the results of laboratory tests, it has been suggested that a correlation between the modulus of elasticity of soil and the undrained shear strength may provide a basis for settlement calculation. The modulus E_s may be expressed as

$$E_s = Ac_u \qquad (6.19)$$

where the value of A for inorganic stiff clay varies from about 500 to 1500 (Bjerrum, 1972), and c_u is the undrained cohesion. It may generally be assumed that highly plastic clays give lower values for A, and low plasticity give higher values for A. For organic or soft clays the value of A may vary from 100 to 500. The undrained cohesion c_u can be obtained from any one of the field tests mentioned below and also discussed in Chapter 3.

Field methods

Field methods are increasingly used to determine the soil strength parameters. They have been found to be more reliable than the ones obtained from laboratory tests. The field tests that are normally used for this purpose are:

1. Plate load tests (PLT).
2. Standard penetration test (SPT).
3. Static cone penetration test (CPT).
4. Pressuremeter test (PMT).
5. Flat dilatometer test (DMT).

Plate load tests, if conducted at levels at which E_s is required, give quite reliable values as compared to laboratory tests. Since these tests are too expensive to carry out, they are rarely used except in major projects.

Many investigators have obtained correlations between E_s and field tests such as SPT, CPT and PMT. The correlations between E_s and SPT or CPT are applicable mostly to cohesionless soils and in some cases cohesive soils under undrained conditions. PMT can be used for cohesive soils to determine both the immediate and consolidation settlements together.

Some of the correlations of E_s with N and q_c are given in Table 6.2. These correlations have been collected from various sources.

Table 6.2 Equations for computing E_s by making use of SPT and CPT values (in kPa)

Soil	SPT	CPT
Sand (normally consolidated)	$500 (N_{cor} + 15)$	2 to $4\, q_c$
	$(35{,}000 \text{ to } 50{,}000) \log N_{cor}$	$(1 + D_r^2)\, q_c$
	(USSR Practice)	
Sand (saturated)	$250 (N_{cor} + 15)$	
Sand (overconsolidated)	–	6 to $30\, q_c$
Gravelly sand and gravel	$1{,}200 (N_{cor} + 6)$	
Clayey sand	$320 (N_{cor} + 15)$	3 to $6\, q_c$
Silty sand	$300 (N_{cor} + 6)$	1 to $2\, q_c$
Soft clay	–	3 to $8\, q_c$

6.9 METHODS OF COMPUTING SETTLEMENTS

Many methods are available for computing elastic (immediate) and consolidation settlements. Only those methods that are of practical interest are discussed here. The various methods discussed in this chapter are the following:

Computation of Elastic Settlements

1. Elastic settlement based on the theory of elasticity.
2. Janbu *et al* (1956) method.
3. Schmertmann's method.
4. Pressuremeter method.

Computation of Consolidation Settlement

1. *e*-log *p* method by making use of oedometer test data.
2. Skempton-Bjerrum method.

6.10 ELASTIC SETTLEMENT BENEATH THE CORNER OF A UNIFORMLY LOADED FLEXIBLE AREA BASED ON THE THEORY OF ELASTICITY

The net elastic settlement equation for a flexible surface footing may be written as,

$$S_e = q_n B \frac{\left(1 - \mu^2\right)}{E_s} I_f \tag{6.20a}$$

where S_e = elastic settlement,
  B = width of foundation,
  E_s = modulus of elasticity of soil,
  μ = Poisson's ratio,
  q_n = net foundation pressure,
  I_f = influence factor.

In Eq. (6.20a), for saturated clays, $\mu = 0.5$, and E_s is to be obtained under undrained conditions as discussed earlier. For soils other than clays, the value of μ has to be chosen suitably and the corresponding value of E_s has to be determined. Table 6.3 gives typical values for μ as suggested by Bowles (1996).

Table 6.3 Typical range of values for Poisson's ratio (Bowles, 1996)

Type of soil	m
Clay, saturated	0.4−0.5
Clay, unsaturated	0.1−0.3
Sandy clay	0.2−0.3
Silt	0.3−0.35
Sand (dense)	0.2−0.4
Coarse (void ratio 0.4 to 0.7)	0.15
Fine grained (void ratio = 0.4 to 0.7)	0.25
Rock	0.1−0.4

I_f is a function of the L/B ratio of the foundation, and the thickness H of the compressible layer. Terzaghi has a given a method of calculating I_f from curves derived by Steinbrenner (1934),

for Poisson's ratio of 0.5, $I_f = F_1$,

for Poisson's ratio zero, $I_f = F_1 + F_2$

where F_1 and F_2 are factors which depend upon the ratios of H/B and L/B.

For intermediate values of μ, the value of I_f can be computed by means of interpolation or by the equation

$$I_f = \left[F_1 + \frac{\left(1 - \mu - 2\mu^2\right)F_2}{1 - \mu^2} \right] \qquad (6.20b)$$

The values of F_1 and F_2 are given in Fig. (6.7a). The elastic settlement at any point N Fig. (6.7b) is given by

$$S_e \text{ at point } N = \frac{q_n \left(1 - \mu^2\right)}{E_s} \left[I_{f1} B_1 + I_{f2} B_2 + I_{f3} B_3 + I_{f4} B_4 \right] \qquad (6.20c)$$

To obtain the settlement at the centre of the loaded area, the principle of superposition is followed. In such a case N in Fig. 6.7 (b) will be at the centre of the area when $B_1 = B_4 = L_2 = B_3$ and $B_2 = L_1$. Then the settlement at the centre is equal to four times the settlement at any one corner. The curves in Fig. 6.7 (a) are based on the assumption that the modulus of deformation is constant with depth.

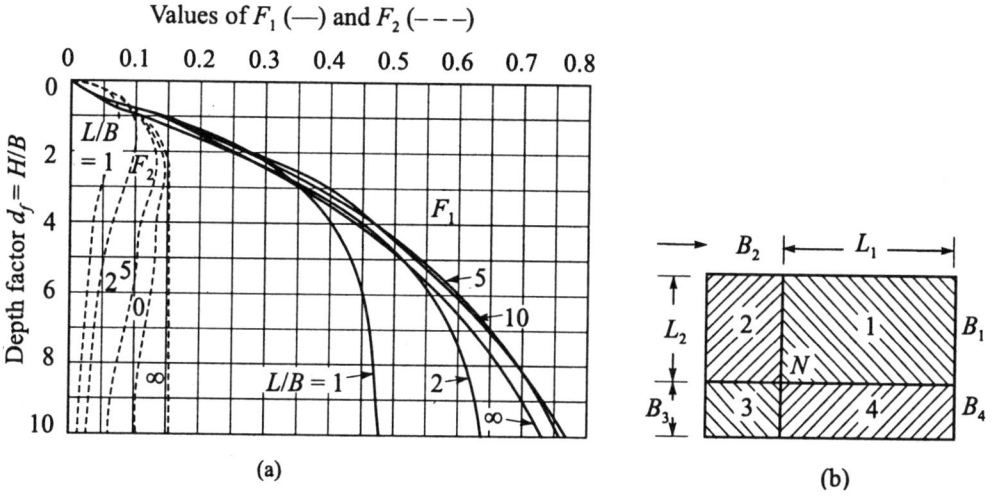

Values of F_1 (—) and F_2 (– – –)

Depth factor $d_f = H/B$

(a)   (b)

Fig. 6.7 Settlement due to load on surface of elastic layer: (a) F_1 and F_2 *versus* H/B, (b) method of estimating settlement (After Steinbrenner, 1934)

In the case of a rigid foundation, the immediate settlement at the centre is approximately 0.8 times that obtained for a flexible foundation at the centre. A correction factor is applied to the immediate settlement to allow for the depth of foundation by means of the depth factor d_f. Figure 6.8 gives Fox's (1948), correction curve for depth factor. The final elastic settlement is

$$S_{ef} = C_r d_f s_e \qquad (6.21)$$

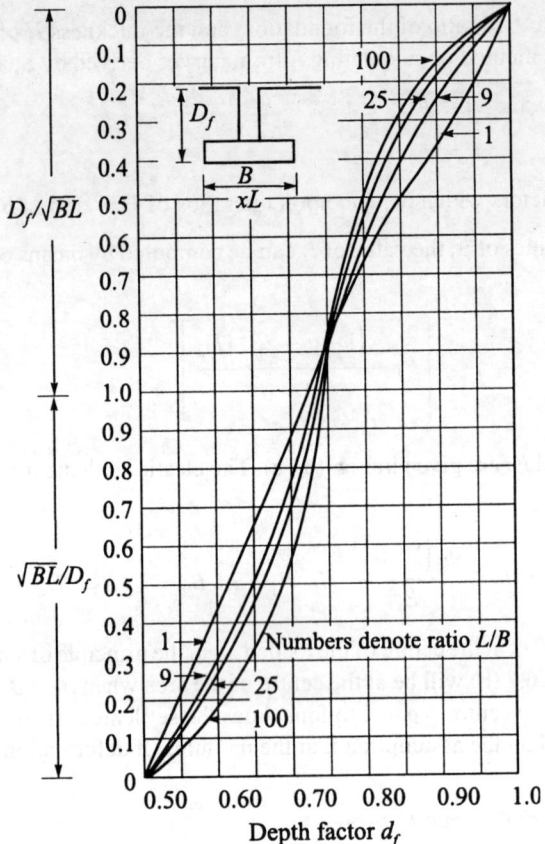

Fig. 6.8 Correction curves for elastic settlement of flexible rectangular foundations at depth (Fox, 1948)

where S_{ef} = final elastic settlement,

  C_r = rigidity factor taken as equal to 0.8 for a highly rigid foundation,

  d_f = depth factor from Fig. 6.8,

$$\text{Depth factor, } d_f = \frac{\text{Corrected settlement for foundation of depth } D_f}{\text{Calculated settlement for foundation at surface}}$$

  S_e = settlement for a surface flexible footing.

Bowles (1996) has given the influence factor for various shapes of rigid and flexible footings as shown in Table 6.4.

6.11 JANBU, BJERRUM AND KJAERNSLI'S METHOD OF DETERMINING ELASTIC SETTLEMENT UNDER UNDRAINED CONDITIONS

Probably the most useful chart is that given by Janbu *et al*, (1956) as modified by Christian and Carrier (1978) for the case of a constant E_s with respect to depth. Figure 6.9 provides estimates of the average immediate settlement of uniformly loaded, flexible strip, rectangular, square or circular footings on homogeneous isotropic saturated clay. The equation for computing the settlement may be expressed as

Table 6.4 Influence factor, I_f (Bowles, 1988)

Shape	I_f (average values)	
	Flexible footing	*Rigid footing*
Circle	0.85	0.88
Square	0.95	0.82
Rectangle	1.20	1.06
$L/B =$ 1.5	1.20	1.06
2.0	1.31	1.20
5.0	1.83	1.70
10.0	2.25	2.10
100.0	2.96	3.40

$$S_e = \frac{\mu_0 \mu_1 \, q_n B}{E} \tag{6.22}$$

In Eq. (6.20), Poisson's ratio is assumed equal to 0.5. The factors μ_0 and μ_1 are related to the D_f/B and H/B ratios of the foundation as shown in Fig. 6.9. Values of μ_1 are given for various L/B ratios. Rigidity and depth factors are required to be applied to Eq. (6.22) as per Eq. (6.21). In Fig. 6.9 the thickness of compressible strata is taken as equal to H below the base of the foundation where a hard stratum is met with.

Generally, real soil profiles which are deposited naturally consist of layers of soils of different properties underlain ultimately by a hard stratum. Within these layers, strength and moduli generally increase with depth. The chart given in Fig. 6.9 may be used for the case of E_s increasing with depth by replacing the multilayered system with one hypothetical layer on a rigid base. The depth of this hypothetical layer is successively extended to incorporate each real layer, the corresponding values of E_s being ascribed in each case and settlements calculated. By subtracting the effect of the hypothetical layer above each real layer the separate compression of each layer may be found and summed to give the overall total settlement.

6.12 SCHMERTMANN'S METHOD OF CALCULATING SETTLEMENT IN GRANULAR SOILS BY USING CPT VALUES

It is normally taken for granted that the distribution of vertical strain under the centre of a footing over uniform sand is qualitatively similar to the distribution of the increase in vertical stress. If true, the greatest strain would occur immediately under the footing, which is the position of the greatest stress increase. The detailed investigations of Schmertmann (1970), Eggestad, (1963), and others, indicate that the greatest strain would occur at a depth equal to half the width for a square or circular footing. The strain is assumed to increase from a minimum at the base to a maximum at $B/2$, then decrease and reaches zero at a depth equal to $2B$. For strip footings of $L/B > 10$, the maximum strain is found to occur at a depth equal to the width and reaches zero at a depth equal to $4B$. The modified triangular vertical strain influence factor distribution diagram as proposed by Schmertmann (1978) is shown in Fig. 6.10. The area of this diagram is related to the settlement. The equation (for square as will as circular footings) is

$$S = C_1 C_2 \, q_n \sum_0^{2B} \frac{I_z}{E_s} \Delta z \tag{6.23}$$

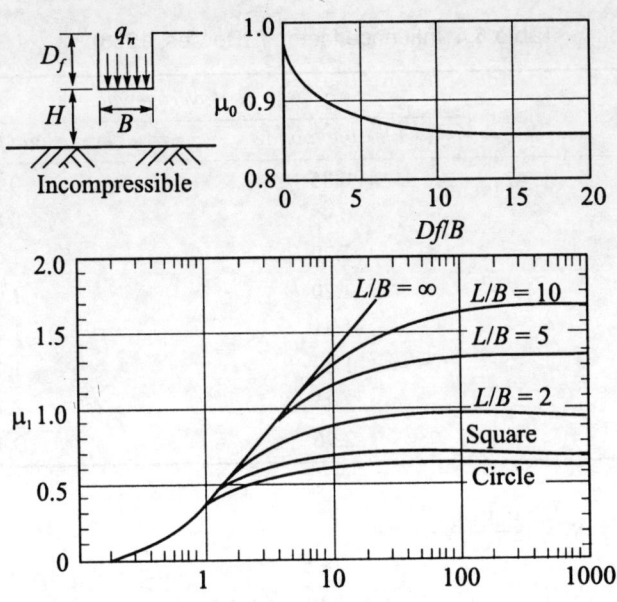

Fig. 6.9 Factors for calculating the average immediate settlement of a loaded area (after Christian and Carrier, 1978)

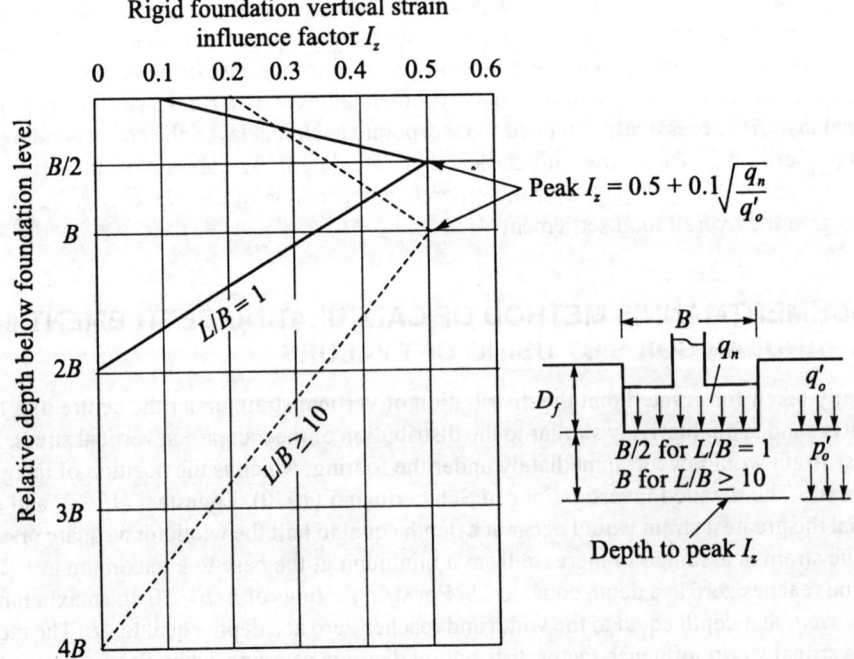

Fig. 6.10 Vertical strain influence factor diagrams (after Schmertmann *et al*, 1978)

where, S = total settlement,

q_n = net foundation base pressure = $(q - q_o')$,

q = total foundation pressure,

q'_o = effective overburden pressure at foundation level,

Δz = thickness of elemental layer,

I_z = vertical strain influence factor,

C_1 = depth correction factor,

C_2 = creep factor.

The equations for C_1 and C_2 are

$$C_1 = 1 - 0.5 \left(\frac{q_o'}{q_n} \right) \tag{6.24}$$

$$C_2 = 1 + 0.2 \log_{10} \left(\frac{t}{0.1} \right) \tag{6.25}$$

where t is time in years for which period settlement is required.

Equation (6.25) is also applicable for $L/B \geq 10$ except that the summation is from 0 to $4B$.

The modulus of elasticity to be used in Eq. (6.23) depends upon the type of foundation as follows: For a square footing,

$$E_s = 2.5q_c \tag{6.26}$$

For a strip footing, $L/B \geq 10$,

$$E_s = 3.5q_c \tag{6.27}$$

Figure 6.10 gives the vertical strain influence factor I_z distribution for both square and strip foundations if the ratio $L/B \geq 10$. Values for rectangular foundations for $L/B < 10$ can be obtained by interpolation. The depths at which the maximum I_z occurs may be calculated as follows (Fig. 6.10),

$$I_z = 0.5 + 0.1 \sqrt{\frac{q_n}{p_o'}} \tag{6.28}$$

where p'_o = effective overburden pressure at depths $B/2$ and B for square and strip foundations respectively.

Further, I_z is equal to 0.1 at the base and zero at depth $2B$ below the base for square footing; whereas for a strip foundation it is 0.2 at the base and zero at depth $4B$.

Values of E_s given in Eqs (6.26) and (6.27) are suggested by Schmertmann (1978). Lunne and Christoffersen (1985) proposed the use of the tangent modulus on the basis of a comprehensive review of field and laboratory tests as follows:

For normally consolidated sands,

$$E_s = 4q_c \text{ for } q_c < 10 \tag{6.29}$$

$$E_s = (2q_c + 20) \text{ for } 10 < q_c < 50 \tag{6.30}$$

$$E_s = 120 \text{ for } q_c > 50 \tag{6.31}$$

For overconsolidated sands with an overconsolidation ratio greater than 2,

$$E_s = 5\,q_c \text{ for } q_c < 50 \tag{6.32a}$$

$$E_s = 250 \text{ for } q_c > 50 \tag{6.32b}$$

where E_s and q_c are expressed in MPa.

The cone resistance diagram is divided into layers of approximately constant values of q_c and the strain influence factor diagram is placed alongside this diagram beneath the foundation which is drawn to the same scale. The settlements of each layer resulting from the net contact pressure q_n are then calculated using the values of E_s and I_z appropriate to each layer. The sum of the settlements in each layer is then corrected for the depth and creep factors using Eqs (6.24) and (6.25) respectively.

Example 6.8

Estimate the immediate settlement of a concrete footing 1.5×1.5 m in size founded at a depth of 1 m in silty soil whose modulus of elasticity is 90 kg/cm². The footing is expected to transmit a unit pressure of 200 kN/m².

Solution

Use Eq. (6.20a)

Immediate settlement,

$$S_e = qB \frac{\left(1 - \mu^2\right)}{E_s} I_f$$

Assume $\mu = 0.35$, $I_f = 0.82$ for a rigid footing.

Given: $q = 200 \text{ kN/m}^2$, $B = 1.5$ m, $E_s = 90 \text{ kg/cm}^2 \approx 9,000 \text{ kN/m}^2$.

By substituting the known values, we have

$$S_e = 200 \times 1.5 \times \frac{1 - 0.35^2}{9,000} \times 0.82 = 0.024 \text{ m} = 2.4 \text{ cm}$$

Example 6.9

A square footing of size 8×8 m is founded at a depth of 2 m below the ground surface in loose to medium dense sand with $q_n = 120 \text{ kN/m}^2$. Standard penetration tests conducted at the site gave the following corrected N_{60} values.

Depth below GL (m)	N_{cor}	Depth below GL	N_{cor}
2	8	10	11
4	8	12	16
6	12	14	18
8	12	16	17
		18	20

The water table is at the base of the foundation. Above the water table $\gamma = 16.5 \text{ kN/m}^3$, and submerged $\gamma_b = 8.5 \text{ kN/m}^3$.

Compute the elastic settlement by Eq. (6.20a). Use the equation $E_s = 250 (N_{cor} + 15)$ for computing the modulus of elasticity of the sand. Assume $\mu = 0.3$ and the depth of the compressible layer $= 2B = 16 \text{ m} (= H)$.

Solution

For computing the elastic settlement, it is essential to determine the weighted average value of N_{cor}. The depth of the compressible layer below the base of the foundation is taken as equal to 16 m ($= H$). This depth may be divided into three layers in such a way that N_{cor} is approximately constant in each layer as given below.

Layer no.	Depth (m)		Thickness (m)	N_{cor}
	From	To		
1	2	5	3	9
2	5	11	6	12
3	11	18	7	17

The weighted average

$$N_{cor}(av) = \frac{9 \times 3 + 12 \times 6 + 17 \times 7}{16} = 13.6 \text{ or say } 14$$

From equation $E_s = 250 (N_{cor} + 15)$, we have

$$E_s = 250 (14 + 15) = 7{,}250 \text{ kN/m}^2$$

The total settlement of the centre of the footing of size 8×8 m is equal to four times the settlement of a corner of a footing of size 4×4 m.

In the Eq. (6.20a), $B = 4$ m, $q_n = 120$ kN/m^2, $\mu = 0.3$.

Now from Fig. 6.7,

for $\qquad H/B = 16/4 = 4, L/B = 1$

$$F_1 = I_f = 0.4 \text{ for } \mu = 0.5$$
$$F_2 = 0.03 \text{ for } \mu = 0.5$$

Now from Eq. (6.20b), I_f for $\mu = 0.3$ is

$$I_f = F_1 + \frac{\left(1 - \mu - 2\mu^2\right) F_2}{1 - \mu^2} = 0.40 + \frac{\left(1 - 0.3 - 2 \times 0.3^2\right)}{1 - 0.3^2} \times 0.03 = 0.42$$

From Eq. (6.20a), we have settlement of a corner of a footing of size 4×4 m as

$$S_e = q_n B \frac{\left(1 - \mu^2\right)}{E_s} I_f = \frac{120 \times 4 \left(1 - 0.3^2\right)}{7{,}250} \times 0.42 \times 100 = 2.53 \text{ cm}$$

With the correction factor, the final elastic settlement from Eq. (6.21) is

$$S_{ef} = C_r d_f S_e$$

where C_r = rigidity factor = 1 for flexible footing d_f = depth factor

From Fig. 6.8 for

$$\frac{D_f}{\sqrt{BL}} = \frac{2}{\sqrt{4 \times 4}} = 0.5, \frac{L}{B} = \frac{4}{4} = 1, \text{ we have } d_f = 0.85$$

Now $$S_{ef} = 1 \times 0.85 \times 2.53 = 2.15 \text{ cm}$$

The total elastic settlement of the centre of the footing is

$$S_e = 4 \times 2.15 = 8.6 \text{ cm} = 86 \text{ mm}$$

Per Table 6.1 (a), the maximum permissible settlement for a raft foundation in sand is 62.5 mm. Since the calculated value is higher, the contact pressure q_n has to be reduced.

Example 6.10

It is proposed to construct an overhead tank at a site on a raft foundation of size 8×12 m with the footing at a depth of 2 m below ground level. The soil investigation conducted at the site indicates that the soil to a depth of 20 m is normally consolidated insensitive inorganic clay with the water table 2 m below ground level. Static cone penetration tests were conducted at the site using a mechanical cone penetrometer. The average value of cone penetration resistance $\bar{q}_c$ was found to be 1,540 kN/m² and the average saturated unit weight of the soil = 18 kN/m³. Determine the immediate settlement of the foundation using Eq. (6.22). The contact pressure $q_n = 100$ kN/m² (= 0.1 MPa). Assume that the stratum below 20 m is incompressible.

Solution

Computation of the modulus of elasticity

Use Eq. (6.19) with $A = 500$

$$E_s = 500 \, c_u$$

where c_u = the undrained shear strength of the soil

From Eq. (Chapter 3)

$$c_u = \frac{\bar{q}_c - p_o}{N_k}$$

where $\bar{q}_c$ = average static cone penetration resistance = 1,540 kN/m²,

p_o = average total overburden pressure = $10 \times 18 = 180$ kN/m²,

N_k = 20 (assumed).

Therefore $$c_u = \frac{1,540 - 180}{20} = 68 \text{ kN/m}^2$$

$$E_s = 500 \times 68 = 34,000 \text{ kN/m}^2 = 34 \text{ MPa}$$

Eq. (6.22) for S_e is

$$S_e = \frac{\mu_0 \mu_1 \, q_n B}{E_s}$$

From Fig. 6.9 for $D_f/B = 2/8 = 0.25$, $\mu_0 = 0.95$, for $H/B = 16/8 = 2$ and $L/B = 12/8 = 1.5$, $\mu_1 = 0.6$.

Substituting

$$S_e \text{ (average)} = \frac{0.95 \times 0.6 \times 0.1 \times 8}{34} = 0.0134 \text{ m} = 13.4 \text{ mm}$$

From Fig. 6.8 for $D_f/\sqrt{BL} = 2/\sqrt{8 \times 12} = 0.2$, $L/B = 1.5$ the depth factor $d_f = 0.94$.

The corrected settlement S_{ef} is

$$S_{ef} = 0.94 \times 13.4 = 12.6 \, \text{mm}$$

Example 6.11

Refer to Example 6.9. Estimate the elastic settlement by Schmertmann's method by making use of the relationship $q_c = 4 \, N_{cor} \, \text{kg/cm}^2$, where q_c = static cone penetration value in kg/cm^2. Assume settlement is required at the end of a period of 3 years.

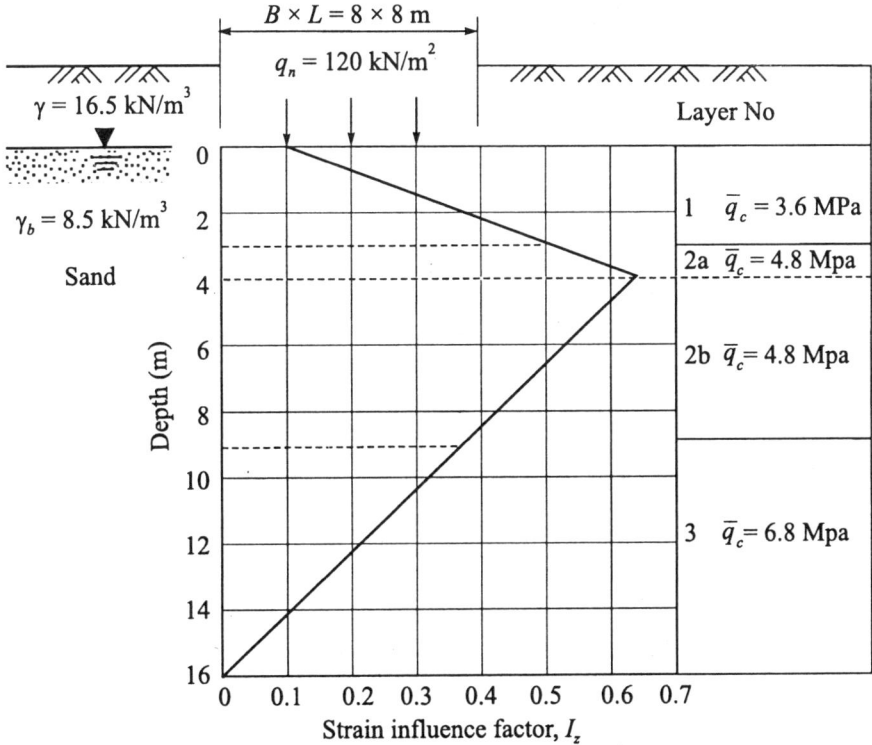

Fig. Ex. 6.11

Solution

The average value of for N_{cor} each layer given in Ex. 6.9 is given below:

Layer no	Average N_{cor}	Average q_c kg/cm^2	Average q_c MPa
1	9	36	3.6
2	12	48	4.8
3	17	68	6.8

The vertical strain factor I_z with respect to depth is calculated by making use of Fig. 6.10.

At the base of the foundation $I_z = 0.1$.

At depth $B/2$, $\qquad I_z = 0.5 + 0.1 \sqrt{\dfrac{q_n}{p_o'}}$,

where $\quad q_n = 120$ kPa

p_o' = effective average overburden pressure at depth = $(2 + B/2) = 6$ m below ground level.

$= 2 \times 16.5 + 4 \times 8.5 = 67$ kN/m^2.

$$I_z \text{(max)} = 0.5 + 0.1 \sqrt{\frac{120}{67}} = 0.63$$

$I_z = 0$ at $z = H = 16$ m below base level of the foundation. The distribution of I_z is given in Fig. Ex. 6.11. The equation for settlement is

$$S = C_1 C_2 q_n \sum_0^{2B} \frac{I_z}{E_s} \Delta z$$

where $\qquad C_1 = 1 - 0.5 \left(\dfrac{q_o'}{q_n}\right) = 1 - 0.5 \left(\dfrac{2 \times 16.5}{120}\right) = 0.86$

$$C_2 = 1 + 0.2 \log \left(\frac{t}{0.1}\right) = 1 + 0.2 \log \left(\frac{3}{0.1}\right) = 1.3$$

where, $t = 3$ years.

The elastic modulus E_s for normally consolidated sands may be calculated by Eq. (6.29).

$$E_s = 4\bar{q}_c \text{ for } q_c < 10 \text{ MPa}$$

where $\bar{q}_c$ is the average for each layer.

Layer 2 is divided into sub-layers 2 (a) and 2 (b) for computing I_z. The average of the influence factors for each of the layers given in Fig. Ex. 6.11 are tabulated along with the other calculations.

Layer no.	Δz (cm)	$\bar{q}_c$ (MPa)	E_s (MPa)	I_z (av)	$\dfrac{I_z \Delta z}{E_s}$
1	300	3.6	14.4	0.3	6.25
2 (a)	100	4.8	19.2	0.56	2.92
2 (b)	500	4.8	19.2	0.50	13.02
3	700	6.8	27.2	0.18	4.63
				Total	26.82

Substituting in the equation for settlement S, we have

$$S = 0.86 \times 1.3 \times 0.12 \times 26.82 = 3.6 \text{ cm} = 36 \text{ mm}$$

6.13 PRESSUREMETER METHOD OF ESTIMATING SETTLEMENT IN COHESIONLESS AND COHESIVE SOILS

Introduction

Pressuremeter method is one of the *in-situ* methods for estimating the ultimate bearing capacity and settlement of footings in both the cohesionless and cohesive soils. It is becoming more and more popular in many countries of the world. The basic data required for computing the settlement are:

1. Pressuremeter tests at equal intervals of one metre in prebored holes.
2. The net limit pressure $\bar{p}_l$ (corrected) where $\bar{p}_l = p_l - p_{oh}$ (See Section 3.10).
3. Pressuremeter modulus E_m (Eq. 3.27).
4. The soil profile with respect to depth from the bore hole records.
5. The rheological factor α from Table 3.9 (a) for the type of soil met at the location of the test.

With the data available as mentioned above, the settlement of footings, circular or rectangular, founded at depth D below the surface can be found out. The method of approach varies a little depending on the type of strata met with. Three types of strata are considered. They are:

(a) Homogeneous soil,

(b) Heterogeneous soil,

(c) Layered soil,

In the case of homogeneous soil, the pressuremeter modulus E_m is assumed as constant with depth. In the case of heterogeneous soil, E_m is assumed to vary with depth (provided the variations are not too great).

In the case of layered strata, a softer strata is considered either just below the footing or sandwiched between two hard stratum below the footing.

Settlement in Homogeneous Soil

Settlement prediction is based on the pressuremeter modulus E_m using a semiempirical formula which was proposed originally by Menard and Roussean (1962), and explained by Baguelin *et al* (1978) as

$$S = \frac{2}{9E_m} q_n B_0 \left(\lambda_d \frac{B}{B_0}\right)^{\alpha} + \frac{\alpha}{9E_m} q_n \lambda_c \frac{B}{c} \qquad (6.33a)$$

Where, E_m = the pressuremeter modulus,

q_n = the net applied bearing pressure = $q - q_o$,

q = the total bearing pressure,

q_o = the overburden surcharge at the foundation level,

B_0 = a reference width, usually 60 cm,

B = width of footing which is supposed to be larger than B_0,

α = rheological factor which depends on the soil type and the ratio of $E_m / \bar{p}_l$ (Table 3.9a),

λ_d, λ_c = shape factors, which depend on the length to width ratio L/B of the foundation as shown in Table 6.5.

Equation (6.33a) is applicable to footings embedded at depths $D_f \geq B$. When the depth of embedment is less than B, Eq. (6.33a) is to be multiplied by a correction factor, C_f, given as

$$C_f = \left(1.2 - 0.2 \frac{D_f}{B}\right) \qquad (6.33b)$$

The total settlement is

$$S_t = C_f S \qquad (6.33c)$$

Table 6.5 Shape factors of λ_d and λ_c for foundation

$L/B =$	1		2	3	5	20
	Circle	*Square*				
λ_d	1	1.12	1.53	1.78	2.14	2.65
λ_c	1	1.10	1.20	1.30	1.40	1.50

where, D_f is less than or equal to B. When $D = 0$ (surface footing), the settlement of the footing given by Eq. (6.33a) will be increased by 20 percent.

In Eq. (6.33a), the first term on the RHS represent the immediate settlement, S_e and the second term represent the consolidation settlement, S_c. In the cohesionless soils there is no dividing line between S_e and S_c as both the settlements take place simultaneously. Whereas, in the case of cohesive soils, S_e occurs under undrained condition, and the final consolidation settlement S_c occurs after a lapse of several years under drained condition.

In the case of cohesionless soils, the net imposed bearing pressure, q_n, to be used in Eq. (6.33a) will be the maximum one that may occur during the life of the structure. Further Eq. (6.33a) is meant to be used with the range of bearing stresses q_n greater than $q_{nu}/10$ where q_{nu} is the net ultimate bearing capacity of the soil.

Equation (6.33a) holds good as long as the foundation width B is larger than $B_0 = 60$ cm. When B is smaller than B_0, the following formula applies

$$S = \frac{q_n B}{9 E_m} (2 \lambda_d^\alpha + \alpha \lambda_c) \qquad (6.34)$$

Settlement in Heterogeneous Soil

In most soils, the pressuremeter modulus varies with depth. If the variation in the values is not very significant a harmonic mean of the values is considered for the settlement calculation. The method recommended for computing the total settlement S in cohesionless soil is (consider Fig. 6.11):

1. Divide the soil below the foundation base into a series fictitious layers of thickness $R = B/2$.

2. Let E_1 be the average modulus of deformation of the first layer. The first layer is supposed to contribute the maximum of the total settlement. In Eq. (6.33a), the second term on RHS contributes to this settlement. The modulus of deformation of this layer may be called for the sake of clarity as E_c. Therefore,

   $$E_c = E_1$$

   This applies to both the cohesionless and cohesive soils.

3. The harmonic mean of the modulus of deformation from layer 1 to 16 is taken as E_d to be used in the first term of Eq. (6.33a). This term is supposed to give the immediate settlement.

The harmonic mean of layers 1 to 16 may be found out as follows,

$$\frac{1}{E_d} = \frac{1}{4}\left(\frac{1}{E_1} + \frac{1}{0.85 E_2} + \frac{1}{E_{3/4/5}} + \frac{1}{2.5 E_{6/7/8}} + \frac{1}{2.5 E_{9/16}} \right) \qquad (6.35a)$$

where, $E_{3/4/5} = $ Harmonic mean of layers 3, 4 and 5 or

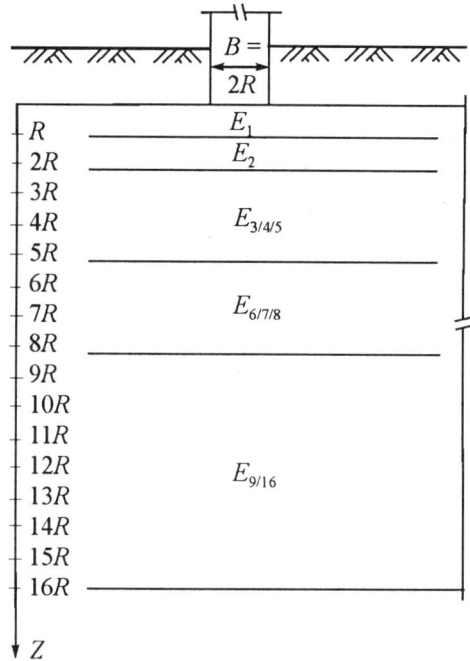

Fig. 6.11 Fictitious layers for settlement calculations in heterogeneous soil (Baguelin *et al*, 1978)

$$\frac{1}{E_{3/4/5}} = \frac{1}{3}\left(\frac{1}{E_3} + \frac{1}{E_4} + \frac{1}{E_5}\right) \tag{6.35b}$$

and $\quad E_{p/q}$ = Harmonic mean of the moduli of layers p to q

If no PMT data is available beyond a certain depth, z, the moduli of the layers deeper than z are assumed to be equal to the deepest measured modulus

Thus, the settlement is,

$$S = \frac{2}{9\,E_d}\,q_n N B_0 \left(\lambda_d\,\frac{B}{B_0}\right)^{\alpha} + \frac{\alpha}{9\,E_c}\,q_n\lambda_c B \tag{6.36}$$

Settlement when the Base of the Foundation Rests on a Soft Layer of Thickness *Z* which is Less than *B*/2 [Fig. 6.12 (a)]

This situation is often encountered beneath raft foundations or embankments. The settlement due to the underlying stiffer material can be neglected since presumably the bearing capacity is low because of the softness of the shallow layer. Only the consolidation settlement of the soft layer is computed. The equation is (Baguelin *et al*, 1978),

$$S = \sum_{1}^{n} \frac{\alpha_i\,\beta_i\,\Delta\,p_i}{E_i}\,\Delta z_i \tag{6.37a}$$

where, $\quad n$ = number of layers constituting the soft layer,

Δp_i = average increase in vertical pressure in the ith layer computed by elastic theory,

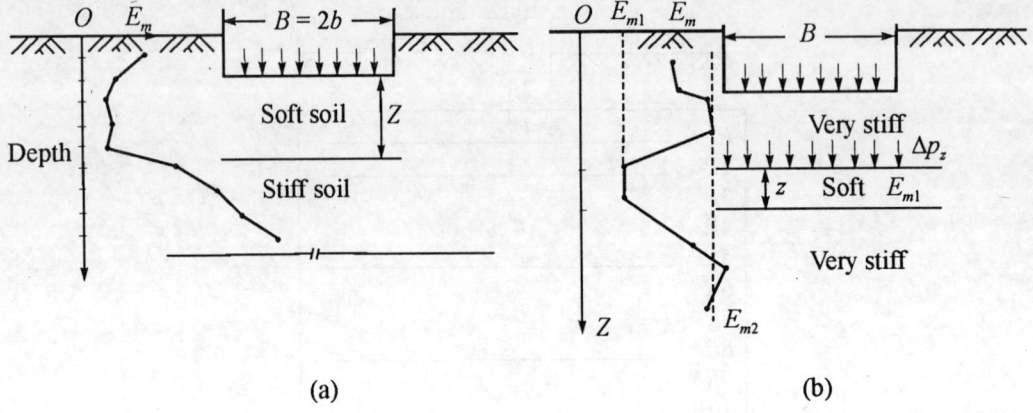

(a)   (b)

Fig. 6.12 Pressuremeter method of settlement calculation in layered soils: (a) Soft layer just below the foundation, (b) soft layer at depth

Δz_i = thickness of the ith layer,

α_i = rheological factor for the ith layer,

E_i = pressuremeter modulus for the ith layer,

β_i = a coefficient which is a function of the safety factor F (the ratio of the ultimate bearing pressure to the increase in the ith layer).

The coefficient β_i is expressed as,

$$\beta_i = \frac{2}{3} \frac{F_s}{F_s - 1} \tag{6.37b}$$

where, F_s = Factor of safety.

The coefficient β tends to take into account the increase in compressibility beyond the preconsolidation pressure and is explained as follows:

1. If the factor of safety is three, the bearing pressure is likely to be close to or smaller than the preconsolidation pressure, and β is 1 in this case.

2. If the factor of safety is less than 3, the bearing pressure is likely to exceed the preconsolidation pressure and β increases accordingly.

Settlement when A Soft Layer of Thickness z Lies at Some Depth below the Base of the Foundation

Figure 6.12 (b) illustrates the type of strata,

1. First compute the settlement S_1 which would occur if the soft layer were the same as the surrounding soil. The value of modulus of deformation of the hard soil is substituted for the compressible layer and the general procedure as explained earlier is followed for computing settlement by dividing the soil below the foundation into fictitious layers.

2. Compute the additional settlement, S_2 of the soft layer due to the difference in the moduli assumed in step 1 above and the actual value. The equation is,

$$S_2 = (m_{v1} - m_{v2}) \Delta p_z \tag{6.38}$$

where, m_{v1} = coefficient of volume compressibility of the soft layer,

m_{v2} = coefficient of compressibility of the adjacent layer,

Δp_z = excess vertical pressure on the top of soft layer due to the foundation pressure q_n calculated by elastic theory,

z = thickness of the soft layer.

Since $m_v = \dfrac{1}{E}$, we may write Eq. (6.38) as

$$S_2 = \left(\frac{1}{E_1} - \frac{1}{E_2} \right) \Delta p_z z \tag{6.39}$$

From Eq. (3.46), $E = \dfrac{E_m}{\alpha}$, substituting in Eq. (6.36), we have

$$S_2 = \left(\frac{1}{E_{m1}} - \frac{1}{E_{m2}} \right) \alpha \Delta p_z z, \tag{6.40}$$

where, E = Young's modulus,

E_{m1} = pressuremeter modulus of the soft layer,

E_{m2} = pressuremeter modulus of the adjacent hard layer,

α = rheological factor of the soft layer.

6.14 ESTIMATION OF CONSOLIDATION SETTLEMENT BY USING OEDOMETER TEST DATA

Equations for Computing Settlement

Settlement calculation from e-log p curves

A general equation for computing oedometer consolidation settlement may be written as follows.

Normally, consolidated clays

$$S_c = H \frac{C_c}{1 + e_0} \log \frac{p_0 + \Delta p}{p_0} \tag{6.41}$$

Overconsolidated clays

for $p_0 + \Delta p < p_c$

$$S_c = H \frac{C_s}{1 + e_0} \log \frac{p_0 + \Delta p}{p_0} \tag{6.42}$$

for $p_0 < p_c < p_0 + \Delta p$

$$S_c = \frac{H}{1 + e_0} \left(C_s \log \frac{p_c}{p_0} + C_c \log \frac{p_0 + \Delta p}{p_c} \right) \tag{6.43}$$

where C_s = swell index, and C_c = compression index.

If the thickness of the clay stratum is more than 3 m the stratum has to be divided into layers of thickness less than 3 m. Further, e_0 is the initial void ratio and p_0, the effective overburden pressure corresponding to the particular layer; Δp is the increase in the effective stress at the middle of the layer due to foundation loading which is calculated by elastic theory. The compression index, and the swell index may be the same for the entire depth or may vary from layer to layer.

Settlement calculation from e-p curve

Equation (6.43) can be expressed in a different form as follows:

$$S_c = \Sigma H m_v \Delta p \tag{6.44}$$

where m_v = coefficient of volume compressibility.

6.15 SKEMPTON-BJERRUM METHOD OF CALCULATING CONSOLIDATION SETTLEMENT (1957)

Calculation of consolidation settlement is based on one dimensional test results obtained from oedometer tests on representative samples of clay. These tests do not allow any lateral yield during the test and as such the ratio of the minor to major principal stresses, K_0, remain constant. In practice, the condition of zero lateral strain is satisfied only in cases where the thickness of the clay layer is small in comparison with the loaded area. In many practical solutions, however, significant lateral strain will occur and the initial pore water pressure will depend on the *in-situ* stress condition and the value of the pore pressure coefficient A, which will not be equal to unity as in the case of a one-dimensional consolidation test. In view of the lateral yield, the ratios of the minor and major principal stresses due to a given loading condition at a given point in a clay layer do not maintain a constant K_0.

The initial excess pore water pressure at a point P (Fig. 6.13) in the clay layer is given by the expression

$$\Delta u = \Delta\sigma_3 + A\,(\Delta\sigma_1 - \Delta\sigma_3)$$

$$= \Delta\sigma_1 \left[A + \frac{\Delta\sigma_3}{\Delta\sigma_1}(1 - A) \right] \tag{6.45}$$

where $\Delta\sigma_1$ and $\Delta\sigma_3$ are the total principal stress increments due to surface loading. It can be seen from Eq. (6.45), $\Delta u > \Delta\sigma_3$ if A is positive and $\Delta u = \Delta\sigma_1$ if $A = 1$.

The value of A depends on the type of clay, the stress levels and the stress system.

Figure 6.13 (a) presents the loading condition at a point in a clay layer below the central line of circular footing. Figures 6.13 (b), (c) and (d) show the condition before loading, immediately after loading and after consolidation respectively.

By the one-dimensional method, consolidation settlement S_{oc} is expressed as

$$S_{oc} = \int_0^H m_v \, \Delta\sigma_1 \, dz \tag{6.46}$$

By the Skempton-Bjerrum method, consolidation settlement is expressed as

$$S_{oc} = \int_0^H m_v \, \Delta u \, dz \tag{6.47}$$

or

$$S_{oc} = \int_0^H m_v \, \Delta\sigma_1 \left[A + \frac{\Delta\sigma_3}{\Delta\sigma_1}(1 - A) \right] \tag{6.48a}$$

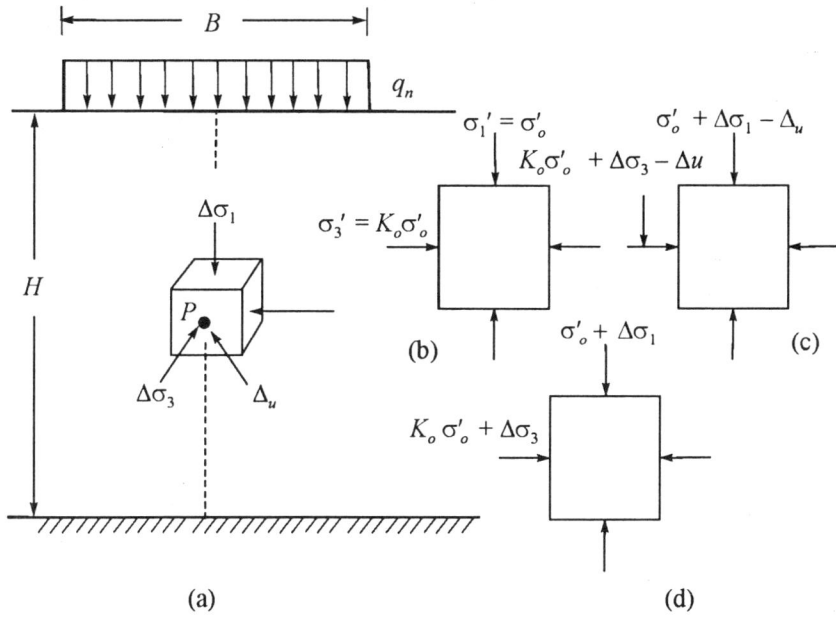

Fig. 6.13 *In-situ* effective stresses: (a) Physical plane, (b) initial conditions, (c) immediately after loading, (d) after consolidation

A settlement coefficient β is used, such that $S_c = \beta \, S_{oc}$

The expression for β is

$$\beta = \frac{S_c}{S_{oc}} = \frac{\int\limits_0^H m_v \, \Delta\sigma_1 \left[A + \dfrac{\Delta\sigma_3}{\Delta\sigma_1} (1 - A) \right] dz}{\int\limits_0^H m_v \, \Delta\sigma_1 \, dz} \qquad (6.48b)$$

or
$$S_c = \beta S_{oc} \qquad (6.49)$$

where β is called the *settlement coefficient*.

If it can be assumed that m_v and A are constant with depth (sub-layers can be used in the analysis), then β can be expressed as

$$\beta = A + (1 - A)\alpha \qquad (6.50a)$$

where
$$\alpha = \frac{\int\limits_0^H \Delta\sigma_3 \, dz}{\int\limits_0^H \Delta\sigma_1 \, dz} \qquad (6.50b)$$

Taking Poisson's ratio μ as 0.5 for a saturated clay during loading under untrained conditions, the value of β depends only on the shape of the loaded area and the thickness of the clay layer in relation to the dimensions of the loaded area and thus β can be estimated from elastic theory.

The value of initial excess pore water pressure (Δu) should, in general, correspond to the *in-situ* stress conditions. The use of a value of pore pressure coefficient A obtained from the results of a triaxial test on a cylindrical clay specimen is strictly applicable only for the condition of axial symmetry, i.e. for the case of settlement under the centre of a circular footing. However, the value of A so obtained will serve as a good approximation for the case of settlement under the centre of a square footing (using the circular footing of the same area).

Under a strip footing plane strain conditions prevail. Scott (1963), has shown that the value of Δu appropriate in the case of a strip footing can be obtained by using a pore pressure coefficient A_s as

$$A_s = 0.866 A + 0.211 \tag{6.51}$$

The coefficient A_s replaces A (the coefficient for the condition of axial symmetry) in Eq. (6.42) for the case of a strip footing, the expression for a being unchanged.

Values of the settlement coefficient β for circular and strip footings, in terms of A and ratios H/B, are given in Fig. 6.14.

Typical values of β are given in Table 6.6 for various types of clay soils.

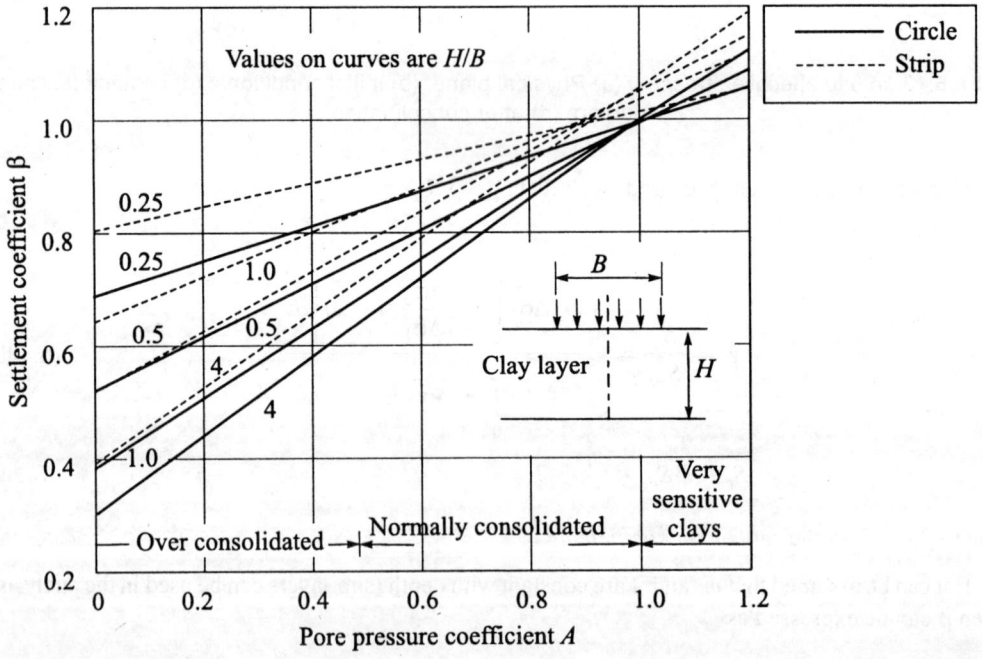

Fig. 6.14 Settlement coefficient *versus* pore-pressure coefficient for circular and strip footings (After Skempton and Bjerrum, 1957)

Table 6.6 Values of settlement coefficient β

Type of clay	β
Very sensitive clays (soft alluvial and marine clays)	1.0 to 1.2
Normally consolidated clays	0.7 to 1.0
Overconsolidated clays	0.5 to 0.7
Heavily overconsolidated clays	0.2 to 0.5

Example 6.12

For the problem given in Ex. 6.10, compute the consolidation settlement by the Skempton-Bjerrum method. The compressible layer of depth 16 m below the base of the foundation is divided into four layers and the soil properties of each layer are given in Fig. Ex. 6.12. The net contact pressure $q_n = 100 \text{ kN/m}^2$.

Solution

From Eq. (6.41), the oedometer settlement for the entire clay layer system may be expressed as

$$S_{oe} = \Sigma H_i \frac{C_c}{1 + e_o} \log \frac{p_o + \Delta p}{p_o}$$

From Eq. (6.49), the consolidation settlement as per Skempton-Bjerrum may be expressed as

$$S_c = \beta S_{oe}$$

where β = settlement coefficient which can be obtained from Fig. 6.14 for various values of A and H/B,

   p_o = effective overburden pressure at the middle of each layer (Fig. Ex. 6.12),

   C_c = compression index of each layer,

   H_i = thickness of ith layer,

   e_o = initial void ratio of each layer,

   Δp = the excess pressure at the middle of each layer obtained from elastic theory.

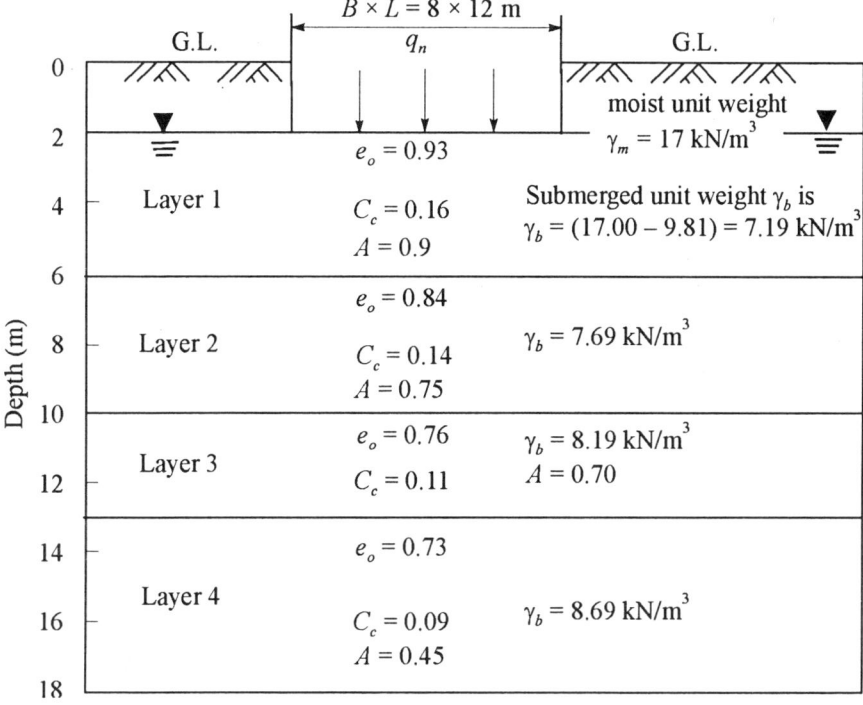

Fig. Ex. 6.12

The average pore pressure coefficient is

$$A = \frac{0.9 + 0.75 + 0.70 + 0.45}{4} = 0.7$$

The details of the calculations are tabulated below.

Layer no.	H_i (cm)	p_o (kN/m^2)	Δp (kN/m^2)	C_c	e_o	$\log \dfrac{p_0 + \Delta p}{p_o}$	S_{oed} (cm)
1	400	48.4	75	0.16	0.93	0.407	13.50
2	400	78.1	43	0.14	0.84	0.191	5.81
3	300	105.8	22	0.11	0.76	0.082	1.54
4	500	139.8	14	0.09	0.73	0.041	1.07
						Total	21.92

For $H/B = 16/8 = 2$, $A = 0.7$, from Fig. 6.14, we have $\beta = 0.8$

The consolidation settlement S_c is

$$Sc = 0.8 \times 21.92 = 17.536 \text{ cm} = 175.36 \text{ mm}$$

6.16 CONSOLIDATION SETTLEMENT BY LAMBE'S STRESS PATH METHOD

Introduction

The principal characteristics of a stress path, the method of constructing effective stress paths (ESP) and the stress–strain contours have been explained in Chpater 2.

The Steps for Computing Settlement

The method is explained with respect to a field problem. Consider Fig. 6.15 (a). Consolidation settlement of a normally consolidated clay layer of thickness H, which is acted by a surface load q_n per unit area on a circular footing of diameter B is required. The various steps required to be taken for computing settlement are:

1. Preparation of stress–strain contours from a series of laboratory tests [Fig. 6.15 (b)].
2. Superposition on the stress–strain contours, the predicted effective stress paths for the field loading [Fig. 6.15 (b)].
3. Estimating the vertical strain from the plot in step 2.
4. Multiplying the vertical strain by the thickness of the layer under consideration to get the total settlement.

If the clay strata of thickness H [Fig. 6.15 (a)] is too great or non-homogeneous, the strata may be divided into convenient layers of suitable thickness each. The stress path method is applied to each layer to calculate the total settlement. For convenience, H is considered here as one single layer.

The first step involves the construction of stress–strain contours as given in Fig. 6.15 (b). Here the figure shows three effective stress paths (ESP) with constant water contents w_1, w_2 and w_3. The stress–strain contours are designated as ε_1, ε_2, ε_3, etc. The stress paths are constructed from the consolidated-undrained triaxial tests conducted on undisturbed soil samples extracted at the middle of the layer of thickness H and lying on the central line of the footing Fig. 6.15 (a).

Superposition of ESP due to Surface Loading

Stress-condition before loading

Let point P in Fig. 6.15 (a) represent the average point which lies on the central line of the footing. The stress condition at point P before the application of foundation load is,

vertical effective stress, $\sigma'_v = \sigma'_0$

horizontal effective stress, $\sigma'_h = K_0 \sigma'_0$

The vertical stress σ'_0 ($= \sigma'_v$) can be calculated by knowing the unit weight of soil. The at-rest earth pressure coefficient, K_0, can be calculated by making use of equation (Jaky, 1944),

$$K_0 = 1 - \sin \phi$$

For the soil at *in-situ* condition, we may write,

$$p' = \frac{\sigma_1' + \sigma_3'}{2} = \frac{\sigma_v' + \sigma_h'}{2} = \frac{\sigma_v'\left(1 + K_0\right)}{2} \tag{6.52a}$$

$$q' = \frac{\sigma_1' - \sigma_3'}{2} = \frac{\sigma_v' - \sigma_h'}{2} = \frac{\sigma_v'\left(1 - K_0\right)}{2} \tag{6.52b}$$

The slope of the K_0-line β is,

$$\beta = \tan^{-1} \frac{q'}{p'} = \tan^{-1}\left(\frac{1 - K_0}{1 + K_0}\right) \tag{6.53}$$

The point b, plotted in Fig. 6.15 (b) represent the *in-situ* condition with coordinates q' and p' as given by Eq. (6.52). This point lies on the K_0-line. An effective stress path $a_1 a_2$ can be sketched passing through b and geometrically similar to the adjoining stress paths with constant water contents w_1 and w_2.

Stress Condition after Loading

If the surface load q_n is imposed on the surface instantaneously, there will be no change in the water content in the compressible layer due to the excess stresses $\Delta\sigma_1$ and $\Delta\sigma_3$ imposed at the point P in Fig. 6.15 (a). As such the stress condition in the soil soon after loading will be represented by a point say c, which lies on the ESP passing through the point b which represent the initial condition before loading. The coordinates of point c, may be written as follows:

Immediately after loading

$$\sigma_1 = \sigma'_0 + \Delta\sigma_1, \; \sigma_3 = K_0 \sigma'_0 + \Delta\sigma_3 \tag{6.54a}$$

$$q' = q = \frac{\sigma_1 - \sigma_3}{2} = \frac{\sigma_1' - \sigma_3'}{2} = \frac{\sigma_0'\left(1 - K_0\right) + \left(\Delta\sigma_1 - \Delta\sigma_3\right)}{2} \tag{6.54b}$$

In Eq. (6.54b), the excess stresses $\Delta\sigma_1$ and $\Delta\sigma_3$ developed due to the surface loading, q_n, have to be found out from elastic theory. When once these are known, the vertical coordinate q' of point c can be found out. Since the stress path $a_1 a_2$ is known, q' of point c is known, the point c on the stress path $a_1 a_2$ can be plotted [Fig. 6.15 (b)].

Stress condition at the end of consolidation

Soon after loading, consolidation of the soil strata of thickness H starts taking place. After some period of time (may be a few months or years), equilibrium condition will be attained. At this stage there will be no further expulsion of water from the pores of the soil. During all this period the vertical coordinate q' of point c [Fig. 6.15 (b)] remains un-altered, only the horizontal coordinate p' goes on increasing and reaches a final value represented by point d in [Fig. 6.15 (b)]. This means the line cd is the effective stress path from the point c soon after loading, and this line is parallel to the abscissa. The abscissa of point d may now be written (from Eq. 6.54a) as follows.

At the end of consolidation, $\sigma_1 = \sigma_1'$, $\sigma_3 = \sigma_3'$. Therefore,

$$p' = \frac{\sigma_1' + \sigma_3'}{2} = \frac{\sigma_0'(1 + K_0) + \Delta\sigma_1 + \Delta\sigma_3}{2} \tag{6.55}$$

Now the complete effective stress path is represented by the broken line bcd, wherein bc is the path due to elastic settlement (without change of water content) and cd represent the consolidation settlement. Sketch an effective stress path $g_1\,g_2$ passing through d, geometrically similar to the adjoining stress paths with constant water contents w_2 and w_3.

Computation of Settlement

Elastic settlement, S_e

Let point c [Fig. 6.15 (b)] lie on the strain contour ε_6, and point b on the strain contour ε_5 (which is close to the K_0-line). The equation for elastic settlement S_e may be written as

$$S_e = (\varepsilon_6 - \varepsilon_5)\,H = \Delta\varepsilon H \tag{6.56}$$

Consolidation settlement, S_c

Let the ESP passing through point d cut the K_0-line at point t. According to the characteristics of *ESP, the volumetric strain between c and d is the same as that between b and t along the K_0-line,* which represent the volumetric strain in one-dimensional consolidation.

For the point t on the ESP $g_1\,g_2$, the principal stresses are σ_{1t}' and σ_{3t}' [Fig. 6.15 (b)].

As per Eq. (6.41), we have

$$\frac{S_c}{H} = \frac{C_c}{1 + e_0}\,\log_{10}\frac{p_0' + \Delta p}{p_0'}$$

or

$$\frac{S_c}{H} = \frac{C_c}{1 + e_0}\,\log_{10}\frac{\sigma_{1t}'}{\sigma_0'} = \varepsilon_v, \tag{6.57}$$

where, $p_0' + \Delta p = \sigma_{1t}'$, and $p_0' = \sigma_0'$, C_c = compression index, and e_0 = initial void ratio.

Equation (6.57), therefore, represents the volumetric strain ε_v. For a horizontal stress path cd [Fig. 6.15 (b)], the relationship between axial strain ε_a and the volumetric strain ε_v may be expressed from elastic theory as

$$\varepsilon_a = \frac{\varepsilon_v}{3} \tag{6.58}$$

Now, the equation for consolidation settlement may be written as

$$S_c = \varepsilon_a H = \frac{1}{3}\,\varepsilon_v H \tag{6.59}$$

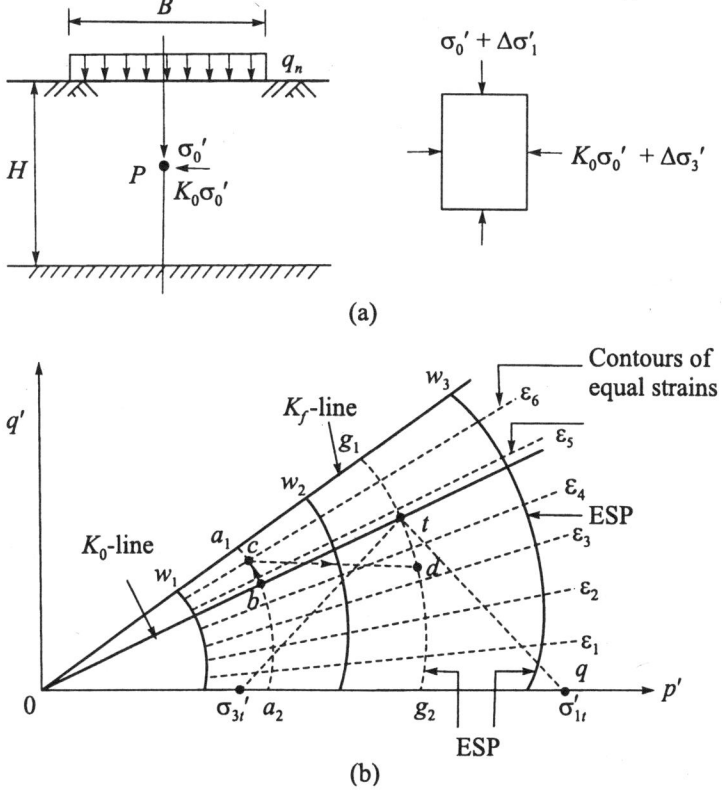

Fig. 6.15 Stress path method of computing settlement: (a) Settlement of compressible layer of thickness H, (b) stress path for computing settlement

Equation (6.57), therefore implies that consolidation tests are required to be carried out on undisturbed samples for computing the value of volumetric strain ε_v.

Total Settlement, *S*

The total settlement is a sum of S_e and S_c or

$$S = S_e + S_c = \Delta \varepsilon H + \varepsilon_a H$$

or

$$S = (\Delta \varepsilon + \varepsilon_a) H. \tag{6.60}$$

Validity of Stress Path Method

1. The effective stress path method is a powerful tool for solving deformation problems in soil engineering. It offers considerable insight into the settlement problem.

2. However, there are very many uncertainties involved in the proper application of this method. They are:

    (*a*) Difficulties in getting an accurate soil profile.

    (*b*) Difficulties in determining the magnitude and distribution of stress in the sub-soil due to the applied surface loadings.

    (*c*) The influence of sample disturbance on stress–strain data obtained in the laboratory.

3. The process is laborious and time consuming. Possibly, it is a very useful tool in major projects where other methods are not adequate to solve the problem.

6.17 PROBLEMS

6.1 A plate load test was conducted in a medium dense sand at a depth of 5 ft below ground level in a test pit. The size of the plate used was 12 × 12 in. The data obtained from the test are plotted in Fig. Prob. 6.1 as a load-settlement curve. Determine from the curve the net safe bearing pressure for footings of size (*a*) 10 × 10 ft, and (*b*) 15 × 15 ft. Assume the permissible settlement for the foundation is 25 mm.

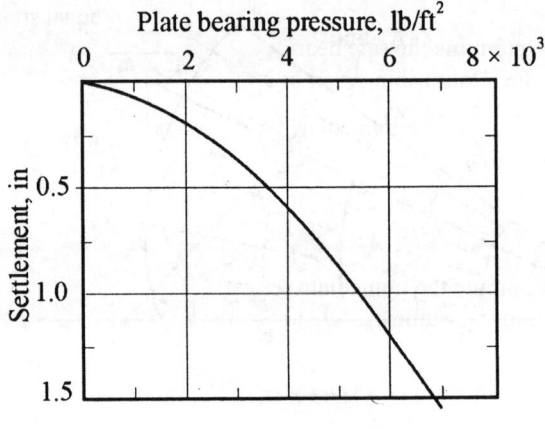

Fig. Prob. 6.1

6.2 Refer to Prob. 6.1, determine the settlements of the footings given in Prob 6.1. Assume the settlement of the plate as equal to 0.5 in. What is the net bearing pressure from Fig. Prob. 6.1 for the computed settlements of the foundation?

6.3 For Problem 6.2, determine the safe bearing pressure of the footings if the settlement is limited to 2 in.

6.4 Refer to Prob. 6.1. if the curve given in Fig. Prob. 6.1 applies to a plate test of 12 × 12 in. conducted in a clay stratum, determine the safe bearing pressures of the footings for a settlement of 2 in.

6.5 Two plate load tests were conducted in a *c*-φ soil as given below.

Size of plates (m)	Load kN	Settlement (mm)
0.3 × 0.3	40	30
0.6 × 0.6	100	30

Determine the required size of a footing to carry a load of 1250 kN for the same settlement of 30 mm.

6.6 A rectangular footing of size 4 × 8 m is founded at a depth of 2 m below the ground surface in dense sand and the water table is at the base of the foundation. $N_{cor} = 30$ (Fig. Prob. 6.6). Compute the safe bearing pressure q_s using the chart given in Fig. 6.5.

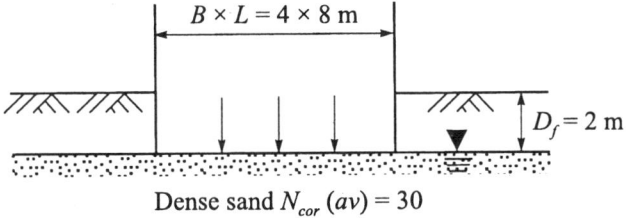

Dense sand N_{cor} (av) = 30

Fig. Prob. 6.6

6.7 Refer to Prob. 6.6. Compute q_s by using modified (*a*) Teng's formula, and (*b*) Meyerhof's formula.

6.8 Refer to Prob. 6.6. Determine the safe bearing pressure based on the static cone penetration test value based on the relationship given in Eq. (6.17b) for $q_c = 120$ kN/m².

6.9 Refer to Prob. 6.6. Estimate the immediate settlement of the footing by using Eq. (6.20a). The additional data available are:

$\mu = 0.30$, $I_f = 0.82$ for rigid footing and $E_s = 11,000$ kN/m². Assume $q_n = q_s$ as obtained from Prob. 6.6.

6.10 Refer to Prob 6.6. Compute the immediate settlement for a flexible footing, given $\mu = 0.30$ and $E_s = 11,000$ kN/m². Assume $q_n = q_s$.

6.11 If the footing given in Prob. 6.6 rests on normally consolidated saturated clay, compute the immediate settlement using Eq. (6.22). Use the following relationships.

$$\bar{q}_c = 120 \text{ kN/m}^2$$

$$c_u = \frac{\bar{q}_c - p_0}{N_k} \text{ and } N_k = 20$$

$$E_s = 600 c_u \text{ kN/m}^2$$

Given: $\gamma_{sat} = 18.5$ kN/m³, $q_n = 150$ kN/m². Assume that the incompressible stratum lies at depth of 10 m below the base of the foundation.

6.12 A footing of size 6 × 6 m rests in medium dense sand at a depth of 1.5 below ground level. The contact pressure $q_n = 175$ kN/m². The compressible stratum below the foundation base is divided into three layers. The corrected N_{cor} values for each layer is given in Fig. Prob. 6.12 with other data. Compute the immediate settlement using Eq. (6.23). Use the relationship $q_c = 400 N_{cor}$ kN/m².

6.13 It is proposed to construct an overhead tank on a raft foundation of size 8 × 16 m with the foundation at a depth of 2 m below ground level. The subsoil at the site is a stiff homogeneous clay with the water table at the base of the foundation. The subsoil is divided into 3 layers and the properties of each layer are given in Fig. Prob. 6.13. Estimate the consolidation settlement by the Skempton-Bjerrum method.

6.14 A footing of size 10 × 10 m is founded at a depth of 2.5 m below ground level on a sand deposit. The water table is at the base of the foundation. The saturated unit weight of soil from ground level to a depth of 22.5 m is 20 kN/m³. The compressible stratum of 20 m below the foundation base is divided into three layers with corrected SPT values (*N*) and CPT values (q_c) constant in each layer as given below.

Layer no.	Depth from (m) foundation level		N_{cor} (av)	q_c (av) MPa
	From	To		
1	0	5	20	8.0
2	5	11.0	25	10.0
3	11.0	20.0	30	12.0

Compute the settlements by Schmertmann's method.

Assume the net contact pressure at the base of the foundation is equal to 70 kPa, and $t = 10$ years.

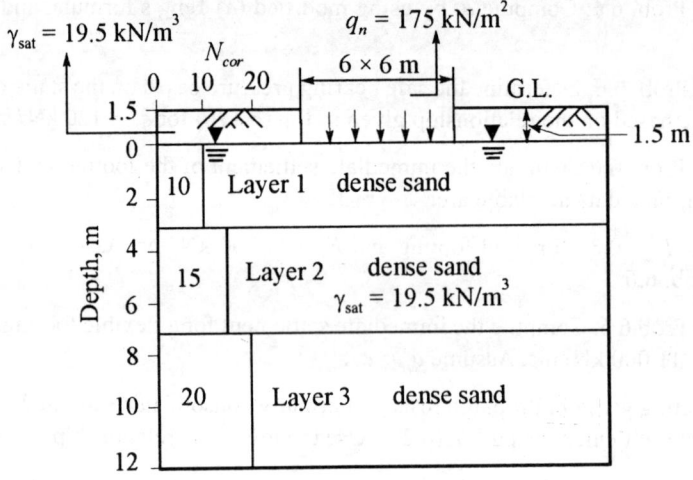

Fig. Prob. 6.12

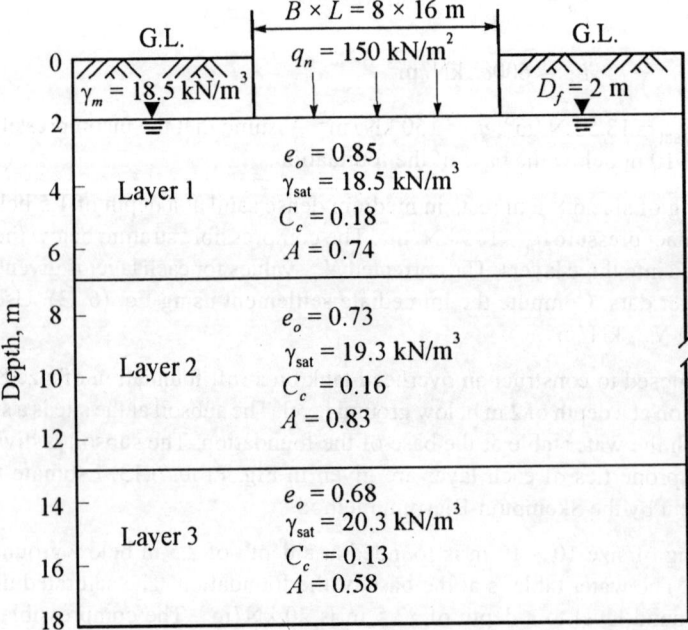

Fig. Prob. 6.13

6.15 A square rigid footing of size 10×10 m is founded at a depth of 2.0 m below ground level. The type of strata met at the site is

Depth below GL (m)	Type of soil
0 to 5	Sand
5 to 7 m	Clay
Below 7 m	Sand

The water table is at the base level of the foundation. The saturated unit weight of soil above the foundation base is 20 kN/m³. The coefficient of volume compressibility of clay, m_v, is 0.0001 m²/kN, and the coefficient of consolidation c_v, is 1 m²/year. The total contact pressure $q_n = 100$ kN/m². Water table is at the base level of foundation.

Compute primary consolidation settlement.

6.16 A circular tank of diameter 3 m is founded at a depth of 1 m below ground surface on a 6 m thick normally consolidated clay. The water table is at the base of the foundation. The saturated unit weight of soil is 19.5 kN/m³, and the *in-situ* void ratio e_0 is 1.08. Laboratory tests on representative undisturbed samples of the clay gave a value of 0.6 for the pore pressure coefficient A and a value of 0.2 for the compression index C_c. Compute the consolidation settlement of the foundation for a total contact pressure of 95 kPa. Use $2:1$ method for computing Δp.

6.17 A raft foundation of size 10×40 m is founded at a depth of 3 m below ground surface and is uniformly loaded with a net pressure of 50 kN/m². The subsoil is normally consolidated saturated clay to a depth of 20 m below the base of the foundation with variable elastic moduli with respect to depth. For the purpose of analysis, the stratum is divided into three layers with constant modulus as given below:

Layer no.	Depth below ground (m)		Elastic modulus
	From	To	Es (MPa)
1	3	8	20
2	8	18	25
3	18	23	30

Compute the immediate settlements by using Eqs (6.20a). Assume the footing is flexible.

Shallow Foundation 4
Combined Footings and Mat Foundations

7

7.1 INTRODUCTION

Chapter 5 has considered the common methods of transmitting loads to subsoil through spread footings carrying single column loads. This chapter considers the following types of foundations:

1. Cantilever footings
2. Combined footings
3. Mat foundations

When a column is near or right next to a property limit, a square or rectangular footing concentrically loaded under the column would extend into the adjoining property. If the adjoining property is a public side walk or alley, local building codes may permit such footings to project into public property. But when the adjoining property is privately owned, the footings must be constructed within the property. In such cases, there are three alternatives which are illustrated in Fig. 7.1 (a). These are:

1. *Cantilever footing:* A cantilever or strap footing normally comprises two footings connected by a beam called a strap. A strap footing is a special case of a combined footing.

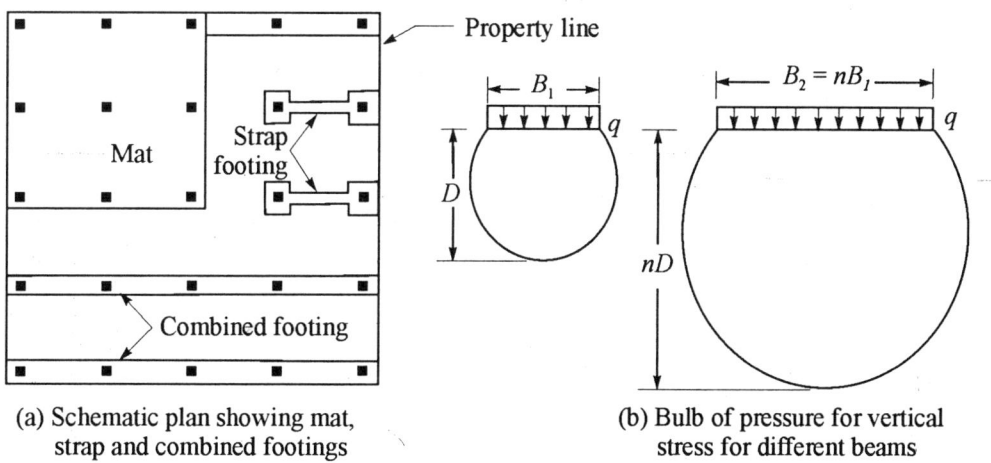

| (a) Schematic plan showing mat, strap and combined footings | (b) Bulb of pressure for vertical stress for different beams |

Fig. 7.1 (a) Types of footings, (b) beams on compressible subgrade

2. *Combined footing:* A combined footing is a long footing supporting two or more columns in one row.

3. *Mat or raft foundations:* A mat or raft foundation is a large footing, usually supporting several columns in two or more rows.

The choice between these types depends primarily upon the relative cost. In the majority of cases, mat foundations are normally used where the soil has low bearing capacity and where the total area occupied by an individual footing is not less than 50 percent of the loaded area of the building.

When the distances between the columns and the loads carried by each column are not equal, there will be eccentric loading. The effect of eccentricity is to increase the base pressure on the side of eccentricity and decrease it on the opposite side. The effect of eccentricity on the base pressure of rigid footings is also considered here.

Mat Foundation in Sand

A foundation is generally termed as a mat if the least width is more than 6 metres. Experience indicates that the ultimate bearing capacity of a mat foundation on cohesionless soil is much higher than that of individual footings of lesser width. With the increasing width of the mat, or increasing relative density of the sand, the ultimate bearing capacity increases rapidly. Hence, the danger that a large mat may break into a sand foundation is to remote to require consideration. On account of the large size of mats the stresses in the underlying soil are likely to be relatively high to a considerable depth. Therefore, the influence of local loose pockets distributed at random throughout the sand is likely to be about the same beneath all parts of the mat and differential settlements are likely to be smaller than those of a spread foundation designed for the same soil pressure. The methods of calculating the ultimate bearing capacity dealt with in Chapter 5 are also applicable to mat foundations.

Mat Foundation in Clay

The net ultimate bearing capacity that can be sustained by the soil at the base of a mat on a deep deposit of clay or plastic silt may be obtained in the same manner as for footings on clay discussed in Chapter 5. However, by using the principle of *flotation*, the pressure on the base of the mat that induces settlement can be reduced by increasing the depth of the foundation. A brief discussion on the principle of flotation is dealt with in this chapter.

Rigid and Elastic Foundation

The conventional method of design of combined footings and mat foundations is to assume the foundation as infinitely rigid and the contact pressure is assumed to have a planar distribution. In the case of an elastic foundation, the soil is assumed to be a truly elastic solid obeying Hooke's law in all directions. The design of an elastic foundation requires a knowledge of the subgrade reaction which is briefly discussed here. However, the elastic method does not readily lend itself to engineering applications because it is extremely difficult and solutions are available for only a few extremely simple cases.

7.2   SAFE BEARING PRESSURES FOR MAT FOUNDATIONS ON SAND AND CLAY

Mats on Sand

Because the differential settlements of a mat foundation are less than those of a spread foundation designed for the same soil pressure, it is reasonable to permit larger safe soil pressures on a raft foundation. Experience has shown that a pressure approximately twice as great as that allowed for individual footings may be used because it does not lead to detrimental differential settlements. The maximum settlement of a mat may be about 50 mm (2 in) instead of 25 mm as for a spread foundation.

The shape of the curve in Fig. 6.3 (a) shows that the net soil pressure corresponding to a given settlement is practically independent of the width of the footing or mat when the width becomes large. The safe soil pressure for design may with sufficient accuracy be taken as twice the pressure indicated in Fig. 6.5. (Peck *et al*, 1974, now modified) recommend the following equation for computing net safe pressure,

$$q_s = 21 \, N_{cor} \, \text{kPa} \qquad (7.1)$$

$$\text{for } 5 < N_{cor} < 50$$

where N_{cor} is the SPT value corrected for energy, overburden pressure and field procedures.

Equation (7.1.), gives q_s values above the water table. A correction factor should be used for the presence of a water table as explained in Chapter 5.

Peck *et al*, (1974), also recommend that the q_s values as given by Eq. (7.1) may be increased somewhat if bedrock is encountered at a depth less than about one-half the width of the raft.

The value of N to be considered is the average of the values obtained up to a depth equal to the least width of the raft. If the average value of N after correction for the influence of overburden pressure and dilatancy is less than about 5, Peck *et al*, say that the sand is generally considered to be too loose for the successful use of a raft foundation. Either the sand should be compacted or else the foundation should be established on piles or piers.

The minimum depth of foundation recommended for a raft is about 2.5 m below the surrounding ground surface. Experience has shown that if the surcharge is less than this amount, the edges of the raft settle appreciably more than the interior because of a lack of confinement of the sand.

Safe Bearing Pressures of Mats on Clay

The quantity in Eq. (5.22b), is the net bearing capacity q_{nu} at the elevation of the base of the raft in excess of that exerted by the surrounding surcharge. Likewise, in Eq. (5.22c), q_{na} is the net allowable soil pressure. By increasing the depth of excavation, the pressure that can safely be exerted by the building is correspondingly increased. This aspect of the problem is considered further in Section 7.10 in *floating foundation.*

As for footings on clay, the factor of safety against failure of the soil beneath a mat on clay should not be less than 3 under normal loads, or less than 2 under the most extreme loads.

The settlement of the mat under the given loading condition should be calculated as per the procedures explained in Chapter 6. The net safe pressure should be decided on the basis of the permissible settlement.

7.3 ECCENTRIC LOADING

When the resultant of loads on a footing does not pass through the centre of the footing, the footing is subjected to what is called *eccentric loading*. The loads on the footing may be vertical or inclined. If the loads are inclined it may be assumed that the horizontal component is resisted by the frictional resistance offered by the base of the footing. The vertical component in such a case is the only factor for the design of the footing. The effects of eccentricity on bearing pressure of the footings have been discussed in Chapter 5.

7.4 THE COEFFICIENT OF SUBGRADE REACTION

The coefficient of subgrade reaction is defined as the ratio between the pressure against the footing or mat and the settlement at a given point expressed as

$$k_s = \frac{q}{S} \tag{7.2}$$

where, k_s = ceofficient of subgrade reaction expressed as force/length^3 (FL^{-3}),

q = pressure on the footing or mat at a given point expressed as force/length^2 (FL^{-2}),

S = settlement of the same point of the footing or mat in the corresponding unit of length.

In other words the coefficient of subgrade reaction is the unit pressure required to produce a unit settlement. In clayey soils, settlement under the load takes place over a long period of time and the coefficient should be determined on the basis of the final settlement. On purely granular soils, settlement takes place shortly after load application. Equation (7.2) is based on two simplifying assumptions:

1. The value of k_s is independent of the magnitude of pressure.
2. The value of k_s has the same value for every point on the surface of the footing.

Both the assumptions are strictly not accurate. The value of k_s decreases with the increase of the magnitude of the pressure and it is not the same for every point of the surface of the footing as the settlement of a flexible footing varies from point to point. However, the method is supposed to give realistic values for contact pressures and is suitable for beam or mat design when only a low order of settlement is required.

Factors Affecting the Value of k_s

Terzaghi (1955), discussed the various factors that affect the value of k_s. A brief description of his arguments is given below.

Consider two foundation beams of widths B_1 and B_2 such that $B_2 = nB_1$ resting on a compressible subgrade and each loaded so that the pressure against the footing is uniform and equal to q for both the beams (Fig. 7.1b). Consider the same points on each beam and, let

$$y_1 = \text{settlement of beam of width } B_1$$
$$y_2 = \text{settlement of beam of width } B_2$$

Hence $k_{s1} = \dfrac{q}{y_1}$ and $k_{s2} = \dfrac{q}{y_2}$

If the beams are resting on a subgrade whose deformation properties are more or less independent of depth (such as a stiff clay) then it can be assumed that the settlement increases in simple proportion to the depth of the pressure bulb.

Then $y_2 = ny_1$

and $$k_{s2} = \frac{q}{ny_1} = \frac{q}{y_1}\frac{B_1}{B_2} = k_{s1}\frac{B_1}{B_2} \tag{7.3}$$

A general expression for k_s can now be obtained if we consider B_1 as being of unit width (Terzaghi used a unit width of one foot which converted to metric units may be taken as equal to 0.30 m).

Hence by putting $B_1 = 0.30$ m, $k_s = k_{s2}$, $B = B_2$, we obtain

$$k_s = 0.3 \, \frac{k_{s1}}{B} \qquad (7.4)$$

where k_s is the coefficient of subgrade reaction of a long footing of width B metres and resting on stiff clay; k_{s1} is the coefficient of subgrade reaction of a long footing of width 0.30 m (approximately), resting on the same clay. It is to be noted here that the value of k_{s1} is derived from ultimate settlement values, that is, after consolidation settlement is completed.

If the beams are resting on clean sand, the final settlement values are obtained almost instantaneously. Since the modulus of elasticity of sand increases with depth, the deformation characteristic of the sand change and become less compressible with depth. Because of this characteristic of sand, the lower portion of the bulb of pressure for beam B_2 is less compressible than that of the sand enclosed in the bulb of pressure of beam B_1.

The settlement value y_2 lies somewhere between y_1 and ny_1. It has been shown experimentally (Terzaghi and Peck, 1948) that the settlement, y, of a beam of width B resting on sand is given by the expression

$$y = y_1 \left(\frac{2B}{B + 0.3} \right)^2 \qquad (7.5)$$

where y_1 = settlement of a beam of width 0.30 m and subjected to the same reactive pressure as the beam of width B metres.

Hence, the coefficient of subgrade reaction k_s of a beam of width B metres can be obtained from the following equation

$$k_s = \frac{q}{y} = \frac{q}{y_1} \left(\frac{B + 0.30}{2B} \right)^2 = k_{s1} \left(\frac{B + 0.30}{2B} \right)^2 \qquad (7.6)$$

where, k_{s1} = coefficient of subgrade reaction of a beam of width 0.30 m resting on the same sand.

Measurement of k_{s1}

A value for k_{s1} for a particular subgrade can be obtained by carrying out plate load tests. The standard size of plate used for this purpose is 0.30×0.30 m size. Let k_1 be the subgrade reaction for a plate of size 0.30×0.30 m size.

From experiments it has been found that $k_{s1} \approx k_1$ for sand subgrades, but for clays k_{s1} varies with the length of the beam. Terzaghi (1955) gives the following formula for clays

$$k_{s1} = k_1 \left(\frac{L + 0.152}{1.5L} \right) \qquad (7.7a)$$

where, L = length of the beam in metres and the width of the beam = 0.30 m. For a very long beam on clay subgrade, we may write

$$k_{s1} = \frac{k_1}{1.5} \qquad (7.7b)$$

Procedure to Find k_s
For sand

1. Determine k_1 from plate load test or from estimation.

2. Since $k_{s1} \approx k_1$, use Eq. (7.6) to determine k_s for sand for any given width B metre.

For clay

1. Determine k_1 from plate load test or from estimation.
2. Determine k_{s1} from Eq. (7.7a) as the length of beam is known.
3. Determine k_s from Eq. (7.4) for the given width B metres.

When plate load tests are used, k_1 may be found by one of the two ways,

1. A bearing pressure equal to not more than the ultimate pressure and the corresponding settlement is used for computing k_1.
2. Consider the bearing pressure corresponding to a settlement of 1.3 mm for computing k_1.

Estimation of k_1 Values

Plate load tests are both costly and time consuming. Generally, a designer requires only the values of the bending moments and shear forces within the foundation. With even a relatively large error in the estimation of k_1, moments and shear forces can be calculated with little error (Terzaghi, 1955); an error of 100 percent in the estimation of k_s may change the structural behaviour of the foundation by up to 15 percent only.

In the absence of plate load tests, estimated values of k_1 and hence k_s are used. The values suggested by Terzaghi for k_1 (converted into SI units) are given in Table 7.1.

Table 7.1 (a) k_1 values for foundations on sand (MN/m³)

Relative density	Loose	Medium	Dense
SPT values (Uncorrected)	< 10	10–30	> 30
Soil, dry or moist	15	45	175
Soil submerged	10	30	100

Table 7.1 (b) k_1 values for foundation on clay

Consistency	Stiff	Very stiff	Hard
c_u (kN/m²)	50–100	100–200	> 200
k_1 (MN/m³)	25	50	100

Source: Terzaghi (1955).

7.5 PROPORTIONING OF CANTILEVER FOOTING

Strap or cantilever footings are designed on the basis of the following assumptions:

1. The strap is infinitely stiff. It serves to transfer the column loads to the soil with equal and uniform soil pressure under both the footings.
2. The strap is a pure flexural member and does not take soil reaction. To avoid bearing on the bottom of the strap a few centimeters of the underlying soil may be loosened prior to the placement of concrete.

A strap footing is used to connect an eccentrically loaded column footing close to the property line to an interior column as shown in Fig. 7.2.

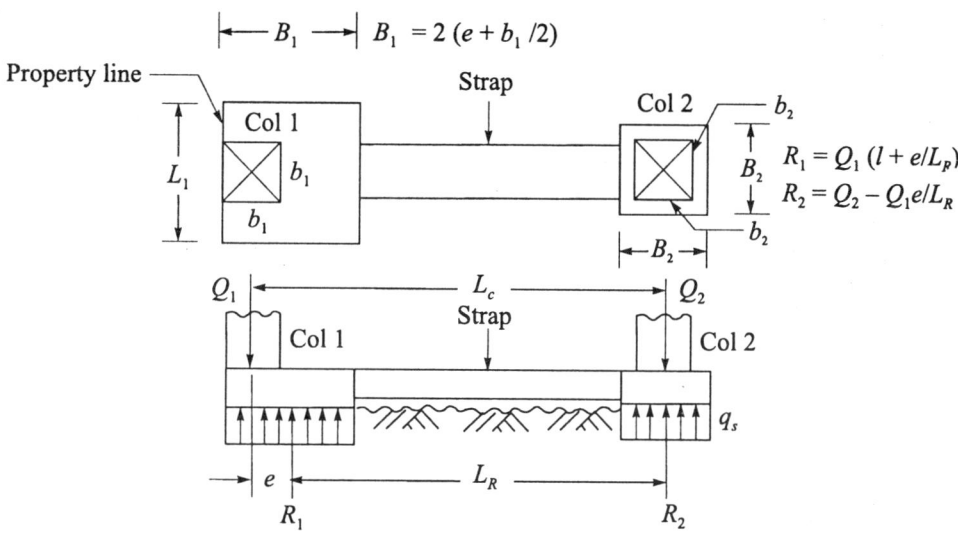

Fig. 7.2 Principles of cantilever or strap footing design

With the above assumptions, the design of a strap footing is a simple procedure. It starts with a trial value of e, Fig. 7.2. Then the reactions R_1 and R_2 are computed by the principle of statics. The tentative footing areas are equal to the reactions R_1 and R_2 divided by the safe bearing pressure q_s. With tentative footing sizes, the value of e is computed. These steps are repeated until the trial value of e is identical with the final one. The shears and moments in the strap are determined, and the straps designed to withstand the shear and moments. The footings are assumed to be subjected to uniform soil pressure and designed as simple spread footings. Under the assumptions given above the resultants of the column loads Q_1 and Q_2 would coincide with the centre of gravity of the two footing areas. Theoretically, the bearing pressure would be uniform under both the footings. However, it is possible that sometimes the full design live load acts upon one of the columns while the other may be subjected to little live load. In such a case, the full reduction of column load from Q_2 to R_2 may not be realised. It seems justified then that in designing the footing under column Q_2, only the dead load or dead load plus reduced live load should be used on column Q_1.

The equations for determining the position of the reactions (Fig. 7.2) are

$$R_1 = Q_1 \left(1 + \frac{e}{L_R}\right), \qquad R_2 = Q_2 - \frac{Q_1 e}{L_R} \tag{7.8}$$

where R_1 and R_2 = reactions for the column loads Q_1 and Q_2 respectively, e = distance of R_1 from Q_1, L_R = distance between R_1 and R_2.

7.6 DESIGN OF COMBINED FOOTINGS BY RIGID METHOD (CONVENTIONAL METHOD)

The rigid method of design of combined footings assumes that:

1. The footing or mat is infinitely rigid, and therefore, the deflection of the footing or mat does not influence the pressure distribution.

2. The soil pressure is distributed in a straight line or a plane surface such that the centroid of the soil pressure coincides with the line of action of the resultant force of all the loads acting on the foundation.

Design of Combined Footings

Two or more columns in a row joined together by a stiff continuous footing form a combined footing as shown in Fig. 7.3 (a). The procedure of design for a combined footing is as follows:

1. Determine the total column loads $\Sigma Q = Q_1 + Q_2 + Q_3 + \cdots$ and location of the line of action of the resultant ΣQ. If any column is subjected to bending moment, the effect of the moment should be taken into account.

2. Determine the pressure distribution q per lineal length of footing.

3. Determine the width, B, of the footing.

4. Draw the shear diagram along the length of the footing. By definition, the shear at any section along the beam is equal to the summation of all vertical forces to the left or right of the section. For example, the shear at a section immediately to the left of Q_1 is equal to the area $abcd$, and immediately to the right of Q_1 is equal to $(abcd - Q_1)$ as shown in Fig. 7.3 (a).

5. Draw the moment diagram along the length of the footing. By definition the bending moment at any section is equal to the summation of moment due to all the forces and reaction to the left (or right) of the section. It is also equal to the area under the shear diagram to the left (or right) of the section.

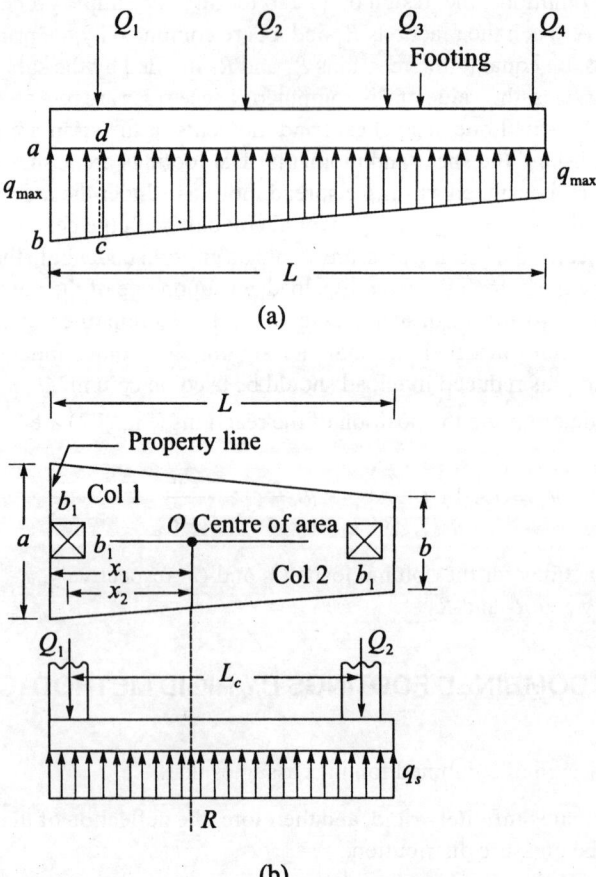

(a)

(b)

Fig. 7.3 Combined or trapezoidal footing design: (a) Combined footing, (b) trapezoidal combined footing

6. Design the footing as a continuous beam to resist the shear and moment.

7. Design the footing for transverse bending in the same manner as for spread footings.

It should be noted here that the end column along the property line may be connected to the interior column by a rectangular or trapezoidal footing. In such a case no strap is required and both the columns together will be a combined footing as shown in Fig. 7.3 (b). It is necessary that the centre of area of the footing must coincide with the centre of loading for the pressure to remain uniform.

7.7 DESIGN OF MAT FOUNDATION BY RIGID METHOD

In the conventional rigid method the mat is assumed to be infinitely rigid and the bearing pressure against the bottom of the mat follows a planar distribution, where the centroid of the bearing pressure coincides with the line of action of the resultant force of all loads acting on the mat. The procedure of design is as follows:

1. The column loads of all the columns coming from the superstructure are calculated as per standard practice. The loads include live and deal loads.

2. Determine the line of action of the resultant of all the loads. However, the weight of the mat is not included in the structural design of the mat because every point of the mat is supported by the soil under it, causing no flexural stresses.

3. Calculate the soil pressure at desired locations by the use of Eq. (5.37a)

$$q = \frac{Q_t}{A} \pm \frac{Q_t e_x}{I_y} x \pm \frac{Q_t e_y}{I_x} y \qquad (7.9)$$

where, $Q_t = \Sigma Q =$ total load on the mat,

$A =$ total area of the mat,

$x, y =$ coordinates of any given point on the mat with respect to the x and y axes passing through the centroid of the area of the mat,

$e_x, e_y =$ eccentricities of the resultant force,

$I_x, I_y =$ moments of inertia of the mat with respect to the x and y axes respectively.

4. The mat is treated as a whole in each of two perpendicular directions. Thus, the total shear force acting on any section cutting across the entire mat is equal to the arithmetic sum of all forces and reactions (bearing pressure) to the left (or right) of the section. The total bending moment acting on such a section is equal to the sum of all the moments to the left (or right) of the section.

7.8 DESIGN OF COMBINED FOOTINGS BY ELASTIC LINE METHOD

The relationship between deflection, y, at any point on an elastic beam and the corresponding bending moment M may be expressed by the equation

$$EI \frac{d^2 y}{dx^2} = M \qquad (7.10)$$

The equations for shear V and reaction q at the same point may be expressed as

$$EI \frac{d^3 y}{dx^3} = V \qquad (7.11)$$

$$EI \frac{d^4 y}{dx^4} = q \tag{7.12}$$

where x is the coordinate along the length of the beam.

From the basic assumption of an elastic foundation

$$q = - yBk_s$$

where, B = width of footing, k_s = coefficient of subgrade reaction.

Substituting for q, Eq. (7.12), may be written as

$$EI \frac{d^4 y}{dx^4} = - yBk_s \tag{7.13}$$

The classical solutions of Eq. (7.13) being of closed form, are not general in their application. Hetenyi (1946), developed equations for a load at any point along a beam. The development of solutions is based on the concept that the beam lies on a bed of elastic springs which is based on Winkler's hypothesis. As per this hypothesis, the reaction at any point on the beam depends only on the deflection at that point.

Methods are also available for solving the beam-problem on an elastic foundation by the method of finite differences (Malter, 1958). The finite element method has been found to be the most efficient of the methods for solving beam-elastic foundation problem. Computer programmes are available for solving the problem.

Since all the methods mentioned above are quite involved, they are not dealt with here. Interested readers may refer to Bowles (1996).

7.9   DESIGN OF MAT FOUNDATIONS BY ELASTIC PLATE METHOD

Many methods are available for the design of mat-foundations. The one that is very much in use is the finite difference method. This method is based on the assumption that the subgrade can be substituted by a bed of uniformly distributed coil springs with a spring constant k_s which is called the *coefficient of subgrade reaction*. The finite difference method uses the fourth order differential equation

$$\nabla^4 w = \frac{q - k_s w}{D}$$

$$\nabla^4 w = \frac{\partial^4 w}{\partial x^4} + 2 \frac{\partial^4 w}{\partial x^2 \partial y^2} + \frac{\partial^4 w}{\partial y^4} \tag{7.14}$$

where, q = subgrade reaction per unit area,

k_s = coefficient of subgrade reaction,

w = deflection,

D = rigidity of the mat = $\dfrac{Et^3}{12\left(1 - \mu^2\right)}$,

E = modulus of elasticity of the material of the footing,

t = thickness of mat,

μ = Poisson's ratio.

Equation (7.14) may be solved by dividing the mat into suitable square grid elements, and writing difference equations for each of the grid points. By solving the simultaneous equations so obtained the deflections at all the grid points are obtained. The equations can be solved rapidly with an electronic computer. After the deflections are known, the bending moments are calculated using the relevant difference equations.

Interested readers may refer to Teng (1969) or Bowles (1996) for a detailed discussion of the method.

7.10 FLOATING FOUNDATION

General Consideration

A *floating foundation* for a building is defined as a foundation, in which the weight of the building is approximately equal to the full weight including water of the soil removed from the site of the building. This principle of flotation may be explained with reference to Figs 7.4. and 7.4 (a) shows a horizontal ground surface with a horizontal water table at a depth d_w below the ground surface. Figure 7.4 (b) shows an excavation made in the ground to a depth D where $D > d_w$, and Fig. 7.4 (c) shows a structure built in the excavation and completely filling it.

If the weight of the building is equal to the weight of the soil and water removed from the excavation, then it is evident that the total vertical pressure in the soil below depth D in Fig. 7.4 (c) is the same as in Fig. 7.4 (a) before excavation.

Since the water level has not changed, the neutral pressure and the effective pressure are therefore unchanged. Since settlements are caused by an increase in effective vertical pressure, if we could move from Fig. 7.4 (a) to Fig. 7.4 (c) without the intermediate case of 7.4 (b), the building in Fig. 7.4 (c) would not settle at all.

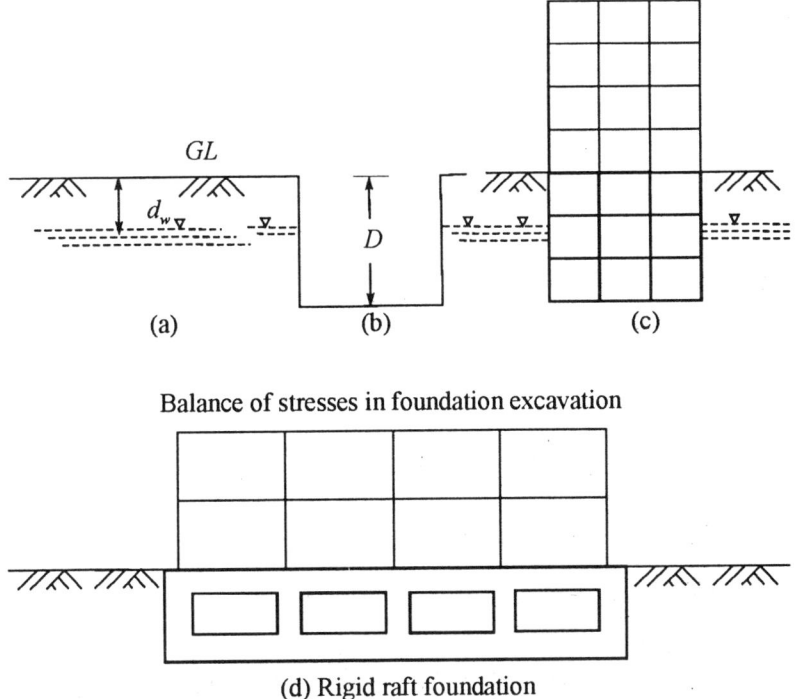

Balance of stresses in foundation excavation

(d) Rigid raft foundation

Fig. 7.4 Principles of floating foundation and a typical rigid raft foundation

This is the principle of a floating foundation, an exact balance of weight removed against weight imposed. The result is zero settlement of the building.

However, it may be noted, that we cannot jump from the stage shown in Fig. 7.4 (a) to the stage in Fig. 7.4 (c) without passing through stage 7.4 (b). The excavation stage of the building is the critical stage.

Cases may arise where we cannot have a fully floating foundation. The foundations of this type are sometimes called *partly compensated foundations* (as against *fully compensated* or *fully floating foundations*).

While dealing with floating foundations, we have to consider the following two types of soils. They are:

Type 1: The foundation soils are of such a strength that shear failure of soil will not occur under the building load but the settlements and particularly differential settlements, will be too large and will constitute *failure* of the structure. A floating foundation is used to reduce settlements to an acceptable value.

Type 2: The shear strength of the foundation soil is so low that rupture of the soil would occur if the building were to be founded at ground level. In the absence of a strong layer at a reasonable depth, the building can only be built on a floating foundation which reduces the shear stresses to an acceptable value. Solving this problem solves the settlement problem.

In both the cases, a rigid raft or box type of foundation is required for the floating foundation [Fig. 7.4 (d)].

Problems to be Considered in the Design of a Floating Foundation

The following problems are to be considered during the design and construction stage of a floating foundation.

1. Excavation

The excavation for the foundation has to be done with care. The sides of the excavation should suitably be supported by sheet piling, soldier piles and timber or some other standard method.

2. Dewatering

Dewatering will be necessary when excavation has to be taken below the water table level. Care has to be taken to see that the adjoining structures are not affected due to the lowering of the water table.

3. Critical depth

In Type 2 foundations the shear strength of the soil is low and there is a theoretical limit to the depth, to which an excavation can be made. Terzaghi (1943), has proposed the following equation for computing the critical depth D_c,

$$D_c = \frac{5.7 s}{\gamma - (s/B)\sqrt{2}} \qquad (7.15)$$

for an excavation which is long compared to its width

where γ = unit weight of soil,

s = shear strength of soil = $q_u/2$,

B = width of foundation,

L = length of foundation.

Skempton (1951) proposes the following equation for D_c, which is based on actual failures in excavations

$$D_c = N_c \frac{s}{\gamma} \tag{7.16}$$

or the factor of safety F_s against bottom failure for an excavation of depth D is

$$F_s = N_c \frac{s}{\gamma D + p}$$

where N_c is the bearing capacity factor as given by Skempton, and p is the surcharge load. The values of N_c may be obtained from Fig. 5.8. The above equations may be used to determine the maximum depth of excavation.

4. Bottom heave

Excavation for foundations reduces the pressure in the soil below the founding depth which results in the heaving of the bottom of the excavation. Any heave which occurs will be reversed and appear as settlement during the construction of the foundation and the building. Though heaving of the bottom of the excavation cannot be avoided it can be minimised to a certain extent. There are three possible causes of heave:

1. Elastic movement of the soil as the existing overburden pressure is removed.
2. A gradual swelling of soil due to the intake of water if there is some delay for placing the foundation on the excavated bottom of the foundation.
3. Plastic inward movement of the surrounding soil.

The last movement of the soil can be avoided by providing proper lateral support to the excavated sides of the trench.

Heaving can be minimised by phasing out excavation in narrow trenches and placing the foundation soon after excavation. It can be minimised by lowering the water table during the excavation process. Friction piles can also be used to minimise the heave. The piles are driven either before excavation commences or when the excavation is at half depth and the pile tops are pushed down to below foundation level. As excavation proceeds, the soil starts to expand but this movement is resisted by the upper part of the piles which go into tension. This heave is prevented or very much reduced.

It is only a *practical and pragmatic approach* that would lead to a safe and sound settlement free floating (or partly floating) foundation.

Example 7.1

A beam of length 4 m and width 0.75 m rests on stiff clay. A plate load test carried out at the site with the use of a square plate of size 0.30 m gives a coefficient of subgrade reaction k_1 equal to 25 MN/m^3. Determine the coefficient of subgrade reaction k_s for the beam.

Solution

First determine k_{s1} from Eq. (7.7a) for a beam of 0.30 m wide and length 4 m. Next determine k_s from Eq. (7.4) for the same beam of width 0.75 m.

$$k_{s1} = k_1 \left(\frac{L + 0.152}{1.5 L} \right) = 25 \left(\frac{4 + 0.152}{1.5 \times 4} \right) = 17.3 \text{ MN/m}^3$$

$$k_s = 0.3 \frac{k_{s1}}{B} = \frac{0.3 \times 17.3}{0.75} \approx 7 \text{ MN/m}^3$$

Example 7.2

A beam of length 4 m and width 0.75 m rests in dry medium dense sand. A plate load test carried out at the same site and at the same level gave a coefficient of subgrade reaction k_1 equal to 47 MN/m³. Determine the coefficient of subgrade reaction for the beam.

Solution

For sand the coefficient of subgrade reaction, k_{s1}, for a long beam of width 0.3 m is the same as that for a square plate of size 0.3 × 0.3 m that is $k_{s1} = k_s$. k_s now can be found from Eq. (7.6) as

$$k_s = k_1 \left(\frac{B + 0.3}{2B} \right)^2 = 47 \left(\frac{0.75 + 0.30}{1.5} \right)^2 = 23 \text{ MN/m}^3$$

Example 7.3

The following information is given for proportioning a cantilever footing with reference to Fig. 7.2.

Column loads: $Q_1 = 1,455$ kN, $Q_2 = 1,500$ kN

Size of column: 0.5 × 0.5 m.

$$L_c = 6.2 \text{ m}, \ q_s = 384 \text{ kN/m}^2$$

It is required to determine the size of the footings for columns 1 and 2.

Solution

Assume the width of the footing for column $1 = B_1 = 2$ m.

First trial

Try $\qquad\qquad e = 0.5$ m.

Now, $\qquad\qquad L_R = 6.2 - 0.5 = 5.7$ m.

Reactions

$$R_1 = Q_1 \left(1 + \frac{e}{L_R} \right) = 1455 \left(1 + \frac{0.5}{5.7} \right) = 1,583 \text{ kN}$$

$$R_2 = \left(Q_2 - \frac{Q_1 e}{L_R} \right) = \left(1,500 - \frac{1,455 \times 0.5}{5.7} \right) = 1,372 \text{ kN}$$

Size of footings – First trial

Col. 1. Area of footing

$$A_1 = \frac{1,583}{384} = 4.122 \text{ sq. m}$$

Col. 2. Area of footing

$$A_2 = \frac{1,372}{384} = 3.57 \text{ sq. m}$$

Try 1.9 × 1.9 m

Second trial

New value of $e = \dfrac{B_1}{2} - \dfrac{b_1}{2} = \dfrac{2}{2} - \dfrac{0.5}{2} = 0.75$ m

New $L_R = 6.20 - 0.75 = 5.45$ m

$$R_1 = 1455 \left(1 + \dfrac{0.75}{5.45}\right) = 1,655 \text{ kN}$$

$$R_2 = \left(1,500 - \dfrac{1,455 \times 0.75}{5.45}\right) = 1,300 \text{ kN}$$

$$A_1 = \dfrac{1,655}{384} = 4.31 \text{ sq.m or } 2.08 \times 2.08 \text{ m}$$

$$A_2 = \dfrac{1,300}{384} = 3.88 \text{ sq.m or } 1.84 \times 1.84 \text{ m}$$

Check $\quad e = \dfrac{B_1}{2} - \dfrac{b_1}{2} = 1.04 - 0.25 = 0.79 \approx 0.75$ m

Use 2.08×2.08 m for Col. 1 and 1.90×1.90 m for Col. 2.

Note: Rectangular footings may be used for both the columns.

Example 7.4

Figure Ex. 7.4 gives a foundation beam with the vertical loads and moment acting thereon. The width of the beam is 0.70 m and depth 0.50 m. A uniform load of 16 kN/m (including the weight of the beam) is imposed on the beam. Draw (*a*) the base pressure distribution, (*b*) the shear force diagram, and (*c*) the bending moment diagram. The length of the beam is 8 m.

Solution

The steps to be followed are:

1. Determine the resultant vertical force R of the applied loadings and its eccentricity with respect to the centre of the beam.
2. Determine the maximum and minimum base pressures.
3. Draw the shear and bending moment diagrams.

$$R = 320 + 400 + 16 \times 8 = 848 \text{ kN}$$

Taking the moment about the right hand edge of the beam, we have

$$Rx = 848x = 320 \times 7 + 400 \times 1 + 16 \times \dfrac{8^2}{2} - 160 = 2992$$

or $\qquad x = \dfrac{2992}{848} = 3.528$ m

$e = 4.0 - 3.528 = 0.472$ m to the right of centre of the beam. Now from Eqs (5.39a and b), using $e_y = 0$,

$$q_{min}^{mix} = \dfrac{\Sigma Q}{A}\left(1 + \pm 6 \dfrac{e_x}{L}\right) = \dfrac{848}{8 \times 0.7}\left(1 \pm \dfrac{6 \times 0.472}{8}\right) = 205.02 \text{ or } 97.83 \text{ kN/m}^2$$

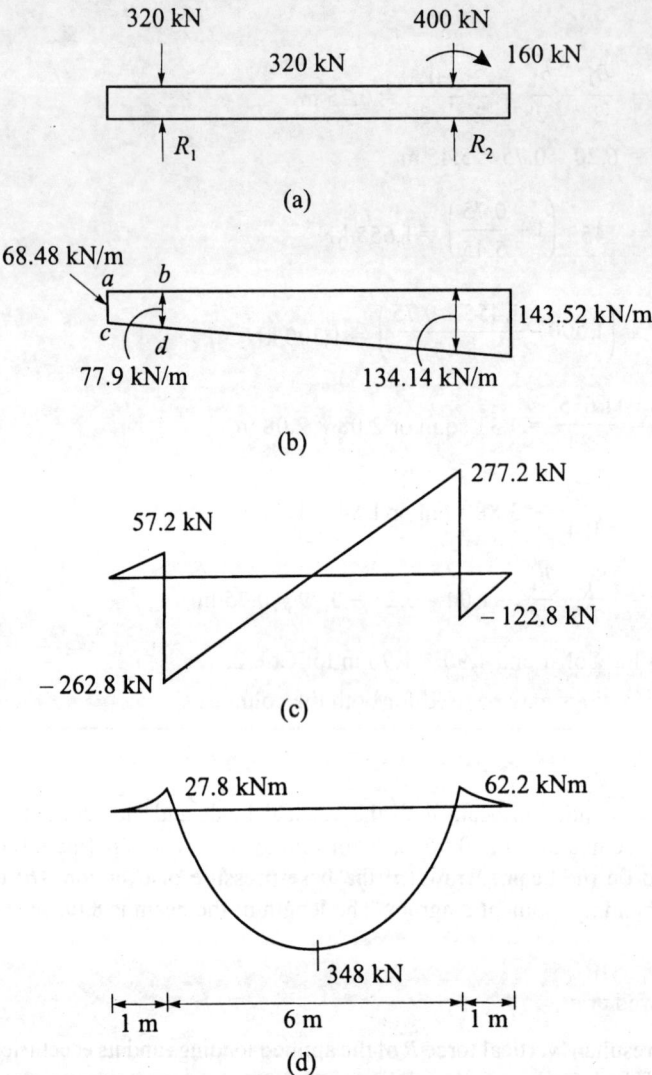

Fig. Ex. 7.4 (a) Applied load, (b) base reaction, (c) shear force diagram, (d) bending moment diagram

Convert the base pressures per unit area to load per unit length of beam.

The maximum vertical load $= 0.7 \times 205.02 = 143.52$ kN/m.

The minimum vertical load $= 0.7 \times 97.83 = 68.48$ kN/m.

The reactive loading distribution is given in Fig. Ex. 7.4 (b).

Shear force diagram

Calculation of shear for a typical point such as the reaction point R_1 [Fig. Ex. 7.4 (a)] is explained below.

Consider forces to the left of R_1 (without 320 kN).

Shear force

$V = $ upward shear force equal to the area $ab\,dc$ – downward force due to distributed load on beam ab

$$= \frac{68.48 + 77.9}{2} - 16 \times 1 = 57.2 \text{ kN}$$

Consider to the right of reaction point R_1 (with 320 kN).

$$V = -320 + 57.2 = -262.8 \text{ kN}.$$

In the same way, the shear at other points can be calculated. Figure Ex. 7.4 (c) gives the complete shear force diagram.

Bending moment diagram

Bending moment at the reaction point R_1 = moment due to force equal to the area *ab dc* + moment due to distributed load on beam *ab*

$$= 68.48 \times \frac{1}{2} + \frac{9.42}{2} \times \frac{1}{3} - 16 \times \frac{1}{2}$$

$$= 27.8 \text{ kN} - \text{m}$$

The moments at other points can be calculated in the same way. The complete moment diagram is given in Fig. Ex. 7.4 (d).

Example 7.5

The end column along a property line is connected to an interior column by a trapezoidal footing. The following data are given with reference to Fig. 7.3 (b):

Column loads:

$$Q_1 = 2,016 \text{ kN}, \ Q_2 = 1,560 \text{ kN}$$

Size of columns: 0.46 × 0.46 m

$$L_c = 5.48 \text{ m}$$

Determine the dimensions *a* and *b* of the trapezoidal footing. The net allowable bearing pressure $q_{na} = 190$ kPa.

Solution

Determine the centre of bearing pressure x_2 from the centre of Column 1. Taking moments of all the loads about the centre of Column 1, we have

$$(2,016 + 1,560) x_2 = 1,560 \times 5.48$$

$$x_2 = \frac{1,560 \times 5.48}{3,576} = 2.39 \text{ m}$$

Now

$$x_1 = 2.39 + \frac{0.46}{2} = 2.62 \text{ m}$$

Point O in Fig. 7.3 (b) is the centre of the area coinciding with the centre of pressure. From the allowable pressure $q_a = 190$ kPa, the area of the combined footing required is

$$A = \frac{3,576}{190} = 18.82 \text{ sq m}$$

From geometry, the area of the trapezoidal footing [Fig. 7.3 (b)] is

$$A = \frac{(a+b)L}{2} = \frac{(a+b)}{2} (5.94) = 18.82$$

or $\quad\quad\quad (a+b) = 6.34$ m

where, $\quad\quad\quad L = L_c + b_1 = 5.48 + 0.46 = 5.94$ m

From the geometry of the Fig. 7.3 (b), the distance of the centre of area x_1 can be written in terms of a, b and L as

$$x_1 = \frac{L}{3} \frac{2a+b}{a+b}$$

or $\quad\quad\quad \dfrac{2a+b}{a+b} = \dfrac{3x_1}{L} = \dfrac{3 \times 2.62}{5.94} = 1.32$ m

but $a + b = 6.32$ m or $b = 6.32 - a$. Now substituting for b, we have

$$\frac{2a + 6.34 - a}{6.34} = 1.32$$

and solving, $a = 2.03$ m, from which, $b = 6.34 - 2.03 = 4.31$ m.

7.11 PROBLEMS

7.1 A beam of length 6 m and width 0.80 m is founded on dense sand under submerged conditions. A plate load test with a plate of 0.30×0.30 m conducted at the site gave a value for the coefficient of subgrade reaction for the plate equal to 95 MN/m^3. Determine the coefficient of subgrade reaction for the beam.

7.2 If the beam in Prob. 7.1 is founded in very stiff clay with the value for k_1 equal to 45 MN/m^3, what is the coefficient of subgrade reaction for the beam?

7.3 Proportion a strap footing given the following data with reference to Fig. 7.2:

$$Q_1 = 580 \text{ kN}, Q_2 = 900 \text{ kN}$$

$$L_c = 6.2 \text{ m}, b_1 = 0.40 \text{ m}, q_s = 120 \text{ kPa}.$$

7.4 Proportion a rectangular combined footing given the following data with reference to Fig. 7.3 (the footing is rectangular instead of trapezoidal):

$$Q_1 = 535 \text{ kN}, Q_2 = 900 \text{ kN}, b_1 = 0.40 \text{ m},$$

$$L_c = 4.75 \text{ m}, q_s = 100 \text{ kPa}.$$

When a pile is driven into granular soil, the soil so displaced, equal to the volume of the driven pile, compacts the soil around the sides since the displaced soil particles enter the soil spaces of the adjacent mass which leads to densification of the mass. The pile that compacts the soil adjacent to it is sometimes called a _compaction pile._ The compaction of the soil mass around a pile increases its bearing capacity.

If a pile is driven into saturated silty or cohesive soil, the soil around the pile cannot be densified because of its poor drainage qualities. The displaced soil particles cannot enter the void space unless the water in the pores is pushed out. The stresses developed in the soil mass adjacent to the pile due to the driving of the pile have to be borne by the pore water only. This results in the development of pore water pressure and a consequent decrease in the bearing capacity of the soil. The soil adjacent to the piles is remolded and loses to a certain extent its structural strength. The immediate effect of driving a pile in a soil with poor drainage qualities is, therefore, to decrease its bearing strength. However, with the passage of time, the remolded soil regains part of its lost strength due to the reorientation of the disturbed particles (which is termed _thixotropy_) and due to consolidation of the mass. The advantages and disadvantages of driven piles are:

Advantages

1. Piles can be precast to the required specifications.
2. Piles of any size, length and shape can be made in advance and used at the site. As a result, the progress of the work will be rapid.
3. A pile driven into granular soil compacts the adjacent soil mass and as a result the bearing capacity of the pile is increased.
4. The work is neat and clean. The supervision of work at the site can be reduced to a minimum. The storage space required is very much less.
5. Driven piles may conveniently be used in places where it is advisable not to drill holes for fear of meeting ground water under pressure.
6. Driven pile are the most favoured for works over water such as piles in wharf structures or jetties.

Disadvantages

1. Precast or prestressed concrete piles must be properly reinforced to withstand handling stresses during transportation and driving.
2. Advance planning is required for handling and driving.
3. Requires heavy equipment for handling and driving.
4. Since the exact length required at the site cannot be determined in advance, the method involves cutting off extra lengths or adding more lengths. This increases the cost of the project.
5. Driven piles are not suitable in soils of poor drainage qualities. If the driving of piles is not properly phased and arranged, there is every possibility of heaving of the soil or the lifting of the driven piles during the driving of a new pile.
6. Where the foundations of adjacent structures are likely to be affected due to the vibrations generated by the driving of piles, driven piles should not be used.

Cast-in-situ Piles

Cast-in-situ piles are concrete piles. These piles are distinguished from drilled piers as small diameter piles. They are constructed by making holes in the ground to the required depth and then filling the hole with concrete. Straight bored piles or piles with one or more bulbs at intervals may be cast at the site. The latter type are called *under-reamed piles*. Reinforcement may be used as per the requirements. *Cast-in-situ* piles have advantages as well as disadvantages.

Advantages

1. Piles of any size and length may be constructed at the site.
2. Damage due to driving and handling that is common in precast piles is eliminated in this case.
3. These piles are ideally suited in places where vibrations of any type are required to be avoided to preserve the safety of the adjoining structure.
4. They are suitable in soils of poor drainage qualities since *cast-in-situ* piles do not significantly disturb the surrounding soil.

Disadvantages

1. Installation of *cast-in-situ* piles requires careful supervision and quality control of all the materials used in the construction.
2. The method is quite cumbersome. It needs sufficient storage space for all the materials used in the construction.
3. The advantage of increased bearing capacity due to compaction in granular soil that could be obtained by a driven pile is not produced by a *cast-in-situ* pile.
4. Construction of piles in holes where there is heavy current of ground water flow or artesian pressure is very difficult.

A straight bored pile is shown in Fig. 8.1 (a).

Driven and *Cast-in-situ* Piles

This type has the advantages and disadvantages of both the driven and the *cast-in-situ* piles. The procedure of installing a *driven* and *cast-in-situ* pile is as follows:

A steel shell is driven into the ground with the aid of a mandrel inserted into the shell. The mandrel is withdrawn and concrete is placed in the shell. The shell is made of corrugated and reinforced thin sheet steel (mono-tube piles) or pipes (Armco welded pipes or common seamless pipes). The piles of this type are called a shell type. The shell-less type is formed by withdrawing the shell while the concrete is being placed. In both the types of piles the bottom of the shell is closed with a conical tip which can be separated from the shell. By driving the concrete out of the shell an enlarged bulb may be formed in both the types of piles. Franki piles are of this type. The common types of driven and *cast-in-situ* piles are given in Fig. 8.1. In some cases the shell will be left in place and the tube is concreted. This type of pile is very much used in piling over water.

8.4 USES OF PILES

The major uses of piles are:

1. To carry vertical compression load.
2. To resist uplift load.
3. To resist horizontal or inclined loads.

Normally, vertical piles are used to carry vertical compression loads coming from superstructures such as buildings, bridges, etc. The piles are used in groups joined together by pile caps. The loads carried by the piles are transferred to the adjacent soil. If all the loads coming on the tops of piles are transferred to the tips, such piles are called *end-bearing* or *point-bearing piles.* However, if all the

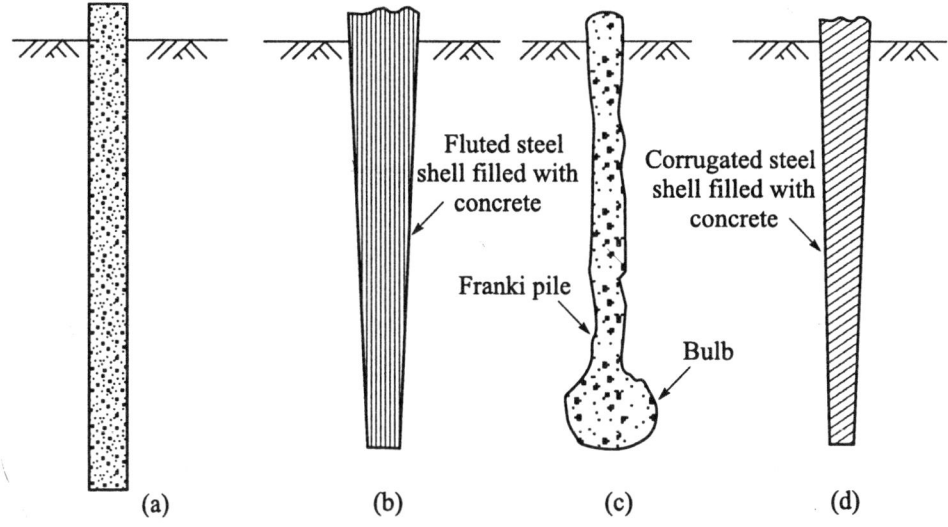

Fig. 8.1 Types of *cast-in-situ* and riven *cast-in-situ* concrete piles

load is transferred to the soil along the length of the pile such piles are called *friction piles*. If, in the course of driving a pile into granular soils, the soil around the pile gets compacted, such piles are called *compaction piles*. Figure 8.2 (a) shows piles used for the foundation of a multistoried building to carry loads from the superstructure.

Piles are also used to resist uplift loads. Piles used for this purpose are called *tension piles* or *uplift piles* or *anchor piles*. Uplift loads are developed due to hydrostatic pressure or overturning movement as shown in Fig. 8.2 (a).

Piles are also used to resist horizontal or inclined forces. Batter piles are normally used to resist large horizontal loads. Figure 8.2 (b) shows the use of piles to resist lateral loads.

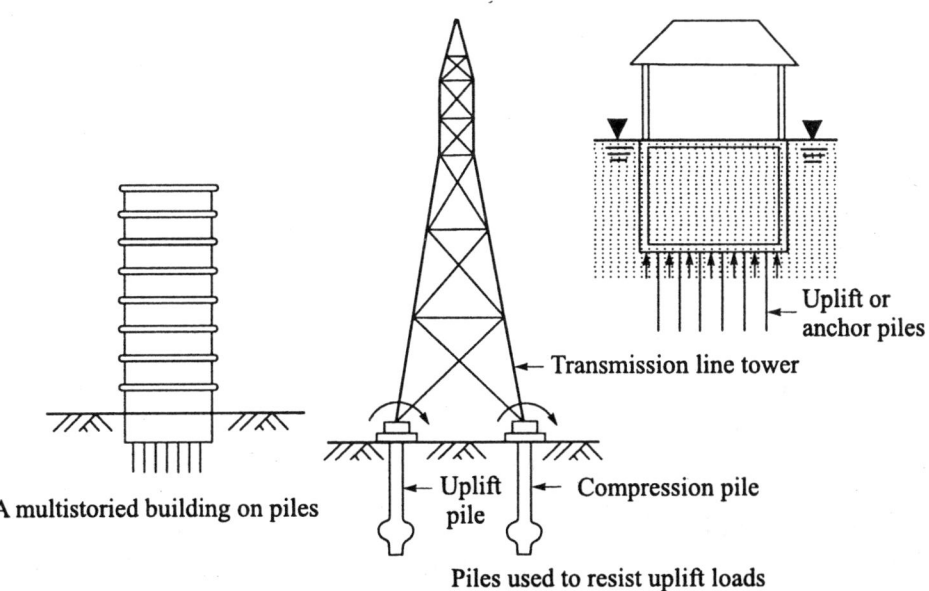

Fig. 8.2 (a) Piles for multistoried buildings and to resist up lift loads

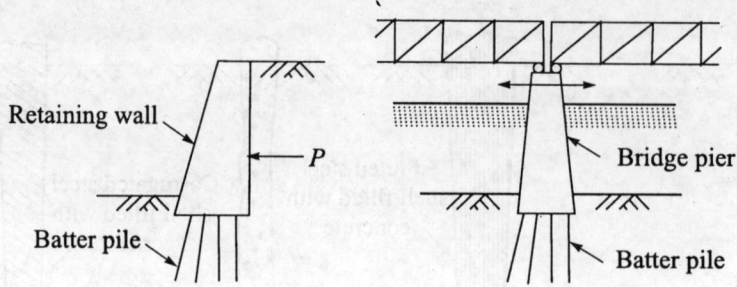

Fig. 8.2 (b) Piles used to resist lateral loads

8.5 SELECTION OF PILE

The selection of the type, length and capacity is usually made from estimation based on the soil conditions and the magnitude of the load. In large cities, where the soil conditions are well-known and where a large number of pile foundations have been constructed, the experience gained in the past is extremely useful. Generally, the foundation design is made on the preliminary estimated values. Before the actual construction begins, pile load tests must be conducted to verify the design values. The foundation design must be revised according to the test results. The factors that govern the selection of piles are:

1. Length of pile in relation to the load and type of soil,
2. Character of structure,
3. Availability of materials,
4. Type of loading,
5. Factors causing deterioration,
6. Ease of maintenance,
7. Estimated costs of types of piles, taking into account the initial cost, life expectancy and cost of maintenance, and
8. Availability of funds.

All the above factors have to be largely analysed before deciding up on a particular type.

8.6 INSTALLATION OF PILES

The method of installing a pile at a site depends upon the type of pile. The equipment required for this purpose varies. The following types of piles are normally considered for the purpose of installation.

1. Driven piles

The piles that come under this category are:

(*a*) Timber piles,
(*b*) Steel piles, *H*-section and pipe piles,
(*c*) Precast concrete or prestressed concrete piles, either solid or hollow sections.

2. Driven *cast-in-situ* piles

This involves driving of a steel tube to the required depth with the end closed by a detachable conical tip. The tube is next concreted and the shell is simultaneously withdrawn. In some cases the shell will not be withdrawn.

3. Bored *cast-in-situ* piles

Boring is done either by auguring or by percussion drilling. After boring is completed, the bore is concreted with or without reinforcement.

Pile Driving Equipment for Driven and Driven *Cast-in-situ* Piles

Pile driving equipment contains three parts. They are:

1. A pile frame,
2. Piling winch,
3. Impact hammers.

Pile frame

Pile driving equipment is required for *driven* piles or driven *cast-in-situ* piles. The driving pile frame must be such that it can be mounted on a standard tracked crane base machine for mobility on land sites or on framed bases for mounting on stagings or pontoons in offshore construction. Figure 8.3 gives a typical pile frame for both onshore and offshore construction. Both the types must be capable

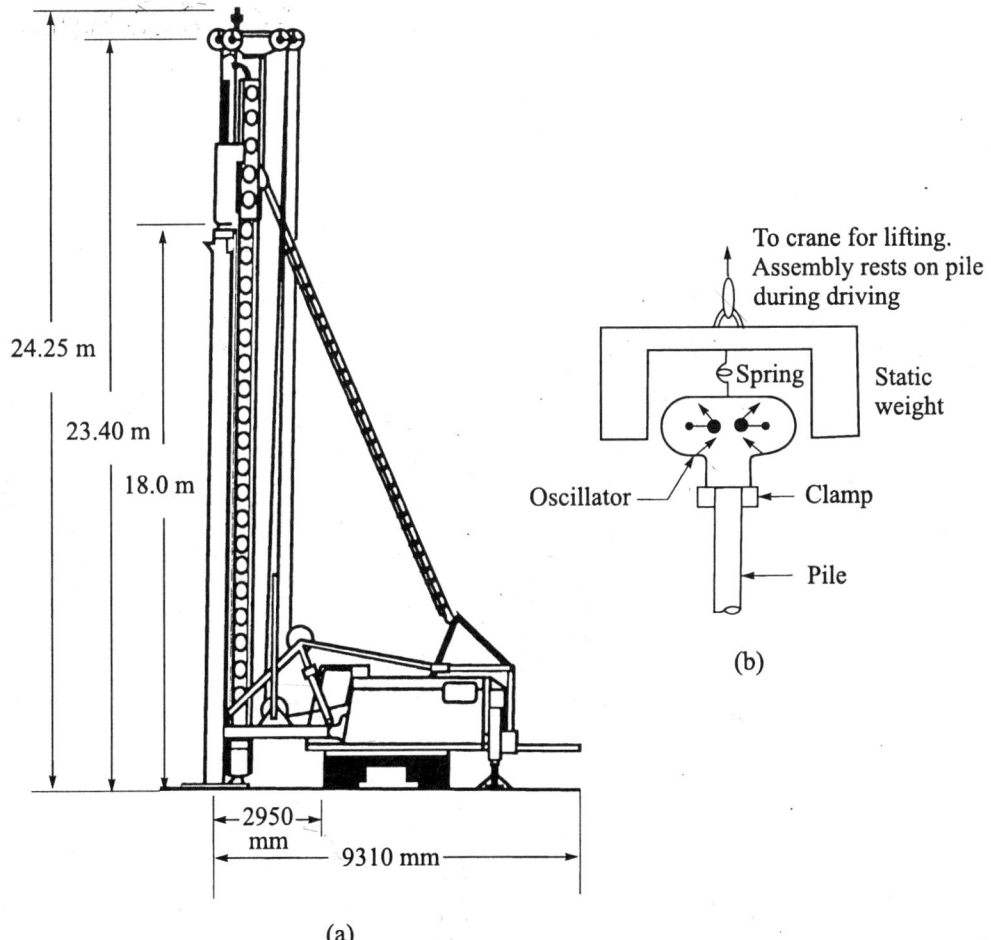

(a)

Fig. 8.3 Pile driving equipment and vibratory pile driver: (a) The ackermanns M14-5P pile frame, (b) diagrammatic sketch of vibratory pile driver

of full rotation and backward or forward raking. All types of frames consist essentially of *leaders,* which are a pair of steel members extending for the full height of the frame and which guide the hammer and pile as it is driven into the ground. Where long piles have to be driven the leaders can be extended at the top by a telescopic boom.

The base frame may be mounted on swivel wheels fitted with self-contained jacking screws for levelling the frame or it may be carried on steel rollers. The rollers run on steel girders or long timbers and the frame is moved along by winching from a deadman set on the roller track, or by turning the rollers by a tommy-bar placed in holes at the ends of the rollers. Movements parallel to the rollers are achieved by winding in a wire rope terminating in hooks on the ends of rollers; the frame then skids in either direction along the rollers. It is important to ensure that the pile frame remains in its correct position throughout the driving of a pile.

Piling winches

Piling winches are mounted on the base. Winches may be powered by steam, diesel or gasoline engines, or electric motors. Steam-powered winches are commonly used where steam is used for the piling hammer. Diesel or gasoline engines, or electric motors (rarely) are used in conjunction with drop hammers or where compressed air is used to operate the hammers.

Impact hammers

The impact energy for driving piles may be obtained by any one of the following types of hammers. They are:

1. Drop hammers,
2. Single-acting steam hammer,
3. Double-acting steam hammers,
4. Diesel hammer,
5. Vibratory hammer.

Drop hammers are at present used for small jobs. The weight is raised and allowed to fall freely on the top of the pile. The impact drives the pile into the ground.

In the case of a *Single-acting steam hammer* steam or air raises the moveable part of the hammer which then drops by gravity along. The blows in this case are much more rapidly delivered than for a drop hammer. The weights of hammers vary from about 1,500 to 10,000 kg with the length of stroke being about 90 cm. In general the ratio of ram weight to pile weight may vary from 0.5 to 1.0.

In the case of a *double-acting hammer* steam or air is used to raise the moveable part of the hammer and also to impart additional energy during the down stroke. The downward acceleration of the ram owing to gravity is increased by the acceleration due to steam pressure. The weights of hammers vary from about 350 to 2,500 kg. The length of stroke varies from about 20 to 90 cm. The rate of driving ranges from 300 blows per minute for the light types, to 100 blows per minute for the heaviest types.

Diesel or internal combustion hammers utilise diesel-fuel explosions to provide the impact energy to the pile. Diesel hammers have considerable advantage over steam hammers because they are lighter, more mobile and use a smaller amount of fuel. The weight of the hammer varies from about 1,000 to 2,500 kg.

The advantage of the power-hammer type of driving is that the blows fall in rapid succession (50 to 150 blows per minute) keeping the pile in continuous motion. Since the pile is continuously moving, the effects of the blows tend to convert to pressure rather than impact, thus reducing damage to the pile.

The *vibration method* of driving piles is now coming into prominence. Driving is quiet and does not generate local vibrations. Vibration driving utilises a variable speed oscillator attached to the top of the pile [Fig. 8.3 (b)]. It consists of two counter rotating eccentric weights which are in phase twice per cycle (180° apart) in the vertical direction. This introduces vibration through the pile which can be made to coincide with the resonant frequency of the pile. As a result, a push–pull effect is created at the pile tip which breaks-up the soil structure allowing easy pile penetration into the ground with a relatively small driving effort. Pile driving by the vibration method is quite common in Russia.

Jetting piles

Water jetting may be used to aid the penetration of a pile into dense sand or dense sandy gravel. Jetting is ineffective in firm to stiff clays or any soil containing much coarse to stiff cobbles or boulders.

Where jetting is required for pile penetration a stream of water is discharged near the pile point or along the sides of the pile through a pipe 5 to 7.5 cm in diameter. An adequate quantity of water is essential for jetting. Suitable quantities of water for jetting a 250 to 350 mm pile are

Fine sand	15–25 litres/second,
Coarse sand	25–40 litres/second,
Sandy gravels	45–600 litres/second.

A pressure of at least 5 kg/cm^2 or more is required.

8.7 LOAD TRANSFER MECHANISM

Statement of the Problem

Figure 8.4 (a) gives a single pile of uniform diameter d (circular or any other shape) and length L driven into a homogeneous mass of soil of known physical properties. A static vertical load is applied on the top. It is required to determine the ultimate bearing capacity Q_u of the pile.

When the ultimate load applied on the top of the pile is Q_u, a part of the load is transmitted to the soil along the length of the pile and the balance is transmitted to the pile base. The load transmitted to the soil along the length of the pile is called the *ultimate friction load* or *skin load* Q_f and that transmitted to the base is called the *base* or *point load* Q_b. The total ultimate load Q_u is expressed as the sum of these two, that is,

$$Q_u = Q_b + Q_f = q_b A_b + f_s A_s \tag{8.1}$$

where, Q_u = ultimate load applied on the top of the pile,

q_b = ultimate unit bearing capacity of the pile at the base,

A_b = bearing area of the base of the pile,

A_s = total surface area of pile embedded below ground surface,

f_s = unit skin friction (ultimate).

Load Transfer and Types of Failure

Consider the pile shown in Fig. 8.4 (b) is loaded to failure by gradually increasing the load on the top. If settlement of the top of the pile is measured at every stage of loading after an equilibrium condition is attained, a load settlement curve as shown in Fig. 8.4 (c) can be obtained.

If the pile is instrumented, the load distribution along the pile can be determined at different stages of loading and plotted as shown in Fig. 8.4 (b).

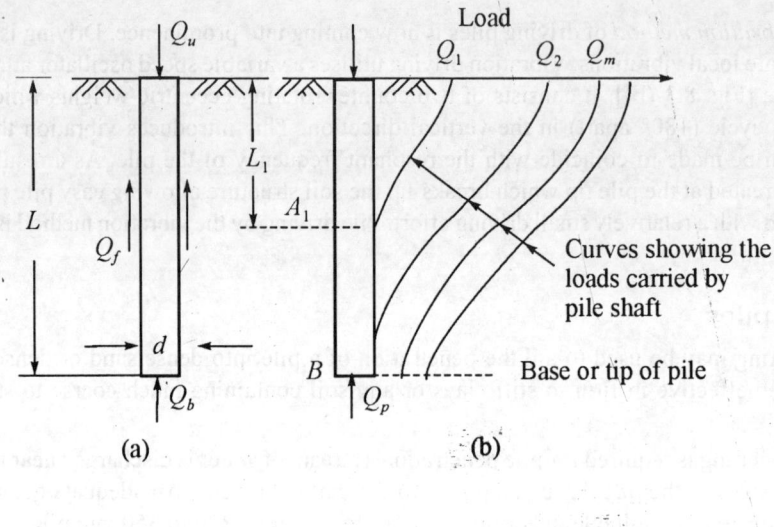

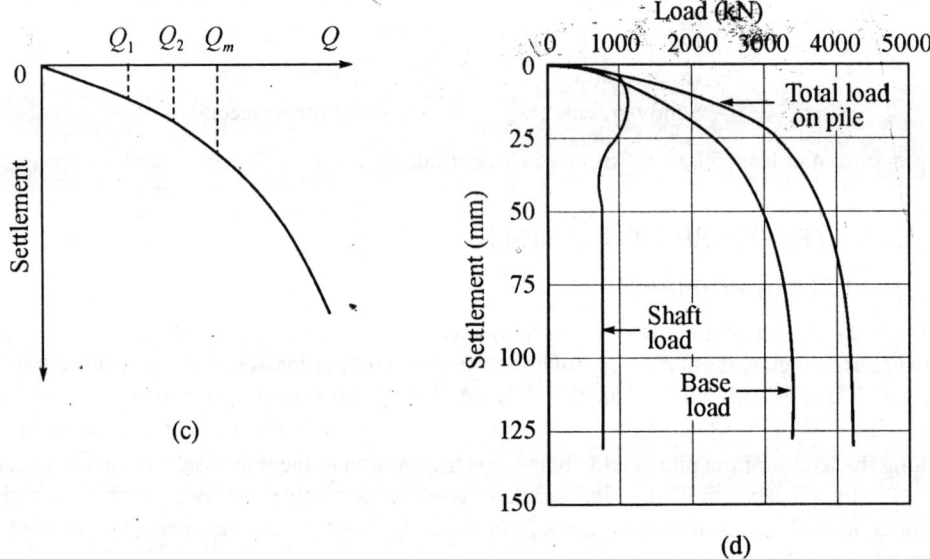

Fig. 8.4 Load transfer mechanism: (a) Single pile, (b) load-transfer curves, (c) load-settlement curve, (d) load-settlement relationships for large-diameter bored and cast-in-place piles (after Tomlinson, 1986)

When a load Q_1 acts on the pile head, the axial load at ground level is also Q_1, but at level A_1 [Fig. 8.4 (b)], the axial load is zero. The total load Q_1 is distributed as friction load within a length of pile L_1. The lower section A_1B of pile will not be affected by this load. As the load at the top is increased to Q_2, the axial load at the bottom of the pile is just zero. The total load Q_2 is distributed as friction load along the whole length of pile L. The friction load distribution curves along the pile shaft may be as shown in the figure. If the load put on the pile is greater than Q_2, a part of this load is transferred to the soil at the base as point load and the rest is transferred to the soil surrounding the pile. With the increase of load Q on the top, both the friction and point loads continue to increase. The friction load attains an ultimate value Q_f at a particular load level, say Q_m, at the top, and any further increment of load added to Q_m will not increase the value of Q_f. However, the point load, Q_p, still goes on increasing till the soil fails by punching shear failure. It has been determined by Van Wheele (1957), that the point load Q_p increases linearly with the elastic compression of the soil at the base.

The relative proportions of the loads carried by skin friction and base resistance depend on the shear strength and elasticity of the soil. Generally, the vertical movement of the pile which is required to mobilise full end resistance is much greater than that required to mobilise full skin friction. Experience indicates that in bored *cast-in-situ* piles full frictional load is normally mobilised at a settlement equal to 0.5 to 1 percent of pile diameter and the full base load Q_b at 10 to 20 percent of the diameter. But, if this ultimate load criterion is applied to piles of large diameter in clay, the settlement at the working load (with a factor of safety of 2 on the ultimate load) may be excessive. A typical load-settlement relationship of friction load and base load is shown in Fig. 8.4 (d) (Tomlinson, 1986) for a large diameter bored and *cast-in-situ* pile in clay. It may be seen from this figure that the full shaft resistance is mobilised at a settlement of only 15 mm whereas the full base resistance, and the ultimate resistance of the entire pile, is mobilised at a settlement of 120 mm. The shaft load at a settlement of 15 mm is only 1,000 kN which is about 25 percent of the base resistance. If a working load of 2,000 kN at a settlement of 15 mm is used for the design, at this working load, the full shaft resistance will have been mobilised whereas only about 50 percent of the base resistance has been mobilised. This means if piles are designed to carry a working load equal to one-third to one-half the total failure load, there is every likelihood of the shaft resistance being fully mobilised at the working load. This has an important bearing on the design.

The type of load-settlement curve for a pile depends on the relative strength values of the surrounding and underlying soil. Figure 8.5 gives the types of failure (Kézdi, 1975). They are as follows:

Figure 8.5 (a) represents a driven pile (wooden or reinforced concrete), whose tip bears on a very hard stratum (rock). The soil around the shaft is too weak to exert any confining pressure or lateral resistance. In such cases, the pile fails like a compressed, slender column of the same material; after a more or less elastic compression buckling occurs. The curve shows a definite failure load.

Figure 8.5 (b) is the type normally met in practice. The pile penetrates through layers of soil having low shear strength down to a layer having a high strength and the layer extending sufficiently below the tip of the pile. At ultimate load Q_u, there will be a base general shear failure at the tip of the pile, since the upper layer does not prevent the formation of a failure surface. The effect of the shaft friction is rather less, since the lower dense layer prevents the occurrence of excessive settlements. Therefore, the degree of mobilisation of shear stresses along the shaft will be low. The load settlement diagram is of the shape typical for a shallow footing on dense soil.

Figure 8.5 (c) shows the case where the shear strength of the surrounding soil is fairly uniform; therefore, a punching failure is likely to occur. The load-settlement diagram does not have a vertical tangent, and there is no definite failure load. The load will be carried by point resistance as well as by skin friction.

Figure 8.5 (d) is a rare case where the lower layer is weaker. In such cases, the load will be carried mainly by shaft friction, and the point resistance is almost zero. The load-settlement curve shows a vertical tangent, which represents the load when the shaft friction has been fully mobilised.

Figure 8.5 (e) is a case when a pull, $-Q$, acts on the pile. Since the point resistance is again zero the same diagram, as in Fig. 8.5 (d), will characterise the behaviour, but heaving occurs.

Definition of Failure Load

The methods of determining failure loads based on load-settlement curves are described in subsequent sections. However, in the absence of a load settlement curve, a failure load may be defined as that which causes a settlement equal to 10 percent of the pile diameter or width (as per the suggestion of Terzaghi) which is widely accepted by engineers. However, if this criterion is applied to piles of

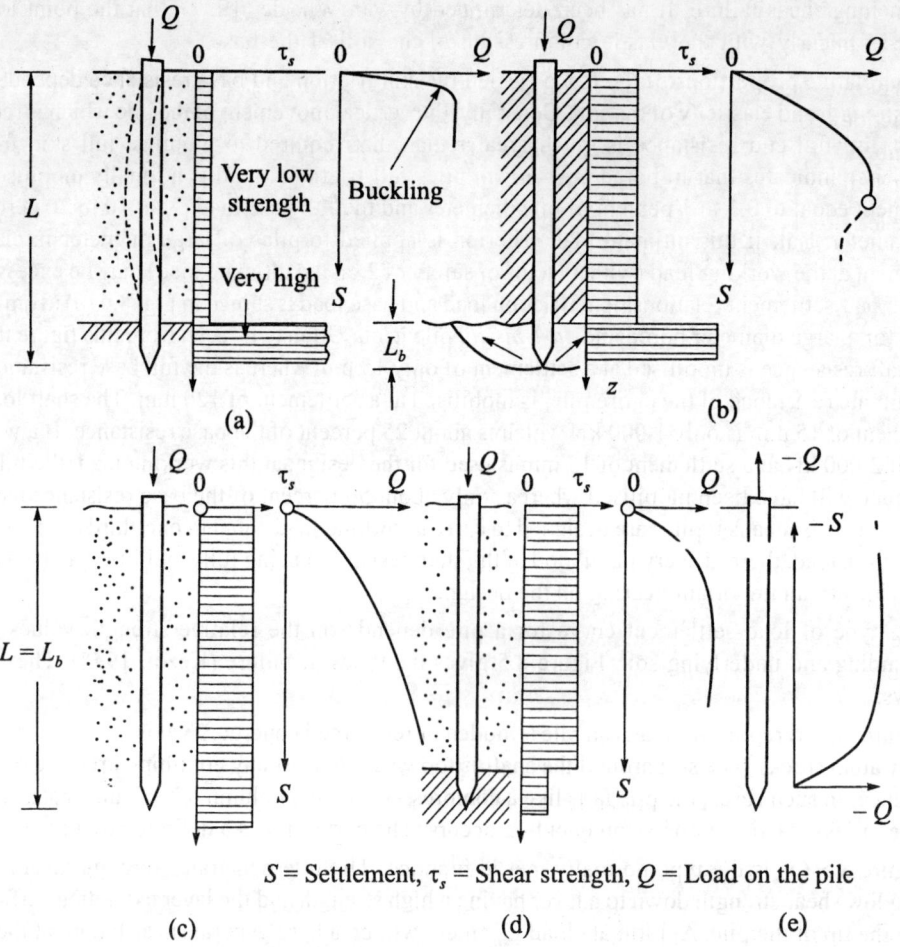

S = Settlement, τ_s = Shear strength, Q = Load on the pile

Fig. 8.5 Types of failure of piles; indicate how strength of soil determines the type of failure: (a) Buckling in very weak surrounding soil, (b) general shear failure in the strong lower soil, (c) soil of uniform strength, (d) low strength soil in the lower layer, skin friction predominant, (e) skin friction in tension (Kézdi, 1975)

large diameter in clay and a nominal factor of safety of 2 is used to obtain the working load, then the settlement at the working load may be excessive.

Factor of Safety

In almost all cases where piles are acting as structural foundations, the allowable load is governed solely from considerations of tolerable settlement at the working load.

The working load for all pile types in all types of soil may be taken as equal to the sum of the base resistance and shaft friction divided by a suitable factor of safety. A safety factor of 2.5 is normally used. Therefore, we may write

$$Q_a = \frac{Q_b + Q_f}{2.5} \tag{8.2}$$

In case where the values of Q_b and Q_f can be obtained independently, the allowable load can be written as

$$Q_a = \frac{Q_b}{3} + \frac{Q_f}{1.5} \tag{8.3}$$

It is permissible to take a safety factor equal to 1.5 for the skin friction because the peak value of skin friction on a pile occurs at a settlement of only 3–8 mm (relatively independent of shaft diameter and embedded length but may depend on soil parameters), whereas the base resistance requires a greater settlement for full mobilisation.

The least of the allowable loads given by Eqs (8.2) and (8.3) is taken as the design working load.

8.8 METHODS OF DETERMINING ULTIMATE LOAD BEARING CAPACITY OF A SINGLE VERTICAL PILE

The ultimate bearing capacity, Q_u, of a single vertical pile may be determined by any of the following methods.

1. By the use of static bearing capacity equations.
2. By the use of SPT and CPT values.
3. By field load tests.
4. By dynamic method.

The determination of the ultimate point bearing capacity, q_b, of a deep foundation on the basis of theory is a very complex one since there are many factors which cannot be accounted for in the theory. The theory assumes that the soil is homogeneous and isotropic which is normally not the case. All the theoretical equations are obtained based on plane strain conditions. Only shape factors are applied to take care of the three-dimensional nature of the problem. Compressibility characteristics of the soil complicate the problem further. Experience and judgement are therefore very essential in applying any theory to a specific problem. The skin load Q_f depends on the nature of the surface of the pile, the method of installation of the pile and the type of soil. An exact evaluation of Q_f is a difficult job even if the soil is homogeneous over the whole length of the pile. The problem becomes all the more complicated if the pile passes through soils of variable characteristics.

8.9 GENERAL THEORY FOR ULTIMATE BEARING CAPACITY

According to Vesic (1967), only punching shear failure occurs in deep foundations irrespective of the density of the soil so long as the depth–width ratio L/d is greater than 4 where L = length of pile and d = diameter (or width of pile). The types of failure surfaces assumed by different investigators are shown in Fig. 8.6 for the general shear failure condition. The detailed experimental study of Vesic indicates that the failure surfaces do not revert back to the shaft as shown in Fig. 8.6 (b).

The total failure load $\bar{Q}_u$ may be written as follows

$$\bar{Q}_u = Q_u + W_p = Q_b + Q_f + W_p \tag{8.4}$$

where, Q_u = load at failure applied to the pile,

Q_b = base resistance,

Q_f = shaft resistance,

W_p = weight of the pile.

The general equation for the base resistance may be written as

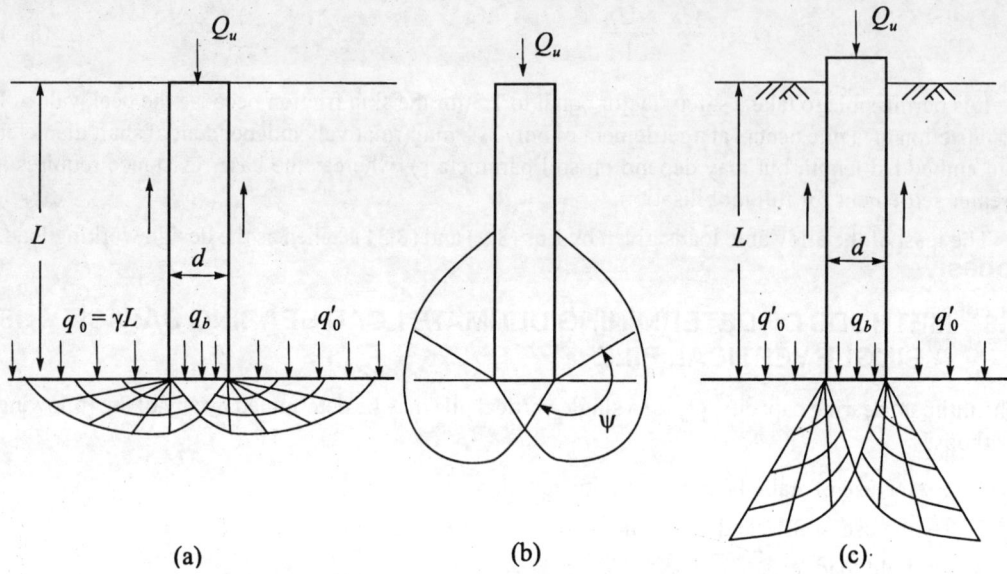

Fig. 8.6 The shapes of failure surfaces at the tips of piles as assumed by: (a) Terzaghi, (b) Meyerhof, and (c) Vesic

$$Q_b = \left(cN_c + q_o' N_q + \frac{1}{2}\gamma\, dN_\gamma \right) A_b \tag{8.5}$$

where d = width or diameter of the shaft at base level,

   q_o' = effective overburden pressure at the base level of the pile,

   A_b = base area of pile,

   c = cohesion of soil,

   γ = effective unit weight of soil,

N_c, N_q, N_γ = bearing capacity factors which take into account the shape factor.

Cohesionless soils

For cohesionless soils, $c = 0$ and the term $1/2\gamma dN_\gamma$ becomes insignificant in comparison with he term $q_o N_q$ for deep foundations. Therefore, Eq. (8.5) reduces to

$$Q_b = q_o' N_q A_b = q_b A_b \tag{8.6}$$

Equation (8.4) may now be written as

$$\bar{Q}_b = Q_u + W_p = q_o' N_q A_b + W_p + Q_f \tag{8.7}$$

The net ultimate load in excess of the overburden pressure load $q_o A_b$ is

$$Q_u + W_p - q_o' A_b = q_o' N_q A_b + W_p - q_o' A_b + Q_f \tag{8.8}$$

If we assume, for all practical purposes, W_p and $q_o' A_b$ are roughly equal for straight side or moderately tapered piles, Eq. (8.8) reduces to

$$Q_u = q_o' N_q A_b + Q_f$$

or
$$Q_u = q'_o N_q A_b + A_s \bar{q}'_o \bar{K}_s \tan \delta \tag{8.9}$$

where, A_s = surface area of the embedded length of the pile,

$\bar{q}'_o$ = average effective overburden pressure over the embedded depth of the pile,

$\bar{K}_s$ = average lateral earth pressure coefficient,

δ = angle of wall friction.

Cohesive soils

For cohesive soils such as saturated clays (normally consolidated), we have for $\phi = 0$, $N_q = 1$, and $N_\gamma = 0$. The ultimate base load from Eq. (8.5) is

$$\bar{Q}_b = (c_b N_c + q'_o) A_b \tag{8.10}$$

The net ultimate base load is

$$(\bar{Q}_b - q'_o A_b) = Q_b = c_b N_c A_b \tag{8.11}$$

Therefore, the net ultimate load capacity of the pile, Q_u, is

$$Q_u = c_b N_c A_b + Q_f$$

or
$$Q_u = c_b N_c A_b + A_s \, \alpha \, \bar{c}_u \tag{8.12}$$

where α = adhesion factor,

$\bar{c}_u$ = average undrained shear strength of clay along the shaft,

c_b = Undrained shear strength of clay at the base level,

N_c = bearing capacity factor.

Equations (8.9) and (8.12) are used for analysing the net ultimate load capacity of piles in cohesionless and cohesive soils respectively. In each case the following types of piles are considered.

1. Driven piles.
2. Driven and *cast-in-situ* piles.
3. Bored piles.

8.10 ULTIMATE BEARING CAPACITY IN COHESIONLESS SOILS

Effect of Pile Installation on the Value of the Angle of Friction

When a pile is driven into loose sand its density is increased (Meyerhof, 1959), and the horizontal extent of the compacted zone has a width of about 6 to 8 times the pile diameter. However, in dense sand, pile driving decreases the relative density because of the dilatancy of the sand and the loosened sand along the shaft has a width of about 5 times the pile diameter (Kerisel, 1961). On the basis of field and model test results, Kishida (1967) proposed that the angle of internal friction decreases linearly from a maximum value of ϕ_2 at the pile tip to a low value of ϕ_1 at a distance of $3.5d$ from the tip where d is the diameter of the pile, ϕ_1 is the angle of friction before the installation of the pile and ϕ_2 after the installation as shown in Fig. 8.7. Based on the field data, the relationship between ϕ_1 and ϕ_2 in sands may be written as

$$\phi_2 = \frac{\phi_1 + 40}{2} \tag{8.13}$$

An angle of $\phi_1 = \phi_2 = 40°$ in Eq. (8.13) means no change of relative density due to pile driving. Values of ϕ_1 are obtained from *in-situ* penetration tests with no correction due to overburden pressure, but corrected for field procedure by using the relationships established between ϕ and SPT or CPT values. Kishida (1967) has suggested the following relationship between ϕ and the SPT value N_{cor} as

$$\phi° = \sqrt{20 N_{cor}} + 15° \qquad (8.14)$$

However, Tomlinson (1986) is of the opinion that it is unwise to use higher values for ϕ due to pile driving. His argument is that the sand may not get compacted, as for example, when piles are driven into loose sand, the resistance is so low and little compaction is given to the soil. He suggests that the value of ϕ used for the design should represent the *in-situ* condition that existed before driving.

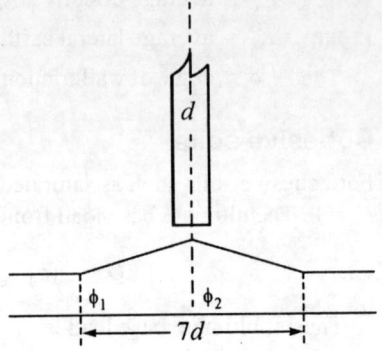

Fig. 8.7 The effect of driving a pile on ϕ

With regard to driven and *cast-in-situ* piles, there is no suggestion by any investigator as to what value of ϕ should be used for calculating the base resistance. However, it is safer to assume the *in-situ* ϕ value for computing the base resistance.

With regard to bored and *cast-in-situ* piles, the soil gets loosened during boring. Tomlinson (1986) suggests that the ϕ value for calculating both the base and skin resistance should represent the loose state. However, Poulos *et al* (1980) suggests that for bored piles, the value of ϕ be taken as

$$\phi = \phi_1 - 3 \qquad (8.15)$$

where ϕ_1 = angle of internal friction prior to installation of the pile.

8.11 CRITICAL DEPTH

The ultimate bearing capacity Q_u in cohesionless soils as per Eq. (8.9) is

$$Q_u = q'_o N_q A_b + \bar{q}'_o \bar{K}_s \tan \delta A_s \qquad (8.16a)$$

or $$Q_u = q_b A_b + f_s A_s \qquad (8.16b)$$

Equation (8.16b) implies that both the point resistance q_b and the skin resistance f_s are functions of the effective overburden pressure q_o in cohesionless soils and increase linearly with the depth of embedment, L, of the pile. However, extensive research work carried out by Vesic (1967) has revealed that the base and frictional resistances remain almost constant beyond a certain depth of embedment which is a function of ϕ. This phenomenon was attributed to arching by Vesic. One conclusion from the investigation of Vesic is that in cohesionless soils, the bearing capacity factor, N_q, is not a constant depending on ϕ only, but also on the ratio L/d (where L = length of embedment of pile, d = diameter or width of pile). In a similar way, the frictional resistance, f_s, increases with the L/d ratio and remains constant beyond a particular depth. Let L_c be the depth, which may be called the *critical depth*, beyond which both q_b and f_s remain constant. Experiments of Vesic have indicated that L_c is a function of ϕ. The L_c/d ratio as a function of ϕ may be expressed as follows (Poulos and Davis, 1980)

For $28° < \phi < 36.5°$

$$L_c/d = 5 + 0.24 (\phi° - 28°) \qquad (8.17a)$$

For $\qquad$ $36.5° < \phi < 42°$

$$L_c/d = 7 + 2.35\,(\phi° - 36.5°) \qquad\qquad (8.17b)$$

The above expressions have been developed based on the curve given by Poulos and Davis, (1980), giving the relationship between L_c/d and $\phi°$

The Eq. (8.17) indicate

$L_c/d = 5$ at $\phi = 28°$

$L_c/d = 7$ at $\phi = 36.5°$

$L_c/d = 20$ at $\phi = 42°$

The ϕ values to be used for obtaining L_c/d are as follows (Poulos and Davis, 1980)

for driven piles $\qquad\qquad \phi = 0.75\,\phi_1 + 10° \qquad\qquad\qquad (8.18a)$

for bored piles: $\qquad\qquad \phi = \phi_1 - 3° \qquad\qquad\qquad\qquad (8.18b)$

where, ϕ_1 = angle of internal friction prior to the installation of the pile.

8.12 TOMLINSON'S SOLUTION FOR Q_b IN SAND

Driven Piles

The theoretical N_q factor in Eq. (8.9) is a function of ϕ. There is great variation in the values of N_q derived by different investigators as shown in Fig. 8.8. Comparison of observed base resistances of piles by Nordlund (1963) and Vesic (1964) have shown (Tomlinson, 1986) that N_q values established by Berezantsev *et al*, (1961), which take into account the depth to width ratio of the pile, most nearly conform to practical criteria of pile failure. Berezantsev's values of N_q as adopted by Tomlinson (1986) are given in Fig. 8.9.

It may be seen from Fig. 8.9 that there is a rapid increase in N_q for high values of ϕ, giving thereby high values of base resistance. As a general rule (Tomlinson, 1986), the allowable working load on an isolated pile driven to virtual refusal, using normal driving equipment, in a dense sand or gravel consisting predominantly of quartz particles, is given by the allowable load on the pile considered as a structural member rather than by consideration of failure of the supporting soil, or if the permissible working stress on the material of the pile is not exceeded, then the pile will not fail.

As per Tomlinson, the maximum base resistance q_b is normally limited to 11,000 kN/m^2 (110 t/ft^2) whatever might be the penetration depth of the pile.

Bored and *Cast-in-situ* Piles in Cohesionless Soils

Bored piles are formed in cohesionless soils by drilling with rigs. The sides of the holes might be supported by the use of casing pipes. When casing is used, the concrete is placed in the drilled hole and the casing is gradually withdrawn. In all the cases the sides and bottom if the hole will be loosened as a result of the boring operations, even though it may be initially be in a dense or medium dense state. Tomlinson suggests that the values of the parameters in Eq. (8.9) must be calculated by assuming that the ϕ value will represent the loose condition.

However, when piles are installed by rotary drilling under a bentonite slurry for stabilising the sides, it may be assumed that the ϕ value used to calculate both the skin friction and base resistance will correspond to the undisturbed soil condition (Tomlinson, 1986).

The assumption of loose conditions for calculating skin friction and base resistance means that the ultimate carrying capacity of a bored pile in a cohesionless soil will be considerably lower than

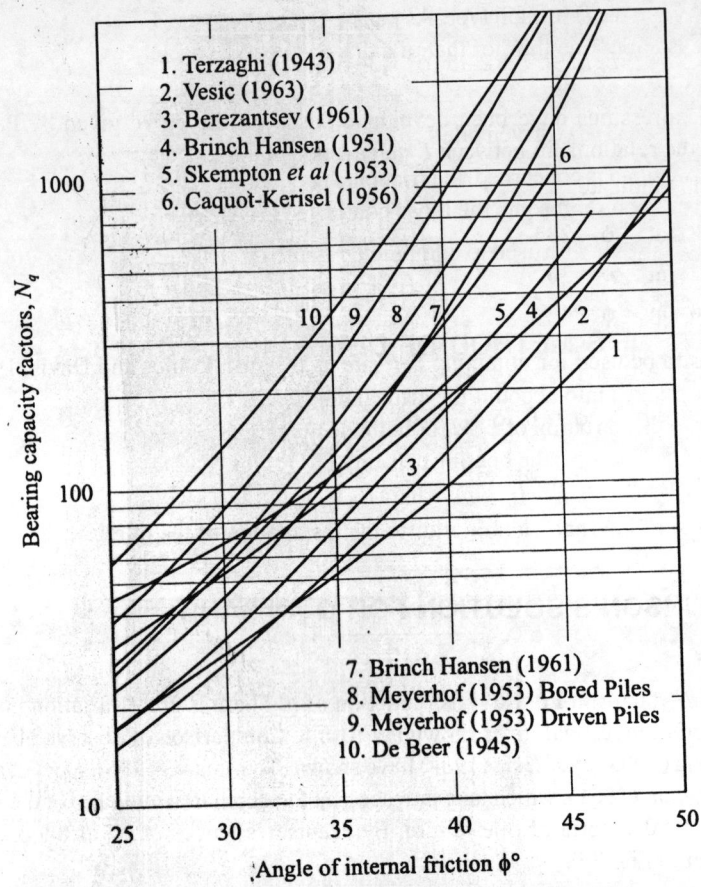

Fig. 8.8 Bearing capacity factors for circular deep foundations (after Kézdi, 1975)

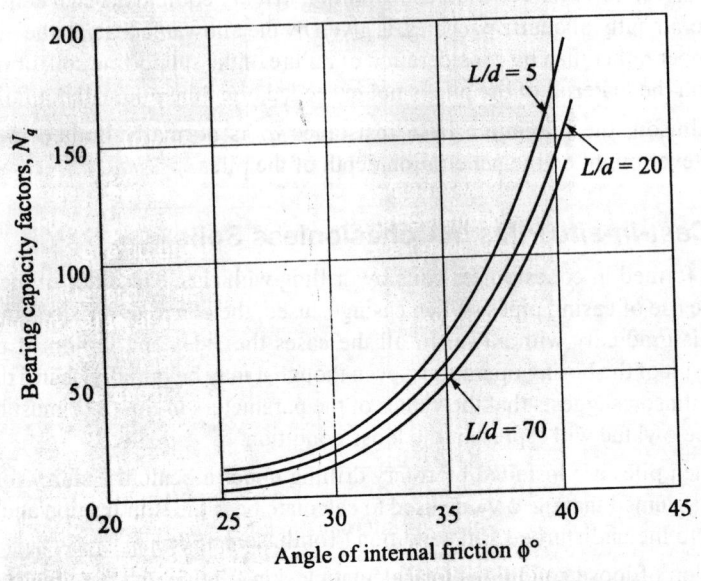

Fig. 8.9 Berezantsev's bearing capacity factor, N_q (after Tomlinson, 1986)

that of a pile driven in the same soil type. As per De Beer (1965), the base resistance q_b of a bored and *cast-in-situ* pile is about one third of that of a driven pile.

We may write,

q_b (bored pile) = $(1/3) \, q_b$ (driven pile)

So far as friction load is concerned, the frictional parameter may be calculated by assuming a value of ϕ equal to 28° which represents the loose condition of the soil.

The same Eq. (8.9) may be used to compute Q_u based on the modifications explained above.

8.13 MEYERHOF'S METHOD OF DETERMINING Q_b FOR PILES IN SAND

Meyerhof (1976), takes into account the critical depth ratio (L_c/d) for estimating the value of Q_b. Figure 8.10 shows the variation of L_c/d for both the bearing capacity factors N_c and N_q as a function of ϕ. According to Meyerhof, the bearing capacity factors increase with L_b/d and reach a maximum value at L_b/d equal to about 0.5 (L_c/d), where L_b is the actual thickness of the bearing stratum. For

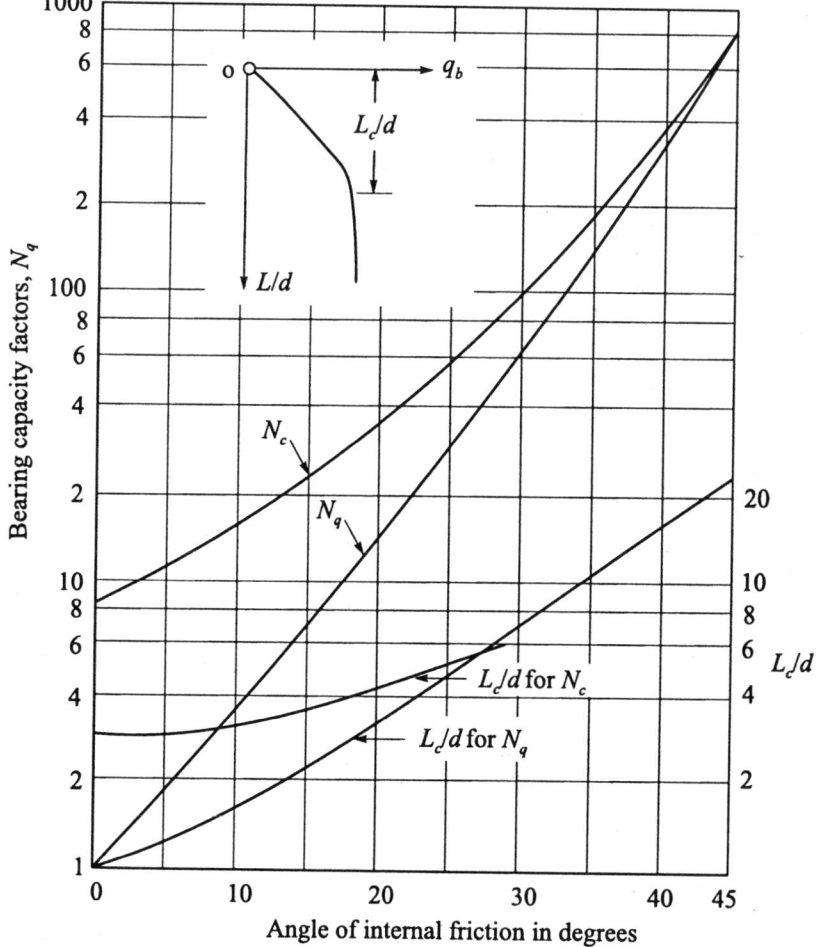

Fig. 8.10 Bearing capacity factors and critical depth ratios L_c/d for driven piles (after Meyerhof, 1976)

example, in a homogeneous soil Fig. 8.5 (c), L_b is equal to L, the actual embedded length of pile; whereas in Fig. 8.5 (b), L_b is less than L.

As per Fig. 8.10, the value of L_c/d is about 25 for ϕ equal to 45° and it decreases with a decrease in the angle of friction ϕ. Normally, the magnitude of L_b/d for piles is greater than 0.5 (L_c/d) so that maximum values of N_c and N_q may apply for the calculation of q_b, the unit bearing pressure of the pile. Meyerhof prescribes a limiting value for q_b, based on his findings on static cone penetration resistance. The expression for the limiting value, q_{bl}, is

for dense sand: $\qquad q_{bl} = 50\, N_q \tan \phi \text{ kN/m}^2$ $\qquad\qquad\qquad\qquad\qquad$ (8.19a)

for loose sand: $\qquad q_{bl} = 25\, N_q \tan \phi \text{ kN/m}^2$ $\qquad\qquad\qquad\qquad\qquad$ (8.19b)

where ϕ is the angle of shearing resistance of the bearing stratum. The limiting q_{bl} values given by Eqs (8.19a and b) remain practically independent of the effective overburden pressure and groundwater conditions beyond the critical depth.

The equation for base resistance in sand may now be expressed as

$$Q_b = q_o' N_q A_b \leq q_{bl} A_b \qquad\qquad\qquad\qquad (8.20)$$

where q_o' = effective overburden pressure at the tip of the pile L_c/d and N_q = bearing capacity factor (Fig. 8.10).

Equation (8.20) is applicable only for driven piles in sand. For bored *cast-in-situ* piles the value of q_b is to be reduced by one third to one-half.

Clay Soil ($\phi = 0$)

The base resistance Q_b for piles in saturated clay soil may be expressed as

$$Q_b = N_c c_u A_b = 9 c_u A_b \qquad\qquad\qquad\qquad (8.21)$$

where $N_c = 9$, and c_u = undrained shear strength of the soil at the base level of the pile.

8.14 VESIC'S METHOD OF DETERMINING Q_b

The unit base resistance of a pile in a $(c - \phi)$ soil may be expressed as (Vesic, 1977)

$$q_b = c N_c^* + q_o' N_q^* \qquad\qquad\qquad\qquad (8.22)$$

where $\qquad c$ = unit cohesion,

$\qquad\qquad q_o'$ = effective vertical pressure at the base level of the pile

N_c^* and N_q^* = bearing capacity factors related to each other by the equation

$$N_c^* = (N_q^* - 1) \cot \phi \qquad\qquad\qquad\qquad (8.23)$$

As per Vesic, the base resistance is not governed by the vertical ground pressure q_o' but by the mean effective normal ground stress σ_m expressed as

$$\sigma_m = \left(\frac{1 + 2 K_o}{3}\right) q_o' \qquad\qquad\qquad\qquad (8.24)$$

in which K_o = coefficient of earth pressure for the at rest condition = $1 - \sin \phi$.

Now the bearing capacity in Eq. (8.22) may be expressed as

$$q_b = cN_c^* + \sigma_m N_\sigma^*$$ (8.25)

An equation for N_σ^* can be obtained from Eqs (8.22), (8.24) and (8.25) as

$$N_\sigma^* = \frac{3 N_q^*}{1 + 2 K_o}$$ (8.26)

Vesic has developed an expression for N_σ^* based on the ultimate pressure needed to expand a spherical cavity in an infinite soil mass as

$$N_\sigma^* = \alpha_1 e^{\alpha_2} N_\phi (I_{rr})^{\alpha_3}$$ (8.27)

where $\alpha_1 = \dfrac{3}{3 - \sin \phi}$, $\alpha_2 = \left(\dfrac{\pi}{2} - \phi\right) \tan \phi$, $\alpha_3 = \dfrac{1.33 \sin \phi}{(1 + \sin \phi)}$, and $N_\phi = \tan^2 (45° + \phi/2)$

According to Vesic

$$I_{rr} = \frac{I_r}{1 + I_r \Delta}$$ (8.28)

where I_r = rigidity index = $\dfrac{E_s}{2 (1 + \mu) (c + q_o' \tan \phi)} = \dfrac{G}{(c + q_o' \tan \phi)}$ (8.29)

where I_{rr} = reduced rigidity index for the soil,

   Δ = average volumetric strain in the plastic zone below the pile point,

   E_s = modulus of elasticity of soil,

   G = shear modulus of soil,

   μ = Poisson's ratio of soil.

Figures 8.11 (a) and (b) given plots of N_σ^* versus ϕ, and N_c^* versus ϕ for various values of I_{rr} respectively.

The values of rigidity index can be computed knowing the values of shear modulus G and the shear strength s $(= c + q_o' \tan \phi)$.

When an undrained condition exists in the saturated clay soil or the soil is cohesionless and is in a dense state, we have $\Delta = 0$ and in such a case $I_r = I_{rr}$.

For $\phi = 0$ (undrained condition), we have

$$N_c^* = 1.33 (\ln I_{rr} + 1) + \frac{\pi}{2} + 1$$ (8.30)

The value of I_r depends upon the soil state, (*a*) for sand, loose or dense, and (*b*) for clay low, medium or high plasticity. For preliminary estimates the following values of I_r may be used.

Soil type	I_r
Sand ($D_r = 0.5 - 0.8$)	75 – 150
Silt	50 – 75
Clay	150 – 250

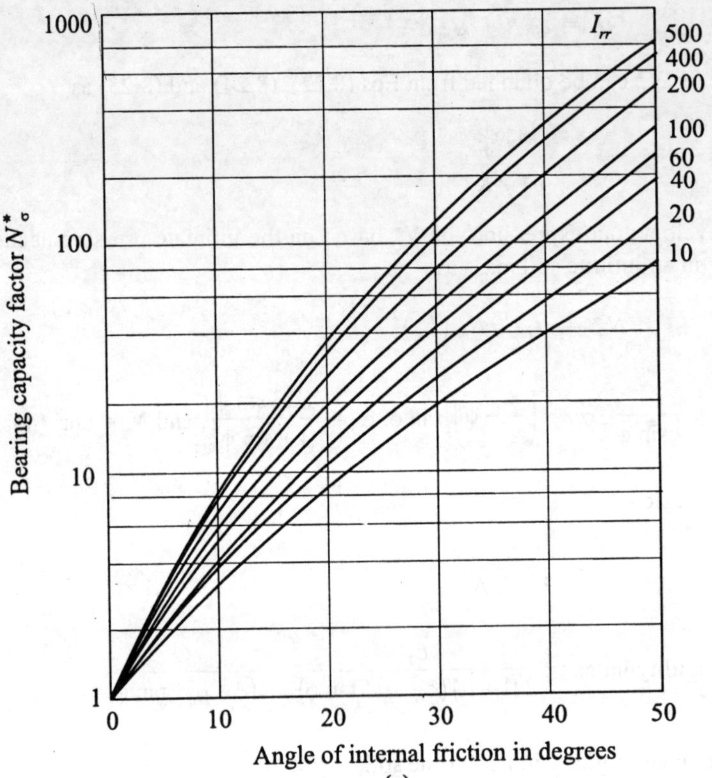

(a)

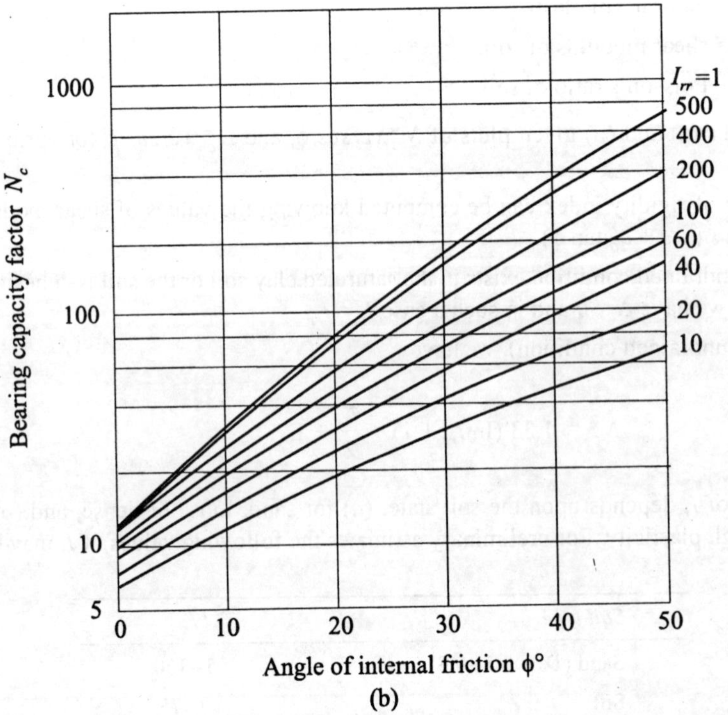

(b)

Fig. 8.11 (a) Bearing capacity factor N_σ^*, (b) bearing capacity factor N_c (Vesic, 1977)

8.15   JANBU'S METHOD OF DETERMINING Q_b

The bearing capacity equation of Janbu (1976) is the same as Eq. (8.22) and is expressed as

$$Q_b = (cN_c^* + q_o' N_q^*) A_b \tag{8.31}$$

The shape of the failure surface as assumed by Janbu is similar to that given in Fig. 8.6 (b). Janbu's equation for N_q^* is

$$N_q^* = \left(\tan\phi + \sqrt{1 + \tan^2\phi}\right)^2 e^{2\psi \tan\phi} \tag{8.32}$$

where ψ = angle as shown in Fig. 8.6 (b). This angle varies from 60° in soft compressible soil to 105° in dense sand. The values for N_c^* used by Janbu are the same as those given by Vesic (Eq. 8.23). Table 8.1 gives the bearing capacity factors of Janbu.

Table 8.1 Bearing capacity factors N_q^* and N_c^* by Janbu

$\psi =$	75°		90°		105°	
$\phi°$	N_q^*	N_c^*	N_q^*	N_c^*	N_q^*	N_c^*
0	1.00	5.74	1.00	5.74	1.00	5.74
5	1.50	6.25	1.57	6.49	1.64	7.33
10	2.25	7.11	2.47	8.34	2.71	9.70
20	5.29	11.78	6.40	14.83	7.74	18.53
30	13.60	21.82	18.40	30.14	24.90	41.39
35	23.08	31.53	33.30	46.12	48.04	67.18
40	41.37	48.11	64.20	75.31	99.61	117.52
45	79.90	78.90	134.87	133.87	227.68	226.68

Since Janbu's bearing capacity factor N_q^* depends on the angle ψ, there are two uncertainties involved in this procedure. They are:

1. The difficulty in determining the values of ψ for different situations at base level.
2. The settlement required at the base level of the pile for the full development of a plastic zone.

For full base load Q_b to develop, at least a settlement of about 10 to 20 percent of the pile diameter is required which is considerable for larger diameter piles.

8.16   COYLE AND CASTELLO'S METHOD OF ESTIMATING Q_b IN SAND

Coyle and Castello (1981) made use of the results of 24 full scale pile load tests driven in sand for evaluating the bearing capacity factors. The form of equation used by them is the same as Eq. (8.6) which may be expressed as

$$Q_b = q_o' N_q^* A_b \tag{8.33}$$

where q_o' = effective overburden pressure at the base level of the pile,

N_q^* = bearing capacity factor.

Coyle and Castello collected data from the instrumented piles, and separated from the total load, the base load, and friction load. The total force at the top of the pile was applied by means of a jack.

The soil at the site was generally fine sand with some percentage of silt. The lowest and the highest relative densities were 40 to 100 percent respectively. The pile diameter was generally around 1.5 ft and pile penetration was about 50 ft. Closed end steel pipe was used for the tests in same places and precast square piles or steel H piles were used at other places.

The bearing capacity factor N_q^* was evaluated with respect to depth ratio L/d in Fig. 8.12 for various values of ϕ.

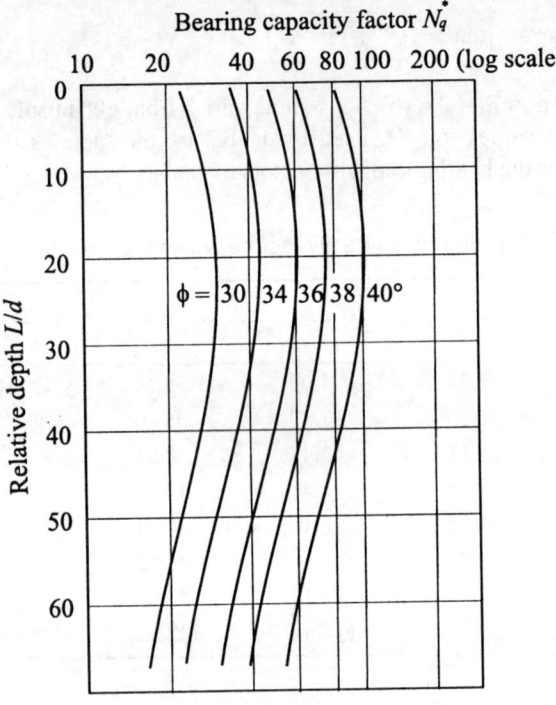

Fig. 8.12 N_q^* *versus L/d* (after Coyle and Castello, 1981)

8.17   THE ULTIMATE SKIN RESISTANCE OF A SINGLE PILE IN COHESIONLESS SOIL

Skin Resistance (Straight Shaft)

The ultimate skin resistance in a homogeneous soil as per Eq. (8.9) is expressed as

$$Q_f = A_s \, \overline{q}_o{}' \, \overline{K}_s \tan \delta \tag{8.34a}$$

In a layered system of soil $\overline{q}_o{}'$, $\overline{K}_s$ and δ vary with respect to depth. Equation (8.34a) may then be expressed as

$$Q_f = \int_0^L P \overline{q}_o{}' \, \overline{K}_s \, \tan \delta dz \tag{8.34b}$$

where, $\overline{q}_o{}'$, $\overline{K}_s$ and δ refer to thickness dz of each layer and P is the perimeter of the pile.

As explained in Section 8.10 the effective overburden pressure does not increase linearly with depth and reaches a constant value beyond a particular depth L_c, called the *critical depth*, which is a function of ϕ. It is therefore natural to expect the skin resistance f_s also to remain constant beyond depth L_c. The magnitude of L_c may be taken as equal to $20d$.

Equation (8.17) can be used for determining the critical length L_c for any given set of values of ϕ and d. Q_f can be calculated from Eq. (8.34) if $\overline{K}_s$ and δ are known.

The values of $\overline{K}_s$ and δ vary not only with the relative density and pile material but also with the method of installation of the pile.

Broms (1966) has related the values of $\overline{K}_s$ and δ to the effective angle of internal friction ϕ of cohesionless soils for various pile materials and relative densities (D_r) as shown in Table 8.2. The values are applicable to driven piles. As per the present state of knowledge, the maximum skin friction is limited to 110 kN/m^2 (Tomlinson, 1986).

Table 8.2 Values of $\overline{K}_s$ and δ (Broms, 1966)

Pile material	δ	Values of $\overline{K}_s$	
		Low D_r	High D_r
Steel	20°	0.5	1.0
Concrete	3/4 ϕ	1.0	2.0
Wood	2/3 ϕ	1.5	4.0

Equation (8.34) may also be written as

$$Q_f = \int_0^L P\overline{q}_o' \beta \, dz \tag{8.35}$$

where, $\beta = \overline{K}_s \tan \delta$.

Poulos and Davis, (1980) have given a curve giving the relationship between β and $\phi°$ which is applicable for driven piles and all types of material surfaces. According to them there is not sufficient evidence to show that β would vary with the pile material. The relationship between β and ϕ is given in Fig. 8.13 (a). For bored piles, Poulos *et al*, recommend the relationship given by Meyerhof (1976) between ϕ and β [Fig. 8.13 (b)].

Skin Resistance on Tapered Piles

Nordlund (1963) has shown that even a small taper of 1° on the shaft gives a four fold increase in unit friction in medium dense sand under compression loading. Based on Nordlund's analysis, curves have been developed (Poulos and Davis, 1980) giving a relationship between taper angle $\omega°$ and a taper correction factor F_ω, which can be used in Eq. (8.35) as

$$Q_f = \int_0^L F_\omega P\overline{q}_o' \beta \, dz \tag{8.36}$$

Equation (8.36) gives the ultimate skin load for tapered piles. The correction factor F_ω can be obtained from Fig. 8.13 (c). The value of ϕ to be used for obtaining F_ω is as per Eq. (8.18a) for driven piles.

8.18 SKIN RESISTANCE Q_f BY COYLE AND CASTELLO METHOD (1981)

For evaluating frictional resistance, Q_f, for piles in sand, Coyle and Castello (1981), made use of the results obtained from 24 field tests on piles. The expression for Q_f is the one given in Eq. (8.34a).

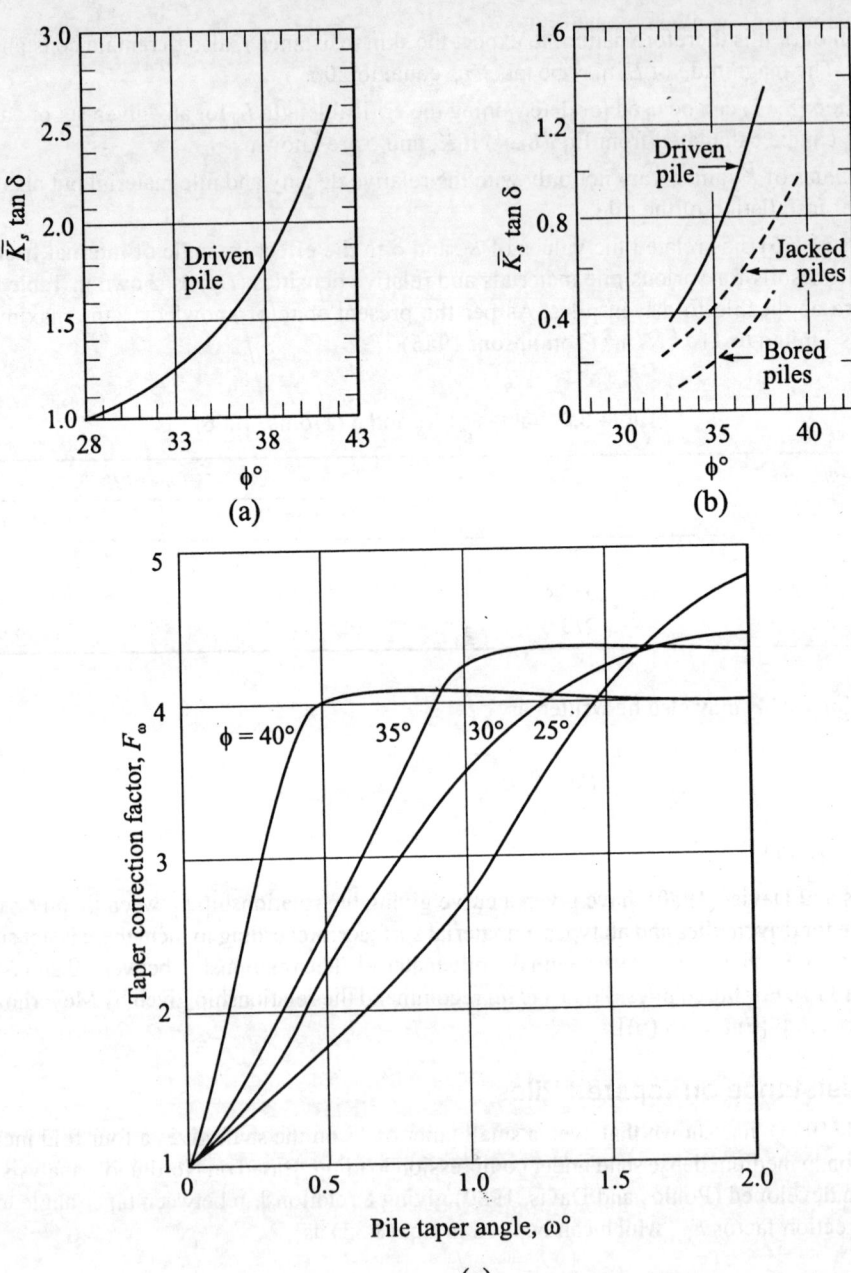

Fig. 8.13 Values of $K_s \tan \delta$ in sand as per: (a) Poulos and Davis 1980, (b) Meyerhof, 1976, and (c) taper factor F_ω (after Nordlund, 1963)

They developed a chart (Fig. 8.14) giving relationships between $\bar{K}_s$ and ϕ for various L/d ratios. The angle of wall friction δ is assumed equal to $0.8\,\phi$. The expression for Q_f is Eq. (8.34a)

$$Q_f = A_s \, \bar{q}'_o \, \bar{K}_s \, \tan \delta$$

where $\bar{q}'_o$ = average effective overburden pressure and δ = angle of wall friction = 0.8ϕ.

The value of $\bar{K}_s$ can be obtained Fig. 8.14.

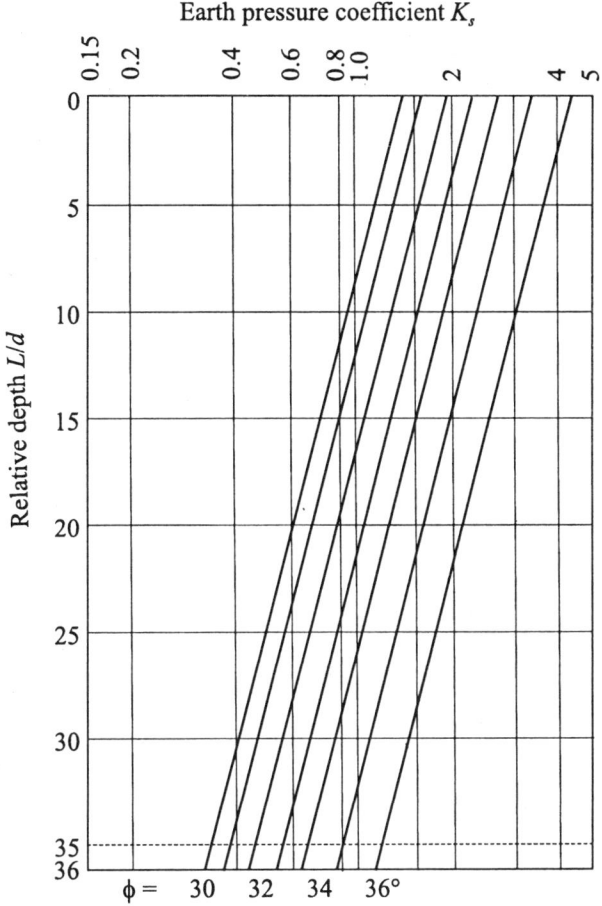

Fig. 8.14 Coefficient $\bar{K}_s$ versus L/d, $\delta = 0.8\phi$ (after Coyle and Castello, 1981)

8.19 STATIC BEARING CAPACITY OF PILES IN CLAY SOIL

Equation for Ultimate Bearing Capacity

The static ultimate bearing capacity of piles in clay as per Eq. (8.12) is

$$Q_u = Q_b + Q_f = c_b N_c A_b + \alpha \bar{c}_u A_s \tag{8.37}$$

For layered clay soils where the cohesive strength varies along the shaft, Eq. (8.37) may be written as

$$Q_u = c_b N_c A_b + \sum_0^L \alpha \bar{c}_u A_s \tag{8.38}$$

Bearing Capacity Factor N_c

The value of the bearing capacity factor N_c that is generally accepted is 9 which is the value proposed by Skempton (1951) for circular foundations for a L/B ratio greater than 4. The base capacity of a pile in clay soil may now be expressed as

$$Q_b = 9c_b A_b \qquad (8.39)$$

The value of c_b may be obtained either from laboratory tests on undisturbed samples or from the relationships established between c_u and field penetration tests. Equation (8.39) is applicable for all types of pile installations.

Skin Resistance by α-Method

Tomlinson (1986) has given some empirical correlations for evaluating α in Eq. (8.37) for different types of soil conditions and L/d ratios. His procedure requires a great deal of judgement of the soil conditions in the field and may lead to different interpretations by different geotechnical engineers. A simplified approach for such problems would be needed. Dennis and Olson (1983b) made use of the information provided by Tomlinson and developed a single curve giving the relationship between α and the undrained shear strength c_u of clay as shown in Fig. 8.15.

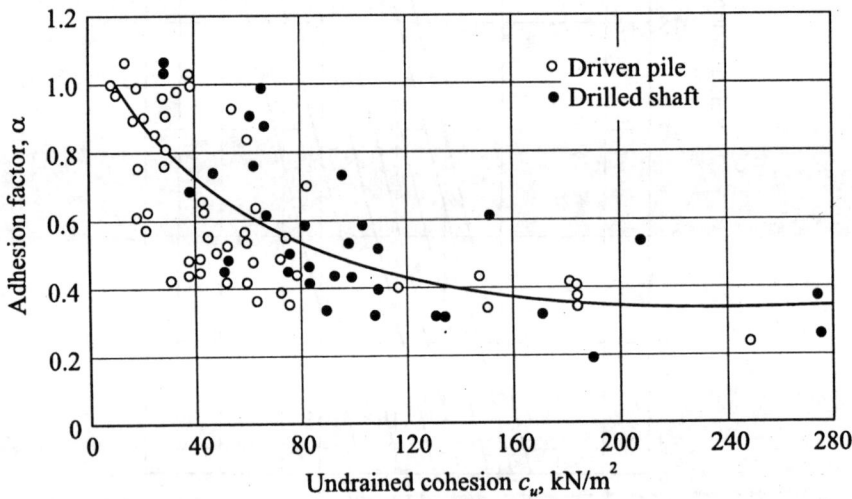

Fig. 8.15 Adhesion factor α for piles with penetration length less than 50 m in clay. (Data from Dennis and Olson 1983a, b; Stas and Kulhawy, 1984)

This curve can be used to estimate the values of α for piles with penetration lengths less than 30 m. As the length of the embedment increases beyond 30 m, the value of α decreases. Piles of such great length experience elastic shortening that results in small shear strain or slip at great depth as compared to that at shallow depth. Investigation indicates that for embedment greater than about 50 m the value of α from Fig. 8.15 should be multiplied by a factor 0.56. For embedments between 30 and 50 m, the reduction factor may be considered to vary linearly from 1.0 to 0.56 (Dennis and Olson, 1893a, b).

Skin Resistance by λ-Method

Vijayvergiya and Focht (1972) have suggested a different approach for computing skin load Q_f for steel-pipe piles on the basis of examination of load test results on such piles. The equation is of the form

$$Q_f = \lambda \, (\bar{q}_o' + 2\bar{c}_u) \, A_s \qquad (8.40)$$

where λ = frictional capacity coefficient,

$\bar{q}'_o$ = mean effective vertical stress between the ground surface and pile tip.

The other terms are already defined. λ is plotted against pile penetration as shown in Fig. 8.16.

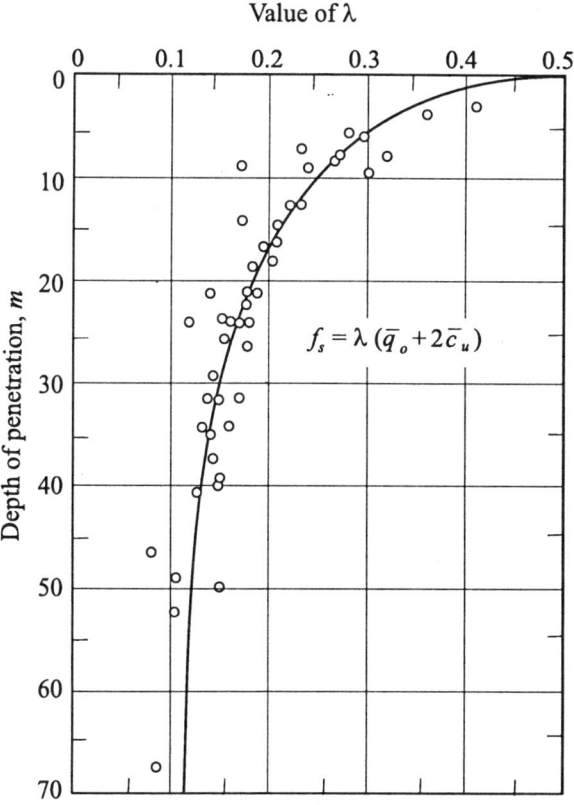

Fig. 8.16 Frictional capacity coefficient λ *versus* pile penetration (Vijayvergiya and Focht, 1972)

Equation (8.40) has been found very useful for the design of heavily loaded pipe piles for offshore structures.

β-Method or the Effective Stress Method of Computing Skin Resistance

In this method, the unit skin friction f_s is defined as

$$f_s = \bar{K}_s \tan \delta \bar{q}'_o = \beta \bar{q}'_o \tag{8.41}$$

$$\beta = \text{the skin factor} = \bar{K}_s \tan \delta, \tag{8.42a}$$

where $\bar{K}_s$ = lateral earth pressure coefficient,

δ = angle of wall friction,

$\bar{q}'_o$ = average effective overburden pressure.

Burland (1973), discusses the values to be used for β and demonstrates that a lower limit for this factor for normally consolidated clay can be written as

$$\beta_o = \bar{K}_o \tan \phi' \tag{8.42b}$$

As per Jaky (1944)

$$K_o = (1 - \sin \phi') \tag{8.42c}$$

Therefore

$$\beta = (1 - \sin \phi') \tan \phi' \tag{8.42d}$$

where ϕ' = effective angle of internal friction.

Since the concept of this method is based on effective stresses, the cohesion intercept on a Mohr circle is equal to zero. For driven piles in stiff overconsolidated clay, $\bar{K}_s$ is roughly 1.5 times greater than $\bar{K}_o$. For overconsolidated clays $\bar{K}_o$ may be found from the expression

$$\bar{K}_o = (1 - \sin \phi') \sqrt{R_{oc}} \tag{8.42e}$$

where R_{oc} = overconsolidation ratio of clay.

For clays, ϕ' may be taken in the range of 20 to 30 degrees. In such a case the value of β in Eq. (8.42d) varies between 10.24 and 0.29.

Meyerhof's method (1976)

Meyerhof has suggested a semi-empirical relationship for estimating skin friction in clays.

For driven piles:

$$f_s = 1.5c_u \tan \phi^\circ \tag{8.43}$$

For bored piles:

$$f_s = c_u \tan \phi' \tag{8.44}$$

By utilising a value of 20° for ϕ' for the stiff to very stiff clays, the expressions reduce to

For driven piles:

$$f_s = 0.55c_u \tag{8.45}$$

For bored piles:

$$f_s = 0.36c_u \tag{8.46}$$

In practice the maximum value of unit friction for bored piles is restricted to 100 kPa.

8.20   BEARING CAPACITY OF PILES IN GRANULAR SOILS BASED ON SPT VALUE

Meyerhof (1956) suggests the following equations for single piles in granular soils based on SPT values.

For displacement piles:

$$Q_u = Q_b + Q_f = 40N_{cor} (L/d) A_b + 2\bar{N}_{cor} A_s \tag{8.47a}$$

For H-piles:

$$Q_u = 40N_{cor} (L/d) A_b + \bar{N}_{cor} A_s \tag{8.47b}$$

where, $q_b = 40N_{cor} (L/d) \leq 400 \, N_{cor}$

For bored piles:

$$Q_u = 133 \, N_{cor} A_b + 0.67\bar{N}_{cor} A_s \tag{8.48}$$

where Q_u = ultimate total load in kN,

     N_{cor} = average corrected SPT value below pile tip,

     $\bar{N}_{cor}$ = corrected average SPT value along the pile shaft,

     A_b = base area of pile in m^2 (for H-piles including the soil between the flanges),

     A_s = Shaft surface area in m^2.

In English units Q_u for a displacement pile is

$$Q_u \text{ (kip)} = Q_b + Q_f = 0.80 N_{cor} (L/d) A_b + 0.04 \bar{N}_{cor} A_s \tag{8.49a}$$

where, A_b = base area in ft^2 and A_s = surface area in ft^2

and

$$0.80 \, N_{cor} \left(\frac{L}{d}\right) A_b \le 8 N_{cor} A_b \text{ (kip)} \tag{8.49b}$$

A minimum factor of safety of 4 is recommended. The allowable load Q_a is

$$Q_a = \frac{Q_u}{4} \tag{8.50}$$

Example 8.1

A concrete pile of 45 cm diameter was driven into sand of loose to medium density to a depth of 15 m. The following properties are known:

(a) Average unit weight of soil along the length of the pile, $\bar{\gamma}$ = 17.5 kN/m^3, average ϕ = 30°, (b) average $\bar{K}_s$ = 1.0 and δ = 0.75ϕ.

Calculate (a) the ultimate bearing capacity of the pile, and (b) the allowable load with F_s = 2.5. Assume the water table is at great depth. Use Berezantsev's method.

Solution

From Eq. (8.9)

$$Q_u = Q_b + Q_f = q'_o A_b N_q + \bar{q}'_o A_s \bar{K}_s \tan \delta$$

where $q'_o = \bar{\gamma} L = 17.5 \times 15 = 262.5 \text{ kN/m}^2$

$$\bar{q}'_o = \frac{1}{2} \bar{\gamma} L = \frac{262.5}{2} = 131.25 \text{ kN/m}^2$$

$$A_b = \frac{3.14}{4} \times 0.45^2 = 0.159 \text{ m}^2$$

$$A_s = 3.14 \times 0.45 \times 15 = 21.195 \text{ m}^2$$

$$\delta = 0.75\phi = 0.75 \times 30 = 22.5°$$

$$\tan \delta = 0.4142$$

From Fig. 8.9,

$$N_q \text{ for } \frac{L}{d} = \frac{15}{0.45} = 33.3$$

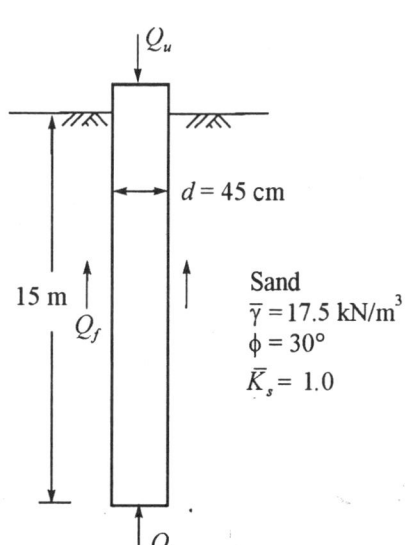

Fig. Ex. 8.1

and $\phi = 30°$ is equal to 16.5.

Substituting the known values, we have

$$Q_u = Q_b + Q_f = 262.5 \times 0.159 \times 16.5 + 131.25 \times 21.195 \times 1.0 \times 0.4142$$

$$= 689 + 1152 = 1841 \text{ kN}$$

$$Q_a = \frac{1841}{2.5} = 736 \text{ kN}$$

Example 8.2

Assume in Ex. 8.1 that the water table is at the ground surface and $\gamma_{sat} = 18.5 \text{ kN/m}^3$. All the other data remain the same. Calculate Q_u and Q_a.

Solution

Water table at the ground surface $\gamma_{sat} = 18.5 \text{ kN/m}^3$

$$\gamma_b = \gamma_{sat} - \gamma_w = 18.5 - 9.81 = 8.69 \text{ kN/m}^3$$

$$q'_o = 8.69 \times 15 = 130.35 \text{ kN/m}^2$$

$$\bar{q}'_o = \frac{1}{2} \times 130.35 = 65.18 \text{ kN/m}^2$$

Substituting the known values

$$Q_u = 130.35 \times 0.159 \times 16.5 \times 65.18 \times 21.195 \times 1.0 \times 0.4142$$

$$= 342 + 572 = 914 \text{ kN}$$

$$Q_a = \frac{914}{2.5} = 366 \text{ kN}$$

Note: It may be noted here that the presence of a water table at the ground surface in cohesionless soil reduces the ultimate load capacity of pile by about 50 percent.

Example 8.3

A concrete pile of 45 cm diameter is driven to a depth of 16 m through a layered system of sandy soil ($c = 0$). The following data are available.

Top layer 1: Thickness = 8 m, $\gamma_d = 16.5 \text{ kN/m}^3$, $e = 0.60$ and $\phi = 30°$.

Layer 2: Thickness = 6 m, $\gamma_d = 15.5 \text{ kN/m}^3$, $e = 0.65$ and $\phi = 35°$.

Layer 3: Extends to a great depth, $\gamma_d = 16.00 \text{ kN/m}^3$, $e = 0.65$ and $\phi = 38°$.

Assume that the value of δ in all the layers of sand is equal to 0.75ϕ. The value of $\bar{K}_s$ for each layer as equal to half of the passive earth pressure coefficient. The water table is at ground level.

Calculate the values of Q_u and Q_a with $F_s = 2.5$ by the conventional method for Q_f and Berezantsev's method for Q_b.

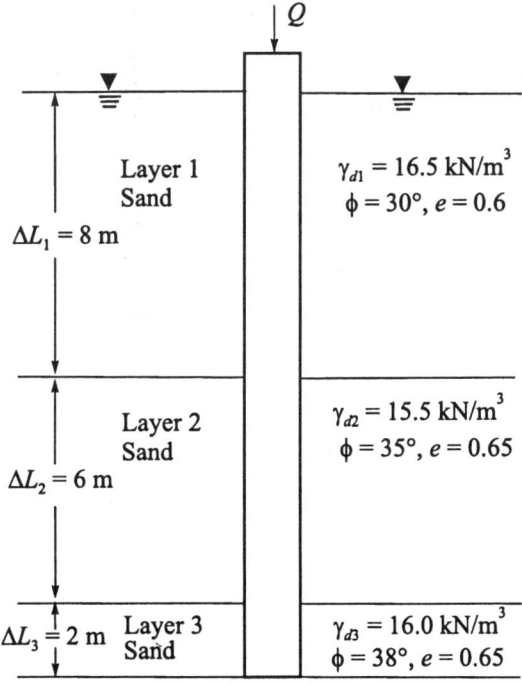

Fig. Ex. 8.3

Solution

The soil is submerged throughout the soil profile. The specific gravity G is required for calculating γ_{sat}.

(a) Using the equation $\gamma_d = \dfrac{\gamma_w G}{1+e}$, calculate G for each layer since γ_d, γ_w and e are known.

(b) Using the equation $\gamma_{sat} = \dfrac{\gamma_w (G+e)}{1+e}$, calculate γ_{sat} for each layer and then $\gamma_b = \gamma_{sat} - \gamma_w$ for each layer.

(c) For a layered system of soil, the ultimate load can be determined by making use of Eq. (8.9). Now

$$Q_u = Q_b + Q_f = q'_o N_q A_b + P \sum_0^L \bar{q}_o{}' K_s \tan \delta \, \Delta L$$

(d) q'_o at the tip of the pile is

$$q'_o = \gamma_{b1}\Delta L_1 + \gamma_{b2}\Delta L_2 + \gamma_{b3}\Delta L_3$$

(e) $\bar{q}'_o$ at the middle of each layer is

$$\bar{q}'_{o1} = \frac{1}{2} \Delta L_1 \gamma_{b1}$$

$$\bar{q}'_{o2} = \Delta L_1 \gamma_{b1} + \frac{1}{2} \Delta L_2 \gamma_{b2}$$

$$\bar{q}'_{o3} = \Delta L_1 \gamma_{b1} + \Delta L_2 \gamma_{b2} + \frac{1}{2} \Delta L_3 \gamma_{b3}$$

(f) $N_q = 95$ for $\phi = 38°$ and $\dfrac{L}{d} = \dfrac{15}{0.45} = 33.33$ from Fig. 8.9.

(g) $A_b = 0.159$ m^2, $P = 1.413$ m.

(h) $\bar{K}_s = \dfrac{1}{2} \tan^2 \left(45° + \dfrac{\phi}{2} \right) = \dfrac{1}{2} K_p$. $\bar{K}_s$ for each layer can be calculated.

(i) $\delta = 0.75\phi$. The values of tan δ can be calculated for each layer.

The computed values for all the layers are given below in a tabular form.

Layer no.	G	γ_b	$\bar{q}'_0$	$\bar{K}_s$	tan δ	ΔL
		kN/m^3	kN/m^2			m
1	2.69	10.36	41.44	1.5	0.414	8
2	2.61	9.57	111.59	1.845	0.493	6
3	2.69	10.05	150.35	2.10	0.543	2
	From middle of layer 3 to tip of pile = 10.05					
	At the tip of pile		$q'_0 = 160.40$ kN/m^2			

$$Q_u = 160.4 \times 95 \times 0.159 + 1.413 \, (41.44 \times 1.5 \times 0.414 \times 8 + 111.59 \times 1.845 \times 0.493 \times 6$$
$$+ 150.35 \times 2.10 \times 0.543 \times 2)$$

$$= 2423 + 1636 \approx 4059 \text{ kN}$$

$$Q_a = \frac{Q_u}{2.5} = \frac{4059}{2.5} = 1624 \text{ kN}$$

Example 8.4

If the pile Ex. 8.2 is a bored and *cast-in-situ*, compute Q_u and Q_a. All the other data remain the same. Water table is close to the ground surface.

Solution

Per Tomlinson (1986), the ultimate bearing capacity of a bored and *cast-in-situ*-pile in cohesionless soil is reduced considerably due to disturbance of the soil. Per Section 8.12, calculate the base resistance for a driven pile and take one-third of this as the ultimate resistance for a bored and *cast-in-situ* pile.

For computing δ, take $\phi = 28°$ and $\bar{K}_s = 1.0$ from Table 8.2 for a concrete pile.

Base resistance for driven pile

For $\phi = 30°$, $N_q = 16.5$ from Fig. 8.9.

$$A_b = 0.159 \text{ m}^2$$

$$q'_o = 130.35 \text{ kN/m}^2 \text{ (From Ex. 8.2)}$$

$$Q_b = 130.35 \times 0.159 \times 16.5 = 342 \text{ kN}$$

For bored pile

$$Q_b = \frac{1}{3} \times 342 = 114 \text{ kN}$$

Skin load

$$Q_f = A_s \bar{q}' \bar{K}_s \tan \delta$$

For $\phi = 28°$, $\delta = 0.75 \times 28 = 21°$, $\tan \delta = 0.384$

$A_s = 21.195 \text{ m}^2$ (from Ex. 8.2)

$\bar{q}'_o = 65.18 \text{ kN/m}^2$ (Ex. 8.2)

Substituting the known values,

$$Q_f = 21.195 \times 65.18 \times 1.0 \times 0.384 = 530 \text{ kN}$$

Therefore,

$$Q_u = 114 + 530 = 644 \text{ kN}$$

$$Q_a = \frac{644}{2.5} \approx 258 \text{ kN}$$

Example 8.5

Solve the problem given in Ex. 8.1 by Meyerhof's method. All the other data remain the same.

Solution

Per Table 3.5 (a), the sand *in-situ* may be considered in a loose state for $\phi = 30°$. The corrected SPT value $N_{cor} = 10$.

Point bearing capacity

From Eq. (8.20)

$$q_b = q'_o Nq \le q_{b1}$$

From Eq. (8.19b)

$$q_{bl} = 25 \, N_q \tan \phi \text{ kN/m}^2$$

Now From Fig. 8.10 $N_q = 60$ for $\phi = 30°$

$$q'_o = \gamma L = 17.5 \times 15 = 262.5 \text{ kN/m}^2$$

$$q_b = 262.5 \times 60 = 15{,}750 \text{ kN/m}^2$$

$$q_{bl} = 25 \times 60 \times \tan 30° = 866 \text{ kN/m}^2$$

Hence, the limiting value for $q_b = 866 \text{ kN/m}^2$

Now, $Q_b = A_b q_b = \dfrac{3.14}{4} \times (0.45)^2 \times 866 = 138 \text{ kN}$

Frictional resistance

Per Section 8.17, the unit skin resistance f_s is assumed to increase from 0 at ground level to a limiting value of f_{sl} at $L_c = 20d$, where L_c = critical depth and d = diameter. Therefore, $L_c = 20 \times 0.45 = 9 \text{ m}$

Now
$$f_{sl} = q'_o \bar{K}_s \tan \delta = \gamma L_c \bar{K}_s \tan \delta$$

Given:
$$\gamma = 17.5 \text{ kN/m}^3, L_c = 9 \text{ m}, \bar{K}_s = 1.0 \text{ and } \delta = 22.5° \text{ m.}$$

Substituting and simplifying, we have

$$f_{sl} = 17.5 \times 9 \times 1.0 \times \tan 22.5 = 65 \text{ kN/m}^2$$

The skin load
$$Q_f = Q_{f1} + Q_{f2} = \frac{1}{2} f_{sl} PL_c + Pf_{sl} (L - L_c)$$

Substituting
$$Q_f = \frac{1}{2} 65 \times 3.14 \times 0.45 \times 9 + 3.14 \times 0.45 \times 65 (15 - 9)$$

$$= 413 + 551 = 964 \text{ kN}$$

The failure load Q_u is

$$Q_u = Q_b + Q_f = 138 + 964 = 1,102 \text{ kN}$$

with
$$F_s = 2.5,$$

$$Q_a = \frac{1,102}{2.5} = 440 \text{ kN}$$

Example 8.6

Determine the base load of the problem in Example 8.5 by Vesic's method. Assume $I_r = I_{rr} = 50$. Determine Q_a for $F_s = 2.5$ using the value of Q_f in Ex. 8.5.

Solution

From Eq. (8.25) for $c = 0$, we have

$$q_b = \sigma_m N^*_\sigma$$

From Eq. (8.24)

$$\sigma_m = \left(\frac{1 + 2K_0}{3} \right) q'_o = \left[\frac{1 + 2(1 - \sin \phi)}{3} \right] q'_o$$

$$q'_o = 15 \times 17.5 = 262.5 \text{ kN/m}^2$$

$$\sigma_m = \left[\frac{1 + 2(1 - \sin 30°)}{3} \right] \times 262.5 = 175 \text{ kN/m}^2$$

From Fig. 8.11 (a), $N^*_\sigma = 36$ for $\phi = 30$ and $I_r = 50$

Substituting

$$q_b = 175 \times 36 = 6,300 \text{ kN/m}^2$$

$$Q_b = A_b q_b = \frac{3.14}{4} \times (0.45)^2 \times 6,300 = 1,001 \text{ kN}$$

$$Q_u = Q_b + Q_f = 1,001 + 964 = 1,965 \text{ kN}$$

$$Q_a = \frac{1,965}{2.5} = 786 \text{ kN}$$

Example 8.7

Determine the base load of the problem in Ex. 8.1 by Janbu's method. Use $\psi = 90°$. Determine Q_a for $F_s = 2.5$ using the Q_f estimated in Ex. 8.5.

Solution

From Eq. (8.31), for $c = 0$, we have

$$q_b = q_0' N_q^*$$

For $\phi = 30°$ and $\psi = 90°$, we have $N_q^* = 18.4$ from Table 8.1 $q_0' = 262.5 \ kN/m^2$ as in Ex. 8.5.

Therefore $\qquad q_b = 262.5 \times 18.4 = 4,830 \ kN/m^2$

$$Q_b = A_b q_b = 0.159 \times 4,830 = 768 \ kN$$

$$Q_u = Q_b + Q_f = 768 + 964 = 1,732 \ kN$$

$$Q_a = \frac{1,732}{2.5} = 693 \ kN$$

Example 8.8

Estimate Q_b, Q_f, Q_u and Q_a by the Coyle and Castello method using the data given in Ex. 8.1.

Solution

Base load Q_b from Eq. (8.33)

$$q_b = q_0' N_q^*$$

From Fig. 8.12, $\qquad N_q^* = 29$ for $\phi = 30°$ and $L/d = 33.3$

$\qquad q_0' = 262.5 \ kN/m^2$ as in Ex. 8.5

Therefore $\qquad q_b = 262.5 \times 29 = 7,612 \ kN/m^2$

$$Q_b = A_b q_b = 0.159 \times 7,612 = 1,210 \ kN$$

From Eq. (8.34a) $\qquad Q_f = A_s \bar{q}_0' \bar{K}_s \tan \delta$

where $\qquad A_s = 3.14 \times 0.45 \times 15 = 21.2 \ m^2$

$$\bar{q}_0' = \frac{1}{2} \times 262.5 = 131.25 \ kN/m^2$$

$$\delta = 0.8\phi = 0.8 \times 30° = 24°$$

From Fig. 8.14, $\qquad \bar{K}_s = 0.35$ for $\phi = 30°$ and $L/d = 33.3$

Therefore $\qquad Q_f = 21.2 \times 131.25 \times 0.35 \tan 24° = 434 \ kN$

$$Q_u = Q_b + Q_f = 1,210 + 434 = 1,644 \ kN$$

$$Q_a = \frac{1,644}{2.5} = 658 \ kN$$

Example 8.9

Determine Q_b, Q_f, Q_u and Q_a by using the SPT value for $\phi = 30°$ from Fig. 5.7.

Solution

From Fig. 5.7, $\quad N_{cor} = 10$ for $\phi = 30°$. Use Eq. (8.47a) for Q_u

$$Q_u = Q_b + Q_f = 40 N_{cor} \left(\frac{L}{d}\right) A_b + 2\bar{N}_{cor} A_s$$

where $\quad Q_b \leq Q_{bl} = 400\, N_{cor}\, A_b$

Given: $\quad L = 15$ m, $d = 0.45$ m, $A_b = 0.159$ m^2, $A_s = 21.2$ m^2

$$Q_b = 40 \times 10 \times \left(\frac{15}{0.45}\right) \times 0.159 = 2{,}120 \text{ kN}$$

$$Q_{bl} = 400 \times 10 \times 0.159 = 636 \text{ kN}$$

Since $Q_b \geq Q_{bl}$, use Q_{bl}

$$Q_f = 2 \times 10 \times 21.2 = 424 \text{ kN}$$

Now $\quad Q_u = 636 + 424 = 1{,}060 \text{ kN}$

$$Q_a = \frac{1{,}060}{2.5} = 424 \text{ kN}$$

Example 8.10

Compare the values of Q_b, Q_f and Q_a obtained by the different methods in Ex. 8.1, and 8.5 through 8.9 and make appropriate comments.

Comparison

The values obtained by different methods are tabulated below.

Method no.	Example no.	Investigator	Q_b kN	Q_f kN	Q_u kN	$Q_a (F_s = 2.5)$ kN
1	8.1	Berezantsev	689	1,152	1,841	736
2	8.5	Meyerhof	138	964	1,102	440
3	8.6	Vesic	1,001	964	1,965	786
4	8.7	Janbu	768	964	1,732	693
5	8.8	Coyle and Castello	1,210	434	1,644	658
6	8.9	Meyerhof (Based on SPT)	636	424	1,060	424

Comments

It may be seen from the table above that there are wide variations in the values of Q_b and Q_f between the different methods.

Method 1 Tomlinson (1986) recommends Berezantsev's method for computing Q_b as this method conforms to the practical criteria of pile failure. Tomlinson does not recommend the critical depth concept.

Method 2 Meyerhof's method takes into account the critical depth concept. Equation (8.19) is based on this concept. The Eq. (8.20) $q_b = q'_o N_q$ does not consider the critical depth concept where q'_o = effective overburden pressure at the pile tip level of the pile. The value of Q_b per this equation is

$$Q_b = q_b A_b = 15,750 \times 0.159 = 2,504 \text{ kN}$$

which is very high and this is close to the value of Q_b (= 2,120 kN) by the SPT method. However, Eq. (8.19b) gives a limiting value for $Q_b = 138$ kN (here the sand is considered lose for $\phi = 30$).

Q_f is computed by assuming Q_f increases linearly with depth from 0 at $L = 0$ to Q_{fl} at depth $L_c = 20d$ and then remains constant to the end of the pile.

Though some investigators have accepted the critical depth concept for computing Q_b and Q_f, it is difficult to generalise this concept as applicable to all types of conditions prevailing in the field.

Method 3 Vesic's method is based on many assumptions for determining the values of I_r, I_{rr}, σ_m, N_σ^*, etc. There are many assumptions in this method. Are these assumptions valid for the field conditions? Designers have to answer this question.

Method 4 The uncertainties involved in Janbu's method are given in Section 8.15 and as such difficult to assess the validity of this method.

Method 5 Coyle and Castello's method is based on full scale field tests on a number of driven piles. Their bearing capacity factors vary with depth. Of the first five methods listed above, the value of Q_b obtained by them is much higher than the other four methods whereas the value of Q_f is very much lower. But on the whole the value of Q_u is lower than the other methods.

Method 6 This method was developed by Meyerhof based on SPT values. The Q_b value (= 2,120 kN) by this method is very higher than the preceding methods, whereas the value of Q_f is the lowest of all the methods. In the table given above the limiting value of $Q_{bl} = 636$ kN is considered. In all the six methods $F_s = 2.5$ has been taken to evaluate Q_a whereas in method 6, $F_s = 4$ is recommended by Meyerhof.

Which method to use

There are wide variations in the values of Q_b, Q_f, Q_u and Q_a between the different methods. The relative proportions of loads carried by skin friction and base resistance also vary between the methods. The order of preference of the methods may be listed as follows:

Preference no.	Method no.	Name of the investigator
1	1	Berezantsev for Q_b
2	2	Meyerhof
3	5	Coyle and Castello
4	6	Meyerhof (SPT)
5	3	Vesic
6	4	Janbu

Example 8.11

A concrete pile 18 in. in diameter and 50 ft long is driven into a homogeneous mass of clay soil of medium consistency. The water table is at the ground surface. The unit cohesion of the soil under undrained condition is 1,050 lb/ft² and the adhesion factor $\alpha = 0.75$. Compute Q_u and Q_a with $F_s = 2.5$.

Solution

Given: $L = 50$ ft, $d = 1.5$ ft, $c_u = 1,050$ lb/ft², $\alpha = 0.75$

From Eq. (8.37), we have

$$Q_u = Q_b + Q_f = c_b\,N_c\,A_b + A_s\alpha\bar{c}_u$$

where, $c_b = \bar{c}_u = 1,050$ lb/ft²; $N_c = 9$; $A_b = 1.766$ ft²; $A_s = 235.5$ ft²

Substituting the known values, we have

$$Q_u = \frac{1,050 \times 9 \times 1.766}{1,000} + \frac{235.5 \times 0.75 \times 1,050}{1,000}$$

$$= 16.69 + 185.46 = 202.15 \text{ kips}$$

$$Q_a = \frac{202.15}{2.5} = 81 \text{ kips}$$

Example 8.12

A concrete pile of 45 cm diameter is driven through a system of layered cohesive soils. The length of the pile is 16 m. The following data are available. The water table is close to the ground surface.

Top layer 1 : Soft clay, thickness = 8 m, unit cohesion $\bar{c}_u = 30$ kN/m² and adhesion factor $\alpha = 0.90$.

Layer 2 : Medium stiff, thickness = 6 m, unit cohesion $\bar{c}_u = 50$ kN/m² and $\alpha = 0.75$.

Layer 3 : Stiff stratum extends to a great depth, unit cohesion $\bar{c}_u = 105$ kN/m² and $\alpha = 0.50$.

Compute Q_u and Q_a with $F_s = 2.5$.

Solution

Here, the pile is driven through clay soils of different consistencies.

The equations for Q_u expressed as (Eq. 8.38) yield

$$Q_u = 9c_b\,A_b + P\sum_0^L \alpha\bar{c}_u\,\Delta L$$

Here, $c_b = \bar{c}_u$ of layer 3, $P = 1.413$ m, $A_b = 0.159$ m².

Substituting the known values, we have

$$Q_u = 9 \times 105 \times 0.159 + 1.413\,(0.90 \times 30 \times 8 + 0.75 \times 50 \times 6 + 0.50 \times 105 \times 2)$$

$$= 150.25 + 771.5 = 921.75 \text{ kN}$$

$$Q_a = \frac{921.75}{2.5} \approx 369 \text{ kN}$$

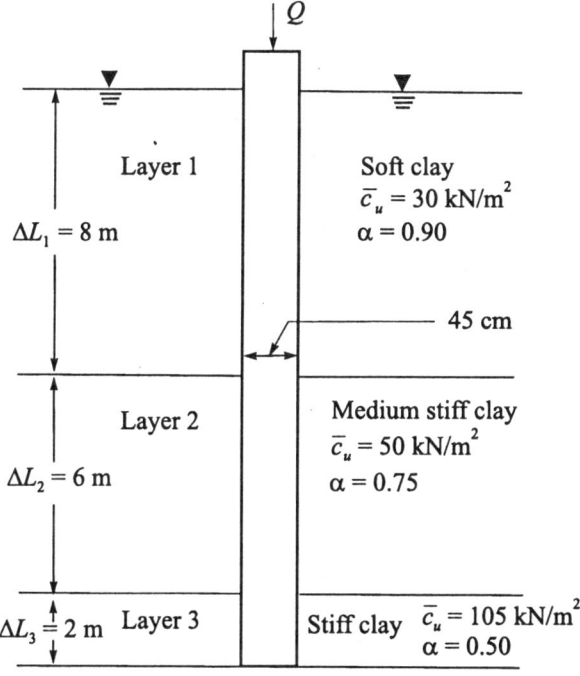

Fig. Ex. 8.12

Example 8.13

A precast concrete pile of size 18×18 in is driven into stiff clay. The unconfined compressive strength of the clay is 4.2 kips/ft². Determine the length of pile required to carry a safe working load of 90 kips with $F_s = 2.5$.

Solution

The equation for Q_u is

$$Q_u = N_c c_u A_b + \alpha \bar{c}_u A_s$$

we have

$$Q_u = 2.5 \times 90 = 225 \text{ kips},$$

$$N_c = 9,\ c_u = 2.1 \text{ kips/ft}^2$$

$$\alpha = 0.48 \text{ from Fig. 8.15,}$$

$$\bar{c}_u = c_u = 2.1 \text{ kip/ft}^2,\ A_b = 2.25 \text{ ft}^2$$

Assume the length of pile $= L$ ft

Now, $A_s = 4 \times 1.5L = 6L$

Substituting the known values, we have

$$225 = 9 \times 2.1 \times 2.25 + 0.48 \times 2.1 \times 6L$$

or $225 = 42.525 + 6.05L$

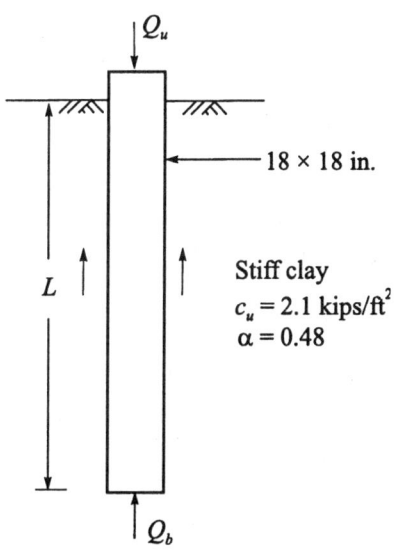

Fig. Ex. 8.13

Simplifying, we have

$$L = \frac{225 - 42.525}{6.05} = 30.2 \text{ ft}$$

Example 8.14

For the problem given in Example 8.11 determine the skin friction load by the λ-method. All the other data remain the same. Assume the average unit weight of the soil is 110 lb/ft^3. Use Q_b given in Ex. 8.11 and determine Q_a for $F_s = 2.5$.

Solution

Per Eq. (8.40)

$$Q_f = \lambda\, (\bar{q}'_o + 2c_u)\, A_s$$

$$\bar{q}'_o = \frac{1}{2} \times 50 \times 110 = 2{,}750 \text{ lb/ft}^2$$

Depth = 50 ft = 15.24 m

From Fig. 8.16, $\lambda = 0.2$ for depth $L = 15.24$ m

Now $Q_f = \dfrac{0.2\,(2{,}750 + 2 \times 1{,}050) \times 235.5}{1{,}000} = 228.44 \text{ kips}$

Now $Q_u = Q_b + Q_f = 16.69 + 228.44 = 245$ kips

$$Q_a = \frac{245}{2.5} = 98 \text{ kips}$$

Example 8.15

A reinforced concrete pile of size 30 × 30 cm and 10 m long is driven into coarse sand extending to a great depth. The average total unit weight of the soil is 18 kN/m^3 and the average N_{cor} value is 15. Determine the allowable load on the pile by the static formula. Use $F_s = 2.5$. The water table is close to the ground surface.

Solution

In this example only the N-value is given. The corresponding ϕ value can be found from Fig. 5.7 which is equal to 32°.

Now from Fig. 8.9, for $\phi = 32°$, and $\dfrac{L}{d} = \dfrac{10}{0.3} = 33.33$, the value of $N_q = 25$.

$$A_b = 0.3 \times 0.3 = 0.09 \text{ m}^2,$$

$$A_s = 10 \times 4 \times 0.3 = 12 \text{ m}^2$$

$$\delta = 0.75 \times 32 = 24°, \tan \delta = 0.445$$

The relative density is loose to medium dense. From Table 8.2, we may take

$$\bar{K}_s = 1 + \frac{1}{3}\,(2 - 1) = 1.33$$

Now,

$$Q_u = q_o' N_q A_b + \overline{q}_o' K_s \tan \delta A_s$$

$$\gamma_b = \gamma_{sat} - \gamma_w = 18.0 - 9.81 = 8.19 \text{ kN/m}^2$$

$$q_o' = \gamma_b L = 8.19 \times 10 = 81.9 \text{ kN/m}^2$$

$$\overline{q}_o' = \frac{q_o'}{2} = \frac{81.9}{2} = 40.905 \text{ kN/m}^2$$

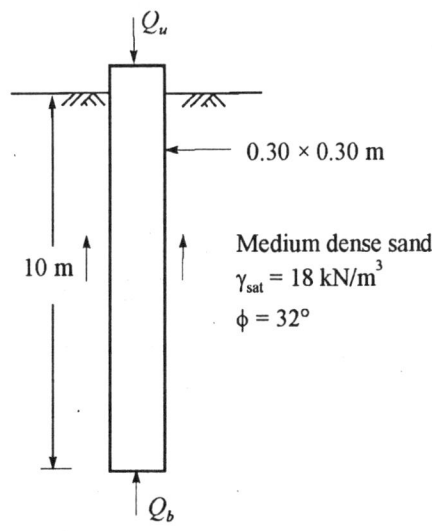

0.30 × 0.30 m

10 m

Medium dense sand
γ_{sat} = 18 kN/m³
ϕ = 32°

Substituting the known values, we have

$$Q_u = 81.9 \times 25 \times 0.09 + 40.95 \times 1.33 \times 0.445 \times 12$$

$$= 184 + 291 = 475 \text{ kN}$$

$$Q_a = \frac{475}{2.5} = 190 \text{ kN}$$

Fig. Ex. 8.15

Example 8.16

Determine the allowable load on the pile given in Ex. 8.15 by making use of the SPT approach by Meyerhof.

Solution

Per Ex. 8.15, $\qquad N_{cor} = 15$

Expression for Q_u is (Eq. 8.47a)

$$Q_u = Q_b + Q_f = 40 \, N_{cor} \, (L/d) A_b + 2 \, \overline{N}_{cor} \, A_s$$

Here, we have to assume

$$N_{cor} = \overline{N}_{cor} = 15$$

$$A_b = 0.3 \times 0.3 = 0.09 \text{ m}^2,$$

$$A_s = 4 \times 0.3 \times 10 = 12 \text{ m}^2$$

Substituting, we have

$$Q_b = 40 \times 15 \left(\frac{10}{0.3}\right) \times 0.09 = 1,800 \text{ kN}$$

$$Q_{b1} = 400 \, N_{cor} \times A_b = 400 \times 15 \times 0.09 = 540 \text{ kN} < Q_b$$

Hence, Q_{b1} governs.

$$Q_f = 2 \, \overline{N}_{cor} A_s = 2 \times 15 \times 12 = 360 \text{ kN}$$

A minimum $F_s = 4$ is recommended, thus

$$Q_a = \frac{Q_b + Q_f}{4} = \frac{540 + 360}{4} = 225 \text{ kN}$$

Example 8.17

Precast concrete piles 16 in. in diameter are required to be driven for a building foundation. The design load on a single pile is 100 kips. Determine the length of the pile if the soil is loose to medium dense sand with an average N_{cor} value of 15 along the pile and 21 at the tip of the pile. The water table may be taken at the ground level. The average saturated unit weight of soil is equal to 120 lb/ft³. Use the static formula and $F_s = 2.5$.

Solution

It is required to determine the length of a pile to carry an ultimate load of $Q_u = 2.5 \times 100 = 250$ kips.

The equation for Q_u is

$$Q_u = q'_o N_q A_b + \bar{q}'_o \bar{K}_s \tan \delta A_s$$

The average value of ϕ along the pile and the value at the tip may be determined from Fig. 5.7.

For $N_{cor} = 15$, $\phi = 32°$; for $N_{cor} = 21$, $\phi = 33.5°$.

Since the soil is submerged

$$\gamma_b = 120 - 62.4 = 57.6 \text{ lb/ft}^3$$

Now

$$q'_o = \gamma_b L = 57.6L \text{ lb/ft}^2$$

$$\bar{q}'_o = \frac{57.6\,L}{2} = 28.8L \text{ lb/ft}^2$$

$$A_b = \frac{3.14}{4} \times (1.33)^2 = 1.39 \text{ ft}^2$$

$$A_s = 3.14 \times 1.33 \times L = 4.176L \text{ ft}^2$$

From Fig. 8.9,

$$N_q = 40 \text{ for } L/d = 20 \text{ (assumed)}$$

From Table 8.2,

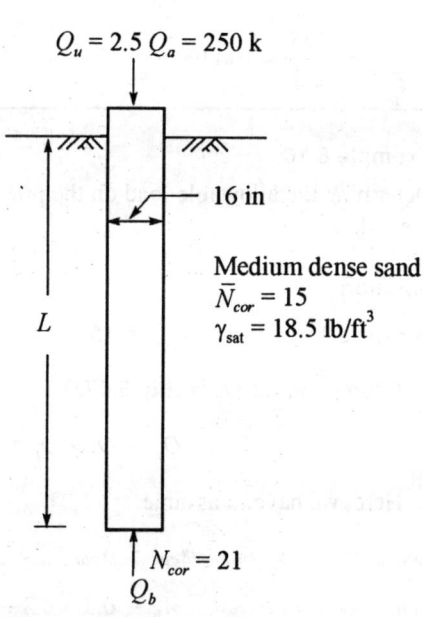

$Q_u = 2.5\, Q_a = 250\text{ k}$

16 in

Medium dense sand
$\bar{N}_{cor} = 15$
$\gamma_{sat} = 18.5 \text{ lb/ft}^3$

L

$N_{cor} = 21$
Q_b

Fig. Ex. 8.17

$$\bar{K}_s = 1.33 \text{ for the lower side of medium dense sand}$$

$$\delta = \frac{3}{4} \times 33.5 = 25.1°, \tan \delta = 0.469$$

Now by substituting the known values, we have

$$250 = \frac{(57.6\,L) \times 40 \times 1.39}{1,000} + \frac{(28.8\,L) \times 0.469 \times 1.33 \times (4.176\,L)}{1,000}$$

$$= 3.203L + 0.075L^2$$

or $L^2 + 42.71L - 3,333 = 0$

Solving this equation gives a value of $L = 40.2$ ft or say 41 ft.

Example 8.18

Refer to the problem in Ex. 8.17. Determine directly the ultimate and the allowable loads using N_{cor}. All the other data remain the same.

Solution

Use Eq. (8.49a)

$$Q_u = Q_b + Q_f = 0.8N_{cor}(L/d)A_b + 0.04\overline{N}_{cor}A_s \text{ (kips)}$$

Given: $\quad N_{cor} = 21, \overline{N}_{cor} = 15, d = 16 \text{ in}, L = 41 \text{ ft}$

$$A_b = \frac{3.14}{4}\left(\frac{16}{12}\right)^2 = 1.39\text{ft}^2; A_s = 3.14 \times \left(\frac{16}{12}\right) \times 41 = 171.65 \text{ ft}^2$$

Substituting

$$Q_b = 0.80 \times 21 \left(\frac{41}{1.33}\right) \times 1.39 = 720 \leq 8 \, N_{cor}A_b \text{ kips}$$

$$Q_{bl} = 8 \times 21 \times 1.39 = 234 \text{ kips}$$

The limiting value of $Q_b = 234$ kips

$$Q_f = 0.04N_{cor}A_s = 0.04 \times 15 \times 171.65 = 103 \text{ kips}$$

$$Q_u = Q_b + Q_f = 234 + 103 = 337 \text{ kips}$$

$$Q_a = \frac{337}{4} = 84 \text{ kips}$$

8.21 BEARING CAPACITY OF PILES BASED ON STATIC CONE PENETRATION TESTS (CPT)

Methods of Determining Pile Capacity

The cone penetration test may be considered as a small scale pile load test. As such the results of this test yield the necessary parameters for the design of piles subjected to vertical load. The types of static cone penetrometers and the methods of conducting the tests have been discussed in detail in Chapter 3. Various methods for using CPT results to predict vertical pile capacity have been proposed. The following methods will be discussed:

1. Vander Veen's method.
2. Schmertmann's method.

Vander Veen's Method for Piles in Cohesionless Soils

In the Vander Veen *et al*, (1957) method, the ultimate end-bearing resistance of a pile is taken, equal to the point resistance of the cone. To allow for the variation of cone resistance which normally occurs, the method considers average cone resistance over a depth equal to three times the diameter of the pile above the pile point level and one pile diameter below point level as shown in Fig. 8.17 (a). Experience has shown that if a safety factor of 2.5 is applied to the ultimate end resistance as determined from cone resistance, the pile is unlikely to settle more than 15 mm under the working load (Tomlinson, 1986). The equations for ultimate bearing capacity and allowable load may be written as,

pile base resistance, $\quad q_b = q_p$ (cone) $\hspace{4cm}$ (8.51a)

ultimate base capacity, $Q_b = A_b q_p$ $\hspace{4cm}$ (8.51b)

allowable base load, $\quad Q_a = \dfrac{A_b q_p}{F_s}$ $\hspace{3.5cm}$ (8.51c)

where, q_p = average cone resistance over a depth $4d$ as shown in Fig. 8.17 (a) and F_s = factor of safety.

The skin friction on the pile shaft in cohesionless soils is obtained from the relationships established by Meyerhof (1956) as follows.

For displacement piles, the ultimate skin friction, f_s, is given by

$$f_s = \frac{\bar{q}_c}{2} \text{ (kPa)} \hspace{4cm} (8.52a)$$

and for H-section piles, the ultimate limiting skin friction is given by

$$f_s = \frac{\bar{q}_c}{4} \text{ (kPa)} \hspace{4cm} (8.52b)$$

where $\bar{q}_c$ = average cone resistance in kg/cm^2 over the length of the pile shaft under consideration.

Meyerhof states that for straight sided displacement piles, the ultimate unit skin friction, f_s, has a maximum value of 107 kPa and for H-sections, a maximum of 54 kPa (calculated on all faces of flanges and web). The ultimate skin load is

$$Q_f = A_s f_s \hspace{4cm} (8.53a)$$

The ultimate load capacity of a pile is

$$Q_u = Q_b' + Q_f \hspace{4cm} (8.53b)$$

The allowable load is

$$Q_a = \frac{Q_b + Q_f}{2.5} \hspace{4cm} (8.53c)$$

If the working load, Q_a, obtained for a particular position of pile in Fig. 8.17 (a), is less than that required for the structural designer's loading conditions, then the pile must be taken to a greater depth to increase the skin friction f_s or the base resistance q_b.

Schmertmann's Method for Cohesionless and Cohesive Soils

Schmertmann (1978) recommends one procedure for all types of soil for computing the point bearing capacity of piles. However, for computing side friction, Schmertmann gives two different approaches, one for sand and one for clay soils.

Point bearing capacity Q_b in all types of soil

The method suggested by Schmertmann (1978) is similar to the procedures developed by De Ruiter and Beringen (1979) for sand. The principle of this method is based on the one suggested by Vander Veen (1957) and explained earlier. The procedure used in this case involves determining a representative cone point penetration value, q_p, within a depth between 0.7 to $4d$ below the tip level of the pile and $8d$ above the tip level as shown in Fig. 8.17 (b) and (c). The value of q_p may be expressed as

$$q_p = \frac{(q_{c1} + q_{c2})/2 + q_{c3}}{2} \tag{8.54}$$

where q_{c1} = average cone resistance below the tip of the pile over a depth which may vary between 0.7d and 4d, where d = diameter of pile,

q_{c2} = minimum cone resistance recorded below the pile tip over the same depth 0.7d to 4d,

q_{c3} = average of the envelope of minimum cone resistance recorded above the pile tip to a height of 8d.

Now, the unit point resistance of the pile, q_b, is

$$q_b \text{ (pile)} = q_p \text{ (cone)} \tag{8.55a}$$

The ultimate base resistance, Q_b, of a pile is

$$Q_b = A_b q_p \tag{8.55b}$$

The allowable base load, Q_a is

$$Q_a = \frac{A_b q_p}{F_s} \tag{8.55c}$$

Method of computing the average cone point resistance q_p

The method of computing q_{c1}, q_{c2} and q_{c3} with respect to a typical q_c-plot shown in Fig. 8.17 (b) and (c) is explained below.

Case 1: When the cone point resistance q_c below the tip of a pile is lower than that at the tip [Fig. 8.17 (b)] within depth $4d$.

$$q_{c1} = \frac{d_3 (q_o + q_b)/2 + d_2 (q_b + q_c)/2 + d_1 (q_d + q_c)/2}{4d} \tag{8.56a}$$

where q_o, q_b, etc. refer to the points o, b, etc. on the q_c-profile, $q_{c2} = q_c$ = minimum value below tip within a depth of $4d$ at a point c on the q_c-profile.

The envelope of minimum cone resistance above the pile tip is as shown by the arrow mark along (8.17b) *aefghk*.

$$q_{c3} = \frac{d_4 q_e + d_5 (q_e + q_f)/2 + d_6 q_f + d_7 (q_g + q_h)/2 + d_8 q_h}{8d} \tag{8.56b}$$

where $q_a = q_e$, $q_f = q_g$, $q_h = q_k$.

Case 2: When the cone resistance q_c below the pile tip is greater than that at the tip within a depth 4d. [Fig. 8.17 (c)].

In this case q_p is found within a total depth of 0.7d as shown in Fig. 8.17 (c).

$$q_{c1} = \frac{q_o + q_b}{2}$$

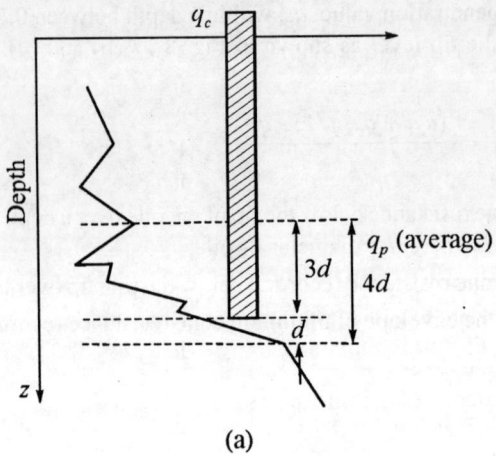

(a)

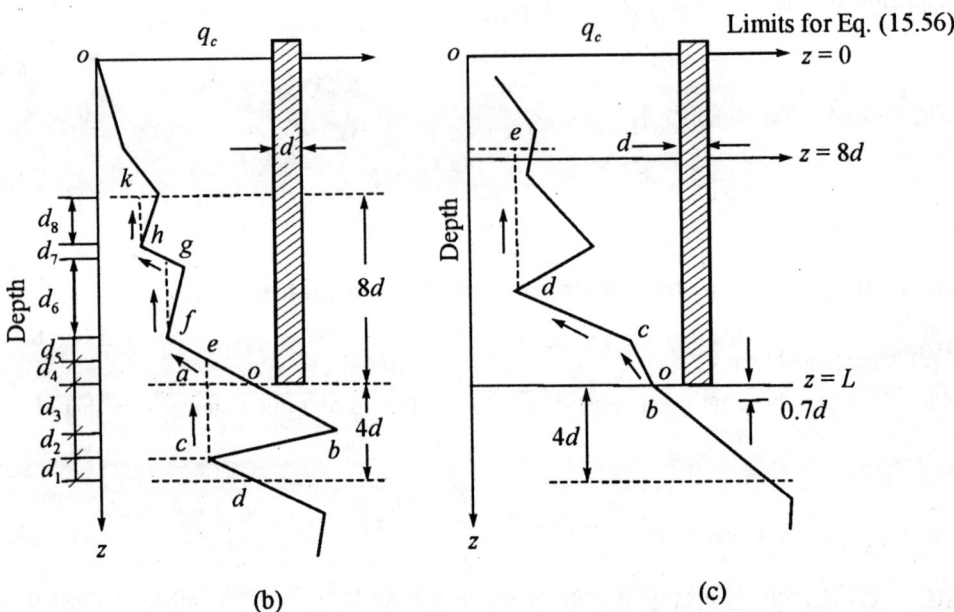

(b)   (c)

Fig. 8.17 Pile capacity by use of CPT values: (a) Vander Veen's method, (b) resistance below pile tip lower than that at pile tip within depth 4 d, (c) Schmertmann's method, resistance below pile tip greater than that at pile tip within 0.7 d depth

$q_{c2} = q_o$ = minimum value at the pile tip itself, q_{c3} = average of the minimum values along the envelope $ocde$ as before.

In determining the average q_c above, the minimum values q_{c2} selected under Case 1 or 2 are to be disregarded.

Effect of overconsolidation ratio in sand

Reduction factors have been developed that should be applied to the theoretical end bearing of a pile as determined from the CPT if the bearing layer consists of overconsolidated sand. The problem in many cases will be to make a reasonable estimate of the overconsolidation ratio in sand. In sands with a high q_c, some conservation in this respect is desirable, in particular for shallow foundations. The influence of overconsolidation on pile end bearing is one of the reasons for applying a limiting value to pile end bearing, irrespective of the cone resistances recorded in the bearing layer. A limit pile end bearing of 15 MN/m^2 is generally accepted (De Ruiter and Beringen, 1979), although in dense sands cone resistance may be greater than 50 MN/m^2. It is unlikely that in dense normally consolidated sand ultimate end bearing values higher than 15 MN/m^2 can occur but this has not been adequately confirmed by load tests.

Design CPT values for sand and clay

The application of CPT in evaluating the design values for skin friction and bearing as recommended by De Ruiter and Beringen (1979) is summarised in Table 8.3.

Table 8.3 Application of CPT in pile design (After De Ruiter and Beringen, 1979)

Item	Sand	Clay	Legend
Unit friction f_s	Minimum of $f_1 = 012$ MPa $f_2 =$ CPT sleeve friction $f_3 = q_c/300$ (compression) or $f_3 = q_c/400$ (tension)	$f_s = \alpha' c_u$, where $\alpha' = 1$ in NC clay $= 0.5$ in OC clay	$q_c =$ cone resistance below pile tip $c_u = q_c/N_k$
Unit end bearing q_p	Minimum of q_p from Fig. 8.17 (b) and (c)	$q_p = N_c c_u$ $N_c = 9$	$q_p =$ ultimate resistance of pile

Ultimate skin load Q_f in cohesionless soils

For the computation of skin load, Q_f, Schmertmann (1978) presents the following equation

$$Q_f = K \left(\sum_{z=0}^{z=8d} \frac{z}{8d} f_c A_s + \sum_{z=8d}^{z=L} f_c A_s \right) \tag{8.56c}$$

where, $K = f_s/f_c =$ correction factor for f_c,
$f_s =$ unit pile friction,
$f_c =$ unit sleeve friction measured by the friction jacket,
$z =$ depth of f_c value considered from ground surface,
$d =$ pile diameter or width,
$A_s =$ pile-soil contact area per f_c depth interval,
$L =$ embedded depth of pile.

When f_c does not vary significantly with depth, Eq. (8.56c) can be written in a simplified form as

$$Q_f = K \left[\frac{1}{2} \left(\overline{f}_c A_s \right)_{o-8d} + \left(\overline{f}_c A_s \right)_{8d-L} \right] \tag{8.56d}$$

where $\overline{f}_c$ is the average value within the depths specified. The correction factor K is given in Fig. 8.18 (a).

Ultimate skin load Q_f for piles in clay soil

For piles in clay Schmertmann gives the expression

$$Q_f = \alpha' \overline{f}_c A_s \qquad (8.57)$$

where, α' = ratio of pile to penetrometer sleeve friction,

f_c = average sleeve friction,

A_s = pile to soil contact area.

Figure 8.18 (b) gives values of α'.

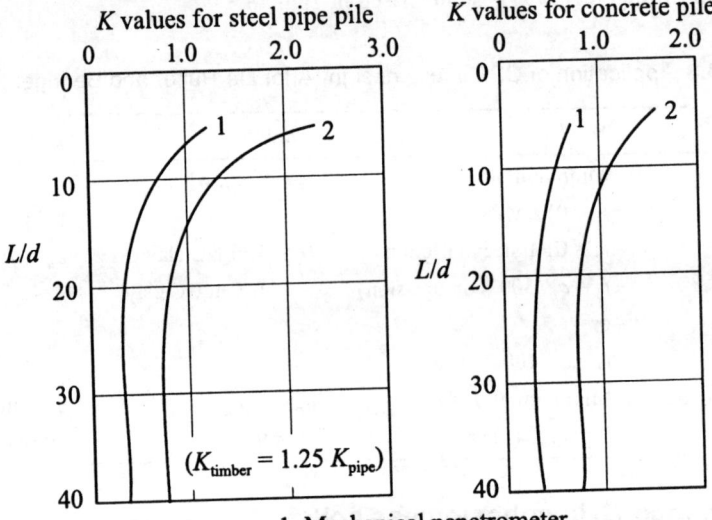

1. Mechanical penetrometer
2. Electrical penetrometer

(a)

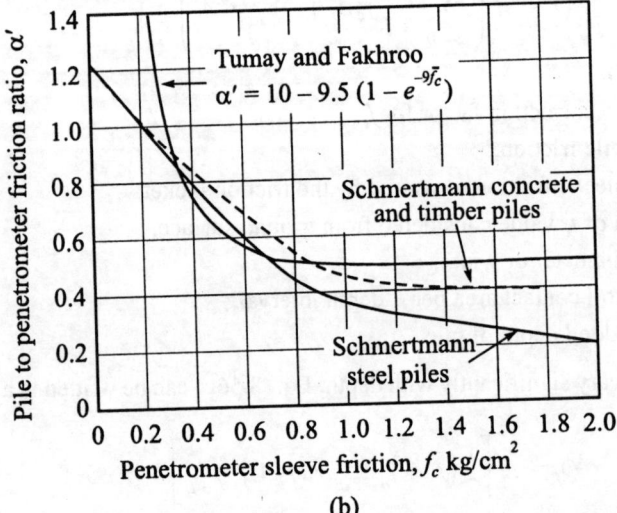

(b)

Fig. 8.18 Penetrometer design curves: (a) For pile side friction in sand (Schmertmann, 1978), and (b) for pile side friction in clay (Schmertmann, 1978)

Example 8.19

A concrete pile of 40 cm diameter is driven into a homogeneous mass of cohesionless soil. The pile carries a safe load of 650 kN. A static cone penetration test conducted at the site indicates an average value of $q_c = 40$ kg/cm^2 along the pile and 120 kg/cm^2 below the pile tip. Compute the length of the pile with $F_s = 2.5$. (Fig. Ex. 8.19).

Solution

From Eq. (8.51a)

$$q_b \text{ (pile)} = q_p \text{ (cone)}$$

Given, $q_p = 120$ kg/cm^2, therefore,

$$q_b = 120 \text{ kg/cm}^2 = 120 \times 100 = 12,000 \text{ kN/m}^2$$

Per Section 8.12, q_b is restricted to 11,000 kN/m^2.

Therefore,

$$Q_b = A_b\, q_b = \frac{3.14}{4} \times 0.4^2 \times 11,000 = 1,382 \text{ kN}$$

Assume the length of the pile $= L$m

The average,

$$\bar{q}_c = 40 \text{ kg/cm}^2$$

Per Eq. (8.52a),

$$f_s = \frac{\bar{q}_c}{2} \text{ kN/m}^2 = \frac{40}{2} = 20 \text{ kN/m}^2$$

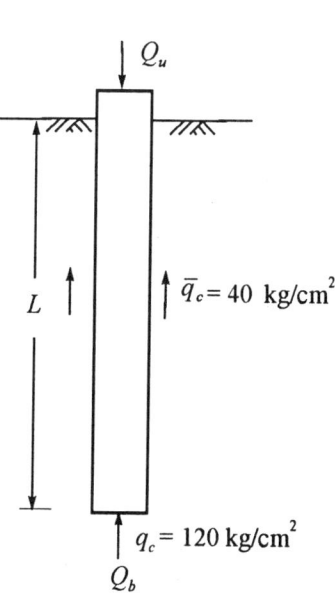

Fig. Ex. 8.19

Now, $Q_f = f_s A_s = 20 \times 3.14 \times 0.4 \times L = 25.12L$ kN

Given $Q_a = 650$ kN. With $F_s = 2.5$, $Q_u = 650 \times 2.5 = 1,625$ kN.

Now, $1,625 = Q_b + Q_f = (1,382 + 25.12L)$ kN

or $$L = \frac{1,625 - 1,382}{15.12} = 9.67 \text{ m or say 10 m}$$

The pile has to be driven to a depth of 10 m to carry a safe load of 650 kN with $F_s = 2.5$.

Example 8.20

A concrete pile of size 0.4 × 0.4 m is driven to a depth of 12 m into medium dense sand. The water table is close to the ground surface. Static cone penetration tests were carried out at this site by using an electric cone penetrometer. The values of q_c and f_c as obtained from the test have been plotted against depth and shown in Fig. Ex. 8.20. Determine the safe load on this pile with $F_s = 2.5$ by Schmertmann's method (Section 8.21).

Solution

First determine the representative cone penetration value q_p by using Eq. (8.54)

$$q_p = \frac{(q_{c1} + q_{c2}) + q_{c3}}{2}$$

Now from Fig. Ex. 8.20 and Eq. (8.56a)

$$q_{c1} = \frac{d_3(q_o + q_b)/2 + d_2(\dot{q}_b + q_d)/2 + d_1(q_d + q_c)/2}{4d}$$

$$= \frac{0.7(76 + 85)/2 + 0.3(85 + 71)/2 + 0.6(71 + 80)/2}{4 \times 0.4}$$

$$= 78 \text{ kg/cm}^2$$

$q_{c2} = q_d =$ the minimum value below the tip of the pile within $4d$ depth $= 71 \text{ kg/cm}^2$.

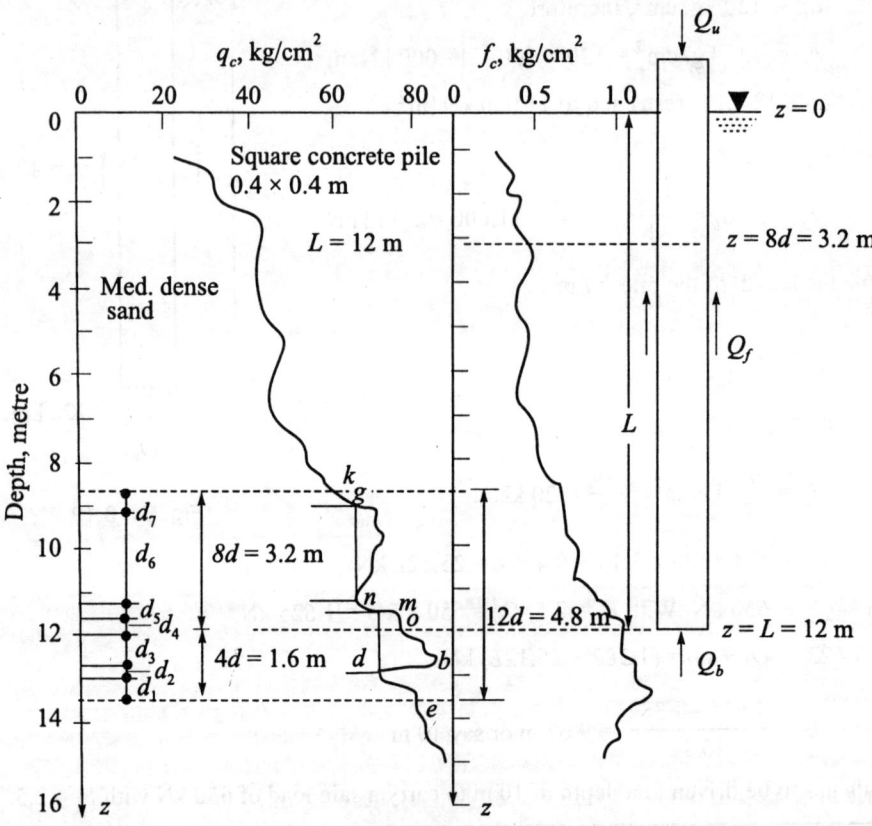

Fig. Ex. 8.20

From Eq. (8.56b)

$$q_{c3} = \frac{d_4 q_m + d_5(q_m + q_n)/2 + d_6 q_n + d_7(q_g + q_k)/2}{8d}$$

$$= \frac{0.4 \times 71 + 0.3(71 + 65)/2 + 2.1 \times 65 + 0.4(65 + 60)/2}{8 \times 0.4}$$

$$= 66 \text{ kg/cm}^2 = 660 \text{ t/m}^2 \text{ (metric)}$$

From Eq. (8.54)

$$q_p = \frac{(78+71)/2+66}{2} = 70 \text{ kg/cm}^2 \approx 700 \ t/m^2 \text{ (metric)}$$

Ultimate base load

$$Q_b = q_b A_b = q_p A_b = 700 \times 0.4^2 = 112 \ t \text{ (metric)}$$

Frictional load Q_f

From Eq. (8.56d)

$$Q_f = K\left[\frac{1}{2}\left(\bar{f}_c A_s\right)_{0-8d} + \left(\bar{f}_c A_s\right)_{8d-L}\right]$$

where K = correction factor from Fig. 8.18 (a) for electrical penetrometer.

For $\dfrac{L}{d} = \dfrac{12}{0.4} = 30$, $K = 0.75$ for concrete pile. It is now necessary to determine the average sleeve friction $\bar{f}_c$ between depths $z = 0$ and $z = 8d$, and $z = 8d$ and $z = L$ from the top of pile from f_c profile given in Fig. Ex. 8.20.

$$Q_f = 0.75 \left(\frac{1}{2} \times 0.34 \times 10 \times 4 \times 0.4 \times 3.2 + 0.71 \times 10 \times 4 \times 0.4 \times 8.8\right)$$

$$= 0.75 \ (8.7 + 99.97) = 81.5t \text{ (metric)}$$

$$Q_u = Q_b + Q_f = 112 + 81.5 = 193.5t$$

$$Q_a = \frac{Q_b + Q_f}{2.5} = \frac{193.5}{2.5} = 77.4t \text{ (metric)} = 759 \text{ kN}$$

Example 8.21

A concrete pile of section 0.4 × 0.4 m is driven into normally consolidated clay to a depth of 10 m. The water table is at ground level. A static cone penetration test (CPT) was conducted at the site with an electric cone penetrometer. Figure Ex. 8.21 gives a profile of q_c and f_c values with respect to depth. Determine safe loads on the pile by the following methods:

(a) α-method, (b) λ-method, given: $\gamma_b = 8.5 \text{ kN/m}^3$, and (c) Schmertmann's method. Use a factor of safety of 2.5.

Solution

(a) α-method

The α-method requires the undrained shear strength of the soil. Since this is not given, it has to be determined by using the relation between q_c and c_u given in Eq. (3.18).

$$c_u = \frac{q_c}{N_k} \text{ by neglecting the overburden effect,}$$

where N_k = cone factor = 20.

It is necessary to determine the average $\bar{c}_u$ along the pile shaft and $\bar{c}_b$ at the base level of the pile. For this purpose find the corresponding f_c (sleeve friction) value from Fig. Ex. 8.21.

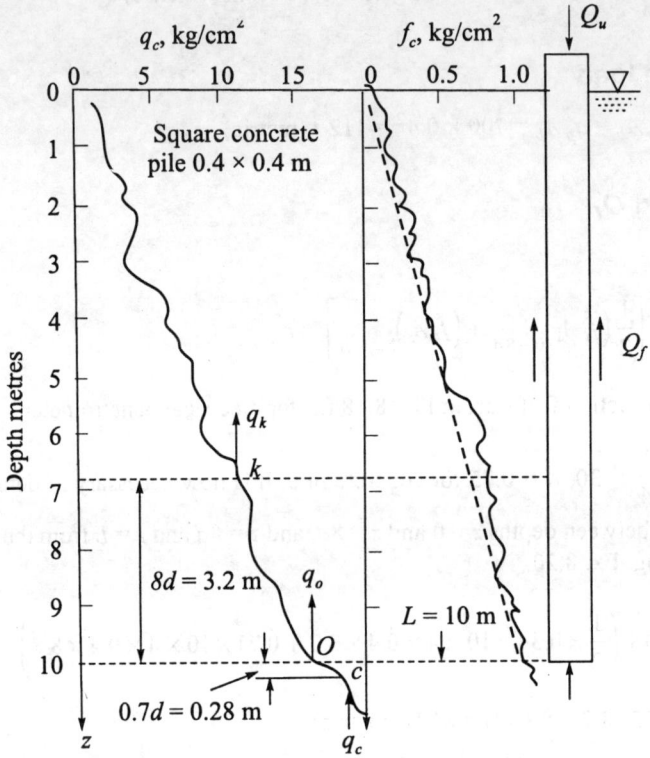

Fig. Ex. 8.21

Average $\bar{q}_c$ along the shaft,

$$\bar{q}_c = \frac{1+16}{2} = 8.5 \, \text{kg/cm}^2.$$

Average of q_c within a depth $3d$ above the base and d below the base of the pile [Refer to Fig. 8.17 (a)]

$$q_p = \frac{15 + 18.5}{2} = 17 \, \text{kg/cm}^2$$

$$\bar{c}_u = \frac{8.5}{20} = 0.43 \, \text{kg/cm}^2 \approx 43 \, \text{kN/m}^2$$

$$c_b = \frac{17}{20} = 0.85 \, \text{kg/cm}^2 \approx 85 \, \text{kN/m}^2.$$

Ultimate Base Load, Q_b

From Eq. (8.39)

$$Q_b = 9c_b A_b = 9 \times 85 \times 0.40^2 = 122 \, \text{kN}.$$

Ultimate Friction Load, Q_f

From Eq. (8.37)

$$Q_f = \alpha \bar{c}_u A_s$$

From Fig. 8.15 for $\bar{c}_u = 43$ kN/m^2, $\alpha = 0.70$

$$Q_f = 0.70 \times 43 \times 10 \times 4 \times 0.4 = 481.6 \text{ kN or say } 482 \text{ kN}$$

$$Q_u = 122 + 482 = 604 \text{ kN}$$

$$Q_a = \frac{604}{2.5} = 241.6 \text{ kN or say } 242 \text{ kN}$$

(b) λ-method

Base Load Q_b

In this method the base load is the same as in (a) above. That is

$$Q_b = 122 \text{ kN}$$

Friction Load

From Fig. 8.17, $\lambda = 0.25$ for $L = 10$ m (= 32.4 ft). From Eq. (8.40)

$$f_s = \lambda (\bar{q}_o + 2\bar{c}_u)$$

$$\bar{q}_o = \frac{1}{2} \times 10 \times 8.5 = 42.5 \text{ kN/m}^2$$

$$f_s = 0.25 (42.5 + 2 \times 43) = 32 \text{ kN/m}^2$$

$$Q_f = f_s A_s = 32 \times 10 \times 4 \times 0.4 = 512 \text{ kN}$$

$$Q_u = 122 + 512 = 634 \text{ kN}$$

$$Q_a = \frac{634}{2.5} = 254 \text{ kN.}$$

(c) Schmertmann's Method

Base load Q_b

Use Eq. (8.54) for determining the representative value for q_p. Here, the minimum value for q_c is at point O on the q_c-profile in Fig. Ex. 8.21 which is the base level of the pile. Now q_{c1} is the average q_c at the base and $0.7d$ below the base of the pile, that is,

$$q_{c1} = \frac{q_o + q_e}{2} = \frac{16 + 18.5}{2} = 17.25 \text{ kg/cm}^2$$

$$q_{c2} = q_o = 16 \text{ kg/cm}^2$$

$q_{c3} = $ The average of q_c within a depth $8d$ above the base level

$$= \frac{q_o + q_k}{2} = \frac{16 + 11}{2} = 13.5 \text{ kg/cm}^2$$

From Eq. (8.45),

$$q_p = \frac{(17.25 + 16)/2 + 13.5}{2} = 15 \text{ kg/cm}^2.$$

From Eq. (8.55a)

$$q_b \text{ (pile)} = q_p \text{ (cone)} = 15 \text{ kg/cm}^2 \approx 1{,}500 \text{ kN/m}^2$$

$$Q_b = q_b A_b = 1{,}500 \times (0.4)^2 = 240 \text{ kN}$$

Friction Load Q_f

Use Eq. (8.57)

$$Q_f = \alpha' \bar{f_c} A_s$$

where α' = ratio of pile to penetrometer sleeve friction.

From Fig. Ex. 8.21 for $\bar{f_c} = \dfrac{0 + 1.15}{2} = 0.58 \text{ kg/cm}^2 \approx 58 \text{ kN/m}^2.$

From Fig. 8.18 (b) for $f_c = 58 \text{ kN/m}^2$, $\alpha' = 0.70$

$$Q_f = 0.70 \times 58 \times 10 \times 4 \times 0.4 = 650 \text{ kN}$$

$$Q_u = 240 + 650 = 890 \text{ kN}$$

$$Q_a = \frac{890}{2.5} = 356 \text{ kN}$$

Note: The values given in the examples are only illustrative and not factual.

8.22 BEARING CAPACITY OF A SINGLE PILE BY LOAD TEST

A pile load test is the most acceptable method to determine the load carrying capacity of a pile. The load test may be carried out either on a driven pile or a *cast-in-situ* pile. Load tests may be made either on a single pile or a group of piles. Load tests on a pile group are very costly and may be undertaken only in very important projects.

Pile load tests on a single pile or a group of piles are conducted for the determination of

1. Vertical load bearing capacity,
2. Uplift load capacity,
3. Lateral load capacity.

Generally, load tests are made to determine the bearing capacity and to establish the load settlement relationship under a compressive load. The other two types of tests may be carried out only when piles are required to resist large uplift or lateral forces.

Usually, pile foundations are designed with an estimated capacity which is determined from a thorough study of the site conditions. At the beginning of construction, load tests are made for the purpose of verifying the adequacy of the design capacity. If the test results show an inadequate factor of safety or excessive settlement, the design must be revised before construction is under way.

Load tests may be carried out either on

1. A working pile or
2. A test pile.

A *working pile* is a pile driven or *cast-in-situ* along with the other piles to carry the loads from the superstructure. The maximum test load on such piles should not exceed one and a half times the design load.

A *test pile* is a pile which does not carry the loads coming from the structure. The maximum load that can be put on such piles may be about 2½ times the design load or the load imposed must be such as to give a total settlement not less than one-tenth the pile diameter.

Method of Carrying Out Vertical Pile Load Test

A vertical pile load test assembly is shown in Fig. 8.19 (a). It consists of

1. An arrangement to take the reaction of the load applied on the pile head.
2. An hydraulic jack of sufficient capacity to apply load on the pile head.
3. A set of three dial gauges to measure settlement of the pile head.

Load application

A load test may be of two types:

1. Continuous load test.
2. Cyclic load test.

In the case of a continuous load test, continuous increments of load are applied to the pile head. Settlement of the pile head is recorded at each load level.

In the case of the cyclic load test, the load is raised to a particular level, then reduced to zero, again raised to a higher level and reduced to zero. Settlements are recorded at each increment or decrement of load. Cyclic load tests help to separate frictional load from point load.

The total elastic recovery or settlement S_e, is due to

1. The total plastic recovery of the pile material,
2. Elastic recovery of the soil at the tip of the pile, $\overline{S}_e$.

The total settlement S due to any load can be separated into elastic and plastic settlements by carrying out cyclic load tests as shown in Fig. 8.19 (b).

A pile loaded to Q_1 gives a total settlement S_1. When this load is reduced to zero, there is an elastic recovery which is equal to S_{e1}. This elastic recovery is due to the elastic compression of the pile material and the soil. The net settlement or plastic compression is S_{p1}. The pile is loaded again from zero to the next higher load Q_2 and reduced to zero thereafter. The corresponding settlements may be found as before. The method of loading and unloading may be repeated as before.

Allowable Load from Single Pile Load Test Data

There are many methods by which allowable loads on a single pile may be determined by making use of load test data. If the ultimate load can be determined from load-settlement curves, allowable loads are found by dividing the ultimate load by a suitable factor of safety which varies from 2 to 3. A factor of safety of 2.5 is normally recommended. A few of the methods that are useful for the determination of ultimate or allowable loads on a single pile are given below:

1. The ultimate load, Q_u, can be determined as the abscissa of the point where the curved part of the load-settlement curve changes to a falling straight line, Fig. 8.20 (a).
2. Q_u is the abscissa of the point of intersection of the initial and final tangents of the load-settlement curve, Fig. 8.20 (b).

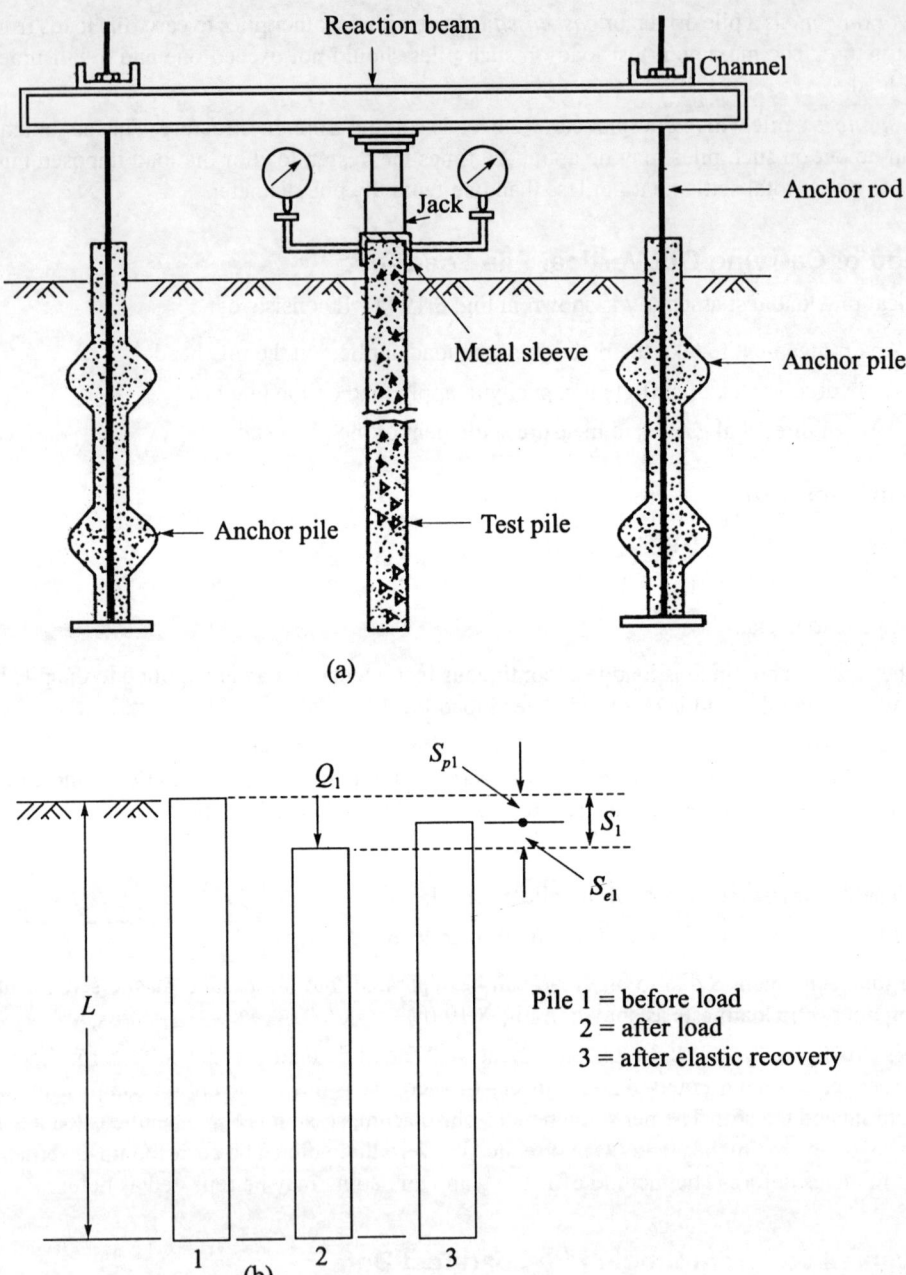

Fig. 8.19 (a) Vertical pile load test assembly, and (b) elastic compression at the base of the pile

3. The allowable load Q_a is 50 percent of the ultimate load at which the total settlement amounts to one-tenth of the diameter of the pile for uniform diameter piles.

4. The allowable load Q_a is sometimes taken as equal to two-thirds of the load which causes a total settlement of 12 mm.

5. The allowable load Q_a is sometimes taken as equal to two-thirds of the load which causes a net (plastic) settlement of 6 mm.

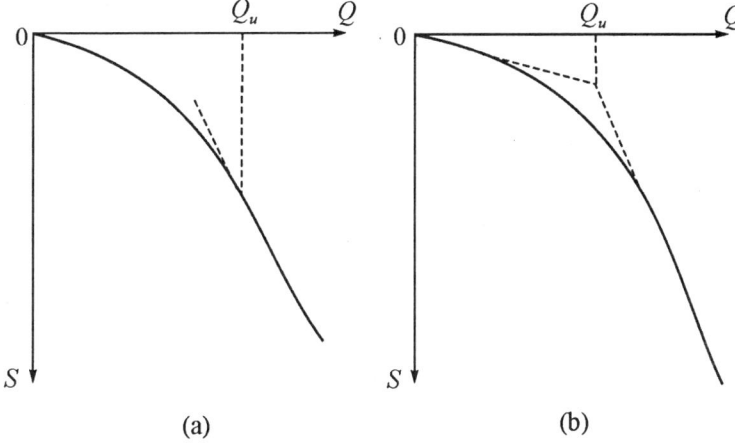

Fig. 8.20 Determination of ultimate load from load-settlement curves

If pile groups are loaded to failure, the ultimate load of the group, Q_{gu}, may be found by any one of the first two methods mentioned above for single piles. However, if the groups are subjected to only to one and a half-times the design load of the group, the allowable load on the group cannot be found on the basis of 12 or 6 mm settlement criteria applicable to single piles. In the case of a group with piles spaced at less than 6 to 8 times the pile diameter, the stress interaction of the adjacent piles affects the settlement considerably. The settlement criteria applicable to pile groups should be the same as that applicable to shallow foundations at design loads.

8.23   PILE BEARING CAPACITY FROM DYNAMIC PILE DRIVING FORMULAS

The resistance offered by a soil to penetration of a pile during driving gives an indication of its bearing capacity. Qualitatively speaking, a pile which meets greater resistance during driving is capable of carrying a greater load. A number of dynamic formulae have been developed which equate pile capacity in terms of driving energy.

The basis of all these formulae is the simple energy relationship which may be stated by the following equation (Fig. 8.21).

$$Wh = Q_u s$$

or
$$Q_u = \frac{Wh}{s} \tag{8.58}$$

where W = weight of the driving hammer,
h = height of fall of hammer,
Wh = energy of hammer blow,
Q_u = ultimate resistance to penetration,
s = pile penetration under one hammer blow,
$Q_u s$ = resisting energy of the pile.

Hiley Formula

Equation (8.58) holds only if the system is 100 percent efficient. Since the driving of a pile involves many losses, the energy of the system may be written as

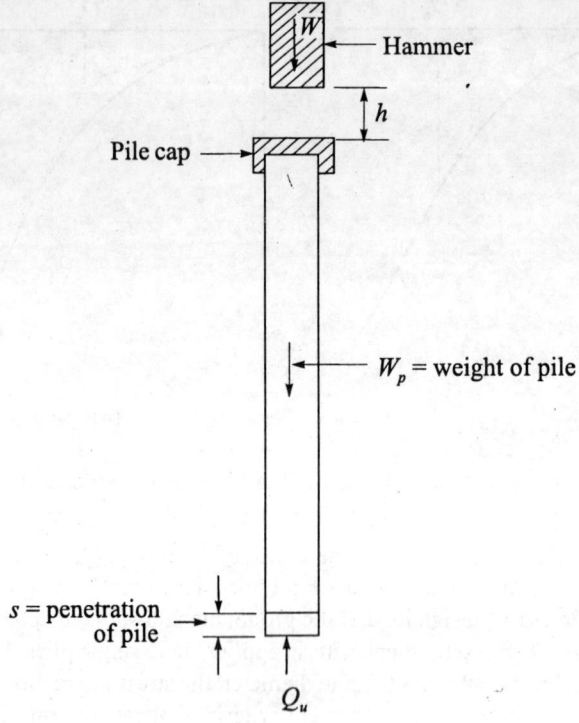

Fig. 8.21 Basic energy relationship

Energy input = Energy used + Energy losses

or   Energy used = Energy input – Energy losses.

The expressions for the various energy terms used are:

1. Energy used = $Q_u s$.
2. Energy input = $\eta_h W h$, where η_h is the efficiency of the hammer.
3. The energy losses are due to the following:

   (i) The energy loss E_1 due to the elastic compressions of the pile cap, pile material and the soil surrounding the pile. The expression for E_1 may be written as

$$E_1 = \frac{1}{2} Q_u (c_1 + c_2 + c_3) = Q_u C$$

   where c_1 = elastic compression of the pile cap,
             c_2 = elastic compression of the pile,
             c_3 = elastic compression of the soil.

   (ii) The energy loss E_2 due to the interaction of the pile hammer system (impact of two bodies). The expression for E_2 may be written as

$$E_2 = W h W_p \frac{1 - C^2}{W + W_p}$$

   where W_p = weight of pile,
           C_r = coefficient of restitution.

Substituting the various expressions in the energy equation and simplifying, we have

$$Q_u = \frac{\eta_h \, Wh}{s + C} \times \frac{1 + C_r^2 R}{1 + R} \qquad (8.59)$$

where $\quad R = \dfrac{W_p}{W}$

Equation (8.59) is called the Hiley formula. The allowable load Q_a may be obtained by dividing Q_u by a suitable factor of safety.

If the pile tip rests on rock or relatively impenetrable material, Eq. (8.59) is not valid. Chellis (1961), suggests for this condition that the use of $W_p/2$ instead of W_p may be more correct. The various coefficients used in the Eq. (8.59) are as given below:

1. *Elastic compression c_1 of cap and pile head*

Pile Material	Range of Driving Stress kg/cm^2	Range of c_1
Precast concrete pile with packing inside cap	30–150	0.12–0.50
Timber pile without cap	30–150	0.05–0.20
Steel *H*-pile	30–150	0.04–0.16

2. *Elastic compression c_2 of pile*
   This may be computed using the equation

$$C_2 = \frac{Q_u L}{AE}$$

   where $\quad L =$ embedded length of the pile,
   $\qquad\quad A =$ average cross-sectional area of the pile,
   $\qquad\quad E =$ Young's modulus.

3. *Elastic compression c_3 of soil*
   The average value of c_3 may be taken as 0.1 (the value ranges from 0.0 for hard soil to 0.2 for resilient soils).

4. *Pile-hammer efficiency*

Hamer type	η_h
Drop	1.00
Single acting	0.75–0.85
Double acting	0.85
Diesel	1.00

5. *Coefficient of restitution C_r*

Material	C_r
Wood pile	0.25
Compact wood cushion on steel pile	0.32
Cast iron hammer on concrete pile without cap	0.40
Cast iron hammer on steel pipe without cushion	0.55

Engineering News Record (ENR) Formula

The general form of the Engineering News Record Formula for the allowable load Q_a may be obtained from Eq. (8.59) by putting.

$\eta_h = 1$ and $C_r = 1$ and a factor of safety equal to 6. The formula proposed by AM Wellington, editor of the Engineering News, in 1886, is

$$Q_a = \frac{Wh}{6(s+C)} \tag{8.60}$$

where Q_a = allowable load in kg,

W = weight of hammer in kg,

h = height of fall of hammer in cm,

s = final penetration in cm per blow (which is termed as set). The set is taken as the average penetration per blow for the last 5 blows of a drop hammer or 20 blows of a steam hammer,

C = empirical constant,

= 2.5 cm for a drop hammer,

= 0.25 cm for single and double acting hammers.

The equations for the various types of hammers may be written as :

1. Drop hammer

$$Q_a = \frac{Wh}{6(s+2.5)} \tag{8.61}$$

2. Single-acting hammer

$$Q_a = \frac{Wh}{6(s+0.25)} \tag{8.62}$$

3. Double-acting hammer

$$Q_a = \frac{(W+ap)}{6(s+0.25)} \tag{8.63}$$

a = effective area of the piston in sq. cm,

p = mean effective steam pressure in kg/cm^2.

Comments on the use of dynamic formulae

1. Detailed investigations carried out by Vesic (1967) on deep foundations in granular soils indicate that the Engineering News Record Formula applicable to drop hammers, Eq. (8.61), gives pile loads as low as 44% of the actual loads. To obtain better agreement between the one computed and observed loads, Vesic suggests the following values for the coefficient C in Eq. (8.60).

   For steel pipe piles, $C = 1$ cm.

   For precast concrete piles $C = 1.5$ cm.

2. The tests carried out by Vesic in granular soils indicate that Hiley's formula does not give consistent results. The values computed from Eq. (8.59) are sometimes higher and sometimes lower than the observed values.

3. Dynamic formulae in general have limited value in pile foundation work mainly because the dynamic resistance of soil does not represent the static resistance, and because often the results obtained from the use of dynamic equations are of questionable dependability. However, engineers prefer to use the Engineering News Record Formula because of its simplicity.

4. Dynamic formulae could be used with more confidence in freely draining materials such as coarse sand. If the pile is driven to saturated loose fine sand and silt, there is every possibility of development of liquefaction which reduces the bearing capacity of the pile.

5. Dynamic formulae are not recommended for computing allowable loads of piles driven into cohesive soils. In cohesive soils, the resistance to driving increases through the sudden increase in stress in pore water and decreases because of the decreased value of the internal friction between soil and pile because of pore water. These two oppositely directed forces do not lend themselves to analytical treatment and as such the dynamic penetration resistance to pile driving has no relationship to static bearing capacity.

There is another effect of pile driving in cohesive soils. During driving the soil becomes remolded and the shear strength of the soil is reduced considerably. Though there will be a regaining of shear strength after a lapse of some days after the driving operation, this will not be reflected in the resistance value obtained from the dynamic formulae.

Example 8.22

A 40 × 40 cm reinforced concrete pile 20 m long is driven through loose sand and then into dense gravel to a final set of 3 mm/blow, using a 30 kN single-acting hammer with a stroke of 1.5 m. Determine the ultimate driving resistance of the pile if it is fitted with a helmet, plastic dolly and 50 mm packing on the top of the pile. The weight of the helmet and dolly is 4 kN. The other details are: weight of pile = 74 kN; weight of hammer = 30 kN; pile hammer efficiency η_h = 0.80 and coefficient of restitution C_r = 0.40.

Use the Hiley formula. The sum of the elastic compression $\bar{C}$ is

$$\bar{C} = c_1 + c_2 + c_3 = 19.6 \text{ mm.}$$

Solution

Hiley formula

Use Eq. (8.59)

$$Q_u = \frac{\eta_h Wh}{s + C} \times \frac{1 + C_r^2 R}{1 + R}$$

where η_h = 0.80, W = 30 kN, h = 1.5 m, $R = \dfrac{W_p}{W} = \dfrac{(74 + 4)}{30} = 2.6$, C_r = 0.40, s = 0.30 cm.

Substituting, we have

$$Q_u = \frac{0.8 \times 30 \times 150}{0.3 + 1.96/2} \times \frac{1 + 0.4^2 \times 2.6}{1 + 2.6} = 2{,}813 \times 0.393 = 1{,}105 \text{ kN}$$

8.24 BEARING CAPACITY OF PILES FOUNDED ON A ROCKY BED

Piles are at times required to be driven through weak layers of soil until the tips meet a hard strata for bearing. If the bearing strata happens to be rock, the piles are to be driven to refusal to obtain the maximum carrying capacity from the piles. If the rock is strong at its surface, the pile will refuse further driving at a negligible penetration. In such cases the carrying capacity of the piles is governed by the strength of the pile shaft regarded as a column as shown in Fig. 8.6 (a). If the soil mass through which the piles are driven happens to be stiff clays or sands, the piles can be regarded as being supported on all sides from buckling as a strut. In such cases, the carrying capacity of a pile is calculated from the safe load on the material of the pile at the point of minimum cross-section. In practice, it is necessary to limit the safe load on piles regarded as short columns because of the likely deviations from the vertical and the possibility of damage to the pile during driving.

If piles are driven to weak rocks, working loads as determined by the available stress on the material of the pile shaft may not be possible. In such cases the frictional resistance developed over the penetration into the rock and the end bearing resistance are required to be calculated. Tomlinson (1986) suggests an equation for computing the end bearing resistance of piles resting on rocky strata as

$$q_u = 2N_\phi q_{ur} \tag{8.64}$$

where $N_\phi = \tan^2 (45 + \phi/2)$,

q_{ur} = unconfined compressive strength of the rock.

Boring of a hole in rocky strata for constructing bored piles may weaken the bearing strata of some types of rock. In such cases low values of skin friction should be used and normally may not be more than 20 kN/m² (Tomlinson, 1986), when the boring is through friable chalk or mud stone. In the case of moderately weak to strong rocks where it is possible to obtain core samples for unconfined compression tests, the end bearing resistance can be calculated by making use of Eq. (8.64).

8.25 UPLIFT RESISTANCE OF PILES

Piles are also used to resist uplift loads. Piles used for this purpose are called *tension piles*, uplift piles or anchor piles. Uplift forces are developed due to hydrostatic pressure or overturning moments as shown in Fig. 8.22.

Figure 8.22 shows a straight edged pile subjected to uplift force. The equation for the uplift force P_{ul} may be written as

$$P_{ul} = W_p + A_s f_r \tag{8.65}$$

where P_{ul} = uplift capacity of pile,

W_p = weight of pile,

f_r = unit resisting force,

A_s = effective area of the embedded length of pile.

Uplift Resistance of Pile in Clay

For piles embedded in clay, Eq. (8.65), may written as

$$P_{ul} = W_p + A_s \alpha \bar{c}_u \tag{8.66}$$

where, $\bar{c}_u$ = average undrained shear strength of clay along the pile shaft,

α = adhesion factor (= c_a/c_u),

c_a = average adhesion.

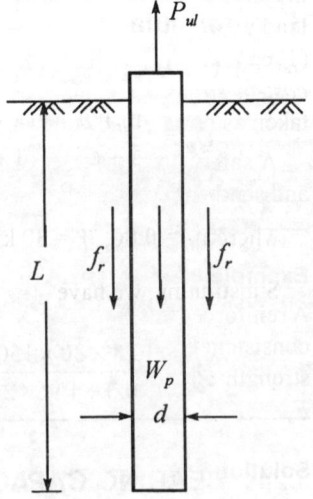

Fig. 8.22 Single pile subjected to uplift

Figure 8.23 gives the relationship between α and c_u based on pull out test results as collected by Sowa (1970). As per Sowa, the values of c_a agree reasonably well with the values for piles subjected to compression loadings.

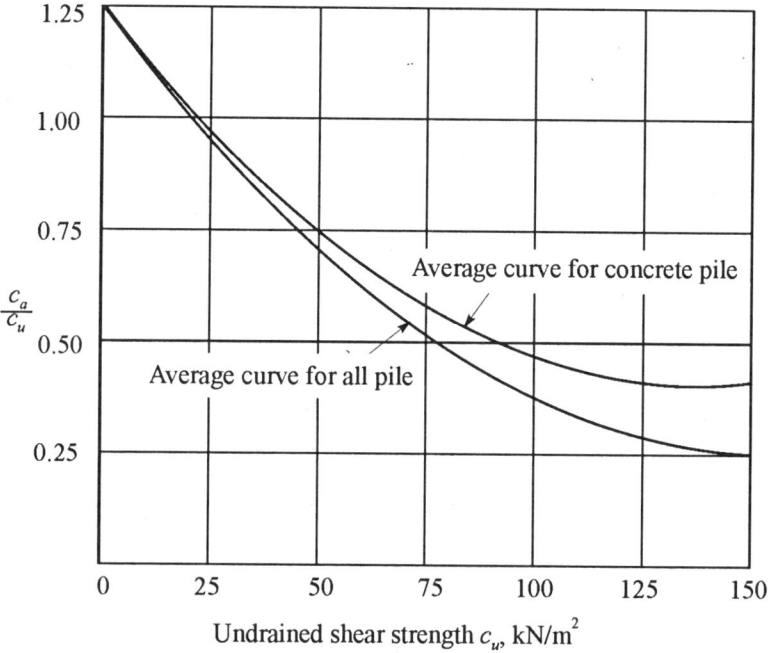

Fig. 8.23 Relationship between adhesion factor α and undrained shear strength c_u
(*Source:* Poulos and Davis, 1980)

Uplift Resistance of Pile in Sand

Adequate confirmatory data are not available for evaluating the uplift resistance of piles embedded in cohesionless soils. Ireland (1957) reports that the average skin friction for piles under compression loading and uplift loading are equal, but data collected by Sowa (1970) and Downs and Chieurzzi (1966) indicate lower values for upward loading as compared to downward loading especially for *cast-in-situ* piles. Poulos and Davis (1980) suggest that the skin friction of upward loading may be taken as two-thirds of the calculated shaft resistance for downward loading.

A safety factor of 3 is normally assumed for calculating the safe uplift load for both piles in clay and sand.

Example 8.23

A reinforced concrete pile 30 ft long and 15 in. diameter is embedded in a saturated clay of very stiff consistency. Laboratory tests on samples of undistorbed soil gave an average undrained cohesive strength $c_u = 2,500$ lb/ft^2. Determine the net pullout capacity and the allowable pullout load with $F_s = 3$.

Solution

Given: $L = 30$ ft, $d = 15$ in. diameter, $c_u = 2,500$ lb/ft^2, $F_s = 3$.

From Fig. 8.23, $c_a/c_u = 0.41$ for $c_u = 2,500 \times 0.0479 \approx 120$ kN/m^2 for concrete pile.

From Eq. (8.66)

$$P_{ul} \text{ (net)} = \alpha \bar{c}_u A_s$$

where $\alpha = c_a/c_u = 0.41$, $c_u = 2,500$ lb/ft^2

$$A_s = 3.14 \times \frac{15}{12} \times 30 = 117.75 \text{ ft}^2$$

Substituting

$$P_{ul} \text{ (net)} = \frac{0.41 \times 2,500 \times 117.75}{1,000} = 120.69 \text{ kips}$$

$$P_{ul} \text{ (allowed)} = \frac{120.69}{3} \approx 40 \text{ kips}$$

Example 8.24

Refer to Ex. 8.23. If the pile is embedded in medium dense sand, determine the net pullout capacity and the net allowable pullout load with $F_s = 3$.

Given: $L = 30$ ft, $\phi = 38°$, $\bar{K}_s = 1.5$, and $\delta = 25°$, γ (average) $= 110$ lb/ft^3.

The water table is at great depth. Refer to Section 8.25.

Solution

Downward skin resistance Q_f

$$Q_f = \bar{q}'_o \bar{K}_s \tan \delta A_s$$

where

$$\bar{q}'_o = \frac{1}{2} \times 30 \times 110 = 1,650 \text{ lb/ft}^2$$

$$A_s = 3.14 \times 1.25 \times 30 = 117.75 \text{ ft}^2$$

$$Q_f \text{ (down)} = \frac{1,650 \times 1.5 \tan 25° \times 117.75}{1,000} = 136 \text{ kips}$$

Based on the recommendations of Poulos and Davis (1980)

$$Q_f(\text{up}) = \frac{2}{3} Q_f \text{ (down)} = \frac{2}{3} \times 136 = 91 \text{ kips}$$

8.26 PROBLEMS

8.1 A 45 cm diameter pipe pile of length 12 m with closed end is driven into a cohesionless soil having $\phi = 35°$. The void ratio of the soil is 0.48 and $G = 2.65$. The water table is at the ground surface. Estimate: (a) The ultimate base load Q_b, (b) the frictional load Q_f, and (c) the allowable load Q_a with $F_s = 2.5$.

Use the Berezantsev method for estimating Q_b. For estimating Q_f, use $\bar{K}_s = 0.75$ and $\delta = 20°$.

8.2 Refer to Problem 8.1. Compute Q_b by Meyerhof's method. Determine Q_f using the critical depth concept, and Q_a with $F_s = 2.5$. All the other data given in Prob. 8.1 remain the same.

8.3 Estimate Q_b by Vesic's method for the pile given in Prob. 8.1. Assume $I_r = I_{rr} = 60$. Determine Q_a for $F_s = 2.5$ and use Q_f obtained in Prob. 8.1.

8.4 For Problem 8.1, estimate the ultimate base resistance Q_b by Janbu's method. Determine Q_a with $F_s = 2.5$. Use Q_f obtained in Prob. 8.1. Use $\psi = 90°$.

8.5 For Problem 8.1, estimate Q_b, Q_f, and Q_a by Coyle and Castello method. All the data given remain the same.

8.6 For problem 8.1, determine Q_b, Q_f, and Q_a by Meyerhof's method using the relationship between N_{cor} and ϕ given in Fig. 5.7.

8.7 A concrete pile 40 cm in diameter is driven into homogeneous sand extending to a great depth. Estimate the ultimate load bearing capacity and the allowable load with $F_s = 3.0$ by Coyle and Castello's method. Given : $L = 15$ m, $\phi = 36°$, $\gamma = 18.5$ kN/m³.

8.8 Refer to Prob. 8.7. Estimate the allowable load by Meyerhof's method using the relationship between ϕ and N_{cor} given in Fig. 5.7.

8.9 A concrete pile of 15 in. diameter, 40 ft long is driven into a homogeneous's stratum of clay with the water table at ground level. The clay is of medium stiff consistency with the undrained shear strength $c_u = 600$ lb/ft². Compute Q_b by Skempton's method and Q_f by the α-method. Determine Q_a for $F_s = 2.5$.

8.10 Refer to Prob. 8.9. Compute Q_f by the λ-method. Determine Q_a by using Q_b computed in Prob. 8.9. Assume $\gamma_{sat} = 120$ lb/ft³.

8.11 A pile of 40 cm diameter and 18.5 m long passes through two layers of clay and is embedded in a third layer. Figure Prob. 8.11 gives the details of the soil system. Compute Q_f by the α-method and Q_b by Skempton's method. Determine Q_a for $F_s = 2.5$.

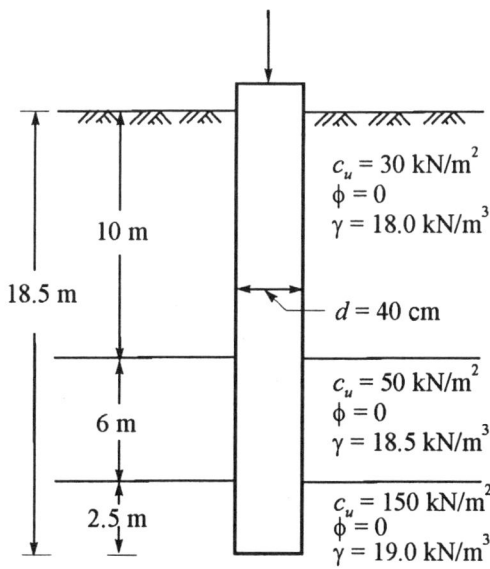

Fig. Prob. 8.11

8.12 A concrete pile of size 16 × 16 in. is driven into a homogeneous clay soil of medium consistency. The water table is at ground level. The undrained shear strength of the soil is 500 lb/ft². Determine the length of pile required to carry a safe load of 50 kips with $F_s = 3$. Use the α-method.

8.13 Refer to Prob. 8.12. Compute the required length of pile by the λ-method. All the other data remain the same. Assume $\gamma_{sat} = 120$ lb/ft³.

8.14 A concrete pile 50 cm in diameter is driven into a homogeneous mass of cohesionless soil. The pile is required to carry a safe load of 700 kN. A static cone penetration test conducted at the site gave an average value of $q_c = 35$ kg/cm^2 along the pile and 60 kg/cm^2 below the base of the pile. Compute the length of the pile with $F_s = 3$.

8.15 Refer to Problem 8.14. If the length of the pile driven is restricted to 12 m, estimate the ultimate load Q_u and safe load Q_a with $F_s = 3$. All the other data remain the same.

8.16 A reinforced concrete pile 20 in. in diameter penetrates 40 ft into a stratum of clay and rests on a medium dense sand stratum. Estimate the ultimate load.

Given: For sand $\phi = 35°$, $\gamma_{sat} = 120$ lb/ft^3.

For clay $\gamma_{sat} = 119$ lb/ft^3, $c_u = 800$ lb/ft^2.

Use (a) the α-method for computing the frictional load, (b) Meyerhof's method for estimating Q_b. The water table is at ground level.

8.17 A ten-storey building is to be constructed at a site where the water table is close to the ground surface. The foundation of the building will be supported on 30 cm diameter pipe piles. The bottom of the pile cap will be at a depth of 1.0 m below ground level. The soil investigation at the site and laboratory tests have provided the saturated unit weights, the shear strength values under undrained conditions (average), the corrected SPT values, and the soil profile of the soil to a depth of about 40 m. The soil profile and the other details are given below.

Depth (m)		Soil	γ_{sat}	N_{cor}	$\phi°$	c (average)
From	To		kN/m^3			kN/m^2
0	6	Sand	19	18	33°	–
6.0	22	Med. stiff clay	18	–	–	60
22	30	Sand	19.6	25	35°	–
30	40	Stiff clay	18.5	–	–	75

Determine the ultimate bearing capacity of a single pile for lengths of (a) 15 m, and (b) 25 m below the bottom of the cap.

Use $\alpha = 0.50$ and $\overline{K}_s = 1.2$. Assume $\delta = 0.8\,\phi$.

8.18 For a pile designed for an allowable load of 400 kN driven by a steam hammer (single acting) with a rated energy of 2,070 kN-cm, what is the approximate terminal set of the pile using the ENR formula ?

8.19 A reinforced concrete pile of 40 cm diameter and 25 m long is driven through medium dense sand to a final set of 2.5 mm, using a 40 kN single-acting hammer with a stroke of 150 cm. Determine the ultimate driving resistance of the pile, if it is fitted with a helmet, plastic dolly and 50 mm packing on the top of the pile. The weight of the helmet, with dolly is 4.5 kN. The other particulars are : weight of pile = 85 kN, weight of hammer 35 kN; pile hammer efficiency $\eta_h = 0.85$; the coefficient restitution $C_r = 0.45$. Use Hiley's formula. The sum of elastic compression $\overline{C} = c_1 + c_2 + c_3 = 20.1$ mm.

8.20 A reinforced concrete pile 45 ft long and 20 in. in diameter is driven into a stratum of homogeneous saturated clay having $c_u = 800$ lb/ft^2. Determine: (a) the ultimate load capacity and the allowable load with $F_s = 3$, (b) the pullout capacity and the allowable pullout load with $F_s = 3$. Use the α-method for estimating the compression load.

8.21 Refer to Prob. 8.20. If the pile is driven to medium dense sand, estimate: (a) the ultimate compression load and the allowable load with $F_s = 3$, and (b) the pullout capacity and the allowable pullout load with $F_s = 3$. Use the Coyle and Castello method for computing Q_b and Q_f. The other data available are: $\phi = 36°$, and $\gamma = 115$ lb/ft^3. Assume the water table is at a great depth.

Deep Foundation 2

9

Behaviour of Single Vertical and Batter Piles Subjected to Lateral Loads

9.1 INTRODUCTION

When a soil of low bearing capacity extends to a considerable depth, piles are generally used to transmit vertical and lateral loads to the surrounding soil media. Piles that are used under tall chimneys, television towers, high-rise buildings, high retaining walls, offshore structures, etc. are normally subjected to high lateral loads. These piles or pile groups should resist not only vertical movements but also lateral movements. The requirements for a satisfactory foundation are:

1. The vertical settlement or the horizontal movement should not exceed an acceptable maximum value.
2. There must not be failure by yield of the surrounding soil or the pile material.

Piles or pile groups may be subjected to static, cyclic or dynamic loadings. The static loadings may be of short duration or sustained loadings. The lateral loads on tall structures, due to wind and on offshore structures due to sea waves or wind are examples of cyclic loadings. Loads due to ship thrust on offshore structures, lateral loads due to earthquakes, bomb blasts, etc. are examples of dynamic loadings. Sustained lateral loads occur on piles used in the foundations of earth retaining structures and other similar types.

Vertical piles are used in foundations to take normally vertical loads and small lateral loads. When the horizontal load per pile exceeds the value suitable for vertical piles, *batter piles* are used in combination with vertical piles. Batter piles are also called as *inclined piles* or *raker piles*. The degree of batter, that is the angle made by the pile with the vertical may go up to 30°. If the lateral load acts on the pile in the direction of batter, it is called as *in batter* or *negative batter* pile. If the lateral load acts in the direction opposite to that of the batter, it is called as an *out batter* or *positive batter* pile. Figure 9.1 (a) shows the two types of batter piles.

As explained earlier, pile foundations may be subjected to combined vertical and lateral loads or horizontal loads only. These loads cause lateral and vertical displacements and rotation of the pile cap. These displacements of the pile cap produce certain movements at each of the pile heads. The movements at the pile heads, in turn, cause an axial load, lateral load and a moment to be applied to the pile cap. The entire system will be in equilibrium when the reactions at the pile heads are consistent with the deformation characteristics of the piles. A exact solution to the problem can only be obtained if the deformation behaviours of each of the piles is taken precisely into account.

When a pile foundation is subjected to lateral loads, the soil directly in contact with the undersurface of the pile cap may carry a part of the lateral load owing to the shearing resistance exerted by the soil on the cap. But, in the design, this is not normally taken into account.

To understand the deformation behaviour of each of the pile in a pile group subjected to lateral loads or a combination of vertical and lateral loads, it is necessary to have a clear idea of the behaviour of single piles of similar batter under lateral loads. This chapter therefore deals with the behaviour of single vertical and batter piles subjected to lateral loads only. Chapter 10 deals with pile groups subjected to vertical and/or lateral loads.

The many investigators in the past have made analytical studies of the design of pile foundations. Mention may be made, in this connection, of a few pioneers such as Culmann (1866), Westergaard (1917), Vetter (1939), Hrennikoff (1949), Vesic (1956), and Vandepitte (1957). The theories of Culmann and Westergaard ignore the lateral resistance of the soil which according to the experimental and theoretical investigation provides a major part for the lateral stability of the pile foundation. Vetter and Vandepitte have introduced dummy piles to take care of soil resistance. Hrennikoff and Vesic take the soil restraint directly into account in their theories.

Extensive theoretical and experimental investigations have been conducted on single vertical piles by many investigators. Generalised solutions for laterally loaded vertical piles are given by Matlock and Reese (1960). The effect of vertical loads in addition to lateral loads has been evaluated by Davisson (1960) in terms of non-dimensional parameters. Broms (1964, 1965), Poulos and Davis (1980) have given different approaches for solving laterally loaded pile problems. Brom's method is ingenious and is based primarily on the use of limiting values of soil resistance. The method of Poulos and Davis is based on the theory of elasticity. Both these methods have had considerable use in practice.

Finite difference method of solving the differential equation for a laterally loaded pile is also very much in use where computer facility is available. Reese *et al.* (1974, 75) and Matlock (1970) have developed the concept of (*p-y*) curves for solving laterally loaded pile problems. This method is quite popular in USA and in some other countries.

However, the work on batter piles is very little as compared to vertical piles. Three series of tests on single 'in' and 'out' batter piles subjected to lateral loads have been reported by Matsuo (1938, 1939). They were run at three scales. The small and medium scale tests were conducted using timber piles embedded in sand in the laboratory under controlled density conditions. Loos and Breth (1949) have reported a few model tests in dry sand on vertical and batter piles. Model tests to determine the effect of batter on pile load capacity have been reported by Tschebotarioff (1953), Yoshimi (1964), and Awad and Petrasovits (1968). The effect of batter on deflections has also been investigated by Kubo (1965) and Awad and Petrasovits (1968) for model piles in sand.

Full-scale field tests on single vertical and batter piles, and also groups of piles have been made from time to time by many investigators in the past. The field test values have been used mostly to check the theories formulated for the behaviour of vertical piles only.

It is therefore obvious that reliable experimental data on batter piles are rather scarce compared to that of vertical piles. Though Kubo (1965) used instrumented model piles to study the deflection behaviour of batter piles, his investigation in this field was quite limited. The work of Awad and Petrasovits was based on non-in-strumented piles and as such does not throw much light on the behaviour of batter piles.

The author (Murthy, 1965) conducted a comprehensive series of model tests on instrumented piles embedded in dry sand. The batter used by the author varied from $-45°$ to $+45°$. The test results in detail have not so far been reported till to-date (1990) due to some unavoidable circumstances. A part of the author's study on the behaviour of batter piles based on his own research work has been included in this Chapter.

9.2 WINKLER'S HYPOTHESIS

Most of the theoretical solutions for laterally loaded piles involve the concept of *modulus of subgrade reaction* or otherwise termed as *soil modulus* which is based on Winkler's (1867), assumption that a soil medium may be approximated by a series of infinitely closely spaced independent elastic springs. Figure 9.1 (b) shows a loaded beam resting on an elastic foundation. The reaction at any point on the base of the beam is actually a function of every point along the beam since soil material exhibit varying degrees of continuity. The beam shown in Fig. 9.1 (b) can be replaced by a beam in Fig. 9.1 (c). In this figure the beam rests on a bed of elastic springs wherein each spring is independent of the other. According to Winkler's hypothesis the reaction at any point on the base of the beam in Fig. 9.1 (c) depends only on the deflection at that point. Vesic (1961) has shown that the error inherent in Winkler's hypothesis is not significant.

The problem of laterally loaded pile embedded in soil is closely related to the beam on an elastic foundation. A beam can be loaded at one or more points along its length whereas in the case of piles the external loads and moments are applied at or above the ground surface only.

The nature of a laterally loaded pile-soil system is illustrated in Fig. 9.1 (d) for a vertical pile. The same principle applies to batter piles also. A series of non-linear springs represent the force-deformation characteristics of the soil. The springs attached to the blocks of different sizes indicate reaction increasing with deflection and then reaching a yield point, or limiting value that depends on depth; the taper on the springs indicate a non-linear variation of load with deflection. The gap between the pile and the springs indicate the moulding away of the soil by repeated loadings and the increasing stiffness of the soil is shown by shortening of the springs as the depth below the surface increases.

9.3 THE DIFFERENTIAL EQUATION

Compatibility

As stated earlier, the problem of the laterally loaded pile is similar to the beam-on-foundation problem. The interaction between the soil and the pile or the beam must be treated quantitatively in the problem solution. The two conditions that must be satisfied for a rational analysis of the problem are:

1. Each element of the structure must be in equilibrium and
2. Compatibility must be maintained between the superstructure, foundation and the supporting soil.

If the assumption is made that structure can be maintained by selecting appropriate boundary conditions at the top of the pile, the remaining problem is to obtain a solution that insures equilibrium and compatibility of each element of the pile, taking into account the soil response along the pile. Such a solution can be made by solving the differential equation that describes the pile behaviour.

The Differential Equation of the Elastic Curve

The standard differential equations for slope, moment, shear and soil reaction for a beam on an elastic foundation are equally applicable to laterally loaded piles.

The deflection of a point on the elastic curve of a pile is given by y in Fig. 9.2 (a). The pile may be vertical or inclined. In both the cases of inclined and vertical piles, the lateral load is taken normal to the pile's axis. The x-axis is along the pile axis and deflection is measured normal to the pile-axis. Deflection to the right is positive. Slopes of the elastic curve at pints 1 and 2 are negative, while slopes at 3 and 4 are positive. However, the moment is positive in both the instances.

The relationships between y, slope moment, shear, and soil reaction at any point on the deflected elastic curve may be written as follows.

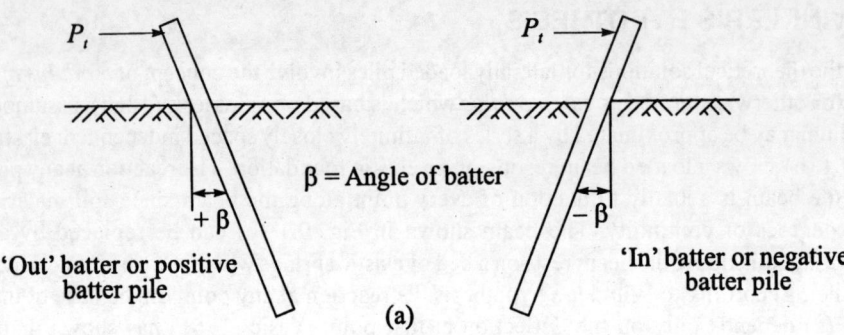

β = Angle of batter

'Out' batter or positive batter pile

'In' batter or negative batter pile

(a)

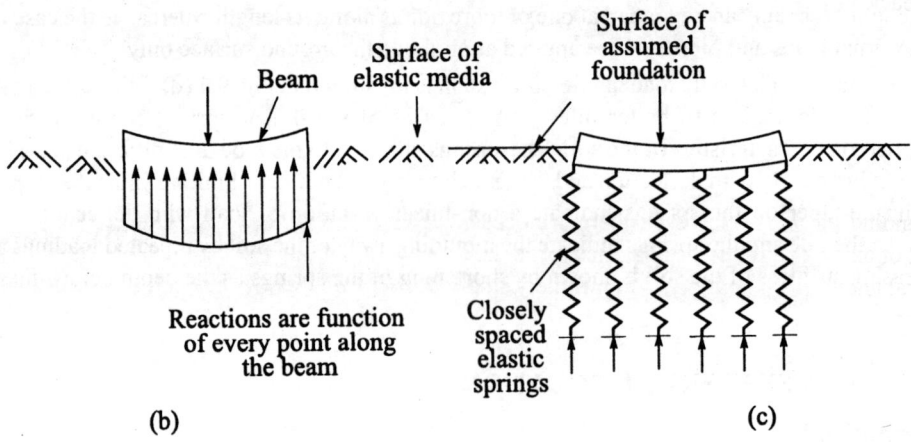

Beam

Surface of elastic media

Surface of assumed foundation

Reactions are function of every point along the beam

Closely spaced elastic springs

(b)   (c)

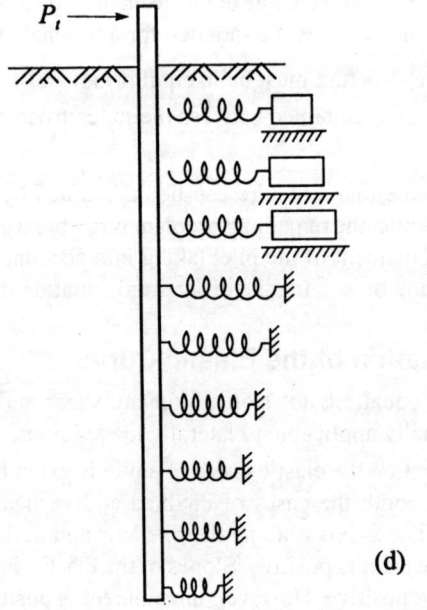

(d)

Fig. 9.1 (a) Batter piles, (b, c) Winkler's hypothesis, (d) The concept of laterally loaded pile-soil system

$$y = \text{deflection of the elastic curve,}$$

slope of the deflected pile $= \dfrac{dy}{dx}$ (9.1)

moment of pile $\quad M = EI \dfrac{d^2 y}{dx^2}$ (9.2)

shear $\quad V = EI \dfrac{d^3 y}{dx^3}$ (9.3)

soil reaction $\quad p = EI \dfrac{d^4 y}{dx^4}$ (9.4)

where, EI is the flexual rigidity of pile material.

Differential Equation for the Beam-Column

In most cases the axial load on a laterally loaded pile is of such magnitude that it has a small influence on bending moment. However, there are occasions when it is necessary to include a term for the effect of axial loading in the analytical process. Consider a pile shown in Fig. 9.2 (a) is subjected to vertical and lateral loadings. If an infinitely small unloaded element, bounded by two horizontals, a distance dx apart, is cut out of this bar [Fig. 9.2 (a)], the equilibrium of moments (ignoring second-order terms) leads to the equation [Fig. 9.2 (b)],

$$(M + dM) - M + Q_x dy - V dx = 0 \quad (9.5)$$

or $\qquad \dfrac{dM}{dx} + Q_x \dfrac{dy}{dx} - V = 0$ (9.6)

Differentiating Eq. (9.6) with respect to x, we have

$$\dfrac{d^2 M}{dx^2} + Q_x \dfrac{d^2 y}{dx^2} - \dfrac{dV}{dx} = 0 \quad (9.7)$$

From Eqs (9.2) through (9.4), we have the following identities

$$\dfrac{d^2 M}{dx^2} = EI \dfrac{d^4 y}{dx^4} \quad (9.8)$$

$$\dfrac{dV}{dx} = p \quad (9.9)$$

Further, we may write for the deflection y at any point x on the pile, the soil reaction, p, per unit length of pile as equal to

$$p = -E_s y \quad (9.10)$$

where, E_s is called as *soil modulus for the point x on the pile*. The negative sign in Eq. (9.10) indicates that the direction of soil reaction is opposite to that of deflection.

Making the relevant substitutions in Eq. (9.7), we have

$$EI \frac{d^4y}{dx^4} + Q_x \frac{d^2y}{dx^2} + E_sy = 0 \tag{9.11}$$

The Eq. (9.11) is valid under the following conditions:

1. The pile is straight and has a uniform cross-section.
2. The pile material is homogeneous.
3. The modulus elasticity of the pile material is the same for tension and compression.
4. The pile is not subjected to dynamic loading.

The sign conventions for a laterally loaded pile-system is shown in Fig. 9.2 (c).

Equation (9.11) can be solved in a closed form for known boundary conditions if the soil modulus E_s is constant with depth. Series type solutions have been developed for the case in which E_s has a linear variation with depth. For any form of variation of E_s with depth the general approach has been to make use of either the difference equation method or the analog computer method.

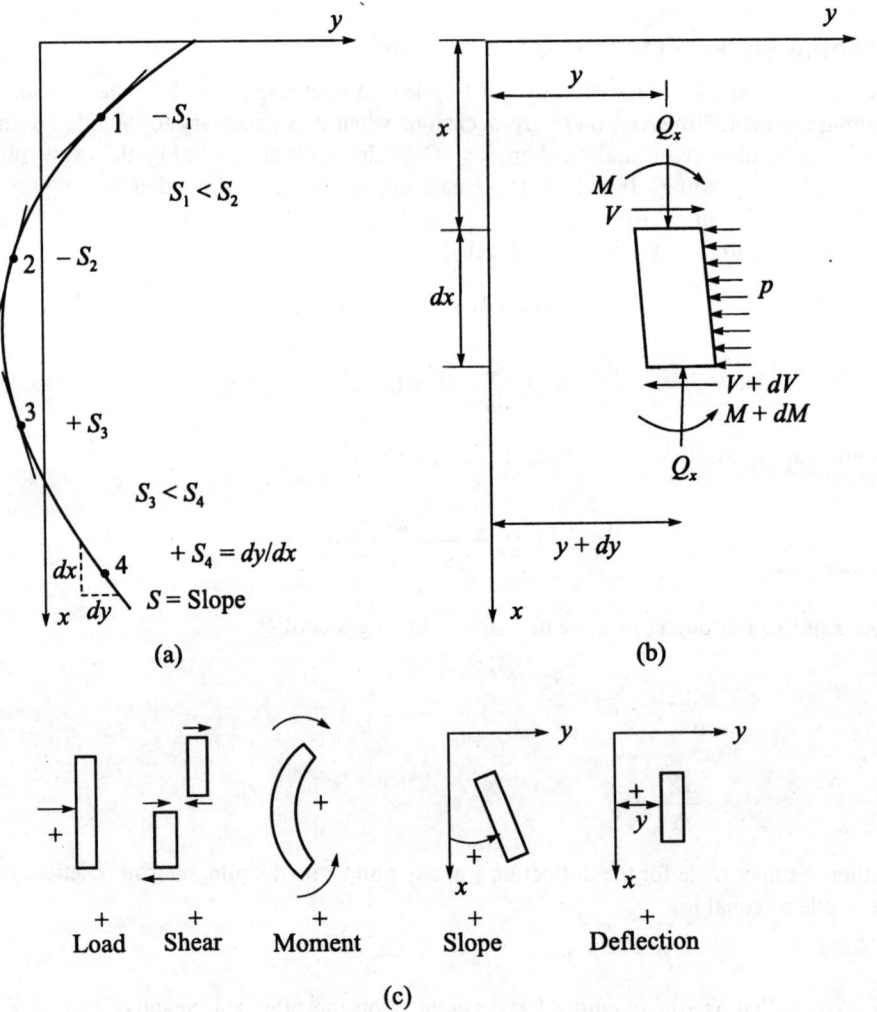

Fig. 9.2 Differential equation concept for a beam-column: (a) Segment of a deflected pile, (b) element from the beam-column, (c) sign conventions

PART A: VERTICAL PILES SUBJECTED TO LATERAL LOADS

9.4 SOLUTION FOR LATERALLY LOADED SINGLE PILES

Introduction

We come across both vertical and batter piles in the foundations of structures subjected to lateral loads and moments. Our objective in this chapter will be to study the behaviour of single piles, vertical or inclined, subjected to lateral loads and moments only. Axial loads are not considered. Most of the theories that have been formulated so far or available, apply to the behaviour of vertical piles only under lateral loads. Very little information is available in published literatures on the behaviour of batter piles subjected to lateral loads. The author (Murthy, 1965) carried out a series of tests on the behaviour of batter piles subjected to lateral loads. A brief account of the author's study along with the other information available in this field is given in this chapter.

The analysis of laterally loaded single piles are based on the following assumptions.

1. The laterally loaded pile behaves as an elastic member and the supporting soil behaves as an ideal elastic material.
2. The theory of subgrade reaction is applicable.
3. There is no axial load.

A departure to the second assumption is the one proposed by Poulos (1980), who considers the continuity of the soil material into account and presents his own approach for solving laterally loaded pile problems. A brief discussion of his approach is presented in this Chapter.

Two types of piles are encountered in practice. They are:

1. Long piles.
2. Short piles.

When a pile is greater than a particular length, the length loses its significance. The behaviour of the pile will not be affected if the length is greater than this particular length. Such piles are called as *long flexible piles*. In the case of short piles, the flexural stiffness *EI* of the material of the pile loses its significance. The pile behaves as a rigid member and rotates as a unit.

When we think of solving laterally loaded pile problems, we have to consider the types of loads applied at the pile heads. Three types of boundary conditions are normally applicable.

1. Free-head pile.
2. Fixed-head pile.
3. Partially-restrained-head pile.

In the case of free-head pile, the lateral load may act at or above ground level and the pile head is free to rotate without any restraint. A fixed-pile is free to move only laterally but rotation is prevented completely, whereas a pile with a partially restrained head moves and rotates under restraint. The partially restrained head is normally encountered in offshore drilling platforms and other similar structures.

Statement of the Problem

If a vertical pile, as shown in Fig. 9.3, is subjected to a lateral load and moment at its head, it is required to determine the curves of (*a*) deflection, (*b*) slope, (*c*) moment, (*d*) shear, and (*e*) soil reaction as a function of depth. Figure 9.3 gives a set of typical curves required for a long flexible pile.

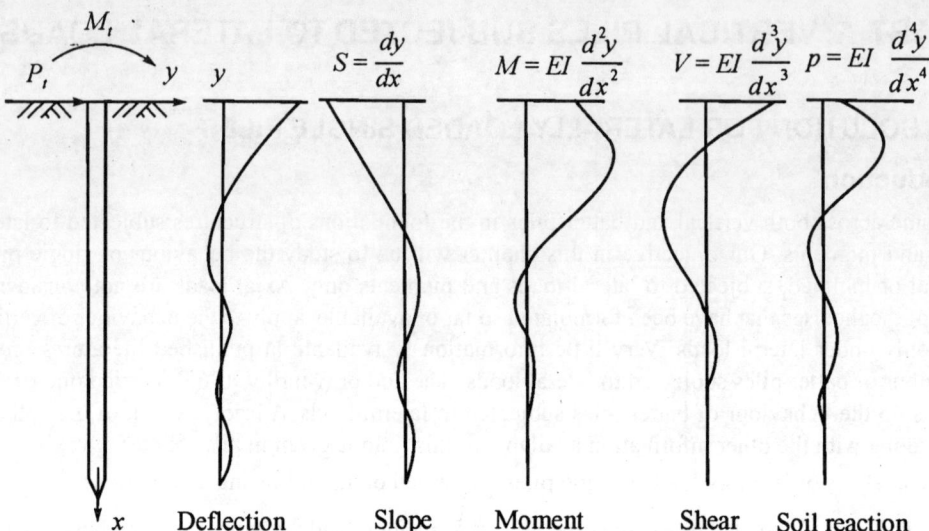

Fig. 9.3 A complete solution to a laterally loaded pile problem

Solutions

There are many methods that are available for solving the problem of laterally loaded piles. Each has its own merits and demerits. Some of the methods that are discussed in this Chapter are:

1. Closed-Form Solution.
2. Difference Equation Method.
3. Non-dimensional Method.
4. A Direct Method.
5. Pressuremeter Method.
6. Broms Method.
7. Poulos Method.

9.5 MODULUS OF SUBGRADE REACTION

The key to the solution of laterally loaded pile problems lies in the determination of the value of the modulus of subgrade reaction with respect to depth along the pile. Figure 9.4 (a) shows a vertical pile subjected to a lateral load P_t at the ground level. The deflected portion of the pile and the corresponding soil reaction curve is also shown in the same figure. The soil modulus, E_s, at any point x below the surface along the pile is defined as (according to Winkler's hypothesis).

$$E_s = -\frac{p}{y} \qquad (9.12)$$

in which y is the deflection of the pile and p the soil reaction expressed as force per unit length of the pile. As the load P_t at the top of the pile increases, the deflection y and the corresponding soil reaction, p, increases. Relationship between p and y can be established along the pile at depths $x = x_1$, $x = x_2$, etc. A family of p-y curves, plotted in the appropriate quadrant, is shown in Fig. 9.4 (b). That the curves plotted in the second and fourth quadrant's are merely an indication that the soil resistance p is opposite in sign to the deflection y. While the p-y curves in Fig. 9.4 (b) are only illustrative, they are typical of many such families of curves in that the initial stiffness and the maximum resistance increases with depth.

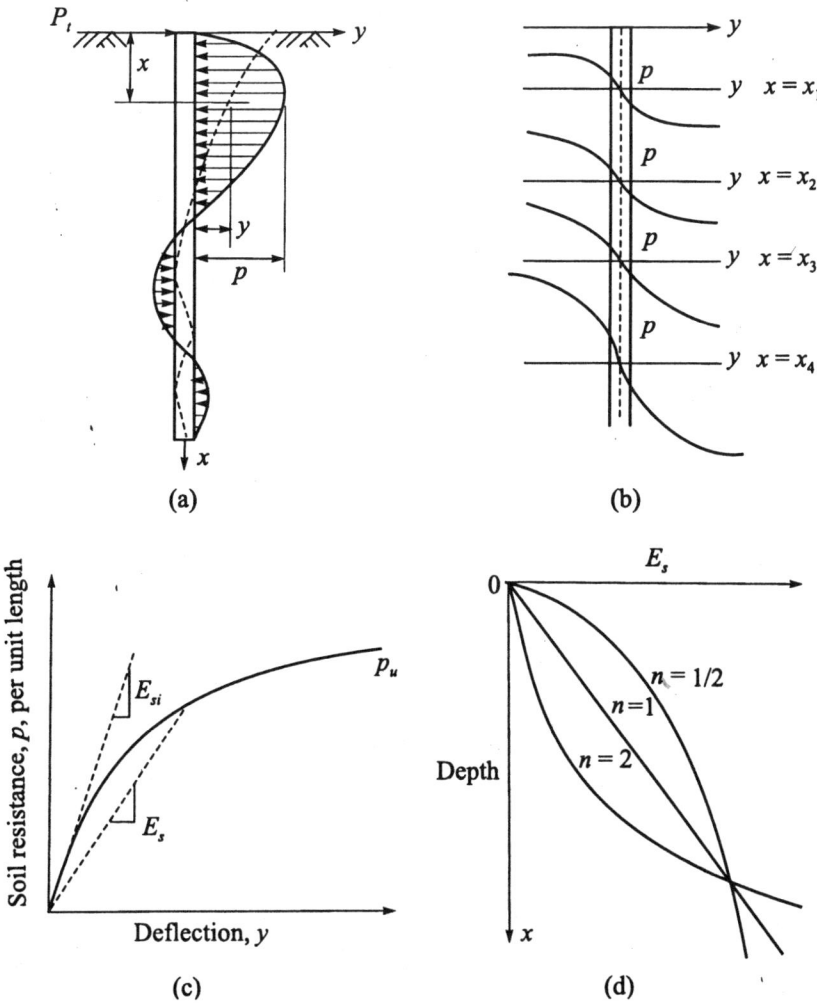

Fig. 9.4 The concept of (*p-y*) curves: (a) Laterally loaded pile, (b) family of (*p-y*) curves, (c) characteristic shape of a curve, (d) form of variation of E_s with depth

A typical *p-y* curve is shown in Fig. 9.4 (c). It is plotted in the first quadrant for convenience. The curve is strongly non-linear, changing from an initial stiffness E_{si} to an ultimate resistance p_u. As is evident the soil modulus E_s is not a constant and changes with deflection.

There are many factors that have an influence on the value of E_s. The most important factors are:

1. The pile width or diameter.
2. The soil properties.
3. The nature and magnitude of loading.
4. The flexural stiffness EI of the pile material.

The present state of art is not yet clear as to how the soil modulus varies with the pile diameter (or width) and flexural stiffness EI of the pile materials. It is well-known qualitatively that E_s varies with soil properties.

But no quantitative relationships established amongst the factors mentioned above are yet available in any published literature till-to-date (1990). The author studied a series of case histories involving all the above factors and has established a relationship which helps to solve the laterally loaded pile problems.

The non-linear-force deformation of the p-y curves at various depths would make E_s a function of both x and y. However, for any particular load level, E_s may be considered a function of x only. The variation of E_s with depth may take many forms depending upon the soil property and the batter of pile. The most popular and useful form of variation of the soil modulus is the power form which is expressed as

$$E_s = n_h x^n \qquad (9.13a)$$

in which n_h is termed as *coefficient of soil modulus variation*. The value of the power n in Eq. (9.13a) depends upon the type of soil and batter of pile. Typical curves for the form of variation of E_s with depth for values of n equal to ½, 1 and 2 are given in Fig. 9.4 (d). Here, x is the distance (depth for vertical piles) along the axis of pile, and E_s is normal to the axis of pile.

The information with regards to n_h and n available in published literatures are applicable to vertical piles only. The author discussed the variation of n_h and n for batter piles in this chapter.

The most common assumptions for vertical piles are that $n = 0$ for stiff clays (that is, the modulus is constant with depth) and $n = 1$ for granular soils and normally consolidated clays which means that the modulus E_s increases linearly with depth.

It was Terzaghi (1955), who recommended $n = 0$ for stiff clays. He based his recommendations on the notion that the deformation characteristics of stiff clays are more or less independent of depth which has been found to be not realistic. The present method of analysis is based on the assumption that E_s increases linearly with depth even in stiff clays. The principal reasons for this are:

1. Soils frequently increase in strength with depth as a result of overburden pressures and of natural deposition and consolidation processes.

2. Pile deflections decrease with depth for any given loading and the corresponding equivalent elastic moduli of soil reaction increase with decreasing deflection.

Table 9.1, however, gives the recommendations of Terzaghi for E_s for stiff clays.

Table 9.1 Soil modulus E_s for laterally loaded piles in stiff clays

Consistency	Stiff	Very stiff	Hard
q_u, kg/cm^2	$1 - 2$	$2 - 4$	> 4
E_s, kg/cm^2	$32 - 65$	$65 - 130$	> 130

Note: q_u = unconfined compressive strength

When the soil modulus increases linearly with depth, Eq. (9.13a) reduces to

$$E_s = n_h x \qquad (9.13b)$$

Terzaghi (1955) also studied the problem of laterally loaded vertical piles in sand and he found that n_h in sand depends only on the overburden pressure and density. He proposed the following equation for determining n_h

$$n_h = \frac{A\gamma}{1.35} \ (\text{tons/ft}^3) \qquad (9.14)$$

Typical values of A and n_h are given in Table 9.2 (a).

Table 9.2 (a) Values of n_h (Ton/ft^3) for sand (After Terzaghi, 1955)

Relative density	*Loose*	*Medium*	*Dense*
Range of value of A	100–300	300–1000	1000–2000
Adopted value of A	200	600	1500
n_h, dry or moist sand	7	21	56
n_h, submerged sand	4	14	34

Table 9.2 (b) gives typical values of n_h for cohesive soils.

Table 9.2 (b) Typical values of n_h for cohesive soils (Taken From Poulos, 1980)

Soil type	n_h *lb/in^3*	*Reference*
Soft NC clay	0.6 to 12.7	Reese and Matloc, 1956
	1.0 to 2.0	Davisson and Prakash, 1963
NC organic clay	0.4 to 1.0	Peck and Davisson, 1962
	0.4 to 3.0	Davisson, 1970
Peat	0.2	Davisson, 1970
	0.1 to 0.4	Wilson and Hills, 1967
Loess	29 to 40	Bowles, 1968

Values of the coefficient of modulus variation n_h were determined directly from lateral loading tests on instrumented piles in submerged sand at Mustang Island, Texas, USA. The tests were made for both static and cyclic loading conditions and the values obtained, as quoted by Reese (1975), were considerably higher than those of Terzaghi. The investigators recommend that Mustang Island values should be used for pile design. The values as obtained by them are given in Fig. 9.5.

It may be noted here that the n_h values as proposed by the investigators are not constants. It varies with the soil properties, (density for granular materials and shear strength for clay soils), the width or diameter of pile, the flexural stiffness of the pile material and the deflection of the pile. It is a very complex phenomenon which does not depend on a single factor.

9.6 CLOSED-FORM SOLUTION FOR PILE OF INFINITE LENGTH

Introduction

A closed-form solution to Eq. (9.11) has been obtained by Hetenyi (1946) for the soil modulus, E_s remaining constant with depth and with zero axial load.

The pile is assumed to be supported along its entire length by a continuous stratum of soil which is capable of exerting a reaction to the pile in a direction opposite to the pile deflection. EI is assumed to remain constant with depth.

It is beyond the scope of this book to give a detailed development of the solution. The final form of equations for free head and fixed piles are as given below.

Free-Head Pile

The expressions for deflections, slope, moment, shear and soil reaction for lateral load P_t and moment M_t acting at the head of the pile are as follows,

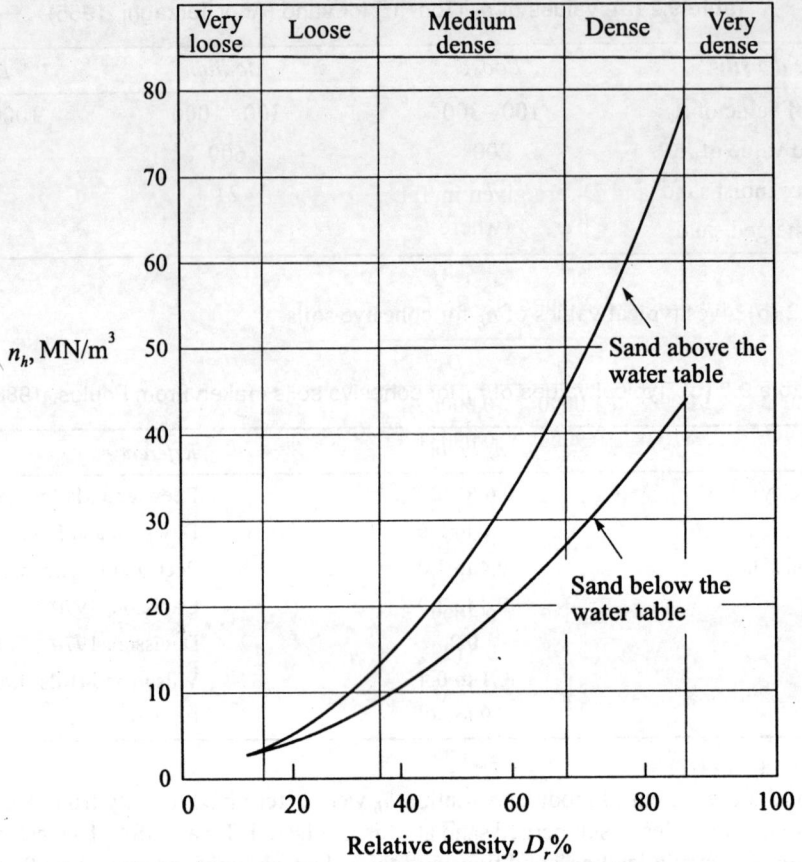

Fig. 9.5 Variation of n_h with relative density (Reese, 1975)

$$y = \frac{2 P_t \beta}{\alpha} C_1 + \frac{M_t}{2 EI \beta^2} B_1 \qquad (9.15)$$

$$S = \frac{-2 P_t \beta}{\alpha} A_1 - \frac{M_t}{EI \beta} C_1 \qquad (9.16)$$

$$M = \frac{P_t}{\beta} D_1 + M_t A_1 \qquad (9.17)$$

$$V = P_t B_1 - 2M_t \beta D_1 \qquad (9.18)$$

$$p = -2 P_t \beta C_1 - 2M_t \beta^2 B_1 \qquad (9.19)$$

where, $\quad A_1 = e^{-\beta x}(\cos \beta x + \sin \beta x) \qquad (9.20)$

$$B_1 = e^{-\beta x}(\cos \beta x - \sin \beta x) \qquad (9.21)$$

$$C_1 = e^{-\beta x} \cos \beta x \qquad (9.22)$$

$$D_1 = e^{-\beta x} \sin \beta x \qquad (9.23)$$

$$\beta = \left(\frac{\alpha}{4\,EI}\right)^{\frac{1}{4}} \tag{9.24a}$$

$$\alpha = -p/y = -E_s \tag{9.24b}$$

Values of A_1, B_1, C_1 and D_1 are given in Table 9.3. Timoshenko (1941) says that the long pile solution is satisfactory where $\beta L \geq 4$ (where L = length of pile).

Table 9.3 Values A_1, B_1, C_1 and D_1 for pile of infinite length

βx	A_1	B_1	C_1	D_1	βx	A_1	B_1	C_1	D_1
0	1.0000	1.0000	1.0000	0.0000	2.4	−0.0056	−0.1282	−0.0669	0.0613
0.1	0.9907	0.8100	0.9003	0.0903	2.6	−0.0254	−0.1019	−0.0636	0.0383
0.2	0.9651	0.6398	0.8024	0.1627	2.8	−0.0369	−0.0777	−0.0573	0.0204
0.3	0.9267	0.4888	0.7077	0.2189	3.2	−0.0431	−0.0383	−0.0407	−0.0024
0.4	0.8784	0.3564	0.6174	0.2610	3.6	−0.0366	−0.0124	−0.0245	−0.0121
0.5	0.8231	0.2415	0.5323	0.2908	4.0	−0.0258	0.0019	−0.0120	−0.0139
0.6	0.7628	0.1431	0.4530	0.3099	4.4	−0.0155	0.0079	−0.0038	−0.0117
0.7	0.6997	0.0599	0.3798	0.3199	4.8	−0.0075	0.0089	0.0007	−0.0082
0.8	0.6354	−0.0093	0.3131	0.3223	5.2	−0.0023	0.0075	0.0026	−0.0049
0.9	0.5712	−0.0657	0.2527	0.3185	5.6	0.0005	0.0052	0.0029	−0.0023
1.0	0.5083	−0.1108	0.1988	0.3096	6.0	0.0017	0.0031	0.0024	−0.0007
1.1	0.4476	−0.1457	0.1510	0.2967	6.4	0.0018	0.0015	0.0017	0.0003
1.2	0.3899	−0.1716	0.1091	0.2807	6.8	0.0015	0.0004	0.0010	0.0006
1.3	0.3355	−0.1897	0.0729	0.2626	7.2	0.0011	−0.00014	0.00045	0.00060
1.4	0.2849	−0.2011	0.0419	0.2430	7.6	0.00061	−0.00036	0.00012	0.00049
1.5	0.2384	−0.2068	0.0158	0.2226	8.0	0.00028	−0.00038	−0.0005	0.00033
1.6	0.1959	−0.2077	−0.0059	0.2018	8.4	0.00007	−0.00031	−0.00012	0.00019
1.7	0.1576	−0.2047	−0.0235	0.1812	8.8	−0.00003	−0.00021	−0.00012	0.00009
1.8	0.1234	−0.1985	−0.0376	0.1610	9.2	−0.00008	−0.00012	−0.00010	0.00002
1.9	0.0932	−0.1899	−0.0484	0.1415	9.6	−0.00008	−0.00005	−0.00007	−0.00001
2.0	0.0667	−0.1794	−0.0563	0.1230	10.0	−0.00006	−0.00001	−0.00004	−0.00002
2.2	0.0244	−0.1548	−0.0652	0.0895					

Fixed-Head Pile

For a pile whose head is fixed against rotation, the final expressions for long piles are as follows:

$$y = \frac{P_t \beta}{\alpha} A_1 \tag{9.25}$$

$$S = -\frac{P_t}{2\,EI\,\beta^2} D_1 \tag{9.26}$$

$$M = -\frac{P_t}{2B} B_1 \tag{9.27}$$

$$V = P_t C_1 \tag{9.28}$$

$$p = -P_t \beta A_1 \qquad (9.29)$$

The values of factors A_1, B_1, C_1 and D_1 are the same as given in Table 9.3.

9.7 FINITE DIFFERENCE METHOD OF SOLVING THE DIFFERENTIAL EQUATION FOR A LATERALLY LOADED LONG PILE (GLESSER, 1953)

Difference Equations

The differential Eq. (9.11) can be solved in a closed form if the soil modulus E_s is constant with depth as explained in Section 9.6. For any form of variation of E_s with depth, the numerical finite difference method as suggested by Palmer and Thompson (1948), is the most convenient method. A convenient way of solving the difference equation has been suggested by Glesser (1953). The differential equation required to be solved is of the form (with axial load $Q_x = 0$).

$$\frac{d^4y}{dx^4} + \frac{E_s}{EI} y = 0 \qquad (9.30)$$

The basic forms of difference relationships may be explained with reference to a laterally loaded deflected pile shown in Fig. 9.6 (a). Figure 9.6 (b) shows the method of dividing the pile. The pile of length L is divided into t equal parts each of length h. Two imaginary points are shown below the tip of the pile and two above the top of the pile.

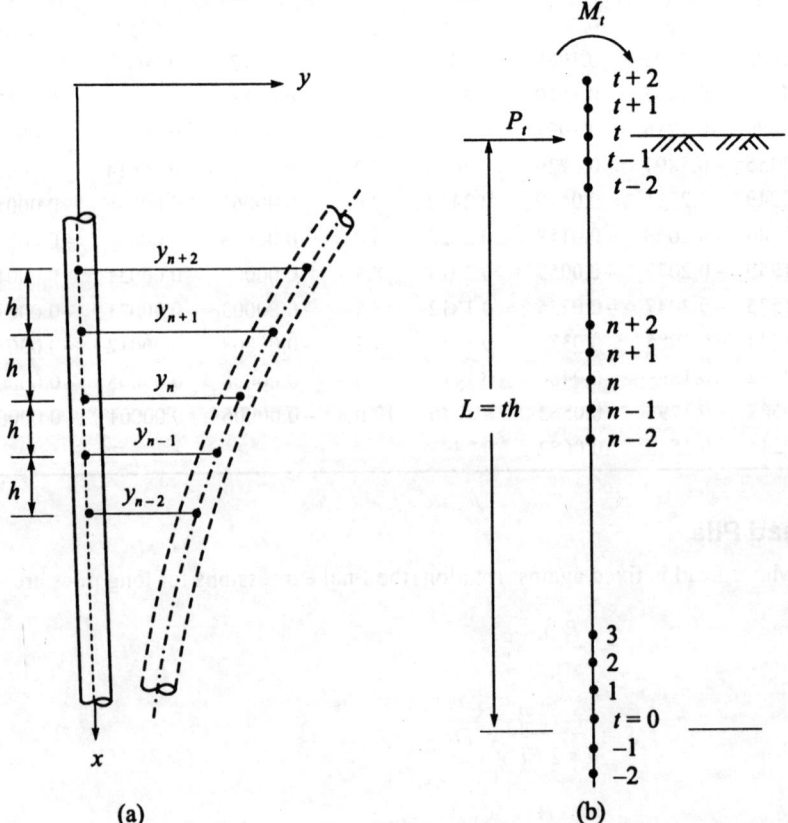

Fig. 9.6 Finite difference method of analysis of laterally loaded piles: (a) Representation of deflected pile, (b) method of sub-dividing pile

$$\left(\frac{dy}{dx}\right)_{x=n} = \frac{y_{n-1} - y_{n+1}}{2h} \tag{9.31}$$

$$\left(\frac{d^2 y}{dx^2}\right)_{x=n} = \frac{\frac{y_{n-1} - y_n}{h} - \frac{y_n - y_{n+1}}{h}}{h} = \frac{\left(y_{n-1} - 2y_n + y_{n+1}\right)}{h^2} \tag{9.32}$$

In the same way,

$$\left(\frac{d^3 y}{dx^3}\right)_{x=n} = \frac{\left(y_{n-2} - 2y_{n-1} + 2y_{n+1} - y_{n+2}\right)}{2h^3} \tag{9.33}$$

$$\left(\frac{d^4 y}{dx^4}\right)_{x=n} = \frac{y_{n-2} - 4y_{n-1} + 6y_n - 4y_{n+1} + y_{n+2}}{h^4} \tag{9.34}$$

If Eq. (9.34) is substituted into Eq. (9.30), we have

$$y_{n-2} - 4y_{n-1} + 6y_n - 4y_{n+1} + y_{n+2} = \frac{-E_{sn}h^4}{EI} y_n \tag{9.35}$$

Since E_s is supposed to have been known for all points along the pile, it is possible to write $t + 1$ algebraic equations similar to Eq. (9.35) for points 0 through t. Two boundary conditions at the tip of the pile and two at the top of the pile yield four additional equations, giving a total of $t + 5$ simultaneous equations. When solved, these equations give the deflection of the pile from point -2 through points $t + 2$. A solution can be obtained for any number of subdivisions of the pile.

Glesser (1953) has streamlined a method of solving the $t + 5$ simultaneous equations by successive elimination of the unknowns beginning with the equations at the bottom of the pile and progressively working upward. At the top of the pile the boundary conditions are used to solve the deflections y_t, y_{t+1} and y_{t+2}. These deflections can then be used to work back down the pile and solve for deflections, slopes, moments, shears and soil reactions for all points along the pile.

The solution of $t + 5$ equations requires a knowledge of four boundary conditions. The boundary conditions at the pile top may be of three forms. They are:

1. Lateral load P_t and moment M_t.
2. Lateral load P_t and slope S_t.
3. Lateral load P_t and rotational restraint constant M_t/S_t.

The other two boundary conditions are that the shear and moment are zero at the bottom of pile.

The method of Glesser can suitably be adopted for computer solution. A computer programme can handle different boundary conditions and interactive methods can be employed to account for the non-linear behaviour of pile. The number of divisions of the pile can be increased to account for the accuracy required.

Glesser's method of solving the simultaneous equations is given in below.

Generalised Equations for Free-Head Pile Subjected to Lateral Loads

No.	Equation
(1)	$y_1 - 2y_0 + y_{-1} = 0$

(2) $\quad -y_2 + 2y_1 - 2y_{-1} + y_{-2} = 0$

(3) $\quad y_2 - 4y_1 + 6y_0 - 4y_{-1} + y_{-2} = -A_0 y_0$

(n + 3) $\quad y_{n+2} - 4y_{n+1} + 6y_n - 4y_{n-1} + y_{n-2} = -A_n y_n$

(t + 3) $\quad y_{t+2} - 4y_{t+1} + 6y_t - 4y_{t-1} + y_{t-2} = -A_t y_t$

(t + 4) $\quad -y_{t+2} + 2y_{t+1} - 2y_{t-1} + y_{t-2} = \dfrac{-2h^3 P_t}{EI}$

(t + 5) $\quad y_{t+1} - 2y_t + y_{t-1} = \dfrac{-h^2 M_t}{EI}$

Solution

From Eqs (1), (2) and (3)

(t + 6) $\quad y_0 = \dfrac{-2y_2 + 4y_1}{2 + A_0} = -B_1 y_2 + 2B_1 y_1$

From Eqs (t + 6) and (4)

(t + 7) $\quad y_1 = \dfrac{-y_3 + y_2(4 - 2B_1)}{5 + A_1 - 5B_1} = -B_2 y_3 + B_3 y_2$

From Eqs (t + 7), (t + 6) and 5

(t + 8) $\quad y_2 = \dfrac{-y_4 + y_3\left[4 - B_2(4 - 2B_1)\right]}{6 + A_2 - B_1 - B_3(4 - 2B_1)} = -B_4 y_4 + B_5 y_3$

In general, thereafter

(t + 9) $\quad y_n = \dfrac{-y_{n+2} + y_{n+1}\left[4 - B_{2n-2}(4 - B_{2n-3})\right]}{6 + A_n - B_{2n-4} - B_{2n-1}(4 - B_{2n-3})}$

$\qquad = -B_{2n} y_{n+2} + B_{2n+1} y_{n+1}$

until a solution is obtained for y_t. Substituting this and the solution for y_{t-1} into Eq. (t + 5),

(t + 10) $\quad y_{t+2} = \dfrac{1}{B_{2t}(2 - B_{2t-1})}\left\{\dfrac{-h^2 Mt}{EI} + \left[B_{2t-2} - 1 + B_{2t+1}(2 - B_{2t-1})\right]y_{t-1}\right\}$

$\qquad = -B_{2t+2} + B_{2t+3} y_{t+1}$

Substituting into Eq. (t + 4)

(t + 11) $\quad y_{t+1}\{-B_{2t-3} + 2 + [(2 - B_{2t-3})B_{2t-1} + B_{2t-4}] \times (B_{2t}B_{2t+3} - B_{2t+1})$

$\qquad + (2 - B_{2t-3})B_{2t-2}\}$

$$= -\frac{2 P_t h^3}{EI} - B_{2t+2} + B_{2t+2} B_{2t} (2B_{2t-1} + B_{2t-4} - B_{2t-3} B_{2t-1})$$

To obtain deflections substitute value for y_{t+1} from Eq. $(t+11)$ into $(t+10)$ to obtain y_{t-2}. Substitute these values into Eq. $(t+9)$ to obtain y_t, etc. until y_0, y_{-1} and y_{-2} are obtained.

Constants

$$B_1 = \frac{2}{2+A_1} ; B_2 = \frac{1}{5+A_1 - 4B_1}$$

$$B_3 = B_2 (4 - 2B_1); B_4 = \frac{1}{6 + A_2 - B_1 - B_3 (4 - 2B_1)}$$

$$B_5 = B_4 (4 - B_3)$$

For all even constants B_6 through B_{2t}, inclusive,

$$B_{2n} = \frac{1}{6 + A_n - B_{2n-4} - B_{2n-1} (4 - B_{2n-3})}$$

For all odd constants B_7 through B_{2t+1}, inclusive,

$$B_{2n+1} = B_{2n} (4 - B_{2n-1})$$

$$B_{2t-2} = \frac{-h^2 M_t}{EI (2 - B_{2t-1}) B_{2t}}$$

$$B_{2t+3} = \frac{B_{2t-2} - 1 + B_{2t+1} (2 - B_{2t-1})}{B_{2t} (2 - B_{2t-1})}$$

Generalised Equations for Fixed-Head Pile Subjected to Lateral Loads

No.	Equation
	Equations $(1')$ through $(t'+4')$ for fixed head piles are identical with those for the free head pile, as are the solutions in equations $(t'+6')$ through $(t'+9')$ inclusive.
$(t'+5')$	$y'_{t-1} - y'_{t+1} = 2\,h$ times the slope of pile at ground $= 0$ or $y'_{t-1} = y'_{t+1}$

Solution

Substituting the value of $(t'+5')$ and the results of equations $(t'+6')$ through $(t'+9')$ into Eq. $(t'+4')$:

$(t'+10')$ $\quad y'_{t+2} = -2h^3 P_t (1 + B'_{2t-2} - B'_{2t-1} B'_{2t+1}) / EI (C + D)$

where, $\quad C = -1 + B'_{2t} B'_{2t-4} + B'_{2t-4} B'_{2t-2} B'_{2t}$

$\quad D = -B'_{2t} B'_{2t-1} B'_{2t-3} - B'_{2t-2} + B'_{2t+1} B'_{2t-1}$

$(t'+11')$ $\quad y'_{t+1} = y'_{t-1} = \dfrac{(-B'_{2t-1} B'_{2t}) y'_{t+2}}{1 + B'_{2t-2} - B'_{2t+1} B'_{2t-1}}$

$(t' + 12')$ $y'_{2t} y'_{t+2} + B'_{2t+1} y'_{t-1}$

To obtain remaining deflections, substitute values of y'_{t+2}, y'_{t+1} and y'_t into Eq. $(t' + 9')$ for y'_{t-1}, etc. until value for y'_3 is obtained, after which use Eqs $(t' + 8')$, $(t' + 7')$, $(t' + 6')$, $(3')$, $(2')$ and $(1')$ to obtain y'_2, y'_1, y'_0, y'_{-1} and y'_{-2}.

Constants

For fixed piles, all constants are the same as for free head pile to, and including B'_{2t} and B'_{2t+1}. Thereafter, use Eqs $(t' + 10')$ and $(t' + 11')$.

Nomenclature

In the above equations

y_n = deflection at point n,

t = number of equal units into which L is divided,

P_t = applied lateral load,

M_t = applied moment,

E = Young's modulus for the pile material,

I = moment of inertia of the pile,

$$A_n = \frac{E_s h^4}{EI}$$

E_s = soil modulus = $n_h x$,

n_h = coefficient of soil modulus variation,

x = distance of point n on the pile from ground level.

In the generalised equations for free head pile, Eqs (1) and (2) represent the boundary conditions at the tip of the pile ($n = 0$) and express the fact that the moment and shear at that point are zero. Equations $(t + 4)$ and $(t + 5)$ express the moment and shear at the ground surface ($n = t$) and are derived from basic equations (9.2), (9.3), (9.32) and (9.33).

For the fixed pile Eq. $(t' + 5')$ expresses the fact that the head of the pile at the ground surface ($n = t$) is restrained to remain in a vertical position, that is, that its slope relative to the vertical is zero and is derived from Eq. (9.31).

It may be noted here that the number of constants involved in the computation is equal to $(2t + 3)$ that is for example, if a pile is divided into 10 equal parts, the number constants is 23 (B_1 to B_{23}).

Computation of Slope, Moment, Shear and Soil Reaction

When once the deflections are known at all the n points, the slopes, moments, shears and soil reactions at all these points may be calculated by making use of the following equations.

Slope, $$S_n = \frac{y_{n-1} - y_{n+1}}{2h}$$

Moment, $$M_n = \frac{EI}{h^2} (y_{n-1} - 2y_n + y_{n+1})$$

Shear, $$V_n = \frac{EI}{2h^3} (y_{n-2} - 2y_{n-1} + y_{n+1} - y_{n+2})$$

Soil reaction, $$p_n = -E_{sn} y_n$$

Example 9.1

A steel pipe pile of 61 cm outside diameter with a wall thickness of 2.5 cm is driven into loose sand ($D_r = 30\%$) under submerged condition to a depth of 20 m. The submerged unit weight of the soil is 8.75 kN/m^3 and the angle of internal friction is 33°. The EI value of the pile is 4.35×10^{11} kg-cm^2 (4.35×10^2 MN – m^2). Compute the ground line deflection of the pile under a lateral load of 268 kN at ground level under free head condition by difference equation method (Refer Section 9.7, Fig. 9.6).

Solution

Determine n_h value from Fig. 10.5 for $D_r = 30\%$ under submerged condition which is equal to 6 MN/m^3 (≈ 0.6 kg/cm^3). The number of increments t is taken as equal to 5 to facilitate calculation. For accurate results the number of increments must be at least 20. The increment value $h = 4$ m.

Calculation of A-values

$$A_n = \frac{n_h x h^4}{EI} = \frac{6 \times 4^4 \, x}{4.35 \times 10^2} = 3.531x$$

Nodal points (Starting from tip of pile)	*x (m)* (starting from top of pile)	*A*
0	20	71
1	16	57
2	12	42
3	8	28
4	4	14
5	0	0

Computation of B-constants

The number of B-constants = $(2t + 3) = 13$.

$$B_1 = \frac{2}{2 + A_0} = \frac{2}{2 + 71} = 0.0274$$

$$B_2 = \frac{1}{5 + A_1 - 4B_1} = \frac{1}{5 + 57 - 4 \times 0.0274} = 0.0162$$

$$B_3 = B_2 (4 - 2B_1) = 0.0162 (4 - 2 \times 0.0274) = 0.0639$$

$$B_4 = \frac{1}{6 + A_2 - B_1 - B_3 (4 - 2B_1)}$$

$$= \frac{1}{6 + 42 - 0.0274 - 0.0639 (4 - 2 \times 0.0274)} = 0.021$$

$$B_5 = B_4 (4 - B_3) = 0.021 (4 - 0.0639) = 0.0827$$

$$B_6 = B_{2 \times 3} = B_{2n} = \frac{1}{6 + A_n - B_{2n-4} - B_{2n-1}\left(4 - B_{2n-3}\right)}$$

$$= \frac{1}{6 + A_3 - B_2 - B_5\left(4 - B_3\right)}$$

$$= \frac{1}{6 + 28 - 0.0162 - 0.0827\left(4 - 0.0639\right)} = 0.0297$$

$$B_7 = B_{2n+1} = B_{2n}\left(4 - B_{2n-1}\right) = B_6\left(4 - B_5\right)$$

$$= 0.0297\left(4 - 0.0827\right) = 0.1163$$

$$B_8 = \frac{1}{6 + A_4 - B_4 - B_7\left(4 - B_5\right)}$$

$$= \frac{1}{6 + 14 - 0.021 - 0.1163\left(4 - 0827\right)} = 0.0512$$

$$B_9 = B_8\left(4 - B_7\right) = 0.0512\left(4 - 0.1163\right) = 0.1988$$

$$B_{10} = \frac{1}{6 + A_5 - B_6 - B_9\left(4 - B_7\right)}$$

$$= \frac{1}{6 + 0 - 0.0297 - 0.1988\left(4 - 0.1163\right)} = 0.1924$$

$$B_{11} = B_{10}\left(4 - B_9\right) = 0.1924\left(4 - 0.1988\right) = 0.7314$$

As per equation $(t + 10)$, $B_{12} = B_{2t+2} = 0$. Since $M_t = 0$

$$B_{13} = B_{2t+3} = \frac{B_{2t-2} - 1 + B_{2t+1}\left(2 - B_{2t-1}\right)}{B_{2t}\left(2 - B_{2t-1}\right)}$$

$$= \frac{B_8 - 1 + B_{11}\left(2 - B_9\right)}{B_{10}\left(2 - B_9\right)}$$

$$= \frac{0.0512 - 1 + 0.7314\left(2 - 0.1988\right)}{0.1924\left(2 - 0.1988\right)} = 1.0635$$

Computation of Deflection

From equation $(t + 11)$, Section 9.7, for $t = 5$, we have

$$y_6\left[- B_7 + 2 + \{(2 - 0.1163)\,0.1988 + 0.0297\} \times \{0.1924 \times 1.0635 - 0.7314\}\right.$$
$$\left. + \{2 - 0.1163\}\,0.0572\right]$$

$$= \frac{- 2 \times 0.268 \times 4^3}{4.35 \times 10^2} = 1.7785 y_6 = - 0.0789$$

$$y_6 = -0.0443 \text{ m} = -4.43 \text{ cm.}$$

From equation $(t + 10)$,

$$y_{t+2} = -B_{2t+2} + B_{2t+3} y_{t+1}$$

or

$$y_7 = -B_{12} + B_{13} y_6$$

$$= 0 + 1.0635 \times (-4.43) = -4.7113 \text{ cm.}$$

From equation $(t + 9)$

$$y_n = -B_{2n} y_{n+2} + B_{2n+1} y_{n+1}$$

For $n = 5$,

$$y_5 = -B_{10} y_7 + B_{11} y_6 = -0.1924 \times (-4.7113) + 0.7314 (-4.43)$$

$$= -2.3336 \text{ cm.}$$

For $n = 4$,

$$y_4 = -B_8 y_6 + B_9 y_5$$

$$= -0.0512 \times (-4.43) + 0.1988 \times (-2.3336)$$

$$= -0.2371 \text{ cm.}$$

For $n = 3$,

$$y_3 = -B_6 y_5 + B_n y_4$$

$$= -0.0297 \times (-2.336) + 0.1163 (-0.2371)$$

$$= +0.0417 \text{ cm.}$$

For $n = 2$,

$$y_2 = -B_4 y_4 + B_5 y_3$$

$$= -0.021 \times (-0.2371) + 0.0827 \times (+0.04717)$$

$$= 0.0044 \text{ cm.}$$

For $n = 1$,

$$y_1 = -B_2 y_3 + B_3 y_2$$

$$= -0.0162 \times 0.0417 + 0.0639 \times 0.0044$$

$$= -0.0004 \text{ cm.}$$

From equation $(t + 6)$,

$$y_0 = -B_1 y_2 + 2B_1 y_1$$

$$= -0.0274 \times 0.0044 + 2 \times 0.0274 \times (-0.0004)$$

$$= -0.0001 \text{ cm.}$$

The deflection at ground line is $y_5 = 2.3336$ cm.

9.8 NON-DIMENSIONAL METHOD OF ANALYSIS OF VERTICAL PILES SUBJECTED TO LATERAL LOADS

Dimensional Analysis for Elastic Piles

The non-dimensional method for the analysis of laterally loaded piles presented in this section is based on the paper published by Reese and Matlock (1956).

The principle of dimensional analysis may be used to establish the form of non-dimensional relations for the laterally loaded pile. With the use of model theory, the necessary relations can be determined between a "prototype" having any given set of dimensions, and a similar "model" for which solution may be available.

For very long piles, the length L loses its significance because the deflection may be nearly zero for much of the length of the pile. It is convenient to introduce some characteristic length as a substitute. A linear dimension T is therefore included in the analysis. Since T in the analysis expresses a relation between the stiffness of the soil and the flexural stiffness of the pile material, it is called the *relative stiffness factor.*

For the case of an applied lateral load P_t and moment M_t, the solution for deflections of the elastic curve will include the relative stiffness factor and be expressed as

$$y = f_y (x, T, L, E_s, EI, P_t, M_t) \tag{9.36}$$

Other boundary conditions can be substituted for P_t and M_t.

It the assumption of elastic behaviour is introduced for the pile, and if deflections remain small relative to the pile dimensions, the principle of superposition may be employed. Thus, the effects of an imposed lateral load P_t and imposed moment M_t, if considered independently give rise to deflection

$$y = y_A + y_B \tag{9.37}$$

where, y_A is the deflection due to P_t, and y_B is the deflection caused by the moment M_t (Fig. 9.7). The solution for the two cases of deflections may be expressed in the following forms.

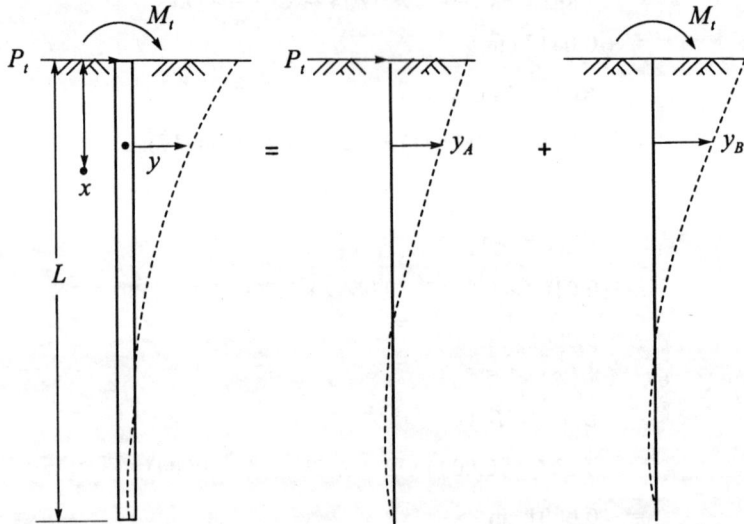

Fig. 9.7 Principle of superposition for the deflection of laterally loaded piles

For case A

$$\frac{y_A}{P_t} = f_A (x, T, L, E_s, EI) \tag{9.38}$$

For case B

$$\frac{y_B}{M_t} = f_B (x, T, L, E_s, EI) \tag{9.39}$$

where, f_A and f_B represent two different functions of the same terms. In each case there are six terms and two dimensions (force and length). There are therefore four independent non-dimensional groups which can be formed as follows.

For case A

$$\frac{y_A EI}{P_t T^3}, \frac{x}{T}, \frac{L}{T}, \frac{E_s T^4}{EI} \qquad (9.40)$$

For case B

$$\frac{y_B EI}{M_t T^2}, \frac{x}{T}, \frac{L}{T}, \frac{E_s T^4}{EI} \qquad (9.41)$$

To satisfy the conditions of similarity, each of these groups must be equal for both the model and prototype. A group of non-dimensional parameters may be defined which will have the same numerical value for any model and prototype. They are:

Depth coefficient, $\quad Z = \dfrac{x}{T}$ $\qquad (9.42)$

Maximum depth coefficient,

$$Z_{max} = \frac{L}{T} \qquad (9.43)$$

Soil modulus function

$$\phi(Z) = \frac{E_s T^4}{EI} \qquad (9.44)$$

Case A, deflection coefficient,

$$A_y = \frac{y_A EI}{P_t T^3} \qquad (9.45)$$

Case B, deflection coefficient,

$$B_y = \frac{y_B EI}{M_t T^2} \qquad (9.46)$$

The general form of variation of E_s with depth that is normally used is the power form expressed as per Eq. (9.13a) as

$$E_s = n_h x^n \qquad (9.13a)$$

If Eq. (9.13a) is substituted in Eq. (9.44), the result is

$$\phi(Z) = \frac{n_h}{EI} x^n T^4 \qquad (9.47)$$

For the elastic pile case, it is convenient to define the relative stiffness factor, T, by the following expression

$$T^{n+4} = \frac{EI}{n_h} \qquad (9.48)$$

Substituting Eq. (9.48) into Eq. (9.47) gives

$$\phi(Z) = \frac{x^n T^4}{T^{n+4}} = \left(\frac{x}{T}\right)^n = Z^n \tag{9.49}$$

where, $x/T = Z$.

As per definition, the equation for the relative stiffness factor is

$$T = \left(\frac{EI}{n_h}\right)^{\frac{1}{n+4}} \tag{9.50}$$

The total deflection y as per Eqs (9.37), (9.45), and (9.46) is

$$y = \left(\frac{P_t T^3}{EI}\right) A_y + \left(\frac{M_t T^2}{EI}\right) B_y \tag{9.51}$$

By carrying out dimensional analysis in the same way as for deflection, the solutions for slope, moment, shear and soil reaction may be expressed as:

Slope,
$$S = S_A + S_B = \left(\frac{P_t T^2}{EI}\right) A_s + \left[\frac{M_t T}{EI}\right] B_s \tag{9.52}$$

Moment,
$$M = M_A + M_B = (P_t T) A_m + (M_t) B_m \tag{9.53}$$

Shear,
$$V = V_A + V_B = (P_t) A_v + \left(\frac{M_t}{T}\right) B_v \tag{9.54}$$

Soil reaction,
$$p = p_A + p_B = \left(\frac{P_t}{T}\right) A_p + \left(\frac{M_t}{T^2}\right) B_p \tag{9.55}$$

In the Eqs (9.51) through (9.55), A and B are the sets of non-dimensional coefficients, whose values are to be determined as a function of depth coefficient.

Determination of *A* and *B* coefficients

The basic differential equations for the deflections y_A and y_B may be written separately as [see Eq. (9.30)]

$$\frac{d^4 y_A}{dx^4} + \frac{E_s}{EI} y_A = 0 \tag{9.56}$$

$$\frac{d^4 y_B}{dx^4} + \frac{E_s}{EI} y_B = 0 \tag{9.57}$$

Equations (9.56) and (9.57) may now be written in terms of non-dimensional parameters by making use of the Eqs (9.42) through (9.46) as follows:

For deflection y_A,

$$\frac{d^4 A_y}{dZ^4} = \left(\frac{dA_y}{dy_A}\right)\left(\frac{dx}{dZ}\right)^4 \frac{d^4 y_A}{dx^4} \tag{9.58}$$

From Eq. (9.45), we have

$$\frac{dA_y}{dy_A} = \frac{EI}{P_t T^3} \tag{9.59}$$

From Eq. (9.42), we have

$$\frac{dx}{dZ} = T \tag{9.60}$$

substituting for $\dfrac{dA_y}{dy_A}$ and $\dfrac{dx}{dZ}$ in Eq. (9.58), and simplifying, we obtain

$$\frac{d^4 y_A}{dx^4} = \frac{P_t}{TEI} \frac{d^4 A_y}{dZ^4} \tag{9.61}$$

Now by substituting in Eq. (9.56) for $d^4 y_A/dx^4$ from Eq. (9.61), for E_s from Eq. (9.44), for y_A from Eq. (9.45), and $\phi\,(Z)$ from Eq. (9.49) and simplifying, we get the differential equation in non-dimensional form for deflection y_A as

$$\frac{d^4 A_y}{dZ^4} + Z^n A_y = 0 \tag{9.62}$$

Similarly, for deflection y_B we may write the nondimensional equation as

$$\frac{d^4 B_y}{dZ^4} + Z^n B_y = 0 \tag{9.63}$$

The Eqs (9.62) and (9.63) can be solved for A and B coefficients in a closed form for the boundary conditions if the soil modulus E_s is constant with depth. For any form of variation of E_s with depth, that is for any value of n, the general approach has been to make use of the difference equation method as explained in Section 9.7. Non-dimensional solutions have been developed by Reese and Matlock (1956) for $E_s = n_h x$ by making use of the difference equation method. When once the A_y and B_y coefficients are obtained as a function of depth coefficient Z, the other sets of A and B coefficients may be obtained by making use of the following differential equations by the use of the difference equation method

$$A_s = \frac{dA_y}{dZ}, \; B_s = \frac{d B_y}{dZ} \tag{9.64}$$

$$A_m = \frac{d^2 A_y}{dZ^2}, \; B_m = \frac{d^2 B_y}{dZ^2} \tag{9.65}$$

$$A_v = \frac{d^3 A_y}{dZ^3}, \; B_v = \frac{d^3 B_y}{dZ^3} \tag{9.66}$$

$$A_p = \frac{d^4 A_y}{dZ^4}, \; B_p = \frac{d^4 B_y}{dZ^4} \tag{9.67}$$

A and B coefficients as obtained by Reese and Matlock (1956) for vertical piles on the assumption of linear increase of soil modulus with depth is given in Table 9.4. The computations have indicated that the values of the A and B coefficients remain almost the same so long the maximum depth coefficient Z_{max} is equal to or greater than $4T$, where

$$T = \sqrt[5]{\frac{EI}{n_h}} \tag{9.68}$$

This definition of length is applicable to vertical piles but not necessarily for inclined piles.

Boundary Conditions

Formulae can be presented for three sets of boundary conditions at the top of the pile. They are:

1. The pile head free to rotate.
2. The pile head fixed against rotation.
3. Pile head restricted against rotation.

Case I: Pile head free to rotate

When the pile head is free to rotate the Eqs (9.51) to (9.55) are applicable for computing the values of deflection, slope, moment, shear and soil reaction respectively. The corresponding A and B coefficients may be obtained from Table 9.4. However, we require often the deflection and slope at ground level. The corresponding equations for these two cases are

$$y_g = 2.43 \frac{P_t T^2}{EI} + 1.62 \frac{M_t T^2}{EI} \tag{9.69}$$

$$S_g = 1.62 \frac{P_t T^2}{EI} + 1.75 \frac{M_t T}{EI} \tag{9.70}$$

Case II: Pile head fixed against rotation

Case II may be used to obtain a solution for the case where the superstructure translates under load but does not rotate. This happens in cases where the superstructure is very stiff in relation to the pile. In such cases equations for deflection and moment can be developed from the corresponding equation for a free head pile. When the pile cap does not rotate, a resisting moment acts on the pile cap which neutralises the actuating moment. For the rotation to be zero, the condition is [from Eq. (9.52)],

$$\frac{P_t T^2}{EI} A_s = \frac{M_t T}{EI} B_s$$

or
$$\frac{M_t}{P_t T} = \frac{A_s}{B_s} \tag{9.71}$$

Since the direction of M_t is negative, Eq. (9.51) for deflection may be written as (for fixed pile head)

$$y = \left(\frac{P_t T^3}{EI}\right) A_y - \left(\frac{M_t T^2}{EI}\right) B_y$$

or
$$y = \frac{P_t T^3}{EI} \left(A_y - \frac{M_t}{P_t T} B_y\right)$$

Table 9.4 The A and B coefficients as obtained by Reese and Matlock for long vertical piles on the assumption $E_s = n_h x$

Z	A_y	A_s	A_m	A_v	A_p
0.0	2.435	− 1.623	0.000	1.000	0.000
0.1	2.273	− 1.618	0.100	0.989	− 0.277
0.2	2.112	− 1.603	0.198	0.966	− 0.422
0.3	1.952	− 1.578	0.291	0.906	− 0.586
0.4	1.796	− 1.545	0.379	0.840	− 0.718
0.5	1.644	− 1.503	0.459	0.764	− 0.822
0.6	1.496	− 1.454	0.532	0.677	− 0.897
0.7	1.353	− 1.397	0.595	0.585	− 0.947
0.8	1.216	− 1.335	0.649	0.489	− 0.973
0.9	1.086	− 1.268	0.693	0.392	− 0.977
1.0	0.962	− 1.197	0.727	0.295	− 0.962
1.2	0.738	− 1.047	0.767	0.109	− 0.885
1.4	0.544	− 0.893	0.772	− 0.056	− 0.761
1.6	0.381	− 0.741	0.746	− 0.193	− 0.609
1.8	0.247	− 0.596	0.696	− 0.298	− 0.445
2.0	0.142	− 0.464	0.628	− 0.371	− 0.283
3.0	− 0.075	− 0.040	0.225	− 0.349	0.226
4.0	− 0.050	0.052	0.000	− 0.016	0.201
5.0	− 0.009	0.025	− 0.033	0.013	0.046
Z	B_y	B_s	B_m	B_v	B_p
0.0	1.623	− 1.750	1.000	0.000	0.000
0.1	1.453	− 1.650	1.000	− 0.007	− 0.145
0.2	1.293	− 1.550	0.999	− 0.028	− 0.259
0.3	1.143	− 1.450	0.994	− 0.058	− 0.343
0.4	1.003	− 1.351	0.987	− 0.095	− 0.401
0.5	0.873	− 1.253	0.976	− 0.137	− 0.436
0.6	0.752	− 1.156	0.960	− 0.181	− 0.451
0.7	0.642	− 1.061	0.939	− 0.226	− 0.449
0.8	0.540	− 0.968	0.914	− 0.270	− 0.432
0.9	0.448	− 0.878	0.885	− 0.312	− 0.403
1.0	0.364	− 0.792	0.852	− 0.350	− 0.364
1.2	0.223	− 0.629	0.775	− 0.414	− 0.268
1.4	0.112	− 0.482	0.668	− 0.456	− 0.157
1.6	0.029	− 0.354	0.594	− 0.477	− 0.047
1.8	− 0.030	− 0.245	0.498	− 0.476	0.054
2.0	− 0.070	− 0.155	0.404	− 0.456	0.140
3.0	− 0.089	0.057	0.059	− 0.0213	0.268
4.0	− 0.028	0.049	0.042	0.017	0.112
5.0	0.000	0.011	0.026	0.029	− 0.002

Substituting for $\dfrac{M_t}{P_t T}$, we have

or

$$y = \frac{P_t T^3}{EI} \left(A_y - \frac{A_s}{B_s} B_y \right) = F_y \frac{P_t T^3}{EI} \qquad (9.72)$$

where,

$$F_y = \left(A_y - \frac{A_s}{B_s} B_y \right)$$

The values of F_y for various depth coefficients, Z, can be computed by taking the corresponding A and B factors from Table 9.4. The deflection at ground level may be expressed as

$$y_0 = 0.93 \frac{P_t T^3}{EI} \qquad (9.73)$$

Similarly, the moment equation for fixed pile may be written as [from Eq. (9.53)]

$$M = (P_t T) A_m - (M_t) B_m$$

or

$$M = P_t T \left(A_m - \frac{M_t}{P_t T} B_m \right)$$

Substituting for $\dfrac{M_t}{P_t T}$ from Eq. (9.71), we have

$$M = P_t T \left(A_m - \frac{A_s}{B_s} B_m \right) = F_m P_t T \qquad (9.74)$$

where,

$$F_m = \left(A_m - \frac{A_s}{B_s} B_m \right)$$

The values of F_m with respect to the depth coefficient Z can be calculated as before. For moment at ground level, Eq. (9.74) reduces to

$$M_t = -0.93 P_t T \qquad (9.75)$$

Case III: Pile head restricted against rotation

We come across pile heads restrained against rotation in the case of offshore pile supported structures and the like where the superstructure translates under angular restraint provided by the framework of the superstructure. A typical example of the offshore pile supported structure with the movement of a single pile column is shown in Fig. 9.8.

From the analysis of superstructure, it is possible to obtain the ratio of M_t and S_t where M_t is the resisting moment acting on the top of the pile under a lateral load of P_t, and S_t the deflected slope of the top of the pile. The ratio M_t/S_t is called as the *Spring Stiffness factor*, K_θ that is

$$K_\theta = \frac{M_t}{S_t} \qquad (9.76)$$

which is expressed as force per radian of rotation.

The slope at the top of the pile is also expressed as

$$S_t = A_{st} \frac{P_t T^2}{EI} + B_{st} \frac{M_t T}{EI} \qquad (9.77)$$

Combining Eqs (9.76) and (9.77), and rearranging, we have

$$\frac{M_t}{P_t T} = \frac{A_{st} T K_\theta}{(EI - K_\theta B_{st} T)} \qquad (9.78)$$

The ratio of $\dfrac{M_t}{P_t T}$ in Eq. (9.78) can be found out if T is known since all the other constants on the right hand side of the equation are known. If $\dfrac{M_t}{P_t T}$ is known, the deflection of the pile head can be computed by making use of a new set of non-dimensional deflection coefficients as explained below.

The general equation for deflection is (Eq. 9.51)

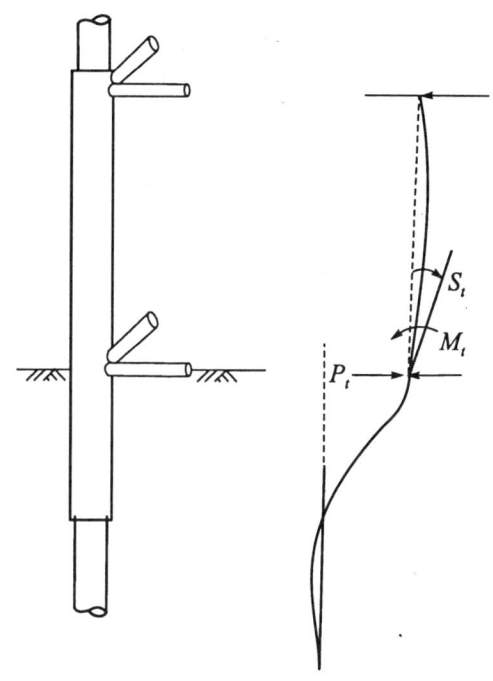

Fig. 9.8 Pile head restrained against rotation

$$y = A_y \frac{p_t T^3}{EI} + B_y \frac{M_t T^2}{EI}$$

Equation (9.51) may be written as

$$y = C_y \frac{p_t T^3}{EI} \qquad (9.79)$$

where, $\quad C_y = A_y + \dfrac{M_t}{P_t T} B_y \qquad (9.80)$

The values of C_y at different depth coefficients $Z (= x/T)$ can be calculated from Eq. (9.80) for known values of $\dfrac{M_t}{P_t T}$. The value of $M_t/P_t T$ will range from zero for pinned case to -0.93 for the case where the structure prevents any rotation of the pile head. Values of C_y have been developed by Matlock and Reese (1961), and are given in Fig. 9.9 (a). While getting the value of C_y, the direction of M_t should be taken in its proper sign. Values of C_y can be calculated and plotted for $\dfrac{M_t}{P_t T}$ greater than zero also if required. However, Fig. 9.9 (a) gives values of C_y for $\dfrac{M_t}{P_t T}$ lying between -1.0 and 0 only.

In the same way the moment equation (9.53) can be written in a different form as follows.

$$M = P_t T A_m + M_t B_m$$

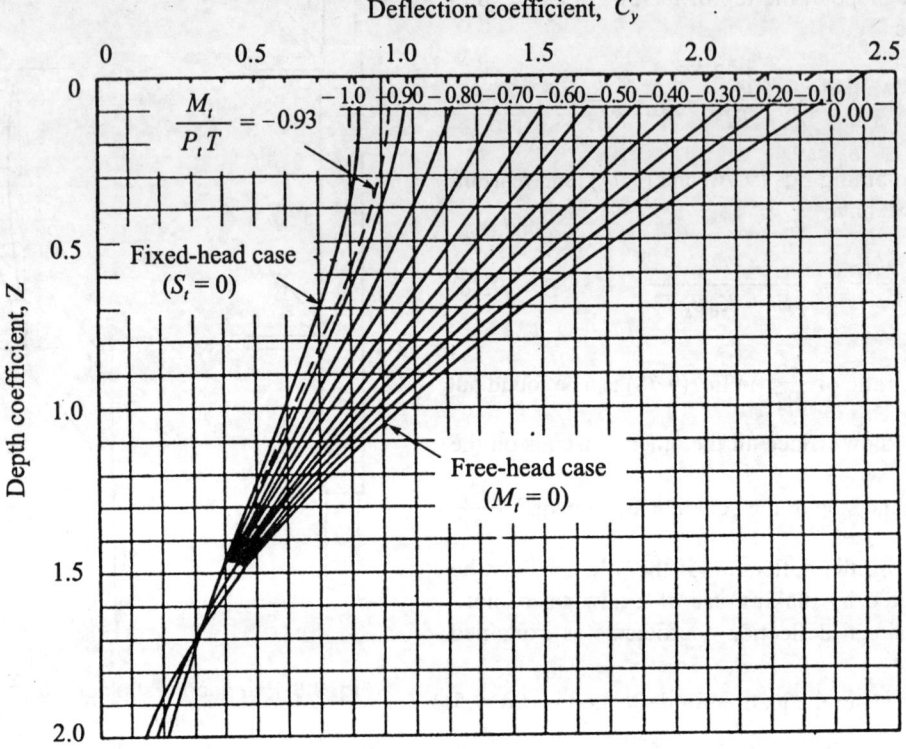

Fig. 9.9 (a) Non-dimensional coefficients for lateral deflection of a pile for $E_s = n_h x$ (Matlock and Reese, 1961)

$$= P_t T \left(A_m + \frac{M_t}{P_t T} B_m \right) = C_m P_t T \qquad (9.81)$$

where,
$$C_m = A_m + \frac{M_t}{P_t T} B_m \qquad (9.82)$$

A set of curves for C_m can be developed for various values of $\frac{M_t}{P_t T}$ as shown in Fig. 9.9 (b).

The values of $\frac{M_t}{P_t T}$ may range from zero to -0.93 for a restrained pile head. Values of C_m can be

obtained for $\frac{M_t}{P_t T}$ greater than zero also. Figure 9.9 (b) gives values of C_m for $M_t/P_t T$ lying between -1.0 and $+1.0$.

Example 9.2

For the problem given in Ex. 9.1, compute the ground line deflection by the use of Reese and Matlock (1956) non-dimensional parameters.

Solution

Use Eq. (9.69)

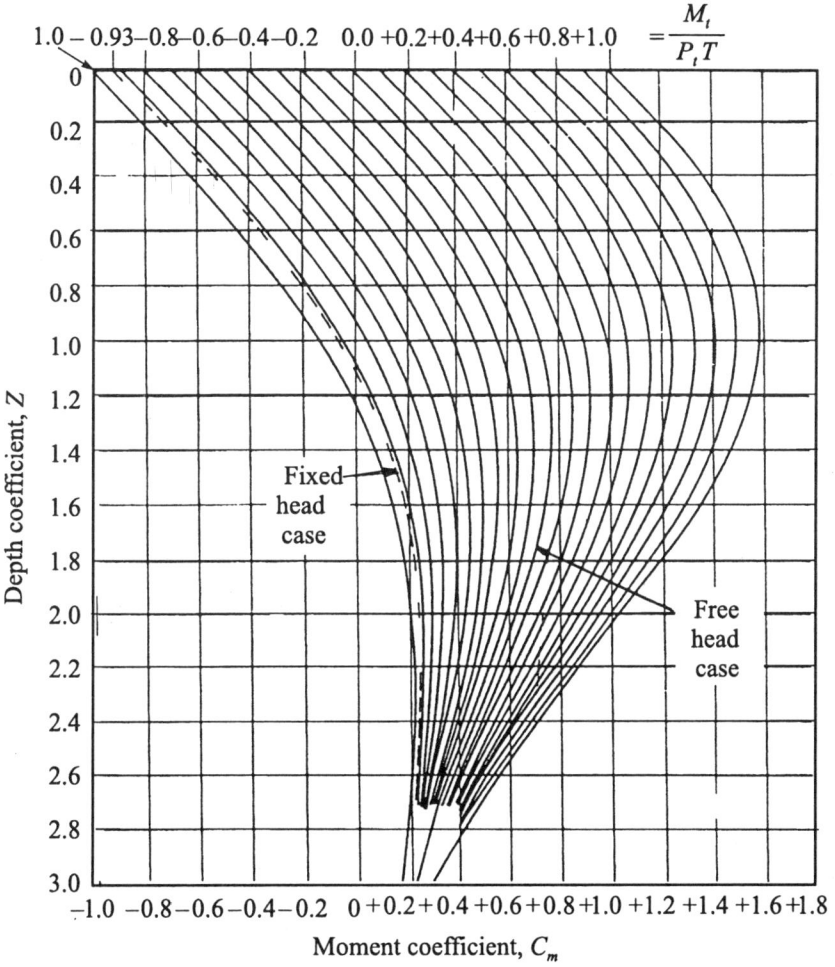

Fig. 9.9 (b) Curves of moment coefficient C_m for long piles (Matlock and Reese, 1961)

$$y_g = 2.43 \frac{P_t T^3}{EI} \text{ for } M_t = 0$$

From Eq. (9.68), $T = \left(\dfrac{EI}{n_h}\right)^{\frac{1}{5}}$

where, P_t = 0.268 MN,

$EI = 4.35 \times 10^2$ MN – m^2,

$n_h = 6$ MN/m^3,

$$T = \left(\frac{4.35 \times 10^2}{6}\right)^{\frac{1}{5}} = 2.3554 \text{ m}$$

Now, $y_g = \dfrac{2.43 \times 0.268 \times (2.3554)^3}{4.35 \times 10^2} = 0.0196 \text{ m} = 1.96 \text{ cm}$

The deflection by this method is quite close to the value in Ex. 9.1.

Example 9.3

If the pile in Ex. 9.1 is subjected to a lateral load at a height 2 m above ground level, what will be the ground line deflection?

Solution

Use Eq. (9.69)

$$y_g = 2.43 \frac{P_t T^3}{EI} + 1.62 \frac{M_t T^2}{EI}$$

As in Ex. 9.2, $T = 2.3554$ m, $M_t = 0.268 \times 2 = 0.536$ MN – m.

Substituting,

$$y_g = \frac{2.43 \times 0.268 \times (2.3554)^3}{4.35 \times 10^2} + \frac{1.62 \times 0.536 \times (2.3554)^2}{4.35 \times 10^2}$$

$$= 0.0196 + 0.0111 = 0.0307 \text{ m} = 3.07 \text{ cm.}$$

Example 9.4

If the pile in Ex. 9.1 is fixed against rotation, calculate the deflection at the ground line level.

Solution

Use Eq. (9.73),

$$y_g = \frac{0.93 \, P_t T^3}{EI}$$

The values of P_t, T and EI are as given in Ex. 9.1 substituting

$$y_g = \frac{0.93}{2.43} \times 1.96 = 0.75 \text{ cm.}$$

Example 9.5

If the pile head given in Ex. 9.1 is partially fixed having a spring stiffness $K_\theta = 50$ MN/radian, compute the ground line deflection.

Solution

From Eq. (9.78),

$$\frac{M_t}{P_t T} = \frac{A_{st} T K_\theta}{(EI - k_\theta B_{st} T)}$$

where, $T = 2.3554$ m, $EI = 4.35 \times 10^2$ MN – m^2,

$P_t = 268$ kN.

From Table 9.4, $A_{st} = 1.623$, $B_{st} = 1.75$.

Substituting,

$$\frac{M_t}{P_t T} = \frac{1.623 \times 2.3554 \times 50}{4.35 \times 10^2 - 50 \times 1.75 \times 2.3554} = 0.83$$

From Fig. 9.9 (a),

$$C_y = 1.2 \text{ for } \frac{M_t}{P_t T} = 0.83$$

From Eq. (9.79),

$$y_g = \frac{C_y P_t T^3}{EI}$$

$$= \frac{1.2 \times 0.268 \times (2.3554)^3}{4.35 \times 10^2} = 0.0097 \text{ m}$$

$$= 0.97 \text{ cm}$$

9.9 BROMS METHOD FOR THE ANALYSIS OF LATERALLY LOADED PILES (1964a, 1964b)

Broms theory of laterally loaded piles given here is based on his papers published in the year 1964. His theory deals with the following:

1. Lateral deflections of piles at working loads.
2. Ultimate soil resistance of soil.

He has considered short and long piles embedded in both cohesive and cohesionless soils. His theory is considered under the following headings:

1. Lateral deflections at working loads in saturated cohesive soils (1964a).
2. Ultimate lateral resistance of piles in cohesive soils (1964a).
3. Lateral deflections at working loads in cohesionless soils (1964b).
4. Ultimate lateral resistance of piles in cohesionless soils (1964b).

9.10 LATERAL DEFLECTIONS AT WORKING LOADS IN SATURATED COHESIVE SOILS (BROMS, 1964a)

Introduction

Broms has developed methods for calculating deflections of laterally loaded piles driven into saturated cohesive soils. He has considered short and long piles either fixed or free to rotate at the head. Lateral deflections at working loads have been calculated using the concept of subgrade reaction. It is assumed that the deflections increase approximately linearly with the applied loads when the loads applied are less than one-half to one-third the ultimate lateral resistance of the pile.

The deflections, bending moments and soil reactions depend primarily on the dimensionless length β where,

$$\beta = \sqrt[4]{\frac{kd}{4EI}} \qquad (9.83)$$

EI = stiffness of the pile section,
k = coefficient of subgrade reaction,
d = width or diameter of pile,
L = length of pile.

A pile is considered long or short on the following conditions :

Free-head pile

For long piles, $\beta L > 2.50$.

For short piles, $\beta L < 2.50$.

Fixed-head pile

For long pile, $\beta L > 1.5$.

For short pile, $\beta L < 1.5$.

Calculation of Deflection

Broms has given a set curves giving the relationship between βL and a dimensionless quantity $y_0 k d L / P_t$ for various values of e/L as shown in Fig. 9.10 (a). Equations for calculating lateral deflections, y_0, at ground level are given for two cases, that is, when the pile is fully free or fully fixed at the ground surface.

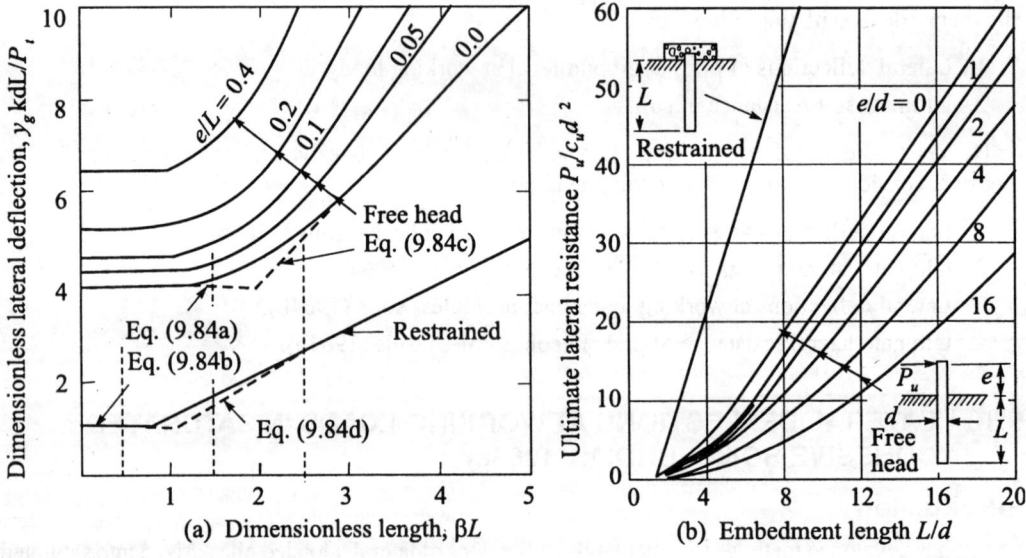

Fig. 9.10 (a) Lateral deflections at ground surface in cohesive soils, (b) ultimate lateral resistance of short piles in cohesive soils related to embedded length (Broms, 1964)

y_0 for infinitely stiff pile when $\beta L < 1.5$ – free head

$$y_0 = \frac{4 P_t \left(1 + 1.5\, e / L\right)}{k\, d\, L} \qquad (9.84a)$$

where, e is the height above the ground level where the lateral load P_t is applied.

y_0 for a restrained pile with $\beta L < 0.5$

$$y_0 = \frac{P_t}{k d L} \qquad (9.84b)$$

The deflection of restrained piles is theoretically one-fourth or less than that of corresponding free-head piles.

y_0 for long piles

Free head: $$y_0 = \frac{2 P_t \beta (e\beta + 1)}{k_\infty d} \qquad (9.84c)$$

Fixed head: $$y_0 = \frac{P_t \beta}{k_\infty d} \qquad (9.84d)$$

where, k_∞ = coefficient of subgrade reaction for long piles.

Figure 9.10 (a) gives the relationship between $y_0 kdL/P_t$ and βL for free-head and restrained piles.

Coefficient of subgrade reaction for long piles

The coefficient of subgrade reaction k_∞ for infinitely long piles is calculated from

$$k_\infty = \frac{\alpha K_0}{d} \qquad (9.84e)$$

where, $$\alpha = 0.52 \sqrt[12]{\frac{K_0 d^4}{EI}} \qquad (9.84f)$$

K_0 = coefficient of subgrade reaction of a plate with a diameter of 12 inches.

According to Broms, α in Eq. (9.84f) may be written as

$$\alpha = n_1 n_2 \qquad (9.84g)$$

in which n_1 and n_2 are functions of the unconfined compressive strength of the supporting soil and of the pile material respectively.

Table 9.5 (a) Values of coefficient n_1

Undrained shearing strength c_u kPa	Value of n_1
< 27	0.32
27–107	0.36
> 107	0.40

Table 9.5 (b) Values of coefficient n_2

Pile material	Value of n_2
Steel	1.00
Concrete	1.15
Wood	1.30

Coefficient of subgrade reaction *k* for short piles

The method proposed by Broms for computing k is quite involved. Tomlinson suggests that it is sufficiently accurate to take k as k_1 (the modulus for 300 mm plate) for the case of soil with constant modulus. Values of k_1 for cohesive soils are given in Table 7.1 (b).

9.11   ULTIMATE LATERAL RESISTANCE OF PILES IN SATURATED COHESIVE SOILS (BROMS, 1964a)

Introduction

The ultimate soil resistance for piles in cohesive soils increases with depth from $2c_u$ at the surface (c_u = undrained shear strength) to 8 to 12 c_u at a depth of about three pile diameters ($3d$) below the surface. Broms suggests a simplified distribution of soil resistance as being zero from ground surface to a depth of $1.5d$ and constant value of $9c_u$ below this depth. The mechanism of failure of soil under ultimate lateral load P_u for the following types of piles are discussed.

1. Short piles, free head and restrained.
2. Long piles, free head and restrained.

Short Pile

Free-head

The distribution of soil reactions and bending moments is shown in Fig. 9.11 (a). Failure takes place when the soil yields along the total length of the pile and the pile rotates as a unit. The maximum moment M_{max} at a depth $(f + 1.5d)$ below the ground surface and at this depth the shear force is equal to zero. We may write for f as

$$f = \frac{P_u}{9c_u d} \tag{9.85a}$$

By taking moments about the maximum moment location, we have

$$M_{max} = P_u (e + 1.5d + f) - \left(\frac{f}{2}\right)(9c_u df)$$

or

$$M_{max} = P_u (e + 1.5d + f) - \frac{P_u f}{2}$$

or

$$M_{max} = P_u (e + 1.5d + 0.5f) \tag{9.85b}$$

Integration of the lower part of the shear diagram yields

$$M_{max} = 2.25 c_u dg^2 \tag{9.85c}$$

Since $L = (1.5d + f + g)$, Eqs (9.85a) and (9.85b) can be solved for the ultimate load P_u that will produce a soil failure. The solution is plotted in Fig. 9.17 (b) in terms of dimensionless parameters L/D and $P_u/c_u d^2$.

Short Fixed Head Pile

Broms considers two types of short piles for fixed-head conditions. They are:

1. Very short pile.
2. Intermediate length of pile.

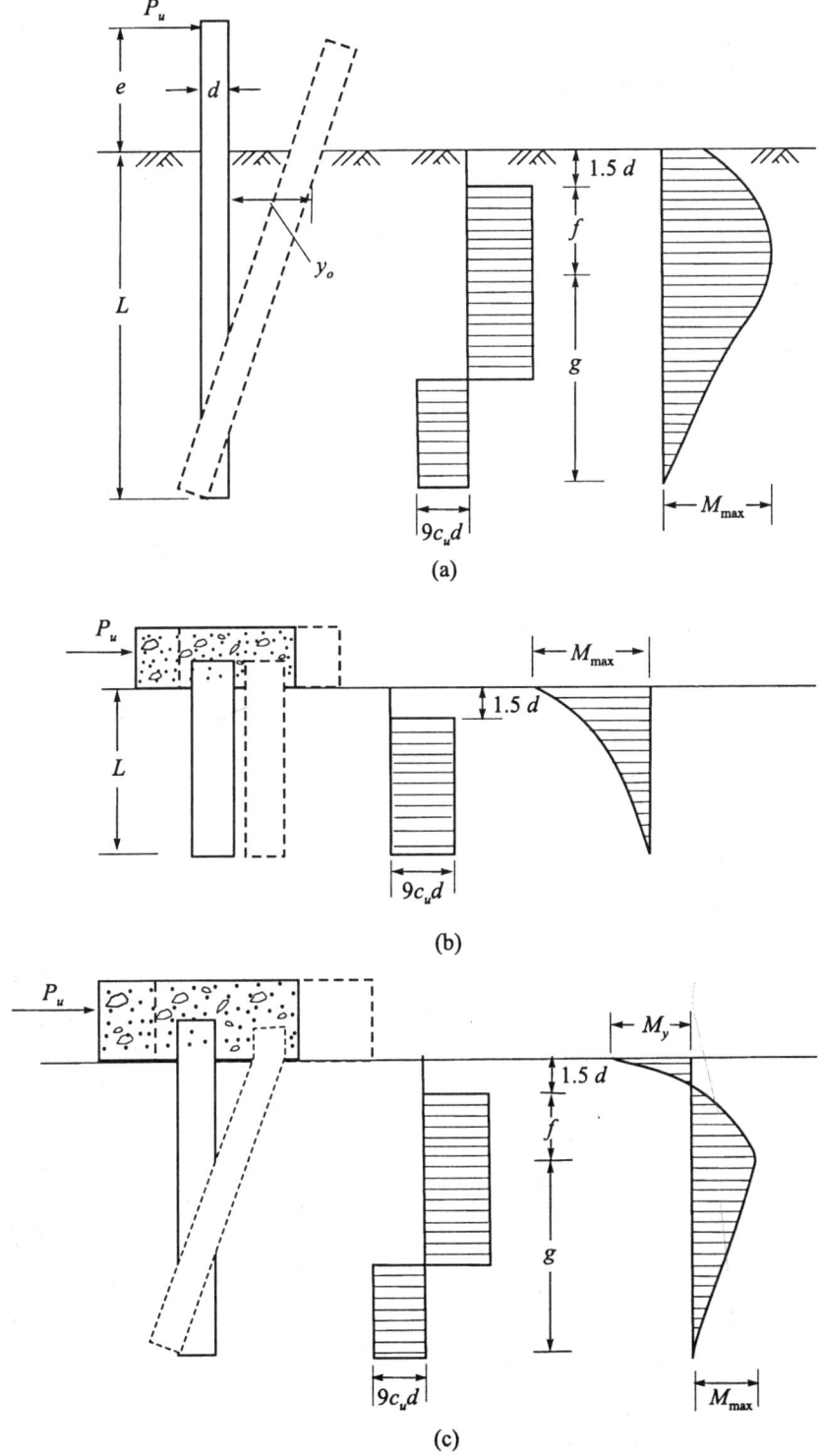

Fig. 9.11 Deflection, soil reaction and bending moment distribution along short piles in cohesive soils: (a) Short pile-free-head, (b) very short pile-fixed head, (c) intermediate length-fixed pile (Broms, 1964)

The failure mechanisms for both the types of piles are shown in Fig. 9.11 (b) and (c). In the case of a very short pile, failure takes place when the applied lateral load P_u is equal to the ultimate lateral resistance of the soil or when

$$P_u = 9c_u d (L - 1.5d)$$
(9.85d)

In the case of an intermediate pile [Fig. 9.11 (c)], the first yield of the pile occurs at the head. The equation for moment equilibrium for the point where the shear is zero is

$$M^+_{max} = P_u (1.5d + f) - f(9c_u d)(f/2) - M^-_{max}$$
(9.85e)

where, M^+_{max} is the maximum positive moment at depth f and M^-_{max} is the yield moment of the pile section at the bottom of the pile cap.

Simplifying Eq. (9.85e) by writing $9c_u df = P_u$, we have

$$M^+_{max} = P_u (1.5d + 0.5f) - M^-_{max}$$
(9.85f)

Employing the shear diagram [Fig. 9.11 (c)] for the lower part of the pile,

$$M^+_{max} = 2.25c_u dg^2$$
(9.85g)

Since $L = 1.5d + f + g$, $f = P_u/9c_u d$, Eq. (9.85f) can be solved for P_u. The ultimate lateral resistance can also be obtained from [Fig. 9.10 (b)], in which the dimensionless quantity $P_u/c_u d^2$ is plotted as a function of L/d.

Long Piles – Free Head

The failure mechanism of long pile under ultimate lateral load condition is shown in Fig. 9.12 (a). In this case a plastic hinge, forms in the pile section at a depth of $1.5d + f$. Figure 9.12 (c) gives a plot of non-dimensional quantity $P_u/c_u d^2$ as a function of $M_y/c_u d^3$.

Fixed head

The mode of failure of a long fixed pile is shown in Fig. 9.12 (b). Failure takes place when two plastic hinges form along the pile. The first hinge is located at the bottom of the pile cap and the second at the section of the maximum positive moment at the depth $(1.5d + f)$ below ground surface. The ultimate lateral resistance can be obtained from Eqs (9.85a) and (9.85f) and it is assumed that the maximum positive bending moment M^+_{max} is equal to the yield resistance of the section M^+_y. The ultimate lateral resistance is equal to

$$P_u = \frac{2M_y}{(1.5d + 0.5f)}$$
(9.85h)

Figure 9.12 (c) gives a plot of $P_u/c_u d^2$ as a function of $M_y/c_u d^3$.

9.12   LATERAL DEFLECTIONS AT WORKING LOADS IN COHESIONLESS SOILS (BROMS, 1964b)

Modulus of Subgrade Reaction

Broms also assumes that the modulus of subgrade reaction, E_s, increases linearly with depth as per Eq. (9.13b)

$$E_s = n_h x$$

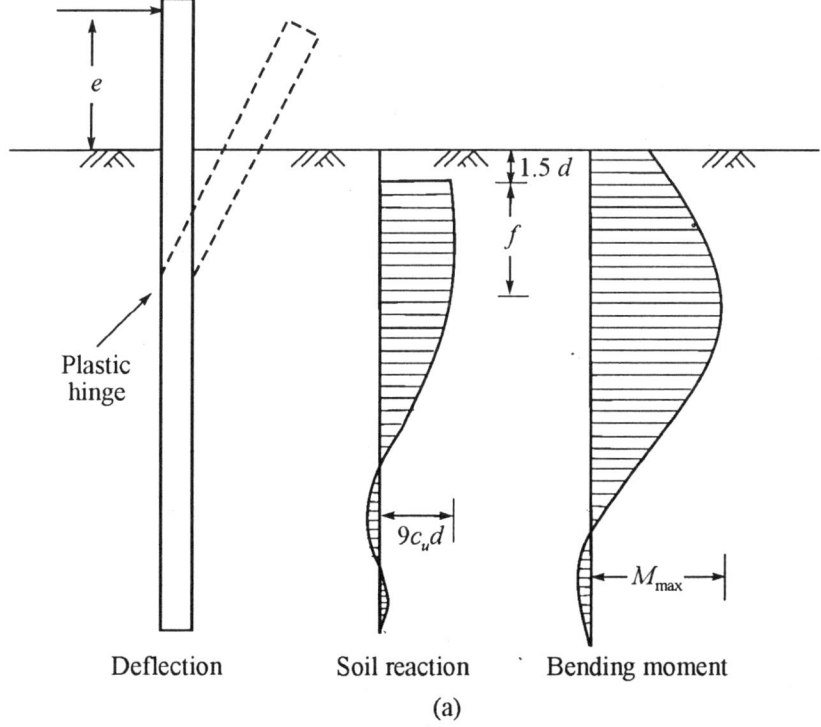

Deflection   Soil reaction   Bending moment

(a)

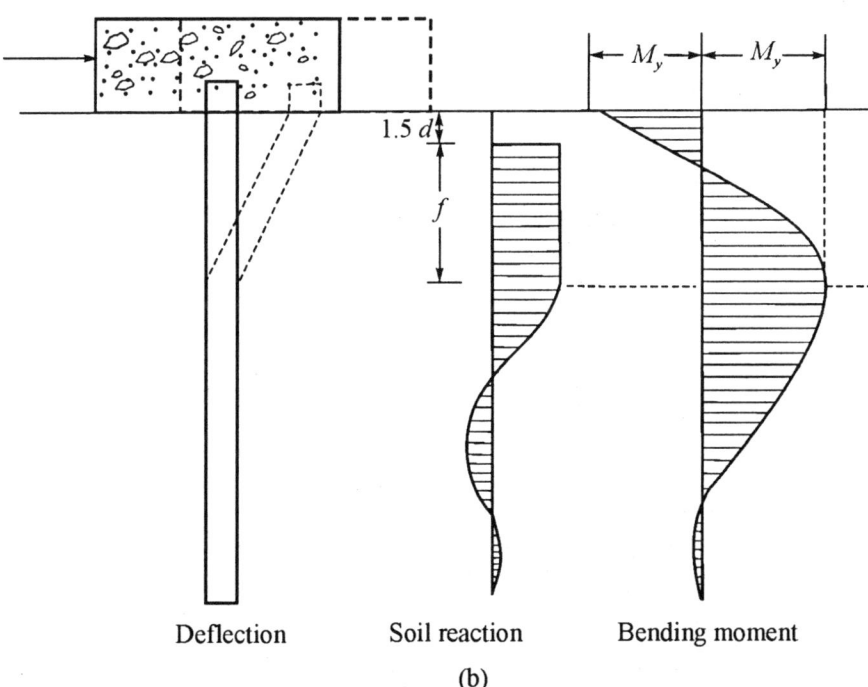

Deflection   Soil reaction   Bending moment

(b)

Fig. 9.12 Deflection, soil reaction and bending moment distribution along piles in cohesive soils: (a) Long pile–free head, (b) long piles–fixed head (Broms, 1964)

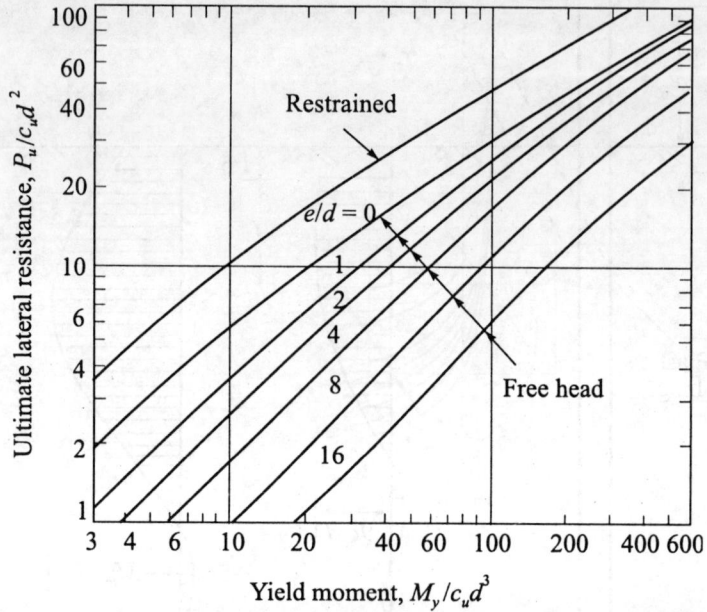

Fig. 9.12 (c) Ultimate lateral resistance of long piles in cohesive soils (Broms, 1964)

The values for n_h assumed by Broms are the same as recommended by Terzaghi and are given in Table 9.2.

Lateral deflections at ground level

Broms gives equation for computing lateral deflections at ground level and he has given a set of curves as shown in Fig. 9.13 (a) for computing deflections at ground level. He has plotted the non-dimensional deflection factor $y_g (EI)^{3/5} (n_h)^{2/5}/P_t L$ as a function of ηL for various values of e/L

where, y_g = deflection at ground level,

$$\eta = \left(\frac{n_h}{EI}\right)^{\frac{1}{5}},\qquad(9.86a)$$

n_h = coefficient of soil modulus variation,

E_I = flexural stiffness of pile material,

P_t = lateral load applied at or above ground level,

L = length of pile.

It can be seen from [Fig. 9.13 (a)] that a laterally loaded pile behaves as an infinitely stiff member when the dimensionless length ηL is less than about 2 and an infinitely long member when ηL exceeds about 4. For short piles ($\eta L < 2$), an increase of the embedment length decreases the lateral deflections at the ground surface, while for long piles ($\eta L > 4$), the lateral deflection at the ground surface are unaffected by the change of the embedment length. Broms has also given some equations for computing deflections as follows.

1. Long free head pile, with lateral load at ground level

$$y_g = \frac{2.4 P_t}{n_h^{3/5} (EI)^{2/5}}\qquad(9.86b)$$

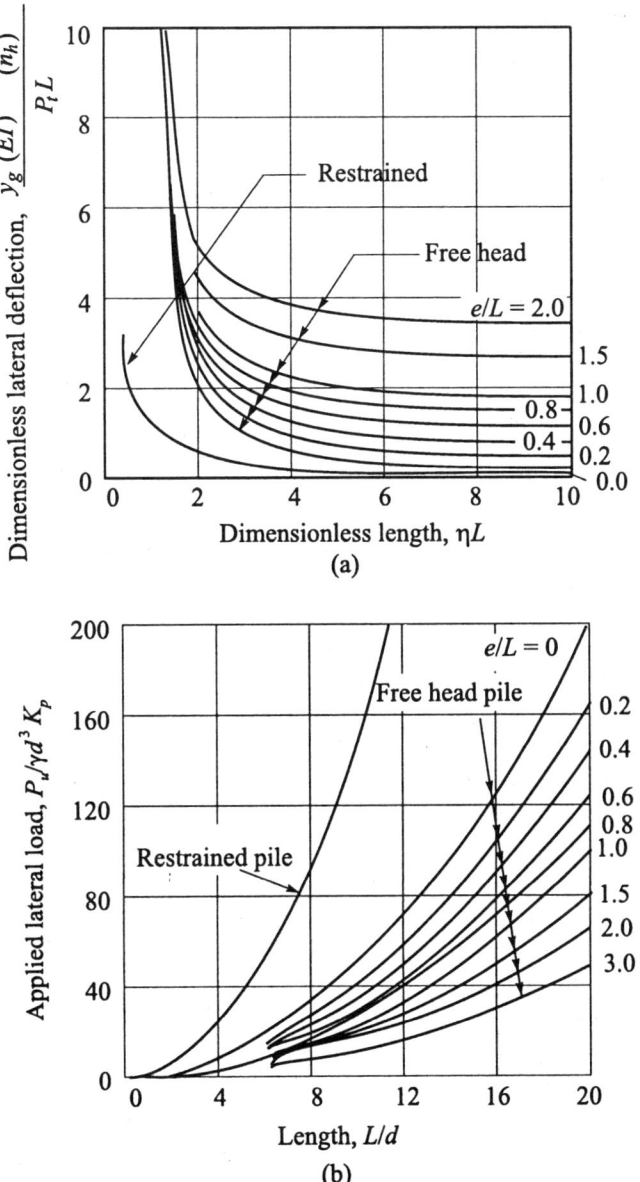

Fig. 9.13 Lateral deflections and ultimate lateral resistance (short piles) in cohesionless soils (Broms, 1964)

2. Long restrained pile

$$y_g = \frac{0.93 P_t}{n_h^{3/5} (EI)^{3/5}}$$ (9.86c)

3. Short piles – free head

$$y_g = \frac{18 P_t \left(1 + 1.33 \dfrac{e}{L}\right)}{L^2 n_h}$$ (9.86d)

4. Fixed head short pile

$$y_g = \frac{2P_t}{L^2 n_h}$$

(9.86e)

9.13 ULTIMATE LATERAL RESISTANCE OF PILES IN COHESIONLESS SOILS (BROMS, 1964b)

Introduction

Broms considers two types of failures of piles in cohesionless soils. They are:

1. Failure of soil.
2. Failure of the pile by the formation of plastic hinges in the pile.

With regards to soil failure in cohesionless soils, Broms assumed that the ultimate lateral resistance is equal; to three times the Rankine passive earth pressure. Thus, at a depth x below the ground surface, the ultimate soil resistance per unit length, p_u can be obtained from

$$p_u = 3d\gamma x K_P$$

(9.87a)

where, $K_P = \tan^2(45° + \phi/2) =$ Rankine passive earth pressure coefficient,

$\gamma =$ effective unit weight of soil,

$\phi =$ angle of internal friction of soil,

$d =$ diameter or width of pile.

Equations for computing ultimate lateral resistance and movement are given below.

Short Free-Head Pile

The mode of failure for the type of loading for a free-head pile is shown in Fig. 9.14 (a). Let P_u is the ultimate applied lateral load at the top of pile at an eccentricity e. The following equation results after taking moments about the bottom of the pile.

$$P_u(e + L) = (3\gamma dLK_p)\left(\frac{L}{2}\right)\left(\frac{L}{3}\right)$$

Solving for P_u

$$P_u = \frac{\gamma dL^3 K_P}{2(e + L)}$$

(9.87b)

Equation (9.87b) may be written in a dimensionless form as

$$\frac{P_u}{K_P \gamma d^3} = \frac{0.5(L/d)^2}{(1 + e/L)}$$

(9.87c)

The dimensionless ultimate lateral resistance $P_u/K_P\gamma d^3$ has been plotted as a function of the dimensionless embedment length L/d for various eccentricity ratios e/L and is shown in Fig. 9.13 (b).

Long Free-Head Pile

The mode of failure of a long free-head pile is shown in Fig. 9.15 (a). Failure takes place when a plastic hinge forms at a distance f below the ground surface (at the location of the maximum moment where the shear will be zero). The distance f can be computed by equating the ultimate load P_u to the total resistance of soil within depth f or

$$P_u - \left(\frac{f}{2}\right)(3\gamma d f K_P) = 0$$

or
$$f = 0.82 \sqrt{\frac{P_u}{\gamma d K_P}} \qquad (9.87d)$$

The corresponding maximum positive bending moment M_{max}^+ can then be determined from

$$M_{max}^+ = P_u(e + 0.67f) \qquad (9.87e)$$

Failure takes place when the M_{max}^+ is equal to the yield resistance of pile section, M_y

Substituting in Eq. (9.87e) for f and writing $M_{max}^+ = M_y$, we have an equation for the ultimate lateral resistance P_u as

$$P_u = \frac{M_y}{\left[e + 0.54\sqrt{\dfrac{P_u}{\gamma d K_P}}\right]} \qquad (9.87f)$$

Equation (9.87f) can be written in a non-dimensional form as

$$\frac{M_y}{\gamma d^4 K_P} = \frac{P_u}{\gamma d^3 K_P}\left[\frac{\iota}{d} + 0.54\sqrt{\frac{P_u}{\gamma d^3 K_P}}\right] \qquad (9.87g)$$

The dimensionless ultimate lateral resistance $P_u/\gamma d^3 K_p$ has been plotted in Fig. 9.16 as a function of dimensionless yield resistance $M_y/\gamma d^4 K_P$ and the eccentricity ratio e/d.

Short Restrained Pile

The lateral deflections and distribution of soil pressure and bending moments for short restrained pile is shown in Fig. 9.14 (b). Failure takes place when the load applied to the pile is equal to the ultimate lateral resistance of the soil as expressed by

$$P_u = 1.5\gamma L^2 d K_P \qquad (9.87h)$$

The dimensionless ultimate lateral resistance $P_u/\gamma d^3 K_p$ as determined from Eq. (9.87h) has been plotted in Fig. 9.13 (b) as a function of embedment length L/d.

Intermediate Length–Restrained Pile

The mode of failure for a restrained pile with soil reaction and moment distribution is shown in Fig. 9.14 (c). Taking moments about the bottom end of the pile, we have

$$P_u L + M_y = \left(\frac{L}{2}\right)(3\gamma d L K_P)\left(\frac{L}{3}\right)$$

Simplifying, the ultimate lateral resistance P_u may be obtained from

$$P_u = 0.5\,\gamma d L^2 K_P - M_y/L \qquad (9.87i)$$

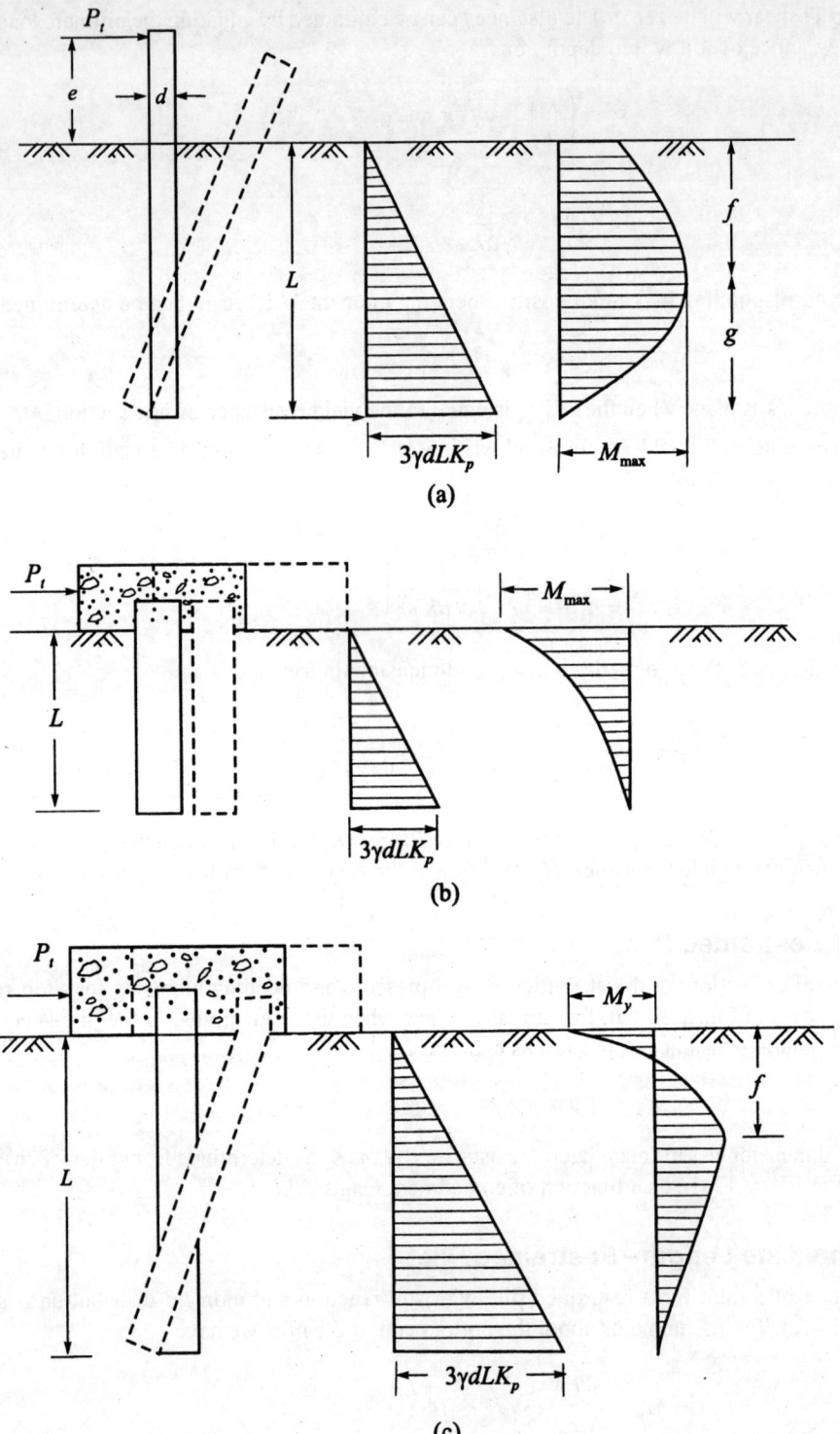

Fig. 9.14 Deflection, soil reaction and moment distribution in cohesionless soils along short piles: (a) Short pile-free head, (b) very short-pile free head, and (c) intermediate length-fixed head pile (Broms, 1964)

Long Restrained Pile

The mode of failure of a long restrained pile is shown in Fig. 9.15 (b). Failure takes place when two plastic things form as shown in the figure and the maximum bending moment at the bottom of the

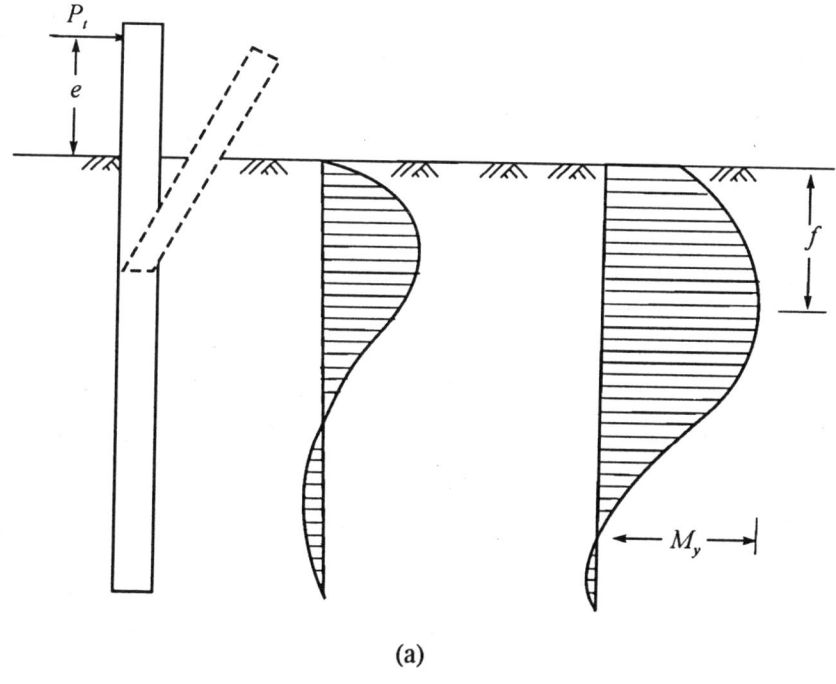

(a)

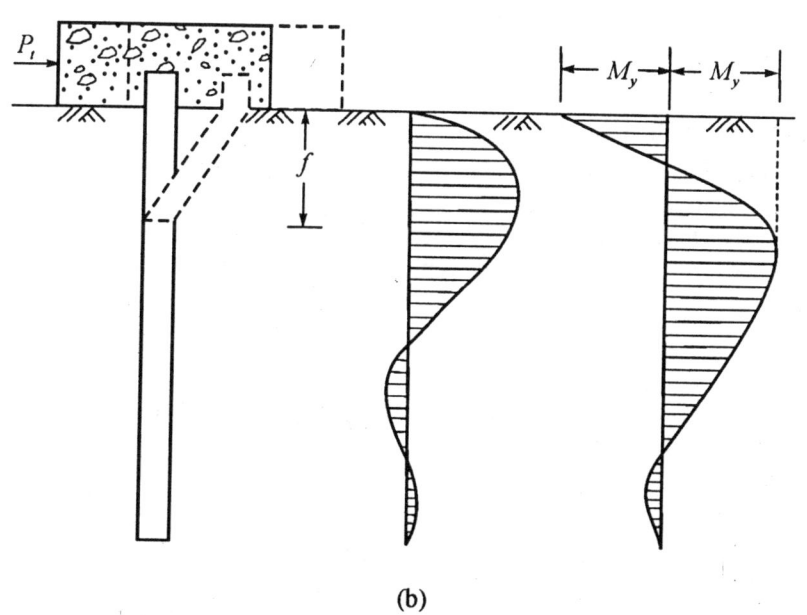

(b)

Fig. 9.15 Deflection, soil reaction and moment distribution along piles in cohesionless soils: (a) Long pile-free head, (b) long pile-fixed head (Broms, 1964)

pile cap reach the yield resistance of the pile section. The ultimate lateral resistance can be calculated by

$$P_u = \frac{2M_y}{e + 0.54 \sqrt{\dfrac{P_u}{\gamma\, dK_P}}} \tag{9.87j}$$

where, it is assumed that the yield resistance of the pile section at the bottom of the cap M_y^- is equal to the maximum positive bending movement M_{max}^+ at depth f.

The ultimate lateral resistance as determined from Eq. (9.87j) is shown in Fig. 9.16 as a function of the dimensionless yield resistance $M_y / \gamma d^4 K_P$.

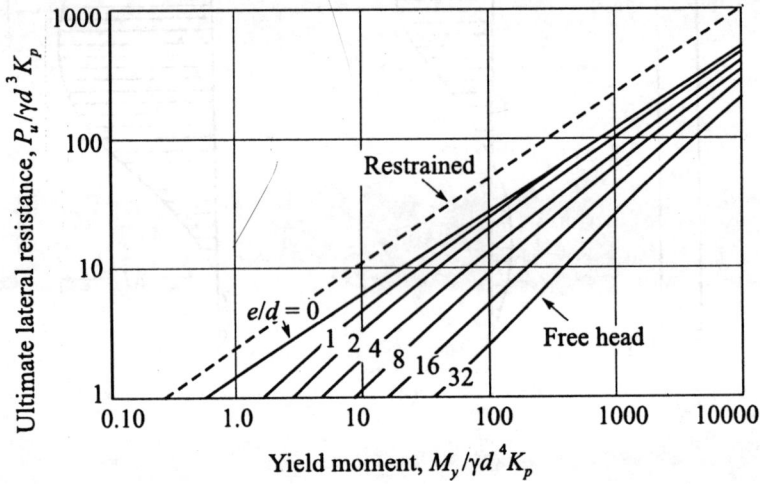

Fig. 9.16 Ultimate lateral resistance of long piles in cohesionless soils (Broms, 1964)

Yield Resistance of Pile Section

Plastic hinges form in steel piles when the stress at the section of maximum bending moment reaches yield strength of the material of pile section. The corresponding plastic resistance of the pile section M_y can be calculated on the basis of an ultimate strength analysis. For a cylindrical steel pipe section, the plastic moment can be estimated when the applied axial load is small from

$$M_y = 1.3\, f_y Z \tag{9.87k}$$

in which f_y is the yield strength of the pile material and Z is the section modulus of the pile section. The coefficient 1.3 is the *plastic moment shape factor* for a circular cross-section. The plastic moment for an *H*-section can be calculated from

$$M_y = 1.1\, f_y Z_{max} \tag{9.87l}$$

when the applied lateral load is in the direction of the largest moment resistance of the pile, and from

$$M_y = 1.5\, f_y Z_{min} \tag{9.87m}$$

when the applied load is in the direction of the minimum resistance of the pile. The ultimate strength of reinforced concrete pile sections can be calculated in a similar manner.

Example 9.6

A steel pipe pile of 61 cm outside diameter with 2.5 cm wall thickness is driven into saturated cohesive soil up to a depth of 20 m. The undrained cohesive strength of the soil is 85 kPa. Calculate the ultimate lateral resistance of the pile by Broms method with the load applied at ground level.

Solution

The pile is considered as a long pile. Use Fig. 9.12 (c) to obtain the ultimate lateral resistance P_u of the pile.

$$\text{The non-dimensional yield moment} = \frac{M_y}{c_u d^3}$$

where, M_y = yield resistance of the pile section

$\quad\quad = 1.3 f_y Z,$

$\quad f_y$ = yield strength of the pile material

$\quad\quad = 2800 \text{ kg/cm}^2 \text{ (assumed)},$

$\quad Z$ = section modulus $= \dfrac{1}{I} = \dfrac{\pi}{64\,R}\,(d_o^4 - d_i^4),$

$\quad I$ = moment of inertia,

$\quad d_o$ = outside diameter = 61 cm,

$\quad d_i$ = inside diameter = 56 cm,

$\quad R$ = radius = 30.5 cm.

$$Z = \frac{3.14}{64 \times 30.5}\,(61^4 - 56^4) = 6452.6 \text{ cm}^3$$

$$M_y = 1.3 \times 2800 \times 6452.6 = 23.487 \times 10^6 \text{ kg-cm}$$

$$\frac{M_y}{c_u d^3} = \frac{23.487 \times 10^6}{0.85 \times 61^3} = 122$$

From [Fig. 9.12 (c)] For $e/d = 0$, $\dfrac{M_y}{c_u d^3} = 122$, $P_u/c_u d^2 \approx 35,$

$$P_u = 35\,c_u d^2 = 35 \times 85 \times 0.61^2 = 1107 \text{ kN}$$

or say 1100 kN.

Example 9.7

If the pile given in Ex. 9.6 is restrained against rotation, calculate the ultimate lateral resistance P_u.

Solution

As per Ex. 9.6, $\dfrac{M_y}{c_u d^3} = 122.$

From [Fig. 9.12 (c)], for $\dfrac{M_y}{c_u d^3} = 122$, for restrained pile $\dfrac{P_u}{c_u d^2} \approx 50$

Therefore, $\quad P_u = \dfrac{50}{35} \times 1107 = 1581 \text{ kN}$

Example 9.8

A steel pipe pile of outside diameter 61 cm, and inside diameter 56 cm is driven into medium dense sand under submerged condition, which is having a relative density 60% and an angle of internal friction of 38°. Compute the ultimate lateral resistance of the pile by Broms method. Assume that the yield resistance of the pile section is the same as that given in Ex. 9.6. The submerged unit weight of the soil $\gamma_b = 8.75$ kN/m³.

Solution

Use Fig. 9.16.

$$\text{Non-dimensional yield moment} = \frac{M_y}{\gamma d^4 K_P}$$

where, $K_P = \tan^2(45° + \phi/2) = \tan^2 64° = 4.20$,

$M_y = 23.487 \times 10^6$ kg-cm,

$\gamma = 8.75$ kN/m³ $\approx 8.75 \times 10^{-4}$ kg/cm³,

$d = 61$ cm.

Substituting,

$$\frac{M_y}{\gamma d^4 K_P} = \frac{23.487 \times 10^6 \times 10^4}{8.75 \times 61^4 \times 4.2} = 462$$

From Fig. 9.16, for $\dfrac{M_y}{\gamma d^4 K_P} = 462$, for $e/d = 0$, we have $\dfrac{P_u}{\gamma d^3 K_P} \approx 80$

Therefore $P_u = 80 \, \gamma d^3 K_P = 80 \times 8.75 \times 0.61^3 \times 4.2 = 667$ kN

Example 9.9

If the pile in Ex. 9.8 is restrained, what is the ultimate lateral resistance of the pile?

Solution

From Fig. 9.16, for $\dfrac{My}{\gamma d^4 K_P} = 462$, the value $\dfrac{P_u}{\gamma d^3 K_P} = 135$.

$$P_u = 136 \, \gamma d^3 K_P = 136 \times 8.75 \times 0.61^3 \times 472 = 1134 \text{ kN}.$$

Exampe 9.10

Compute deflections at ground level by Broms method for the pile given in Ex. 9.2.

Solution

From Eq. (9.86a)

$$\eta = \left(\frac{n_h}{EI}\right)^{\frac{1}{5}} = \left(\frac{6}{4.35 \times 10^2}\right)^{\frac{1}{5}} = 0.424$$

$$\eta L = 0.42 \times 20 = 8.5$$

From Fig. 9.13 (a), for

$$\eta L = 8.5, \ e/L = 0, \text{ we have}$$

$$\frac{y_g \, (EI)^{3/5} \, (n_h)^{2/5}}{P_t L} = 0.2$$

$$y_g = \frac{0.2 \, P_t L}{(EI)^{3/5} \, (n_h)^{2/5}} = \frac{0.2 \times .268 \times 20}{\left(4.35 \times 10^2\right)^{3/5} \, (6)^{2/5}}$$

$$= 0.014 \text{ m} = 1.4 \text{ cm}$$

Example 9.11

If the pile given in Ex. 9.1 is only 4 m long, compute the ultimate lateral resistance of the pile by Broms method.

Solution

From Eq. (9.86a)

$$\eta = \left(\frac{n_h}{EI}\right)^{\frac{1}{5}} = \left(\frac{6}{4.35 \times 10^2}\right)^{\frac{1}{5}} = 0.424$$

$$\eta L = 0.424 \times 4 = 1.696$$

The pile behaves as an infinitely stiff member since $\eta L < 2.0$. $L/d = 4/0.61 = 6.6$
From Fig. 9.13 (b), for $L/d = 6.6$, $e/L = 0$, we have

$$P_u / \gamma d^3 K_p = 25$$

$$\phi = 33°, \gamma = 8.75 \text{ kN/m}^3, \ d = 61 \text{ cm}, \ K_p = \tan^2 (45° + \phi/2) = 3.4$$

Now $\qquad P_u = 20 \ \gamma d^3 K_p = 25 \times 8.75 \times 0.61 \times 3.4 = 454 \text{ kN}.$

If the sand is medium dense as given in Ex. 9.8, then $K_p = 4.20$, the ultimate lateral resistance P_u is

$$P_u = \frac{4.2}{3.4} \times 454 = 561 \text{ kN}$$

As per Ex. 9.8, P_u for long pile = 667 kN, which indicates that the ultimate lateral resistance increases with the length of the pile and remains constant for a long pile.

9.14 A DIRECT METHOD FOR SOLVING THE NON-LINEAR BEHAVIOUR OF LATERALLY LOADED FLEXIBLE PILE PROBLEMS

Key to the Solution

The key to the solution of a laterally loaded vertical pile problem is the development of an equation for n_h. The present state of the art does not indicate any definite relationship between n_h, the properties of the soil, the pile material, and the lateral loads. However, it has been recognised that n_h depends on the relative density of soil for piles in sand and undrained shear strength c for piles in clay. It is well-known that the value of n_h decreases with an increase in the deflection of the pile. It was Palmer et al (1948), who first showed that a change of width d of a pile will have an effect on deflection, moment and soil reaction even while EI is kept constant for all the widths. The selection of an initial value for n_h for a particular problem is still difficult and many times quite arbitrary. The available recommendations in this regard (Terzaghi 1955, and Reese 1975) are widely varying.

The author has been working on this problem since a long time (Murthy, 1965). An explicit relationship between n_h and the other variable soil and pile properties has been developed on the principles of dimensional analysis (Murthy and Subba Rao, 1995).

Development of Expressions for n_h

The term n_h may be expressed as a function of the following parameters for piles in sand and clay.

(a) Piles in sand

$$n_h = f_s (EI, d, P_e, \gamma, \phi) \tag{9.88a}$$

(b) Piles in clay

$$n_h = f_c (EI, d, P_e, \gamma, c) \tag{9.88b}$$

The symbols used in the above expressions have been defined earlier.

In Eqs (9.88a) and (9.88b), an equivalent lateral load P_e at ground level is used in place of P_t acting at a height e above ground level. An expression for P_e may be written from Eq. (9.69) as follows.

$$p_e = P_t (1 + 0.67 \frac{e}{T}) \tag{9.88c}$$

Now the equation for computing groundline deflection y_g is

$$y_g = \frac{2.43 \, P_e T^3}{EI} \tag{9.88d}$$

Based on dimensional analysis the following non-dimensional groups have been established for piles in sand and clay.

Piles in Sand

$$F_n = \frac{n_h P_e^{1/3}}{C_\phi d \, \gamma^{4/3}} \text{ and } F_\varrho = \frac{(EI) \gamma^{1/3}}{d P_e^{4/3}} \tag{9.89a}$$

where C_ϕ = correction factor for the angle of friction ϕ. The expression for C_ϕ has been found separately based on a critical study of the available data. The expression for C_ϕ is

$$C_\phi = 3 \times 10^{-5} (1.316)^{\phi°} \tag{9.89b}$$

Figure 9.17 gives a plot of C_ϕ vs. ϕ.

Piles in Clay

The nondimensional groups developed for piles in clay are

$$F_n = \frac{n_h \sqrt{P_e} \left(1 + e/d\right)^{1.5}}{c^{1.5}} \; ; F_p = \frac{\sqrt{EI\gamma d}}{P_e} \qquad (9.90)$$

In any lateral load test in the field or laboratory, the values of EI, γ, ϕ (for sand) and c (for clay) are known in advance. From the lateral load tests, the ground line deflection curve P_t vs. y_g is known, that is, for any applied load P_t, the corresponding measured y_g is known. The values of $\bar{T}$, n_h and P_e can be obtained from Eqs (9.68), (9.69) and (9.88c) respectively. C_ϕ is obtained from Eq. (9.89b) for piles in sand or from Fig. 9.17. Thus, the right hand side of functions F_n and F_p are known at each load level.

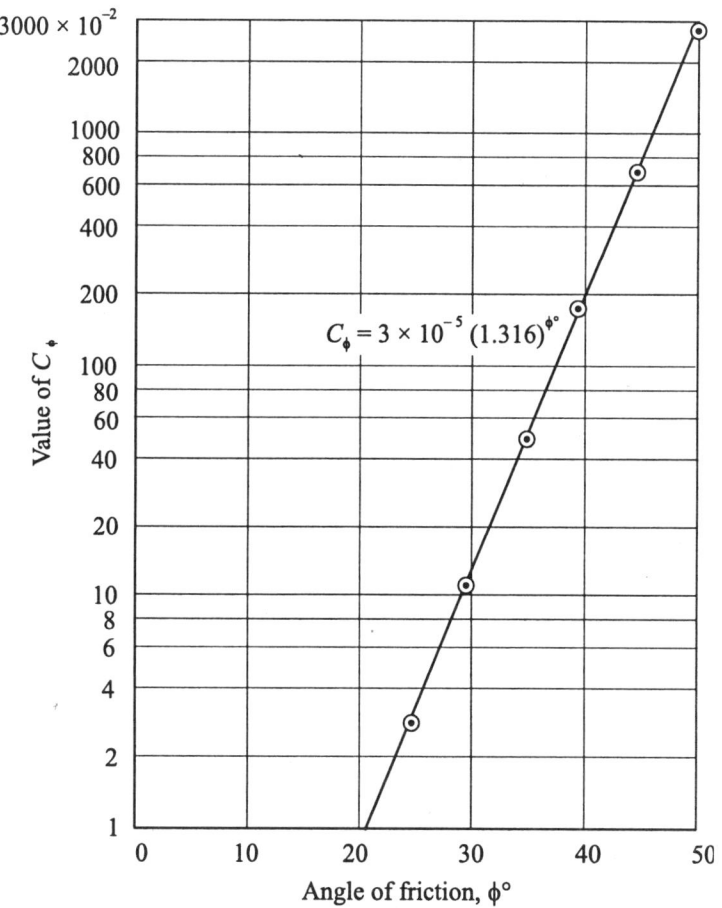

Fig. 9.17 C_ϕ vs. $\phi°$

A large number of pile test data were analysed and plots of $\sqrt{F_n}$ vs. F_p were made on log scale for piles in sand, Fig. 9.18 (a) and F_n vs. F_p for piles in clay, Fig. 9.18 (b). The method of least squares was used to determine the linear trend. The equations obtained are as given below.

Piles in Sand

$$F_n = 150 \sqrt{F_p} \qquad (9.91a)$$

Piles in Clay

$$F_n = 125 F_p \tag{9.91b}$$

Explicit Equations for n_h

By substituting for F_n and F_p, and simplifying, the expressions for n_h for piles in sand and clay are obtained as

for piles in sand,

$$n_h = \frac{150 C_\phi \, \gamma^{1.5} \sqrt{EId}}{P_e} \tag{9.92a}$$

for piles in clay,

$$n_h = \frac{125 c^{1.5} \sqrt{EI\gamma d} \,/\, (1 + e/d)^{1.5}}{P_e^{1.5}} \tag{9.92b}$$

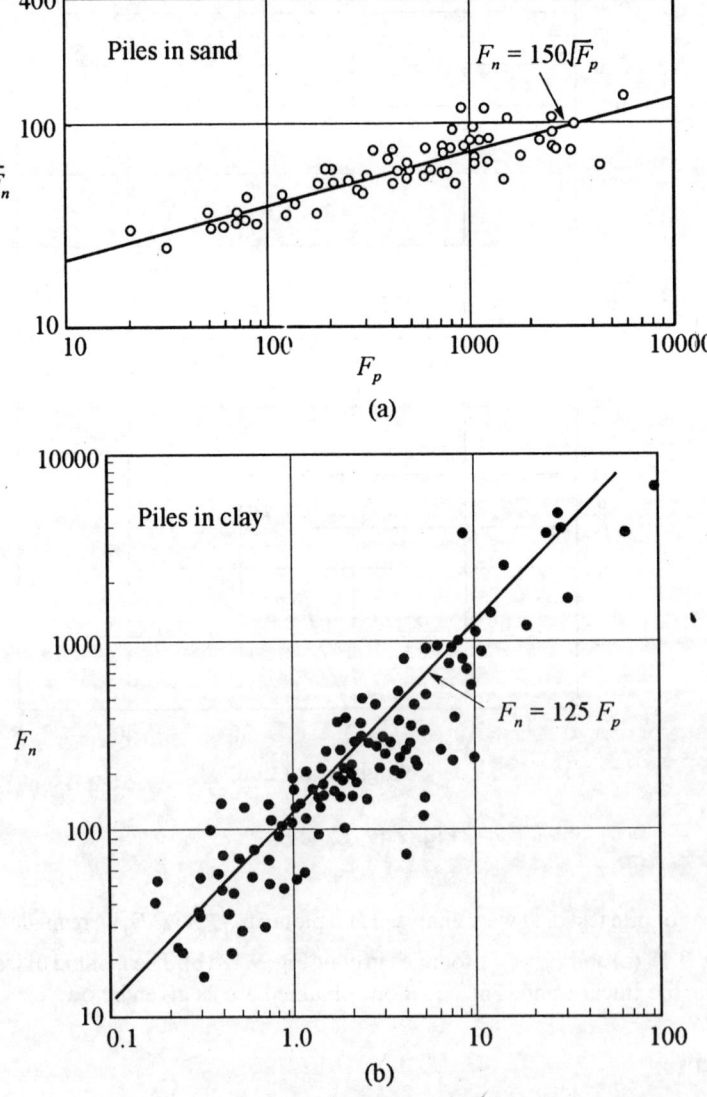

Fig. 9.18 (a) Nondimensional plot for piles in sand, (b) nondimensional plot for piles in clay

It can be seen in the above equations that the numerators in both cases are constants for any given set of pile and soil properties.

The above two equations can be used to predict the non-linear behaviour of piles subjected to lateral loads very accurately.

Equation for P_t at Ground Level where y_g is Known and Vice Versa

For sand

We have the following equations

1. $y_g = \dfrac{2.43\, P_t T^3}{EI}$ ... (a)

2. $n_h = \dfrac{150\, C_\theta\, \gamma^{1.5} \sqrt{EId}}{P_t}$ ... (b)

3. $T = \left(\dfrac{EI}{n_h} \right)^{\frac{1}{5}}$ ... (c)

From the above equations, equations for P_t and y_g may be obtained as given below

$$P_t = 3.65\, C_\phi^{0.4}\, \gamma^{0.6} (EI)^{0.43} d^{0.2}\, y_g^{0.6} \qquad (9.92\text{c})$$

From Eq. (9.92c), we have

$$y_g = \frac{P_t}{8.7\, C_\phi^{0.7}\, \gamma (EI)^{0.72}\, d^{0.3}} \qquad (9.92\text{d})$$

P_t or y_g may be determined from Eqs (9.92c) and (9.92d) without the knowledge of n_h.

Example 9.12

Solve the problem in Ex. 9.1 by the direct method. The soil is loose sand in a submerged condition.

Given: $EI = 4.35 \times 10^{11}$ kg-cm^2 = 4.35×10^5 kN-m^2

                 $d = 61$ cm, $L = 20$ m, $\gamma_b = 8.75$ kN/m^3

                 $\phi = 33°$, $P_t = 268$ kN (since $e = 0$)

Required y_g at ground level.

Solution

For a pile in sand for the case of $e = 0$, use Eq. (9.92a)

$$n_h = \frac{150\, C_\phi\, \gamma^{1.5} \sqrt{EId}}{P_e}$$

For $\phi = 33°$, $C_\phi = 3 \times 10^{-5} (1.316)^{33} = 0.26$ from Eq. (9.89b)

$$n_h = \frac{150 \times 0.26 \times (8.75)^{1.5} \sqrt{4.65 \times 10^5 \times 0.61}}{P_e}$$

$$= \frac{54 \times 10^4}{P_e} = \frac{54 \times 10^4}{268} = 2,015 \text{ kN/m}^3$$

$$T = \left(\frac{EI}{n_h}\right)^{\frac{1}{5}} = \left(\frac{43.5 \times 10^4}{2,015}\right)^{\frac{1}{5}} = 2.93 \text{ m}$$

Now using Eq. (9.88d)

$$y_g = \frac{2.43 \times 268 \times (2.93)^3}{4.35 \times 10^5} = 0.0377 \text{ m} = 3.77 \text{ cm}$$

It may be noted that the direct method gives a greater ground line deflection (= 3.77 cm) as compared to the 1.96 cm in Ex. 9.2.

Example 9.13

Solve the problem in Ex. 9.2 by the direct method. In this case P_t is applied at a height 2 m above ground level. All the other data remain the same.

Solution

From Ex. 9.12

$$n_h = \frac{54 \times 10^4}{P_e}$$

For $P_e = P_t = 268$ kN, we have $n_h = 2,015$ kN/m^3, and $T = 2.93$ m

From Eq. (9.88c)

$$P_e = P_t\left(1 + 0.67\frac{e}{T}\right) = 268\left(1 + 0.67 \times \frac{2}{2.93}\right) = 391 \text{ kN}$$

For $\qquad P_e = 391$ kN, $n_h = \dfrac{54 \times 10^4}{391} = 1,381$ kN/m^3

Now $\qquad T = \left(\dfrac{43.5 \times 10^4}{1,381}\right)^{\frac{1}{5}} = 3.16$ m

As before $\qquad P_e = 268\left(1 + 0.67 \times \dfrac{2}{3.16}\right) = 382$ kN

For $\qquad P_e = 382$ kN, $n_h = 1,414$ kN/m^3, $T = 3.14$ m

Convergence will be reached after a few trials. The final values are

$$P_e = 387 \text{ kN}, \ n_h = 1,718 \text{ kN/m}^3, \ T = 3.025 \text{ m}$$

Now from Eq. (9.88d)

$$y_g = \frac{2.43 \, P_e T^3}{EI} = \frac{2.43 \times 382 \times (3.14)^3}{4.35 \times 10^5} = 0.066 \text{ m} = 6.6 \text{ cm}$$

The n_h value from the direct method is 1,414 kN/m³ whereas from Fig. 9.5 it is 6,000 kN/m³. The n_h from Fig. 9.5 gives y_g which is 50 percent of the probable value and is on the unsafe side.

Example 9.14

Compute the ultimate lateral resistance for the pile given in Ex. 9.6 by the direct method. All the other data given in the example remain the same.

Given: $EI = 4.35 \times 10^5$ kN-m², $d = 61$ cm, $L = 20$ m

$c_u = 85$ kN/m³, $\gamma_b = 10$ kN/m³ (assumed for clay)

$M_y = 2,349$ kN-m; $e = 0$

Required: The ultimate lateral resistance P_u.

Solution

Use Eqs (9.92b) and (9.68)

$$n_h = \frac{125 \, c^{1.5} \sqrt{EI \, \gamma \, d}}{P_t^{1.5}} \quad \text{for } e = 0$$

$$T = \left(\frac{EI}{n_h}\right)^{0.2} \qquad \qquad \text{... (a)}$$

Substituting the known values and simplifying

$$n_h = \frac{1,600 \times 10^5}{P_t^{1.5}} \qquad \qquad \text{... (b)}$$

Step 1

Let $P_t = 1,000$ kN, $n_h = \dfrac{1,600 \times 10^5}{(1,000)^{1.5}} = 5,060$ kN/m³

$$T = \left(\frac{4.35 \times 10^5}{5,060}\right)^{0.2} = 2.437 \text{ m}$$

For $e = 0$, from Table 9.4 and Eq. (9.4) we may write

$M_{\max} = 0.77 \, (P_t T)$

where $A_m = 0.77$ (max) correct to two decimal places.

For $P_t = 1,000$ kN, and $T = 2.437$ m

$M_{\max} = 0.77 \times 1,000 \times 2.437 = 1,876$ kN-m $< M_y$.

Step 2

Let $\quad\quad P_t = 1,500\,\text{kN}$

$\quad\quad\quad n_h = 2,754\,\text{kN/m}^3$ from Eq. (b)

and $\quad\quad T = 2.75\,\text{m}$ from Eq. (a)

Now $\quad\quad M_{max} = 0.77 \times 1,500 \times 2.75 = 3,179\,\text{kN-m} > M_y.$

P_u for $M_y = 2,349$ kN-m can be determined as

$$P_u = 1,000 + (1,500 - 1,000) \times \frac{(2,349 - 1,876)}{(3,179 - 1,876)} = 1,182\,\text{kN}$$

$P_u = 1,100$ kN by Brom's method which agrees with the direct method.

9.15 CASE STUDIES FOR LATERALLY LOADED VERTICAL PILES IN SAND BY DIRECT METHOD

Case 1: Mustang Island Pile Load Test (Reese *et al*, 1974)

Data

Pile diameter, $\quad d = 24$ in, steel pipe (driven pile)

$\quad\quad\quad EI = 4.854 \times 10^{10}\,\text{lb-in}^2$

$\quad\quad\quad L = 69\,\text{ft.}$

$\quad\quad\quad e = 12\,\text{in.}$

$\quad\quad\quad \phi = 39°$

$\quad\quad\quad \gamma = 66\,\text{lb/ft}^3\,(= 0.0382\,\text{lb/in}^3)$

$\quad\quad\quad M_y = 7 \times 10^6\,\text{in-lbs}$

The soil was fine silty sand with WT at ground level.

Required

(a) Load-deflection curve (P_t vs. y_g) and n_h vs. y_g curve

(b) Load-max moment curve (P_t vs. M_{max})

(c) Ultimate load P_u

Solutions

For pile in sand, $\quad n_h = \dfrac{150\,C_\phi\,\gamma^{1.5}\,\sqrt{EId}}{P_e}$

For $\phi = 39°$, $\quad C_\phi = 3 \times 10^{-5}\,(1.316)^{39°} = 1.34$

After substitution and simplifying

$$n_h = \frac{1,631 \times 10^3}{P_e} \quad\quad\quad\quad \text{... (a)}$$

We have $\quad P_e = P_t\left(1 + 0.67\dfrac{e}{T}\right)$ $\quad\quad\quad\quad$ ... (b)

$$T = \left(\frac{EI}{n_h} \right)^{\frac{1}{5}} \qquad \dots (c)$$

(a) Calculation of groundline deflection, y_g

Step 1

Since T is not known to start with, assume $e = 0$, and $P_e = P_t = 10,000$ lbs

Now, from Eq. (a), $n_h = \dfrac{1,631 \times 10^3}{10 \times 10^3} = 163$ lb/in^3

From Eq. (c), $\qquad T = \left(\dfrac{4.854 \times 10^{10}}{163} \right)^{\frac{1}{5}} = 49.5$ in

From Eq. (b) $\qquad P_e = 10 \times 10^3 \left(1 + 0.67 \times \dfrac{12}{49.5} \right) = 11.624 \times 10^3$ lbs

Step 2

For $\qquad P_e = 11.62 \times 10^3$ lb, $n_h = \dfrac{1,631 \times 10^3}{11.624 \times 10^3} = 140$ lb/in^3

As in Step 1 $T = 51$ ins, $P_e = 12.32 \times 10^3$ lbs

Step 3

Continue Step 1 and Step 2 until convergence is reached in the values of T and P_e. The final values obtained for $P_t = 10 \times 10^3$ lb are $T = 51.6$ in, and $P_e = 12.32 \times 10^3$ lbs.

Step 4

The ground line deflection may be obtained from

$$y_g = \frac{2.43 \, P_e T^3}{EI} = \frac{2.43 \times 12.32 \times 10^3 \times (51.6)^3}{4.84 \times 10^{10}} = 0.0845 \text{ in}$$

This deflection is for $P_t = 10 \times 10^3$ lbs. In the same way the values of y_g can be obtained for different stages of loadings. Figure 9.19 (a) gives a plot P_t vs. y_g. Since n_h is known at each stage of loading, a curve of n_h vs. y_g can be plotted as shown in the same figure.

(b) Maximum moment

The calculation under (a) above give the values of T for various loads P_t. By making use of Eq. (9.53) and Table 9.4, moment distribution along the pile for various loads P_t can be calculated. From these curves the maximum moments may be obtained and a curve of P_t vs. $M_{\max}$ may be plotted as shown in Fig. 9.19 (b).

(c) Ultimate load P_u

Figure 9.19 (b) is a plot of $M_{\max}$ vs. P_t. From this figure, the value of P_u is equal to 100 kips for the ultimate pile moment resistance of 7×10^6 in-lb. The value obtained by Broms' method and by computer (Reese, 1985) are 92 and 102 kips respectively.

Comments

Figure 9.19 (a) gives the computed P_t vs. y curve by the direct method and the observed values. There is an excellent agreement between the two. In the same way the observed and the calculated moments and ultimate loads agree well.

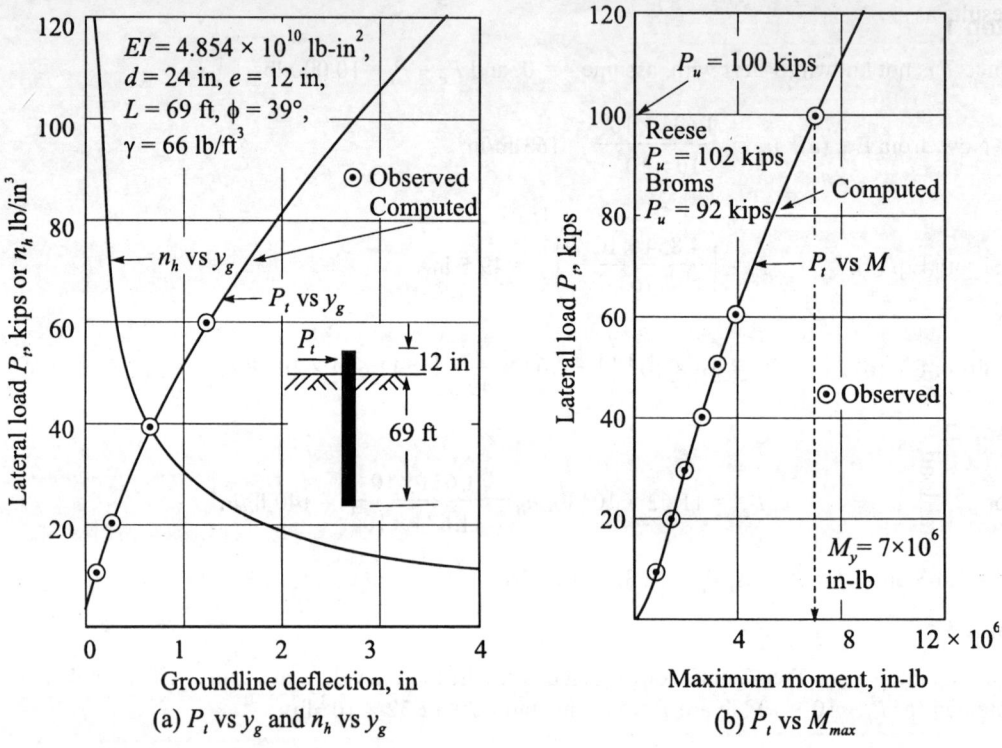

Fig. 9.19 Mustang Island lateral load test: (a) P_t vs. y_g and n_h vs. y_g, (b) P_t vs. M_{max}

Case 2: Florida Pile Load Test (Davis, 1977)

Data

Pile diameter, d = 56 in steel tube filled with concrete,

$\qquad EI$ = 132.5×10^{10} lb-in^2,

$\qquad L$ = 26 ft,

$\qquad e$ = 51 ft,

$\qquad \phi$ = 38°,

$\qquad \gamma$ = 60 lb/ft^3,

$\qquad M_y$ = 4,630 ft-kips.

The soil at the site was medium dense and with water table close to the ground surface.

Required

(a) P_t vs. y_g curve and n_h vs. y_g curve

(b) Ultimate lateral load P_u.

Solution

The same procedure as given for the Mustang Island load test has been followed for calculating the P_t vs. y_g and n_h vs. y_g curves. For getting the ultimate load P_u the P_t vs. M_{max} curve is obtained. The value of P_u obtained is equal to 84 kips which is the same as the ones obtained by Broms (1964) and Reese (1985) methods. There is a very close agreement between the computed and the observed test results as shown in Fig. 9.20.

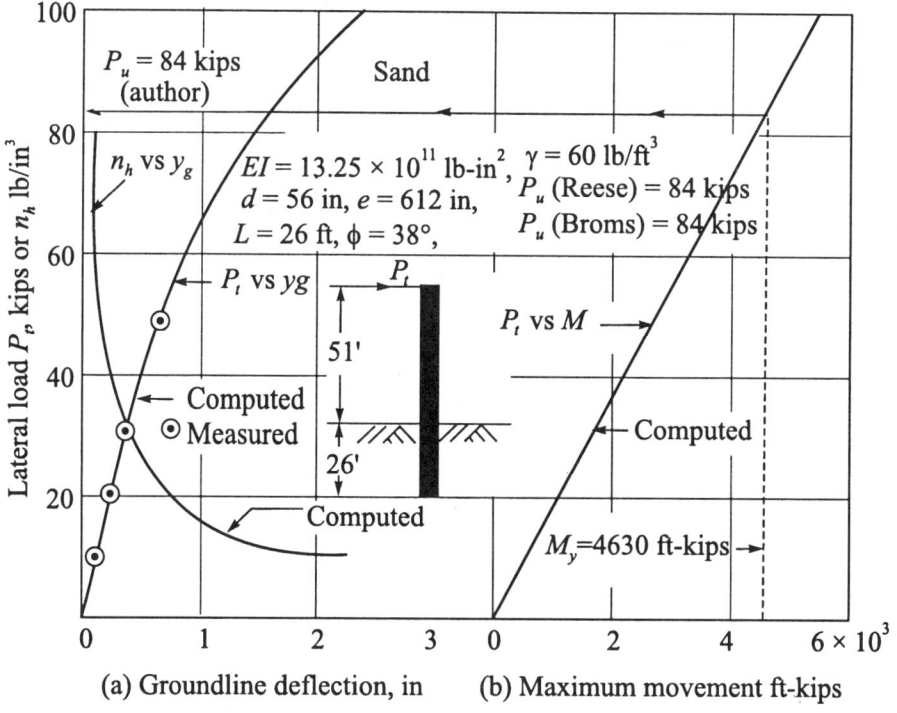

Fig. 9.20 Florida pile test (Davis, 1977)

Case 3: Model Pile Tests in Sand (Murthy, 1965)

Data

Model pile tests were carried out to determine the behaviour of vertical piles subjected to lateral loads. Aluminium alloy tubings, 0.75 in diameter and 0.035 in wall thickness, were used for the test. The test piles were instrumented. Dry clean sand was used for the test at a relative density of 67%. The other details are given in Fig. 9.21.

Solution

Figure 9.21 gives the predicted and observed

(a) Load-ground line deflection curve.

(b) Deflection distribution curves along the pile.

(c) Moment and soil reaction curves along the pile.

There is an excellent agreement between the predicted and the observed values. The direct method has been used.

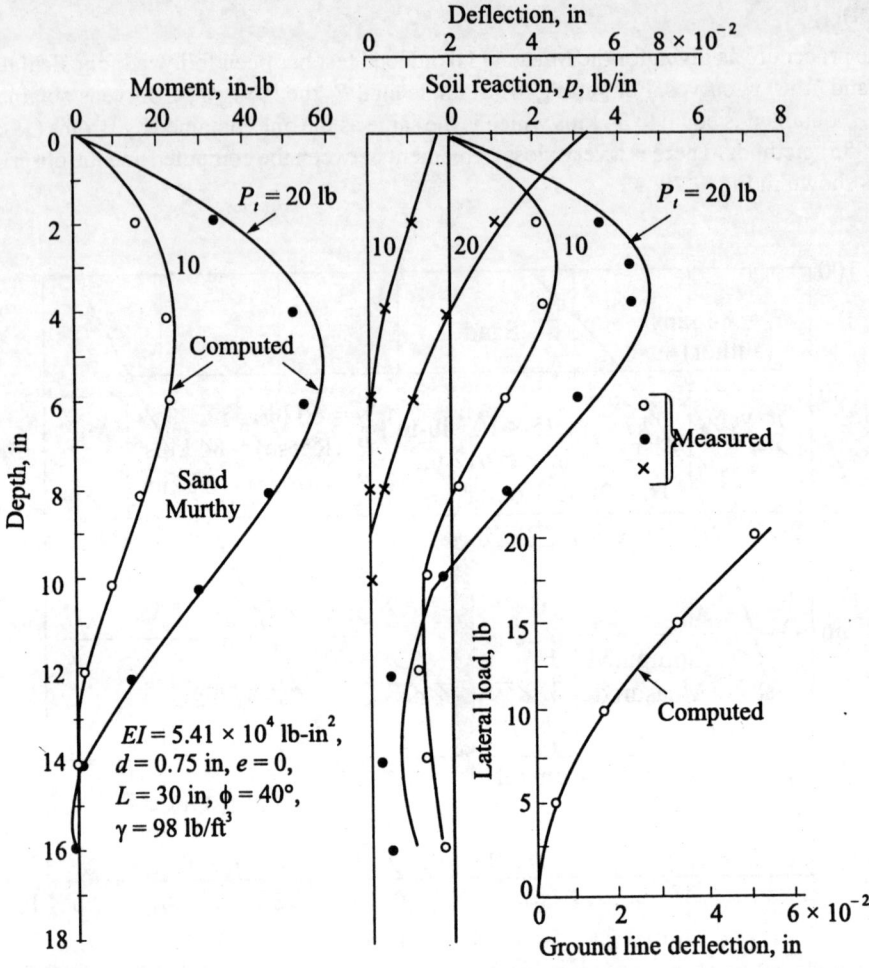

Fig. 9.21 Curves of bending moment, deflection and soil reaction for a model pile in sand
(Murthy, 1965)

9.16 CASE STUDIES FOR LATERALLY LOADED VERTICAL PILES IN CLAY BY DIRECT METHOD

Case 1: Pile Load Test at St. Gabriel (Capazzoli, 1968)

Data

Pile diameter, $d = 10$ in, steel pipe filled with concrete,

$EI = 38 \times 10^8$ lb-in^2,

$L = 115$ ft.,

$e = 12$ in.,

$c = 600$ lb/ft^2,

$\gamma = 110$ lb/ft^3,

$M_y = 116$ ft-kips.

Water table was close to the ground surface.

Required

(a) P_t vs. y_g curve

(b) the ultimate lateral load, P_u

Solution

We have,

$$(a) \; n_h = \frac{125 \, c^{1.5} \, \sqrt{EI \, \gamma \, d}}{(1 + e/d)^{1.5} \, P_e^{1.5}} \qquad (b) \; P_e = P_t \left(1 + 0.67 \frac{e}{T}\right) \qquad (c) \; T = \left(\frac{EI}{n_h}\right)^{\frac{1}{5}}$$

After substituting the known values in Eq. (a) and simplifying, we have

$$n_h = \frac{16,045 \times 10^3}{P_e^{1.5}}$$

(a) Calculation of groundline deflection

1. Let $P_e = P_t = 500$ lbs

   From Eqs (a) and (c), $n_h = 45$ lb/in^3, $T = 38.51$ in

   From Eq. (b), $P_e = 6,044$ lb.

2. For $P_e = 6,044$ lb, $n_h = 34$ lb/in^3 and $T = 41$ in

3. For $T = 41$ in, $P_e = 5,980$ lb, and $n_h = 35$ lb/in^3

4. For $n_h = 35$ lb/in^3, $T = 40.5$ in, $P_e = 5,988$ lb final values.

5. $y_g = \dfrac{2.43 \, P_e T^3}{EI} = \dfrac{2.43 \times 5,988 \times (40.5)^3}{38 \times 10^8} = 0.25$ in

6. Continue steps 1 through 5 for computing y_g for different loads P_t. Figure 9.22 gives a plot of P_t vs. y_g which agrees very well with the measured values.

(b) Ultimate load P_u

A curve of M_{max} vs. P_t is given in Fig. 9.22 following the procedure given for the Mustang Island Test. From this curve $P_u = 23$ k for $M_y = 116$ ft kips. This agrees well with the values obtained by the methods of Reese (1985) and Broms (1964a).

Case 2: Pile Load Test at Ontario (Ismael and Klym, 1977)

Data

Pile diameter, $\quad d = 60$ in, concrete pile (Test pile 38)

$$EI = 93 \times 10^{10} \text{ lb-in}^2,$$

$$L = 38 \text{ ft},$$

$$e = 12 \text{ in.},$$

$$c = 2,000 \text{ lb/ft}^2,$$

$$\gamma = 60 \text{ lb/ft}^3,$$

The soil at the site was heavily overconsolidated.

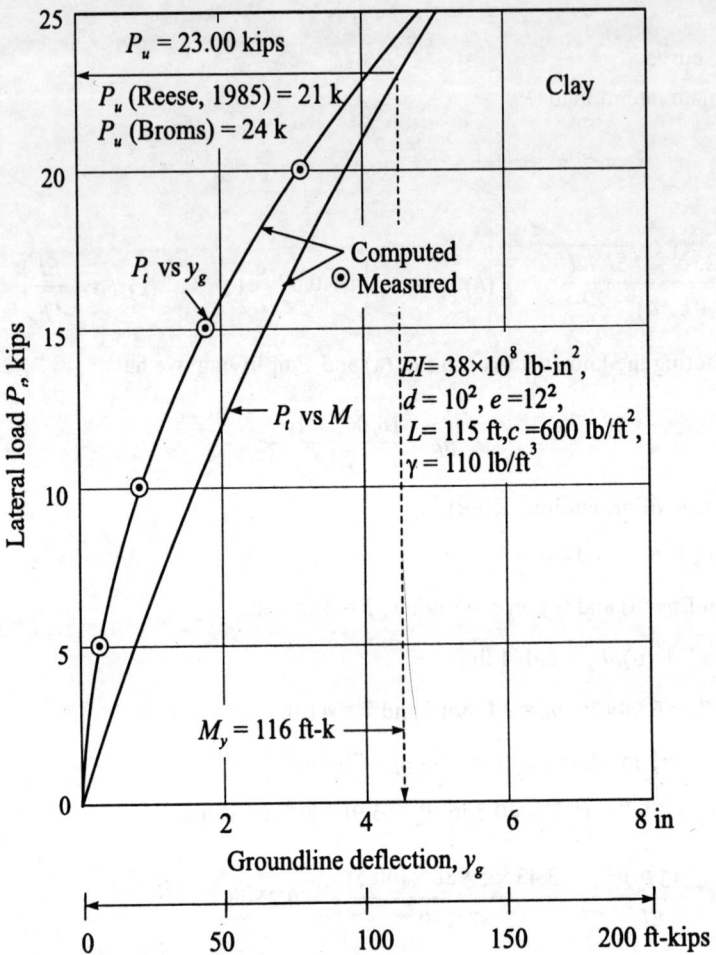

Fig. 9.22 St. Gabriel pile load test in clay

Required

(a) P_t vs. y_g curve.

(b) n_h vs. y_g curve.

Solution

By substituting the known quantities in Eq. (9.92b) and simplifying, we have

$$n_h = \frac{68{,}495 \times 10^5}{P_e^{1.5}}, \quad T = \left(\frac{EI}{n_h}\right)^{\frac{1}{5}}, \quad \text{and } P_e = P_t \left(1 + 0.67 \frac{e}{T}\right)$$

Follow the same procedure as given for Case 1 to obtain values of y_g for the various loads P_t. The load deflection curve can be obtained from the calculated values as shown in Fig. 9.23 measured values are also plotted. It is clear from the curve that there is a very close agreement between the two. The figure also gives the relationship between n_h and y_g.

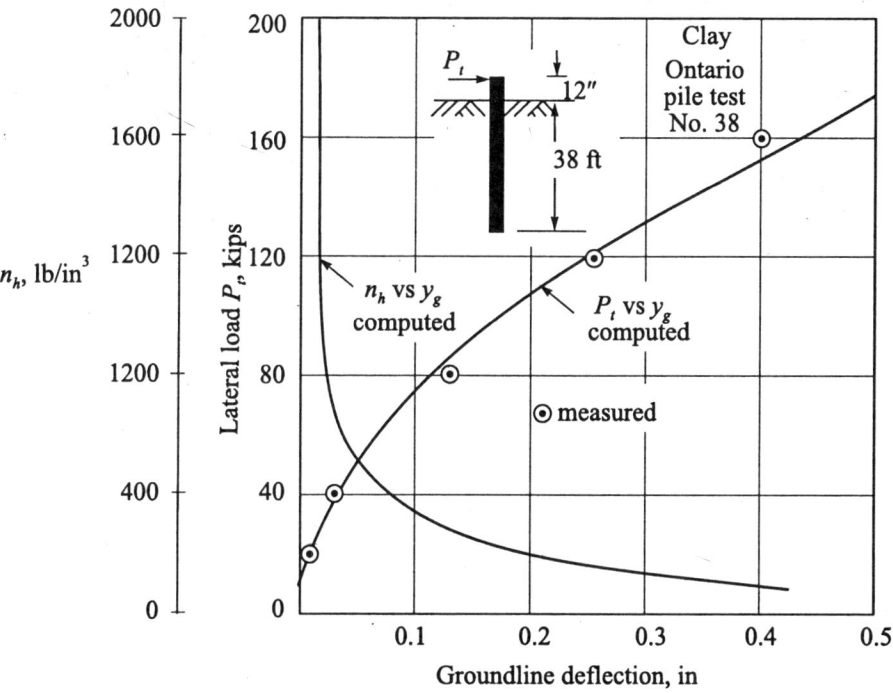

Fig. 9.23 Ontario pile load test (38)

Case 3: Restrained Pile at the Head for Offshore Structure (Matlock and Reese, 1961)

Data

The data for the problem are taken from Matlock and Reese (1961). The pile is restrained at the head by the structure on the top of the pile. The pile considered is below the sea bed. The undrained shear strength c and submerged unit weights are obtained by working back from the known values of n_h and T. The other details are

$$\text{Pile diameter,} \qquad d = 33 \text{ in, pipe pile,}$$

$$EI = 42.35 \times 10^{10} \text{ lb-in}^2,$$

$$c = 500 \text{ lb/ft}^2,$$

$$\gamma = 40 \text{ lb/ft}^3,$$

$$P_t = 150,000 \text{ lbs,}$$

$$(a) \ \frac{M_t}{P_t T} = \frac{-T}{12.25 + 1.078T}, \qquad (b) \ T = \left(\frac{EI}{n_h}\right)^{\frac{1}{5}}, \qquad (c) \ P_e = P_t \left(1 - 0.67\frac{e}{T}\right)$$

Required

(a) Deflection at the pile head.

(b) Moment distribution diagram.

Solution

Substituting the known values in Eq. (9.92b) and simplifying,

$$n_h = \frac{458 \times 10^6}{P_e^{1.5} (1 + e/d)^{1.5}} \qquad \ldots \text{(d)}$$

Calculations

1. Assume $e = 0$, $P_e = P_t = 150{,}000$ lb

   From Eqs (d) and (b) $n_h = 7.9$ lb/in^2, $T = 140$ in

   From Eq. (a) $\quad \dfrac{M_t}{P_t T} = \dfrac{-140}{12.25 + 1.078 \times 140} = -0.858$

   or $\qquad\qquad M_t = -0.858\, P_t T = P_t e$

   Therefore $\qquad e = 0.858 \times 140 = 120$ in

2. $P_e = P_t \left(1 - 0.67 \dfrac{e}{T}\right) = 1.5 \times 10^5 \left(1 - 0.67 \times \dfrac{120}{140}\right) = 63{,}857$ lb

   $$\left(1 + \frac{e}{d}\right)^{1.5} = \left(1 + \frac{120}{33}\right)^{1.5} = 10$$

   Now from Eq. (d), $n_h = 2.84$ lb/in^3, from Eq. (b) $T = 171.64$ in

   After substitution in Eq. (a)

   $$\frac{M_t}{P_t T} = -0.875, \text{ and } e = 0.875 \times 171.64 = 150.2 \text{ in}$$

   $$P_e = \left(1 - 0.67 \times \frac{150.2}{171.64}\right) \times 1.5 \times 10^5 = 62{,}205 \text{ lbs}$$

3. Continuing this process for a few more steps there will be convergence of values of n_h, T and P_e. The final values obtained are

   $$n_h = 2.1 \text{ lb/in}^3, \ T = 182.4 \text{ in, and } P_e = 62{,}246 \text{ lb}$$

   $$M_t = -P_t e = -150{,}000 \times 150.2 = -22.53 \times 10^6 \text{ lb-in}^2$$

   $$y_g = \frac{2.43\, P_e T^3}{EI} = \frac{2.43 \times 62{,}246 \times (182.4)^3}{42.35 \times 10^{10}} = 2.17 \text{ in}$$

Moment distribution along the pile may now be calculated by making use of Eq. (9.53) and Table 9.4. Please note that M_t has a negative sign. The moment distribution curve is given in Fig. 9.24. There is a very close agreement between the computed values by direct method and the Reese and Matlock method. The deflection and the negative bending moment as obtained by Reese and Matlock are

$$y_m = -2.307 \text{ in and } M_t = -24.75 \times 10^6 \text{ lb-in}^2$$

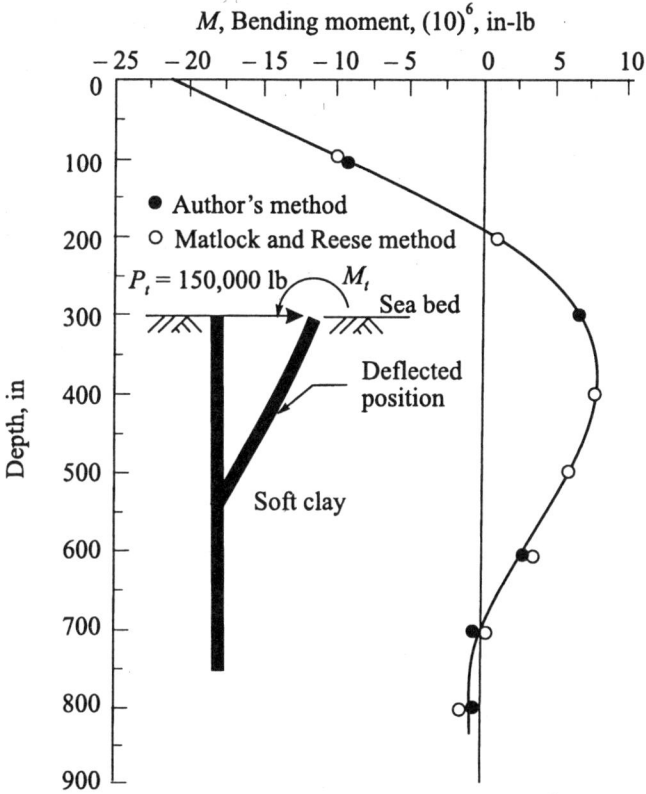

Fig. 9.24 Bending moment distribution for an offshore pile supported structure
(Matlock and Reese, 1961)

9.17 *p-y* CURVES FOR THE SOLUTION OF LATERALLY LOADED PILES

Introduction

Section 9.8 explains the methods of computing deflection, slope, moment shear and soil reaction by making use of equations developed by non-dimensional methods. The prediction of the various curves depend primarily on one single parameter, n_h, which is called as *coefficient of soil modulus variation.* It has been explained earlier that this parameter is a function of the property of soil (γ the unit weight of soil if the soil happens to be sand, c the cohesive strength of soil if the soil is clay), EI of the pile material, width d of the pile and P_t, the lateral load applied on the pile. If it is possible to get the value of n_h independently for each stage of loading P_t, the *p-y* curves at different depths along the pile can be constructed as follows:

1. Determine the value of n_h for a particular stage of loading P_t.

2. Compute T from Eq. (9.68) for the linear variation of E_s with depth.

3. Compute y at specific depths $x = x_1$, $x = x_2$, etc. along the pile by making use of Eq. (9.51), where A and B parameters can be obtained from Table 9.4 for various depth coefficients Z.

4. Compute p by making use of Eq. (9.55) since T is known, for each of the depths $x = x_1$, $x = x_2$, etc.

5. Since the values of p and y are known at each of the depths x_1, x_2, etc. one point on the *p-y* curve at each of these depths is also known.

6. Repeat steps 1 through 5 for different stages of loading and obtain the values of p and y for each stage of loading and plot to get p-y curves at each depth.

The individual p-y curves we get by the above procedure at depths x_1, x_2, etc. can be plotted on a common pair of axes to give a family of curves for the selected depths below the surface. A typical family of p-y curves is given in Fig. 9.25. These curves have been developed by the author for Mustang Island lateral load pile tests (Reese *et al,* 1974). The tests were conducted on 24 in pipe piles embedded in sand. The piles were instrumented along their lengths for the measurement of bending moments. Slopes, moments, and shears along the pile can also be obtained by the same procedure mentioned above for different stages of loading.

The p-y curves shown in Fig. 9.25 are strongly non-linear and these curves can be predicted only if values of n_h are known for each stage of loading. Further, the curves can be extended till the soil reaction, p, reaches ultimate value, p_u, up to a specific depth below the ground surface.

If n_h values are not known to start with at different stages of loading, the above method cannot be followed. Supposing, p-y curves given in Fig. 9.25 can be constructed by some other independent method, then p-y curves are the starting points to obtain the curves of deflection, slope, moment and shear, that means, we are proceeding in the reverse direction in the above method. The methods of constructing p-y curves and predicting the non-linear behaviour of laterally loaded piles are explained in the subsequent sections.

Both cohesive and cohesionless soils are considered. For cohesive soils, the $\phi = 0$ concept is usually employed and deformation of the soil-pile system is assumed to occur under undrained

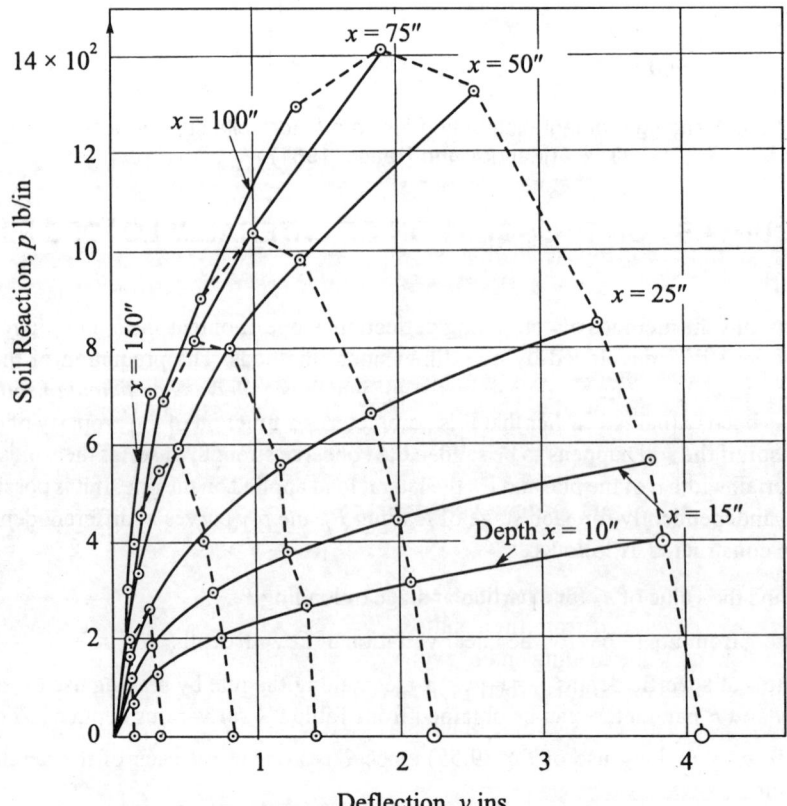

Fig. 9.25 (*p-y*) curves for the lateral load tests at Mustang Island, USA (Murthy, 1990 not published)

conditions. For cohesionless soils, effective strength parameters are used and it is assumed that the soil-pile system deforms under drained condition.

Construction of *p-y* Curves for Clays and Sands

Introduction

Reese and his coworkers have developed methods for constructing *p-y* curves for both clay and sand which can be used for predicting the non-linear behaviour of laterally loaded piles either by the use of difference equation method by making use of the differential Eq. (9.11) or by making use of the non-dimensional solutions discussed in Section 9.8. According to them, the *p-y* curves are independent of shape and stiffness of the pile and represent the deformation of a discrete vertical area of the soil that is unaffected by loading above and below it. Though this assumption is not strictly true, Reese *et al* (1975), confirm as per their experimental investigation that the soil reaction at a point is dependent essentially on the pile deflection at that point, and not on pile deflections above and below. The following types of clays are considered for the construction of *p-y* curves.

1. Soft clays below water table.
2. Stiff clays below water table.
3. Stiff clays above water table.

Both short-term static loads and repeated (cyclic) loads have been considered. Repeated loadings of a clay has pronounced effect on the soil response, particularly when the soil is submerged. The loss of resistance from repeated loading is due to two effects:

1. The breakdown of the structure of the clay (remoulding).
2. The scour.

The remoulding is a result of the repeated strains that occur due to the deflection of the pile. The scour occurs when the pile deflects enough to cause a gap to remain between pile and soil when the load is removed. Water will flow into the gap and will be ejected on the next application of the load. The water in most cases will move out at a high velocity and carry out particles of clay.

If the clay is above the water table only the first of the two effects will be present. Therefore, the position of the water table will have an effect on *p-y* curves.

The criteria for obtaining *p-y* curves for static loading consist mainly of two parts.

1. To obtain an expression to describe the variation of ultimate soil resistance, p_u, with depth.
2. To obtain an expression to describe the variation of soil resistance with deflection at any particular depth along the pile. The methods of construction of *p-y* curves for both static and cyclic loadings have been explained in the subsequent sections of this chapter.

p-y Curves for Soft Clays below Water Table

Static loading

Matlock (1970) developed a procedure for prediction *p-y* curves in a soft, submerged clay deposit. This procedure was developed from the results of tests on fully instrumented flexible pipe piles subjected to short-term static loading and to cyclic loading. Correlations were made with results of field and laboratory tests on "Undisturbed" soil samples obtained from the test sites.

The ultimate soil resistance, p_u can be obtained by using the equation

$$p_u = N_p c_x d \tag{9.93a}$$

where N_p = normalised ultimate soil resistance,

c_x = undrained shear strength at depth x,

d = pile width or diameter.

The value of N_p has been found to be a function of depth below the ground surface. The value of N_p increases with depth until it reaches some limiting value, at which it remains constant at greater depths. For shallow depths, Eq. (9.93a) may be written as

$$p_u = \left(3 + \frac{\gamma' x}{c_x} + \frac{Jx}{d}\right) c_x d \qquad (9.93b)$$

and for deeper depths

$$p_u = 9c_x d \qquad (9.93c)$$

where p_u = ultimate soil resistance per unit length of pile,

γ' = average effective unit weight of soil from ground surface to p-y curve,

x = depth from ground surface to p-y curve,

J = 0.50 for all practical purposes.

The smaller of the values of p_u given by Eqs (9.93b) and (9.93c) has to be used.

Shape of *p-y* curve

To obtain the shape of the p-y curve, a mathematical expression was selected which fitted with the experimental p-y curves. Matlock selected the following equation

$$\frac{p}{p_u} = 0.5 \left(\frac{y}{y_{50}}\right)^{0.33} \qquad (9.93d)$$

where, $y_{50} = 2.5b\varepsilon_{50}$, $\qquad (9.94)$

ε_{50} = strain at 50% of the maximum stress difference, determined from a UU triaxial compression test.

In an undrained triaxial compression test the axial strain is given by the expression

$$\varepsilon = \frac{\sigma_1 - \sigma_3}{E}$$

where, E is the Young's modulus at the stress $(\sigma_1 - \sigma_3)$. If $(\sigma_1 - \sigma_3)f$ is *the failure stress*, ε_{50} is the strain at 50 percent of the failure stress.

The value of p is assumed to remain constant beyond $y = 8y_{50}$. A non-dimensional p-y curve for static loading is shown in Fig. 9.26 (a). A detailed study of unconfined compression tests conducted on undisturbed samples of clay has indicated that the stress-strain relationships of the laboratory test curves are comparable with that of piles under lateral loading.

Cyclic loading

The effects of cyclic loading are to decrease the ultimate soil resistance to $0.72\, p_u$ and to reduce the soil resistance at deflections greater than $3y_{50}$ at depths less than x_r, where

$$x_r = \frac{6d}{\dfrac{\gamma' d}{c_x} + 0.5} \qquad (9.95a)$$

where, x_r = critical depth or depth of transition.

The value of p reduces from $0.72p_u$ at $y = 3y_{50}$ to

$$p = 0.72p_u \left(\frac{x}{x_r}\right) \qquad\qquad (9.95b)$$

at $y = 15y_{50}$. The value of p remains constant beyond $y = 15y_{50}$.

The shape of the cyclic p-y curve is given in Fig. 9.26 (b).

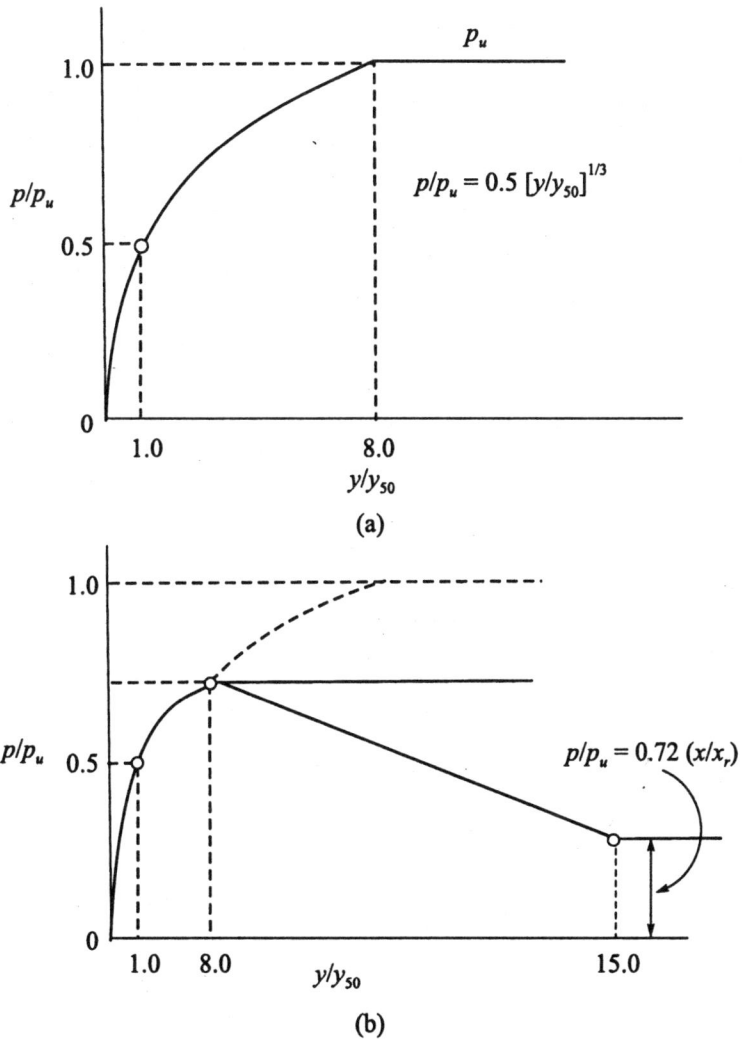

Fig. 9.26 Characteristic shapes of the p-y curves for soft clay below water table: (a) Static loading, (b) cyclic loading (Matlock, 1970)

p-y Curves for Stiff Clays below Water Table

Reese *et al* (1975), performed tests on fully instrumented pipe piles embedded in a submerged, heavily overconsolidated clay deposit. On the basis of test results, they developed criteria for the development of *p-y* curves. For stiff clays, they considered that the clay close to ground surface fail in a wedge shape and the soil moves up. At depths greater than a critical depth the soil moves around the pie. In both the cases the movement of the pile should be adequate to cause failure of soil. The two theoretical expressions developed by them are

$$p_{c1} = 2c_a d + \gamma' dx + 2.83 c_a x \tag{9.96a}$$

$$p_{c2} = 11cd \tag{9.96b}$$

where, p_{c1} = ultimate soil resistance near the surface,

p_{c2} = ultimate soil resistance well below the ground surface,

c_a = average shear strength over the depth *x*,

γ' = submerged unit weight,

d = pile diameter or width.

The least of the two p'_s given by Eqs (9.96a) and (9.96b) is to be used in the computation.

Construction of *p-y* curve under static loading for any depth *x* along the pile

Figure 9.26 (c) shows the shape of the curve as conceived by Reese *et al*. The curve *oabncde* can be constructed as follows.

1. *oa* is a straight line portion which can be established by making use of the relationship

$$p = (n_h x)\, y \tag{9.97a}$$

where, n_h = coefficient of modulus variation which can be selected from Table 9.6 for static loading.

Table 9.6 Representative values of n_h for stiff clays (After Reese et al, 1975)

	Average undrained shear strength		
	50–100	*100–200*	*200–400 kPa*
n_h (static) kN/m^3	1250	2500	5000
n_h (cyclic) kN/m^3	500	1000	2000

Note: The average shear strength be computed from the shear strength of the soil to a depth of 5 pile diameters. It should be defined as half the total maximum principal stress difference in an unconsolidated undrained triaxial test.

2. The parabolic portion *ab* of the curve may be established by the following equation.

$$p = 0.5 p_c \left(\frac{y}{y_{50}} \right)^{0.5} \tag{9.97b}$$

where, $y_{50} = \varepsilon_{50} d$ \hfill (9.97c)

and p_c is the least of the *p* from Eqs (9.96a) and (9.96b).

Use an appropriate value of ε_{50} from results of laboratory tests or from Table 9.7. The abscissa of point *b* is $A_s y_{50}$ where, A_s is an empirical adjustment factor which can be obtained from

Table 9.7 Representative values of ε_{50}

c, lb/ft^2	$\varepsilon_{50}\%$
250 – 500	2
500 – 1000	1
1000 – 2000	0.7
2000 – 4000	0.5
4000 – 8000	0.4

Fig. 9.26 (c) for static loading. Equation (9.97b) defines the portion of the *p-y* curve from the point of intersection with Eq. (9.97a) to a point where $y = A_s y_{50}$.

3. The parabolic curve *bnc* extends from point *b* whose abscissa is $A_s y_{50}$ to point *c* whose abscissa is $6A_s y_{50}$. The curve *bnc* can be constructed by dropping vertical offsets from the extended portion of the parabolic curve *ab* whose equation is the same as Eq. (9.97b). The equation to the curve *bnc* is

$$p = 0.5p_c \left(\frac{y}{y_{50}}\right)^{0.5} - 0.055p_c \left(\frac{y - A_s y_{50}}{A_s y_{50}}\right)^{1.25} \tag{9.97d}$$

At any point *m* on the parabolic curve bmc_1, the length of the offset *mn* is

$$mn = -0.055p_c \left(\frac{y - A_s y_{50}}{A_s y_{50}}\right)^{1.25} \tag{9.97e}$$

4. The straight line portion of the curve *cd* may be established by making use of the equation

$$p = 0.5p_c (6A_s)^{0.5} - 0.411p_c - \frac{0.0625}{y_{50}} \times p_c (y - 6A_s y_{50}) \tag{9.97f}$$

where, the abscissa of the point *d* is equal to $18A_s y_{50}$.

5. The final straight-line portion *de* of the *p-y* curve can be established by making use of the equation

$$p = p_c \left(1.225 \sqrt{A_s} - 0.75 A_s - 0.411\right) \tag{9.97g}$$

Note: The step-by-step procedure is outlined and Fig. 9.26 (a) is drawn as if there is an intersection between Eqs (9.97a) and (9.97b). However, there may be no intersection of Eq. (9.97a) with any of the other equations or if no intersection occurs, Eq. (9.97a) defines the complete *p-y* curve.

Construction of *p-y* curve for cyclic loading

1. For cyclic loading also the least of the *p's* as given by Eqs (9.96a) and (9.96b) is to be used.

2. The initial portion of the curve *oa* is a straight line as for static loading.

3. Choose the appropriate value of A_c from Fig. 9.26 (c) for cyclic loading for the particular non-dimensional depth x/d.

4. Compute $y_p = 4.1\, A_s y_{50}$ (9.97h)

5. The parabolic portion of the curve abc on the p-y curve can be established by making use of the equation

$$p = A_c P_c \left[1 - \left(\frac{y - 0.45\, y_p}{0.45\, y_p} \right)^{2.5} \right]$$ (9.97i)

where, the abscissa of point c is equal to $0.6 y_p$.

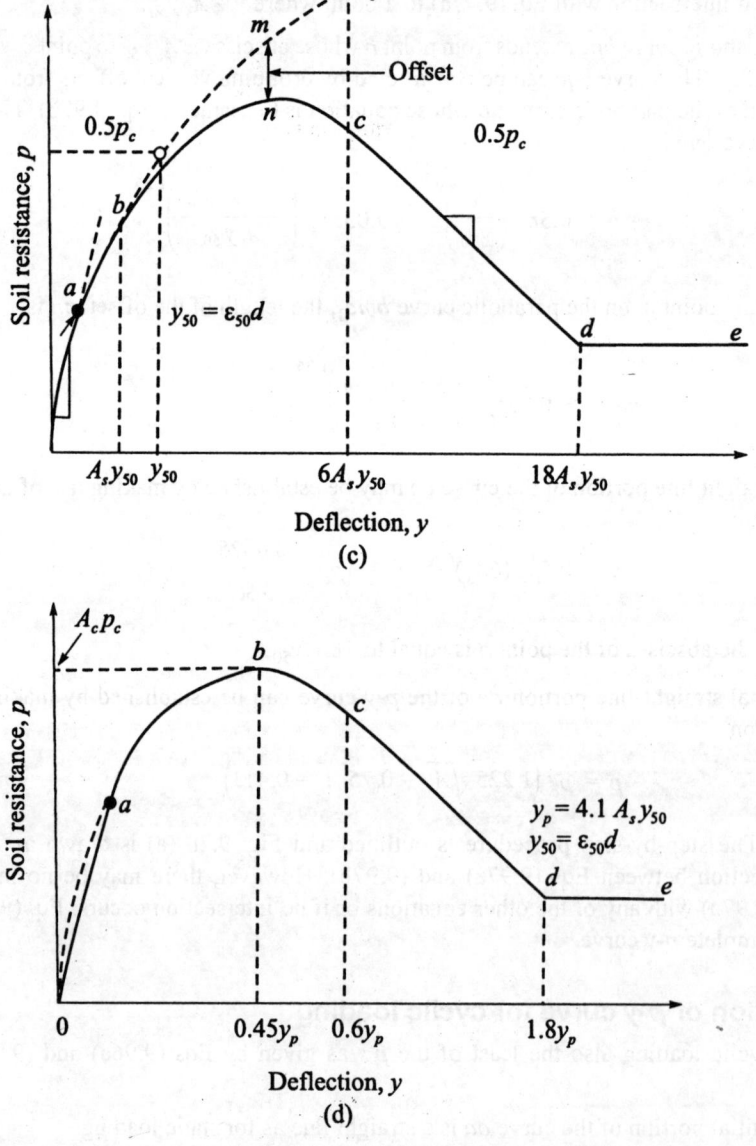

Fig. 9.26 Characteristic shapes of p-y curves for stiff clays below water table: (c) Static loading (d) cyclic loading (Reese *et al*, 1975)

6. Establish the straight line portion *cd* of the curve by

$$p = 0.936 A_c p_c - \frac{0.085}{y_{50}} p_c (y - 0.6 y_p) \qquad (9.97j)$$

where, the abscissa of point *d* is $1.8 y_p$.

7. Establish the final straight-line portion *de* of the curve by

$$p = 0.936 A_c p_c - \frac{0.102}{y_{50}} p_c y_p \qquad (9.97h)$$

Equation (9.97h) applies for all values of *y* of greater than $1.8 y_p$.

Figure 9.26 (b) is drawn on the assumption that there is an intersection between Eqs (9.97a) and (9.97i). If there is no intersection between the two equations or any other equation defining the *p-y* curve with Eq. (9.97a), then the equation should be employed that gives the smallest value of *p* for any value of *y*.

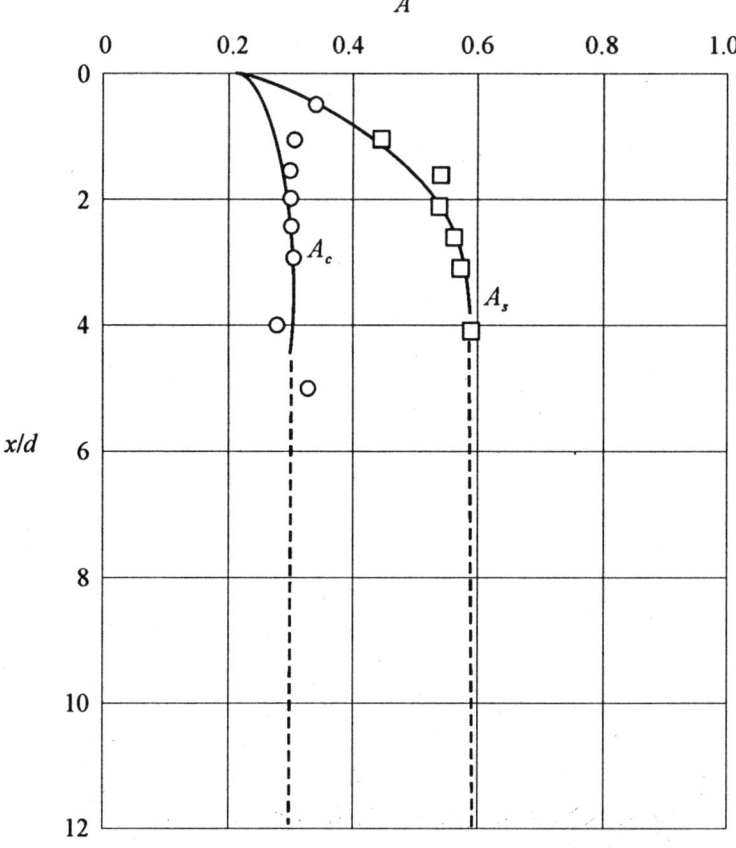

Fig. 9.26 (e) Values of constants A_s and A_c (Reese *et al* 1975)

p-y Curves for Stiff Clays above Water Table

Reese and Welch (1975) proposed criteria for predicting the behaviour of flexible piles in stiff clays above the water table. The following procedure is adopted for constructing p-y curves.

p-y curves for static loading

1. Obtain the values of c, γ, and d as before. Obtain the values of ε_{50} from stress–strain curves or from Table 9.6.
2. Compute p_u from Eqs (9.96a) or (9.96b). The smaller of the two is used.
3. Compute the deflection y_{50} from Eq. (9.94).
4. Points describing the p-y curve may be computed from the relationship

$$\frac{p}{p_u} = 0.5 \left(\frac{y}{y_{50}}\right)^{0.25} \tag{9.98a}$$

5. Beyond $y = 16 y_{50}$, $p = p_u$ for all values y.

Figure 9.27 (a) gives the shape of p-y curve for static loading in stiff clays above water table.

p-y curves for cyclic loading

Figure 9.27 (b) shows the curves under cyclic loading. The procedure for the construction is:

1. First determine the p-y curve for short-term static loading.
2. Determine the number of times the design lateral load will be applied to the pile.
3. For several values of p/p_u obtain the value of C, the parameter describing the effect of repeated loading or deformation from the equation (Welch and Reese, 1972)

$$C = 9.6 \left(\frac{p}{p_u}\right)^4 \tag{9.98b}$$

4. At the value of p corresponding to the value of p/p_u selected in step 3 above, compute new values of y for cyclic loading from equation

$$y_c = y_s + y_{50} \, C \log N \tag{9.99}$$

where, y_c = deflection under N cycles of load,
y_s = deflection under short-term static load,
y_{50} = deflection under short-term static load at one-half the ultimate resistance,
N = number of cycles of load application.

5. The p-y curve defines the soil response after N-cycles of load [Fig. 9.27 (b)].

Construction of p-y Curves for Sand

Reese *et al* (1974) proposed criteria for cohesionless soils for analysing the behaviour of piles under static and cyclic loadings. The procedures were developed from the results of tests at Mustang Island on 24 inch diameter flexible pipe piles embedded in a deposit of submerged, dense fine sand. The following procedure were recommended for the construction of p-y curves under both static and cyclic loading conditions.

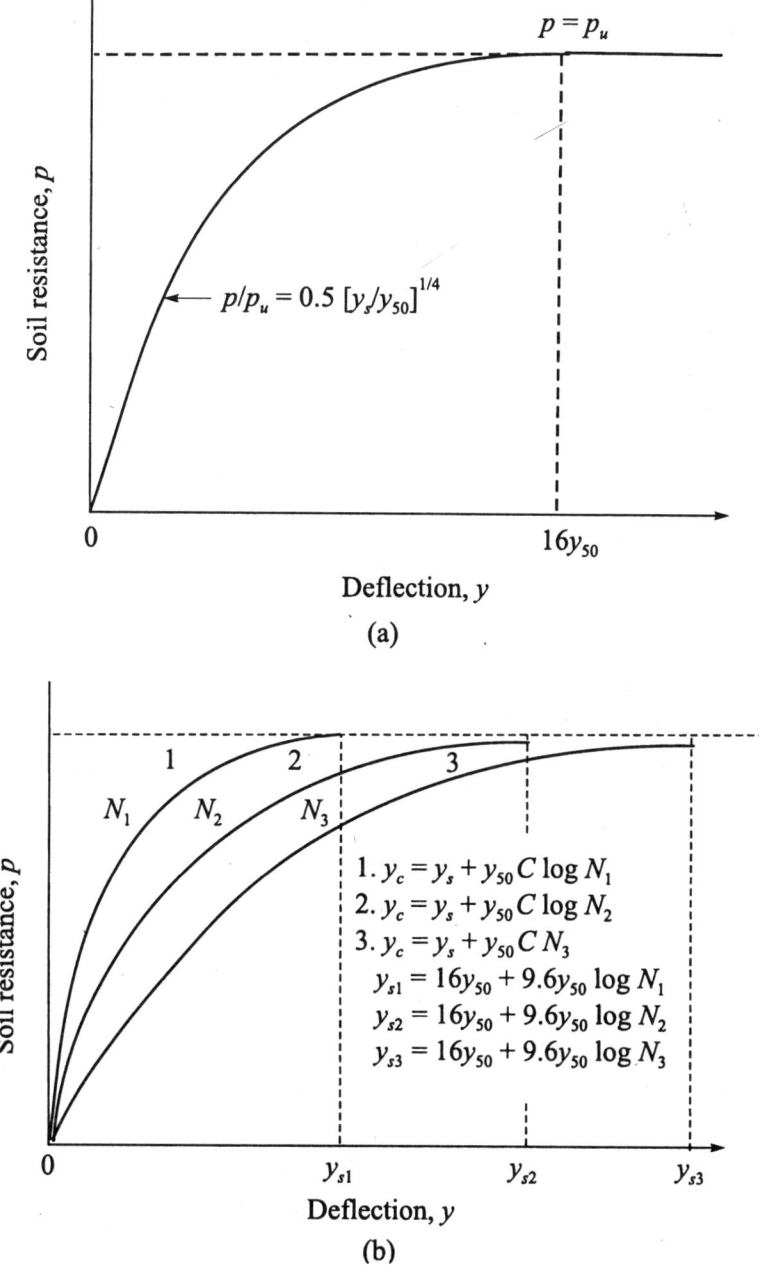

Fig. 9.27 Characteristic shapes of *p-y* curves for stiff clays above water table: (a) Static loading, (b) cyclic loading (Reese and Welch, 1975)

p-y curves under short-term static and cyclic loadings

1. Obtain the values of the angle of internal friction ϕ the effective unit weight γ, and the pile diameter, d.

2. Obtain the ultimate soil resistances per unit length of pile by making use of the following equations:

For soil close to ground surface

$$p_{st} = \gamma x \left[\frac{K_0 x \tan \phi \sin \beta}{\tan (\beta - \phi) \cos \alpha} + \frac{\tan \beta}{\tan (\beta - \phi)} (d + x \tan \beta \tan \alpha) \\ + K_0 x \times \tan \beta (\tan \phi \sin \beta - \tan \alpha) - K_A d \right] \qquad (9.100a)$$

For soil well below ground surface

$$p_{sd} = K_A d \gamma x (\tan^8 \beta - 1) + K_0 d \gamma x \tan \phi \tan^4 \beta \qquad (9.100b)$$

where, $\alpha = \phi/2$, $\beta = 45° + \phi/2$, $K_0 = 0.4$, and

$K_A = \tan^2 (45° - \phi/2)$,

γ = effective unit weight of soil.

Use the smaller of the values given by Eqs (9.100a) and (9.100b).

3. In making the computation in step 2, find the depth x_r at which there is an intersection of Eqs (9.100a) and (9.100b). Above this depth use Eq. (9.100a) and below this depth use Eq. (9.100b).

4. Select a depth x at which p-y curve is required.

5. Establish $y_u = 3d/80$, compute p_u by the following equation

$$p_u = A_s p_s \text{ or } p_u = A_c p_s \qquad (9.100c)$$

Use the appropriate value of A_s or A_c from Fig. 9.28 (a) for the particular non-dimensional depth x/d.

The value of p_s in Eq. (9.100c) is obtained from Eq. (9.100a) or (9.100b).

6. Establish $y_m = d/60$, compute p_m as

$$p_m = B_s p_s \text{ or } p_m = B_c p_s \qquad (9.100d)$$

Use the appropriate value of B_s or B_c from Fig. 9.28 (b) for the particular non-dimensional depth and for either the static or cyclic case. Use the appropriate equation for p_s.

7. The two straight line portions of the curve beyond the point where $y = d/60$ can now be established as shown in Fig. 9.28 (c).

8. Establish the initial portion Ok of the p-y curve by the equation

$$p = (n_h x) y \qquad (9.100e)$$

The values of n_h are given in Table 9.8.

9. Establish the parabolic section km of the p-y curve by the relation

$$p = C y^{1/n} \qquad (9.100f)$$

(a) Fit the parabola between points k and m as follows:

$$m = \frac{p_u - p_m}{y_u - y_m} \qquad (9.101a)$$

(b) Obtain the power of the parabolic section by

$$n = \frac{p_m}{m y_m} \qquad (9.101b)$$

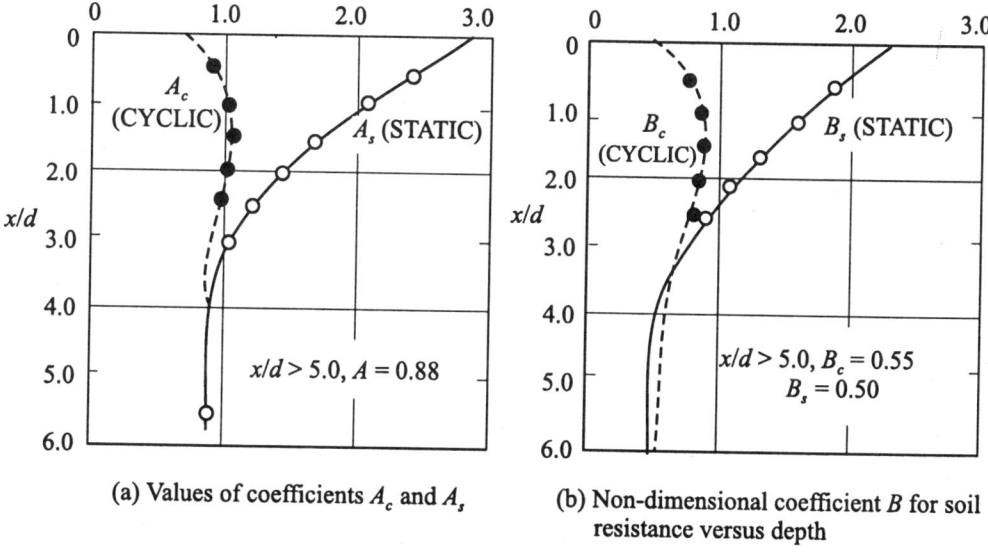

(a) Values of coefficients A_c and A_s

(b) Non-dimensional coefficient B for soil resistance versus depth

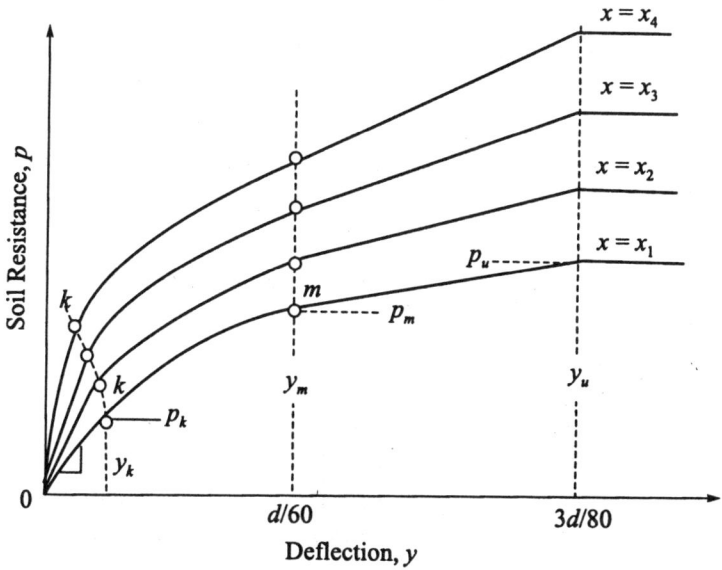

(c) Characteristic shape of a family of p-y curves for static and cyclic loadings in sand

Fig. 9.28 (a) and (b) Non-dimensional coefficients A and B for the development of p-y curves for piles in sand, (c) characteristic shapes of p-y curves for piles in sand (Reese *et al*, 1974)

Table 9.8 Recommended values of n_h for sand MN/m³
(After Reese, 1986)

Relative density	Above water table	Submerged
Loose	7	6
Medium	25	15
Dense	60	30

(c) Obtain the coefficient C as

$$C = \frac{p_m}{y_m^{1/n}}$$ (9.101c)

(d) Determine point k as

$$y_k = \left(\frac{C}{n_h x}\right)^{n/(n-1)}$$ (9.101d)

$$p_k = C y_k^{1/n}$$ (9.101e)

(e) Compute appropriate number of points on the parabola by using Eq. (9.100f).

The soil response curve for other depths can be found out by repeating the above steps for each desired depth. The final p-v curves are given in Fig. 9.28 (c).

Simplified equation for *p-y* curve

Fenske (1981) has simplified the Eqs (9.100a) and (9.100b) and presented the same in the following form

$$p_{st} = \gamma d^2 [S_1 (x/d) + S_2 (x/d)^2]$$ (9.102a)

$$p_{sd} = \gamma d^2 [S_3 (x/d)]$$ (9.102b)

where, S_1, S_2 and S_3 are non-dimensional coefficients which are functions of ϕ and x_r/d. The values are given in Table 9.9.

Table 9.9 Non-dimensional coefficients for *p-y* curves for sand (After Fenske)

ϕ, degree	S_1	S_2	S_3	x_r/d
25.0	2.05805	1.21808	15.68459	11.18690
26.0	2.17061	1.33495	17.68745	11.62351
27.0	2.28742	1.46177	19.95332	12.08526
28.0	2.40879	1.59947	22.52060	12.57407
29.0	2.53509	1.74906	25.43390	13.09204
30.0	2.66667	1.91170	28.74513	13.64147
31.0	2.80391	2.08866	32.51489	14.22489
32.0	2.94733	2.28134	36.81400	14.84507
33.0	3.09732	2.49133	41.72552	15.50508
34.0	3.25442	2.72037	47.34702	16.20830
35.0	3.41918	2.97045	53.79347	16.95848
36.0	3.59222	2.24376	61.20067	17.75976
37.0	3.77421	3.54280	69.72952	18.61673
38.0	3.96586	3.87034	79.57113	19.53452
39.0	4.16799	4.22954	90.95327	20.51883
40.0	4.38147	4.62396	104.14818	21.57604

9.18 SOLUTION FOR THE LATERALLY LOADED PILES BY THE USE OF *p-y* CURVES

The methods used for the construction of *p-y* curves in clay and sand have been explained in the earlier sections. The (*p-y*) method of solution for the problem of the laterally loaded pile is quite popular in USA and in many other countries. The step-by-step approach for the solution of a given laterally loaded pile problem is explained below for the case $E_s = n_h x$.

1. Construct *p-y* curves.
2. To start with assume a suitable value for n_h and compute the corresponding value for T from Eq. (9.68).
3. Compute the deflections y, at depths $x = 0$, $x = x_1$, $x = x_2$, etc. from Eq. (9.51). Suitable boundary conditions may be applied to this equation to get the necessary deflections.
4. For the known deflections, y, at different depths, get the coresponding p values from the constructed *p-y* curves.
5. Compute the E_s values at each depth by using the equation $E_s = p/y$.
6. Plot E_s vs. depth and draw an average straight line passing through the origin of coordinates and most of the points plotted. More importance is to be given for points obtained at depths close to the ground surface. Obtain the value of n_h as

$$n_h = \frac{E_s}{x}$$

Trial plotting of E_s against depth is shown in Fig. 9.29 (a). Compute T by using equation

$$T = \sqrt[5]{\frac{EI}{n_h}}$$

This completes the first iteration of the solution procedure.

7. Continue the second trial from Step 3 onwards till a new value for T is obtained as in Step 6. A plot may be made of the values of T tried and T obtained, and a final T value may be obtained as shown in Fig. 9.29 (b).

Convergence may be achieved within 2 or 3 trials.

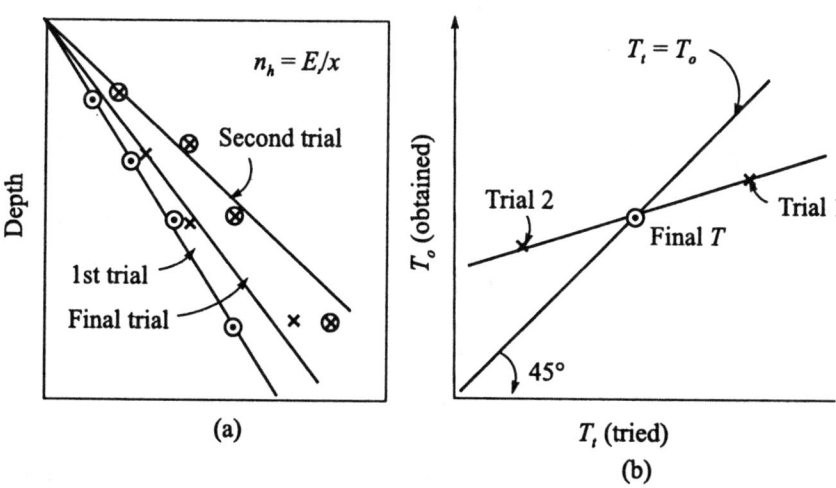

Fig. 9.29 Method of obtaining final n_h and T values by the use of *p-y* curves: (a) Trial plot of E_s values, (b) final value of T

8. Compute the values of deflection, moment and shear by making use of the appropriate equations given in Section 9.8.

Example 9.15

A steel pipe pile of 61 cm outside diameter with 2.5 cm wall thickness is driven into saturated cohesive soil up to a depth of 20 m. The undrained cohesive strength of the soil is 20 kPa. The submerged unit weight of soil is 8.75 kN/m^3. Construct (p-y) curves for static and cyclic loadings at depths of 2, 6 and 10 metres [Refer. Fig. 9.26 (a)].

Solution

This is the case of construction of (p-y) curve for soft clay below water table. The ultimate soil resistance, p_u, per unit length of pile may be expressed as

for shallow depths Eq. (9.93b)

$$p_u = \left(3 + \frac{\gamma' x}{c_x} + \frac{0.5x}{d}\right) c_x d$$

for deeper depths [Eq. (9.93c)]

$$p_u = 9c_x d$$

where, $\gamma' = 8.75$ kN/m^3,

$c_x = 20$ kPa, constant with respect to depth,

$d =$ diameter of pile $= 0.61$ m,

$x =$ depth in metres below ground surface.

Substituting,

$$p_u = \left(3 + \frac{8.75x}{20} + \frac{0.5x}{0.61}\right) \times 20 \times 0.61$$

$$= (36.6 + 9.06x) \text{ kN/m}$$

$$p_u = 9 \times 20 \times 0.61 = 110 \text{ kN/m}$$

Here the smaller of the two values of p_u has to be used for constructing (p-y) curves at each depth. The depth at which both the values of p_u are equal may be found out by equating the two values as

$$36.6 + 9.06x = 110$$

$$x_r = x \approx 8 \text{ m}$$

That is up to depth $x_r = 8$ m Eq. (9.93b) and beyond 8 m, Eq. (9.93c) has to be used. The value of ε_{50} may be taken from Table 9.7 as equal to 0.02 for c between 250 and 500 lb/ft^2 (Here $c \approx 400$ lb/ft$^2 \approx 20$ kPa).

From Eq. (9.94),

$$y_{50} = 2.5d\varepsilon_{50} = 2.5 \times 0.61 \times 0.02 = 0.0305 \text{ m}$$

The value of p_u remains constant beyond a deflection $= 8y_{50} = 0.244$ m ≈ 24.4 cm. The (p-y) curve at any depth may be constructed by making use of Eq. (9.93d).

$$\frac{p}{p_u} = 0.5 \left(\frac{y}{y_{50}}\right)^{0.33}$$

In this equation, y_{50} is known. At $y = 8y_{50}$, $p/p_u \approx 1$. Therefore, for any assumed ratios of $(y/50)$ up to 8 the corresponding ratio of p/p_u can be calculated.

Calculation of p_u

At depth 2 m, $p_u = 36.6 + 9.06 \times 2 = 54.72$ kN/m.

At depth 4 m, $p_u = 36.6 + 9.06 \times 4 = 72.84$ kN/m.

At depth 6 m, $p_u = 35.5 + 9.06 \times 6 = 91$ kN/m.

At depth 8 m and below, $p_u = 110$ kN/m.

Construction of (p-y) curves

y/y_{50}	y(m)	p/p_u	2 m	4 m	6 m	10 m
0.25	0.0076	0.31	16.90	22.5	28.2	34.1
0.50	0.0153	0.40	21.80	29.2	36.4	44.0
1.0	0.0305	0.50	27.36	36.5	45.50	55.0
2.0	0.0610	0.63	34.40	46.0	57.30	69.3
4.0	0.1220	0.79	43.23	58.0	71.9	86.9
8.0	0.2440	1.0	54.72	73.0	91.0	111.0

The (p-y) curves as plotted in shown in Fig. Ex. 9.15 (a)

Cyclic loading [Refer Fig. 9.26 (b)]

Let, p_u = ultimate soil resistance per unit length for static loading,

   p_{uc} = ultimate soil resistance for cyclic loading.

As per Eq. (9.95b)

$$p_{uc} = 0.72 p_u$$

Steps

1. The (p-y) curve up to $0.72\, p_u = p_{uc}$ is to be constructed in the same way as for static load by using Eq. (9.93d) up to $y/y_{50} = 3$ and up to depth $x = x_r$ as for static loading.

2. The (p-y) curve beyond $y = 15\, y_{50}$ may be constructed for depths x less than x_r by using the equation.

$$p = 0.72 p_u \left(\frac{x}{x_r}\right)$$

The value of p remains constant beyond $y = 15 y_{50}$.

3. If the depth to the p-y curve is greater than or equal to x_r then p is equal to $0.72\, p_u$ for all values of y greater than $3y_{50}$.

   Now as for static loading $x_r = 8$ m,

$$y_{50} = 0.0305 \text{ m,}$$

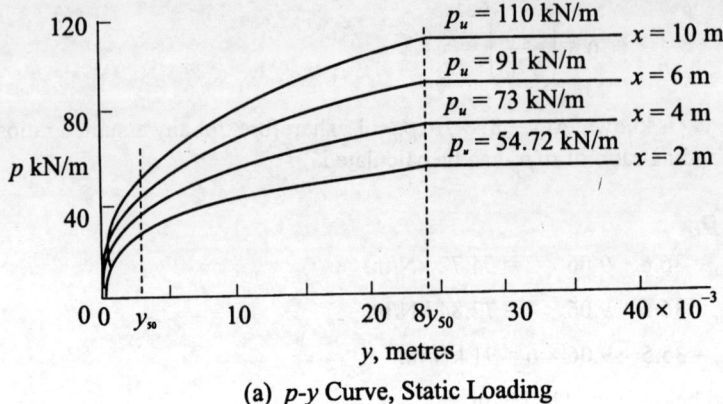

(a) *p-y* Curve, Static Loading

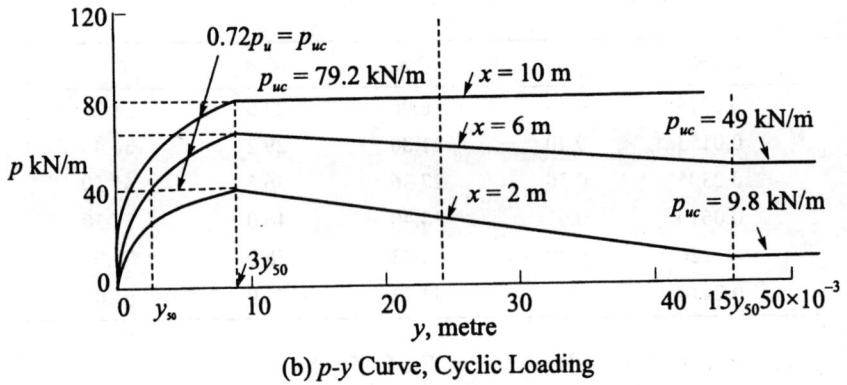

(b) *p-y* Curve, Cyclic Loading

Fig. Ex. 9.15 (a) *p-y* curve, static loading, (b) *p-y* curve, cylic loading

$$3y_{50} = 0.0915\,\text{m},$$
$$15y_{50} = 0.4575\,\text{m}.$$

Construction of (*p-y*) curves

From $y = 0$ to $y = 3y_{50}$

At $x = 2\,\text{m}, p_{uc} = 0.72, p_u = 0.72 \times 54.72 = 39.4\,\text{kN/m}$,

At $x = 6\,\text{m}, p_{uc} = 0.72 \times 91 = 65.5\,\text{kN/m}$,

At $x = 8\,\text{m}, 0.72 \times 110 = 79.2\,\text{kN/m}$.

			Values of p at depths of		
y/y_{50}	y (m)	p/p_u	*2 m*	*6 m*	*10 m*
0.25	0.0076	0.31	16.96	28.2	34.1
0.50	0.0153	0.40	21.8	36.4	44.0
1.0	0.0305	0.50	27.36	45.5	55.0
2.0	0.0610	0.63	34.4	57.3	69.3
3.0	0.0915	0.72	39.4	65.5	79.2

(p-y) *curve from* $y = 3y_{50}$ *to* $y = 15\ y_{50}$

Calculate p_{uc} at $y = 15y_{50}$ as per step 2 above. The values of $p\ (= p_{uc})$ at $y = 15y_{50}$ for the various depths are as given below ($x_r = 8$ m)

		$p = p_{uc}$ (kN/m) at depths of		
y/y_{50}	y (m)	2 m	6 m	10 m
15	0.4575	9.8	49.0	79.2

The p-y curves for both the static and cyclic loadings are given in Fig. Ex. (9.15).

Example 9.16

Determine deflection and moment as a function of depth along the pile in Ex. 9.15 at depths of 0, 2, 4, 6 and 10 m with the horizontal load of 400 kN acting at a height of 2 m above ground level by making use of the p-y curves developed for the static condition. The EI of the pile is 4.35×10^2 MN-m^2 (4.35×10^{11} kg-cm^2).

Solution

The various steps for computing deflections and moments are

1. As a first trial, assume $n_h = 1.51$ lb/in^3 ($= 0.041$ kg/cm^3).

2. $T = \left(\dfrac{EI}{n_h}\right)^{\frac{1}{5}} = \left(\dfrac{4.35 \times 10^{11}}{0.041}\right)^{\frac{1}{5}} = 403$ cm

3. $Z_{max} = \dfrac{L}{T} = \dfrac{20 \times 100}{403} = 4.96$, the pile is a long one.

4. Compute y at depths 0, 2, 6 and 10 m by the use of Eq. (9.51)

$$y = y_A + y_B = \frac{P_t T^3}{EI} A_y + \frac{M_t T^2}{EI} B_y$$

The values of A_y and B_y may be obtained from Table 9.4 for various values of depth coefficient $Z = x/T$. The computed deflections are tabulated below.

Depth x, cm	Z	A_y	B_y	y_A cm	y_B cm	y cm
0	0	2.43	1.62	14.62	4.82	19.44
200	0.496	1.64	0.87	9.88	2.60	12.48
400	0.999	0.96	0.45	5.78	1.34	7.12
600	1.488	0.46	0.07	2.76	0.20	2.96
1000	2.48	0.03	-0.08	0.18	-0.24	-0.06

5. Obtain p from p-y curves given in Fig. Ex. 9.15 (a) for the various values of y given above, and compute E_s by

$$E_s = \frac{p}{y}$$

The values of E_s are as given below:

Depth x cm	y cm	p kg/cm	E_s kg/cm^2
0	19.44	0	0
200	12.48	42	3.36
400	7.12	50	7.0
600	2.96	46	15.54
1000	-0.6		

6. Plot E_s against depth [Fig. Ex. 9.16 (a)], determine n_h by drawing an average line. From Fig. Ex. 19.16 (a), $n_h = 0.0188$ kg/cm^3 as the value obtained.

   The value T obtained is

$$T_0 = \left(\frac{4.35 \times 10^{11}}{0.0188}\right)^{\frac{1}{5}} = 471 \text{ cm}$$

7. For the second trial, use $T = 471$ cm, and calculate y as in Step 4. Obtain p, for the new values of y from the p-y curves in Fig. Ex. 9.15 (a). Calculate new values of E_s. Plot E_s vs. depth for the second trial as shown in Fig. Ex. 9.16 (a).

   From the second trial

$$n_h = 0.013 \text{ kg/cm}^3$$

8. The new value of T for the third trial is

$$T_0 = \left(\frac{4.35 \times 10^{11}}{0.0131}\right)^{\frac{1}{5}} = 506 \text{ cm}$$

9. Compute new values of y as in step 4 above, and new E_s values as in step 5. Plot E_s vs. depth as shown in Fig. Ex. 9.16 (a). The value of n_h is

$$n_h = 0.012 \text{ kg/cm}^3$$

   The new value of T obtained is

$$T_0 = \left(\frac{4.35 \times 10^{11}}{0.012}\right)^{\frac{1}{5}} = 515 \text{ cm}$$

10. Since T obtained from step 9 above is almost close to the one in Step 8, $T = 515$ cm is taken as the final value.

11. With $T = 515$ cm (final), the final deflections are as given below.

Depth x cm	Z	A_y	B_y	y_A cm	y_B cm	y cm
0	0	2.43	1.62	30.52	7.89	38.41
200	0.39	1.81	1.00	22.73	4.87	27.66
400	0.78	1.24	0.56	15.57	2.72	18.29
600	1.17	0.77	0.24	9.67	1.17	10.84
1000	1.94	0.17	-0.06	2.13	-0.29	1.84

12. The calculated moments are tabulated below as per Eq. (9.53)

$$M = P_t T A_m + M_t B_m$$

$$P_t = 400 \text{ kN}, T = 515 \text{ cm} = 5.15 \text{ m}$$

$$M_t = 400 \times 2 = 800 \text{ kN-m}$$

Depth x (m)	Z	A_m	B_m	M_A kN	M_B kN	M kN
0	00	00	1.00	0	800	800
2	0.39	0.37	0.99	762	792	1554
4	0.78	0.64	0.92	1318	736	2054
6	1.17	0.76	0.79	1566	632	2198
10	1.94	0.65	0.43	1339	344	1683
15	2.91	0.26	0.093	536	74	610
20	3.90	0.025	0.04	52	32	84

13. The deflections and moments along the pile are plotted against depth and shown in Fig. Ex. 9.16 (b).

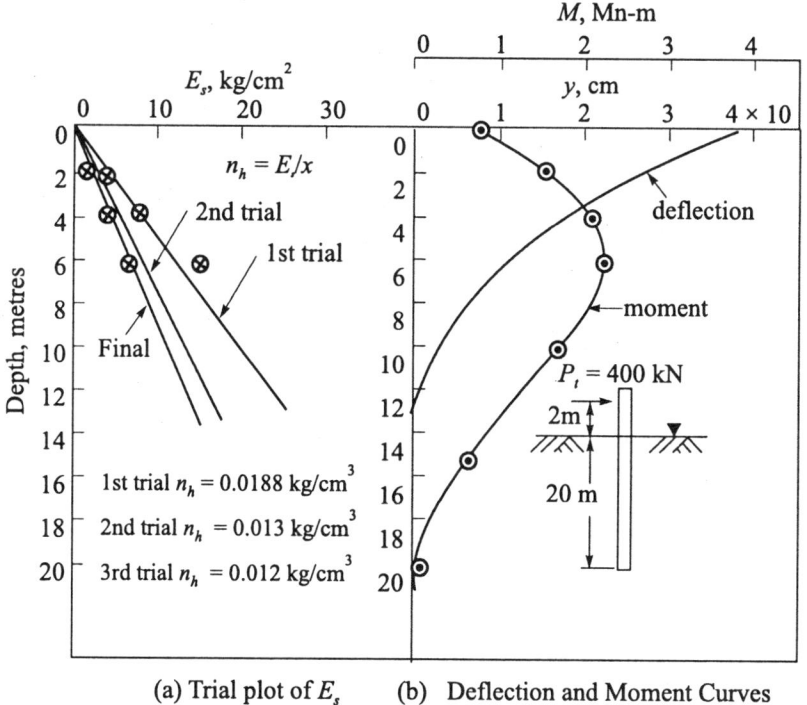

(a) Trial plot of E_s (b) Deflection and Moment Curves

Fig. Ex. 9.16

Example 9.17

Construct *p-y* curves for stiff clay below water table for a pile with a diameter of 61 cm driven to a depth of 20 m. The average undrained cohesive strength of clay is 150 kPa and the effective unit of soil is 8.75 kN/m³. The pile is subjected to static loading. The p-y curves are required to be constructed for the depths of 1, 2, 4, 6, 10 and 15 m below ground level.

Solution

The step-by-step procedure for constructing the p-y curve as explained earlier will not be repeated here. The various calculations are as given below.

1. The ultimate soil resistance will be the least of the following [Eqs (9.96a) and (9.96b)].

$$p_{c1} = 2c_a d + \gamma' dx + 2.83 c_u x$$

$$p_{c2} = 11\ cd$$

$$p_{c1} = 2 \times 150 \times 0.61 + 8.75 \times 0.6x + 2.83 \times 150x$$

$$= 183 + 429.83x\ \text{kN/m}$$

$$p_{c2} = 11 \times 150 \times 0.61 = 1006.5\ \text{kN/m}$$

The depth at which $p_{c1} = p_{c2}$ may be found out by equating the two equations which gives

$$x = x_r = 1.92\ \text{m}$$

Use p_{c1} up to depth 2 m and p_{c2} thereafter for constructing the p-y curves.

2. Compute

$$y_{50} = \varepsilon_{50} d$$

where, $\varepsilon_{50} = 0.5\%$ from Table 9.6 for $c = 150$ kPa (≈ 3000 lb/ft²).

Therefore

$$y_{50} = 0.005 \times 0.61 = 0.0031\ \text{m} = 0.31\ \text{cm}.$$

3. $p = n_h xy$ from Eq. (9.97a),

where, $n_h = 2500$ kN/m³ for $c = 150$ kPa,

y = deflection in metres,

x = distance from ground level in metres.

Therefore, $p = 2500\ xy$ kN/m.

The straight line portion oa of the p-y curve in Fig. 9.26 (a) may now be constructed by the above equation for various values of y at any depth x.

4. The equation for the parabolic portion ab of the curve in Fig. 9.26 (a) is Eq. (9.97b)

$$p = 0.5 p_c \left(\frac{y}{y_{50}}\right)^{0.5}$$

For depth at 1 m

$$p = 0.5\ (183 + 429.83x) \left(\frac{y}{0.0031}\right)^{0.5}$$

For depths at 2 m and below

$$p = 0.5 \times 1006.5 \left(\frac{y}{0.0031}\right)^{0.5} = 503.05 \left(\frac{y}{0.0031}\right)^{0.5}$$

The parabolic portion of the curve ab in Fig. 9.26 (a) can be constructed by making use of the above equation.

The intersection of Eq. (9.97a) and (9.97b) gives the point a.

The abscissa of point b on the p-y curve in Fig. 9.26 (a) is represented by the deflection

$$y = A_s y_{50}$$

From Fig. 9.26 (c), $A_s = 0.5$ for depth 1 m below ground level $(x/d = 1.64)$.

$A_s = 0.6$ for depths greater than 2 m $(x/d \geq 3.3)$.

Therefore, the abscissa of point b is :

For depth 1 m, $y = 0.5 \times 0.0031 = 0.0016$ m.

For depths greater than 2 m, $y = 0.6 \times 0.0031 = 0.0019$ m.

5. The abscissa of point c on the p-y curve in Fig. 9.26 (a) is $y = 6A_s y_{50}$.

For depth 1 m, $y = 6 \times 0.5 \times 0.0031 = 0.0093$ m

For depths greater than 2 m, $y = 6 \times 0.6 \times 0.0031 = 0.0112$ m.

6. The abscissa of point d in Fig. 9.26 (a) is

$$y = 18\, A_s y_{50}$$

For 1 m depth, $y = 18 \times 0.5 \times 0.0031 = 0.0279$ m.

For 2 m depth and above, $y = 18 \times 0.6 \times 0.0031 = 0.0335$ m.

7. Equation for the parabolic portion bnc of p-y curve [Fig. 9.26 (a)] is

$$p = 0.5 p_c \left(\frac{y}{y_{50}}\right)^{0.5} - 0.055\, p_c \left(\frac{y - A_s y_{50}}{A_s y_{50}}\right)^{1.25}$$

For depth 1 m

$$p = (306) \left(\frac{y}{0.0031}\right)^{0.5} - (33.7) \left(\frac{y - 0.0016}{0.0016}\right)^{1.25}$$

At $y = 6A_s y_{50} = 0.0093$ m, $p = 290$ kN/m.

For depth 2 m and above,

$$p = 0.5 \times 1006.5 \left(\frac{y}{0.0031}\right)^{0.5} - 0.055 \times 1006.5 \left(\frac{y - 0.0016}{0.0016}\right)^{1.25}$$

or $\quad p = 503.25 \left(\dfrac{y}{0.0031}\right)^{0.5} - 55.36 \left(\dfrac{y - 0.0016}{0.0016}\right)^{1.25}$

At $y = 6\, A_s y_{50} = 0.0112$ m, $p = 436$ kN/m.

The parabolic portion of the p-y curve bnc may now be constructed by the use of above equations. It may be noted here that the bnc portion of the p-y curve does not change with depth after a depth of 2 m. The abscissa of the curve varies from $A_s y_{50}$ to $6A_s y_{50}$.

8. The ordinate p of the point d on the p-y curve [Fig. 9.26 (a)] may be calculated from Eq. (9.97g).

$$p = p_c \left(1.225 \sqrt{A_s} - 0.75\, A_s - 0.411\right)$$

At 1 m depth,

$$p = (183 + 429.83 \times 1) \times \left(1.225 \sqrt{0.5} - 0.75 \times 0.5 - 0.411\right)$$

$$= 612.83 \times 0.0802 = 49 \text{ kN/m}$$

At depths 2 m and below

$$p = 1006.5 \left(1.225 \sqrt{0.6} - 0.75 \times 0.6 - 0.411\right)$$

$$= 89 \, kN$$

9. The straight line portion cd of the p-y curve [Fig. 9.26 (a)] can be constructed since the values of p at deflections $6A_s y_{50}$ and $18A_s y_{50}$ are known.

10. It may be noted from Fig. Ex. 9.17 that the equation $p = n_h xy$, Eq. (9.97a) cuts only the lower straight line portion of the p-y curve. As such the deflections of the pile has to be calculated by considering the straight line portion of the p-y curve. As such the deflections of the pile has to be calculated by considering the straight line portion of the Eq. (9.97a) till it cuts the p-y curve and then the corresponding p-y curve. There is one p-y curve for all depths below 2 m.

For example the p-y curve to be used at depth 1.0 m is $Os_1 d_1 e_1$, and for depth 4.0 m, the p-y curve is $Os_2 d_2 e_2$. (Fig. Ex. 9.17).

The procedure for calculating deflections, moments, etc. along the pile is the same as that explained in Ex. 9.16.

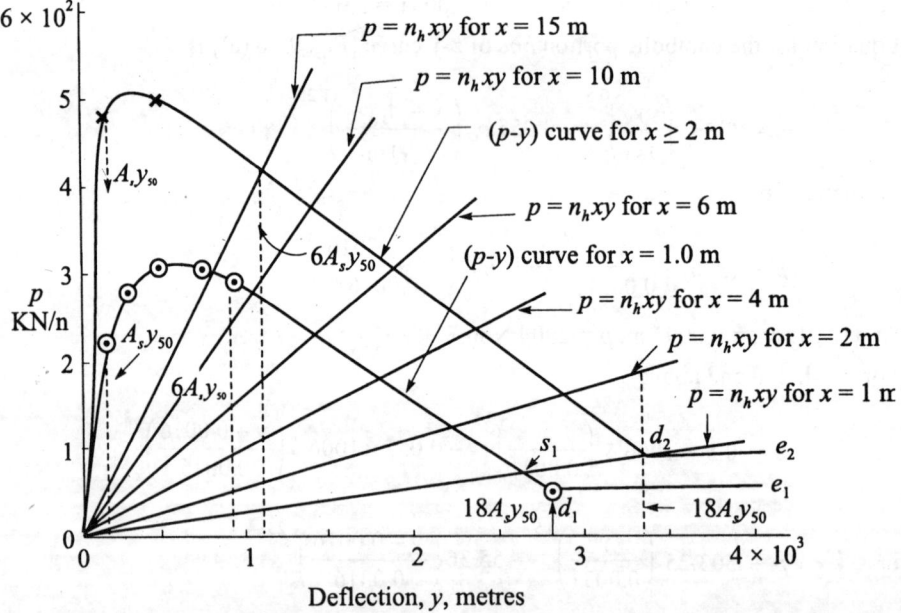

Fig. Ex. 9.17

Example 9.18

A 24 inch outside diameter steel pipe pile was used for lateral load tests at Mustang island, near Corpus Christi, Texas, USA (Reese *et al*, 1974). The pile was instrumented and driven to a depth of 69 ft. The soil at the site consisted of clean fine sand to silty fine sand in dense state under submerged condition. Static load test was carried out with the lateral load acting at a height of 12″ above ground level. The angle of internal friction, ϕ, was found to be 39° and the submerged unit weight 66 lb/ft^3 (0.0382 lb/in^3). Construct p-y curves for the pile at depths of 15, 25, 50, 75 and 100 inches below ground level.

Solution

The ultimate soil resistance per unit length of pile may be calculated by any one of the following methods:

1. By the use of Eqs (9.100a) and (9.100b).
2. By the use of simplified equations (9.102a) and (9.102b).

Only the first method is explained here. The various steps for the computation of p-y curves are as follows.

1. Computation of ultimate lateral resistance of soil for soil close to ground surface. Use Eq. (9.100a)

$$p_{st} = \gamma x \left[\frac{K_0 x \tan \phi \sin \beta}{\tan (\beta - \phi) \cos \alpha} + \frac{\tan \beta}{\tan (\beta + \phi)} (d + x \tan \beta \tan \alpha) + K_0 x \tan \beta (\tan \phi \sin \beta - \tan \alpha) - K_A d \right]$$

For soil at deeper depths. Use Eq. (9.100b). ·

$$p_{sd} = K_A d \gamma x (\tan^8 \beta - 1) + K_0 d \gamma x \tan \phi \tan^4 \beta.$$

we have, $\phi = 39°$, $\gamma = 0.0382$ lb/in^3, $d = 24$ inches,

$$\alpha = \phi/2 = 19.5°, \beta = 45° + \phi/2 = 64°.5, \beta - \phi = 25°.5,$$

$$K_0 = 0.4, \tan \phi = 0.8098, \sin \beta = 0.9026,$$

$$\tan (\beta - \phi) = 0.477, \cos \alpha = 0.9426, \tan \beta = 2.097,$$

$$\tan \alpha = 0.3541, K_A = \tan^2 (45° - \phi/2) = 0.25.$$

Substituting in above equations and simplifying, we have

$$p_{st} = 0.0382 x (4.2308 x + 99.51),$$

$$p_{sd} = 91.21 x.$$

The solution of the equations gives the critical depth x_r

$$x_r = 541 \text{ inches.}$$

The p_{st} values hold good for all depths.

2. Deflections y_u and y_m

$$y_u = \frac{3d}{80} = \frac{3 \times 24}{80} = 0.90 \text{ in}$$

$$y_m = \frac{d}{60} = \frac{24}{60} = 0.40 \text{ in}$$

3. Computation of p_u, p_m, and m

$$p_u = A_s p_{st}$$

$$p_m = B_s p_{st}$$

Get the values of A_s and B_s for various depths x/d from Fig. 9.28.

$$m = \frac{p_u - p_m}{y_u - y_m} = \frac{p_u - p_m}{0.9 - 0.4} = \frac{p_u - p_m}{0.5}$$

The values are tabulated below.

Depth (ins)	A_s	B_s	P_{st} lb/in	P_m lb/in	P_u lb/in	m
15	2.4	1.94	93	180	223	86
25	2.1	1.50	196	294	412	236-
50	1.5	1.0	594	594	891	594
75	1.1	0.7	1194	836	1313	954
100	0.9	0.55	1996	1098	1796	1396

With the coordinates (p_m, y_m) and (p_u, y_u) known, the straight line portions of the p-y curve can be constructed.

4. Compute

$$n = \frac{p_m}{m y_m} = \frac{p_m}{0.4 \text{ m}}, \quad C = \frac{p_m}{y_m^{1/n}} = \frac{p_m}{(0.4)^{1/n}}$$

$$y_k = \left(\frac{C}{n_h x}\right)^{n/(n-1)}, \quad p_k = C y_k^{1/n}$$

For dense sand under submerged conditions

$$n_h = 30 \text{ MN/m}^3 \approx 110 \text{ lb/in}^3 \text{ (from Table 9.8)}$$

The various values are tabulated below.

Depth, (ins)	P_m lb/in	m	n	$1/n$	$n/1-n$	C	$n_h x$ lb/in^2	y_k in	p_k lb/in
15	180	86	5.23	0.19	1.24	214	1650	0.079	132
25	294	236	3.11	0.32	1.46	394	2750	0.059	159
50	594	594	2.50	0.40	1.67	857	5500	0.0448	217
75	836	954	2.19	0.46	1.67	1274	8250	0.032	261
100	1098	1396	1.97	0.51	2.03	1751	11,000	0.024	261

5. Construct the parabolic portion of the curve k_m Fig. 9.28 (c) by using equation

$$p = C_y^{1/n}$$

where, the value of y lies between y_k and y_m.
The values of p for the various assumed values of y are tabulated below.

Depth, x (ins)	C	$1/n$	p (lb/in) at values of y			
			0.1	0.15	0.25	0.35 in
15	214	0.19	138	149	164	175
25	394	0.32	188	215	253	282
50	857	0.40	341	401	492	563
75	1274	0.46	441	532	673	786
100	1751	0.51	541	665	863	1025

Figure Ex. 9.18 gives the various p-y curves developed for different depths. These curves may be compared with the curves given in Fig. 9.10 which are also for Mustang island pile load test.

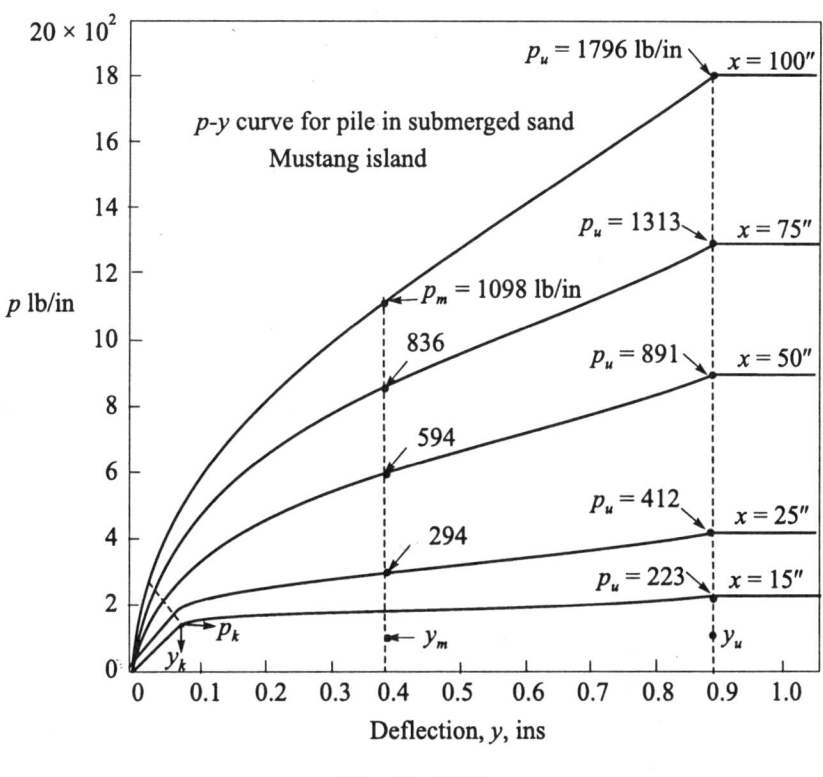

Fig. Ex. 9.18

9.19 PRESSUREMETER METHOD TO SOLVE LATERALLY LOADED PILE PROBLEMS

The pressuremeter test made in a borehole is particularly suitable for use in establishing p-y curves for laterally loaded piles. The test produces a curve as shown in Fig. 9.30 (a). The details of the pressuremeter tests are given in Chapter 3. The initial portion represents a linear relationship between pressure and volume change that is the radial expansion of the walls of the borehole. At the creep pressure, p_f, the pressure volume relationship becomes non-linear indicating plastic yielding of the soil and at the limit pressure, p_l, the volume increases rapidly without increase of pressure as represented by the horizontal portion of the p-y curve. The equation as given by Baguelin *et al*, Eq. (3.25) is

$$E_m = 2.66 \, V_m \, \frac{\Delta p}{\Delta v}$$

where, E_m = pressuremeter modulus,

$\Delta p = p_f - p_{om}$,

$\Delta v = v_f - v_0$,

V_m = mid-point volume,

$\Delta p / \Delta v$ = slope of the curve between v_0 and v_f.

Baguelin *et al* (1978) give two sets of curves relating the response of the soil to lateral loading for the two stages in the pressuremeter tests as shown in Fig. 9.30 (b). The upper curve is for depths below the ground surface equal to or greater than the critical depth x_c at which surface have affects the validity of the calculation method. When there is a pile cap there is no heave, x_c is zero, and the lower curve in Fig. 9.30 (b) applies. The equation for the modulus of subgrade reaction is expressed as

$$E_s = \frac{p}{y} \tag{9.103a}$$

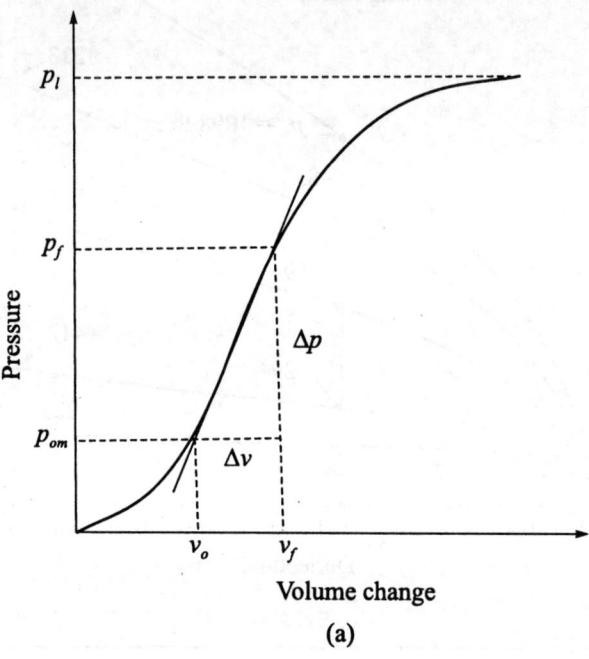

(a)

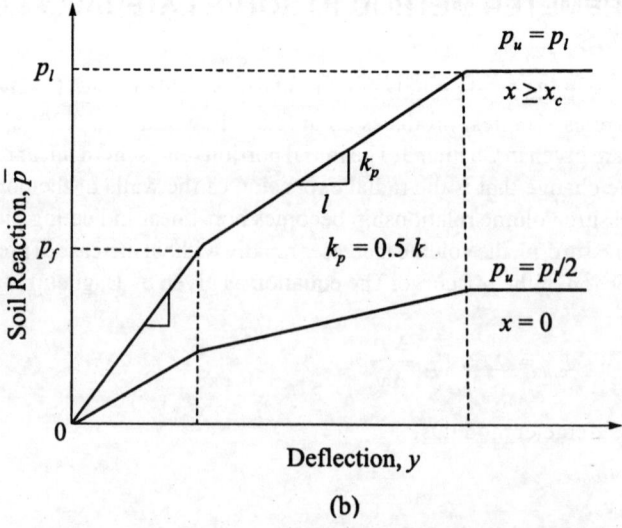

(b)

Fig. 9.30 Pressuremeter method of solving laterally loaded pile problems: (a) Pressuremeter curve, (b) design reaction curve (Baguelin *et al*, 1978)

or $\qquad\qquad p \stackrel{\pm}{=} E_s y$ (9.103b)

where, p represent the force per unit length of pile.

Let $\bar{p} = \dfrac{p}{d}$ where $\bar{p}$ is the soil reaction per unit area. We may now write Eq. (9.103a) as

$$\bar{p} = \frac{E_s}{d}\, y = ky$$ (9.103c)

where, k is called as *modulus of reaction* which has the unit of F/L^3.

The equation for computing the modulus of reaction, k, as given by Baguelin *et al* is for piles of widths greater than 600 mm

$$\frac{1}{k} = \frac{2\,d_0}{9\,E_m}\left(\frac{d}{d_0} \times 2.65\right)^{\alpha} + \frac{\alpha\,d}{6\,E_m}$$ (9.103d)

for piles of widths less than 600 mm

$$\frac{1}{k} = \frac{d}{E_m}\left[\frac{4\,(2.65)^{\alpha} + 3\alpha}{18}\right]$$ (9.103e)

where, E_m = the mean value of the pressuremeter modulus over the characteristic length of the pile,

d_0 = the reference diameter = 600 mm,

d = pile width or diameter,

α = a rheological factor.

It may be noted here that the subgrade modulus E_s is related to the Menard pressuremeter modulus E_m as follows.

$$E_s = C_f E_m$$ (9.103f)

where, C_f is the conversion factor given in Table 9.10 for various rheological factors.

Between the ground surface and critical depth x_c, the value of k should be reduced by the coefficient λ_x given by

$$\lambda_x = \frac{1 + x/x_c}{2}$$ (9.103g)

Table 9.10 Conversion factor C_f for estimating values of E_s from values E_m
(After Baguelin *et al*, 1978)

Type of soil	a	Values of C_f for piles of diameter	
		$d < 0.6$ m	$d = 1.20$ m
Peat	1.0	1.33	1.33
Clay	0.67	1.90	2.25
Silt	0.50	2.30	3.00
Sand	0.33	2.80	4.00
Sand and gravel			

This means that between the ground surface and the critical depth x_c, that is $0 < x < x_c$, the reaction modulus k becomes $\lambda_x k$. For cohesive soils x_c is of the order of $2d$, and for granular soils, it is of the order of $4d$.

Baguelin *et al* give the following equations for calculating deflections, bending moments and shears at any depth x, below the ground surface for conditions of a constant k value with depth.

Deflection
$$y(z_4) = \frac{2P_t}{Rkd} F_1 + \frac{2M_t}{R^2 kd} F_2 \qquad (9.104a)$$

$$M(z) = P_t R F_3 + M_t F_2 \qquad (9.104b)$$

Shear,
$$V(z) = P_t F_4 - \frac{2M_t}{R} F_3 \qquad (9.104c)$$

where, R = the transfer length given by the equation.

$$R = \sqrt[4]{\frac{4EI}{kd}} = \sqrt[4]{\frac{4EI}{E_s}} , \qquad (9.104d)$$

P_t = horizontal load applied to the pile head,

z = dimensionless coefficient $= x/R$,

F_1 to F_4 = non-dimensional factors.

It is generally accepted that if

$$\frac{L}{R} \geq 3$$

the pile is flexible and if

$$\frac{L}{R} \leq 1$$

the pile is rigid,

where, L = length of pile. The equations given here are for flexible piles.

The values of F_1 to F_4 are given in Fig. 9.30 (c). At the ground surface the deflections and slope become,

deflection $y_g = \dfrac{2P_t}{Rkd} + \dfrac{2M_t}{R^2 kd}$ (9.105a)

slope, $S_g = -\dfrac{2P_t}{R^2 kd} - \dfrac{4M_t}{R^3 kd}$ (9.105b)

If the head of the pile is fixed so that if does not rotate the values are

$$y(z) = \frac{P_t}{Rkd} F_2 \qquad (9.106a)$$

$$M(z) = \frac{-P_t R}{2} F_4 \qquad (9.106b)$$

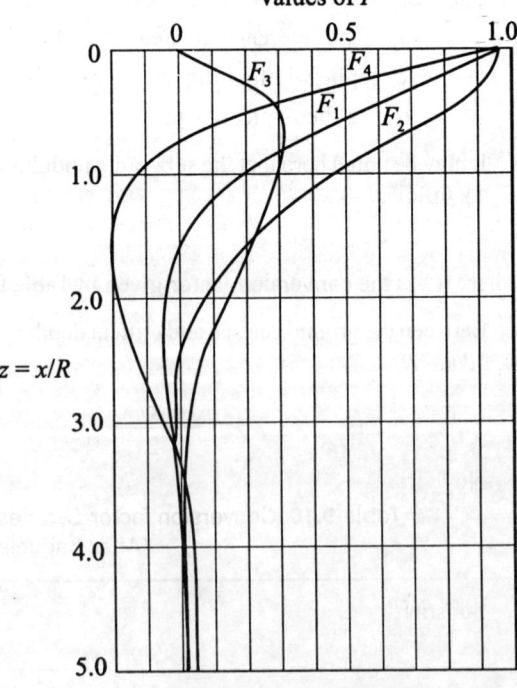

Values of F

$z = x/R$

Values of the coefficients F_1 to F_4
(After Baguelin et al)

Fig. 9.30 (c) Values of the coefficients F_1 to F_4 for solving laterally loaded pile problems by pressuremeter method (Baguelin et al, 1978)

$$V(z) = P_t F_3 \tag{9.106c}$$

$$y_0 = \frac{P_t}{Rkd} \tag{9.106d}$$

$$M_t = -\frac{P_t R}{2} \tag{9.106e}$$

Example 9.19

A RCC pile of 60 cm diameter is driven to a depth of 9.5 m into homogeneous clay with the following characteristics obtained from pressuremeter tests:

$$\bar{p}_f = 370 \text{ kPa}, \bar{p}_l = 630 \text{ kPa and } E_m = 7000 \text{ kPa}$$

A lateral load, P_t, of 90 kN and a moment M_t of 130 kN were applied at ground level. Required deflection at ground level. The EI of the pile is given as equal to 19×10^4 kN-m^2.

Solution

From Table 9.10, for clay $\alpha = 0.67$, $C_f = 1.9$ for $d = 0.6$ m.

$$E_s = 1.9 \times 7000 = 13,300 \text{ kPa [Eq. (9.103f).}$$

The value of subgrade reaction k is

$$kd = E_s = 13,300 \text{ kPa,}$$

or
$$k = \frac{13,300}{0.6} = 22,170 \text{ kN/m}^3.$$

The transfer length R is equal to [Eq. (9.104d)]

$$R = \left(\frac{4\,EI}{kd}\right)^{\frac{1}{4}} = \left(\frac{4 \times 19 \times 10^4}{13,000}\right)^{\frac{1}{4}} = 2.75 \text{ m}$$

$$\frac{L}{R} = \frac{9.5}{2.75} = 3.45 \text{ The pile is flexible.}$$

Now, from Eq. (9.105a)

$$y_g = \frac{2\,P_t}{Rkd} + \frac{2\,M_t}{R^2 kd}$$

Substituting,

$$y_g = \frac{2 \times 90}{2.75 \times 13,300} + \frac{2 \times 130}{(2.75)^2 \times 13,000}$$

$$= 4.9 \times 10^{-3} + 2.59 \times 10^{-3} = 7.49 \times 10^{-3} \text{ m} = 7.49 \text{ mm}$$

Note: If k is not given, this can be found out from Eqs (9.103d) or (9.103e) according to the diameter of the pile.

9.20 POULOS METHOD OF ELASTIC ANALYSIS FOR LATERALLY LOADED SINGLE PILES

Introduction

Poulos and Davis (1980) analysis considers the soil an elastic continuum since the displacements at a point are influenced by stresses and forces acting at other points within the soil. Since his analysis is more involved, only the final solutions with regards to deflections and moments are given in this Chapter.

The soil has been assumed to be an ideal, homogeneous, isotropic, linear and elastic material of semi-infinite dimensions. Poulos used Mindlin equation for horizontal displacement due to horizontal load within a semi-infinite mass to compute soil displacements. Beam theory has been used to compute pile displacements. The soil and pile displacements are evaluated and equated at element centres along the pile. Poulos *et al* consider in the analysis the following types of soil.

1. Uniform soil with the Yound's modulus $\bar{E}_s$ remaining constant with depth.
2. Soil with linearly increasing soil modulus, E_s, with depth.

His equations take into account the effect of yield of soil on deflection, slope and moments. However, this aspect of the problem is not considered here.

Young's Modulus Constant with Depth
Free-head pile

For a purely elastic soil the ground level displacement y_g and S_g are given by the following equations:

$$y_g = I_{yp}\left(\frac{P_t}{\bar{E}_s L}\right) + I_{ym}\left(\frac{M_t}{\bar{E}_s L^2}\right) \tag{9.107a}$$

$$S_g = I_{sp}\left(\frac{P_t}{\bar{E}_s L^2}\right) + I_{sm}\left(\frac{M_t}{\bar{E}_s L^3}\right) \tag{9.107b}$$

where, P_t = shear load at ground level,

M_t = applied moment at ground level,

$\bar{E}_s$ = Young's modulus of soil,

L = length of pile,

I_{yp} = influence coefficient for computing pile head deflection for applied shear at ground level,

I_{ym} = influence coefficient for computing pile head deflection for applied moment at ground level,

I_{sp} = influence coefficient for computing pile head rotation for applied shear at ground level,

I_{sm} = influence coefficient for computing pile head rotation for applied moment at round level.

Figures 9.31 (a) and (b) give values of I_{yp} and I_{ym} ($= I_{sp}$) and Fig. 9.31 (c) the values of I_{sm} as a function of K_R for various values of L/d. It may be noted here that $I_{ym} = I_{sp}$ from the reciprocal theorem. Poulos *et al* define K_R as a flexibility factor and its values may be computed from equation

$$K_R = \frac{EI}{\overline{E}_s L^4} \tag{9.108}$$

where, EI = flexural stiffness of pile material,

K_R = flexibility factor. $K_R = 0$ for infinitely long piles,

$K_R = \infty$ for infinitely rigid pile.

The maximum moment in a free-head pile subjected to lateral load is given in Fig. 9.31 (d) as a function of K_R and L/d. In all the cases, the Poisson's ratio μ is assumed as equal to 0.5.

Fixed head pile

For a pile that is fixed against rotation at the groundline, the equation for deflection is

$$y_{fg} = I_{yf} \frac{P_t}{\overline{E}_s L} \tag{9.109}$$

where, y_{fg} = ground line deflection,

I_{yf} = influence coefficient.

Figure 9.32 (a) gives the values of I_{yf}. The bending moment at the top of a fixed-head pile is given in Fig. 9.32 (b) as a function of K_R, and L/d.

Solution for Pile with Soil Modulus E_s Increasing Linearly with Depth

Free-head pile

Poulos gives the following equations for ground line deflection, y_g and rotation S_g for free head pile

$$y_g = I'_{yp} \left(\frac{P_t}{n_h L^2} \right) + I'_{ym} \left(\frac{M_t}{n_h L^3} \right) \tag{9.110a}$$

$$S_g = I'_{sp} \left(\frac{P_t}{n_h L^3} \right) + I'_{sm} \left(\frac{M_t}{n_h L^4} \right) \tag{9.110b}$$

where, I'_{yp}, I'_{ym}, I'_{sp} and I'_{sm} are the corresponding influence factors, and n_h is the coefficient of modulus variation.

The elastic influence factors I'_{yp} etc. are given in Figs 9.33 and 9.34 as functions of K_N. Poulos *et al* define K_N also as a flexibility factor for E_s increasing linearly with depth as

$$K_N = \frac{EI}{n_h L^5} \tag{9.111}$$

Fixed pile

For a fixed pile, the ground line deflection is given by

$$y_{fg} = \frac{P_t}{n_h L^2} I'_{yf} \tag{9.112}$$

Figure 9.35 (a) gives the elastic influence factor I'_{yf} as a function of K_N for various values of L/d.

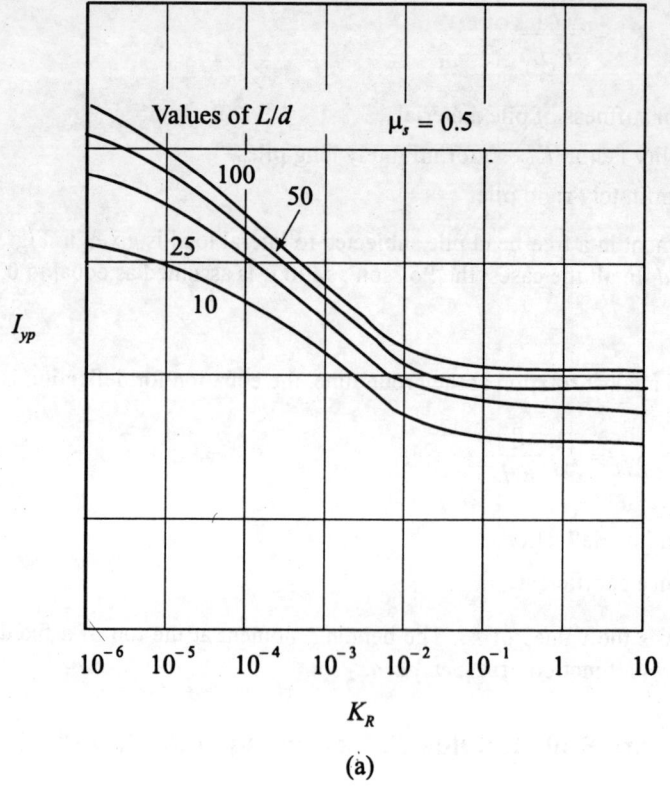

(a)

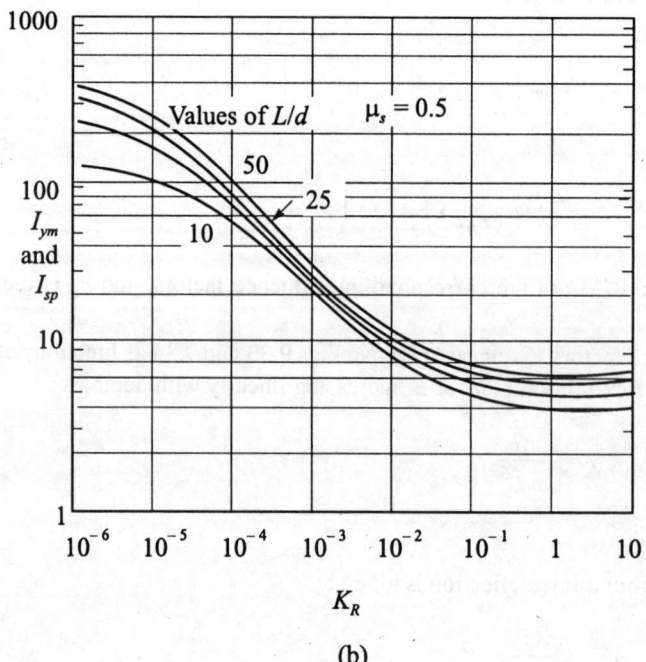

(b)

Fig. 9.31 (a) Values of I_{yp} for free-head floating pile, constant soil modulus, (b) values of I_{ym} and I_{sp} for free-head floaitng pile, constant soil modulus (Poulos and Davis, 1980)

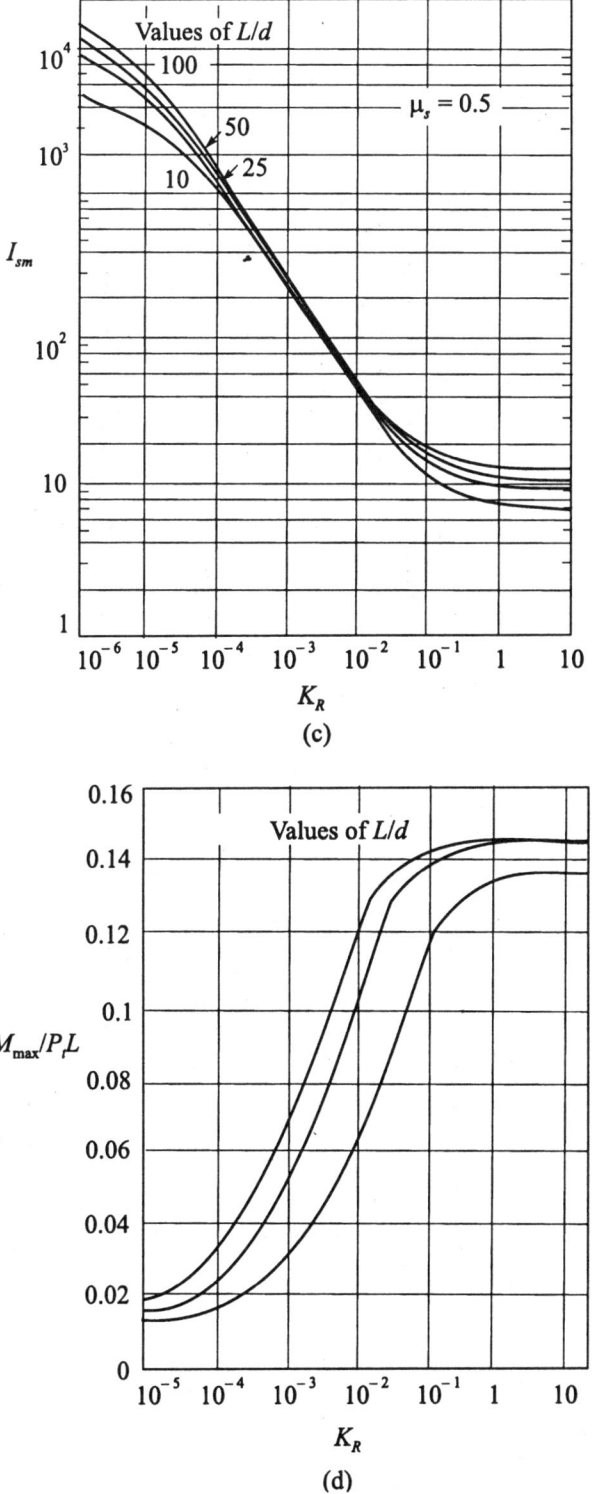

Fig. 9.31 (c) Values of I_{sm} for free-head floating pile, constant soil modulus, (d) maximum moment in free-head pile (Poulos and Davis, 1980)

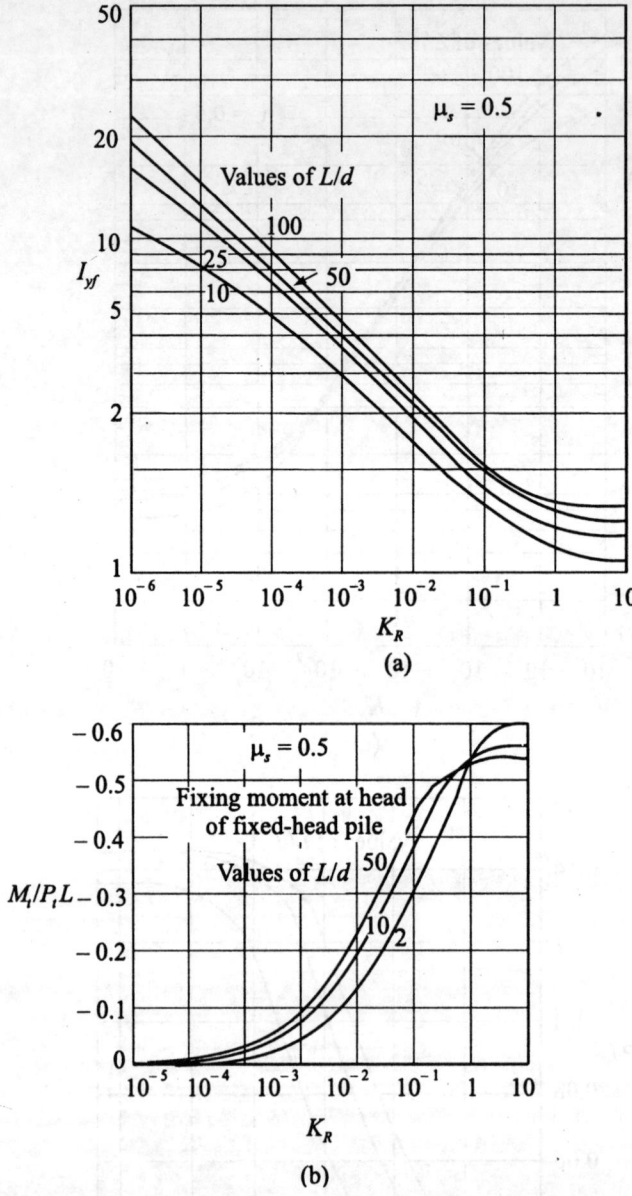

Fig. 9.32 (a) Influence factor I_{yf} for fixed-head floating pile, constant soil modulus, (b) fixing moment at head of fixed-head pile (poulos, 1980)

Moments in Pile

For free-head pile, the maximum moment caused by the horizontal load is given in Fig. 9.34 (a) as a function of L/d and of the flexibiligy factor K_N. For a fixed head pile Fig. 9.35 (b) gives the moment at the head of the pile.

Determination of Soil Modulus

Poulos makes use of two types of soil modulus. They are:

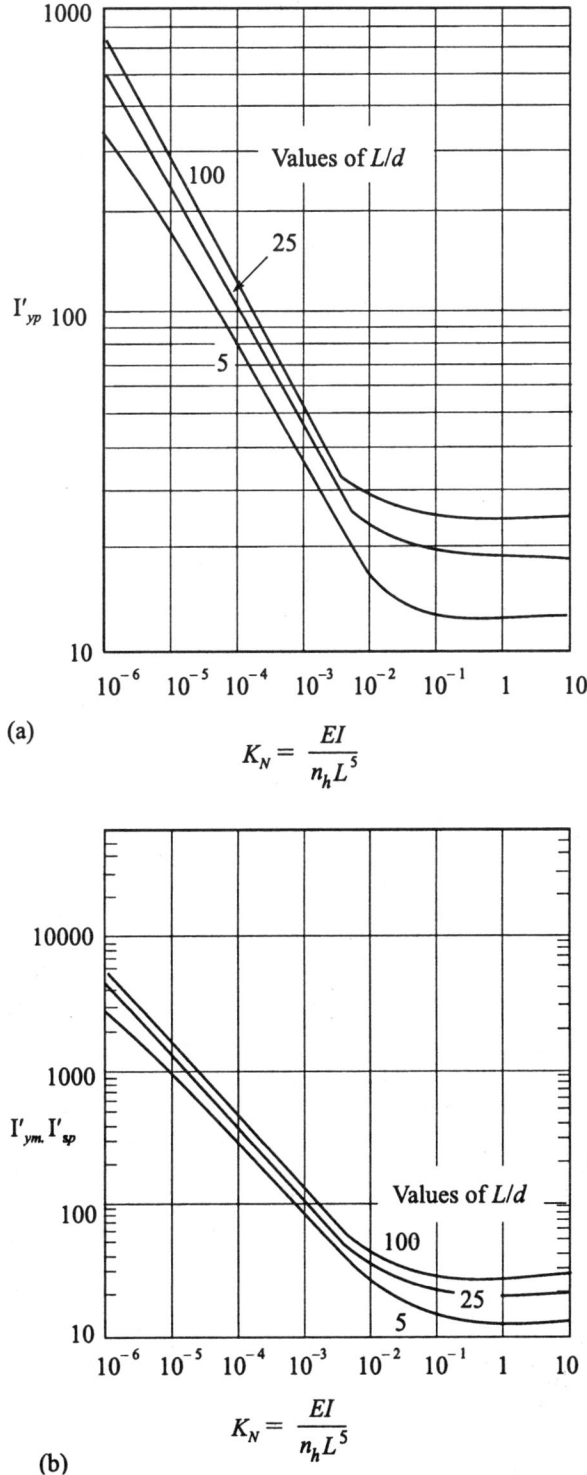

Fig. 9.33 (a) Values of I'_{yp} for free-head floating pile, linearly varying soil modulus (Poulos and Davis, 1980), (b) values of I'_{ym} and I'_{sp} for free-head floating pile, linearly varying soil modulus (Poulos and Davis, 1980)

1. Young's modulus, $\bar{E}_s$.
2. Modulus of subgrade reaction, E_s.

Polous suggests a number of methods for computing Young's modulus $\bar{E}_s$. They are:

1. Laboratory tests in which the stress path of typical elements of soil along the pile are simulated.
2. Plate-bearing tests, preferably on vertical plates, at various depths.
3. Pressuremeter tests.
4. The use of full-scale load tests to backfigure the modulus.
5. Empirical relations with other properties.

According to Poulos, there is not much evidence to indicate whether the first two methods would give satisfactory values of soil modulus. Pressuremeter method has already been discussed earlier. Poulos feels that full-scale lateral load tests are probably the most satisfactory means of determining the soil modulus. Secant value of soil modulus can be back-figured by making use of Eqs (9.107a) or (9.107b) with the known values of ground line deflections and slopes from field tests.

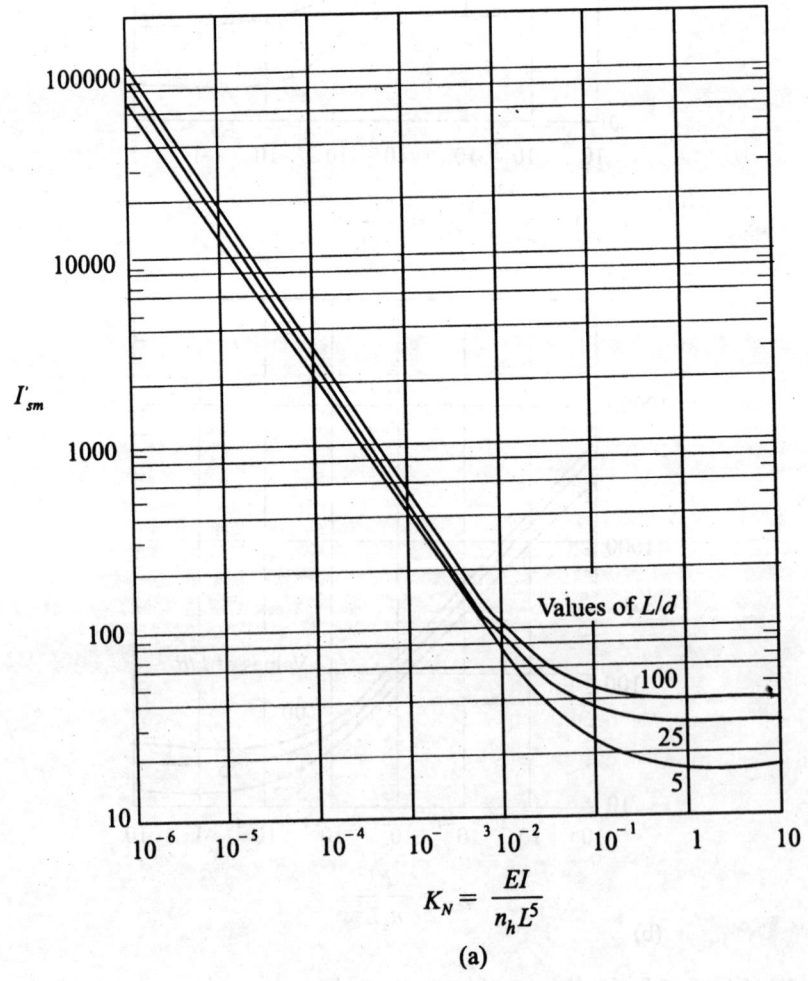

$$K_N = \frac{EI}{n_h L^5}$$

(a)

Fig. 9.34 Values of I'_{sm} for free-head floating pile linearly varying soil modulus
(Poulos and Davis, 1980)

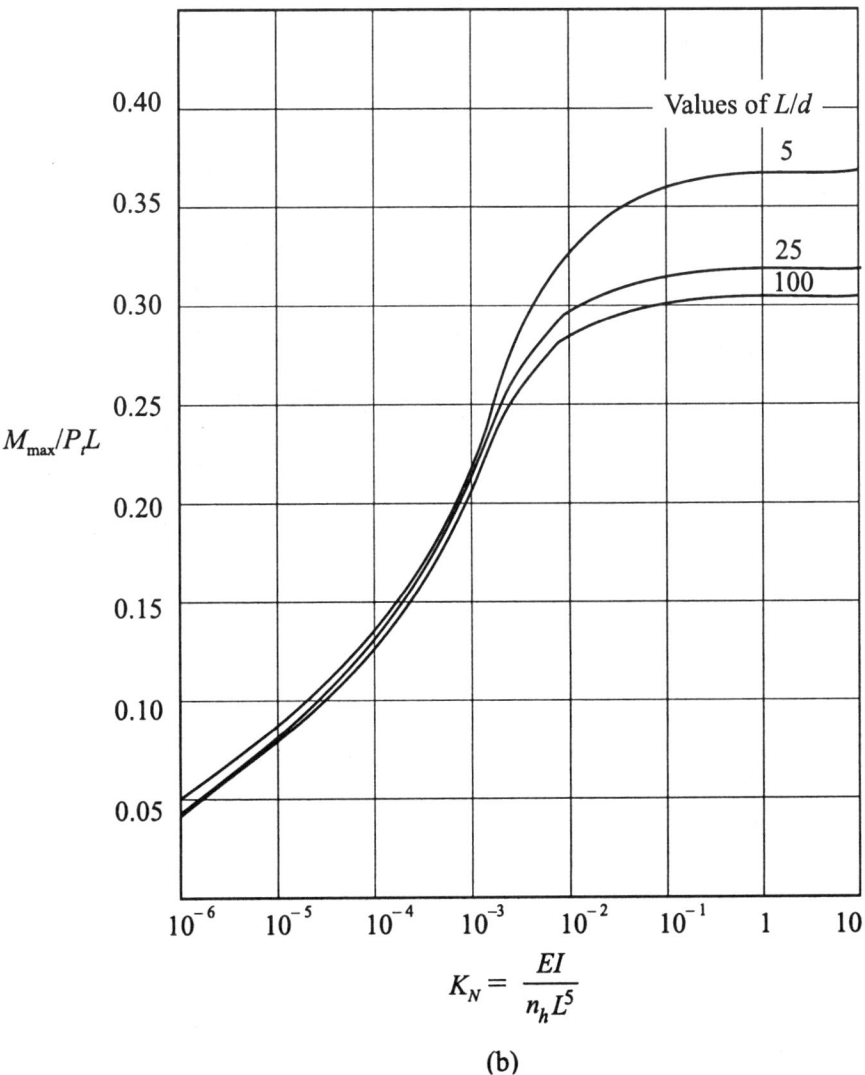

$$K_N = \frac{EI}{n_h L^5}$$

(b)

Fig. 9.34 Maximum moment in free-head pile, linearly varying soil modulus
(Poulos and Davis, 1980)

If the soil modulus E_s values linearly with depth, the methods of obtaining, n_h, the coefficient of soil modulus variation, has been discussed under Section 9.5.

The following empirical correlations have been suggested by Poulos for clay soils.

Secant values of $\bar{E}_s$ from

$$\bar{E}_s = 15c_u \text{ to } 95c_u \tag{9.113a}$$

where, lower value applies to soft clays and higher values to stiff clays.

Tangent values of $\bar{E}_s$ from

$$\bar{E}_s = 250\,c_u \text{ to } 400\,c_u \tag{9.113b}$$

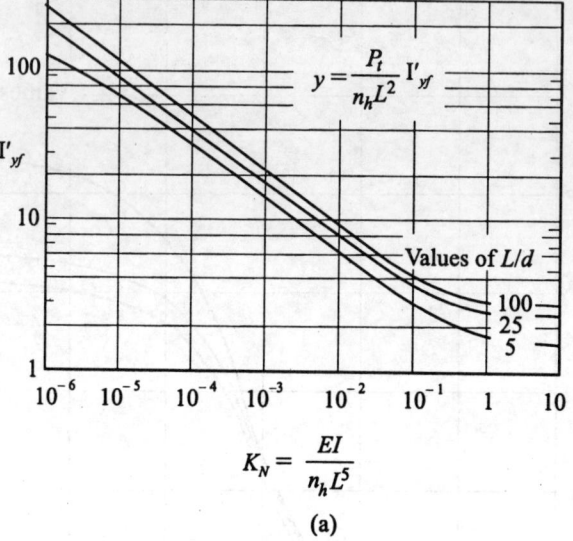

$$y = \frac{P_t}{n_h L^2} I'_{yf}$$

Values of L/d

100
25
5

$$K_N = \frac{EI}{n_h L^5}$$

(a)

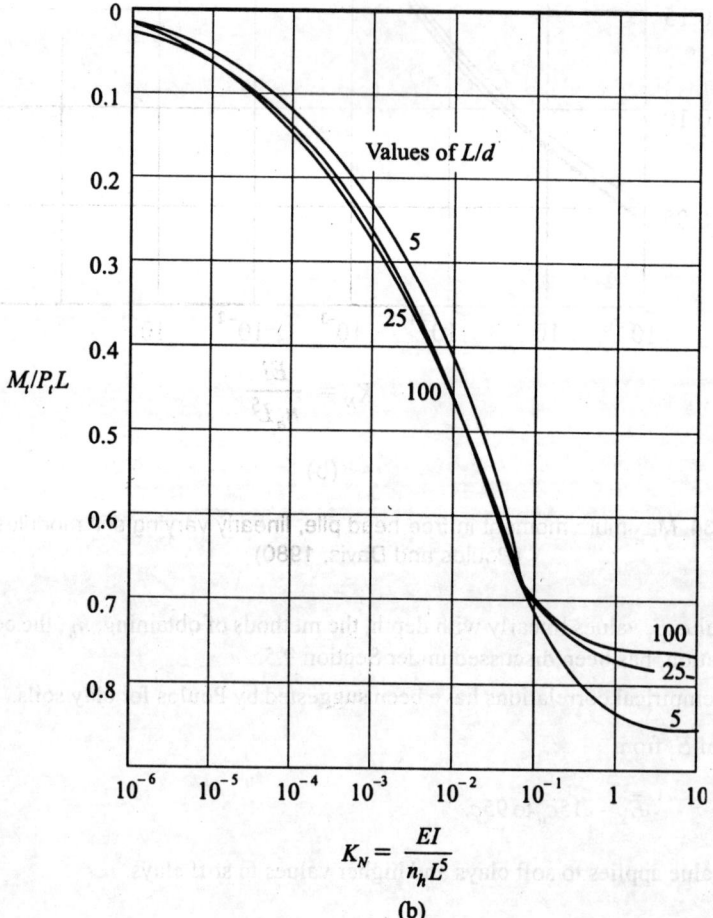

Values of L/d

$M_t/P_t L$

$$K_N = \frac{EI}{n_h L^5}$$

(b)

Fig. 9.35 (a) Values of I_{yf} for fixed-head floating pile, linearly varying soil modulus, (b) fixing moment in fixed-head pile linearly varying soil modulus (Poulos and Davis, 1980)

Example 9.20

A steel pipe pile of outside diameter 61 cm, and wall thickness 2.5 cm is driven into medium dense sand under submerged condition upto a depth of 20 m. The relative density of sand is 30%. The EI of the pile is 4.35×10^{11} kg-cm^2. The coefficient of soil modulus variation n_h as per Fig. 9.5 is 6 MN/m^3. Compute the lateral deflection for a lateral load of 268 kN applied a height of 2 m above the ground level for free head condition by Poulos method.

Solution

Use Eq. (9.110a)

$$y_g = I'_{yp} \left(\frac{P_t}{n_h L^2} \right) + I'_{ym} \left(\frac{M_t}{n_h L^3} \right)$$

From Eq. (9.111)

$$k_N = \frac{EI}{n_h L^5} = \frac{4.35 \times 10^2}{6 \times 20^5} = 2.26 \times 10^{-5}$$

$$L/d = \frac{20}{0.61} \approx 33$$

From Fig. 9.33 (a), for $K_N = 2.26 \times 10^{-5}$, $L/d = 33$, $I'_{yp} \approx 250$.

From Fig. 9.33 (b), $I'_{ym} = 1600$,

$$M_t = 2.268 \times 2 = 0.536 \text{ MN–m},$$

$$y_g = \frac{250 \times 0.268}{6 \times 20^2} + \frac{1600 \times 0.536}{6 \times 20^3}$$

$$= 0.0279 + 0.0179 = 0.0458 \text{ m} \approx 4.58 \text{ cm}$$

Poulos method gives greater deflection as compared to Reese method (Ex. 9.3).

PART B: BATTER PILES IN COHESIONLESS SOILS

9.21 MECHANISM OF FAILURE OF BATTER PILES UNDER LATERAL LOADS IN COHESIONLESS SOILS

First consider an 'in' batter pile subjected to lateral loads for the purpose of analysis as shown in Fig. 9.36 (a). The equilibrium of forces normal to the pile surface requires that

$$P_1 \cos \beta = E'_1 - E''_1 \tag{9.114a}$$

where,

$$E''_1 = f_1 P_1 \cos \beta = \left(\frac{a_1}{b_1} \right) P_1 \cos \beta \text{ and}$$

$$E'_1 = (1 + f_1) P_1 \cos \beta \tag{9.114b}$$

The resistance of the pile to the axial pull depends on frictional forces which (in a cohesionless soil) are proportional to the normal pressures on the pile surface:

$$P_1 \sin \beta = F'_1 + F''_1$$

where,

$$F'_1 = \frac{P_1 \sin \beta (E_1')}{(E_1' + E_1'')} = \frac{P_1 \sin \beta (1 + f_1)}{(1 + 2 f_1)} \tag{9.114c}$$

Failure of soil around a pile subjected to this type of loading starts at the soil surface. The depth of the failure zone depends on the flexibility of the pile. For purposes of this comparative analysis it will be assumed that the pile is rigid so that its stability will depend on the resistance of the soil above the point O_1. The horizontal and vertical components of all the forces above that point are:

$$H'_1 = E'_1 \cos \beta + F'_1 \sin \beta$$

$$= P_1 \left[\cos^2 \beta (1 + f_1) + \frac{\sin^2 \beta (1 + f_1)}{(1 + 2 f_1)} \right] \tag{9.114d}$$

and

$$V'_1 = E'_1 \sin \beta - F'_1 \cos \beta$$

$$= P_1 \left[\sin \beta \cos \beta (1 + f_1) - \frac{\sin \beta \cos \beta (1 + f_1)}{(1 + 2 f_1)} \right]$$

$$= P_1 \sin 2\beta \, (f_1 + f_1^2)/(1 + 2f_1) \tag{9.114e}$$

Similar considerations to the case of an 'out' batter pile [Fig. 9.36 (b)], that is $+ \beta$ angle, show that

$$P_2 \cos \beta = E'_2 - E''_2 \tag{9.115a}$$

where,

$$E''_2 = f_2 P_2 \cos \beta = \left(\frac{a_2}{b_2} \right) \cos \beta \tag{9.115b}$$

$$P_2 \sin \beta = F'_2 + F''_2 + F_p \tag{9.115c}$$

where, F_p is the point resistance of the pile, and

$$F'_2 = \frac{(P_2 \sin \beta - F_p) E_2'}{(E_2' + E_2'')}$$

$$= \frac{(P_2 \sin \beta - F_p)(1 + f_2)}{(1 + 2 f_2)} \tag{9.115d}$$

$$H'_2 = E'_2 \cos \beta + F'_2 \sin \beta$$

$$= P_2 \left[\cos^2 \beta (1 + f_2) + \frac{\sin^2 \beta (1 + f_2)}{1 + 2 f_2} \right] - \sin \beta \, F_p \frac{(1 + f_2)}{(1 + 2 f_2)} \tag{9.115e}$$

$$V'_2 = - E'_2 \sin \beta + F'_2 \cos \beta$$

$$= -\left[\frac{P_2 \sin 2\beta \left(f_2 + f_2^2\right)}{\left(1 + 2 f_2\right)} + \frac{\cos \beta \, F_p \left(1 + f_2\right)}{\left(1 + 2 f_2\right)}\right] \tag{9.115f}$$

If we assume that the component of the pile P_2 which acts along the axis of an 'out' batter pile is resisted only by friction (that is, $F_p = 0$), then a comparison of Eqs (9.114d) and (9.115e) shows that the horizontal components H_1' and H_2' of an 'in' and of an 'out' batter pile are in all respects identical provided $f_1 = f_2$. On the other hand, a comparison of Eqs (9.114e) and (9.115f) shows tha the vertical components V_1' and V_2' are identical if $F_p = 0$ and $f_1 = f_2$ except for opposite signs. In the case of an 'in' batter pile V_1' is directed downwards, and so is $W' = E_1' \tan \delta'$, where δ' is the partially effective angle of wall friction. The slip surfaces are therefore deflected downwards. In the case of an 'out' batter pile V_2' and W' are directed upwards, deflecting thereby, the slip surfaces upwards (Fig. 9.36). The shearing resistance along the slip surfaces of the type shown on Fig. 9.36 for an 'out' batter pile will be smaller than for an 'in' batter pile. Yielding will occur sooner for an 'out' batter pile so that P_2 will be smaller than P_1 and the point O_2 will be located lower than O_1.

The above analysis was advanced by Tschebotarioff (1953) to justify the lateral load tests results on batter piles of Matsuo (1939). It will be seen later on in this chapter that the author's work on model piles also corroborates the above analysis.

The failure mechanism of soil under lateral loads of piles in cohesionless soil is in agreement with the general theory of earth pressure, since walls where the face in contact with the soil has a positive β angle have smaller coefficients of passive lateral earth pressure than walls where the face in contact with the soil has a negative value of the angle β.

It is to be noted here that the component of the horizontal pull which acts along the axis of a batter pile does not change the relationship given in the above analysis.

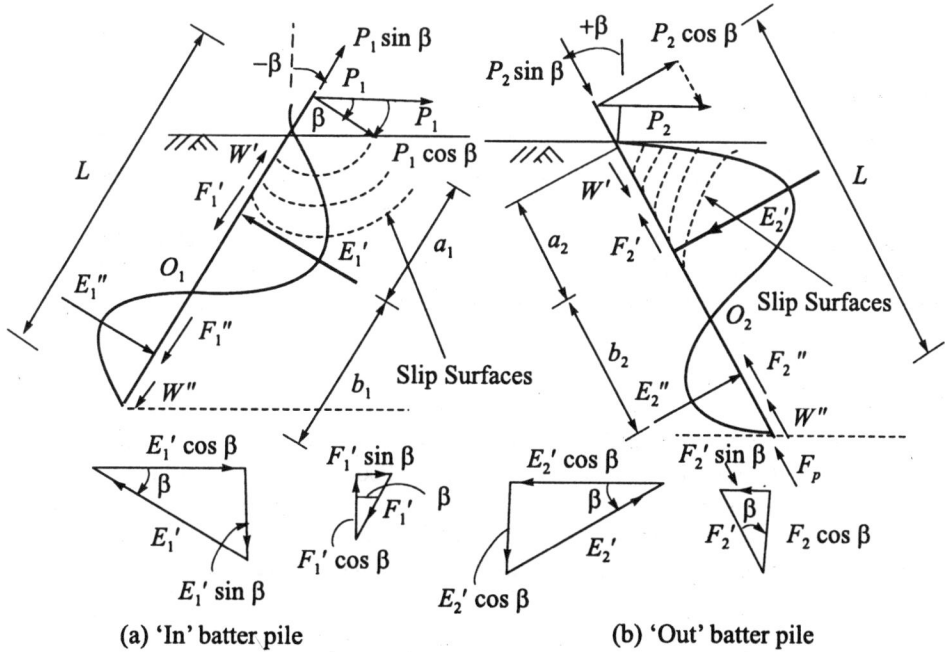

(a) 'In' batter pile (b) 'Out' batter pile

Fig. 9.36 Analysis of forces on batter piles (Tschebotarioff, 1953)

9.22 STATEMENT OF THE PROBLEM OF BATTER PILES SUBJECTED TO LATERAL LOADS

As in the case of a vertical pile subjected to lateral loads, the objective of the study of long flexible batter piles subjected to lateral loads is to obtain the curves of (*a*) deflection, (*b*) slope, (*c*) moment, (*d*) shear and (*e*) soil reaction as a function of length along the pile. As of today (1990), there is no published literature which provides a solution to the problem of laterally loaded batter piles. The author's (Murthy, 1965) detailed experimental investigation on instrumented model piles have provided the necessary non-dimensional solution for the case of lateral loads acting at ground level. Research work is still required for the case where the load acts above the ground level. Necessary relationships have, however, been established to predict the behaviour of batter piles based on the known behaviour of vertical piles.

9.23 MODEL TESTS ON INSTRUMENTED BATTER PILES IN COHESIONLESS SOIL (MURTHY, 1965)

Model Piles and Instrumentation

The model piles used for the tests were aluminium alloy tubings of 0.75″ outside diameter and 0.035 ins wall thickness (ALCOA, 6061-T6). The flexural stiffness, *EI*, as found out from tests was 5.14×10^4 lb-in^2. The length of the pile used for the tests was 30 inches.

Seven piles were instrumented with electric resistance strain gauges for measuring flexural strains along the embedded portion of the pile. The gauges were fixed one inch apart for a length of 24 inches of the embedded portion of the pile. Two gauges were fixed diametrically opposite to each other at each level. Tatnall metal film epoxy strain gauges, type C12-121, 1/8 inch gauge length, were used. The gauges were temperature compensated. The strain gauge data were:

Resistance 120 ± 0.20 ohms.

Gauge factor 2.06 ± 0.5%.

Each strain gauge was calibrated for flexure and calibration constants were determined

Test Apparatus and Soil

The test tank used was a metal tank of length 4.25 ft length, 2.50 ft wide and 3.0 ft deep. Adjustable pulley arrangements were provided on either sides of the tank for applying lateral loads by means of cables tied to the piles and strung over pulleys.

Ennore standard sand in dry condition was used for the work because of its uniformity in size. The specific gravity and the uniformity coefficient of sand were 2.67 and 1.1 respectively. The tests were conducted at a density of 98 lb/ft^3 (Relative density, D_r = 67%) and the value of angle of internal friction as found out was 40°.

The Test

Seven instrumented piles were positioned in the tank at batters of 0°, 15°, 30° and 45°. There were two piles of each batter of 15°, 30° and 45° so that load could be applied in the direction of batter ('in' batter) and against the batter ('out' batter) at ground level. Care was taken to see that adequate distances between piles and, between piles and sides of tank were maintained to avoid interference between pile movements and side effects respectively. Sand was filled in layers and suitably vibrated to obtain uniform density.

Two series of tests on instrumented piles were conducted. One series with the load horizontal at ground level, and the other series with the load normal to the piles' axes. The loads on each of the piles were taken up to failure condition.

Lateral load tests on all th seven piles in each of the seven piles in each of the series were conducted one after the other either with the load horizontal or normal to the piles' axis. Loads were applied in stages in the direction of batter and against the batter on piles of the same batter.

Lateral deflections at ground level were measured by the use of mechanical dial gauges. Tepic Indicator of make Instrument Huggenberger, Zurich, was used for measuring bending strains on instrumented piles. Tinsley, rotary type, switching unit with 100 measuring points was used with the Tepic Indicator for connecting of large number of active gauges to the Bridge.

Data Processing of Instrumented Piles

The tests on each of the instrumented pile for each stage of loading yielded flexural strains. These flexural strains were converted to bending moments by multiplying it by the appropriate calibration constants.

Successive integration of bending moments distributed along the embedded length of a pile with the proper application of boundary conditions led to curves of rotation and deflection. Successive differentiation of a moment curve yielded curves of shear and soil reaction. Integration is a smoothing and averaging process and errors in the experimental data get reduced in the integrated curves. Whereas differentiation is a process that magnifies the experimental errors. There would be a very considerable scattering in the soil reaction values if the moments are not measured to reach a high order of precision and if the local fluctuations are not smoothed out by curve fitting.

Data processing was carried out on a digital computer. Orthogonal polynomials of degree 7 in x (where x is the distance along the pile from the ground level) was used to fit the observed moment data by Least Square approximations.

Distances were measured along the axis and deflections normal to the axis. The moment curves were obtained from fitted moment values. The other curves were obtained either by integration or differentiation of the fitted moment curves. Typical moment and soil reaction curves are given in Fig. 9.37 and 9.38 respectively for a lateral normal load of 10 lbs. For piles of all batter the very fact that the derived soil reaction curves were very smooth indicate the high order of precision obtained for the measured values.

9.24 VARIATION OF SOIL MODULUS ALONG BATTER PILES

Soil modulus, E_s, at any point on a pile at a distance x from the ground surface may be expressed as for a vertical pile as [Eq. (9.12)].

$$E_s = \frac{p}{y} \tag{9.12}$$

where, p = soil reaction per unit length of pile (F/L),

y = deflection normal to the axis of pile.

The values of E_s were calculated from the known values of p and y along the pile and plotted for all the piles tested. Curves were fitted to pass through the plotted points by using the equation of the power form [Eq. (9.13a)].

$$E_s = n_h x^n$$

where, n_h = coefficient of soil modulus variation.

The values of n_h and n were found out for each load level and for all the batters by the method of least squares. The variation of E_s with distance x along the piles for piles of all batter is given in Fig. 9.39 for a lateral load of 10 lbs.

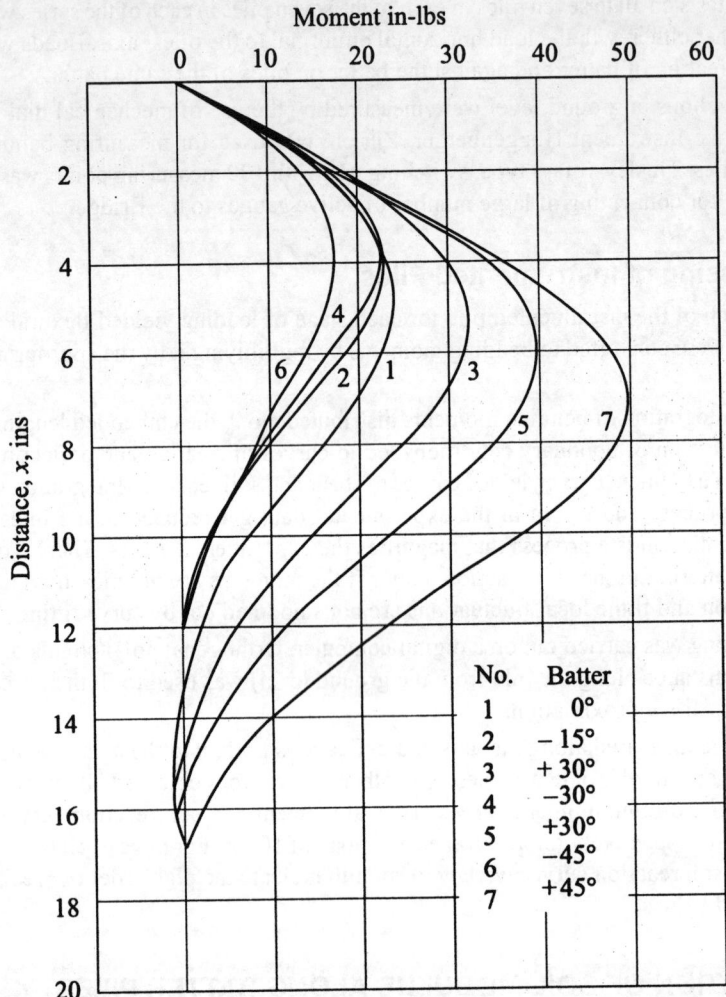

Fig. 9.37 Typical Experimental moment distribution Curves along piles of batter − 45° to + 45° (Murthy, 1965)

9.25 NON-DIMENSIONAL SOLUTIONS FOR LATERALLY LOADED BATTER PILES IN SAND (MURTHY, 1965)

Experimental non-dimensional factors A_y, A_s, A_v and A_p were obtained for all the piles tested ($-45°$, $-30°, -15°, 0°, +15°, +30°, +45°$), by making use of the following equations as applicable for a vertical pile.

$$A_y = \frac{yEI}{P_t T^3} \tag{9.116a}$$

$$A_s = \frac{SEI}{P_t T^2} \tag{9.116b}$$

$$A_m = \frac{M}{P_t T} \tag{9.116c}$$

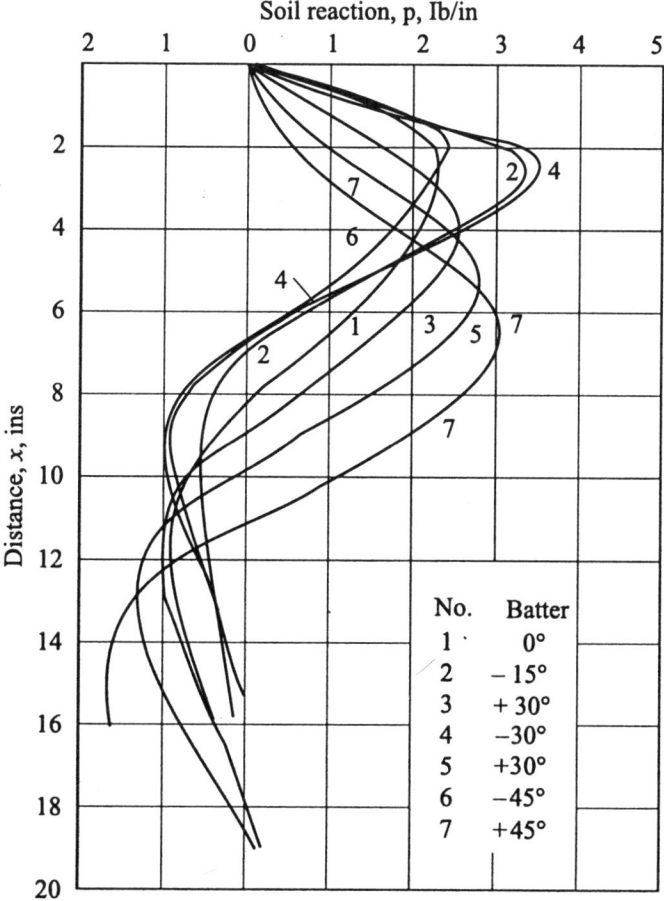

Fig. 9.38 Typical experimental soil reaction distribution curves along piles of batter − 45° to + 45° (Murthy, 1965)

$$A_v = \frac{V}{P_t} \qquad (9.116\text{d})$$

$$A_p = \frac{pT}{P_t} \qquad (9.116\text{e})$$

where, y = deflection normal to pile axis,
EI = flexural stiffness of pile,
P_t = lateral load at ground level normal to pile axis,
S = slope of pile at any point along the pile,
M = moment at any point on the pile axis,
V = shear at any point on the pile axis,
p = soil reaction per unit length of pile at any point on the pile axis.

The non-dimensional distance to any point on the pile axis from ground level is expressed as

$$D_c = \frac{x}{T} \qquad (9.117)$$

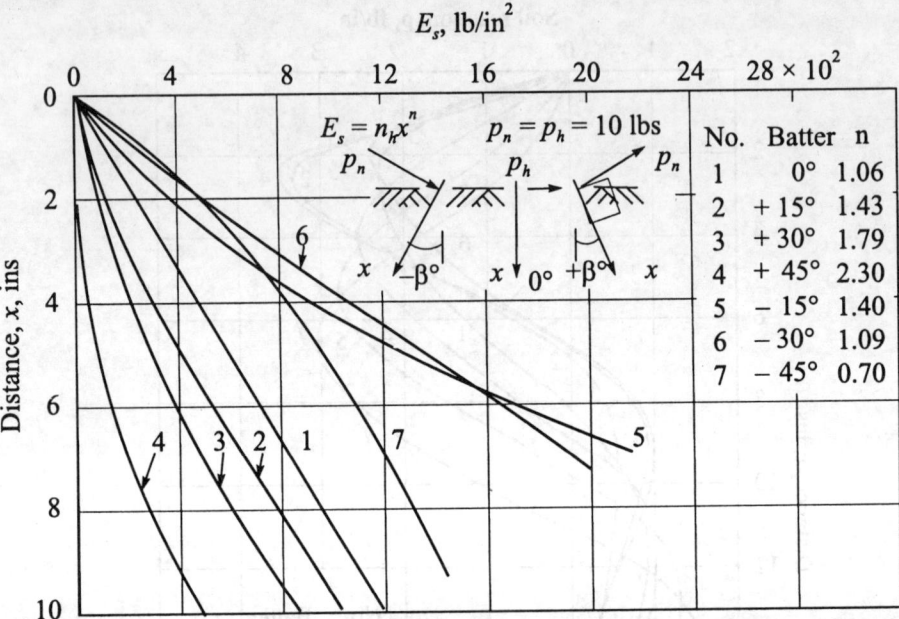

Fig. 9.39 Typical experimental curves showing the variation of E_s with x (distance along the pile) for piles of batter varying firm $-45°$ to $+45°$ (Murthy, 1965)

where, D_c is called as the *distance coefficient* which is the same as [Eq. (9.42)] Z, used as a depth coefficient for a vertical pile. The equation for the relative stiffness factor T is the same as Eq. (9.48).

$$T^{n+4} = \frac{EI}{n_h} \qquad (9.118)$$

where the value of n depends on the batter of pile.

The values of T can be calculated from Eq. (9.118) from the known values of n, EI and n_h for each of the piles tested and for each stage of loading. In Eqs (9.170) all the values on the right hand side are known, and as such the A-factors for different values of distance coefficient D_c can be computed. The values of A factors plotted against the distance coefficient D_c gives the corresponding non-dimensional curves. The average non-dimensional curves as obtained for all the piles tested are given in Figs 9.40 to 9.46.

Maximum values of the non-dimensional A-factors for deflection, moment and soil reaction are given in Table 9.11.

Table 9.11 Maximum values of non-dimensional A-factors for deflection, moment and soil reaction (after, Murthy, 1965)

Batter of pile	0°	+15°	+30°	+45°	−15°	−30°	−45°
A_y	2.27	2.40	2.52	2.76	2.20	2.36	1.76
A_m	0.75	0.78	0.88	0.98	0.72	0.74	0.58
A_p	0.90	1.02	1.14	1.43	0.89	0.98	0.78

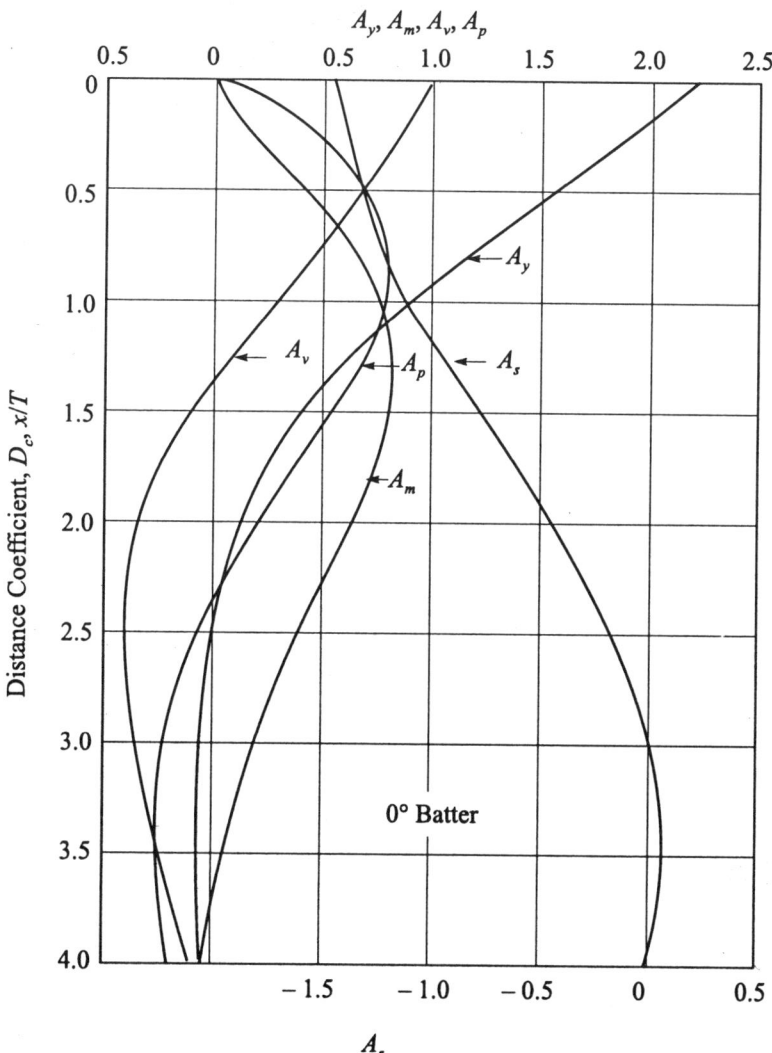

Fig. 9.40 Non-dimensional parameters A_y, A_m, A_v, A_p and A_s for vertical pile (Murthy, 1965)

9.26 RELATIVE STIFFNESS FACTOR FOR BATTER PILES IN SAND (MURTHY, 1965)

As per Eq. (9.118), the relative stiffness factor T is a function of n, n_h and EI. The value of n_h varies with the batter of pile and the load level, whereas the value of n may be taken as constant for any particular batter. From the known values of n, n_h and EI, the value of T can be calculated for any given batter and load level. A relationship can be established between the values of stiffness factors T_b for a pile of batter β and for a vertical pile. From model test results the following relationships have been established for equal normal loads.

For piles of batter $- 22.5°$ to $+ 45°$

$$T_b = T_0 (1 + 7.5\ 5\ 10^{-3}\beta) \tag{9.119a}$$

For piles of batter $- 22.5°$ to $- 45°$

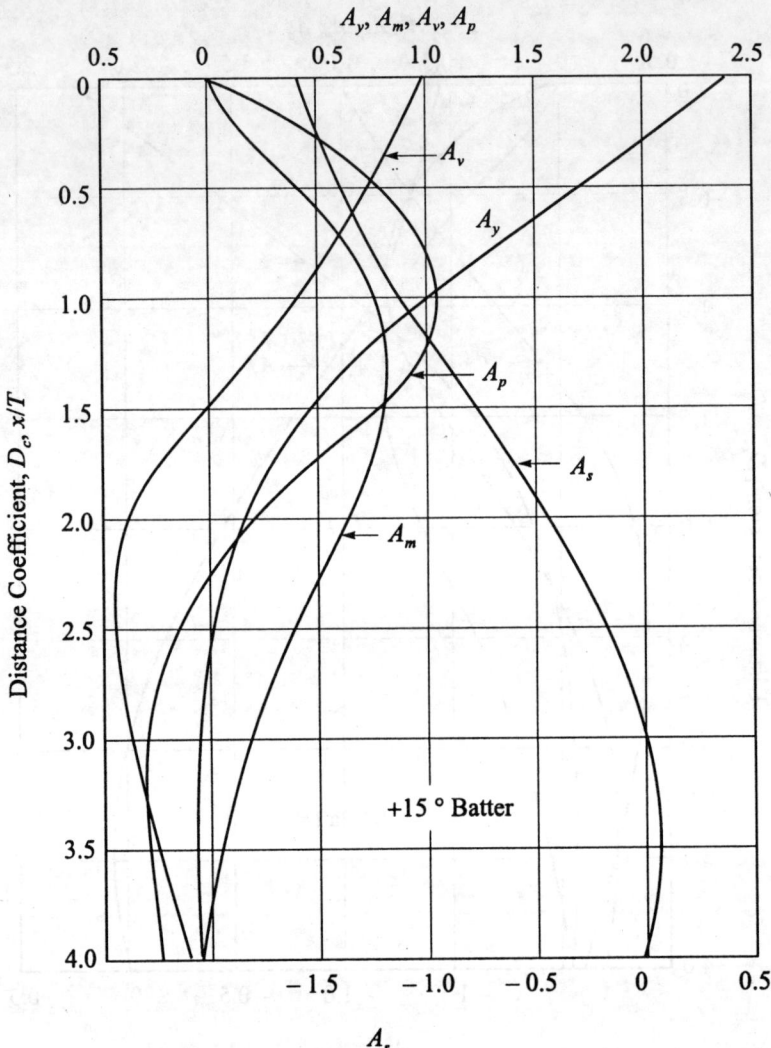

Fig. 9.41 Non-dimensional parameters A_y, A_m, A_v, A_p and A_s for + 15° batter pile (Murthy, 1965)

$$T_b = 0.86\ T_0 \tag{9.119b}$$

Here β is the angle of batter expressed in degrees, negative for 'in' batter and positive for 'out' batter piles. The Eq. (9.119) helps to compute T_b of batter pile from the known values of T_0 of vertical piles.

9.27 ULTIMATE LATERAL BEARING CAPACITY OF BATTER PILES IN SAND (MURTHY, 1965)

One series of tests on non-instrumented model piles with lateral loads horizontal were carried out up to failure. (Murthy, 1965) Fig. 9.47 gives the load-displacement curves as obtained from the test. As can be seen from the curves, the resistance to lateral load increases in the order of the batter of + 45°, + 30°, + 15°, 0°, – 15°, – 30° and – 45°. However at higher lateral loads the resistance of negative batter piles depend on its resistance to pull out under the axial component of the horizontal load. In the case of a −45° batter pile, the pile got pulled out when the lateral load was increased from 50 lb to 55 lbs (Fig. 9.47). It was observed during the tests, that in all the cases of piles with positive batter, the separation of soil grains from the pile at the rear started at loads under 15 lbs, whereas for piles with negative batter, separation took place at much higher loads. The reasons for higher resistances of 'in' batter piles over 'out' batter piles have been analysed earlier.

There is no standard method by which to establish the ultimate lateral bearing capacity of piles. For the purpose of comparison, it is assumed here that the pile has failed at a lateral displacement of 0.3 inch. Ratios of ultimate later resistance of batter piles to that of vertical piles have been worked and plotted in Fig. 9.48. It is clear from this plot that the lateral resistance of negative batter pile increases from unity at 0° batter to a maximum 1.2 at −30° batter and then drops to 0.94 at −45° batter. Where as for the positive batter piles the resistance decreases from unity at 0° batter to 0.70 at + 45° batter. The ratios will be somewhat different if any other lateral displacement is considered. Higher ratios can be obtained at lower displacements.

Awad and Petrasovits (1968) carried out similar tests on vertical and batter piles driven into cohesionless soils. They also used aluminium alloy pipe. They used lengths of 20, 35 and 50 cm and

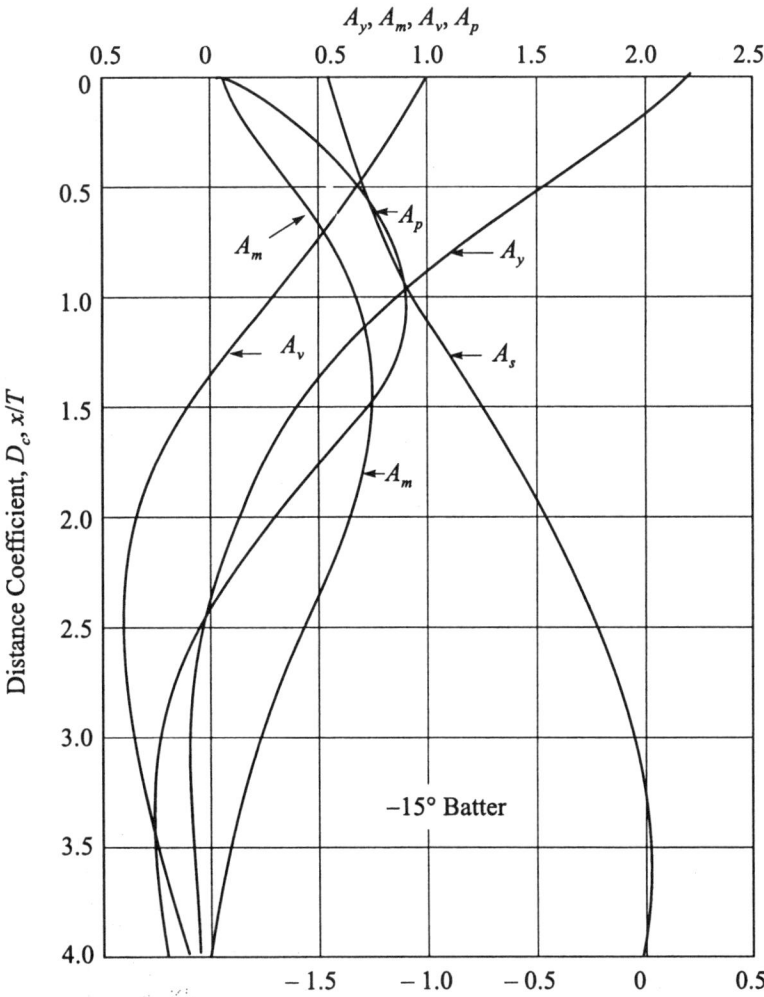

Fig. 9.42 Non-dimensional parameters A_y, A_m, A_v, A_p and A_s for − 15° batter pile (Murthy, 1965)

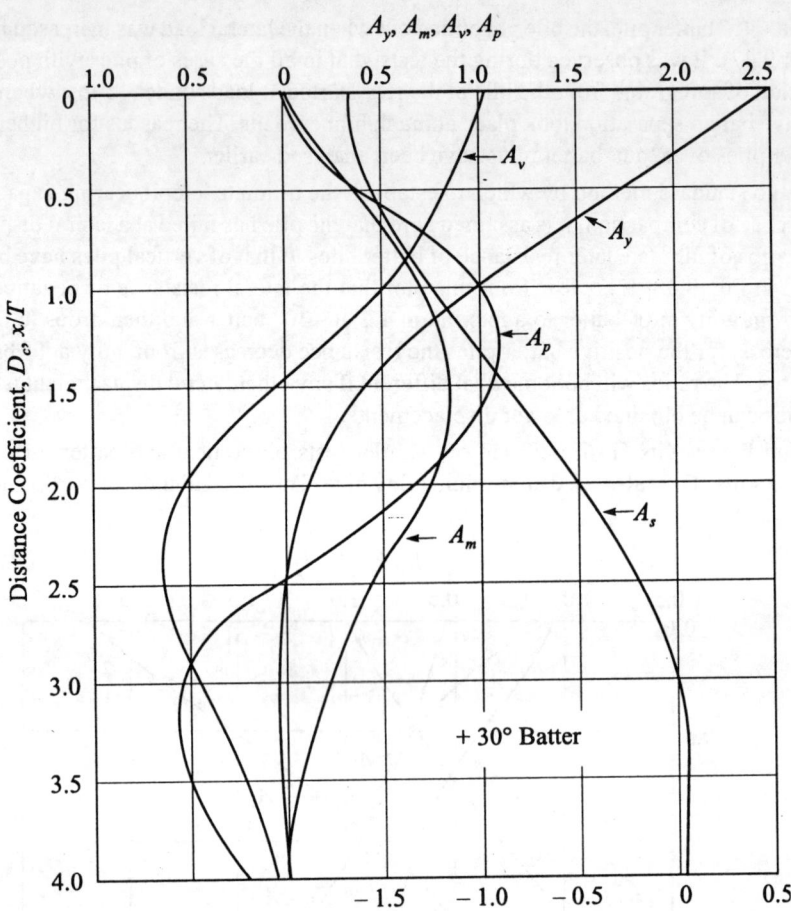

Fig. 9.43 Non-dimensional parameters A_y, A_m, A_v, A_p and A_s for $+30°$ batter pile (Murthy, 1965)

the diameter of all piles was 20 mm. Horizontal load was applied at about 12 cm above ground level. They considered a horizontal displacement $y = 25$ mm at the level of load application as a measure for the bearing capacity. Table 9.12 gives a comparison between the results. Obtained by the author and that of Awad and Petrasovits for the pile of 50 cm length. There is somewhat close agreement between the two investigators.

Table 9.12 Ratio $P^\beta u/P^o{}_u$ for piles of different batter

Batter	$-45°$	$-37.5°$	$-30°$	$-15°$	$0°$	$+15°$	$+22.5°$
1. Values of P_u^β/P_u^o by Awad et al.	0.93	1.42	1.33	1.13	1.0	0.86	0.75
2. Value of P_u^β/P_u^o by the author	0.94	1.16	1.20	1.10	1.0	0.90	0.70

9.28 LATERAL RESISTANCE OF BATTER PILES AS A RATIO TO THAT OF VERTICAL PILE IN SAND (MURTHY, 1965)

Displacements within elastic limits are considered for determining the ratios of lateral displacements of batter piles to that of vertical pile. The load-displacement curves as obtained on instrumented piles subjected to horizontal loads have been used to compute the ratios P_h^β where P_h^β and P_h^o are the horizontal loads of batter and vertical piles for equal horizontal displacements. The results have been plotted in Fig. 9.49. It can be seen from this figure the ratios R_{lh} $(= P_h^\beta / P_h^o)$ is the highest equal to 1.6 for $-45°$ batter pile and least equal to 0.77 for $+45°$ batter pile. Though this curve is obtained for lateral load at ground level, it is assumed that it may be applicable for loads applied above ground level as well. The curve in Fig. 9.49 is quite useful for computing lateral loads of batter piles for any assumed horizontal displacement provided the lateral load of a vertical pile is known for the corresponding displacement.

9.29 COEFFICIENTS OF PASSIVE EARTH PRESSURE FOR BATTER PILES IN SAND

The magnitude of the coefficient of earth pressure for a pile under three-dimensional passive condition is very much more than that for a wall in a two-dimensional state. It is possible to compute the three dimensional passive earth pressure coefficients $\bar{K}_P$ for piles of different batter from the data available from pile load tests. First it is necessary to establish an equation for computing $\bar{K}_P$.

Consider a pile with a positive batter β, and diameter d as shown in Fig. 9.50 (inset). Let point a on the pile at depth f_m from the ground level represent the point of maximum moment at which point the shear is zero under a normal load P_n acting at point O.

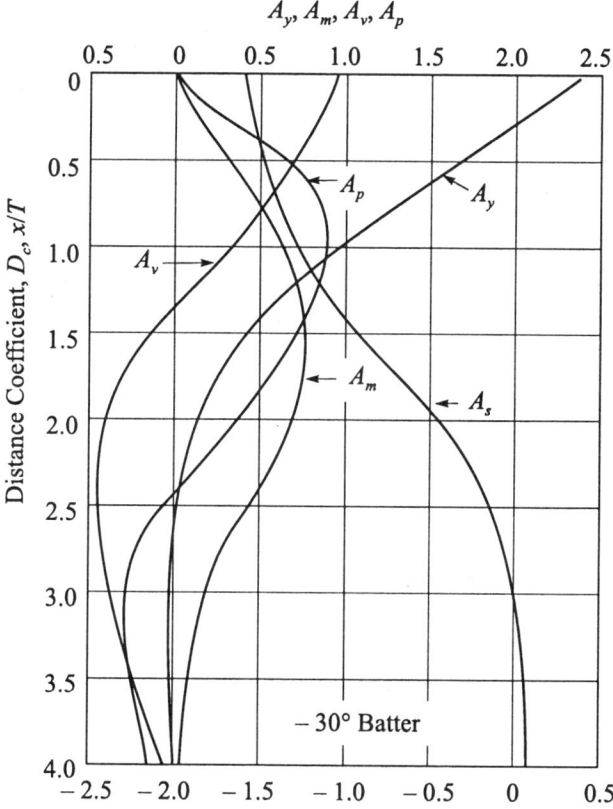

Fig. 9.44 Non-dimensional parameters A_y, A_m, A_v, A_p and A_s for $-30°$ batter pile (Murthy, 1965)

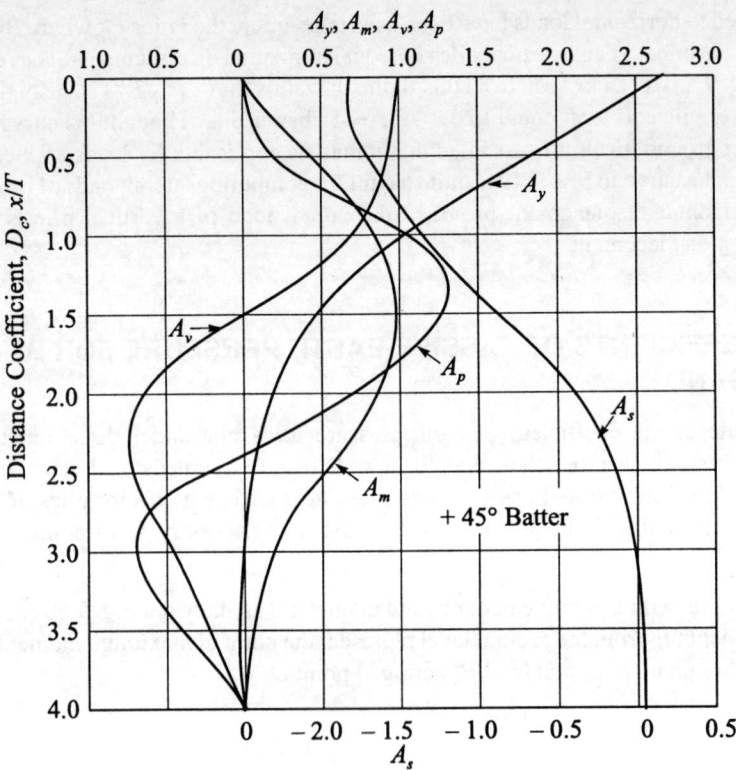

Fig. 9.45 Non-dimensional parameters A_y, A_m, A_v, A_p and A_s for $+45°$ batter pile (Murthy, 1965)

The normal component of the passive earth pressure per unit area of the surface of pile at depth z from ground level may be written as

$$p_p = \gamma z \overline{K}_P \tag{9.120a}$$

where, γ is the unit weight of soil. The passive earth pressure (which is the soil reaction) per unit length of pile of diameter d at point a is

$$p = \gamma d f_m \overline{K}_P \tag{9.120b}$$

Now in Fig. 9.50 (inset), the triangle Oab represents the linear distribution of soil reaction under ultimate limit condition. The area of this triangle must be equal to the load P_n applied normal to the pile axis at point O. Therefore we may write

$$P_n = \frac{1}{2} \gamma d f f_m \overline{K}_P \tag{9.120c}$$

where, $f = Oa$ along the pile axis.

Since $$f = f_m / \cos \beta$$

$$P_n = \frac{1}{2} \frac{\gamma d f_m^2 \overline{K}_P}{\cos \beta} \tag{9.120d}$$

where, β is the batter of pile.

The equation for $\bar{K}_P$ may now be written as

$$\bar{K}_P = \frac{2 P_n \cos \beta}{\gamma \, df_m^2} = \frac{2 P_h \cos^2 \beta}{\gamma \, df_m^2} \qquad (9.120e)$$

where, $P_n = P_h \cos \beta$, P_h = horizontal load.

One series of tests on instrumented piles were carried out with the loads P_n normal to the piles axis at groundlevel. In Eq. (9.120e) the depth to the point of maximum moment f_m can be found out from the fitted moment curves for each stage of loading. For any given batter, β, all the values on the right hand side of Eq. (9.120e) are known $\bar{K}_P$ may be calculated for each stage of loading. The value of $\bar{K}_P$ increases with the increase in the load P_h and remains constant at ultimate load condition. This can be found out by a plot of $\bar{K}_p$ vs. P_h for piles of each batter.

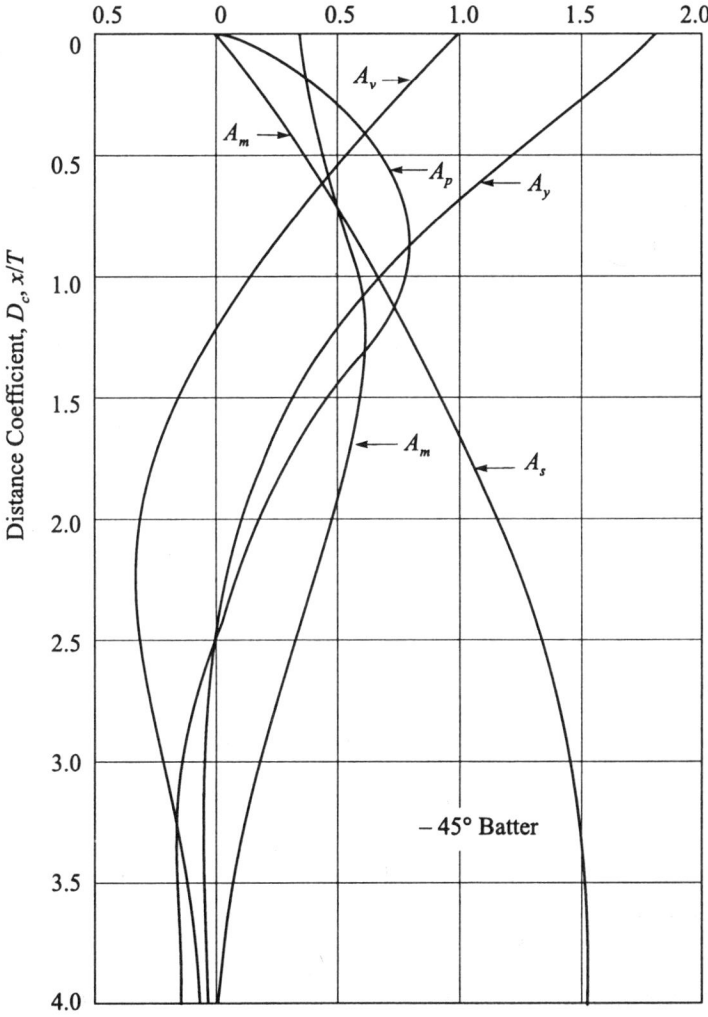

Fig. 9.46 Non-dimensional parameters A_y, A_m, A_v, A_p and A_s for $-45°$ batter pile (Murthy, 1965)

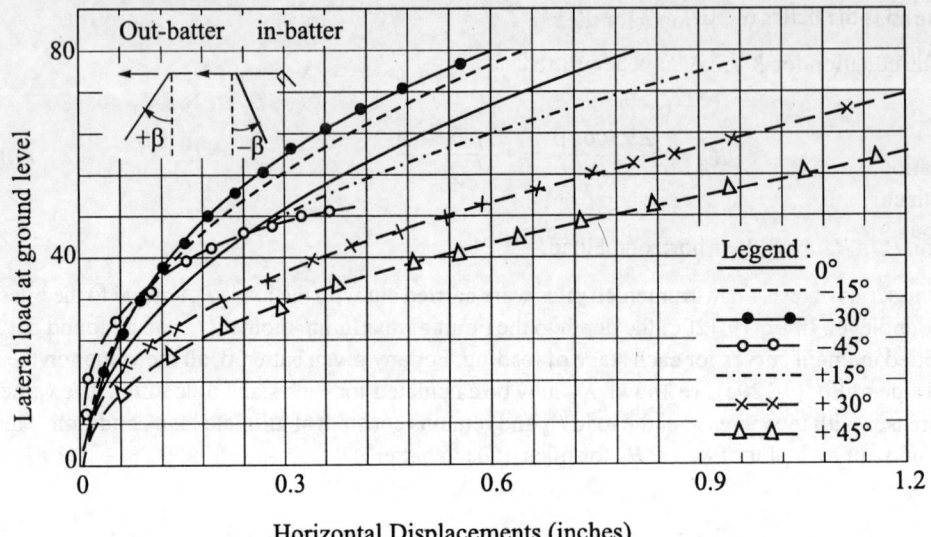

Fig. 9.47 Lateral load-horizontal displacement curves for piles of batter varying from −45° to 45° (Murthy, 1965)

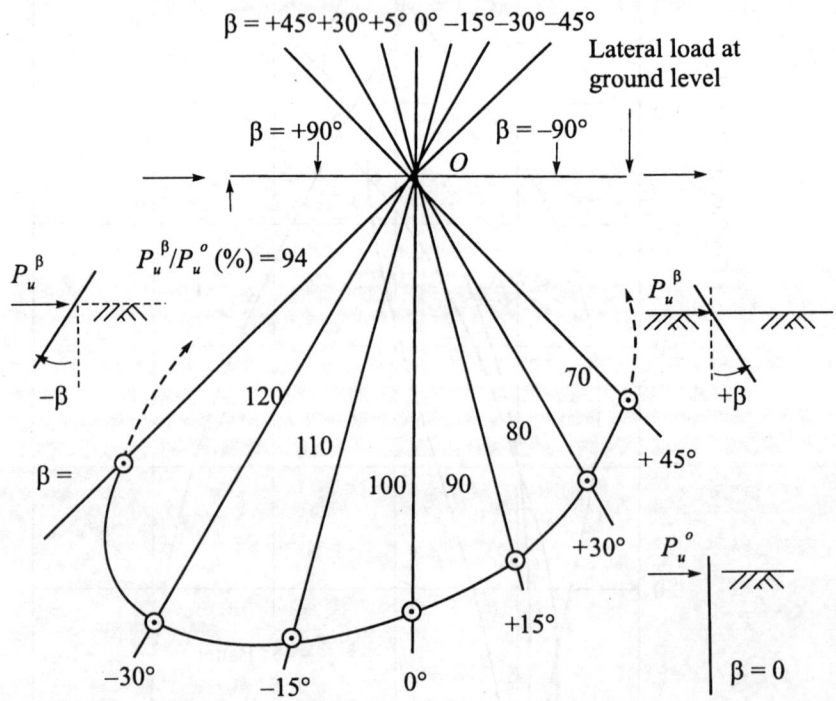

Fig. 9.48 Effect of batter on ultimate lateral resistance of piles (Murthy, 1965)

Figure 9.50 gives a plot of $\bar{K}_P$ *vs.* batter of pile. It can be seen from this plot that $\bar{K}_P$ increases from a minimum of 15 at +45° batter to a maximum of 89 at −37.5° batter and then decreases to 88 at −45° batter. The value of $\bar{K}_P$ for a vertical pile is 48.

The angle of internal friction ϕ for the sand used in the test at a relative density of 67 percent was 40°. The Rankine's passive earth pressure coefficient K_{PR} is

$$K_{PR} = \tan^2 (45° + \phi/2) = \tan^2 65° = 4.6$$

The ratios $\bar{K}_P / \bar{K}_{PR}$ ($= R_P$) as found out for piles of each batter has been plotted in Fig. 9.50 and a smooth curve is drawn. It can be seen from this figure that the ratio $\bar{K}_P / \bar{K}_{PR}$ varies from a minimum of about 3 for a pile of $+45°$ batter to a maximum of about 19 for a $-37.5°$ batter. The ratio is about

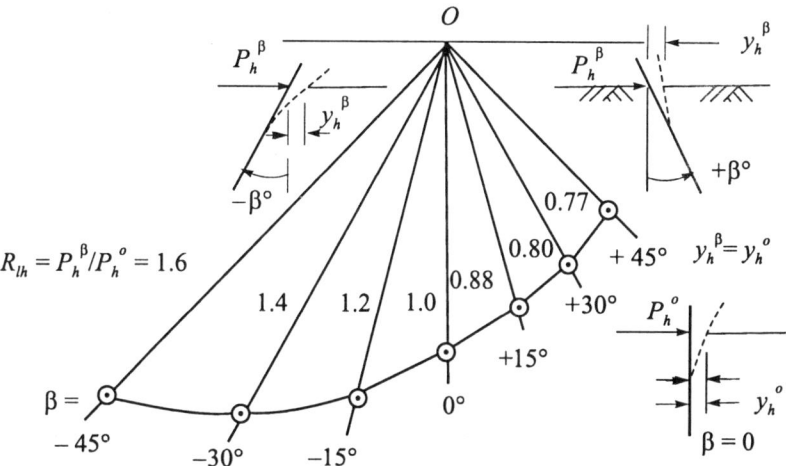

Fig. 9.49 Effect of batter on lateral resistance of pile for equal horizontal displacements (Murthy, 1965)

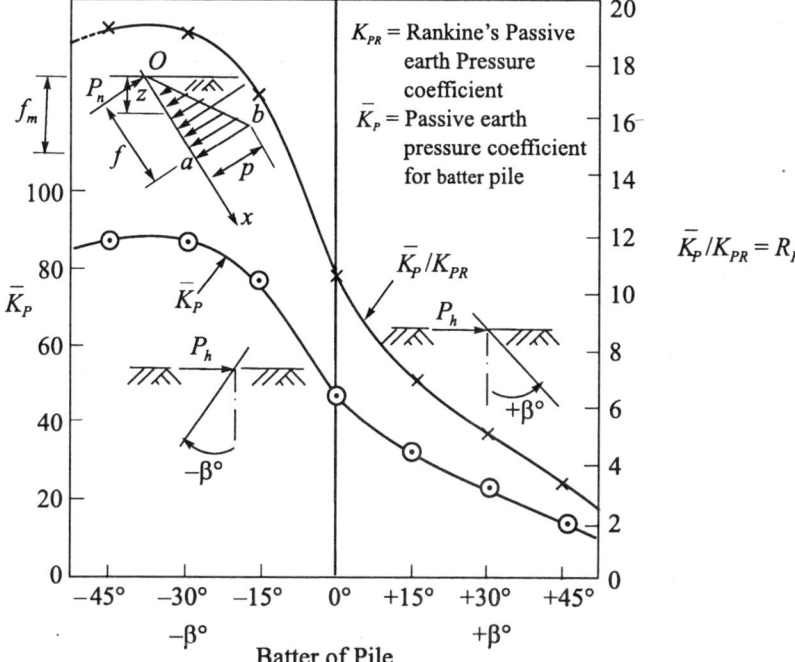

Fig. 9.50 Effect of batter of piles on passive earth pressure coefficients (Murthy, 1965)

10 for a vertical pile. Broms has used a ratio of 3 for vertical piles for computing ultimate lateral resistance which seems to be quite conservative.

9.30 BEHAVIOUR OF LATERALLY LOADED BATTER PILES IN SAND (MURTHY, 1965)

General Considerations

The earlier sections dealt with the behaviour of long vertical piles. The author has so far not come across any rational approach for predicting the behaviour of batter piles subjected to lateral loads. He has been working on this problem for a long time (Murthy, 1965). Based on the work done by the author and others, a method for predicting the behaviour of long batter piles subjected to lateral load has now been developed.

Model Tests on Piles in Sand (Murthy, 1965)

A series of seven instrumented model piles were tested in sand with batters varying from 0 to $\pm 45°$ as has been explained in section 9.23. Based on the test results a relationship was established between the n_h^b values of batter piles and n_h^o values of vertical piles. Figure 9.51 gives this relationship between n_h^b/n_h^o and the angle of batter β. It is clear from this figure that the ratio increases from a minimum of 0.1 for a positive 30° batter pile to a maximum of 2.2 for a negative 30° batter pile. The values obtained by Kubo (1965) are also shown in this figure. There is close agreement between the two.

The other important factor in the prediction is the value of n in Eq. (9.13a). The values obtained from the experimental test results are also given in Fig. 9.51. The values of n are equal to unity for vertical and negative batter piles and increase linearly for positive batter piles up to a maximum of 2.0 at 30° batter. (Average values are shown).

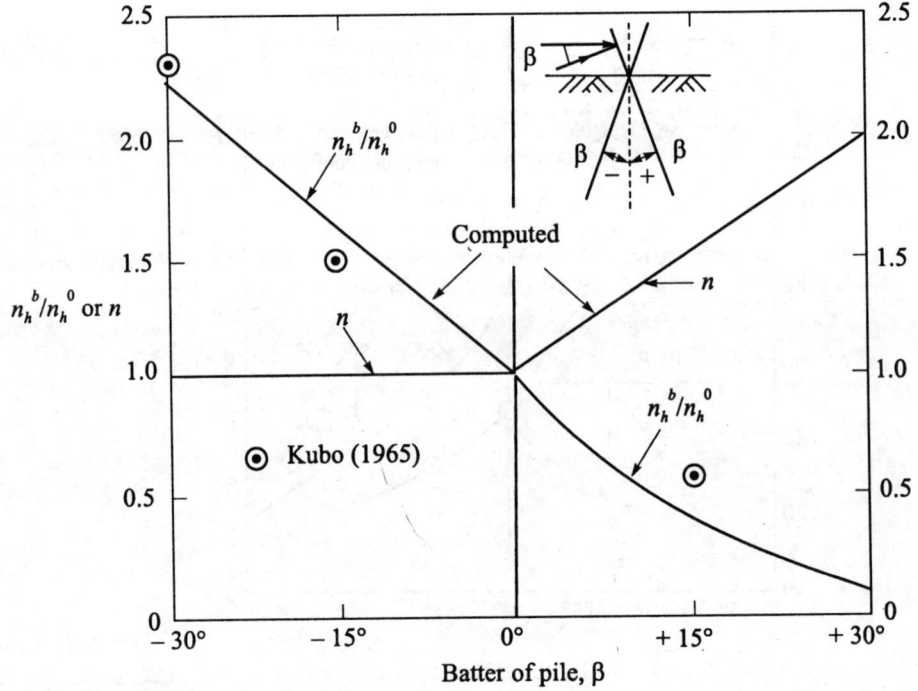

Fig. 9.51 Effect of batter on n_h^b/n_h^o and n (after Murthy, 1965)

In the case of batter piles the loads and deflections are considered normal to the pile axis for the purpose of analysis. The corresponding loads and deflections in the horizontal direction may be written as

$$P_t \, (Hor) = \frac{P_t \, (Nor)}{\cos \beta} \tag{9.121a}$$

$$y_g \, (Hor) = \frac{y_g \, (Nor)}{\cos \beta} \tag{9.121b}$$

where $P_t(Nor)$ and $y_g(Nor)$ are normal to the pile axis; P_t (Hor) and y_g (Hor) are the corresponding horizontal components.

Application of the use of n_h^b/n_h^o and n

It is possible now to predict the non-linear behaviour of laterally loaded batter piles in the same way as for vertical piles by making use of the ratio n_h^b/n_h^o and the value of n. The validity of this method is explained by considering a few case studies.

Case Studies

Case 1: Model pile test (Murthy, 1965)

Piles of + 15° and + 30° batters have been used here to predict the P_t vs. y_g and P_t vs. M_{max} relationships. The properties of the pile and soil are given below.

$$EI = 5.14 \times 10^4 \text{ lb in}^2, \, d = 0.75 \text{ in}, \, L = 30 \text{ in}; \, e = 0$$

$$\gamma = 98 \text{ lb/ft}^3 \text{ and } \phi = 40°$$

From Eq. (9.89b) for $\phi = 40°$, $C_\phi = 1.767 \, [= 3 \times 10^{-5} (1.316)^\phi]$

From Eq. (9.92a), $\quad n_h^o = \dfrac{150 \, C_\phi \, \gamma^{1.5} \sqrt{EId}}{P_t}$

After substituting the known values and simplifying we have

$$n_h^o = \frac{700}{P_t}$$

Solution

+ 15° batter pile

From Fig. 9.51 $\quad n_h^b/n_h^o = 0.4, \, n = 1.5$

From Eq. (9.50), $\quad T_b = \left(\dfrac{EI}{n_h^b} \right)^{\frac{1}{1.5+4}} = 5.33 \left(\dfrac{5.14}{n_h^b} \right)^{0.1818}$

Calculations of deflection y_g

For $P_t = 5$ lbs, $n_h^o = 141$ lbs/in^3, $n_h^b = 141 \times 0.4 = 56$ lb/in^3 and $T_b = 3.5$ in

$$y_g = \frac{2.43 \, P_t^3 \, T_b^3}{5.14 \times 10^4} = 0.97 \times 10^{-2} \text{ in}$$

Similarly, y_g can be calculated for $P_t = 10$, 15 and 20 lbs.

The results are plotted in Fig. 9.52 along with the measured values of y_g. There is a close agreement between the two.

Calculation of maximum moment, M_{max}

For $P_t = 5$ lb $T_b = 3.5$ in, The equation for M is [Eq. (9.53)]

$$M = A_m P_t T_b = 0.77 P_t T_b$$

where $A_m = 0.77$ (max) from Tale 9.4.

By substituting and calculating, we have

$$M_{(max)} = 15 \text{ in-lb}$$

Similarly $M_{(max)}$ can be calculated for other loads. The results are plotted in Fig. 9.52 along with the measured values of $M_{(max)}$. There is very close agreement between the two.

+30° batter pile

From Fig. 9.51, $n_h^b / n_h^o = 0.1$, and $n = 2$; $T_b = \left(\dfrac{EI}{n_h^b} \right)^{\frac{1}{2+4}} = 4.64 \left(\dfrac{5.14}{n_h^b} \right)^{0.1667}$, $n_h^o = \dfrac{700}{P_t}$

For $P_t = 5$ lbs, $n_h^o = 141$ lbs/in³, $r_h^b = 0.1 \times 141 = 14.1$ lb/in³, $T_b = 3.93$ in.

For $P_t = 5$ lbs, $T_b = 3.93$ in, we have, $y_g = 1.43 \times 10^{-2}$ in

As before, $M_{(max)} = 0.77 \times 5 \times 3.93 = 15$ in-lb.

The values of y_g and $M_{(max)}$ for other loads can be calculated in the same way. Figure 9.52 gives P_t vs. y_g and P_t vs. $M_{(max)}$ along with measured values. There is close agreement up to about $P_t = 10$ lb, and beyond this load, the measured values are greater than the predicted by about 25 percent

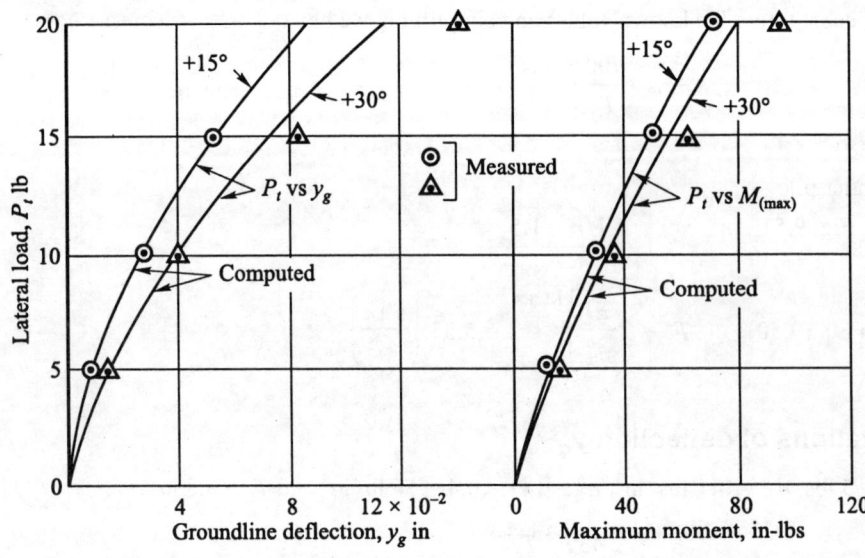

Fig. 9.52 Model piles of batter +15° and +30° (Murthy, 1965)

which is expected since the soil yields at a load higher than 10 lb at this batter and there is a plastic flow beyond this load.

Case 2: Arkansas river project (pile 12) (Alizadeh and Davisson, 1970)

Given: $EI = 278.5 \times 10^8$ lb-in^2, $d = 14$ in, $e = 0$.

$\phi = 41°$, $\gamma = 63$ lb/ft^3, $\beta = 18.4°$ (−ve)

From Fig. 9.17, $C_\phi = 2.33$, from Fig. 9.51 $n_h^b / n_h^o = 1.7$, $n = 1.0$

From Eq. (9.92a), after substituting the known values and simplifying, we have,

$$(a)\ n_h^o = \frac{1528 \times 10^3}{P_t}, \text{ and } (b)\ T_b = 39.8 \left(\frac{278.5}{n_h^b}\right)^{0.2}$$

Calculation for $P_t = 12.6^k$

From Eq. (a), $n_h^o = 121$ lb/in^3; now $n_h^b = 1.7 \times 121 = 206$ lb/in^3

From Eq. (b), $T_b = 42.27$ in

$$y_g = \frac{2.43 \times 12,600\,(42.27)^3}{278.5 \times 10^8} = 0.083 \text{ in}$$

$$M_{(max)} = 0.77\,P_t T = 0.77 \times 12.6 \times 3.52 = 34 \text{ ft-kips.}$$

The values of y_g and $M_{(max)}$ for $P_t = 24.1^k$, 35.5^k, 42.0^k, 53.5^k, 60^k can be calculated in the same way the results are plotted the Fig. 9.53 along with the measured values. There is a very close agreement between the computed and measured values of y but the computed values of M_{max} are

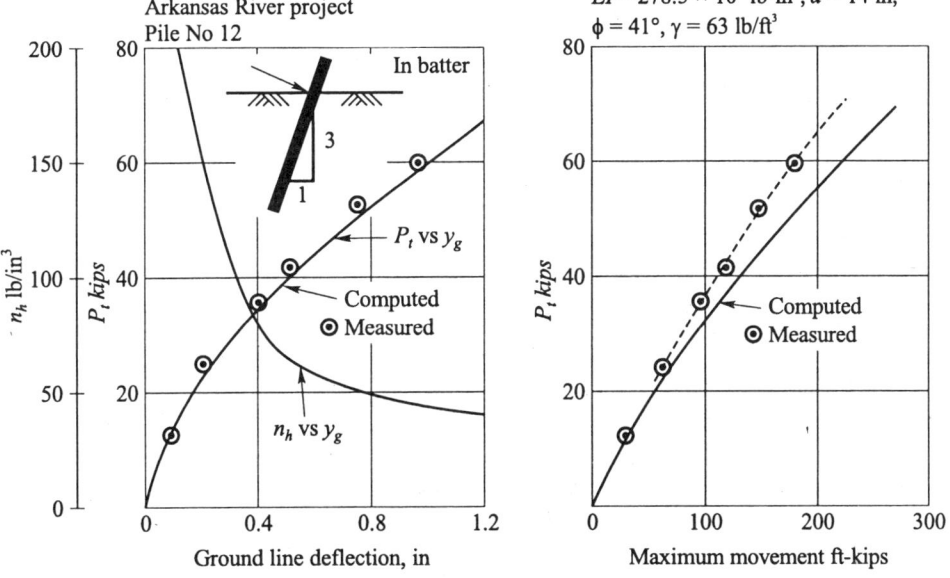

Fig. 9.53 Lateral load test-batter pile 12-arkansas river project (Alizadeh and Davisson, 1970)

higher than the measured values at higher loads. At a load of 60 kips, $M_{(max)}$ is higher than the measured by about 23% which is quite reasonable.

Case 3: Arkansas river project (pile 13) (Alizadeh and Davisson, 1970)

Given: $EI = 288 \times 10^8$ lb-ins, $d = 14"$, $e = 6$ in.

$\gamma = 63$ lbs/ft^3, $\phi = 41°$ ($C_\phi = 2.33$)

$\beta = 18.4°$ (+ vex), $n = 1.6$ $n_h^b/n_h^o = 0.3$

$$T_b = \left(\frac{EI}{n_h^b}\right)^{\frac{1}{1.6+4}} = 27 \left(\frac{288}{n_h^b}\right)^{0.1786} \qquad \text{... (a)}$$

After substituting the known values in the equation for n_h^o [Eq. (9.92a)] and simplifying, we have

$$n_h^o = \frac{1597 \times 10^3}{P_e} \qquad \text{... (b)}$$

Calculations for y_g for $P_t = 141.4k$

1. From Eq. (b), $n_h^o = 39$ lb/in^3, hence $n_h^b = 0.3 \times 39 = 11.7$ lb/in^3

   From Eq. (a), $T_b \approx 48$ in.

2. $P_e = P_t \left(1 + 0.67 \dfrac{e}{T}\right) = 41.4 \left(1 + 0.67 \times \dfrac{6}{48}\right) = 44.86$ kips

3. For $P_e = 44.86$ kips, $n_h^o = 36$ lb/in^3, and $n_h^b = 11$ lb/in^3, $T_b \approx 48$ in

4. Final values: $P_e = 44.86$ kips, $n_h^b = 11$ lb/in^3, and $T_b = 48$ in

5. $y_g = \dfrac{2.43 \, P_e T_b^3}{EI} = \dfrac{2.43 \times 44,860 \times (48)^3}{288 \times 10^8} = 0.42$ in.

6. Follow Steps 1 through 5 for other loads. Computed and measured values of y are plotted in Fig. 9.54 and there is a very close agreement between the two. The n_h values against y_g are also plotted in the same figure.

Calculation of moment distribution

The moment at any distance x along the pile may be calculated by the equation

$$M = (P_t T)A_m + (M_t)B_m$$

As per the calculations shown above, the value of T will be known for any lateral load level P_t. This means $(P_t T)$ will be known. The values of A_m and B_m are functions of the depth coefficient Z which can be taken from Table 9.4 for the distance x ($Z = x/T$). The moment at distance x will be known from the above equation. In the same way moments may be calculated for other distances. The same procedure is followed for other load levels. Figure 9.54 gives the computed moment distribution along the pile axis. The measured values of M are shown for two load levels $P_t = 67.4$ and 80.1 kips. The agreement between the measured and the computed values is very good.

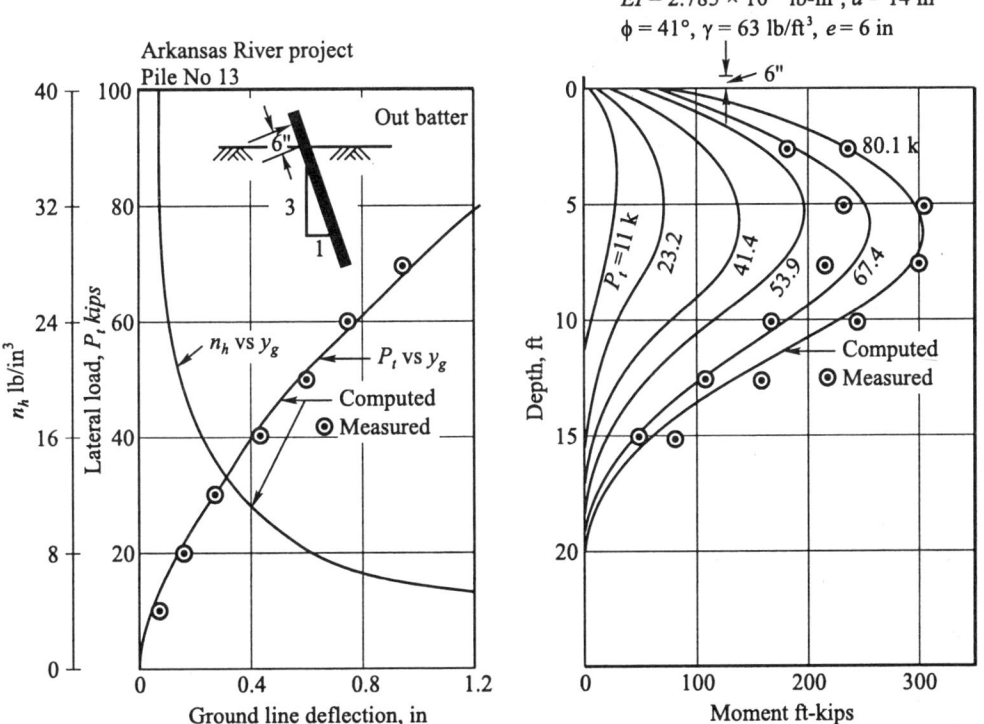

Fig. 9.54 Lateral load test-batter pile 13-arkansas river project (Alizadeh and Davisson, 1970)

Example 9.21

A steel pipe pile of 61 cm diameter is driven vertically into a medium dense sand with the water table close to the ground surface. The following data are available:

Pile: $EI = 43.5 \times 10^4$ kN-m², $L = 20$ m, the yield moment M_y of the pile material $= 2349$ kN-m.

Soil: Submerged unit weight $\gamma_b = 8.75$ kN/m³, $\phi = 38°$.

Lateral load is applied at ground level ($e = 0$)

Determine:

(a) The ultimate lateral resistance P_u of the pile

(b) The groundline deflection yg at the ultimate lateral load level.

Solution

From Eq. (9.92a) the expression for n_h is

$$n_h = \frac{150 \, C_\phi \, \gamma^{1.5} \sqrt{EId}}{P_t} \quad \text{since } P_e = P_t \text{ for } e = 0$$

From Eq. (9.89b) $C_\phi = 3 \times 10^{-5} (1.326)^{38°} = 1.02$

Substituting the known values for n_h we have

$$n_h = \frac{150 \times 1.02 \times (8.75)^{1.5} \sqrt{43.5 \times 10^4 \times 0.61}}{P_t} = \frac{204 \times 10^4}{P_t} \text{ kN/m}^3 \quad \text{... (a)}$$

(a) Ultimate lateral load P_u

Step 1

Assume $\qquad P_u = P_t = 1000$ kN

Now from Eq. (a) $n_h = \dfrac{204 \times 10^4}{1000} = 2040$ kN/m^3

From Eq. (9.50) $\quad T = \left(\dfrac{EI}{n_h}\right)^{\frac{1}{n+4}} = \left(\dfrac{EI}{n_h}\right)^{\frac{1}{1+4}} = \left(\dfrac{EI}{n_h}\right)^{\frac{1}{5}}$

Substituting and simplifying

$$T = \left(\frac{43.5 \times 10^4}{2040}\right)^{\frac{1}{5}} = 2.92 \text{ m}$$

The moment equation for $e = 0$ may be written as (Eq. 9.53)

$$M = A_m (P_t T m)$$

Substituting and simplifying we have [where A_m (max) = 0.77]

$$M_{max} = 0.77 (1000 \times 2.92) = 2248 \text{ kN-m}$$

which is less than $\quad M_y = 2349$ kN-m.

Step 2

Try $\qquad\qquad\qquad P_t = 1050$ kN.

Following the procedure given in Step 1

$$T = 2.95 \text{ m for } P_t = 1050 \text{ kN}$$

Now $\qquad\qquad M_{max} = 0.77 (1050 \times 2.95) = 2385$ kN-m

which is greater than $M_y = 2349$ kN-m.

The actual value P_u lies between 1000 and 1050 kN which can be obtained by proportion as

$$P_u = 1000 + (1050 - 1000) \times \frac{(2349 - 2248)}{(2385 - 2248)} = 1037 \text{ kN}$$

(b) Groundline deflection for $P_u = 1037$ kN

For this the value T is required at $P_u = 1037$ kN. Following the same procedure as in Step 1 we get $T = 2.29$ m.

Now from Eq. (9.69) for $e = 0$

$$y_g = 2.43 \frac{P_t T^3}{EI} = \frac{2.43 \times 1037 \times (2.944)^3}{43.5 \times 10^4} = 0.1478 \text{ m} = 14.78 \text{ cm}$$

Example 9.22

Refer to Ex. 9.21, if the pipe pile is driven at an angle of 30° to the vertical, determine the ultimate lateral resistance and the corresponding groundline deflection for the load applied (*a*) against batter, and (*b*) in the direction of batter.

In both the cases the load is applied normal to the pile axis.

All the other data given in Ex. 9.21 remain the same.

Solution

From Ex. 9.21, the expression for n_h for vertical pile is

$$n_h = n_h^o = \frac{204 \times 10^4}{P_t} \text{ kN/m}^3 \qquad \qquad \text{... (a)}$$

+30° batter pile

From Fig. 9.51
$$\frac{n_h^{+b}}{n_h^o} = 0.1 \text{ and } n = 2 \qquad \qquad \text{... (b)}$$

From Eq. (9.50)
$$T = T_b = \left(\frac{EI}{n_h^b}\right)^{\frac{1}{n+4}} = \left(\frac{EI}{n_h^b}\right)^{\frac{1}{2+4}} = \left(\frac{EI}{n_h^b}\right)^{\frac{1}{6}} \qquad \text{... (c)}$$

Determination of P_u

Step 1

Assume
$$P_e = P_t = 500 \text{ kN.}$$

Following the Step 1 in Ex. 9.21, and using Eq. (a) above

$$n_h^o = 4{,}083 \text{ kN/m}^3, \text{ hence } n_h^{+b} = 4083 \times 0.1 \approx 408 \text{ kN/m}^3$$

Form Eq. (c),
$$T_b = \left(\frac{43.5 \times 10^4}{408}\right)^{\frac{1}{6}} = 3.2 \text{ m}$$

As before,
$$M_{max} = 0.77 P_t T_b = 0.77 \times 500 \times 3.2 = 1232 \text{ kN-m} < M_y$$

Step 2

Try
$$P_t = 1{,}000 \text{ kN}$$

Proceeding in the same way as given in Step 1 we have $T_b = 3.59$ m, $M_{max} = 2764$ kN-m which is more than M_y. The actual P_u is

$$P_u = 500 + (1000 - 500) \times \frac{(2349 - 1232)}{(2764 - 1232)} = 865 \text{ kN}$$

Step 3

As before the corresponding T_b for $P_u = 865$ kN is 3.5 m.

Step 4

The groundline deflection is

$$y_g^{+b} = \frac{2.43 \times 865 \times (3.5)^3}{43.5 \times 10^4} = 0.2072 \text{ m} = 20.72 \text{ cm}$$

−30° batter pile

From Fig. 9.51, $\quad \dfrac{n_h^{-b}}{n_h^o} = 2.2$ and $n = 1.0$ $\hfill$ (d)

and $\quad T_b = \left(\dfrac{EI}{n_h^{-b}}\right)^{\frac{1}{n+4}} = \left(\dfrac{EI}{n_h^{-b}}\right)^{\frac{1}{1+4}} = \left(\dfrac{EI}{n_h^{-b}}\right)^{\frac{1}{5}}$ $\hfill$ (e)

Determination of P_u

Step 1

Try $\qquad\qquad P_t = 1000$ kN

From Eq. (a) $\quad n_h^o = 2040$ kN/m^3 and from Eq. (d) $n_h^{-b} = 2.2 \times 2040 = 4488$ kN/m^3

Now from Eq. (e), $\quad T_b = 2.5$ m

As before $\quad M_{max} = 0.77 \times 1000 \times 2.5 = 1925$ kN-m

which is less than $\quad M_y = 2349$ kN-m

Step 2

Try $\qquad\qquad P_t = 1,500$ kN

Proceeding as in Step 1, $T_b = 2.71$ m, and $M_{max} = 0.77 \times 1500 \times 2.71 = 3130$ kN-m which is greater than M_y.

Step 3

The actual value of P_u is therefore

$$P_u = 1000 + (1500 - 1000) \times \frac{(2349 - 1925)}{(2764 - 1925)} = 1253 \text{ kN}$$

Step 4

Groundline deflection

$$T_b = 2.58 \text{ m for } P_u = 1253 \text{ kN}$$

Now $\qquad y_g^{-b} = \dfrac{2.43 \times 1176 \times (2.58)^3}{43.5 \times 10^4} = 0.1202 \text{ m} = 12.0 \text{ cm}$

The above calculations indicate that the negative batter piles are more resistant to lateral loads than vertical or positive batter piles. Besides, the groundline deflections of the negative batter piles are less than the vertical and corresponding positive batter piles.

Example 9.23

A steel pipe pile of outside diameter 61 cm is driven into medium dense sand under submerged condition which is having a relative density of 60% and an angle of internal friction 38°. The submerged unit weight of soil is 8.75 kN/m³. Compute the ultimate lateral resistance of the pile, if the pile is driven at a batter of 30°, for the following loading conditions.

(*a*) Load applied against the batter, and (*b*) and load applied in the direction of batter. The *EI* of the pile is 4.35×10^{11} kg-cm² (4.35×10^5 kN-m²)

Solution

First it is necessary to known the ultimate lateral resistance of the vertical pile which can be obtained from Ex. 9.8. The ultimate lateral resistance has been worked out for the same pile by Brom's theory for the same pile,

$$P_u = 667 \text{ kN}$$

For computing the ultimate lateral resistance of batter piles, read Section 9.27. As per Fig. 9.48,

for 30° "out" batter pile, $\dfrac{P_u^\beta}{P_u^o} = 0.8$

for 30° "in" batter pile, $\dfrac{P_u^\beta}{P_u^o} = 1.2$

Now, for +30°, $\quad P_u = 0.8 \times 667 = 534 \text{ kN}$

for – 30°, $\quad P_u = 1.2 \times 667 = 800 \text{ kN}$

Example 9.24

For the pile given in Ex. 9.23 determine the following:

(*a*) .The lateral load for the vertical pile for $y_g = 1.5$ cm.

(*b*) For the same lateral load in (*a*) acting horizontally on piles of ± 30° batter, the normal and horizontal deflections.

Solution

(*a*) For $\phi = 38°$, $C_\phi = 1.5$ from Fig. 9.17

Use Eq. (9.92c) for determining P_t

$$P_t = 3.65 \, C_\phi^{0.4 \text{ and } 0.6} \, (EI)^{0.43} \, d^{0.2} \, y_g^{0.60}$$

After substitution and simplifying

$$P_t = 3.65 \times 1.5^{0.4} \times (8.75)^{0.6} \times (4.35 \times 10^5)^{0.43} \times (0.61)^{0.2} \times (0.015)^{0.6}$$

$$= 310 \text{ kN}$$

(*b*) Normal load for pile of ± 30° batter in

$$P_n = P_t \cos \beta = 310 \cos 30° = 268 \text{ kN}.$$

From Fig. 9.51

for – 30° batter, $\quad n = 1, \; n_h^b / n_h^o = 2.25$

for + 30° batter, $n = 2$, $n_h^b/n_h^o = 0.1$

For Vertical pile:

From Eq. (9.92a)

$$n_h = \frac{150\, C_\phi^{1.5} \sqrt{EId}}{P_t} = \frac{150 \times 1.5 \times (8.75)^{1.5} \times \left(4.35 \times 10^5\right)^{\frac{1}{2}}}{310}$$

$$= 9681 \text{ kN/m}^3$$

$$T_o = \left(\frac{EI}{n_h}\right)^{\frac{1}{5}} = \left(\frac{4.35 \times 10^5}{9681}\right)^{\frac{1}{2}} = 2.1 \text{ m}$$

For batter piles

+ 30°, $\qquad n_h^{+b} = 0.1 \times 9681 = 968 \text{ kN/m}^3$

For – 30°, $\qquad n_h^{-b} = 2.25 \times 9681 = 21{,}782 \text{ kN/m}^3$

For + 30°, $\qquad T_b = \left(\frac{4.35 \times 10^5}{968}\right)^{\frac{1}{6}} = 2.8 \text{ m}$

For – 30° m $\qquad T_b = \left(\frac{4.35 \times 10^5}{21{,}782}\right)^{0.2} = 1.82 \text{ m}$

From equation $\qquad y_g = \dfrac{2.43\, P_t\, (T)^3}{EI}$, the expression

for a batter pile is $\quad y_n = \dfrac{2.43\, P_n\, (T_b)^3}{EI}$

Now for + 30°, $\qquad y_n = \dfrac{2.43 \times 268\, (2.8)^3}{4.35 \times 10^5} = 3.30 \text{ cm}$

$$y_h = \frac{y_n}{\cos \beta} = \frac{3.3}{\cos 30°} = 3.8 \text{ cm}$$

for – 30°, $\qquad y_n = \dfrac{2.43 \times 268\, (1.82)^3}{4.35 \times 10^5} = 0.9 \text{ cm}$

$$y_h = \frac{0.90}{\cos 30°} = 1.04 \text{ cm}$$

Example 9.25

For the pile in Ex. 9.24, determine P_t (horizontal) for y_g (horizontal) = 1.5 cm for piles of batter +30° and −30°

Solution

For batter pile $\pm 30°$, $\quad y_n = 1.5 \times \cos 30° = 1.3$ cm.

For $+30°$ pile $\quad y_n = 1.3$ cm $= 0.013$ m $= \dfrac{2.43 \left(P_n\right) \times \left(2.8\right)^3}{4.35 \times 10^5}$

or $\quad P_n = \dfrac{0.013 \times 10^4}{1.23} = 106$ kN

$P_t \text{ (horizontal)} = \dfrac{P_n}{\cos \beta} = \dfrac{106}{\cos 30°} = 122$ kN

In the same way

For $-30°$ pile, 0.013 m $= \dfrac{2.43\ P_n \left(1.82\right)^3}{4.35 \times 10^5}$

Simplifying, $\quad P_n = 386$ kN

$P_t \text{ (horizontal)} = \dfrac{386}{\cos 30°} = 446$ kN

Example 9.26

For the pile given in Ex. 9.23, determine the lateral load at ground level for a horizontal deflection of 1.50 cm. use Fig. 9.49 and Table 9.8

Solution

It is first necessary to determine the lateral load of a vertical pile for the same deflection of 1.50 cm, and then use Fig. 9.49 for determining the lateral loads for batter piles at the same deflection

For lateral deflection at ground level for vertical pile use Eq. (9.69). With $M_t = 0$,

$$y_g = \frac{2.43\ P_t T^3}{EI}$$

In this equation, $\quad y_g = 1.50$ cm,

$EI = 4.35 \times 10^{11}$ kg-cm^2.

We do not know P_t and T. The value of T can be found out by using a suitable value for n_h. From Table 9.8, $n_h = 15$ MN/m^3 ($= 1.5$ kg/cm^3) for medium dense sand under submerged condition

Now, $\quad T = \left(\dfrac{EI}{n_h}\right)^{\frac{1}{5}} = \left(\dfrac{4.35 \times 10^{11}}{1.5}\right)^{\frac{1}{5}} = 196$ cm

Therefore, $\quad P_t = \dfrac{y_g EI}{2.43 T^3} = \dfrac{1.5 \times 4.35 \times 10^{11}}{2.43 \times 196^3} = 35.66$ Tonnes

≈ 357 kN.

From Fig. 9.49, the ratio of lateral resistance of batter pile to vertical pile (R_{lh}) is

$$R_{lh} = \frac{P_h^\beta}{P_h^0} = 0.80 \text{ for } + 30° \text{ batter pile}$$

$$= 1.4 \text{ for } - 30° \text{ batter pile.}$$

Therefore, For $+ 30°$ batter pile,

$$P_t = 357 \times 0.8 = 285.6 \text{ kN,}$$

for $- 30°$ batter pile, $\quad P_t = 357 \times 1.4 = 499.8 \text{ kN.}$

Example 9.27

A lateral load test was carried out on a vertical pile driven to 20 m depth with the load at ground level. The soil and pile properties are as given in Ex. 9.23. If the deflection of the vertical pile is 1.50 cm at a lateral load of 357 kN, compute the horizontal deflections of the same pile inclined at a batter of 30° with the load 357 kN acting (a) against the batter, and (b) in the direction of batter.

Solution

As per Ex. 9.26, the relative stiffness factor T_0 of the vertical pile is 1.96 m. The T_b of batter piles may be calculated from Eq. (9.119).

For $+30°$ pile $\quad T_b = 1.96 (1 + 7.5 \times 10^{-3} \times 30)$

$$= 1.96 \times 1.225 = 2.40 \text{ m.}$$

For $-30°$ pile $\quad T_b = 0.86 \, T_0 = 0.86 \times 1.96 = 1.69 \text{ m}$

The equation for normal deflection of a batter pile may be written as

$$y_n = A_y \frac{P_n T_b^3}{EI}$$

From Table 9.11.

$$A_y = 2.52 \text{ for } + 30° \text{ pile,}$$

Now, $\qquad\quad = 2.36 \text{ for } - 30° \text{ pile,}$

$$P_n = P_h \cos \beta$$

The horizontal deflection

$$y_h = \frac{y_n}{\cos \beta}$$

Therefore $\qquad y_n = y_h \cos \beta$

Substituting for y_h and P_h, the general equation in terms of y_h and P_h is

$$y_h \cos \beta = \frac{A_y P_h \cos \beta T^3}{EI}$$

or $\qquad\qquad y_h = \frac{A_y P_h T^3}{EI}$

Substituting, we have

for $+30°$ pile, $\qquad y_h = \dfrac{2.52 \times 357 \times (2.4)^3 \times 10^6 \times 10^2}{4.35 \times 10^{11}} = 2.859$ cm

for $-30°$ pile, $\qquad y_h = \dfrac{2.36 \times 357 \times (1.69)^3 \times 10^6 \times 10^2}{4735 \times 10^{11}} = 0.93$ cm

The deflection for a vertical pile $y_h^0 = 1.50$ cm. The ratio of deflection of batter pile to that of a vertical pile may be written as

$$y_h^\beta / y_h^0 = 1.91 \text{ for } +30° \text{ pile}$$
$$= 0.62 \text{ for } -30° \text{ pile.}$$

This indicates that a $+30°$ piles deflects about 3 times that of a $-30°$ pile under the same lateral load. When compared to a vertical pile, $+30°$ pile deflects about 2 times and $-30°$ pile about 0.6 times that of a vertical pile.

Example 9.28

Compute the maximum bending moment and soil reactions for the batter piles given in Ex. 9.27 for the same horizontal load.

Solution

As per Ex. 9.27

The relative stiffness factors are

T for vertical pile $= 1.96$ m

for $+30°$ pile $= 2.40$ m

for $-30°$ pile $.= 1.69$ m

Max moment
$$P_h = 357 \text{ kN, } P_n = 357 \cos 30° = 309 \text{ kN}$$

The general equation for moment is

$$M = P_n T A_m$$

From Table 9.11, $\quad A_m = 0.88$ for $+30°$ pile

$$= 0.74 \text{ for } -30° \text{ pile}$$
$$= 0.75 \text{ for } 0° \text{ (as per author's work).}$$

For $+30°$ pile, M (max) $= 0.88 \times 309 \times 2.4 = 653$ kN-m,

$\qquad -30°$ pile M (max) $= 0.74 \times 309 \times 1.69 = 386$ kN-m.

For vertical pile $\qquad P_n = P_h = 357$ kN

$$M \text{ (max)} = 0.75 \times 357 \times 1.96 = 525 \text{ kN-m.}$$

Under the same horizontal load of 357 kN, the ratios of

$$\begin{array}{cc} & +30° \quad -30° \\ \dfrac{M^\beta}{M_o} = & 1.243 \quad 0.735 \end{array}$$

Maximum Soil Reaction, p

The general equation for soil reaction, p is

$$p = \frac{P_t}{T} A_p$$

$P_t = P_h = 357$ kN for vertical pile,

$P_t = P_n = P_h \cos 30° = 309$ kN for 30° batter pile.

From Table 9.11, the values of maximum A_p and hence p are

	0°	+30°	−30°
A_p	0.90	1.14	0.98
p (kg/cm)	164	147	179

The maximum moment as per Reese et al non-dimensional parameters

From Table 9.4,

$$A_m(\max) = 0.772 \text{ for a vertical pile}$$

$$M(\max) = 0.772 \times 357 \times 1.96 = 540 \text{ kN-m.}$$

The value of M (max) by Reese *et al* method is higher than the author's method by about 3% which is negligible.

9.31 PROBLEMS

9.1 A concrete pile 50 cm square in section is driven into medium dense sand upto a depth of 20 m. The sand is in a submerged state. A lateral load of 50 kN is applied on the pile at a height of 5 m above the ground level. Compute the lateral deflection of the pile at ground level. Given: $n_h = 15$ MN/m³, $EI = 115 \times 10^9$ kg-cm². The submerged unit weight of the soil is 8.75 kN/m³.

9.2 If the pile given in Prob. 9.1 is fully restrained at the top, what is the deflection at ground level.

9.3 If the pile given in Prob. 9.1 is only 3 m long below the ground level, what will be the deflection at ground level (*a*) when the top of the pile is free and (*b*) when the top of the pile is restrained?

9.4 Consider the pile given in Prob. 9.1 is inclined at an angle of 30° to the vertical. What would be the horizontal deflection at ground level if the lateral load of 50 kN is applied at ground level normal to the axis of the pile (*a*) in the direction of batter, and (*b*) against batter?

9.5 A precast concrete pile of 30 cm diameter is driven to a depth of 10 m in a vertical direction into medium dense sand which is in a semi-dry state. The value of the coefficient of soil modulus variation (n_h) may be assumed as equal to 0.8 kg/cm³. A lateral load of 40 kN is applied at a height of 3 m above ground level. Compute (*a*) the deflection at ground level, (*b*) the maximum bending moment on the pile and (*c*) the maximum soil reaction (Assume $E = 2.1 \times 10^5$ kg/cm²).

9.6 If the pile in Prob. 9.5 is driven at a batter of 22.5° to the vertical, and lateral load is applied at ground level, compute (*a*) the horizontal deflection at ground level (*b*) the maximum bending moment and (*c*) the maximum soil reaction for the cases of the load acting in the direction of batter and against batter.

9.7 If the soil in Prob. 9.5 has a unit weight of 17.5 kN/m^3, compute the ultimate lateral resistance of pile when the pile is in a vertical position by Broms method. The angle of internal friction of sand, ϕ, may be taken as equal to 38°.

9.8 If the pile given in Prob. 9.1 is driven into saturated normally consolidated clay having an unconfined compressive strength of 70 kPa, what would be the ultimate lateral resistance of the soil under (a) free head condition, and (b) fixed condition. Make necessary assumption for the yield strength of the material.

9.9 Compute the deflection at ground level by difference equation method for the pile given in Prob. 9.1.

9.10 Compute the lateral deflection of the pile in Prob. 9.1 by Poulos method

9.11 Construct *p-y* curves for the pile given in Prob. 9.1 at depths of 2, 4 and 6 metres from ground level for static and cyclic loading conditions. The angle of internal friction of the soil may be taken as equal to 36°.

9.12 Construct (*p-y*) curves for the pile given in Prob. 9.1 if the pile is driven into stiff clay of unconfined compressive strength 275 kPa and the water table is at great depth. Assume γ = 17.5 kN/m^3. Consider both static and cyclic loading conditions.

9.13 Compute deflection and moment as a function of depth for the pile given in Prob. 9.1 by making use of the *p-y* curves developed in Prob. 9.11.

Deep Foundation 3 ⑩
Pile Groups Subjected to Vertical and Lateral Loads

10.1 INTRODUCTION

Chapter 8 has dealt with single vertical piles subjected to vertical loads only and Chapter 9 with the behaviour of single vertical and batter piles subjected to lateral loads only. This chapter deals with the behaviour of pile groups with or without batter piles subjected to vertical/lateral loads.

10.2 NUMBER AND SPACING OF PILES IN A GROUP

Very rarely structures are founded on single piles. Normally, there will be a minimum of three piles under a column or a foundation element because of alignment problems and inadvertent eccentricities. The spacing of piles in a group depends upon many factors such as

1. overlapping of stresses of adjacent piles,
2. cost of foundation,
3. efficiency of pile group.

The pressure isobars of a single pile with load Q acting on the top is shown in Fig. 10.1 (a). When piles are placed in a group, there is a possibility of pressure isobars of adjacent piles overlapping each other as shown in Fig. 10.1 (b). The soil is highly stressed in the zones of overlapping of pressures. With sufficient overlap, either the soil will fail or the pile group will settle excessively since the combined pressure bulb extend to a considerable depth below the base of the piles. It is possible to avoid overlap by installing the piles at considerable distances apart as shown in Fig. 10.1 (c). Large spacings are not recommended sometimes, since this would result in a bigger size of pile cap which would increase the cost of the foundation.

The spacing of pile depends upon the method of installing a pile and the type of soil. The piles can be driven piles or *cast-in-situ* piles. When the piles are driven there will be greater overlapping of stresses due to the displacement of soil. If the displacement of soil compacts the soil in between the piles just as in the case of loose sandy soils, the piles may be placed at closer intervals. But if the piles are driven into saturated clay or silty soils, the displaced soil will not compact the soil between the piles. As a result the soil between the piles may move upwards and in this process lift the pile cup. Greater spacing between piles is required in soils of this type to avoid lifting of piles. When

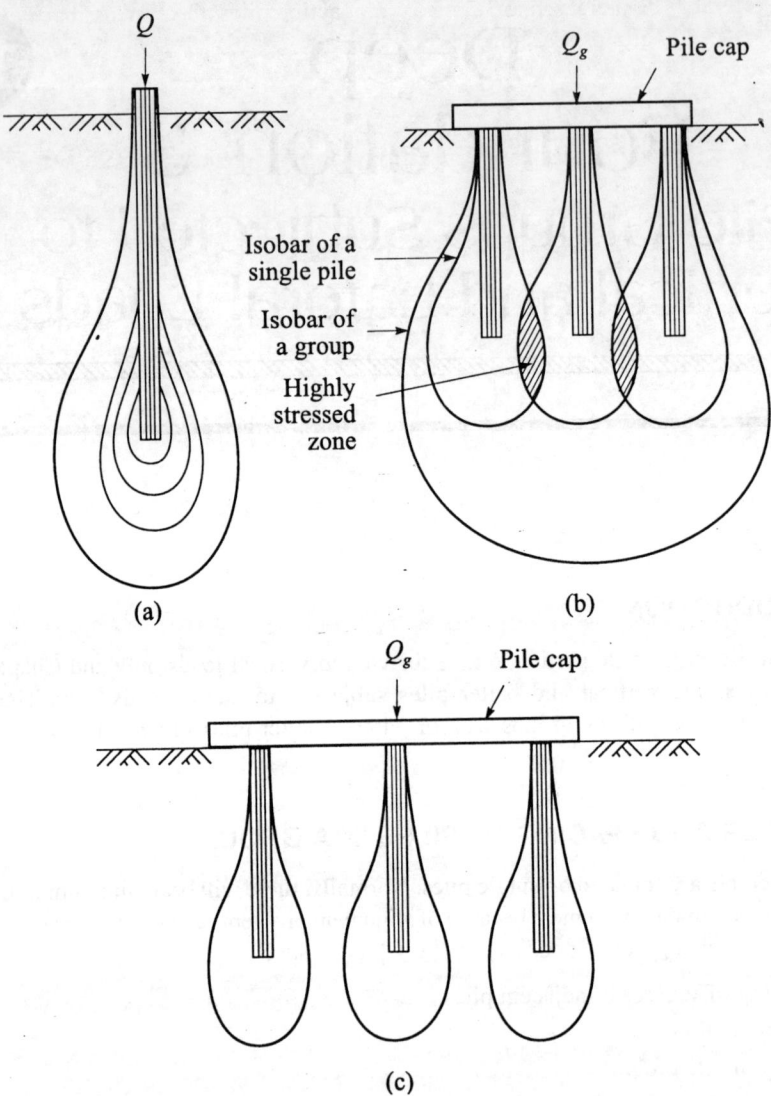

Fig. 10.1 Pressure isobars of (a) single pile, (b) group of piles, closely spaced, and (c) group of piles with piles for apart

piles are *cast-in-situ*, the soils adjacent to the piles are not stressed to that extent and as such smaller spacings are permitted.

Generally, the spacing for point bearing piles, such as piles founded on rock, can be much less than that friction piles since the high-point-bearing stresses and the superposition effect of overlap of the point stresses will most likely not overstress the underlying material nor cause excessive settlements.

The minimum allowable spacing of piles is usually stipulated in building codes. The spacing for straight uniform diameter piles may vary from 2 to 6 times the diameter of the shaft. For friction piles, the minimum spacing recommended is $3d$ where d is the diameter of the pile. For end bearing piles passing through relatively compressible strata the spacing of piles shall not be less than $2.5d$. Whereas for end bearing piles passing through compressible strata and resting in stiff clay, the spacing may preferably be increased to 3.5*d*. For compaction piles, the spacing may be 2*d*. Typical arrangements of piles in groups are shown in Fig. 10.2.

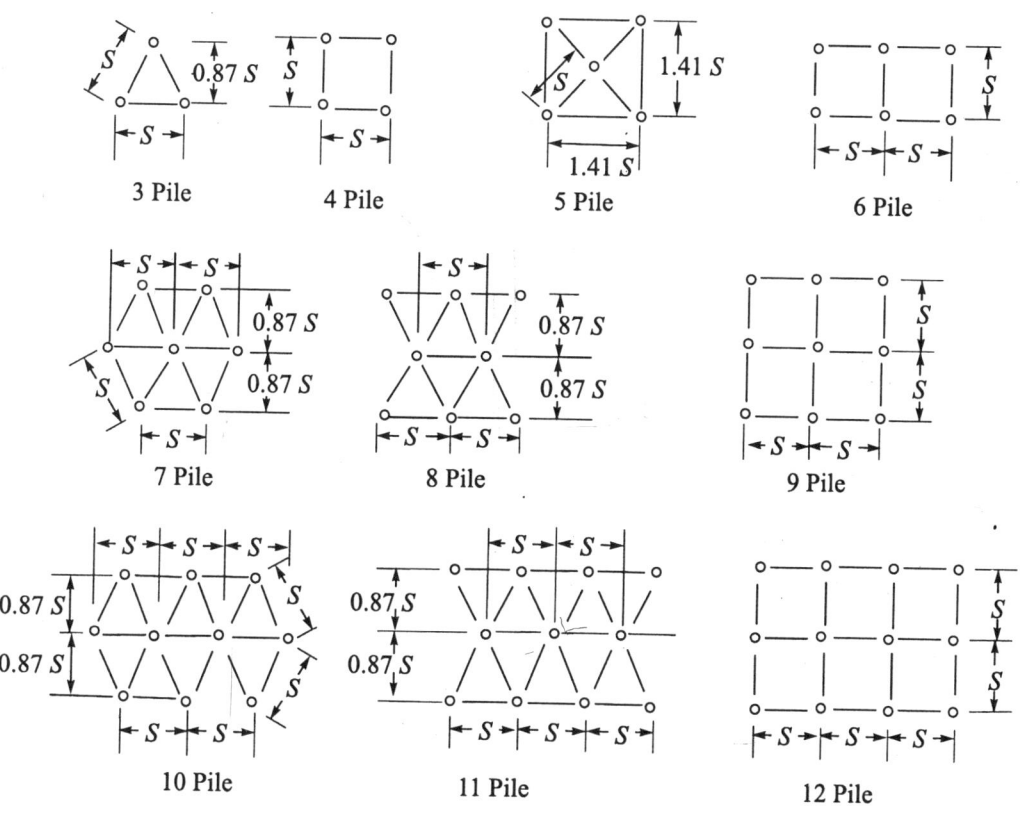

Fig. 10.2 Typical arrangements of piles in groups

10.3 PILE GROUP EFFICIENCY

The spacing of piles is usually predetermined by practical and economical considerations. The design of a pile foundation subjected to vertical loads comprises of:

1. The determination of the ultimate load bearing capacity of the group Q_{gu}.
2. Determination of the settlement of the group, S_g under an allowable load Q_{ga}.

It is well known that the ultimate load of the group is generally different from the sum of the ultimate loads of individual piles Q_a.

The factor

$$E_g = \frac{Q_{gu}}{\Sigma Q_u} \tag{10.1}$$

is called group efficiency which depends on parameters such as

1. type of soil in which the piles are embedded,
2. method of installation of piles, i.e. either driven or cast-in-situ piles, and
3. spacing of piles.

There is no acceptable *efficiency formula* for group bearing capacity. There are a few formulae such as Field rule, Converse-Labarre formula that are sometimes used by engineers. These formulae are empirical and give efficiency factors less than unity. But when piles are installed in sand, efficiency factors greater than unity can be obtained as shown by Vesic by his experimental investigation on groups of piles in sand. There is not sufficient experimental evidence to determine group efficiency for piles embedded in clay soils.

10.4 EFFICIENCY OF PILE GROUPS IN SAND

Vesic (1967) carried out tests on 4 and 9 pile groups driven into sand under controlled conditions. Piles of spacings 2, 3, 4, and 6 times the diameter were used in the tests. The tests were conducted in homogeneous and medium dense sand. His findings are given in Fig. 10.3. The figure gives the following:

1. The efficiencies of 4 and 9 pile groups when the pile caps do not rest on the surface.

2. The efficiencies of 4 and 9 pile groups when the pile caps rest on the surface.

3. The skin efficiency of 4 and 9 pile groups.

4. The average point efficiency of all the pile groups.

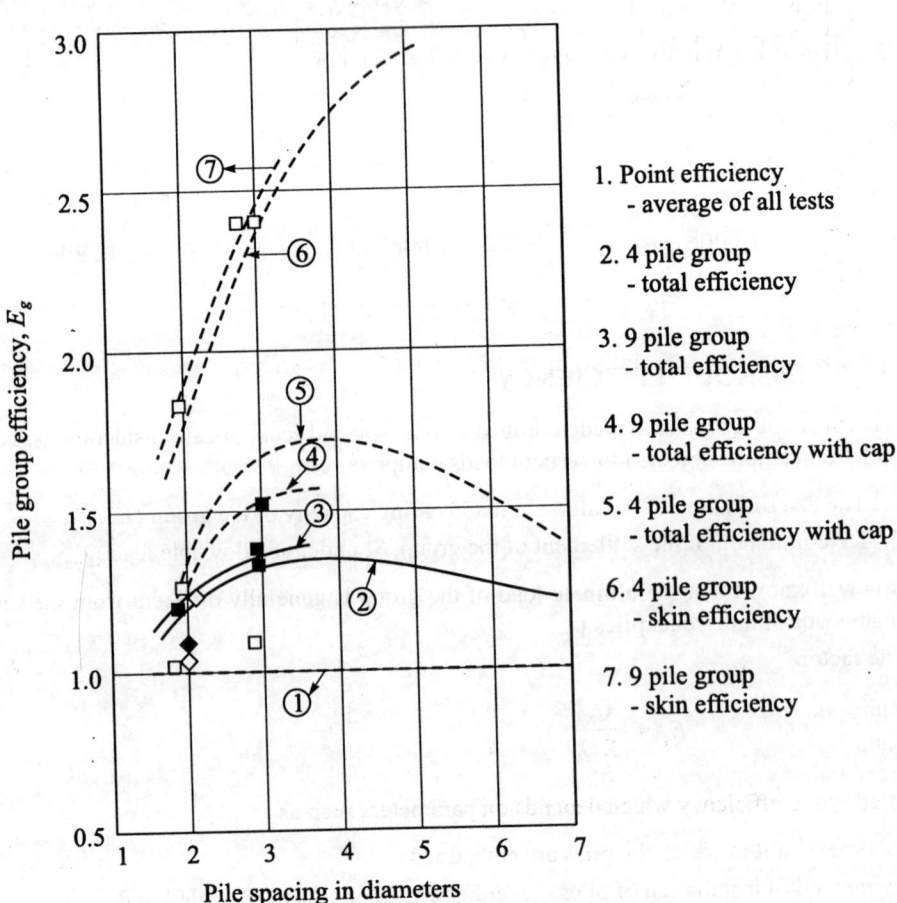

1. Point efficiency
   - average of all tests

2. 4 pile group
   - total efficiency

3. 9 pile group
   - total efficiency

4. 9 pile group
   - total efficiency with cap

5. 4 pile group
   - total efficiency with cap

6. 4 pile group
   - skin efficiency

7. 9 pile group
   - skin efficiency

Fig. 10.3 Efficiency of pile groups in sand (Vesic, 1967)

It may be mentioned here that a pile group with pile cap resting on the surface takes more load than the one with free standing piles above the surface. In the former case, a part of the load is taken by the soil directly under the cap and the rest is taken by the piles. The pile cap behaves just the same way as a shallow foundation of the same size. Though the percentage of load taken by the group is quite considerable, building codes have not so far considered the contribution made by the cap.

It may be seen from the Fig. 10.3 that the overall efficiency of a four pile group with cap resting on the surface increases to a maximum of about 1.7 at pile spacings of 3 to 4 pile diameters, becoming somewhat lower with the further increase of spacing. A sizable part of the increased bearing capacity comes from the caps. If the loads transmitted by the caps are deduced, the group efficiency drops to a maximum of about 1.3.

Very similar results are indicated from tests with 9 pile groups. Since these tests in this case were carried out only up to spacing of 3 diameter of piles, the full picture of the curve is not available. However, it may be seen that the contribution of the cap for the bearing capacity is relatively smaller.

Vesic measured the skin loads of all the piles. The skin efficiencies for both the 4 and 9-pile groups indicate an increasing trend. For the 4-pile group series the efficiency increases from about 1.8 at 2 pile diameter to a maximum of about 3 at 5 piles diameters and beyond. In contrast to this, the average point load efficiency for the groups is about 1.01. Vesic showed for the first time that the increasing bearing capacity of a pile group for piles driven in sand comes primarily from increase in skin loads. The point loads seem to be virtually unaffected by group action.

10.5 PILE GROUP EFFICIENCY EQUATION

There are many pile group equations. These equations are to be used very cautiously, and may in many cases, be no better than a good guess. The Converse-Labarre Formula is one of the most widely used group-efficiency equation which is expressed as

$$E_g = 1 - \frac{\theta(n-1)m + (m-1)n}{90 \, mn} \tag{10.2}$$

where, m = number of columns of piles in a group,

n = number of rows,

$\theta = \tan^{-1}\dfrac{d}{s}$ in degrees,

d = diameter of pile,

s = spacing of pile.

10.6 VERTICAL BEARING CAPACITY OF PILE GROUPS EMBEDDED IN SANDS AND GRAVELS

Driven piles: If piles are driven into loose sands and gravel, the soil around the piles to a radius of at least three times the pile diameter gets computed. When piles are, therefore, driven in a group at close spacing, the soil around and between them becomes highly compacted. When the group is loaded, the piles and the soil between them move together as a unit. Thus, the pile group acts as a pier foundation having a base area equal to the gross plan area contained by the piles. The efficiency of the pile group will be greater than unity as explained earlier. It is normally assumed that the efficiency falls to unity ' when the spacing is increased to five or six diameters. Since the present knowledge is not sufficient to evaluate the efficiency for different spacing of piles, it is quite conservative to assume an efficiency factor of unity for all practical purposes. We may, therefore, write

$$Q_{gn} = nQ_u \tag{10.3}$$

where, n = the number of piles in the group.

The procedure explained above is not applicable if pile tips rest on compressible soil such as silts or clays. When the pile tips rest on compressible soils, the stresses transferred so the compressible soils from the pile group might result in over-stressing or extensive consolidation. The carrying capacity of pile groups under these conditions is governed by the shear strength and compressibility of the soil, rather than by the *efficiency* of the group within the sand or gravel structure.

10.7 BORED PILE GROUPS IN SAND AND GRAVEL

Bored piles are *cast-in-situ* concrete piles. The method of installation involves

1. boring a hole of the required diameter and depth,
2. pouring in of concrete.

There will always be a general loosening of the soil during boring and that too when the boring has to be done below water table. Though benotonite slurry (what is sometimes called as drilling mud) is used for stabilising the sides and bottom of the bores, loosening of the soil cannot be avoided. Cleaning of the bottom of the bore hole prior to concreting is always a problem which will never be achieved quite satisfactorily. Since bored piles do not compact the soil between the piles, the efficiency factor will never be greater than unity. However, for all practical purposes, the efficiency may be taken as unity.

10.8 PILE GROUPS IN COHESIVE SOILS

The effect of driving piles into cohesive soils (clays and silts) is very different from that of cohesionless soils. It has already been explained earlier that when piles are driven into clay soils, particularly when the soil is soft and sensitive, there will be considerable remoulding of the soil. Besides, there will be heaving of the soil between the piles since compaction during driving cannot be achieved in soils of such low permeability. There is every possibility of lifting of the pile also during this process of heaving of the soil. Bored piles are, therefore, preferred to driven piles in cohesive soils. In case driven piles are to be used, the following steps should be kept in view :

1. Piles should be spaced at greater distances apart.
2. Piles should be driven from the centre of the group towards the edges.
3. The rate of driving of each pile should be adjusted as to minimise the development of porewater pressure.

Experimental results have indicated that when a pile group installed in cohesive soils is loaded, it may fail by any one of the following ways:

1. May fail as a block called as *block failure*.
2. Individual piles in the group may fail.

When piles are spaced at closer intervals, the soil contained between the piles move downward with the piles and at failure, piles and soil move together to give the typical *block failure*. Normally this type of failure occurs when piles are placed within 2 to 3 pile diameters. But for wider spacings, the piles fail individually. The efficiency ratio is less than unity at closer spacings and may reach unity at a spacing of about 8 diameters.

The equation for block failure may be written as (Fig. 10.4).

$$Q_{gu} = cN_c A_g + P_g L\bar{c} \qquad (10.4)$$

where, c = cohesive strength of clay beneath the pile group,

$\bar{c}$ = average cohesive strength of clay around the group,

L = length of pile,

P_g = perimeter of pile group,

A_g = sectional Area of group,

N_c = bearing capacity factor which may be assumed as 9 for deep foundation.

Bearing capacity of pile group on the basis of individual pile failure may be written as

$$Q_{gu} = nQ_u \qquad (10.5)$$

where, n = number of piles in the group,

Q_u = bearing capacity of an individual pile.

Terzaghi and Peck recommend that bearing capacity of a pile group may be taken as the smaller of the two given by Eqs (10.4) and (10.5).

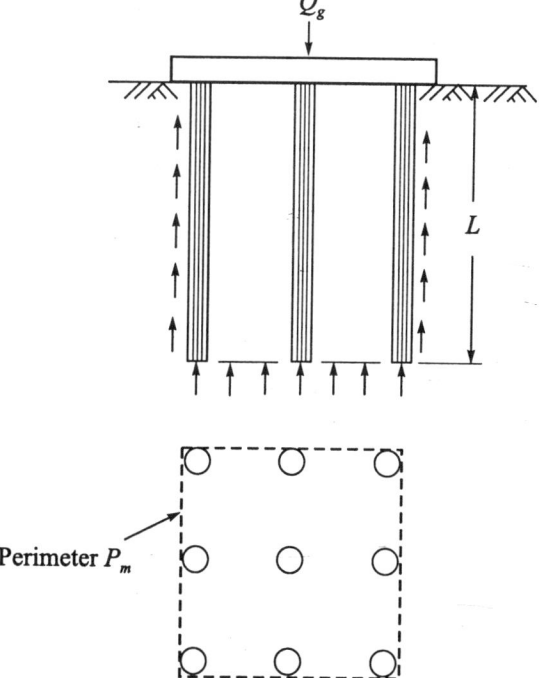

Fig. 10.4 Block failure of a pile group in clay soil

10.9 SETTLEMENT OF PILES AND PILE GROUPS IN SANDS AND GRAVELS

The present knowledge is not sufficient to evaluate the settlements of piles and pile groups. For most engineering structures, the loads to be applied to a pile group will be governed by consideration of consolidation settlement rather than by bearing capacity of the group divided by an arbitrary factor of safety of 2 or 3. It has been found from field observation that the settlement of a pile group is many times the settlement of a single pile at the corresponding working load. The settlement of a group is affected by the shape and size of group, length of piles, method of installation of piles and possibly many other factors.

Vesic has proposed an equation to determine the settlement of a single pile. The equation has been developed on the basis of the experimental results he obtained from tests on piles. Tests on piles of diameters ranging from 2 to 18 inches were carried out in sands of different relative densities D_r. Tests were also carried out on driven piles, jacked piles, and bored piles (jacked piles are those that are pushed into the ground by using a jack). The equation for total settlement of a single pile may be expressed as

$$S = S_p + S_f \qquad (10.6)$$

where, S = total settlement,

S_p = settlement of the pile tip,

S_f = settlement due to the deformation of the pile shaft.

The equation for S_p is

$$S_p = \frac{C_w Q_p}{\left(1 + D_r^2\right) q_{pu}} \tag{10.7}$$

The equation for S_f is

$$S_f = (Q_p + \alpha Q_f)\, \frac{L}{AE} \tag{10.8}$$

where, Q_p = point load,

$\qquad d$ = diameter of pile at the base,

$\qquad q_{pu}$ = ultimate point resistance per unit area,

$\qquad D_r$ = relative density of sand,

$\qquad C_w$ = settlement coefficient = 0.04 for driven piles

$\qquad\qquad\qquad\qquad\qquad\quad$ = 0.05 for jacked piles

$\qquad\qquad\qquad\qquad\qquad\quad$ = 0.18 for bored piles,

$\qquad Q_f$ = friction load,

$\qquad L$ = length of pile,

$\qquad A$ = cross-sectional area of pile,

$\qquad E$ = modulus of deformation of pile shaft,

$\qquad \alpha$ = coefficient which depends on the distribution of skin friction along the shaft and can be taken equal to 0.6.

Settlement of piles cannot be predicted accurately by making use of equations such as the one given here. One should use such equations with caution. It is better to rely on load tests for piles in sand.

Settlement of Pile Groups in Sand

The relation between settlement of a group and a single pile at corresponding working loads may be expressed as

$$F_g = \frac{S_g}{S} \tag{10.9}$$

where, F_g = group settlement factor,

$\qquad S_g$ = settlement of group,

$\qquad S$ = settlement of a single pile.

Vesic has obtained the curve given in Fig. 10.5 which is obtained by plotting F_g against B/d where d is the diameter of the pile and B, the distance between the centre to centre of outer piles in the group (only square pile groups are considered). It should be remembered here that the curve is based on the results obtained from the tests on groups of piles embedded in medium dense sand. It is possible that groups in much looser or much denser deposits might give somewhat different behaviour. Also the group settlement ratio is very likely be affected by the ratio of the pile point settlement S_p to total pile settlement.

Skempton, Yassin and Gibson (1953) have published curves relating F_g with the width of pile groups as shown in Fig. 10.6. These curves can be taken as applying to driven or bored piles.

Since the abscissa for the curve in Fig. 10.6 is not expressed as a ratio, this curve cannot directly be compared with Vesic's curve given in Fig. 10.5. According to Fig. 10.6 a pile group of 3 m wide would settle 5 times that of a single test pile.

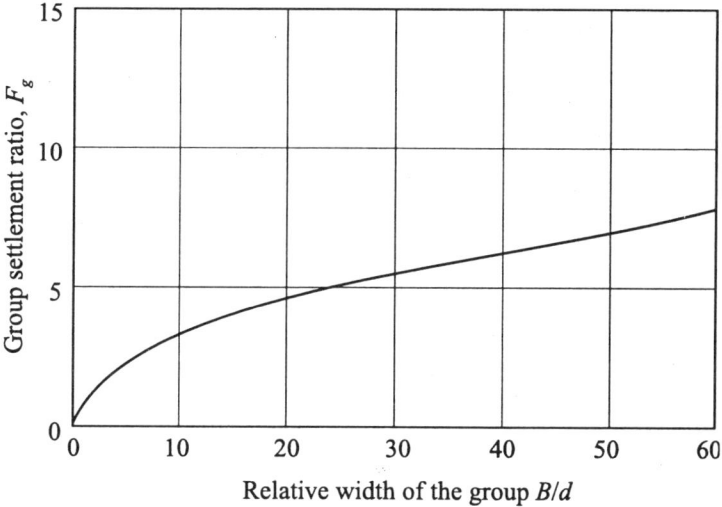

Fig. 10.5 Curve showing the relationship between group settlement ratio and relative widths of pile groups in sand (Vesic, 1967)

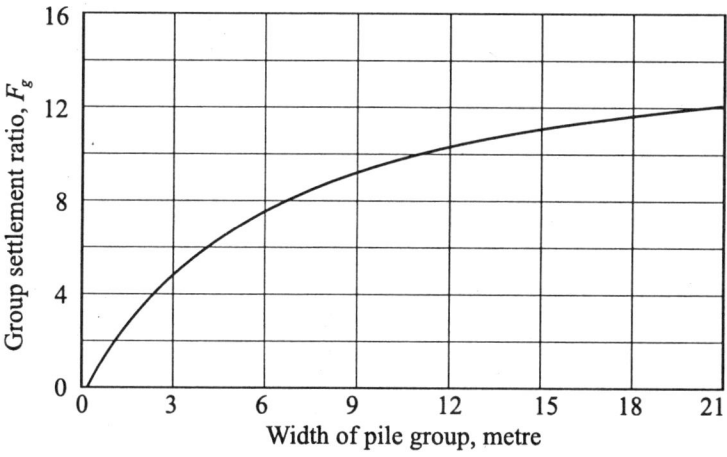

Fig. 10.6 Curve showing relationship between F_g and pile group width (Skempton Yassin and Gibson, 1953)

10.10 SETTLEMENT OF PILE GROUPS IN COHESIVE SOILS

The total settlement of pile groups may be calculated by making use of consolidation settlement equations. The problem involved here is to evaluate the increase in stress Δp beneath a pile group when the group is subjected to a vertical load Q_g. The computation of stresses depends on the type of soil through which the pile passes. The methods of computing the stresses are explained below:

1. The soil in the first group given in (a) of the Fig. 10.7 is homogeneous clay. The load Q_g is assumed to act on a fictitious footing at a depth $2/3L$ from the surface and distributed over the sectional area of the group. The load on the pile group acting at this level is assumed to spread out at 2 : 1 slope. The stress Δp at any depth z below the fictitious footing may be found out as explained in Chapter 2.

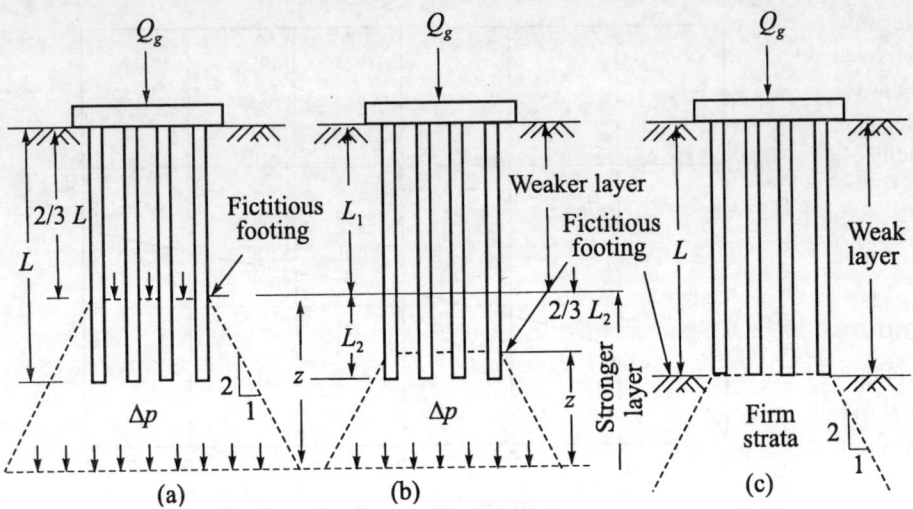

Fig. 10.7 Settlement of pile groups in clay soils

2. In the second group given in (b) of the figure, the pile passes through a very weak layer of depth L_1 and the lower portion of length L_2 is embedded in a strong layer. In this case, the load Q_g is assumed to act at a depth equal to $2/3\,L_2$ below the surface of the strong layer and spreads at 2 : 1 slope as before.

3. In the third case shown in (c) of the figure, the piles are point bearing piles. The load in this case is assumed to act at the level of the firm strata and spreads out at 2 : 1 slope.

10.11 ALLOWABLE LOADS ON GROUPS OF PILES

The basic criterion governing the design of a pile foundation should be the same as that of a shallow foundation, that is, the settlement of the foundation must not exceed some permissible value. The permissible values of settlements assumed for shollow foundations in Chapter 6 are also applicable to pile foundations. The allowable load on a group of piles should be the least of the values computed on the basis of the following two criterions.

1. Shear failure criterion.
2. Settlement criterion.

Procedures have been given in the earlier chapters as to how to compute allowable loads on the basis of shear failure criterion. The settlement of pile groups should not exceed the permissible limits under these loads.

10.12 NEGATIVE FRICTION ON PILES

Figure 10.8 (a) shows a single pile and (b) a group of piles passing through a recently filled cohesive soil. The soil below the fill had completely consolidated under its own overburden pressure.

When the filled up soils starts consolidating under its own overburden pressure, it develops a drag on the surface of the pile. This drag on the surface of the pile is called as *negative friction*. Negative friction may also be developed if the fill material is loose cohesionless soil. Negative friction can also occur when fill is placed over peat or soft clay strata as shown in Fig. 10.8 (c). The superimposed loading on such compressible strata causes heavy settlement of the fill with consequent drag on piles.

Negative friction may also be developed by the lowering of the ground water which increases the effective stress causing consolidation of the soil with the resultant settlement and friction forces being developed on the pile.

Negative friction must be allowed for when considering the factor of safety on the ultimate carrying capacity of pile. The factor of safety, F_s, where negative friction is likely to occur may be written as

$$F_s = \frac{\text{Ultimate carrying capacity of a single or group of piles}}{\text{Working load} + \text{Negative skin friction load}}$$

Computation of Negative Friction on Single Piles

The magnitude of negative friction F_n for a single pile in filled up soils may be taken as [Fig. 10.8 (a)].

(*a*) For cohesive soils

$$F_n = PL_n s \tag{10.10}$$

(*b*) For cohesionless soils

$$F_n = \frac{1}{2} PL_n^2 \gamma K \tan \delta \tag{10.11}$$

where, L_n = length of piles in the compressible material,
s = shear strength of cohesive soils in the filled up zone,
P = perimeter of pile,
K = earth pressure coefficient which lies between the active and the passive earth pressure coefficients,
δ = angle of wall friction which may vary from $\phi/2$ to ϕ.

Negative Friction on Pile Groups

When a group of piles passes through compressible filled up soil, the negative friction, F_{ng}, on the group may be found by any of the following methods [Fig. 10.8 (b)].

(*a*) $F_{ng} = nF_n$ $\tag{10.12}$

(*b*) $F_{ng} = sL_n P_g + \gamma L_n A_g$ $\tag{10.13}$

where, n = Number of piles in the group,
γ = unit weight of soil within the pile group upto depth L_n,
P_g = perimeter of pile group,
A_g = area of pile group within the perimeter P_g,
s = shear strength of soil along the perimeter of the group.

Equation (10.12) gives the negative friction forces of the group as equal to sum of the friction forces of all the single piles.

Equation (10.13) assumes the possibility of block shear failure along the perimeter of the group which includes the volume of the soil $\gamma L_n A_g$ enclosed in the group. The maximum value from Eqs (10.12) or (10.13) should be used.

When the fill is underlain by a compressible stratum as shown in Fig. 10.8 (c), the total negative friction may be found out as follows:

$$F_{ng} = n (F_{n1} + F_{n2}) \tag{10.14}$$

$$F_{ng} = s_1 L_1 P_g + s_2 L_2 P_g + \gamma_1 L_2 A_g + \gamma L_2 A_g$$

$$= P_g (s_1 L_1 + s_2 L_2) + A_g (\gamma_1 L_1 + \gamma_2 L_2) \tag{10.15}$$

Wherein, L_1 = depth of filled up soil,

$\quad\quad L_2$ = depth of compressible natural soil,

$\quad\quad s_1, s_2$ = shear strengths of the fill and compressible soils respectively,

$\quad\quad \gamma_1, \gamma_2$ = unit weights of fill and compressible soils respectively,

$\quad\quad F_{n1}$ = negative friction of a single pile in the fill,

$\quad\quad F_{n2}$ = negative friction of a single pile in the compressible soil.

The maximum value of the negative friction obtained from Eqs (10.14) or (10.15) should be used for the design of pile groups.

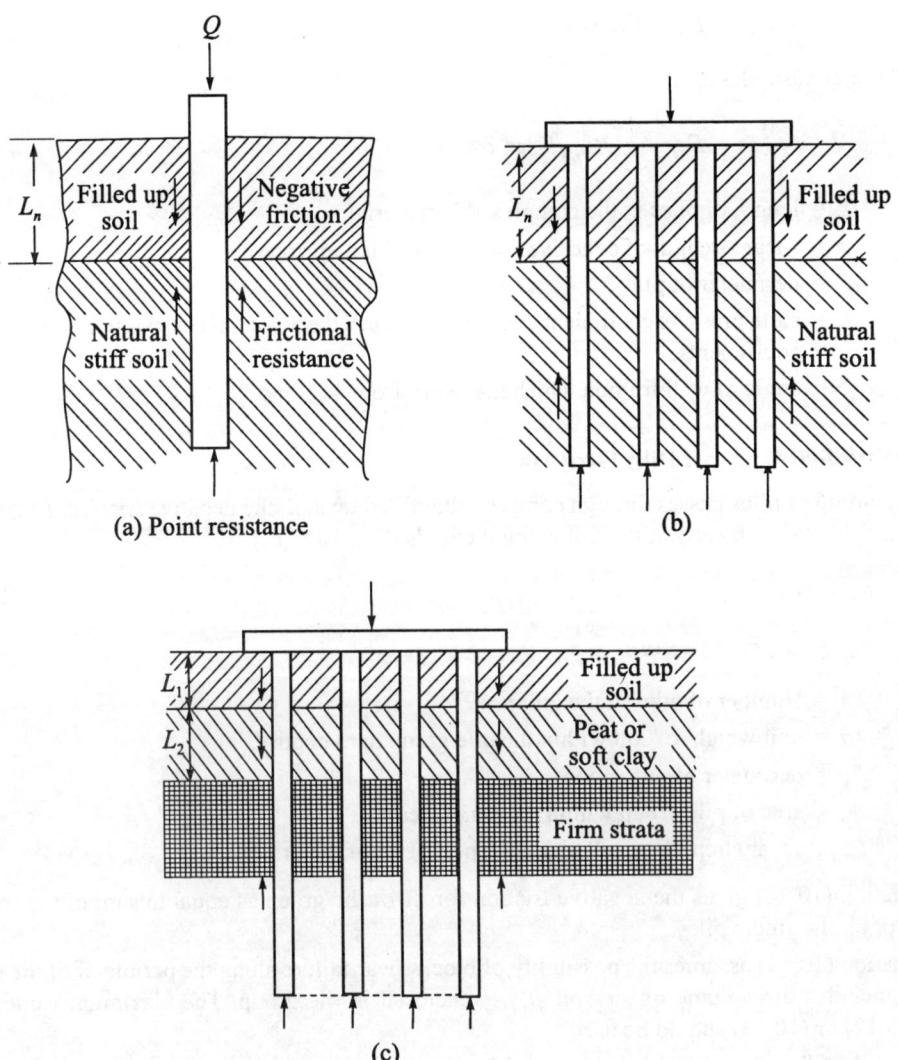

(a) Point resistance (b)

(c)

Fig. 10.8 Negative friction on piles

10.13 ANALYSIS OF PILE FOUNDATIONS COMPRISING VERTICAL AND BATTER PILES AND SUBJECTED TO VERTICAL AND LATERAL LOADS

Introduction

There are quite a few theories that are published in the literature on the design of laterally loaded pile foundations with batter piles. Most of the earlier theories either did not take the soil restraint into account or used it in an arbitrary way. It was Hrennikoff (1949) who was the first to propose a theory for the analysis of pile foundations by taking the soil restraint directly into account. Later on Vesic (1956) proposed a different approach to the same problem though the basic ideas underlying his method are substantially the same as those of the Hrennikoff's method. The theories advanced by these investigators are quite sound and can be used to analyse pile foundations with batter piles provided the pile constants used by them in the theories are correctly estimated or obtained by some means. Hrennikoff assumes the soil modulus as constant with depth for all soils whereas Vesic assumes it as constant for cohesive soils and varying linearly with depth for granular soils. Since Hrennikoff's theory is simple and easy to understand only this theory is dealt with in this chapter.

Hrennikoff's Theory

Statement of the Problem: Figure 10.9 (a) shows a pile foundation comprising of vertical and batter piles subject to external forces such as vertical load V, lateral load H and moment M. Required to develop Hrennikoff's theory which would help to determine,

1. the lateral, vertical and rotational deflections of the pile cap,
2. the axial loads, shear loads and moments at each of the pile heads.

The axes of coordinates for the pile group X and Z are taken as shown in Fig. 10.9 (a). They indicate the positive directions. The origin O is chosen arbitrarily.

Assumptions Made in Hrennikoff's Theory

1. The load carried by each pile is proportional to the displacement of the pile head. In the most general case this displacement consists of three components, namely

   (*a*) the axial displacement of pile, δ_a,

   (*b*) the transverse displacement, y_t,

   (*c*) the rotational displacement, α.

   All the displacements thus defined are proportional to the forces producing them.
2. All piles behave alike with regard to the load deformation relation.
3. The footing, in which the pile heads are embedded, is absolutely rigid.
4. The problem is two-dimensional—that is, the piles, as well as the external forces, are arranged in planes transverse to the length of the foundation and they are symmetrical with regard to the transverse middle plane.
5. The footing movements are small.

The analysis involves two types of constants. They are :

1. Pile Constants.
2. Foundation Constants.

These constants are defined below.

The Pile Constants

The *Pile constants* are defined as the forces with which the pile acts on the foundation when the pile head is given a unit displacement. There are three sets of these constants, corresponding to three different kinds of displacements as shown in Fig. 10.9.

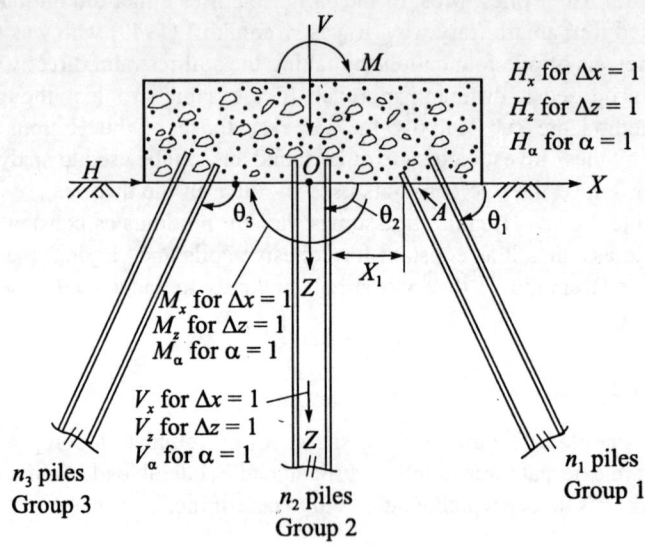

(a) Foundation constants

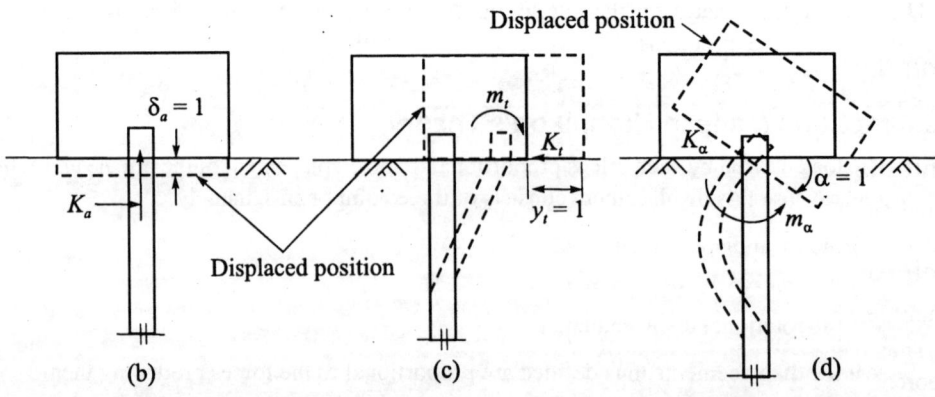

Fig. 10.9 Pile group subjected to combined vertical and horizontal loads, and moment: (a) The definition of foundation constants, (b, c, d) the definition of pile constants

1. Longitudinal displacement

A unit longitudinal displacement [Fig. 10.9 (b)] brings into play a single constant K_a (expressed as the axial load per unit displacement) and acts in the axial direction. This force is called as *coefficient of axial reaction*.

2. Transverse movement

A unit transverse movement $y_t = 1$ of the pile head. Figure 10.9 (c), produces a transverse resistance K_t and a rotational resistance m_t. K_t is defined as the *coefficient of lateral reaction* and is the load required for a pure unit lateral displacement of the pile head with no other load acting on the pile. m_t is defined as the *rotational resistance per unit of transverse displacement of pile head* with rotation equal to zero (expressed in units of force-length per unit of lateral displacement).

3. Rotational displacement

A unit rotation $\alpha = 1$ radian of the pile head, Fig. 10.9 (d), is accompanied by a rotational resistance m_α, and a transverse resistance K_α, measured in units of force-length per radian and force per radian respectively.

It may be observed that no axial resistances accompany either transverse or rotational displacements and no moments or transverse forces are brought into play by the axial displacement. It should also be noted that the pile constants are defined herein represent the effects of the pile on the footing and not *vice-versa*, Therefore, the directions of these constants must be noted.

Since by *Betti's reciprocal theorem*, $K_\alpha = m_t$, there are only four independent constants characterising the load deformation relation of a pile when the pile is deeply embedded in the footing. The four constants are K_a, K_t, m_t and m_α.

When the footing barely engages the piles, their heads may be considered as hinged instead of fixed and $m_t = 0$. Likewise the two constants K_α and K_t need not be considered in this case. The only constants that are to be considered for a hinged case are K_a and K_t.

The Foundation Constants

The foundation constants are defined as the resultant force with which all piles act together on the footing, when the footing is given a unit translation displacement in the positive direction of one of the axes or a unit rotation about the origin in the clockwise direction. Each unit displacement in general brings into play a resisting force, which may be represented by its two components acting along the coordinate axes and the moment about the origin as shown in Fig. 10.9 (a).

Unit displacement along the X-axis

A unit displacement $\Delta X = 1$ along the X-axis gives rise to the constants H_x, V_x, and M_x [Fig. 10.9 (a)].

Unit displacement along the Z-axis

A unit displacement $\Delta Z = 1$ along the Z-axis gives rise to constants H_z, V_z, and M_z.

Unit rotation about the origin O

A unit rotation of $\alpha = 1$ about the origin O produces constants H_α, V_α, and M_α.

Thus there are now nine constants but all are not independent. According to Betti's reciprocal theorem

$$V_x = H_z, \quad M_x = H_\alpha, \text{ and } M_z = V_\alpha.$$

There are therefore only six independent constant which may be taken as H_x, H_z, H_α, V_z, V_α, and M_x. The positive signs of these functions correspond to the positive directions of the axes and to the clockwise direction for the moments.

These foundation constants can be evaluated in terms of the pile constants, K_a, K_t, m_t, and m_α, and the geometry of the foundation. It should be kept in mind that the foundation constants represent the action of the piles on the foundation and not the opposite effect.

Development of Equations for Foundation Constants

Figure 10.9 (a) represents a foundation comprising of vertical and batter piles. Group I comprises n_1 piles making an angle θ_1 with the foundation base; group 2 comprises n_2 piles at an angle θ_2; group

3 comprises n_3 piles at angle θ_3 and so on. Let the total number of pile groups be N. The angle θ is measured from the positive direction of the X-axis clockwise to the given pile.

Displacement of foundation along the *X*-axis through a init distance $\Delta X = 1$ [Fig. 10.10 (a)]

The footing is displaced a unit distance $\Delta X = 1$ parallel to itself tin the positive direction of X-axis. Since the piles are rigidly fixed to the foundation, they retain their original inclinations at the ends. Consider the head of one typical pile A, shown in Fig. 10.10 (a). The forces that would be developed due to the movement of this pile from A to A_1 are shown in the figure.

The total movement of the pile head from A to A_1 along the X-axis equal to $\Delta X = 1$ results in the following deformations.

1. There is a movement of the pile head along the axis of the pile equal to $\cos \theta_1$.

2. There is a lateral displacement of the pile head normal to the axis of the pile equal to $\sin \theta_1$.

These movements along the axis and normal to the axis of the piles give rise to the following forces:

1. An axial force along the pile equal to $K_a \cos \theta_1$.

2. A transverse resisting force equal to $K_t \sin \theta_1$.

3. A moment at the pile head equal to $m_t \sin \theta_1$.

The directions of these forcesare shown in Fig. 10.10 (a). The components of the first two forces in the X and Z directions due to unit displacement $\Delta X = 1$ may be written as

$$h_x = - (K_a \cos^2 \theta_1 + K_t \sin^2 \theta_1)$$

$$v_x = -\frac{1}{2} (K_a - K_t) \sin 2\theta_1$$

The sums of the components of all the induced pile forces of all N groups of piles in the directions of the axes represent the pile constants H_x and V_x. The equations for the corresponding foundation constants may be written as:

$$H_x = \sum_N nh_x = -\sum_N n (K_a \cos^2\theta + K_t \sin^2\theta) \tag{10.16}$$

$$H_z = V_x = \sum_N nv_x = -\frac{1}{2} (K_a - K_t) \sum_N (n \sin 2\theta) \tag{10.17}$$

wherein, the symbol $\sum\limits_N$ covers N terms of the form expressed in the bracket.

The third foundations constant M_x due to unit displacement $\Delta X = 1$ of the foundation may be obtained as follows.

The head of A is at a distance x_1 from the origin O. The moment produced by the resistances of pile A about the origin is

$$m_x = v_x x_1 + m_t \sin \theta_1$$

Substituting for v_x, the expression is

$$m_x = -\frac{1}{2} (K_a - K_t) \sin 2\theta_1 x_1 + m_t \sin \theta_1$$

The moments of all the piles of group 1 about the origin is

$$m_{x1} = -\frac{1}{2} (K_a - K_t) \sin 2\theta_1 (x_1 + x_2 + \cdots x_n) + n_1 m_t \sin \theta_1$$

wherein, x_1, x_2, etc. are the distances of the heads of piles in group 1 from the region O. If $\bar{x}_1$ is the abscissa of the centre of gravity of all the pile heads in group 1, the above equation may be written as

$$m_{x1} = -\frac{1}{2} (K_a - K_t) n_1 \bar{x}_1 \sin 2\theta_1 + n_1 m_t \sin \theta_1$$

Summing up the moments of all N groups, we have

$$H_\infty = M_x = \sum_N m_{x1} = -\frac{1}{2} (K_a - K_t) \sum_N \left(n\bar{x} \sin 2\theta \right) + m_t \sum_N \left(n \sin \theta \right) \quad (10.18)$$

in which $\bar{x}$ is the coordinate of the centre of the group of n piles having the angle θ.

Displacement of foundation parallel to Z-axis

In Fig. 10.10 (b) is shown the head of pile A displaced through a unit distance $\Delta Z = 1$ parallel to the Z-axis. The components of displacements of the pile may be written as

1. the axial movement pile head $= \sin \theta_1$,
2. the lateral movement of pile head $= \cos \theta_1$.

The forces induced by the piles by the axial and lateral displacements are shown in Fig. 10.10 (b) in their proper directions. They are

1. axial force $= K_a \sin \theta_1$,
2. transverse force $= K_t \cos \theta_1$,
3. the rotational moment $= m_t \cos \theta_1$.

As before summing the components of all the pile forces in all the N groups in the X and Z directions gives rise to H_z and V_z constants. Since $H_z = V_x$ is already known, V_z is the only independent constant of the two. The equation for V_z is

$$V_z = -\sum_N \left[n \left(K_a \sin^2\theta + K_t \cos^2 \theta \right) \right] \quad (10.19)$$

As before taking moments of all the pile forces about the origin O gives rise to the second independent constant M_z. The equation is

$$V_\infty = M_z = -\sum_N \left(K_a \sin^2\theta + K_t \cos^2\theta \right) n\bar{x} - m_t \sum_N \left(n \cos \theta \right) \quad (10.20)$$

Rotation of foundation about the origin O

The foundation is rotated through a unit angle $\alpha = 1$ radian about the origin O as shown in Fig. 10.10 (c).

The rotation of the foundation through $\alpha = 1$ radian gives rise to the following displacements of pile ahead.

1. Axial displacement $= x_1 \sin \theta_1$.

2. Lateral displacement $= x_1 \cos \theta_1$.

These displacements give rise to the following forces on the foundation.

1. Axial force $= K_a x_1 \sin \theta_1$.
2. Lateral force $= K_t x_1 \cos \theta_1 + K_\alpha = K_t x_1 \cos \theta_1 + m_t$.
3. Rotational moment $= m_t x_1 \cos \theta_1 + m_\alpha$.

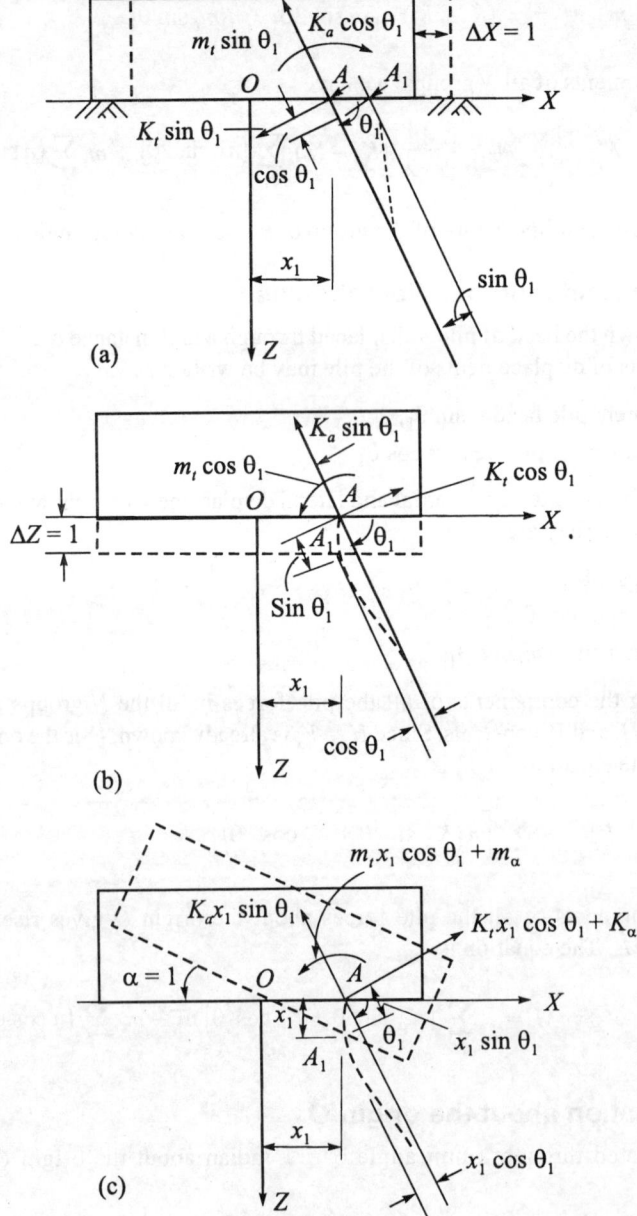

(a)

(b)

(c)

Fig. 10.10 Development of foundation constants. Displacement of foundation: (a) Along X-axis, (b) parallel to Z-axis, and (c) rotation of foundation about the origin O.

The directions of these forces are shown in Fig. 10.10 (c). As before, summing up all the forces of all the piles in the N group in the X and Z directions and summing the moments of all the forces about the origin O gives rise to three foundation constants H_α, V_α and M_α. It is already shown that $H_\alpha = M_x$ and $V_\alpha = M_z$. Expressions for M_x and M_z have already been developed. The only constant left out is M_α. The expression for M_α is

$$M_\alpha = -\sum_N \left[\left(K_a \sin^2\theta + K_t \cos^2\theta \right) \sum_N \left(x^2 \right) \right]$$

$$- 2 m_t \sum_N n\bar{x} \cos\theta - m_\alpha \sum_N (n) \tag{10.21}$$

The foundation constants determined by Eqs (10.16) through (10.21) with their proper signs represent forces and moments exerted by the piles on the footing and not by the footing on piles. Positive signs of these functions indicate forces acting in the positive directions of the coordinates axes or, in the case of moments, in the clockwise direction. The X-coordinates and the trigonometric functions in their expressions must be taken with proper algebraic signs.

The Eqs (10.16) through (10.21) assume full fixity of piles in the footing. If pin-ended conditions exist, the corresponding formula for pile constants may be obtained by substituting $m_t = m_\alpha = 0$ in the equations.

Equations of Equilibrium

In order to determine the pile loads produced by the external forces acting on the footing it is necessary to find the three component displacements of the foundation. The displacements are,

ΔX along the X-axis,

ΔZ along the Z-axis,

and α_1 the angle of rotation about the origin O.

These displacements are found by setting up three simultaneous equations, expressing the equilibrium of the foundation under the action of the external forces and the pile forces induced by these displacements. The equations of equilibrium can be set up by making use of the foundation constants defined earlier. The sum of the forces induced in the X direction due to the displacements ΔX and ΔZ in the X and Z direction and rotation α about the origin O is

$$H_x \Delta X + H_z \Delta Z + H_\alpha \alpha$$

This force is balanced by the horizontal component H of the external force acting on the foundation. Similarly, expression can be written for the forces in the Z-direction and moment about the origin. The equations of equilibrium of the footing may therefore be written as

$$H_x \Delta X + H_z \Delta Z + H_\alpha \alpha + H = 0 \tag{10.22a}$$

$$H_z \Delta X + V_z \Delta Z + V_\alpha \alpha + V = 0 \tag{10.22b}$$

$$H_\alpha \Delta X + V_\alpha \Delta Z + M_\alpha \alpha + M = 0 \tag{10.22c}$$

The component displacements of the footing ΔX, ΔZ and α may be determined from Eq. (10.22) provided values of the foundation constants are known. Once these component displacements are known the displacements of the individual pile heads are found by geometry and the axial, lateral and rotational pile loads may then be computed.

Pile Displacements and Loads

Expressions for pile head displacements may be written as follows with reference to an arbitrary pile A [shown in Fig. 10.9 (a)] situated at a distance of x_1 from the origin O.

The longitudinal displacement δ_a downward,

$$\delta_a = \Delta X \cos \theta_1 + \Delta Z \sin \theta_1 + \alpha x_1 \sin \theta_1 \tag{10.23}$$

The transverse displacement y_t to the right,

$$y_t = \Delta X \sin \theta_1 - \Delta Z \cos \theta_1 - \alpha x_1 \cos \theta_1 \tag{10.24}$$

The rotation of the pile head is α, clockwise. The expressions for the pile loads are as given below. The axial load of pile (compression)

$$Q_g = K_a \delta_a \tag{10.25}$$

The transverse load, acting on the foundation to the right

$$P_g = -K_t y_t + m_t \alpha \tag{10.26}$$

The moment, acting on the foundation clockwise,

$$M_g = m_t y_t - M_\alpha \alpha \tag{10.27}$$

Procedure for Computing the Values of Pile Constants

The values of four pile constants K_a, K_t, m_t and m_α are required for the analysis of pile foundations subjected to vertical and lateral loads. There is no universal method for computing the values of these constants. One should approach this problem with an insight into the deformation characteristics of the pile and the soil and a sound engineering judgement. The following method seems to be reasonable.

Computation of pile constant K_a

Hrennikoff has suggested the following equation for computing K_a.

If the allowable load on the pile in the axial direction is Q_a and the corresponding axial displacement of the pile head is δ_a, then

$$K_a = \frac{Q_a}{\delta_a} \tag{10.28}$$

δ_a may be computed as follows:

For piles driven into very weak soil (i.e. for point bearing piles)

$$\delta_a = \frac{Q_a L}{EA} \tag{10.29}$$

For a pile driven into other soils (friction piles)

$$\delta_a = \frac{Q_a L}{2 EA} \tag{10.30}$$

wherein L = length of pile,
A = average cross-sectional area of pile,
E = modulus of elasticity of pile.

However, Vesic suggests load tests on a vertical pile for determining K_a. A compression test for piles in compression and a pull-out test for piles in tension are needed.

Equations for K_t, m_t and m_α

Using the concept of modulus of subgrade reaction Matlock and Reese have given generalised equations for deflection and slope of laterally loaded piles. As per Eq. (9.51) the displacement of pile head, y_g, at ground level may be expressed as

$$y_g = \frac{P_g T^3}{EI} A_y + \frac{M_g T^2}{EI} B_y$$

As per Eq. (9.52), the corresponding slope or rotation α at ground level is

$$\alpha_g = \frac{P_g T^2}{EI} A_\alpha + \frac{M_g T}{EI} B_\alpha$$

where, $A_\alpha = A_s$, $B_\alpha = B_s$.

T is the relative stiffness factor which is given by Eq. (9.50).

$$T = \left(\frac{EI}{n_h}\right)^{\frac{1}{n+4}}$$

The general expressions for pile constants K_t and m_t may be obtained from Eqs (9.51) and (9.52) by putting $y_g = 1$ and $\alpha_g = 0$ and replacing P_g and M_g by K_t and m_t, respectively, we get

$$K_t = \frac{EI}{T^3} \left(\frac{B_\alpha}{A_y B_\alpha - A_\alpha B_y}\right) = A_1 \frac{EI}{T^3} \tag{10.31}$$

$$m_t = \frac{EI}{T^2} \left(\frac{A_\alpha}{A_\alpha B_y - A_y B_\alpha}\right) = A_2 \frac{EI}{T^2} \tag{10.32}$$

Similarly, if we put $\alpha_g = 1$ and $y_g = 0$ and replace P_g and M_g in Eqs (9.51) and (9.52) by $m_t (= K_\alpha)$ and m_α respectively, we get

$$m_\alpha = \frac{EI}{T} \left(\frac{B_\alpha}{A_y B_\alpha - A_\alpha B_y}\right) = A_1 \frac{EI}{T} \tag{10.33}$$

It can be seen from Eqs (10.31), (10.32) and (10.33) that for a pile of given stiffness EI, the pile constants are the functions of A_y, B_y, A_α, B_α, and T. The A and B coefficients vary with the batter of pile, whereas T is a function of both the batter and the type of soil. A and B coefficients for vertical piles have been computed by Matlock and Reese on the assumption that soil modulus varies linearly with depth (that is for $n = 1$). Hence, pile constants K_t, m_t and m_α may, therefore, be calculated for vertical piles. For batter piles A coefficients have been obtained by the author from model tests and B coefficients are not available. As such the pile constants for batter piles cannot be obtained on the basis of the above equations. However the following equations have been developed by making use of the approach suggested by Vesic

$$K_t = \frac{2.56}{A_y}\frac{EI}{T^3} = A_1\frac{EI}{T^3} \tag{10.34}$$

$$m_t = \frac{2.35}{A_y}\frac{EI}{T^2} = A_2\frac{EI}{T^2} \tag{10.35}$$

$$m_a = \frac{3.54}{A_y}\frac{EI}{T} = A_3\frac{EI}{T} \tag{10.36}$$

The values of A_y as obtained by direct measurements (Murthy, 1965) for different batters are given in Table 10.1.

Table 10.1 Values of A_y

Batter of pile	A_y
0°	2.27
+ 15°	2.40
+ 30°	2.52
+ 45°	2.76
− 15°	2.20
− 30°	2.36
− 45°	1.76

By substituting for A_y in the Eqs (10.34) through (10.36), the values of the coefficients of pile constants are obtained and they are plotted against batter as shown in Fig. 10.11. Average of the plotted points drawn shown that the A_1, A_2 and A_3 lines are almost parallel to each other.

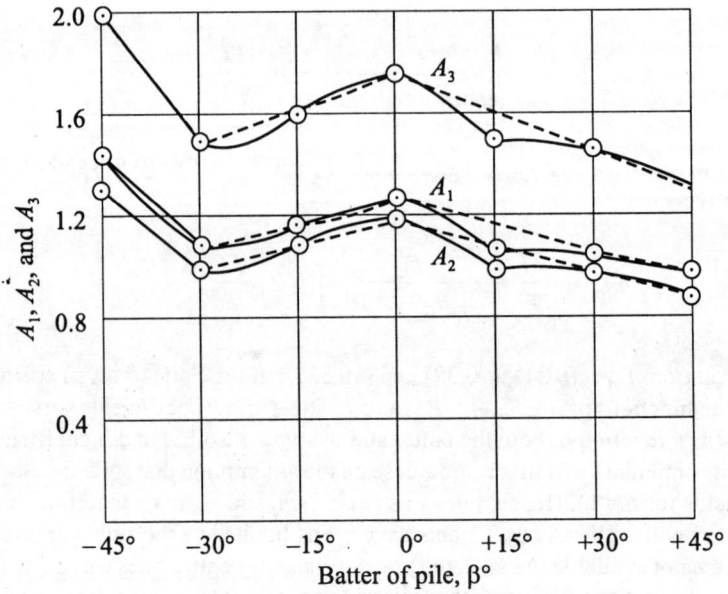

Fig. 10.11 Coefficients of pile constants

It has been found that pile constants are affected if the spacings of the piles are less than six times the diameter of piles. Sufficient data are not available to suggest any change in the pile constants for different spacing of piles.

10.14 PILE GROUPS SUBJECTED TO ECCENTRIC VERTICAL LOADS

The reactions exerted by piles in a group when it is subjected to direct vertical load and moments only may be determined on the following assumptions:

1. The pile cap is a rigid structure.
2. When the pile group is subjected to moment, the reactions exerted by the piles increase linearly with the distance of pile from the centre of gravity of pile group.
3. The resisting moment at pile heads due to the fixity condition between piles and the cap is either negligible or ignored.

Though the above assumptions are not strictly valid, it is considered sufficiently accurate for the purpose of design.

Consider the group of piles shown in Fig. 10.12 (a). O is the centroid of the group. XX and YY are the coordinate axes passing through O. The positive directions of the axes are shown by the arrows. O_x and O_y are points on the X and Y axes with eccentricities e_x and e_y respectively as shown in the figure. O_{xy} is another point with coordinates e_x and e_y.

When a vertical load, V, passes through, O the centroid of the pile group, there will be no moment on the group and V is the vertical load. The reactions of all the piles, in such a case, are equal to each other and is equal to V/n where n is the number of piles in the group. However, when the load V passes through O_x, an eccentric condition develops with one way eccentricity. The pile group is then considered to have been subjected to the vertical load, V, passing through, O, and a moment $M_y = Ve_x$ about the Y-axis as shown in (b) of the figure. If the load passes through O_y instead of O_x similar condition as above develops, but in this case the moment $M_x = Ve_y$ is about the X-axis as shown in (c) of the figure.

When the load V passes through O_{xy}, there will be two way eccentricity. This condition is equivalent to the vertical load, V, passing through O and moments M_x and M_y acting simultaneously about the axes XX and YY respectively.

The reaction developed at pile heads due to V passing through O is equal to V/n and is as shown in (d) of the figure. When only moment M_y (or x) acts on the pile group, the reaction due to this at any pilehead is essumed to vary linearly as shown in (e) of the figure. The combined reactions due to V and moment M_y are as shown in figure (f). If the pile group is subjected to a vertical load with one way eccentricity e_x, the total reaction R at the pile head may be obtained by the equation

$$R = \frac{V}{n} \pm \frac{M_y x}{\Sigma x^2} \tag{10.37}$$

But, if the pile group is subjected to the vertical load V with two way eccentricity, the general equation for determining, R, is

$$R = \frac{V}{n} \pm \frac{M_y x}{\Sigma x^2} \pm \frac{M_x y}{\Sigma y^2} \tag{10.38}$$

where, R = total reaction at the pile head,

V = total vertical load acting on the pile cap,

n = number of piles in the group,

M_x = total moment about X-axis = Ve_y,

M_y = total moment about Y-axis = Ve_x,

x = distance of the pile in question from the Y-axis,

y = distance of the pile in question from the X-axis,

Σx^2 = sum of all the squares of the distances of all the piles from the Y-axis,

Σy^2 = sum of all the squares of the distances of all the piles from the X-axis.

Inspection of Eq. (10.38) indicates that this form is similar to Eq. (8.39) which gives soil pressure at a given point under the base of a shallow foundation with two-way eccentric loading. The number of piles n is substituted for area, the terms Σx^2 and Σy^2 replaces the moments of inertia of the area about YY and XX axes respectively. For this reason Σx^2 or Σy^2 is sometimes called as the moment of inertia of the group of piles. The second or the third term in the Eq. (10.38) may be derived as follows.

Assume the pile group in Fig. 10.12 (a) is subjected to a moment M_y only. The reactions developed at the pile heads give rise to a resisting moment which is equal to the applied moment M_y. Let R_1, R_2, R_3, and R_4 be the reactions of the piles placed at distances of $x_1, x_2 x_3$ and x_4 respectively from the Y-axis [Fig. 10.12 (e)]. We may write,

$$M_y = (4R_1 x_1 + 4R_2 x_2 + 4R_3 x_3 + 4R_4 x_4) \tag{10.39}$$

Since the variation of pile reactions is assumed to be linear, we have

$$R_1/x_1 = R_2/x_2 = R_3/x_3 = R_4/x_4$$

or

$$R_2 = R_1 x_2/x_1$$
$$R_3 = R_1 x_3/x_1$$
$$R_4 = R_1 x_4/x_1$$

Substituting these values of R_2, R_3 and R_4 in Eq. (10.39), we have

$$M_y = (4R_1 x_1^2/x_1 + 4R_1 x_2^2/x_1 + 4R_1 x_3^2/x_1 + 4R_1 x_4^2/x_1)$$

$$= \frac{R_1}{x_1} (4x_1^2 + 4x_2^2 + 4x_3^2 + 4x_4^2)$$

$$= \frac{R_1}{x_1} \Sigma x^2$$

Solving for R_1, we have

$$R_1 = \frac{M_y x_1}{\Sigma x^2} \tag{10.40}$$

Similarly, the reaction at any other pile head may be determined by means of Eq. (10.40) by replacing x_1 by the distance of the pile from the Y-axis.

If the pile group in Fig. 10.12 (a) is subjected to a vertical load passing through O_x only, then all the piles in Col. 1 will carry the maximum load and all the piles in Col. 4 will carry the minimum load. However, if the load V passes through O_{xy}, the pile A will carry the greatest load whereas pile B carries the least. Both M_x and M_y increases the reaction at A and decreases that at B. Thus, it is possible to select by inspection, the proper signs in the application of Eq. (10.38) to any pile.

The determination Σx^2 or Σy^2 for large groups of pile may be considerably simplified by the use of Eq. (10.41), which applies to a single row of piles with equal spacing.

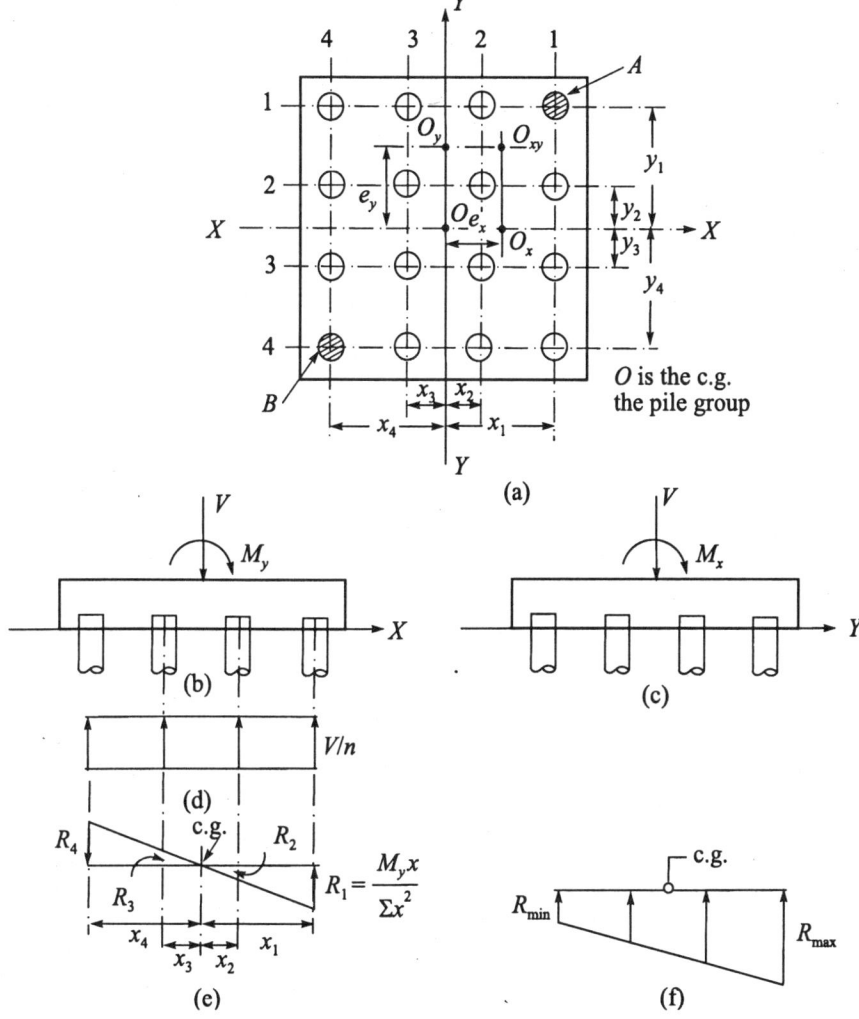

Fig. 10.12 Pile group subjected to eccentric vertical loads

$$\Sigma x^2 \text{ (one row)} = \frac{s^2}{12} \, n_1 \, (n_1^2 - 1), \tag{10.41}$$

where, s = spacings of piles in the row,
n_1 = number of piles in the row.

10.15 ANCHOR PILES

When a pile group is subjected to moment, all the piles in the group may be in compression or some piles in compression and some in tension. If the application of Eq. (10.37) or (10.38) to any pile in the group gives negative value, it indicates that the pile is in tension. When a pile or a group of piles is in tension, the following conditions may develop.

1. The pile or piles may get pulled out of the ground.
2. The pile head may get separated from the pile cap.

By suitably anchoring the pile in the ground it may be prevented from being pulled out. Such piles, are called as *anchor piles* or *tension piles*. If long piles of uniform diameter are used as anchor piles, the length should be sufficient to resist the up-lift force with a suitable factor of safety.

For some of the structures under-reamed piles are quite suitable as anchor piles. Anchor piles are normally required for the foundations of structures such as transmission line towers, gas storage tanks, tall chimneys, etc. The tensile forces beneath such structures are normally caused by moment due to wind. Broken wire condition is also responsible for producing high moments on transmission line towers. Hydrostatic pressures on underground structures also develop uplift forces which may have to be resisted by anchor piles.

10.16   UPLIFT CAPACITY OF A PILE GROUP

The uplift capacity of a pile group, when the vertical piles are arranged in a closely spaced groups may not be equal to the sum of the uplift resistances of the individual piles. This is because, at ultimate load conditions, the block of soil enclosed by the pile group gets lifted. The manner in which the load is transferred from the pile to the soil is quite complex. A simplified way of calculating the uplift capacity of a pile group embedded in cohesionless soil is shown in Fig. 10.13 (a). A spread of load of 1 Horiz : 4 Vert from the pile group base to the ground surface may be taken as the volume of the soil to be lifted by the pile group (Tomlinson, 1977). For simplicity in calculation, the weight of the pile embedded in the ground is assumed to be equal to that of the volume of soil it displaces. If the pile group is partly or fully submerged, the submerged weight of soil below the water table has to be taken.

In the case of cohesive soil, the uplift resistance of the block of soil in undrained shear enclosed by the pile group given in Fig. 10.13 (b) has to be considered. The equation for the total uplift capacity P_{gu} of the group may be expressed by

$$P_{gu} = 2L\,(\bar{L} + \bar{B})\,\bar{c}_u + W \tag{10.42}$$

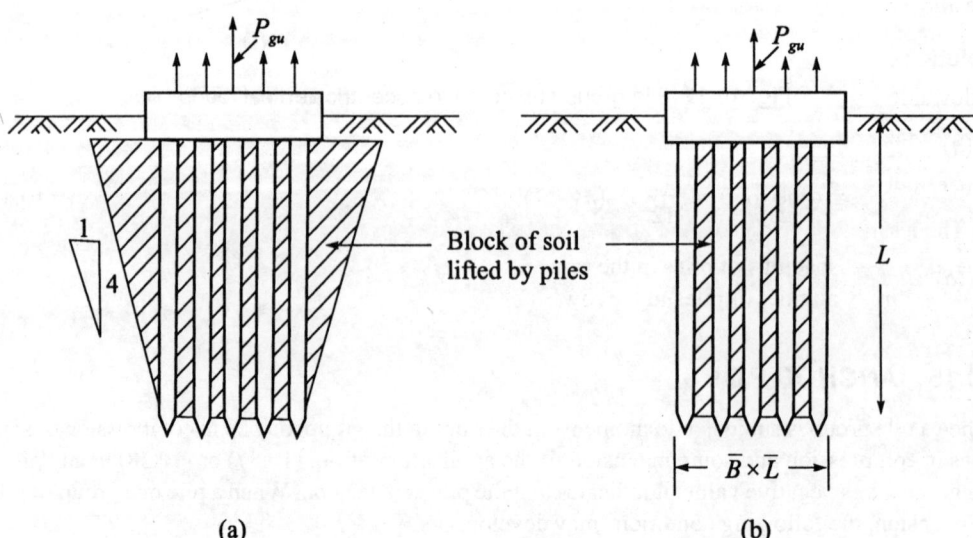

(a)   (b)

Fig. 10.13 Uplift capacity of a pile group: (a) Uplift of a group of closely-spaced piles in cohesonless soils, (b) uplift of a group of piles in cohesive soils

where, L = depth of the pile block,

$\bar{L}$ and $\bar{B}$ = overall length and width of the pile group,

$\bar{c}_u$ = average undrained shear strength of soil around the sides of the group,

W = combined weight of the block of soil enclosed by the pile group plus the weight of the piles and the pile cap.

A factor of safety of 2 may be used in both cases of piles in sand and clay.

The uplift efficiency E_{gu} of a group of piles may be expressed as

$$E_{gu} = \frac{P_{gu}}{nP_{us}}$$ (10.43)

where P_{us} = uplift capacity of a single pile,

n = number of piles in the group.

The efficiency E_{gu} varies with the method of installation of the piles, length and spacing and the type of soil. The available data indicate that E_{gu} increases with the spacing of piles. Meyerhof and Adams (1968) presented some data on uplift efficiency of groups of two and four model circular footings in clay. The results indicate that the uplift efficiency increases with the spacing of the footings or bases and as the depth of embedment decreases, but decreases as the number of footings or bases in the group increases. How far the footings would represent the piles is a debatable point. For uplift loading on pile groups in sand, there appears to be little data from full scale field tests.

10.17 EXAMPLES

Example 10.1

A group of 9 piles with 3 piles in a row were driven into a soft clay extending from ground level to a great depth. The diameter and length of the piles were 30 cm and 10 m respectively. The unconfined compressive strength of clay is 70 kPa. If the piles were spaced at 90 cm centre to centre, compute the allowable load on the pile group on the basis of shear failure criteria for a factor of safety of 2.5.

Solution

Allowable load on the group are to be calculated for two conditions,

(*a*) block failure,

(*b*) individual pile failure.

The least of the two gives the allowable load on the group.

(*a*) *Block failure (Fig. 10.4)*

$$Q_{gu} = cN_c A_g + P_g L\bar{c}$$

$$N_c = 9$$

where, $c = \bar{c} = \dfrac{70}{2} = 35 \text{ kN/m}^2,$

$A_g = 2.1 \times 2.1 = 4.4 \text{ m}^2,$

$P_g = 4 \times 2.1 = 8.4 \text{ m},$

$L = 10 \text{ m},$

$$Q_{gu} = 35 \times 9 \times 4.4 + 8.4 \times 10 \times 35 = 1390 + 2950$$

$$= 4340 \text{ kN},$$

$$Q_a = \frac{4340}{2.5} = 1740 \text{ kN}.$$

(b) *Individual pile failure*

$$Q_u = Q_b + Q_f = q_b A_b + \alpha \bar{c} A_s$$

Assume $\alpha = 1$, Now $q_b = c N_c = 35 \times 9 = 315 \text{ kN/m}^2$

$$A_b = 0.07 \text{ m}^2, A_s = 3.14 \times 0.3 \times 10 = 9.42 \text{ m}^2$$

Substituting, $\quad Q_u = 315 \times 0.07 + 1 \times 35 \times 0.42$

$$= 22 + 330 = 352 \text{ kN}$$

$$Q_{gu} = n Q_u = 9 \times 352 = 3168 \text{ kN}$$

$$Q_a = \frac{3168}{2.5} \approx 1267 \text{ kN}$$

The allowable load is 1267 kN.

Example 10.2

A square pile group similar to the one shown in Fig. 10.4 passes through a recently filled up fill. The depth of fill $L_n = 3$ m. The diameter of pile is 30 cm, and they are spaced at 90 cm apart. If the soil is cohesive with $q_u = 60 \text{ kN/m}^2$, $\gamma = 15 \text{ kN/m}^3$, compute the negative frictional load on the pile group.

Solution

The negative friction on the group is the maximum of the following

(a) $F_{ng} = n F_n$

(b) $F_{ng} = s L_n P_g + \gamma L_n A_g$

where, $P_g = 4 \times 3 = 12$ m, $A_g = 3 \times 3 = 9 \text{ m}^2$

(a) $F_{ng} = 9 \times 3.14 \times 0.3 \times 3.0 \times 30 = 763 \text{ kN}$

(b) $F_{ng} = 30 \times 3 \times 12 + 15 \times 3 \times 9 = 1080 + 405 = 1485 \text{ kN}$

The negative friction of the group is 1485 kN.

Example 10.3

A tall retaining wall is supported on piles comprising vertical and batter piles as shown in Fig. Ex. 10.3. The piles are rigidly fixed to the foundation. The piles are spaced at 1.0 m apart parallel to the face of the wall. The resultant vertical (V), horizontal (H) forces and moment (M) acting on the foundation per metre length of the wall are shown in the figure. The number of piles per metre length of wall is 5 comprising of 3 batter piles all of equal batter of 18° 26′ and two vertical piles and the spacings are shown on the figure. The piles are reinforced concrete piles of 30 cm diameter driven into very loose sand to a depth of 9 m and rests on rocky strata. The coefficient of modulus variation n_h, is assumed as equal to 2.5 MN/m³ ($\approx 0.25 \text{ kg/cm}^3$).

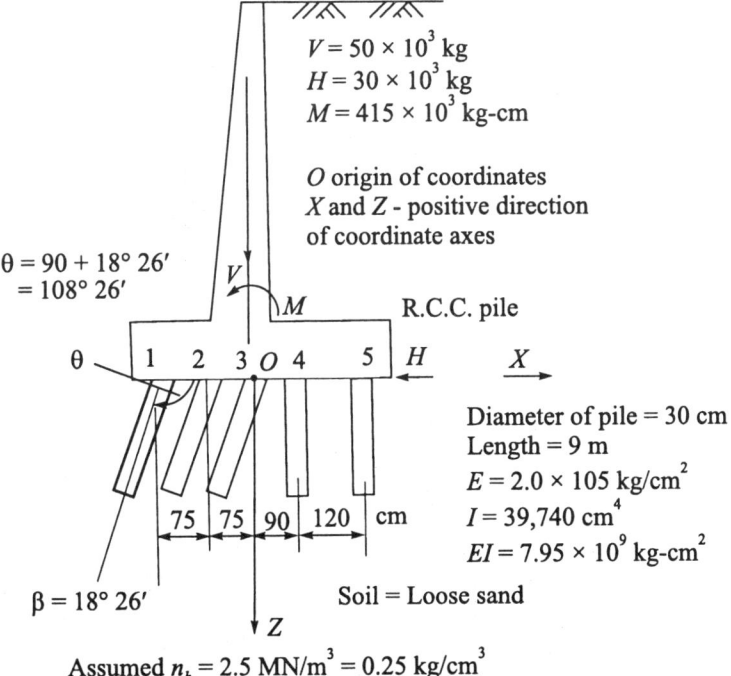

Fig. Ex. 10.3

Determine (a) the lateral, vertical and rotational deflections of the foundation, (b) and the axial loads, shear loads and moments at each of the pile heads.

Solution

The problem is solved by Hrennikoff's theory. The pile constants are determined by the author's method. The pile constants computed for vertical piles are also used for batter piles. The step by step approach of solving the problem is explained below.

1. *Pile constants*

    The pile constants required to be computed are K_a, K_t, m_t, and m_α. K_a is computed here by Eqs (10.28) and (10.29) applicable for weak soil. From these equations

$$K_a = \frac{AE}{L}$$

where, A = sectional area of pile = 707 cm^2,

  E = modulus elasticity of pile = 10.4 × 10^4 kg/cm^2,

  L = length of pile = 900 cm.

Therefore, $K_a = \dfrac{707 \times 2 \times 10^5}{900} = 157 \times 10^3$ kg/cm.

Use Eqs (10.34), (10.35) and (10.36) for computing K_t, m_t and m_α.

$$K_t = \frac{2.56}{A_y} \frac{EI}{T^3}$$

where,

$$T = \left(\frac{EI}{n_h}\right)^{\frac{1}{5}} = \left(\frac{7.95 \times 10^9}{0.25}\right)^{\frac{1}{5}} = 126 \text{ cm}$$

$A_y = 2.43$ as per Reese *et al* non-dimensional factor for deflection at ground level.

Substituting,

$$K_t = \frac{2.56}{2.43} \times \frac{7.95 \times 10^9}{(126)^3} \approx 4.0 \times 10^3 \text{ kg/cm}$$

In the same way,

$$m_t = \frac{2.35}{A_y} \frac{EI}{T^2} = \frac{2.35}{2.43} \times \frac{7.95 \times 10^9}{(126)^2}$$

$$\approx 484 \times 10^3 \text{ (kg-cm)/cm}$$

$$m_\alpha = \frac{3.54}{A_y} \frac{EI}{T} = \frac{3.54}{2.43} \times \frac{7.95 \times 10^9}{126}$$

$$= 91916 \times 10^3 \text{ (kg-cm)/radian.}$$

2. *Foundation constants*

The following foundation constants are required to be computed for solving the simultaneous Eq. (10.22). They are

$$H_x, H_z, H_\alpha, V_z, V_\alpha, \text{ and } M_\alpha.$$

It may be noted here that $V_x = H_z$, $M_x = H_\alpha$ and $M_z = V_\alpha$.

The sign convention may also be noted as per Fig. Ex. 10.3. The positive X and Z are shown in the figure. Clockwise moment is positive. The forces are considered in metric tonnes

For $\theta = 108° 26'$		For $\theta = 90°$
$\sin \theta = 0.94869$	$\sin^2\theta = 0.9$	$\sin \theta = 1.0$
$\cos \theta = -0.31623$	$\cos^2\theta = 0.1$	$\cos \theta = 0$
$\sin 2\theta = -0.6$		$\sin 2\theta = 0.$

$$H_x = -\sum_N n\left(K_a \cos^2\theta + K_t \sin^2\theta\right)$$

$$= -[3 (157 \times 0.1 + 4.0 \times 0.9) + (4) 2] = -65.9T$$

$$\approx -66T.$$

$$H_z = -\frac{1}{2} (K_a - K_t) \sum_N (n \sin 2\theta)$$

$$= -\frac{1}{2} (157 - 4) [3 (-0.6) + 0] = +137.7T \approx +138 \ T.$$

$$H_\alpha = -\frac{1}{2} (K_a - K_t) \sum_N (n\bar{x} \sin 2\theta) + m_t \sum (n \sin \theta)$$

$$= -\frac{1}{2} (157 - 4) (-75 - 150) (-0.6) + 484 (3 \times 0.94869 + 2 \times 1.0)$$

$$= -7983 \text{ T-cm}$$

$$V_z = -\sum_N \left[n \left(K_a \sin^2\theta + K_t \cos^2\theta \right) \right]$$

$$= -[3 \, (157 \times 0.9) + (157 \times 1.0) \, 2 + (4 \times 0.1) \, 3]$$

$$= -739 T$$

$$M_z = V_\alpha = -\sum_N \left(K_a \sin^2\theta + K_t \cos^2\theta \right) n\bar{x} - m_t \sum_N \left(n \cos\theta \right)$$

$$= -[(157 \times 0.9 + 4 \times 0.1)(-150 - 75 + 0) + 157 \, (1)$$

$$\times (90 + 210)] - 489 \, (-0.31623 \times 3) = -15,681 \text{ T-cm}$$

$$M_\alpha = -\sum_N \left(K_a \sin^2\theta + K_t \cos^2\theta \right) \sum_N x^2$$

$$= -2 m_t \sum_N nx \cos\theta - m_\alpha \sum_N \left(n \right)$$

$$= -[(157 \times 0.9 + 4 \times 0.1)(150^2 + 75^2 + 0) + 157 \, (1.0) \, (90^2 + 210^2)]$$

$$-2 \times 484 \, [-0.31623 \, (-150 - 75 + 0)] - 91,916 \times (5)$$

$$= -12,709,168 \text{ T-cm}^2$$

3. *Computation of displacements of footings*

   The equations of equilibrium of the footing [Eq. (10.22)] are

$$H_x \Delta x + H_z \Delta z + H_\alpha \alpha + H = 0$$

$$H_z \Delta x + V_z \Delta z + V_\alpha \alpha + V = 0$$

$$H_\alpha \Delta x + V_\alpha \Delta z + M_\alpha \alpha + M = 0$$

Substituting

$$-66 \Delta x + 138 \Delta z - 7,983\alpha - 30 = 0$$

$$+138 \Delta x - 739 \Delta z - 15,681\alpha + 50 = 0$$

$$-7,983 \Delta x - 15,681 \Delta z - 1,27,09,168\alpha - 4,150 = 0$$

Solving the above equations, we get

$$\Delta x = -0.394 \text{ cm}$$

$$\Delta z = +0.0313 \text{ cm}$$

$$\alpha = +0.000042 \text{ radian}$$

4. *Pile displacements*

   The axial displacement of a pile may be expressed as [Eq. (10.23)]

$$\delta_a = \Delta x \cos\theta_1 + \Delta z \sin\theta_1 + \alpha x_1 \sin\theta_1.$$

Displacement of pile number 1

Axial displacement

$$\delta_a = (-0.394)(-0.31623) + (0.0313)(0.9869) + (0.000042)(-150)(0.94869)$$

$$= 0.1495 \text{ cm}$$

Lateral displacement

$$y_t = \Delta x \sin \theta_1 - \Delta z \cos \theta_1 - \alpha x_1 \cos \theta_1$$

$$= (-0.394)(0.94869) - (0.0313)(-0.31623) - (0.000042)(-150)(-0.31623)$$

$$= -0.366 \text{ cm.}$$

The displacements of other piles may be calculated in the same way.

5. *Pile loads*

Computation of pile loads for pile number 1

Axial load $Q_g = K_a \delta_a = 157 \times 0.1495 = 23.47$ tonnes.

The pile is in compression

The tranverse load,

$$P_g = -K_t y_t + m_t \alpha$$

$$= -4(-0.366) + 484(0.000042)$$

$$= 1.484 \text{ tonnes.}$$

Moment, $M_g = m_t y_t - m_\alpha \alpha$

$$= 484(-0.366) - 91916(0.000042)$$

$$= 181 \text{ T-cm}$$

The computation of pile loads moments of the other piles may be calculated in the same way.

The example worked out is only illustrative. The displacements of the foundation and the magnitudes of the pile loads depends upon the following factors.

(a) The accuracy of the value of n_h.

(b) The accuracy of the pile constant.

(c) The interaction between piles.

More research work is required in this field.

10.18 PROBLEMS

10.1 A group of nine friction piles arranged in a square pattern is to be proportioned in a deposit of medium stiff clay. Assuming that the piles are 30 cm diameter and 10 m long, find the optimum spacing for the piles. Assume $\alpha = 0.8$ and $c_u = 50$ kN/m^2.

10.2 A group of 9 piles with 3 in a row was driven into sand at a site. The diameter and length of the piles are 30 cm and 12 m respectively. The properties of the soil are : $\phi = 30$, $e = 0.7$, and $G = 2.64$.

If the spacing of the piles is 90 cm, compute the allowable load on the pile group on the basis of shear failure for $F_s = 2.0$ with respect to skin resistance, and $F_s = 2.5$ with respect to base resistance. For $\phi = 30°$, assume $N_q = 22.5$ and $N_\gamma = 19.7$. The water table is at ground level.

10.3 Nine RCC piles of diameter 30 cm each are driven in a square pattern at 90 cm centre to centre to a depth of 12 m into a stratum of loose to medium dense sand. The bottom of the pile cap embedding all the piles rests at a depth of 1.5 m below the ground surface. At a depth of 15 m lies a clay stratum of thickness 3 m and below which lies sandy strata. The liquid limit of the clay is 45%. The saturated unit weights of sand and clay are 18.5 kN/m³ and 19.5 kN/m³ respectively. The initial void ratio of the clay is 0.65. Calculate the consolidation settlement of the pile group under the allowable load. The allowable load $Q_a = 120$ kN.

10.4 A square pile group consisting of 16 piles of 40 cm diameter passes through two layers of compressible soils as shown in Fig. 15.32 (c). The thicknesses of the layers are : $L_1 = 2.5$ m and $L_2 = 3$ m. The piles are spaced at 100 cm centre to centre. The properties of the fill material are: top fill $c_u = 25$ kN/m²; the bottom fill (peat), $c_u = 30$ kN/m². Assume $\gamma = 14$ kN/m³ for both the fill materials. Compute the negative frictional load on the pile group.

10.5 9 precast concrete piles of 30 cm diameter were driven three in a row to sandy soil to form a square group. The distance between centre to centre of piles is 75 cm. The piles are connected by a rigid pile cap. The length of pile is 8 m. There is a soft clay strata of 0.5 m thick at a depth of one metre below the tips of the piles (depth to the top edge of the clay strata). Compute the following.

(a) The allowable load on the pile group.

(b) The settlement of the group at this load.

· **Given:** Water table is at ground level. The submerged unit weight of sand = 8.50 kN/m³; Average N-value of sandy strata = 18; liquid limit of clay soil, $w_l = 45\%$, Initial void ratio of clay soil = 0.83; natural moisture content of clay $w_n = 32\%$.

10.6 Assume that the pile group in Prob. 10.1 comprises both vertical and batter piles as follows.

(a) The centre row of three piles are vertical.

(b) The adjacent rows are batter piles inclining away from the centre row at a batter of 30°.

The forces acting on the pile cap level are as follows :

(a) Vertical load 2500 kN.

(b) Horizontal load 1500 kN.

(c) Moment = 2500 kN-m.

Assume $n_h = 15$ MN/m³. $E = 2.1 \times 10^5$ kg-cm².

Compute:

(a) The pile cap movements.

(b) The pile loads.

Deep Foundation 4
Drilled Pier Foundations

11

11.1 INTRODUCTION

Chapter 8 dealt with piles subjected to vertical loads and Chapter 9 with piles subjected to lateral loads. Drilled pier foundations, the subject matter of this chapter, belong to the same category as pile foundations. Because piers and piles serve the same purpose, no sharp deviations can be made between the two. The distinctions are based on the method of installation. A pile is installed by driving, a pier by excavating. Thus, a foundation unit installed in a drill-hole may also be called a bored cast-in-situ concrete pile. Here, distinction is made between a small diameter pile and a large diameter pile. A pile, cast-in-situ, with a diameter less than 0.75 m (or 2.5 ft) is sometimes called a small diameter pile. A pile greater than this size is called a large diameter bored-cast-in-situ pile. The latter definition is used in most non-American countries whereas in the USA, such large diameter bored piles are called *drilled piers*, drilled shafts, and sometimes drilled *caissons*. Chapter 8 deals with small diameter bored-cast-in-situ piles in addition to driven piles.

11.2 TYPES OF DRILLED PIERS

Drilled piers may be described under four types. All four types are similar in construction technique, but differ in their design assumptions and in the mechanism of load transfer to the surrounding earth mass. These types are illustrated in Fig. 11.1.

Straight-shaft end-bearing piers develop their support from end-bearing on strong soil, "hardpan" or rock. The overlying soil is assumed to contribute nothing to the support of the load imposed on the pier [Fig. 11.1 (a)].

Straight-shaft side wall friction piers pass through overburden soils that are assumed to carry none of the load, and penetrate far enough into an assigne bearing stratum to develop design load capacity by side wall friction between the pier and bearing stratum [Fig. 11.1 (b)].

Combination of straight shaft side wall friction and end bearing piers are of the same construction as the two mentioned above, but with both side wall friction and end bearing assigned a role in carrying the design load. When carried into rock, this pier may be referred to as a socketed pier or a "drilled pier with rock socket" [Fig. 11.1 (c)].

Belled or underreamed piers are piers with a bottom bell or underream [Fig. 11.1 (d)]. A greater percentage of the imposed load on the pier top is assumed to be carried by the base.

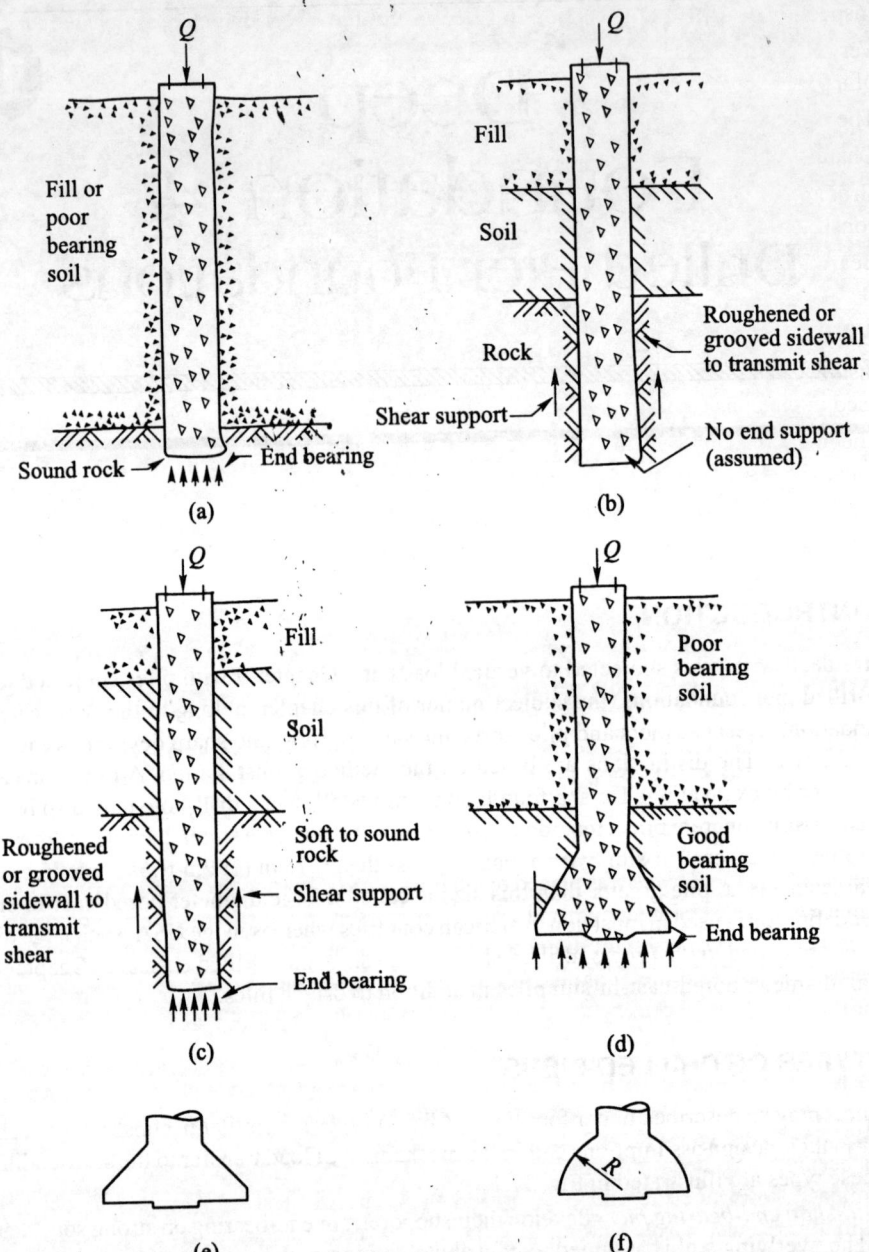

Fig. 11.1 Types of drilled piers and underream shapes: (a) Straight-shaft end-bearing pier, (b) straight-shaft sidewall-shear pier, (c) straight-shaft pier with both sidewall shear and end bearing, (d) underreamed (or belled) pier (30° *bell*), (e) shape of 45° *bell*, (f) shape of domed *bell* (Woodward *et al*, 1972)

11.3   ADVANTAGES AND DISADVANTAGES OF DRILLED PIER FOUNDATIONS

Advantages

1. Pier of any length and size can be constructed at the site.
2. Construction equipment is normally mobile and construction can proceed rapidly.

3. Inspection of drilled holes is possible because of the larger diameter of the shafts.

4. Very large loads can be carried by a single drilled pier foundation thus eliminating the necessity of a pile cap.

5. The drilled pier is applicable to a wide variety of soil conditions.

6. Changes can be made in the design criteria during the progress of a job.

7. Ground vibration that is normally associated with driven piles is absent in drilled pier construction.

8. Bearing capacity can be increased by underreaming the bottom (in non-caving materials).

Disadvantages

1. Installation of drilled piers needs a careful supervision and quality control of all the materials used in the construction.

2. The method is cumbersome. It needs sufficient storage space for all the materials used in the construction.

3. The advantage of increased bearing capacity due to compaction in granular soil that could be obtained in driven piles is not there in drilled pier construction.

4. Construction of drilled piers at places where there is a heavy current of ground water flow due to artesian pressure is very difficult.

11.4 METHODS OF CONSTRUCTION

Earlier Methods

The use of drilled piers for foundations started in the United States during the early part of the twentieth century. The two most common procedures were the Chicago and Gow methods shown in Fig. 11.2. In the Chicago method a circular pit was excavated to a convenient depth and a cylindrical shell of vertical boards or staves was placed by making use of an inside compression ring. Excavation then continued to the next board length and a second tier of staves was set and the procedure continued. The tiers could be set at a constant diameter or stepped in about 50 mm. The Gow method, which used a series of telescopic metal shells, is about the same as the current method of using casing except for the telescoping sections reducing the diameter on successive tiers.

Modern Methods of Construction

Equipment

There has been a phenomenal growth in the manufacture and use of heavy duty drilling equipment in the United States since the end of World War II. The greatest impetus to this development occurred in two states, Texas and California (Woodward *et al*, 1972). Improvements in the machines were made responding to the needs of contractors. Commercially produced drilling rigs of sufficient size and capacity to drill pier holes come in a wide variety of mountings and driving arrangements. Mountings are usually truck crane, tractor or skid. Figure 11.3 shows a tractor mounted rig. Drilling machine ratings as presented in manufacturer's catalogs and technical data sheets are usually expressed as maximum hole diameter, maximum depth, and maximum torque at some particular rpm.

Many drilled pier shafts through soil or soft rock are drilled with the open-helix auger. The tool may be equipped with a knife blade cutting edge for use in most homogeneous soil or with hard-surfaced teeth for cutting stiff or hard soils, stony soils, or soft to moderately hard rock. These augers are available in diameters up to 3 m or more. Figure 11.4 shows commercially available models.

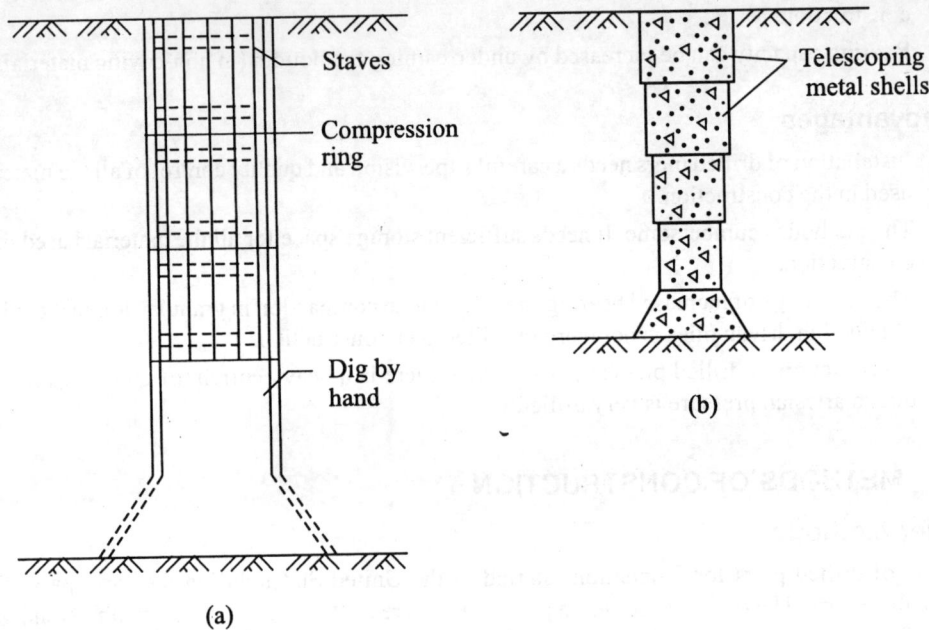

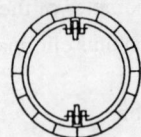

Staves

Compression
ring

Dig by
hand

(a)

Telescoping
metal shells

(b)

Fig. 11.2 Early methods of caisson construction: (a) The chicago method, (b) the gow method

Underreaming tools (or buckets) are available in a variety of designs. Figure 11.5 shows a typical 30° underreamer with blade cutter for soils that can be cut readily. Most such underreaming tools are limited in size to a diameter three times the diameter of the shaft.

When rock becomes too hard to be removed with auger-type tools, it is often necessary to resort to the use of a core barrel. This tool is a simple cylindrical barrel, set with tungsten carbide teeth at the bottom edge. For hard rock which cannot be cut readily with the core barrel set with hard metal teeth, a calyx or shot barrel can be used to cut a core of rock.

General construction methods of drilled pier foundations

The rotary drilling method is the most common method of pier construction in the United States. The methods of drilled pier construction can be classified in three categories as

1. The dry method
2. The casing method
3. The slurry method

Dry method of construction

The dry method is applicable to soil and rock that are above the water table and that will not cave or slump when the hole is drilled to its full depth. The soil that meets this requirement is a homogeneous, stiff clay. The first step in making the hole is to position the equipment at the desired location and to

Fig. 11.3 Tractor mounted hydraulic drilling rig (*Courtesy:* Kelly Tractor Co., USA)

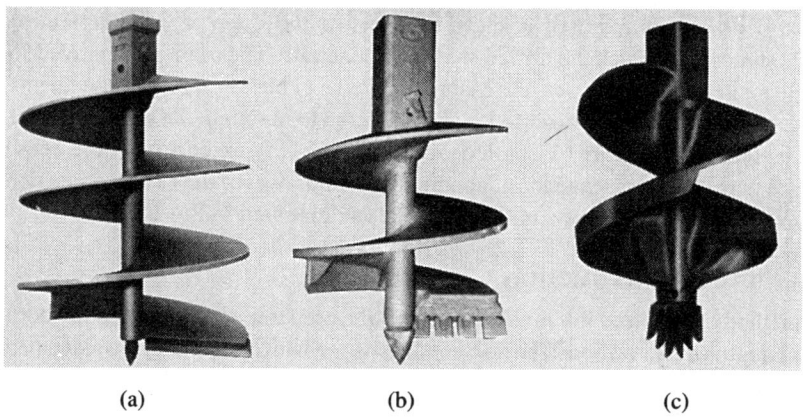

        (a)   (b)   (c)

Fig. 11.4 (a) Single-flight auger bit with cutting blade for soils, (b) single-flight auger bit with hard-metal cutting teeth for hard soils, hardpan, and rock, and (c) cast steel heavy-duty auger bit for hardpan and rock (Source: Woodward *et al*, 1972)

select the appropriate drilling tools. Figure 11.6 (a) gives the initial location. The drilling is next carried out to its fill depth with the spoil from the hole removed simultaneously.

After drilling is complete, the bottom of the hole is underreamed if required. Figure 11.6 (b) and (c) show the next steps of concreting and placing the rebar cage. Figure 11.6 (d) shows the hole completely filled with concrete.

Fig. 11.5. A 30° underreamer with blade cutters for soils that can be cut readily
(*Source:* Woodward *et al*, 1972)

Casing method of construction

The casing method is applicable to sites where the soil conditions are such that caving or excessive soil or rock deformation can occur when a hole is drilled. This can happen when the boring is made in dry soils or rocks which are stable when they are cut but will slough soon afterwards. In such a case, the bore hole is drilled, and a steel pipe casing is quickly set to prevent sloughing. Casing is also required if drilling is required in clean sand below the water table underlain by a layer of impermeable stones into which the drilled shaft will penetrate. The casing is removed soon after the concrete is deposited. In some cases, the casing may have to be left in place permanently. It may be noted here that until the casing is inserted, a slurry is used to maintain the stability of the hole. After the casing is seated, the slurry is bailed out and the shaft extended to the required depth. Figure 11.7 (a) to (h) give the sequence of operations. Withdrawal of the casing, if not done carefully, may lead to voids or soil inclusions in the concrete, as illustrated in Fig. 11.8.

Slurry method of construction

The slurry method of construction involves the use of a prepared slurry to keep the bore hole stable for the entire depth of excavation. The soil conditions for which the slurry displacement method is applicable could be any of the conditions described for the casing method. The slurry method is a viable option at any site where there is a caving soil, and it could be the only feasible option in a permeable, water bearing soil if it is impossible to set a casing into a stratum of soil or rock with low permeability. The various steps in the construction process are shown in Fig. 11.9. It is essential in this method that a sufficient slurry head be available so that the inside pressure is greater than that from the GWT or from the tendency of the soil to cave.

Bentonite is most commonly used with water to produce the slurry. Polymer slurry is also employed. Some experimentation may be required to obtain an optimum percentage for a site, but amounts in the range of 4 to 6 percent by weight of admixture are usually adequate.

The bentonite should be well mixed with water so that the mixture is not lumpy. The slurry should be capable of forming a filter cake on the side of the bore hole. The bore hole is generally not

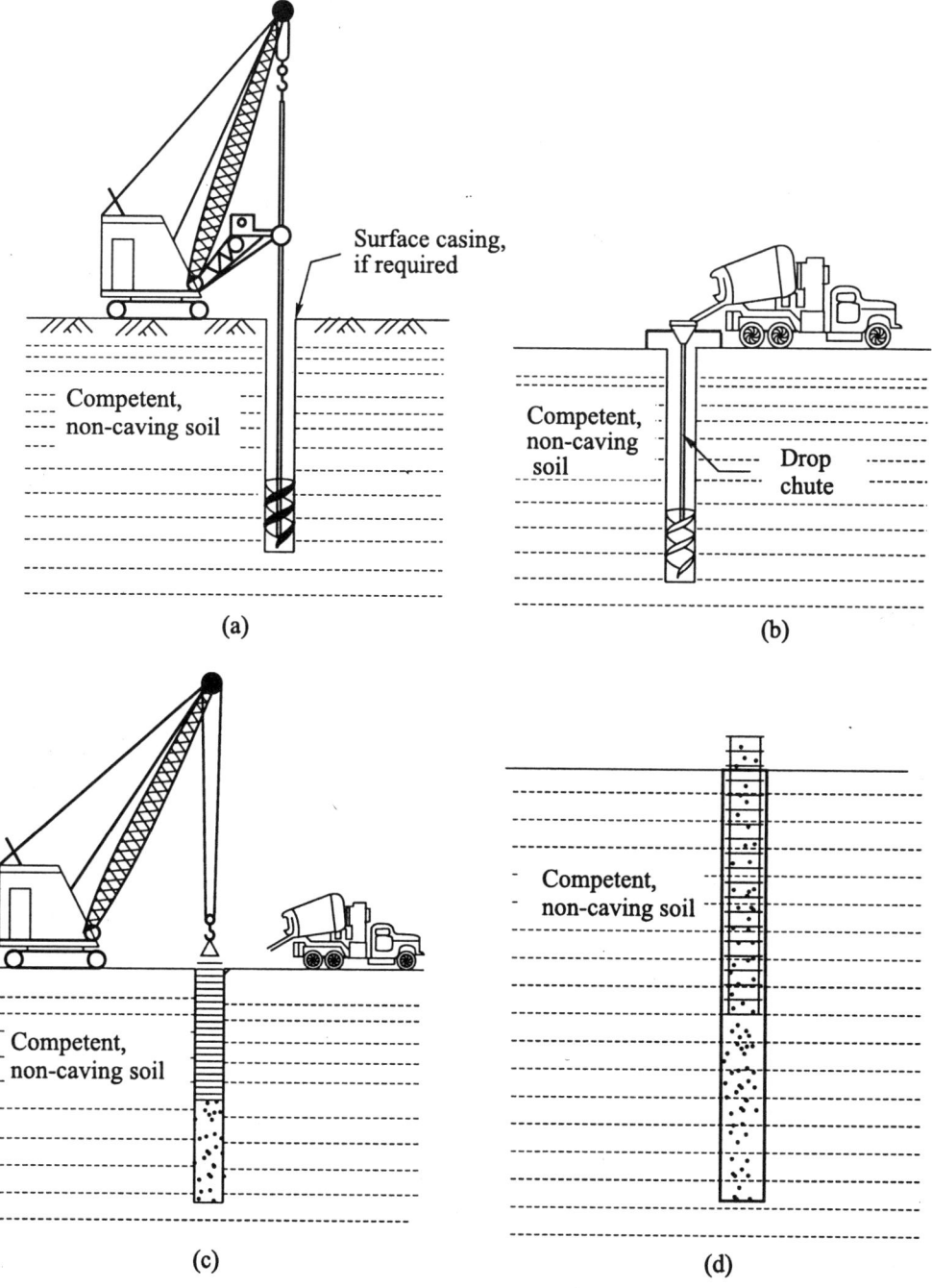

Fig. 11.6 Dry method of construction: (a) Initiating drilling, (b) starting concrete pour, (c) placing rebar cage, and (d) completed shaft (O'Neill and Reese, 1999)

underreamed for a bell since this procedure leaves unconsolidated cuttings on the base and creates a possibility of trapping slurry between the concrete base and the bell roof.

If reinforcing steel is to be used, the rebar cage is placed in the slurry as shown in Fig. 11.9 (b). After the rebar cage has been placed, concrete is placed with a tremie either by gravity feed or by

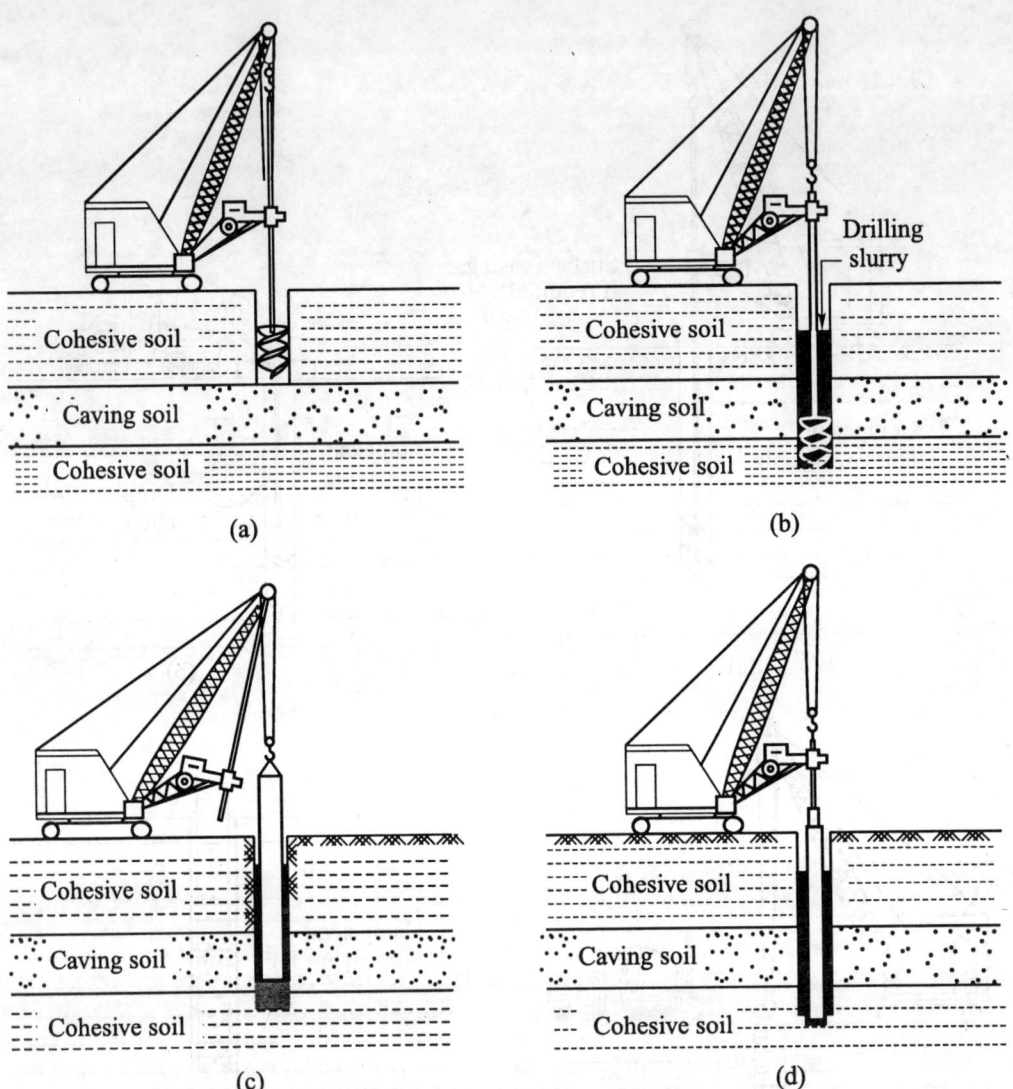

Fig. 11.7 Casing method of construction: (a) Initiating drilling, (b) drilling with slurry; (c) introducing casing, (d) casing is sealed and slurry is being removed from interior of casing

pumping. If a gravity feed is used, the bottom end of the tremie pipe should be closed with a closure plate until the base of the tremie reaches the bottom of the bore hole, in order to prevent contamination of the concrete by the slurry. Filling of the tremie with concrete, followed by subsequent slight lifting of the tremie, will then open the plate, and concreting proceeds. Care must be taken that the bottom of the tremie is buried in concrete at least for a depth of 1.5 m (5 ft). The sequence of operations is shown in Fig. 11.9 (a) to (d).

11.5 DESIGN CONSIDERATIONS

The precess of the design of a drilled pier generally involves the following:

1. The objectives of selecting drilled pier foundations for the project.

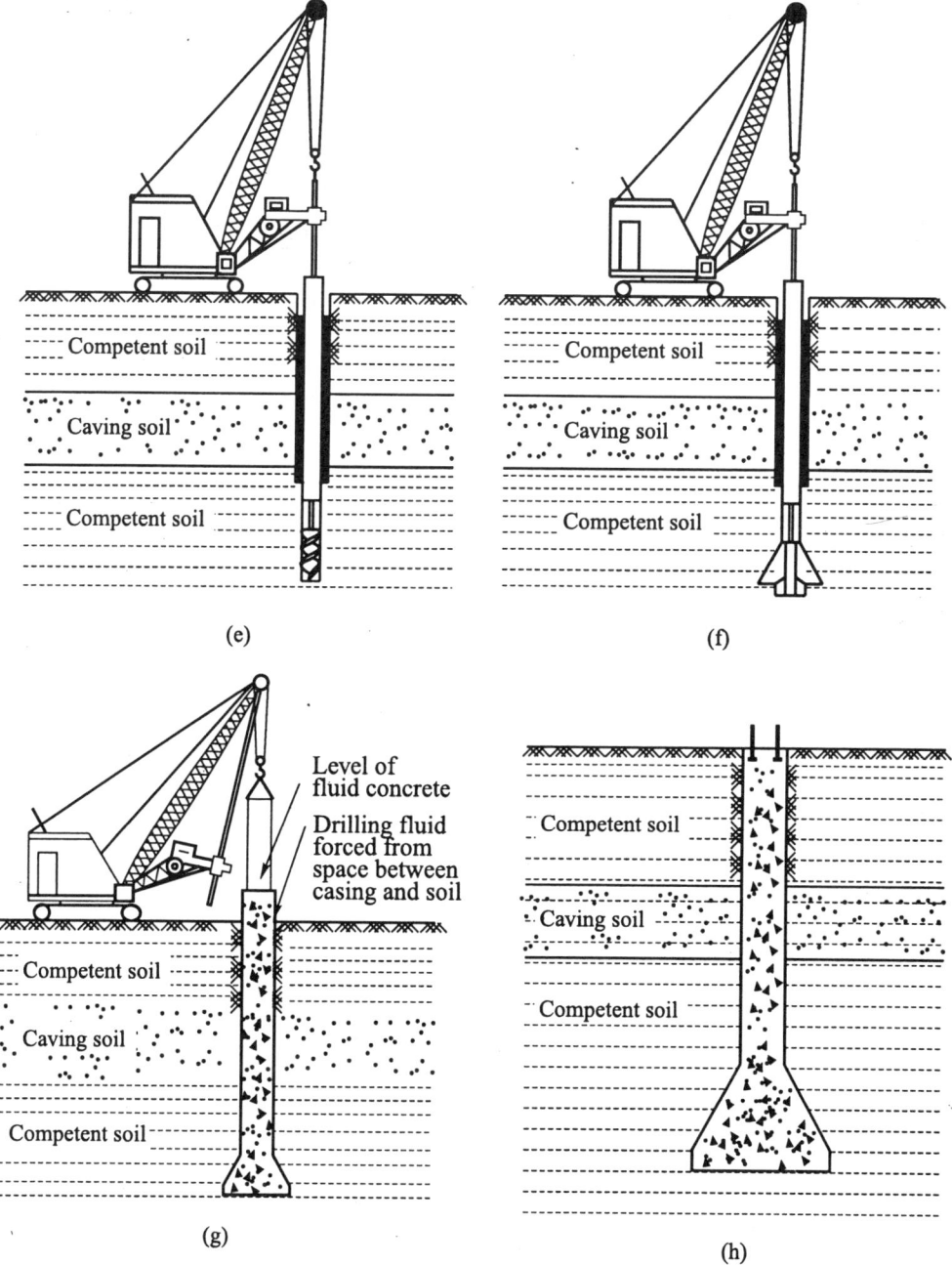

Fig. 11.7 Casing method of construction: (e) Drilling below casing, (f) underreaming, (g) removing casing, and (h) completed shaft (O'Neill and Reese, 1999)

2. Analysis of loads coming on each pier foundation element.
3. A detailed soil investigation and determining the soil parameters for the design.
4. Preparation of plans and specifications which include the methods of design, tolerable settlement, methods of construction of piers, etc.
5. The method of execution of the project.

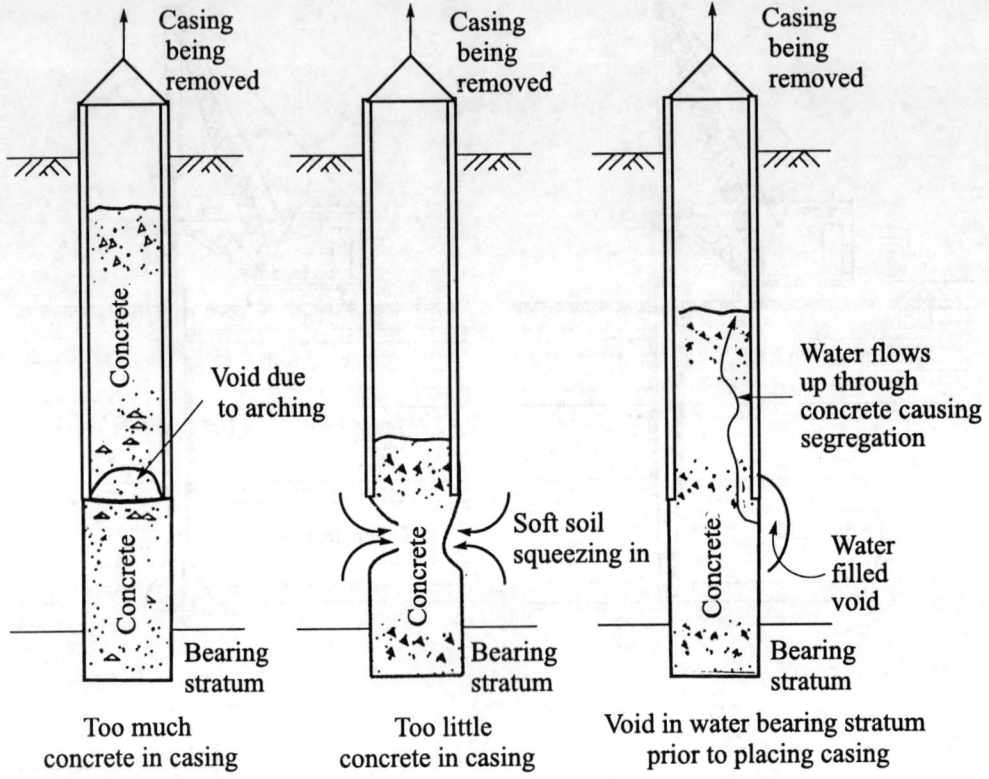

Fig. 11.8 Potential problems leading to inadequate shaft concrete due to removal of temporary casing without care (D'Appolonia, *et al*, 1975)

In general the design of a drilled pier may be studied under the following headings.

1. Allowable loads on the piers based on ultimate bearing capacity theories.
2. Allowable loads based on vertical movement of the piers.
3. Allowable loads based on lateral bearing capacity of the piers.

In addition to the above, the uplift capacity of piers with or without underreams has to be evaluated. The following types of strata are considered.

1. Piers embedded in homogeneous soils, sand or clay.
2. Piers in a layered system of soil.
3. Piers socketed in rocks.

It is better that the designer select shaft diameters that are multiples of 150 mm (6 in) since these are the commonly available drilling tool diameters.

11.6   VERTICAL LOAD TRANSFER MECHANISM

Figure 11.10 (a) shows a single drilled pier of diameter d, and length L constructed in a homogeneous mass of soil of known physical properties. If this pier is loaded to failure under an ultimate load Q_u, a part of this load is transmitted to the soil along the length of the pier and the balance is transmitted to the pier base. The load transmitted to the soil along the pier is

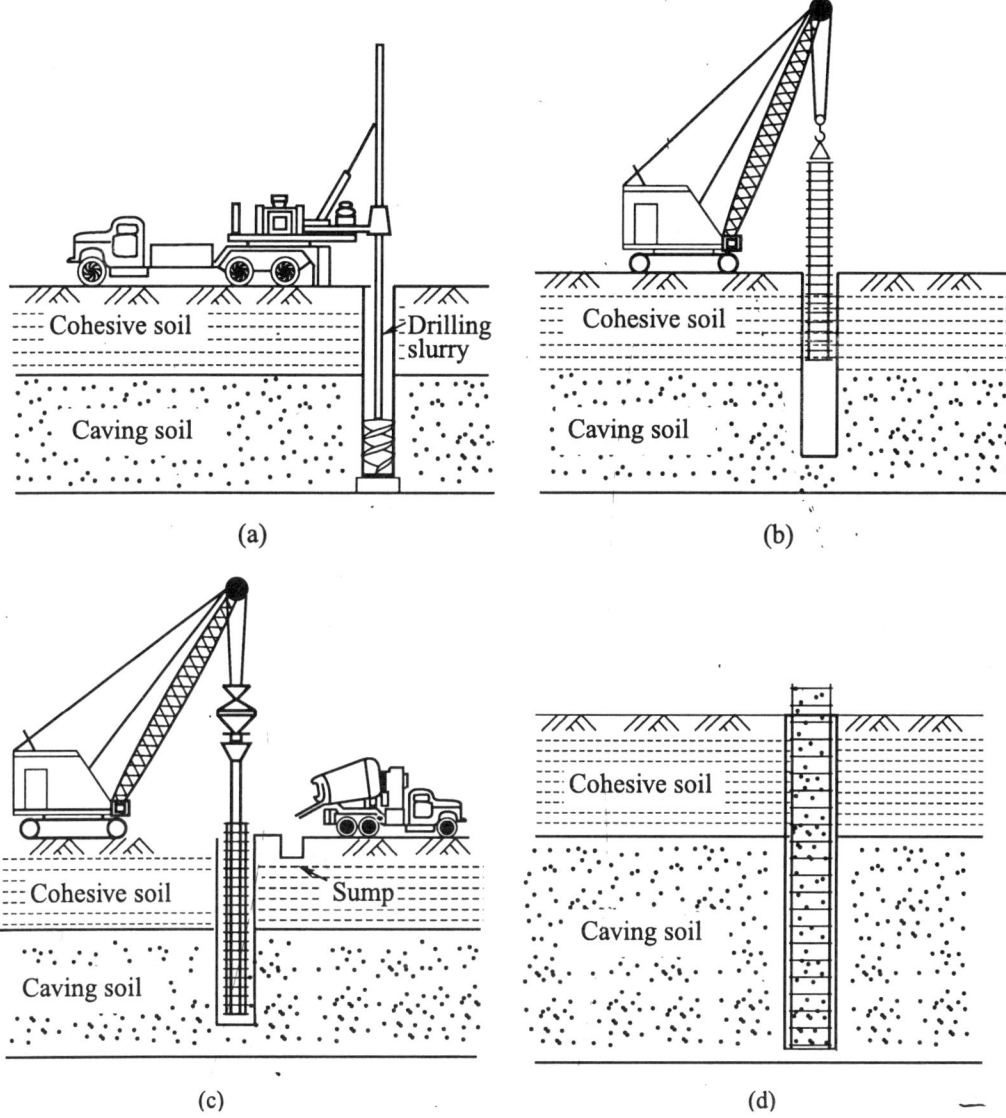

Fig. 11.9 Slurry method of construction: (a) Drilling to full depth with slurry, (b) placing rebar cage, (c) placing concrete, (d) completed shaft (O'Neill and Reese, 1999)

called the ultimate *friction load or skin load,* Q_f and that transmitted to the base is the ultimate base or point load Q_b. The total ultimate load, Q_u, is expressed as (neglecting the weight of the pier)

$$Q_u = Q_b + Q_f = q_b A_b + \sum_{i=1}^{N} f_{si} P_i \Delta z_i \tag{11.1a}$$

where　q_b = net ultimate bearing pressure,

　　　　A_b = base area,

　　　　f_{si} = unit skin resistance (ultimate) of layer i,

P_i = perimeter of pier in layer i,

Δz_i = thickness of layer i,

N = number of layers.

If the pier is instrumented, the load distribution along the pier can be determined at different stages of loading. Typical load distribution curves plotted along a pier are shown in Fig. 11.10 (b) (O'Neill and Reese, 1999). These load distribution curves are similar to the one shown in Fig. 9.5 (b). Since the load transfer mechanism for a pier is the same as that for a pile, no further discussion on this is necessary here. However, it is necessary to study in this context the effect of settlement on the mobilization of side shear and base resistance of a pier. As may be seen from Fig. 11.11, the maximum values of base and side resistance are not mobilised at the same value of displacement. In some soils, and especially in some brittle rocks, the side shear may develop fully at a small value of displacement and then decrease with further displacement while the base resistance is still being mobilised'(O'Neill and Reese, 1999). If the value of the side resistance at point A is added to the value of the base resistance at point B, the total resistance shown at level D is overpredicted. On the other hand, if the designer wants to take advantage primarily of the base resistance, the side resistance at point C should be added to the base resistance at point B to evaluate Q_u. Otherwise, the designer may wish to design for the side resistance at point A and disregard the base resistance entirely.

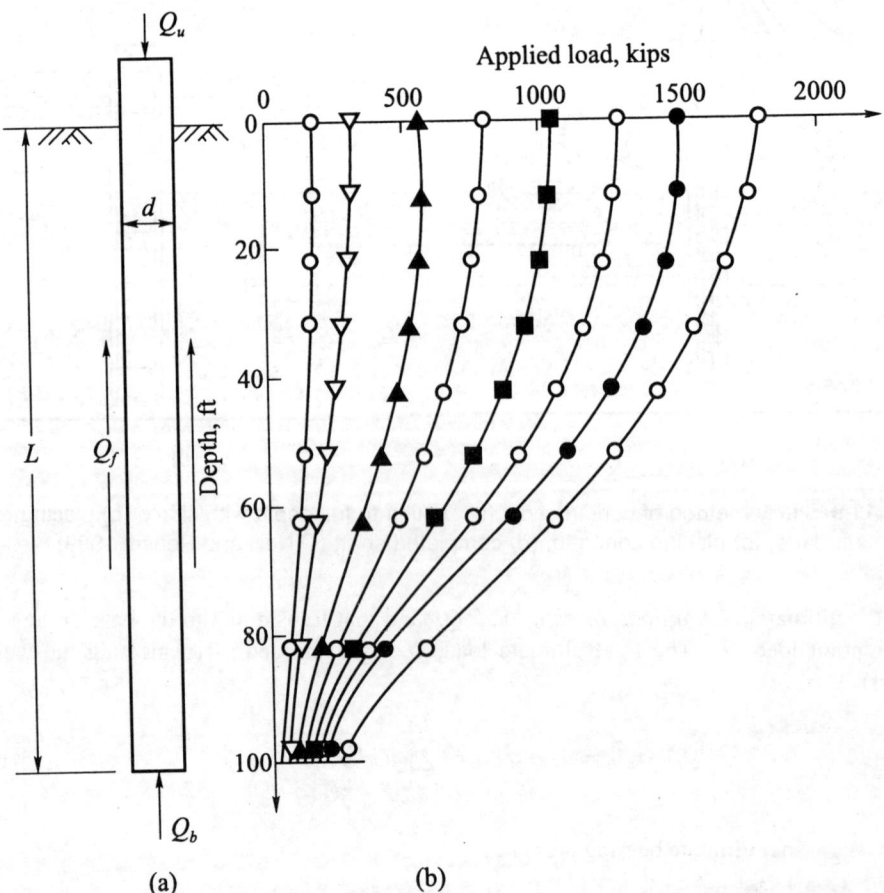

Fig. 11.10 Typical set of load distribution curves (O'Neill and Reese, 1999)

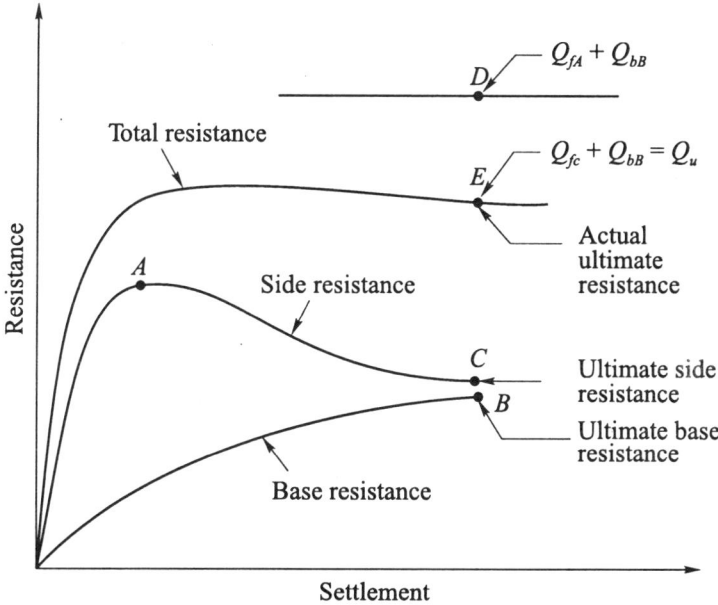

$Q_{fA} + Q_{bB}$

$Q_{fc} + Q_{bB} = Q_u$

Total resistance

A

Side resistance

Actual
ultimate
resistance

C

B

Ultimate side
resistance

Ultimate base
resistance

Base resistance

Settlement

Fig. 11.11 Condition in which ($Q_b + Q_f$) is not equal to actual ultimate resistance

11.7 VERTICAL BEARING CAPACITY OF DRILLED PIERS

For the purpose of estimating the ultimate bearing capacity, the subsoil is divided into layers (Fig. 11.12) based on judgement and experience (O'Neill and Reese, 1999). Each layer is assigned one of four classifications.

1. Cohesive soil [clays and plastic silts with undrained shear strength $c_u \leq 250$ kN/m^2 (2.5 t/ft^2)].
2. Granular soil [cohesionless geomaterial, such as sand, gravel or nonplastic silt with uncorrected SPT (N) values of 50 blows per 0.3/m or less].
3. Intermediate geomaterial [cohesive geomaterial with undrained shear strength c_u between 250 and 2500 kN/m^2 (2.5 and 25 tsf), or cohesionless geomaterial with SPT (N) values > 50 blows per 0.3 m].
4. Rock [highly cemented geomaterial with unconfined compressive strength greater than 5000 kN/m^2 (50 tsf)].

The unit side resistance f_s ($= f_{max}$) is computed in each layer through which the drilled shaft passes, and the unit base resistance q_b ($= q_{max}$) is computed for the layer on or in which the base of the drilled shaft is founded.

The soil along the whole length of the shaft is divided into four layers as shown in Fig. 11.12.

Effective length for computing side resistance in cohesive soil

O'Neill and Reese (1999) suggest that the following effective length of pier is to be considered for computing side resistance in cohesive soil.

Straight shaft: One diameter from the bottom and 1.5 m (5 feet) from the top are to be excluded from the embedded length of pile for computing side resistance as shown in Fig. 11.13 (a).

Belled shaft: The height of the bell plus the diameter of the shaft from the bottom and 1.5 m (5 ft) from the top are to be exclude as shown in Fig. 11.13 (b).

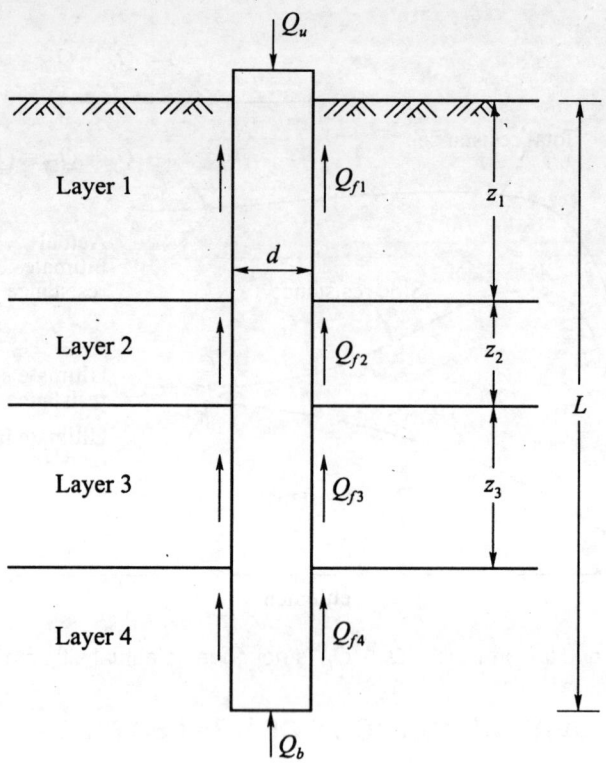

Fig. 11.12 Idealised geomaterial layering for computation of compression load and resistance
(O'Neill and Reese, 1999)

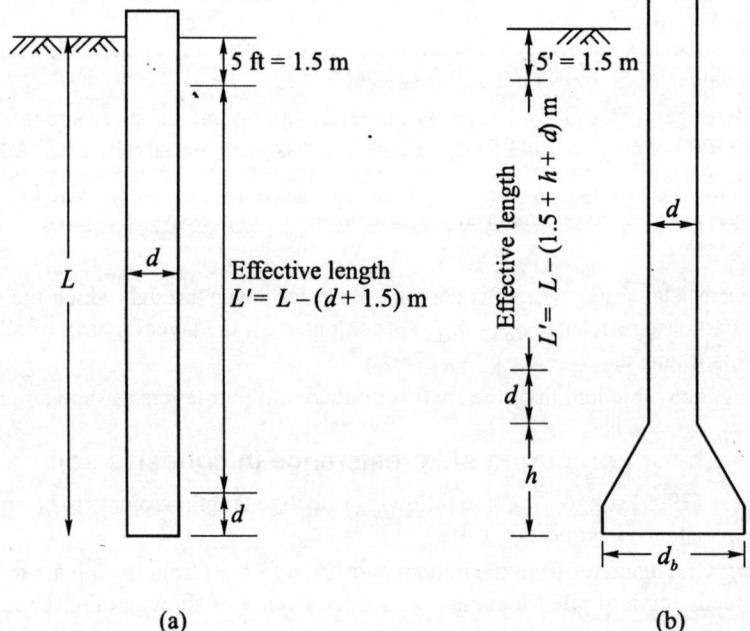

(a) (b)

Fig. 11.13 Exclusion zones for estimating side resistance for drilled shafts in cohesive soils

11.8 THE GENERAL BEARING CAPACITY EQUATION FOR THE BASE RESISTANCE q_b ($= q_{max}$)

The equation for the ultimate base resistance may be expressed as

$$q_b = s_c\, d_c\, N_c c + s_q\, d_q\, (N_q - 1)\, q'_o + \frac{1}{2}\, \gamma d\, s_\gamma\, d_\gamma\, N_\gamma \tag{11.2}$$

where 　N_c, N_q and N_γ = bearing capacity of factors for long footings,
　　　　s_c, s_q and s_γ = shape factors,
　　　　d_c, d_q and d_γ = depth factors,
　　　　q'_o = effective vertical pressure at the base level of the drilled pier,
　　　　γ = effective unit weight of the soil below the bottom of the drilled shaft to a depth = $1.5d$ where d = width or diameter of pier at base level,
　　　　c = average cohesive strength of soil just below the base.

For deep foundations the last term in Eq. (11.2) becomes insignificant and may be ignored. Now Eq. (11.2) may be written as

$$q_b = s_c\, d_c\, N_c c + s_q\, d_q\, (N_q - 1)\, q'_o \tag{11.3}$$

11.9 BEARING CAPACITY EQUATIONS FOR THE BASE IN COHESIVE SOIL

When the Undrained Shear Strength, $c_u \leq 250$ kN/m² (2.5 t/ft²)

For $\phi = 0$, $N_q = 1$ and $(N_q - 1) = 0$, here Eq. (11.3) can be written as (Vesic, 1972)

$$q_b = N_c^* c_u \tag{11.4}$$

in which

$$N_c^* = \frac{4}{3}\, (\ln I_r + 1) \tag{11.5}$$

I_r = rigidity index of the soil

Equation (11.4) is applicable for $c_u \leq 96$ kPa and $L \geq 3d$ (base width)

For $\phi = 0$, I_r may be expressed as (O'Neill and Reese, 1999)

$$I_r = \frac{E_s}{3\, c_u} \tag{11.6}$$

where E_s = Young's modulus of the soil in undrained loading. Refer to Section 6.8 for the methods of evaluating the value of E_s.

Table 11.1 gives the values of I_r and N_c^* as a function of c_u.

Table 11.1 Values of $I_r = E_s/3c_u$ and N_c^*

c_u	I_r	N_c^*
24 kPa (500 lb/ft²)	50	6.5
48 kPa (1000 lb/ft²)	150	8.0
$\geq$ 96 kPa (2000 lb/ft²)	250–300	9.0

If the depth of base (L) < $3d$ (base)

$$q_b (= q_{max}) = \frac{2}{3}\left(1 + \frac{L}{6d}\right) N_c^* c_u \tag{11.7}$$

When $c_u \geq 96$ kPa (2000 lb/ft^2), the equation for q_b may be written as

$$q_b = 9c_u \tag{11.8}$$

for depth of base (= L) $\geq 3d$ (base width).

11.10 BEARING CAPACITY EQUATION FOR THE BASE IN GRANULAR SOIL

Values N_c and N_q in Eq. (11.3) are for strip footings on the surface of rigid soils and are plotted as a function of ϕ in Fig. 11.14. Vesic (1977) explained that during bearing failure, a plastic failure zone develops beneath a circular loaded area that is accompanied by elastic deformation in the surrounding elastic soil mass. The confinement of the elastic soil surrounding the plastic soil has an effect on q_b (= q_{max}). The values of N_c and N_q are therefore dependent not only on ϕ, but also on I_r. They must be corrected for soil rigidity as given below.

$$N_c \text{ (corrected)} = N_c C_c$$
$$N_q \text{ (corrected)} = N_q C_q \tag{11.9}$$

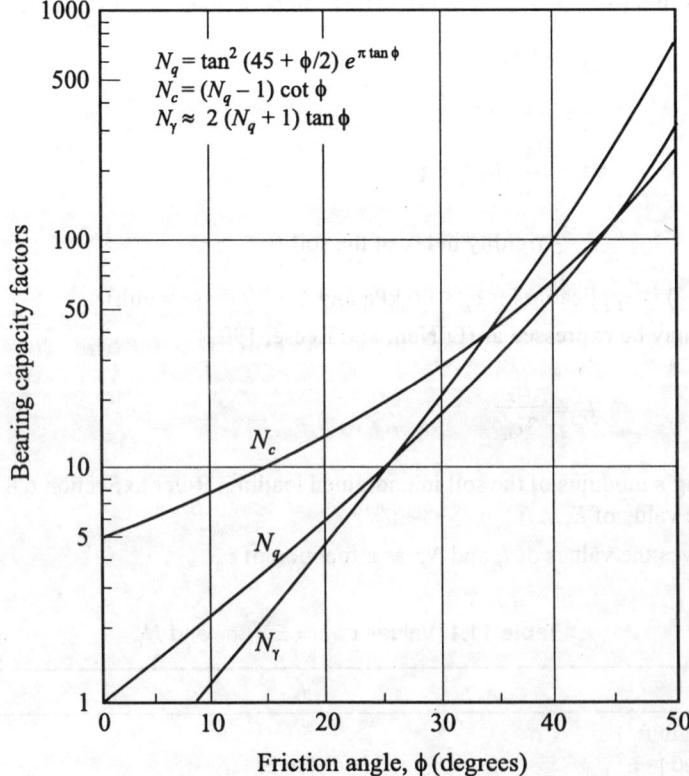

Fig. 11.14 Bearing capacity factors (Chen and Kulhawy, 1994)

where C_c and C_q are the correction factors. As per (Chen and Kulhawy, 1994)

Equation (11.3) may now be expressed as

$$q_b = c\, N_c\, s_c\, d_c\, C_c + (N_q - 1)\, q_o'\, s_q\, d_q\, C_q \tag{11.10}$$

$$C_c = C_q - \frac{1 - C_q}{N_c\, \tan \phi} \tag{11.11a}$$

$$C_q = \exp. \{[(-3.8 \tan \phi) + (3.07 \sin \phi)\, \log_{10} 2I_{rr}]/(1 + \sin \phi)\} \tag{11.11b}$$

where ϕ is an effective angle of internal friction. I_{rr} is the reduced rigidity index expressed as [Eq. (8.28)]

$$I_{rr} = \frac{I_r}{1 + \Delta I_r} \tag{11.12}$$

and

$$I_r = \frac{E_d}{2\,(1 + \mu_d)\, q_o'\, \tan \phi} \tag{11.13}$$

by ignoring cohesion, where,

E_d = drained Young's modulus of the soil,

μ_d = drained Poisson's ratio,

Δ = volumetric strain within the plastic zone during the loading process.

The expressions for μ_d and Δ may be written as (Chen and Kulhawy, 1994)

$$\mu_d = 0.1 + 0.3\phi_{rel} \tag{11.14}$$

$$\Delta = \frac{0.005\,(1 - \phi_{rel})\, q_o'}{p_a} \tag{11.15}$$

where

$$\phi_{rel} = \frac{(\phi° - 25°)}{45° - 25°} \quad \text{for } 25° \le \phi° \le 45° \tag{11.16}$$

= relative friction angle factor, p_a = atmospheric pressure = 101 kPa.

Chen and Kulhawy (1994) suggest that, for granular soils, the following values may be considered.

loose soil, E_d = 100 to $200p_a$

medium dense soil, E_d = 200 to $500p_a$ (11.17)

dense soil, E_d = 500 to $1000p_a$

The correction factors C_c and C_q indicated in Eq. (11.9) need be applied only if I_{rr} is less than the critical rigidity index $(I_r)_{crit}$ expressed as follows

$$(I_r)_{cr} = \frac{1}{2} \exp. \left[2.85 \cot \left(45° - \frac{\phi°}{2} \right) \right] \tag{11.18}$$

The values of critical rigidity index may be obtained from Table 5.4 for piers circular or square in section.

If $I_{rr} > (I_r)_{crit}$, the factors C_c and C_q may be taken as equal to unity.

The shape and depth factors in Eq. (11.3) can be evaluated by making use of the relationships given in Table 11.2.

Table 11.2 Shape and depth factors [Eq. (11.3)] (Chen and Kulhawy, 1994)

Factors	*Value*
s_c	$1 + \dfrac{N_q}{N_c}$
s_d	$d_q - \dfrac{1 - d_q}{N_c \, \tan \phi}$
s_q	$1 + \tan \phi$
d_q	$1 + 2 \tan \phi \, (1 - \sin \phi)^2 \left[\dfrac{\pi}{180} \, \tan^{-1} \dfrac{L}{d} \right]$

Base in Cohesionless Soil

The theoretical approach as outlined above is quite complicated and difficult to apply in practice for drilled piers in granular soils. Direct and simple empirical correlations have been suggested by O'Neill and Reese (1999) between SPT N value and the base bearing capacity as given below for cohesionless soils.

$$q_b (= q_{max}) = 57.5 \, N \text{ kPa} \leq 2900 \text{ kN/m}^2 \tag{11.19a}$$

$$q_b (= q_{max}) = 0.60 \, N \text{ tsf} \leq 30 \text{ tsf} \tag{11.19b}$$

where N = SPT value ≤ 50 blows/0.3 m.

Base in Cohesionless IGM

Cohesionless IGM's are characterized by SPT blow counts if more than 50 per 0.3 m. In such cases, the expression for q_b is

$$q_b (= q_{max}) = 0.60 \left(N_{60} \frac{p_a}{q'_o} \right)^{0.8} q'_o \tag{11.20}$$

where N_{60} = average SPT corrected for 60 percent standard energy within a depth of $2d$ (base) below the base. The value of N_{60} is limited to 100. No correction for overburden pressure,

p_a = atmospheric pressure in the units used for q'_o (= 101 kPa in the SI system),

q'_o = vertical effective stress at the elevation of the base of the drilled shaft.

11.11 BEARING CAPACITY EQUATIONS FOR THE BASE IN COHESIVE IGM OR ROCK (O'NEILL AND REESE, 1999)

Massive rock and cohesive intermediate materials possess common properties. They possess low drainage qualities under normal loadings but drain more rapidly under large loads than cohesive soils. It is for these reasons undrained shear strengths are used for rocks and IGMs.

If the base of the pier lies in cohesive IGM or rock (RQD = 100 percent) and the depth of socket, D_s, in the IGM or rock is equal to or greater than $1.5d$, the bearing capacity may be expressed as

$$q_b (= q_{max}) = 2.5q_u \tag{11.21}$$

where q_u = unconfined compressive strength of IGM or rock below the base

For RQD between 70 and 100 percent,

$$q_b (= q_{max}) = 4.83 \, (q_u) \, 0.51 \text{ MPa} \tag{11.22}$$

For jointed rock or cohesive IGM

$$q_b (= q_{max}) = [s^{0.5} + (ms^{0.5} + s)^{0.5}] \, q_u \tag{11.23}$$

where q_u is measured on intact cores from within $2d$ (base) below the base of the drilled pier. In all the above cases q_b and q_u are expressed in the same units and s and m indicate the properties of the rock or IGM mass that can be estimated from Tables 11.3 and 11.4.

Table 11.3 Descriptions of rock types

Rock type	Description
A	Carbonate rocks with well-developed crystal cleavage (eg., dolostone, limestone, marble)
B	Lithified argillaeous rocks (mudstone, siltstone, shale, slate)
C	Arenaceous rocks (sandstone, quartzite)
D	Fine-grained igneous rocks (andesite, dolerite, diabase, rhyolite)
E	Coarse-grained igneous and metamorphic rocks (amphibole, gabbro, gneiss, granite, norite, quartz-diorite)

Table 11.4 Values of s and m (dimensionless) based on rock classification
(Carter and Kulhawy, 1988)

Quality of rock mass	Joint description and spacing	s	Value of m as function of rock type (A – E) from				
			A	B	C	D	E
Excellent	Intact (closed); spacing > 3 m (10 ft)	1	7	10	15	17	25
Very good	Interlocking; spacing of 1 to 3 m (3 to 10 ft)	0.1	3.5	5	7.5	8.5	12.5
Good	Slightly weathered; spacing of 1 to 3 m (3 to 10 ft)	4×10^{-2}	0.7	1	1.5	1.7	2.5
Fair	Moderately weathered; spacing of 0.3 to 1 m (1 to 3 ft)	10^{-4}	0.14	0.2	0.3	0.34	0.5
Poor	Weathered with gouge (soft material); spacing of 30 to 300 mm (1 in. to 1 ft)	10^{-5}	0.04	0.05	0.08	0.09	0.13
Very poor	Heavily weathered; spacing of less than 50 mm (2 in.)	0	0.007	0.01	0.015	0.017	0.025

11.12 THE ULTIMATE SKIN RESISTANCE OF COHESIVE AND INTERMEDIATE MATERIALS

Cohesive Soil

The process of drilling a borehole for a pier in cohesive soil disturbs the natural condition of the soil all along the side to a certain extent. There is a reduction in the soil strength not only due to boring but also due to stress relief and the time spent between boring and concreting. It is very difficult to quantify the extent of the reduction in strength analytically. In order to take care of the disturbance, the unit frictional resistance on the surface of the pier may be expressed as

$$f_s = \alpha c_u \qquad (11.24)$$

where α = adhesion factor

c_u = undrained shear strength

Relationships have been developed between c_u and α by many investigators based on field load tests. Figure 11.15 gives one such relationship in the form of a curve developed by Chen and Kulhawy (1994). The curve has been developed on the following assumptions (Fig. 11.15).

f_s = 0 up to 1.5 m (= 5 ft) from the ground level.

f_s = 0 up to a height equal $(h + d)$ as per Fig. 11.13

O'Neill and Reese (1999) recommend the chart's trend line given in Fig. 11.15 for designing drilled piers. The suggested relationships are:

$$\alpha = 0.55 \text{ for } c_u/p_a \le 1.5 \qquad (11.25a)$$

$$\text{and } \alpha = 0.55 - 0.1 \left(\frac{c_u}{p_a} - 1.5 \right) \text{ for } 1.5 \le c_u/p_a \le 2.5 \qquad (11.25b)$$

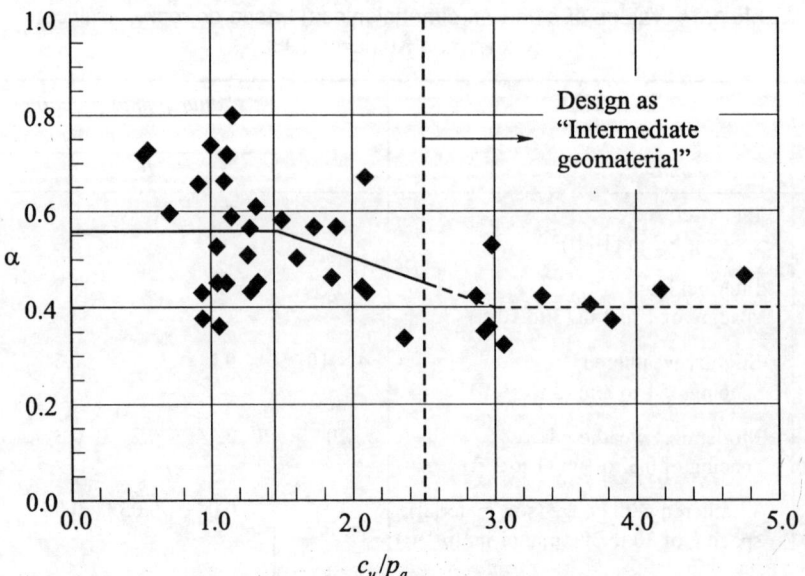

Fig. 11.15 Correlation between α and c_u/p_a

Cohesive Intermediate Geomaterials

Cohesive IGM's are very hard clay-like materials which can also be considered as very soft rock (O'Neill and Reese, 1999). IGM's are ductile and failure may be sudden at peak load. The value of f_a (please note that the term f_a is used instead of f_s for ultimate unit resistance at infinite displacement) depends upon the side condition of the bore hole, that is, whether it is rough or smooth. For design purposes the side is assumed as smooth. The expression for f_a may be written as

$$f_a = \alpha q_u \tag{11.26}$$

where q_u = unconfined compressive strength,

f_a = the value of ultimate unit side resistance which occurs at infinite displacement.

Figure 11.16 gives a chart for evaluating α. The chart is prepared for an effective angle of friction between the concrete and the IGM (assuming that the intersurface is drained) and S_t denotes the settlement of piers at the top of the socket. Further, the chart involves the use of σ_n/p_a where σ_n is the normal effective pressure against the side of the borehole by the drilled pier and p_a is the atmospheric pressure (101 kPa).

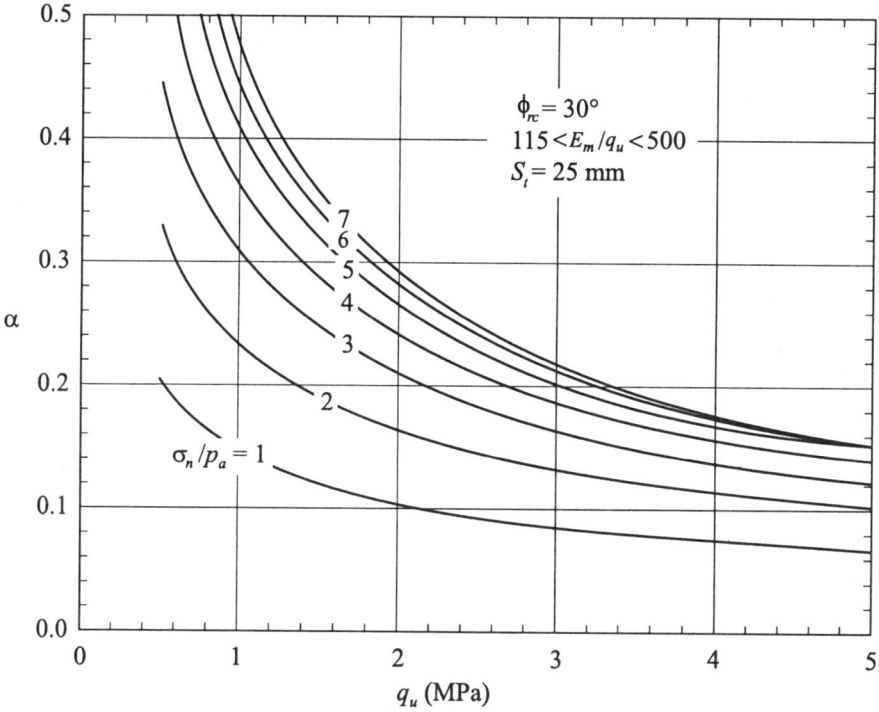

Fig. 11.16 Factor α for cohesive IGM's (O'Neill and Reese, 1999)

O'Neill and Reese (1999) give the following equation for computing σ_n

$$\sigma_n = M \gamma_c z_c \tag{11.27}$$

where γ_c = the unit weight of the fluid concrete used for the construction,

z_c = the depth of the point at which σ_n is required,

M = an empirical factor which depends on the fluidity of the concrete as indexed by the concrete slump.

Figure 11.17 gives the values of M for various slumps.

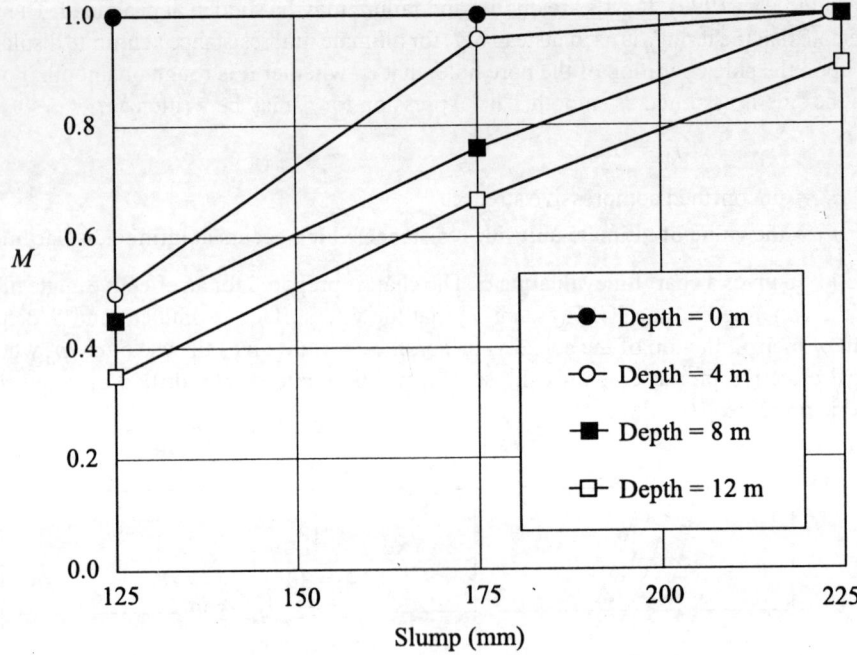

Fig. 11.17 Factor M vs. concrete slump (O'Neill *et al*, 1996)

The mass modulus of elasticity of the IGM (E_m) should be determined before proceeding, in order to verify that the IGM is within the limits of Fig. 11.16. This requires the average Young's modulus of intact IGM core (E_i) which can be determined in the laboratory. Table 11.5 gives the ratios of E_m/E_i for various values of RQD. Values of E_m/E_i less than unity indicate that soft seams and/or joints exist in the IGM. These discontinuities reduce the value of f_a. The reduced value of f_a may be expressed as

$$f_{aa} = f_a R_a \qquad (11.28)$$

where the ratio $R_a = f_{aa}/f_a$ can be determined from Table 11.6.

If the socket is classified as smooth, it is sufficiently accurate to set $f_s = f_{max} = f_{aa}$

Table 11.5 Estimation of E_m/E_i based on RQD (Modified after Carter and Kulhawy, 1988)

RQD (percent)	E_m/E_i	
	Closed joints	*Open joints*
100	1.00	0.60
70	0.70	0.10
50	0.15	0.10
20	0.05	0.05

Note: Values intermediate between tabulated values may be obtained by linear interpolation.

Table 11.6 f_{aa}/f_a based on E_m/E_i (O'Neill *et al*, 1996)

E_m/E_i	f_{aa}/f_a
1.0	1.0
0.5	0.8
0.3	0.7
0.1	0.55
0.05	0.45

11.13 ULTIMATE SKIN RESISTANCE IN COHESIONLESS SOIL AND GRAVELLY SANDS (O'NEILL AND REESE, 1999)

In Sands

A general expression for total skin resistance in cohesionless soil may be written as [Eq. (11.1)]

$$Q_{fi} = \sum_{i=1}^{N} P_i f_{si} \Delta z_i = \sum_{i=1}^{N} P_i q'_{oi} K_{si} \tan \delta_i \Delta z_i \tag{11.29}$$

$$\text{or} \quad Q_{fi} = \sum_{i=1}^{N} P_i \beta_i q'_{oi} \Delta z_i \tag{11.30}$$

where $\quad f_{si} = \beta_i q'_{oi}$ $\tag{11.31}$

$\beta_i = K_{si} \tan \delta_i$,

δ_i = angle of skin friction of the ith layer.

The following equations are provided by O'Neill and Reese (1999) for computing β_i. For SPT N_{60} (uncorrected) ≥ 15 blows/0.3 m

$$\beta_i = 1.5 - 0.245 \, (z_i)^{0.5} \tag{11.32}$$

For SPT N_{60} (uncorrected) < 15 blows/0.3 m

$$\beta_i = \frac{N_{60}}{15} \, [1.5 - 0.245 \, (z_i)^{0.5}] \tag{11.33}$$

In Gravelly Sands or Gravels

For SPT $N_{60} \geq 15$ blows/0.3 m

$$\beta_i = 2.0 - 0.15 \, (z_i)^{0.75} \tag{11.34}$$

In gravelly sands or gravels, use the method for sand if $N_{60} < 15$ blows/0.3 m.

The definitions of various symbols used above are

β_i = dimensionless correlation factor applicable to layer i. Limited to 1.2 in sands and 1.8 in gravelly sands and gravel. Minimum value is 0.25 in both types of soil; f_{si} is limited 200 kN/m^2 (2.1 tsf).

q'_{oi} = vertical effective stress at the middle of each layer.

N_{60} = design value for SPT blow count, uncorrected for depth, saturation or fines corresponding to layer i.

z_i = vertical distance from the ground surface, in meters, to the middle of layer i. The layer thickness Δz_i is limited to 9 m.

11.14 ULTIMATE SIDE AND TOTAL RESISTANCE IN ROCK (O'NEILL AND REESE, 1999)

Ultimate Skin Resistance (for Smooth Socket)

Rock is defined as a cohesive geomaterial with $q_u > 5$ MPa (725 psi). The following equations may be used for computing $f_s (= f_{max})$ when the pier is socketed in rock. Two methods are proposed.

Method 1

$$f_s (= f_{max}) = 0.65 p_a \left(\frac{q_u}{p_a}\right)^{0.5} \leq 0.65 p_a \left(\frac{f_c}{p_a}\right)^{0.5} \qquad (11.35)$$

where, p_a = atmospheric pressure (= 101 kPa),

q_u = unconfined compressive strength of rock mass,

f_c = 28 day compressive cylinder strength of concrete used in the drilled pier.

Method 2

$$f_s (= f_{max}) = 1.42 p_a \left(\frac{q_u}{p_a}\right)^{0.5} \qquad (11.36)$$

Carter and Kulhawy (1988) suggested Eq. (11.36) based on the analysis of 25 drilled shaft socket tests in a very wide variety of soft rock formations, including sandstone, limestone, mudstone, shale and chalk.

Ultimate Total Resistance Q_u

If the base of the drilled pier rests on sound rock, the side resistance can be ignored. In cases where significant penetration of the socket can be made, it is matter of engineering judgement to decide whether Q_f should be added directly to Q_b to obtain the ultimate value Q_u, When the rock is brittle in shear, much side resistance will be lost as the settlement increases to the value required to develop the full value of q_b (= q_{max}). If the rock is ductile in shear, there is no question that the two values can be added directly (O'Neill and Reese, 1999).

11.15 ESTIMATION OF SETTLEMENTS OF DRILLED PIERS AT WORKING LOADS

O'Neill and Reese (1999) suggest the following methods for computing axial settlements for isolated drilled piers:

1. Simple formulas.
2. Normalised load-transfer methods.

The total settlement S_t at the pier head at working loads may be expressed as (Vesic, 1977)

$$S_t = S_e + S_{bb} + S_{bs} \tag{11.37}$$

where, S_e = elastic compression,

S_{bb} = settlement of the base due to the load transferred to the base,

S_{bs} = settlement of the base due to the load transferred into the soil along the sides.

The equations for the settlements are

$$S_e = \frac{L\left(Q_a - 0.5Q_{fm}\right)}{A_b E}$$

$$S_{bb} = C_p\left(\frac{Q_{mb}}{dq_b}\right) \tag{11.38}$$

$$S_{bs} = \left(0.93 + 0.16\sqrt{\frac{L}{d}}\right) C_p\left(\frac{Q_{fm}}{Lq_b}\right) \tag{11.39}$$

where, L = length of the drilled pier,

A_b = base cross-sectional area,

E = Young's modulus of the drilled pier,

Q_a = load applied to the head,

Q_{fm} = mobilised side resistance when Q_a is applied,

Q_{bm} = mobilised base resistance,

d = pier width or diameter,

C_p = soil factor obtained from Table 11.7.

Table 11.7 Values of C_p for various soils (Vesic, 1977)

Soil	C_p
Sand (dense)	0.09
Sand (loose)	0.18
Clay (stiff)	0.03
Clay (soft)	0.06
Silt (dense)	0.09
Silt (loose)	0.12

Normalised Load-Transfer Methods

Reese and O'Neill (1988) analysed a series of compression loading test data obtained from full-sized drilled piers in soil. They developed normalised relations for piers in cohesive and cohesionless soils. Figures 11.18 and 11.19 can be used to predict settlements of piers in cohesive soils and Figs 11.20 and 11.21 in cohesionless soils including soil mixed with gravel.

The boundary limits indicated for gravel in Fig. 11.20 have been found to be approximately appropriate for cemented fine-grained desert IGM's (Walsh et al, 1995). The range of validity of the normalised curves are as follows:

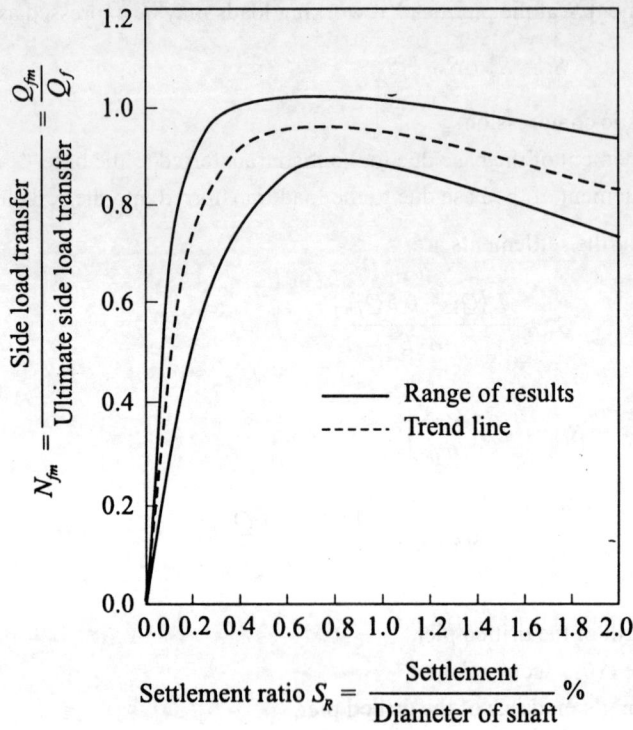

Fig. 11.18 Normalised side load transfer for drilled shaft in cohesive soil (O'Neill and Reese, 1999)

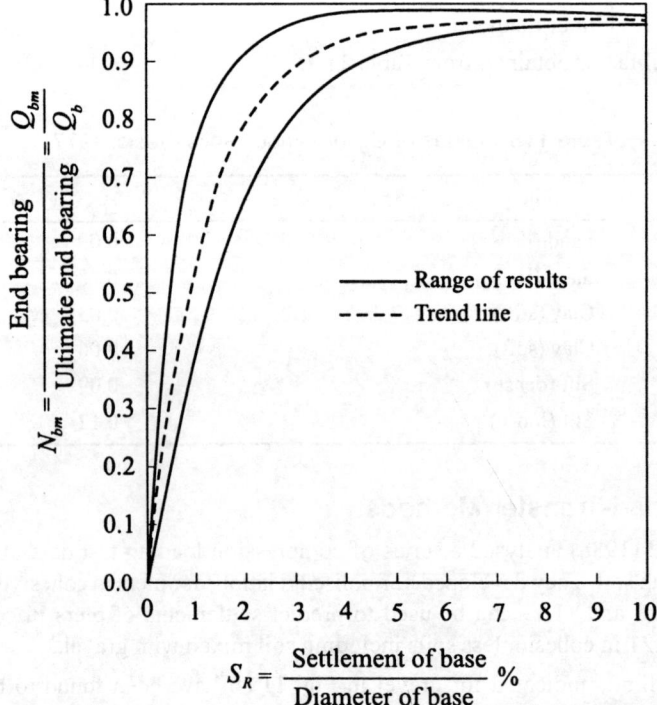

Fig. 11.19 Normalised base load transfer for drilled shaft in cohesive soil (O'Neill and Reese, 1999)

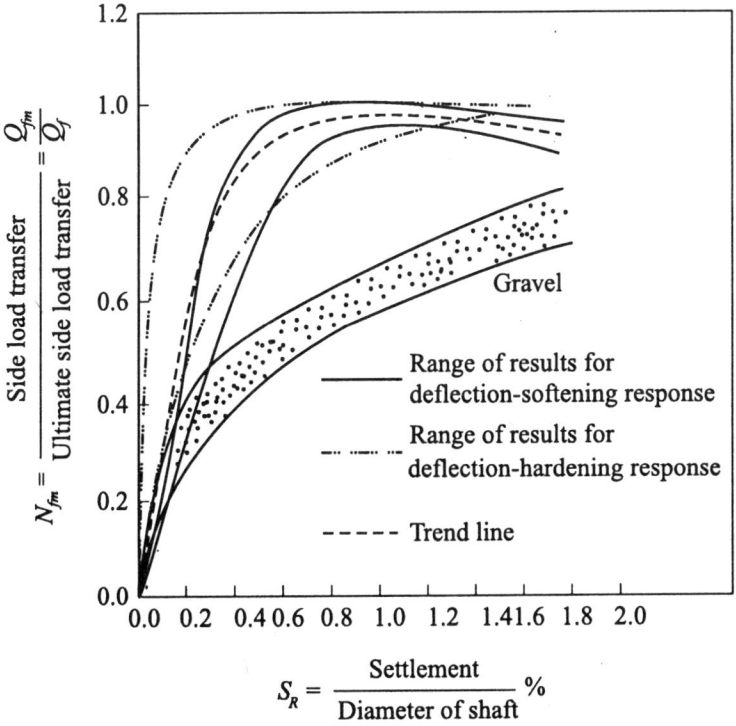

Fig. 11.20 Normalised side load transfer for drilled shaft in cohesionless soil
(O'Neill and Reese, 1999)

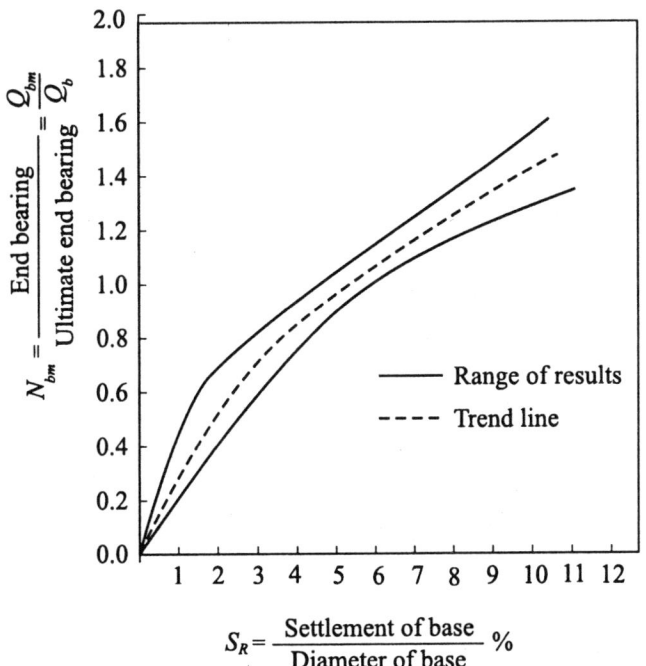

Fig. 11.21 Normalised base load transfer for drilled shaft in cohesionless soil
(O'Neill and Reese, 1999)

Figures 11.18 and 11.19

Normalizing factor = shaft diameter d

Range of d = 0.46 m to 1.53 m

Figures 11.20 and 11.21

Normalizing factor = base diameter

Range of d = 0.46 m to 1.53 m

The following notations are used in the figures:

S_R = Settlement ratio = S_a/d,

S_a = Allowable settlement,

N_{fm} = Normalised side load transfer ratio = Q_{fm}/Q_f,

N_{bm} = Normalize base load transfer ratio = Q_{bm}/Q_b.

Example 11.1

A multistorey building is to be constructed in a stiff to very stiff clay. The soil is homogeneous to a great depth. The average value of undrained shear strength c_u is 150 kN/m². It is proposed to use a drilled pier of length 25 m and diameter 1.5 m. Determine (a) the ultimate load capacity of the pier, and (b) the allowable load on the pier with F_s = 2.5 (Fig.Ex. 11.1).

Solution

Base load

When $c_u \geq 96$ kPa (2000 lb/ft²), use Eq. (11.8) for computing q_b. In this case $c_u > 96$ kPa.

$$q_b = 9c_u = 9 \times 150 = 1350 \text{ kN/m}^2$$

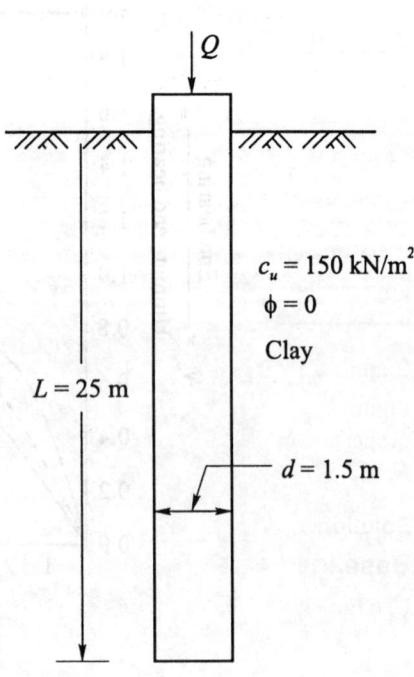

Base load $\quad Q_b = A_b\, q_b = \dfrac{3.14 \times 1.5^2}{4} \times 1350$

$$= 1.766 \times 1350 = 2384 \text{ kN}$$

Frictional load

The unit ultimate frictional resistance f_s is determined using Eq. (11.24)

$$f_s = \alpha c_u$$

From Fig. (11.15),

$$\alpha = 0.55 \text{ for } c_u/p_a = 150/101 = 1.5$$

where p_a is the atmospheric pressure = 101 kPa

Therefore $\quad f_s = 0.55 \times 150 = 82.5$ kN/m²

The effective length of the shaft for computing the frictional load [Fig. 11.13 (a)] is

$$L' = [L - (d + 1.5)] \text{ m}$$

$$= 25 - (1.5 + 1.5) = 22 \text{ m}$$

The effective surface area

$$A_s = \pi\, d\, L' = 3.14 \times 1.5 \times 22$$

$$= 103.62 \text{ m}^2$$

c_u = 150 kN/m²

$\phi = 0$

Clay

L = 25 m

d = 1.5 m

Fig. Ex. 11.1

Therefore $\qquad Q_f = f_s A_s = 82.5 \times 103.62 = 8,549\,\text{kN}$

The total ultimate load is

$$Q_u = Q_f + Q_b = 8,549 + 2,384 = 10,933\,\text{kN}$$

The allowable load may be determined by applying an overall factor of safety to Q_u. Normally $F_s = 2.5$ is sufficient.

$$Q_a = \frac{10,933}{2.5} = 4,373\,\text{kN}$$

Example 11.2

For the problem given in Ex. 11.1, determine the allowable load for a settlement of 10 mm ($= S_a$). All the other data remain the same.

Solution
Allowable skin load

Settlement ratio $\qquad S_R = \dfrac{S_a}{d} = \dfrac{10}{1.5 \times 10^3} \times 100 = 0.67\%$

From Fig. 11.18, for $S_R = 0.67\%$, $N_{fm} = \dfrac{Q_{fm}}{Q_f} = 0.95$ by using the trend line.

$$Q_{fm} = 0.95\,Q_f = 0.95 \times 8,549 = 8,122\,\text{kN}.$$

Allowable base load for S_a = 10 mm

From Fig. 11.19 for $\qquad S_R = 0.67\%$, $N_{bm} = \dfrac{Q_{bm}}{Q_b} = 0.4$

$$Q_{bm} = 0.4\,Q_b = 0.4 \times 2,384 = 954\,\text{kN}$$

Now the allowable load Q_{as} based on settlement consideration is

$$Q_{as} = Q_{fm} + Q_{bm} = 8,122 + 954 = 9,076\,\text{kN}$$

Q_{as} based on settlement consideration is very much higher than Q_a (Ex. 11.1) and as such Q_a governs the criteria for design.

Example 11.3

Figure Ex. 11.3 depicts a drilled pier with a belled bottom. The details of the pile and the soil properties are given in the figure. Estimate (*a*) the ultimate load, and (*b*) the allowable load with $F_s = 2.5$.

Solution
Base load

Use Eq. (11.8) for computing q_b

$$q_b = 9c_u = 9 \times 200 = 1,800\,\text{kN/m}^2$$

Base load $\qquad Q_b = \dfrac{\pi d_b^2}{4} \times q_b = \dfrac{3.14 \times 3^2}{4} \times 1,800 = 12,717\,\text{kN}$

Frictional load

The effective length of shaft

$$L' = 25 - (2.75 + 1.5)$$

$$= 20.75 \, \text{m}$$

From Eq. (11.24)

$$f_s = \alpha c_u$$

For $\dfrac{c_u}{p_a} = \dfrac{100}{101} \approx 1.0$, $\alpha = 0.55$ from Fig. 11.15

Hence $f_s = 0.55 \times 100 = 55 \, \text{kN/m}^2$

$$Q_f = PL'f_s$$

$$= 3.14 \times 1.5 \times 20.75 \times 55$$

$$= 5{,}375 \, \text{kN}$$

$$Q_u = Q_b + Q_f = 12{,}717 + 5{,}375$$

$$= 18{,}092 \, \text{kN}$$

$$Q_a = \frac{18{,}092}{2.5} = 7{,}237 \, \text{kN}$$

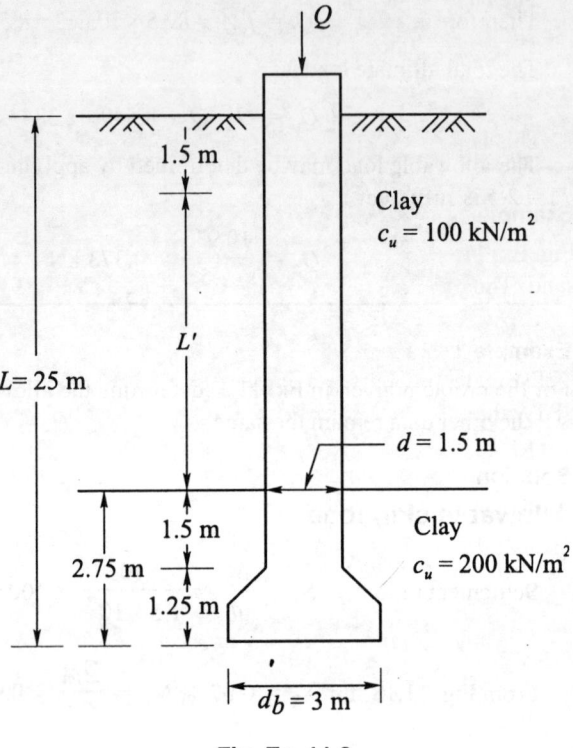

Fig. Ex. 11.3

Example 11.4

For the problem given in Ex. 11.3, determine the allowable load Q_{as} for a settlement $S_a = 10$ mm.

Solution

Skin load Q_{fm} (mobilised)

Settlement ratio $\qquad S_R = \dfrac{10}{1.5 \times 10^3} \times 100 = 0.67\%$

From Fig. 11.18 for $S_R = 0.67$, $N_{fm} = 0.95$ from the trend line.

Therefore $\qquad Q_{fm} = 0.95 \times 5{,}375 = 5{,}106 \, \text{kN}$

Base load Q_{bm} (mobilised)

$$S_R = \frac{10}{3 \times 10^3} \times 100 = 0.33\%$$

From Fig. 11.19 for $S_R = 0.33\%$, $N_{bm} = 0.3$ from the trend line.

Hence $\qquad Q_{bm} = 0.3 \times 12{,}717 = 3815 \, \text{kN}$

$$Q_{as} = Q_{fm} + Q_{bm} = 5{,}106 + 3{,}815 = 8{,}921 \, \text{kN}$$

The factor of safety with respect to Q_u is (from Ex. 11.3)

$$F_s = \frac{18,092}{8,921} = 2.03$$

This is low as compared to the normally accepted value of $F_s = 2.5$. Hence Q_a rules the design.

Example 11.5

Fig. Ex. 11.5 shows a straight shaft drilled pier constructed in homogeneous loose to medium dense sand. The pile and soil properties are:

$$L = 25 \text{ m}, \ d = 1.5 \text{ m}, \ c = 0, \ \phi = 36° \text{ and } \gamma = 17.5 \text{ kN/m}^3$$

Estimate (a) the ultimate load capacity, and (b) the allowable load with $F_s = 2.5$. The average SPT value $N_{cor} = 30$ for $\phi = 36°$.

Use (i) Vesic's method, and (ii) the O'Neill and Reese method.

Solution

(i) Vesic's method

From Eq. (11.10) for $c = 0$

$$q_b = (N_q - 1) \, q'_o \, s_q \, d_q \, C_q \qquad \text{... (a)}$$
$$q'_o = 25 \times 17.5 = 437.5 \text{ kN/m}^2$$

From Eq. (11.16)

$$\phi_{rel} = \frac{\phi° - 25°}{45° - 25°} = 0.55$$

From Eq. (11.15)

$$\Delta = \frac{0.005 \, (1 - 0.55) \times 437.5}{101} \approx 0.01$$

From Eq. (11.14)

$$\mu_d = 0.1 + 0.3\phi_{rel} = 0.1 + 0.3 \times 0.55 = 0.265$$

From Eq. (11.17)

$$E_d = 200p_a = 200 \times 101 = 20,200 \text{ kN/m}^2$$

From Eq. (11.13)

$$I_r = \frac{E_d}{2 \, (1 + \mu_d) \, q'_o \, \tan \phi}$$

$$= \frac{20,200}{2 \, (1 + 0.265) \times 437.5 \tan 36°} = 25$$

From Eq. (11.12)

$$I_{rr} = \frac{I_r}{1 + \Delta I_r} = \frac{25}{1 + 0.01 \times 25} = 20$$

From Eq. (11.11b)

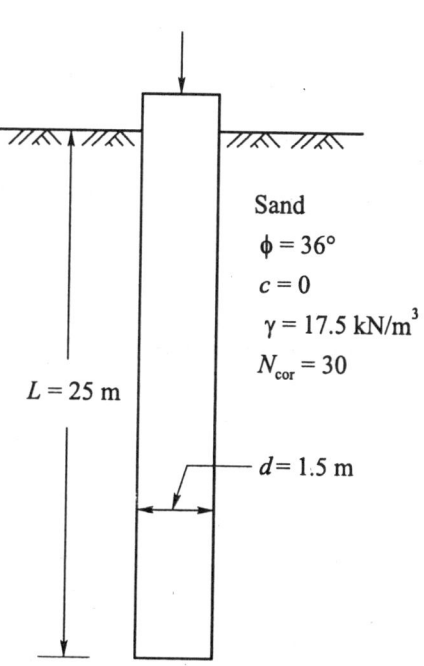

Sand
$\phi = 36°$
$c = 0$
$\gamma = 17.5 \text{ kN/m}^3$
$N_{cor} = 30$

$L = 25 \text{ m}$

$d = 1.5 \text{ m}$

Fig. Ex. 11.5

$$C_q = \exp. \left[(-3.8 \tan 36°) + \left(\frac{3.07 \sin 36° \log_{10} 2 \times 20}{1 + \sin 36°} \right) \right]$$

$$= \exp - (0.9399) = 0.391$$

From Fig. 11.14, $N_q = 30$ for $\phi = 36°$

From Table (11.2) $s_q = 1 + \tan 36° = 1.73$

$$d_q = 1 + 2 \tan 36° (1 - \sin 36°)^2 \left[\frac{3.14}{180} \times \tan^{-1} \frac{25}{1.5} \right] = 1.373$$

Substituting in Eq. (a)

$$q_b = (30 - 1) \times 437.5 \times 1.73 \times 1.373 \times 0.391$$

$$= 11,783 \text{ kN/m}^2 > 11,000 \text{ kN/m}^2$$

As per Tomlinson (1986), the computed q_b should be less than 11,000 kN/m^2.

Hence $Q_b = \dfrac{3.14}{4} \times (1.5)^2 \times 11,000 = 19,429$ kN

Skin load Q_f

From Eqs (11.31) and (11.32)

$$f_s = \beta q'_o, \ \beta = 1.5 - 0.245z^{0.5}, \text{ where } z = \frac{L}{2} = \frac{25}{2} = 12.5$$

Substituting

$$\beta = 1.5 - 0.245 \times (12.5)^{0.5} = 0.63$$

Hence $f_s = 0.63 \times 437.5 = 275.62$ kN/m^2

Per Tomlinson (1986) f_s should be limited to 110 kN/m^2. Hence $f_s = 110$ kN/m^2

Therefore $Q_f = \pi dL f_s = 3.14 \times 1.5 \ 25 \times 110 = 12,953$ kN

Ultimate load $Q_u = 19,429 + 12,953 = 32,382$ kN

$$Q_a = \frac{32,382}{2.5} = 12,953 \text{ kN}$$

O'Neill and Reese method

This method relates q_b to the SPT N value as per Eq. (11.19a)

$$q_b = 57.5N \text{ kN/m}^2 = 57.5 \times 30 = 1,725 \text{ kN/m}^2$$

$$Q_b = A_b q_b = 1.766 \times 1,725 = 3,046 \text{ kN}$$

The method for computing Q_f remains the same as above.

Now $Q_u = 3,406 + 12,953 = 15,999$

$$Q_a = \frac{15,999}{2.5} = 6,400 \text{ kN}$$

Example 11.6

Compute Q_u and Q_a for the pier given in Ex. 11.5 by the following methods.

1. Use the SPT value [Eq. (8.48)] for bored piles
2. Use the Tomlinson method of estimating Q_b and Table 8.2 for estimating Q_f. Compare the results of the various methods.

Solution

Use of the SPT value [Meyerhof Eq. (8.48)]

$$q_b = 133N_{cor} = 133\ 5\ 30 = 3{,}990\ \text{kN/m}^2$$

$$Q_b = \frac{3.14 \times (1.5)^2}{4} \times 3{,}990 = 7{,}047\ \text{kN}$$

$$f_s = 0.67N_{cor} = 0.67 \times 30 = 20\ \text{kN/m}^2$$

$$Q_f = 3.14 \times 1.5 \times 25 \times 20 = 2{,}355\ \text{kN}$$

$$Q_u = 7{,}047 + 2{,}355 = 9{,}402\ \text{kN}$$

$$Q_a = \frac{9{,}402}{2.5} = 3{,}760\ \text{kN}$$

Tomlinson Method for Q_b

For a driven pile

From Fig. 8.9 $\qquad N_q = 65$ for $\phi = 36°$ and $\dfrac{L}{d} = \dfrac{25}{1.5} \approx 17$

Hence $\qquad q_b = q'_o N_q = 437.5 \times 65 = 28{,}438\ \text{kN/m}^2$

For bored pile

$$q_b = \frac{1}{3}\ q_b\ (\text{driven pile}) = \frac{1}{3} \times 28{,}438 = 9{,}479\ \text{kN/m}^2$$

$$Q_b = A_b q_b = 1.766 \times 9{,}479 = 16{,}740\ \text{kN}$$

Q_f from Table 8.2

For $\phi = 36°$, $\qquad \delta = 0.75 \times 36 = 27$, and $\bar{K}_s = 1.5$ (for medium dense sand).

$$f_s = \bar{q}'_o \bar{K}_s \tan\delta = \frac{437.5}{2} \times 1.5 \tan 27° = 167\ \text{kN/m}^2$$

As per Tomlinson (1986) f_s is limited to 110 kN/m². Use $f_s = 110\ \text{kN/m}^2$.

Therefore $\qquad Q_f = 3.14 \times 1.5 \times 25 \times 110 = 12{,}953\ \text{kN}$

$$Q_u = Q_b + Q_f = 16{,}740 + 12953 = 29{,}693\ \text{kN}$$

$$Q_a = \frac{29{,}693}{2.5} = 11{,}877\ \text{kN}$$

Comparison of estimated results ($F_s = 2.5$)

Example No.	Name of method	Q_b (kN)	Q_f (kN)	Q_u (kN)	Q_a (kN)
11.5	Vesic	19,429	12,953	32,382	12,953
11.5	O'Neill and Reese, for Q_b and Vesic for Q_f	3,046	12,953	15,999	6,400
11.6	Meyerhof Eq. (8.49)	7,047	2,355	9,402	3,760
11.6	Tomlinson for Q_b (Fig. 8.9) Table 8.2 for Q_f	16,740	12,953	29,693	11,877

Which method to use

The variation in the values of Q_b and Q_f are very large between the methods. Since the soils encountered in the field are generally heterogeneous in character no theory holds well for all the soil conditions. Designers have to be practical and pragmatic in the selection of any one or combination of the theoretical approaches discussed earlier.

Example 11.7

For the problem given in Example 11.5 determine the allowable load for a settlement of 10 mm. All the other data remain the same. Use (*a*) the values of Q_f and Q_b obtained by Vesic's method, and (*b*) Q_b from the O'Neill and Reese method.

Solution

(a) Vesic's values Q_f and Q_b

Settlement ratio for $S_a = 10$ mm is

$$S_R = \frac{S_a}{d} = \frac{10 \times 10^2}{1.5 \times 10^3} = 0.67\%$$

From Fig. 11.20 for $S_R = 0.67\%$ $N_{fm} = 0.96$ (approx.) using the trend line.

$$Q_{fm} = 0.96 \times Q_f = 0.96 \times 12,953 = 12,435 \text{ kN}$$

From Fig. 11.21 for $S_R = 0.67\%$

$$N_{bm} = 0.20, \text{ or } Q_{bm} = 0.20 \times 19,429 = 3,886 \text{ kN}$$

$$Q_{as} = 12,435 + 3,886 = 16,321 \text{ kN}$$

Shear failure theory give $Q_a = 12,953$ kN which is much lower than Q_{as}. As such Q_a determines the criteria for design.

(b) O'Neill and Reese $Q_b = 3,046$ kN

As above, $Q_{bm} = 0.20 \times 3,046 = 609$ kN

Using Q_{fm} in (*a*) above,

$$Q_{as} = 609 + 12,435 = 13,044 \text{ kN}$$

The value of Q_{as} is closer to Q_a (Vesic) but much higher than Q_a calculated by all the other methods.

Example 11.8

Figure Ex. 11.8 shows a drilled pier penetrating an IGM : clay-shale to a depth of 8 m. Joints exists within the IGM stratum. The following data are available : $L_s = 8$ m ($= z_c$), $d = 1.5$ m, q_u (rock) $= 3 \times 10^3$ kN/m^2, E_i (rock) $= 600 \times 10^3$ kN/m^2, concrete slump $= 175$ mm, unit weight of concrete $\gamma_c = 24$ kN/m^3, E_c (concrete) $= 435 \times 10^6$ kN/m^2, and RQD $= 70$ percent, q_u (concrete) $= 435 \times 10^6$ kN/m^2. Determine the ultimate frictional load Q_f (max).

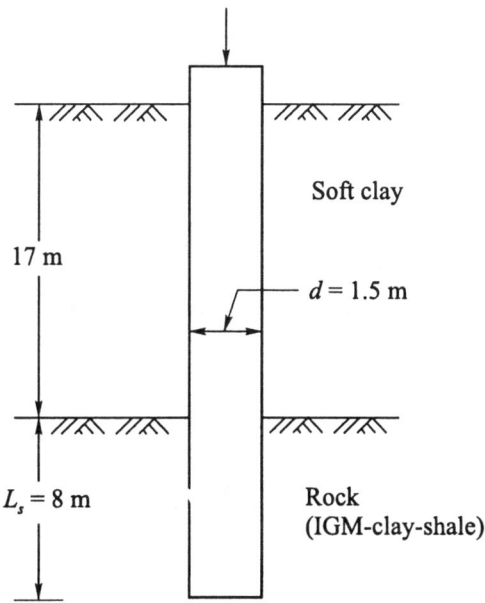

17 m

Soft clay

$d = 1.5$ m

$L_s = 8$ m

Rock
(IGM-clay-shale)

Fig. Ex. 11.8

Solution

(*a*) Determine α in Eq. (11.26)

$$f_a = \alpha q_u \text{ where } q_u = 3 \text{ MPa for rock}$$

For the depth of socket $L_s = 8$ m, and slump $= 175$ mm

$$M = 0.76 \text{ from Fig. 11.17.}$$

From Eq. (11.27)

$$\sigma_n = M\gamma_c z_c = 0.76 \times 24 \times 8 = 146 \text{ kN/m}^2$$

$$p_a = 101 \text{ kN/m}^2, \, \sigma_n/p_a = 146/101 = 1.45$$

From Fig. 11.16 for $q_u = 3$ Mpa and $\sigma_n/p_a = 1.45$, we have $\alpha = 0.11$

(*b*) Determination of f_a

$$f_a = 0.11 \times 3 = 0.33 \text{ MPa}$$

(*c*) Determination f_{aa} in Eq. (11.28)

For RQD $= 70\%$, $E_m/E_i = 0.1$ from Table 11.5 for open joints, and $f_{aa}/f_a \, (= R_a) = 0.55$ from Table 11.6

$$f_{max} = f_{aa} = 0.55 \times 0.33 = 0.182 \text{ MPa} = 182 \text{ kN/m}^2$$

(d) Ultimate friction load Q_f

$$Q_f = PLf_{aa} = 3.14 \times 1.5 \times 8 \times 182 = 6,858 \text{ kN}$$

Example 11.9

For the pier given in Ex. 11.8, determine the ultimate bearing capacity of the base. Neglect the frictional resistance. All the other data remain the same.

Solution

For RQD between 70 and 100 percent

From Eq. (11.22)

$$q_b \ (= q_{max}) = 4.83 \ (q_u) \ 0.5 \text{ MPa} = 4.83 \times (3)^{0.5} = 8.37 \text{ MPa}$$

$$Q_b \ (max) = \frac{3.14}{4} \times 1.5^2 \times 8.37 = 14.78 \text{ MN} = 14,780 \text{ kN}$$

11.16   UPLIFT CAPACITY OF DRILLED PIERS

Structures subjected to large overturning moments can produce uplift loads on drilled piers if they are used for the foundation. The design equation for uplift is similar to that of compression. Figure 11.22 shows the forces acting on the pier under uplift-load Q_{ul}. The equation for Q_{ul} may be expressed as

$$Q_{ul} = Q_{fr} + W_p = A_s f_r + W_p \qquad (11.40)$$

where, Q_{fr} = total side resistance for uplift,

W_p = effective weight of the drilled pier,

A_s = surface area of the pier,

f_r = frictional resistance to uplift.

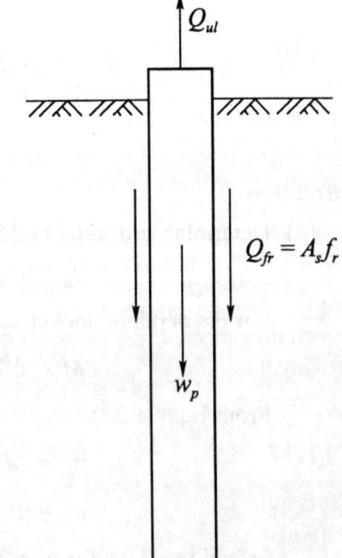

Fig. 11.22 Uplift forces for a straight edged pier

Uplift Capacity of Single Pier (Straight Edge)

For a drilled pier in cohesive soil, the frictional resistance may expressed as (Chen and Kulhawy, 1994)

$$f_r = \alpha c_u \qquad (11.41a)$$

$$\alpha = 0.31 + 0.17 \frac{c_u}{p_a} \qquad (11.41b)$$

where, α = adhesion factor,

c_u = undrained shear strength of cohesive soil,

p_a = atmospheric pressure (101 kPa).

Poulos and Davis (1980) suggest relationships between c_u and α as given in Fig. 11.23. The curves in the figure are based on pull out test data collected by Sowa (1970).

Uplift Resistance of Piers in Sand

There are no confirmatory methods available for evaluating uplift capacity of piers embedded in cohesionless soils. Poulos and Davis, (1980) suggest that the skin frictional resistance for pull out may be taken as equal to two-thirds of the shaft resistance for downward loading.

Uplift Resistance of Piers in Rock

According to Carter and Kulhawy (1988), the frictional resistance offered by the surface of the pier under uplift loading is almost equal to that for downward loading if the drilled pier is rigid relative to the rock. The effective rigidity is defined as $(E_c/E_m)\,(d/D_s)^2$, in which E_c and E_m are the Young's modulus of the drilled pier and rock mass respectively, d is the socket diameter and D_s is the depth of the socket. A socket is rigid when $(E_c/E_m)\,(d/D_s)^2 \geq 4$. When the effective rigidity is less than 4, the frictional resistance f_r for upward loading may be taken as equal to 0.7 times the value for downward loading.

Example 11.10

Determine the uplift capacity of the drilled pier given in Fig. Ex. 11.10. Neglect the weight of the pier.

Solution

From Eq. (11.40)

$$Q_{ul} = A_s f_r$$

From Eq. (11.41a) $f_r = \alpha c_u$

From Eq. (11.41b) $\alpha = 0.31 + 0.17\,\dfrac{c_u}{p_a}$

Given: $L = 25$ m, $d = 1.5$ m, $c_u = 150$ kN/m^2

Hence $\alpha = 0.31 + 0.17 \times \dfrac{150}{101} = 0.56$

$$f_r = 0.56 \times 150 = 84 \text{ kN/m}^2$$

$$Q_{ul} = 3.14 \times 1.5 \times 25 \times 84$$

$$= 9,891 \text{ kN}$$

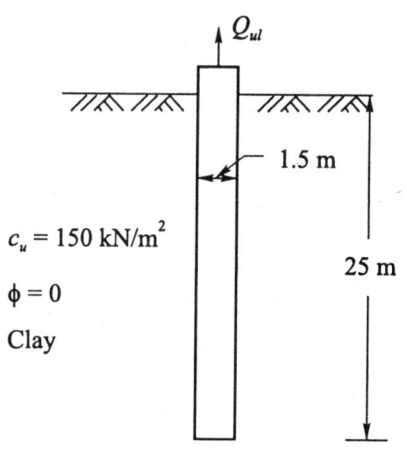

$c_u = 150$ kN/m^2

$\phi = 0$

Clay

1.5 m

25 m

Q_{ul}

Fig. Ex. 11.10

It may be noted here that $f_s = 82.5$ kN/m^2 for downward loading and $f_r = 84$ kN/m^2 for uplift. The two values are very close to each other.

11.17 LATERAL BEARING CAPACITY OF DRILLED PIERS

It is quite common that drilled piers constructed for bridge foundations and other similar structures are also subjected to lateral loads and overturning moments. The methods applicable to piles are applicable to piers also. Chapter 9 deals with such problems. This chapter deals with one more method as recommended by O'Neill and Reese (1999). This method is called **Characteristic load method** and is described below.

Characteristic Load Method (Duncan *et al*, 1994)

The characteristic load method proceeds by defining a characteristic or normalizing shear load (P_c) and a characteristic or normalizing bending moment (M_c) as given below.

For clay

$$P_c = 7.34d^2 \, (ER_I) \left(\frac{c_u}{ER_I} \right)^{0.68} \tag{11.42}$$

$$M_c = 3.86d^3 \, (ER_I) \left(\frac{c_u}{ER_I} \right)^{0.46} \tag{11.43}$$

For sand

$$P_c = 1.57d^2 \, (ER_I) \left(\frac{\gamma' \, d \, \phi' \, K_p}{ER_I} \right)^{0.57} \tag{11.44}$$

$$M_c = 1.33d^3 \, (ER_I) \left(\frac{\gamma' \, d \, \phi' \, K_p}{ER_I} \right)^{0.40} \tag{11.45}$$

where, d = shaft diameter,

E = Young's modulus of the shaft material,

R_I = ratio of moment of inertia of drilled shaft to moment of inertia of solid section (= 1 for a normal uncracked drilled shaft without central voids),

c_u = average value of undrained shear strength of the clay in the top $8d$ below the ground surface,

γ' = average effective unit weight of the sand (total unit weight above the water table, buoyant unit weight below the water table) in the top $8d$ below the ground surface,

ϕ' = average effective stress friction angle for the sand in the top $8d$ below ground surface,

K_p = Rankine's passive earth pressure coefficient = $\tan^2 (45° + \phi'/2)$.

In the design method, the moments and shears are resolved into groundline values, P_t and M_t, and then divided by the appropriate characteristic load values [Eqs (11.42) through (11.45)]. The lateral deflections at the shaft head, y_t are determined from Figs 11.23 and 11.24, considering the conditions of pile-head fixity. The value of the maximum moment in a free or fixed-headed drilled shaft can be determined through the use of Fig. 11.25 if the only load that is applied is ground line shear. If both

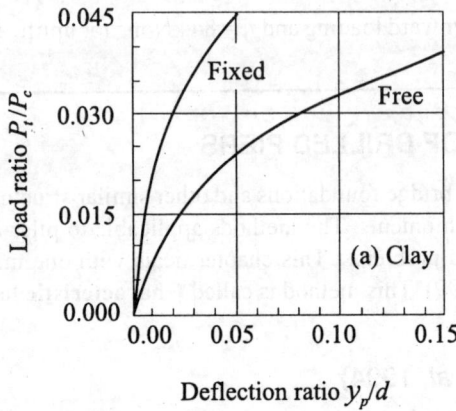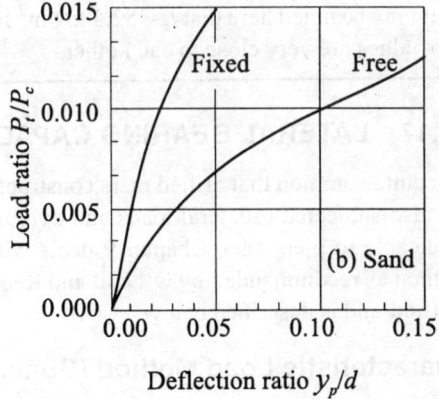

Fig. 11.23 Groundline shear-deflection curves for (a) clay and (b) sand (Duncan *et al,* 1994)

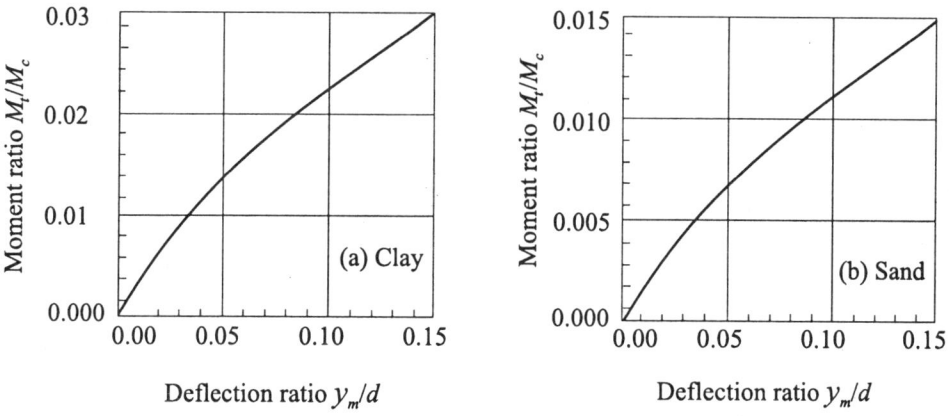

Fig. 11.24 Groundline moment-deflection curves for (a) clay and (b) sand (Duncan *et al*, 1994)

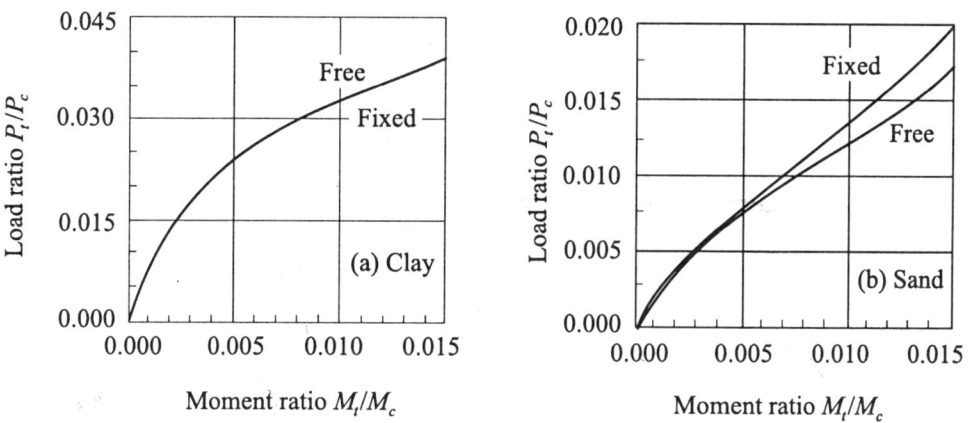

Fig. 11.25 Groundline shear-maximum moment curves for (a) clay and (b) sand
(Duncan *et al*, 1994)

a moment and a shear are applied, one must compute y_t (combined), and then solve Eq. (11.46) for the "characteristic length" T (relative stiffness factor).

$$y_t(\text{combined}) = 2.43 \frac{P_t T^3}{EI} + 1.62 \frac{M_t T^2}{EI} \tag{11.46}$$

where I is the moment of inertia of the cross-section of the drilled shaft.

The principle of superposition is made use of for computing ground line deflections of piers (or piles) subjected to groundline shears and moments at the pier head. The explanation given here applies to a free-head pier. The same principle applies for a fixed head pile also.

Consider a pier shown in Fig. 11.26 (a) subjected to a shear load P_t and moment M_t at the pile head at ground level. The total deflection y_t caused by the combined shear and moment may be written as

$$y_t = y_p + y_m \tag{11.47}$$

where y_p = deflection due to shear load P_t alone with $M_t = 0$,
  y_m = deflection due to moment M_t alone with $P_t = 0$.

Again consider Fig. 11.26 (b). The shear load P_t acting alone at the pile head causes a deflection y_p (as above) which is equal to deflection y_{pm} caused by an equivalent moment M_p acting alone.

In the same way Fig. 11.26 (c) shows a deflection y_m caused by moment M_t at the pile head. An equivalent shear load P_m causes the same deflection y_m which is designated here as y_{mp}. Based on the principles explained above, groundline deflection at the pile head due to a combined shear load and moment may be explained as follows.

1. Use Figs 11.23 and 11.24 to compute groundline deflections y_p and y_m due to shear load and moment respectively.

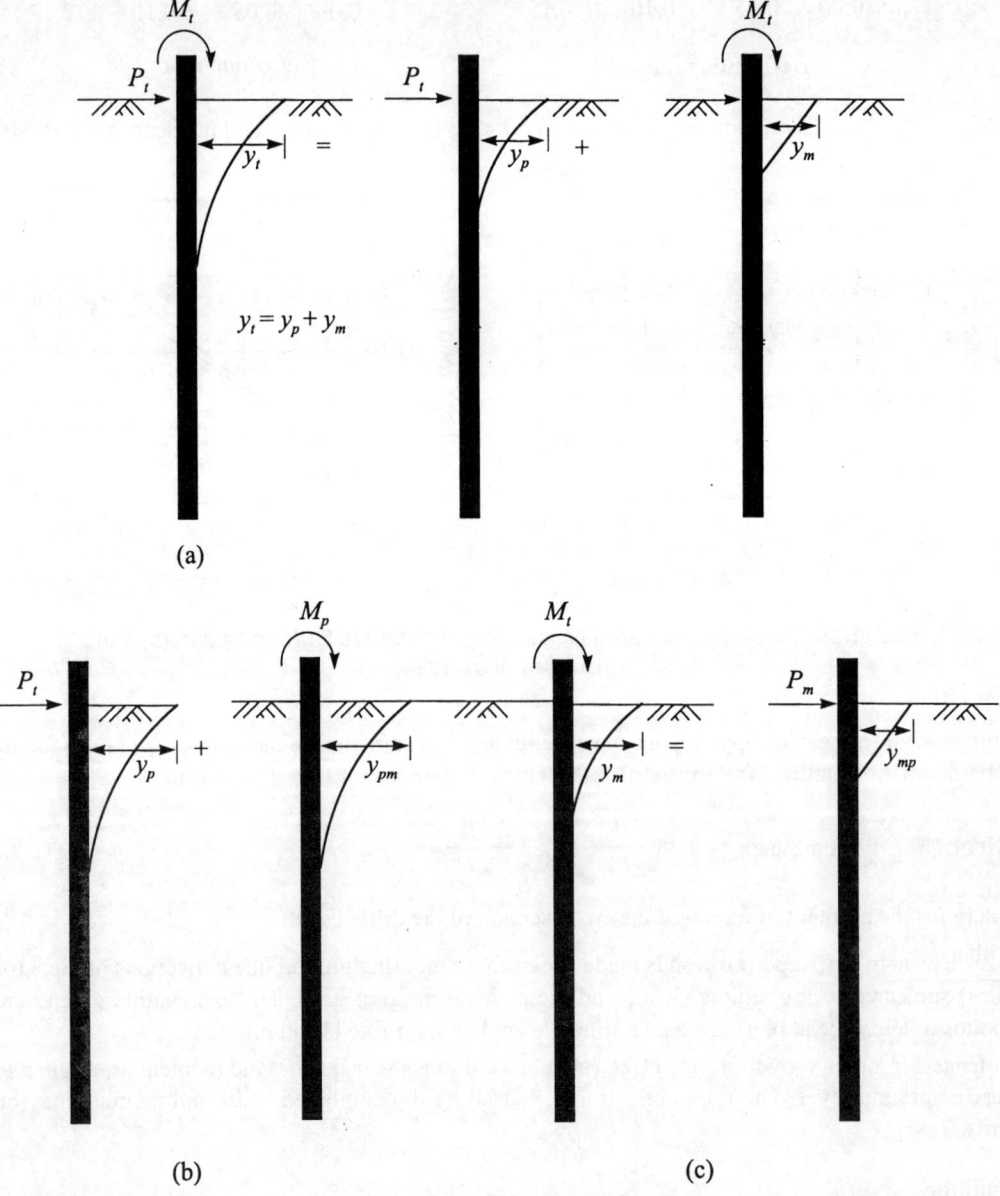

Fig. 11.26 Principle of superposition for computing ground line deflection by Duncan *et al,* (1994) method for a free-head pier

2. Determine the groundline moment M_p that will produce the same deflection as by a shear load P_t [Fig. 11.26 (b)]. In the same way, determine a groundline shear load P_m, that will produce the same deflection as that by the groundline moment M_t [Fig. 11.26 (c)].

3. Now the deflections caused by the shear loads $P_t + P_m$ and that caused by the moments $M_t + M_p$ may be written as follows:

$$y_{tp} = y_p + y_{mp} \tag{11.48}$$

$$y_{tm} = y_m + y_{pm} \tag{11.49}$$

Theoretically, $y_{tp} = y_{tm}$

4. Lastly the total deflection y_t is obtained as

$$y_t = \frac{y_{tp} + y_{tm}}{2} = \frac{\left(y_p + y_{mp}\right) + \left(y_m + y_{pm}\right)}{2} \tag{11.50}$$

The distribution of moment along a pier may be determined using Eq. (9.53) and Table 9.4 or Fig. 11.27.

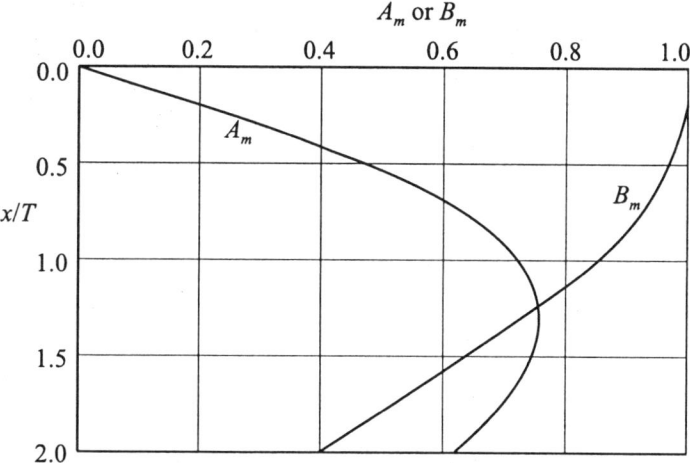

Fig. 11.27 Parameters A_m and B_m (Matlock and Reese, 1961)

Direct Method by Making Use of n_h

The direct method developed by Murthy and Subba Rao (1995) for long laterally loaded piles has been explained in Chapter 9. The application of this method for long drilled piers will be explained with a case study.

Example 11.11 (O'Neill and Reese, 1999)

Refer to Fig. Ex. 11.11. Determine for a free-head pier (*a*) the groundline deflection, and (*b*) the maximum bending moment. Use the Duncan *et al*, (1994) method. Assume $R_I = 1$ in the Eqs (11.44) and (11.45).

Solution

Substituting in Eqs (11.42) and (11.43)

$$P_c = 7.34 \times (0.80)^2 [25 \times 10^3 \times (1)] \left(\frac{0.06}{25 \times 10^3} \right)^{0.68} = 17.72 \, \text{MN}$$

$$M_c = 3.86 (0.80)^3 [25 \times 10^3 \times (1)] \left(\frac{0.06}{25 \times 10^3} \right)^{0.46} = 128.5 \, \text{MN-m}$$

Now
$$\frac{P_t}{P_c} = \frac{0.080}{17.72} = 0.0045, \quad \frac{M_t}{M_c} = \frac{0.4}{128.5} = 0.0031$$

Step 1

From Fig. 11.23 (a) for $\dfrac{P_t}{P_c} = 0.0045$

$$\frac{y_p}{d} = 0.003 \text{ or } y_p = 0.003 \times 0.8 \times 10^3$$

$$= 2.4 \, \text{mm}$$

From Fig. 17.24 (a), $M_t / M_c = 0.0031$

$$\frac{y_m}{d} = 0.006 \text{ or } y_m = 0.006d$$

$$= 0.006 \times 0.8 \times 10^3 = 4.8 \, \text{mm}$$

Step 2

From Fig. 11.23 (a) for $y_m/d = 0.006$, $P_m/P_c = 0.0055$

From Fig. 11.24 (a), for $y_p/d = 0.003$, $M_p/M_c = 0.0015$.

$P_t = 80$ kN $d = 1.8$ m

5 m

9 m

Clay
$c_u = 60$ kPa
$EI = 52.6 \times 10^4$ kN-m^2
$\gamma = 17.5$ kN/m^3 (assumed)
$E = 25 \times 10^6$ kN/m^2

Fig. Ex. 11.11

Step 3

The shear loads P_t and P_m applied at ground level, may be expressed as

$$\frac{P_t}{P_c} + \frac{P_m}{P_c} = 0.0045 + 0.0055 = 0.01$$

From Fig. 11.23,

$$\frac{y_{tp}}{d} = 0.013 \text{ for } \frac{P_t + P_m}{P_c} = 0.01$$

or
$$y_{tp} = 0.013 \times (0.80) \times 10^3 = 10.4 \, \text{mm}$$

Step 4

In the same way as in Step 3

$$\frac{M_t}{M_c} + \frac{M_p}{M_c} = 0.0031 + 0.0015 = 0.0046$$

From Fig. 11.24 (a) $\dfrac{y_{tm}}{d} = \dfrac{y_m + y_{pm}}{d} = 0.011$

Hence $\qquad y_{tm} = 0.011 \times 0.8 \times 10^3 = 8.8$ mm

Step 5

From Eq. (11.50)

$$y_t = \frac{y_{tp} + y_{tm}}{2} = \frac{10.4 + 8.8}{2} = 9.6 \text{ mm}$$

Step 6

The maximum moment for the combined shear load and moment at the pier head may be calculated in the same way as explained in Chapter 9. $M_{(max)}$ as obtained is

$$M_{max} = 470.5 \text{ kN-m}$$

This occurs at a depth of 1.3 m below ground level.

Example 11.12

Solve the problem in Ex. 11.11 by the direct method.

Given: $EI = 52.6 \times 10^4$ kN-m^2, $d = 80$ cm, $\gamma = 17.5$ kN/m^3, $e = 5$ m, $L = 9$ m, $c = 60$ kN/m^2 and $P_t = 80$ kN.

Solution

Groundline deflection

From Eq. (9.92a) for piers in clay

$$n_h = \frac{125 c^{1.5} \sqrt{EI \gamma d}}{\left(1 + \dfrac{e}{d}\right)^{1.5} \times P_e^{1.5}}$$

$$n_h = \frac{125 \times 60^{1.5} \sqrt{52.6 \times 10^4 \times 17.5 \times 0.8}}{\left(1 + \dfrac{5}{0.8}\right)^{1.5} \times P_e^{1.5}} = \frac{808 \times 10^4}{P_e^{1.5}} \text{ kN/m}^3 \qquad \dots \text{(a)}$$

Step 1

Assume $\qquad P_e = P_t = 80$ kN,

From Eq. (a), $n_h = 11,285$ kN/m^3 and

$$T = \left(\frac{EI}{n_h}\right)^{0.2} = \left(\frac{52.6 \times 10^4}{11,285}\right) = 2.16 \text{ m}$$

Step 2

From Eq. (9.88c) $P_e = P_t \times \left(1 + 0.67\dfrac{e}{T}\right) = 80\left[1 + 0.67 \times \dfrac{5}{2.16}\right] = 204 \text{ kN}$

From Eq. (a), $n_h = 2772 \text{ kN/m}^3$, hence $T = 2.86 \text{ m}$

Step 3

Continue the above process till convergence is reached. The final values are

$$P_e = 177 \text{ kN}, \; n_h = 3410 \text{ kN/m}^3 \text{ and } T = 2.74 \text{ m}$$

For $P_e = 190 \text{ kN}$, we have $n_h = 8{,}309 \text{ kN/m}^3$ and $T = 2.29 \text{ m}$

Step 4

From Eq. (11.46)

$$y_t = \frac{2.43 \times 177 \times (2.74)^3}{52.6 \times 10^4} \times 1000 = 16.8 \text{ mm}$$

By Duncan et al, method $y_t = 9.6 \text{ mm}$

Maximum moment from Eq. (9.53)

$$M = (P_t T) A_m + (M'_t) B_m = (80 \times 2.74) A_m + (400) B_m = 219.2 A_m + 400 B_m$$

Depth $x/T = Z$	A_m	B_m	M_1	M_2	M (kN-m)
0	0	1	0	400	400
0.4	0.379	0.99	83	396	479
0.5	0.46	0.98	101	392	493
0.6	0.53	0.96	116	384	500
0.7	0.60	0.94	132	376	**508** (max)
0.8	0.65	0.91	142	364	506

The maximum bending moment occurs at $x/T = 0.7$ or $x = 0.7 \times 2.74 = 1.91 \text{ m}$ (6.26 ft). As per Duncan *et al*, method M (max) = 470.5 kN-m. This occurs at a depth of 1.3 m.

11.18 CASE STUDY OF A DRILLED PIER SUBJECTED TO LATERAL LOADS

Lateral load test was performed on a circular drilled pier by Davisson and Salley (1969). Steel casing pipe was provided for the concrete pier. The details of the pier and the soil properties are given in Fig. 11.28. The pier was instrumented and subjected to cyclic lateral loads. The load deflection curve as obtained by Davison *et al*, is shown in the same figure.

Direct method (Murthy and Subba Rao, 1995) has been used here to predict the load displacement relationship for a continuous load increase by making use of Eq. (9.92a). The predicted curve is also shown in Fig. 11.28. There is an excellent agreement between the predicted and the observed values.

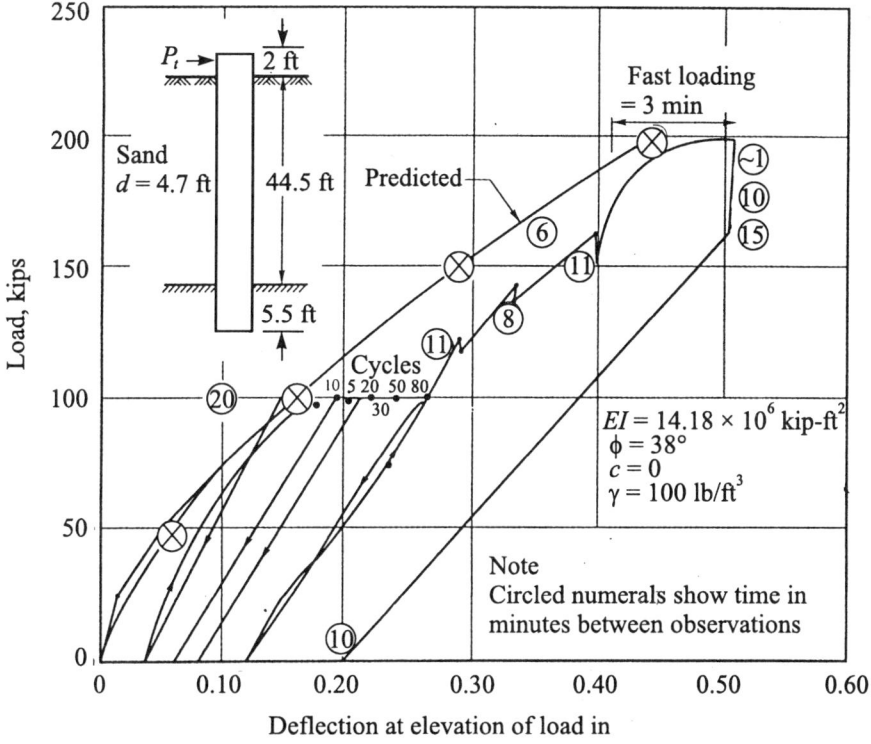

Fig. 11.28 Load deflection relationship, Pier 2S (Davisson *et al*, 1969)

11.19 PROBLEMS

11.1 Figure Prob. 11.1 shows a drilled pier of diameter 1.25 m constructed for the foundation of a bridge. The soil investigation at the site revealed soft to medium stiff clay extending to a great depth. The other details of the pier and the soil are given in the figure. Determine (*a*) the ultimate load capacity, and (*b*) the allowable load for $F_s = 2.5$. Use Vesic's method for base load and α method for the skin load.

11.2 Refer to Prob. 11.1. Given $d = 3$ ft, $L = 30$ ft and $c_u = 1050$ lb/ft². Determine the ultimate (*a*) base load capacity by Vesic method, and (*b*) the frictional load capacity by the α-method.

11.3 Figure Prob. 11.3 shows a drilled pier with a belled bottom constructed for the foundation of a multistory building. The pier passes through two layers of soil. The details of the pier and the properties of the soil are given in the figure. Determine the allowable load Q_a for $F_s = 2.5$. Use (*a*) Vesic's method for the base load, and (*b*) the O'Neill and Reese method for skin load.

11.4 For the drilled pier given in Fig. Prob. 11.1, determine the working load for a settlement of 10 mm. All the other data remain the same. Compare the working load with the allowable load Q_a.

11.5 For the drilled pier given in Prob. 11.2, compute the working load for a settlement of 0.5 in. and compare this with the allowable load Q_a.

11.6 If the drilled pier given in Fig. Prob. 11.6 is to carry a safe load of 2500 kN, determine the length of the pier for $F_s = 2.5$. All the other data are given in the figure.

11.7 Determine the settlement of the pier given in Prob. 11.6 by the O'Neill and Reese method. All the other data remain the same.

11.8 Figure Prob. 11.8 depicts a drilled pier with a belled bottom constructed in homogeneous clay extending to a great depth. Determine the length of the pier to carry an allowable load of 3000 kN with a $F_s = 2.5$. The other details are given in the figure.

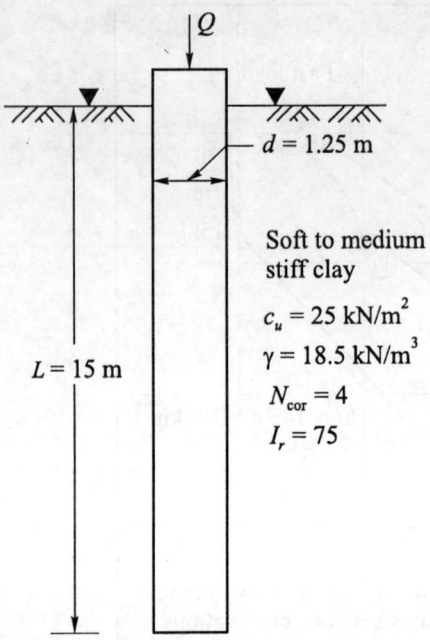

Fig. Prob. 11.1

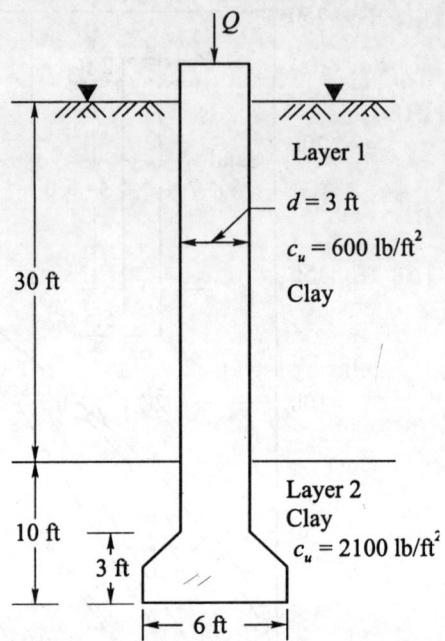

Fig. Prob. 11.3

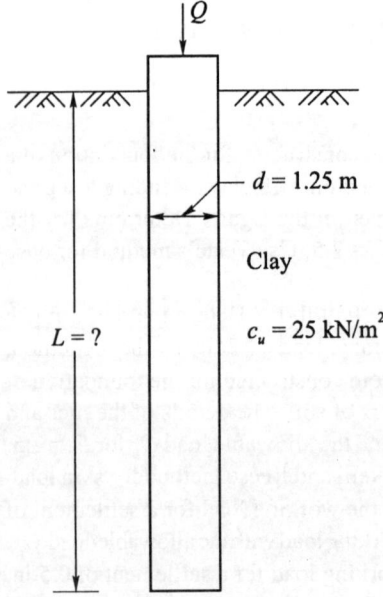

Fig. Prob. 11.6

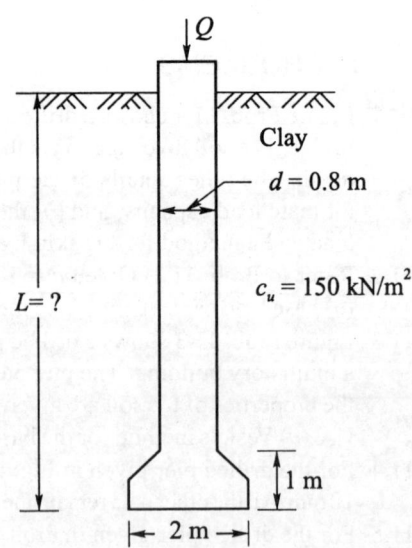

Fig. Prob. 11.8

11.9 Determine the settlement of the pier in Prob. 11.8 for a working load of 3000 kN. All the other data remain the same. Use the length L computed.

11.10 Figure Prob. 11.10 shows a drilled pier. The pier is constructed in homogeneous loose to medium dense sand. The pier details and the properties of the soil are given in the figure. Estimate by Vesic's method the ultimate load bearing capacity of the pier.

11.11 For Prob. 11.10 determine the ultimate base capacity by the O'Neill and Reese method. Compare this value with the one computed in Prob. 11.10.

11.12 Compute the allowable load for the drilled pier given in Fig. 11.10 based on the SPT value. Use Meyerhof's method.

11.13 Compute the ultimate base load of the pier in Fig. Prob. 11.10 by Tomlinson's method.

11.14 A pier is installed in a rocky stratum. Figure Prob. 11.14 gives the details of the pier and the properties of the rock materials. Determine the ultimate frictional load $Q_{f(\max)}$.

11.15 Determine the ultimate base resistance of the drilled pier in Prob. 11.14. All the other data remain the same. What is the allowable load with $F_s = 4$ by taking into account the frictional load Q_f computed in Prob. 11.14 ?

11.16 Determine the ultimate point bearing capacity of the pier given in Prob. 11.14 if the base rests on sound rock with RQD = 100%.

11.17 Determine the uplift capacity of the drilled pier given in Prob. 11.1. Given: $L = 15$ m, $d = 1.25$ m, and $c_u = 25$ kN/m². Neglect the weight of the pile.

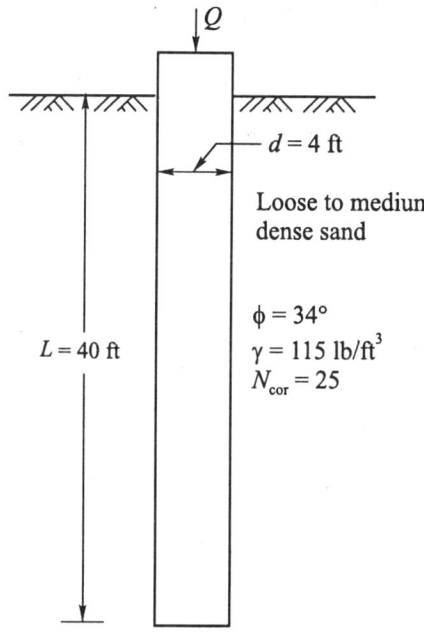

Fig. Prob. 11.10

11.18 The drilled pier given in Fig. Prob. 11.18 is subjected to a lateral load of 120 kips. The soil is homogeneous clay. Given: $d = 60$ in., $EI = 93 \times 10^{10}$ lb-in.², $L = 38$ ft, $e = 12$ in., $c = 2000$ lb ft², and $\gamma_b = 60$ lb/ft³. Determine by the Duncan *et al*, method the groundline deflection.

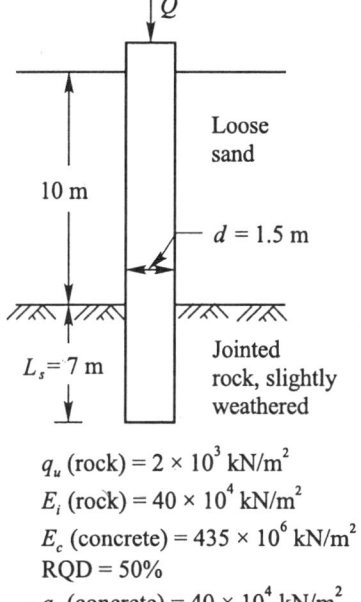

q_u (rock) $= 2 \times 10^3$ kN/m²
E_i (rock) $= 40 \times 10^4$ kN/m²
E_c (concrete) $= 435 \times 10^6$ kN/m²
RQD = 50%
q_u (concrete) $= 40 \times 10^4$ kN/m²
slump = 175 mm, $\gamma_c = 23.5$ kN/m³

Fig. Prob. 11.14

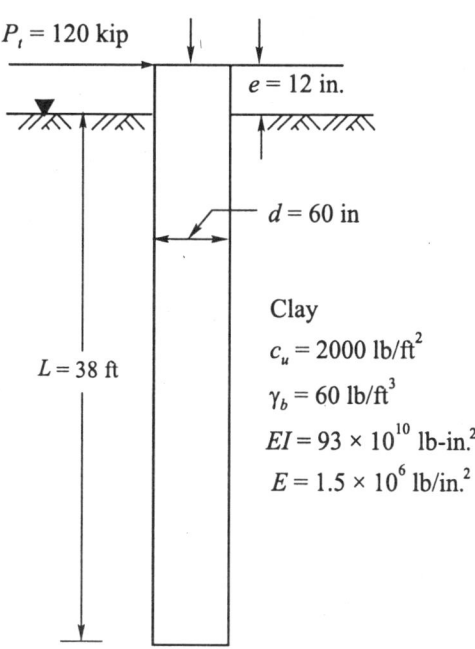

Fig. Prob. 11.18

Deep Foundation 5
Caisson Foundations

12

12.1 INTRODUCTION

Wells, which are also known as caissons, have been in use for foundations of bridges and other important structures since the Roman and Moghul periods in India. Moghuls in particular used wells for the foundations of their monuments, including Taj Mahal which is a standing testimony to the skill of mankind in the earlier days. In modern times, however, one of the earliest use in India is that for an aqueduct for the upper Ganges Canal constructed in the earlier part of the 19th century. With the advent of pneumatic sinking in 1850 AD, and discovery of better materials like reinforced concrete and steel, use of wells as foundations of bridges gained popularity.

Well foundations have been used for most of the major bridges in India. Materials commonly used for construction are reinforced concrete, brick or stone masonry. Use of well or caisson foundations is equally popular in the United States of America and other western countries. The size of caisson used for the San Francisco Oakland Bridge is 29.6 × 60.1 m in section and 74 m depth.

12.2 TYPES OF WELLS OR CAISSONS

There are three types of caissons. They are:

1. Open caissons.
2. Pneumatic caissons.
3. Box caissons.

Open Caissons

The top and bottom of the caisson (Fig. 12.1) is open during construction. They may have any shape in plan as round, oblong, oval, rectangular, etc. They are of cellular construction and the provision of cells reduces the cost of construction. The open-end caisson usually has a cutting edge. The cutting edge is first fabricated at the site and the first segment of the shaft is built on it. The soil inside the shaft is removed by grab buckets and the segment is sunk vertically. Another segment is added to the top and the process of sinking is continued by excavating the soil inside. After the required depth is reached, concrete is placed under water on the open bottom as a seal to a depth that will contain the hydrostatic uplift pressure so as to avoid blowing in of the bottom when the water inside the caisson is pumped out. When the concrete seal is completely cured, the water in the caisson can be pumped out.

Advantages

1. The caisson can be constructed to great depths.

2. The construction cost is relatively low.

Disadvantages

1. The clearing and inspection of bottom of the caisson cannot be done.

2. Concrete seal placed in water will not be satisfactory.

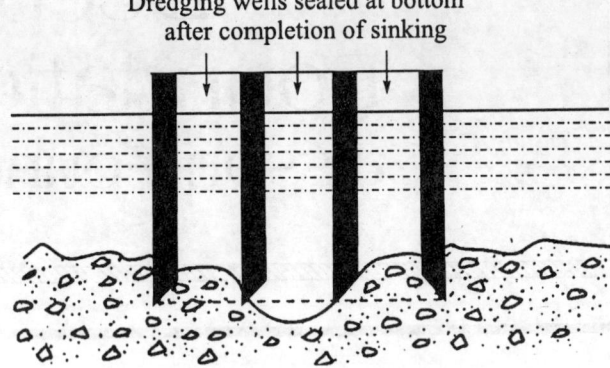

Dredging wells sealed at bottom after completion of sinking

Fig. 12.1 Open caisson

3. The rate of progress will be slowed down if boulders are met during construction.

Pneumatic Caissons

In the case of pneumatic caissons (Fig. 12.2), the working chamber at the bottom of the caisson is kept dry by forcing out water under air pressure. Air locks are provided at the top. The caisson is sunk as the excavation proceeds. Upon reaching its final depth, the working chamber is filled with concrete.

Advantages

1. Control over the work and preparation of foundation for the sinking of caisson are better since the work is done in the dry.

2. The caisson can be sunk vertically as careful supervision is possible.

3. The bottom of the chamber can be sealed effectively with concrete as it can be placed dry.

4. Obstruction to sinking, such as boulders, etc. can be removed easily.

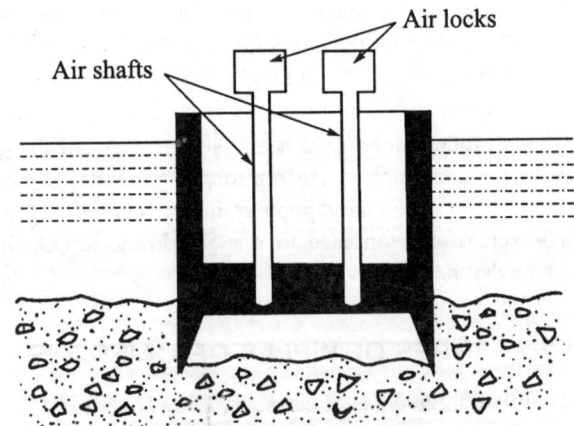

Fig. 12.2 Pneumatic caisson

Disadvantages

1. Construction cost is quite high.

2. The depth of penetration below water is limited to about 35 m (3.5 kg/cm^2). Higher pressures are beyond the endurance of the human body.

Box Caissons

In the case of box caissons (Fig. 12.3) the bottom is closed. This type of caisson is first cast on land and then towed to the site and then sunk on to a previously levelled foundation base. It is sunk by filling inside with sand, gravel, concrete or water. The box type of caisson is also called as *floating caisson*.

Advantages

1. The cost of construction is relatively low.
2. It can be used where the construction of other types of caissons are not possible.

Disadvantages

1. The foundation base shall be prepared in advance of sinking.
2. Deep excavations for seating the caissons at the required depth is very difficult below water level.
3. Due care has to be taken to protect the foundation from scour.
4. The bearing capacity of the base should be assessed in advance.

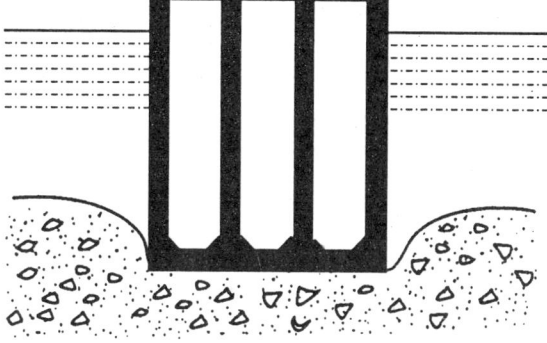

Dredged bed

Fig. 12.3 Box caisson

12.3 STABILITY ANALYSIS OF WELL FOUNDATIONS

Introduction

Two types of soils are considered in the stability analysis of well foundations. They are cohesionless and cohesive soils. This chapter deals with the lateral stability of well foundation only. The vertical bearing capacity of deep foundations dealt with in Chapter 9 is also applicable to well foundations, and as such this aspect of the problem is not considered here.

Statement of the Problem

A well foundation used for a bridge pier shall carry both vertical and lateral loads. Vertical loads comprise of dead and live loads. The dead loads include the weights of superstructure and substructure. The vertical line loads are brought on to the structure due to the passing of vehicles over the bridge. The lateral loads are caused due to braking or traction of vehicles, water current, wind, earth quakes, etc. The lateral forces might act at different points on a pier, but their effect can be simulated by considering an equivalent force acting at bearing level.

Figure 12.4 shows a typical rectangular well foundation with all the external load and the resisting forces acting on the well in cohesionless soil. The external loads are,

W_T = the vertical load at the bearing level of pier which includes loads of superstructure (excluding the pier) and the live loads acting on it,

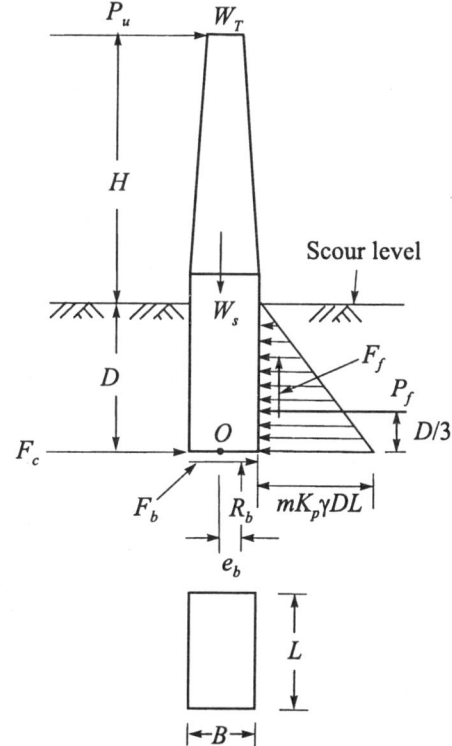

Fig. 12.4 Stability analysis of well foundation in cohesionless soils

W_s = weight of pier and well (considering relief due to buoyancy),

P_u = equivalent lateral load acting at the bearing level at height H above the maximum scour level under ultimate lateral load.

[$P_u = F_l P_t$ where, F_l = load factor, P_t = design lateral load]

The external forces are resisted by the soil surrounding the well. Since the well is a massive one with depth / width ration (D/B) normally not exceeding a value of 3, it is assumed to rotate as a rigid body about a point O lying on the base of the well on the axis passing through well. When the well rotates as a unit passive pressure develops in the front and active pressure at the back, and the active pressure is normally neglected in the analysis since it is quite small compared to the magnitude of the passive pressure. The high lateral pressure that develops at the bottom of the back of the wall is assumed to be resisted by a line load F_c at the bottom of the well.

Two cases are considered:

1. Stability analysis of wells in cohesionless soils.
2. Stability analysis of wells in cohesive soils.

The principle of analysis in both the cases are based on the principles enunciated by Broms (1964) for short piles.

In both the cases it is required to determine the depth of embedment of the well (grip length) under ultimate lateral load conditions. With a suitable factor of safety, the movement or rotation of the well at the bearing level should be within the permissible limits.

The analysis is based on limit equilibrium conditions. Some experimental work has been done by Kapur on model wells (1971) in cohesionless soil on the stability analysis of wells. But there is no way of checking the validity of these methods since field tests are not available to verify his findings. However, some of his findings are briefly mentioned in this chapter. He has also developed equations based on dimensional analysis to compute slopes of wells at working loads in cohesionless soils.

12.4 LIMIT EQUILIBRIUM METHOD OF DETERMINING THE GRIP LENGTH OF WELLS IN COHESIONLESS SOILS

Experimental works of Kapur (1971) on instrumented model wells indicate that the passive pressure distribution on the front face of the well is parabolic, however, under limit equilibrium condition, it is reasonable to assume that the passive pressure increases linearly with depth. It has also been found out that the passive pressure increases linearly with depth. It has also been found out that the passive earth pressure coefficients for wells are greater than those applicable for plane strain conditions.

In order to simplify the procedure, Broms (1964) recommends that the maximum earth pressure which develops at failure may be taken as equal to three times the passive earth pressure calculated by Rankine's earth pressure theory. He has used this approach for the computation of ultimate lateral resistance of piles with D/B ratio greater than 3 (D = depth, B = diameter of piles). For rigid foundations with D/B ratio less than 3, Broms (1987), recommends as follows.

Let, $\bar{K}_P$ = three dimensional passive earth pressure coefficient,

K_P = Rankines passive earth pressure coefficient.

We, may write

$$\bar{K}_P = mK_P \qquad (12.1)$$

The value of m depends according to Broms (1987) on the D/B ratio, and his recommendations are

D/B ratio	Value of m
≤ 1	1.0
$1 < D/B < 3$	D/B
≥ 3	3

The author feels that a value of 1 for m for heavy wells and a maximum of 1.5 for smaller diameter wells might be reasonable.

Figure 12.4 shows a typical rectangular well foundation with all the external loads and resisting forces acting on it

The resisting forces are

P_f = passive earth pressure on the front face of the well,
F_f = vertical frictional force on the front face = $P_f \tan \delta$,
δ = angle of wall friction,
R_b = base reaction acting at an eccentricity at e_b,
F_b = base frictional resistance,
F_c = concentrated lateral force acting at the back of the well at base level.

The maximum passive earth pressure, p_D, at the base of the well per unit depth may be expressed as

$$p_D = m\, K_P \gamma\, DL \tag{12.2}$$

The total passive pressure P_f is

$$P_f = \frac{1}{2}\, \gamma\, Lm\, K_P D^2 \tag{12.3}$$

The total frictional force F_f on the front face is

$$F_f = \frac{1}{2}\, \gamma\, Lm\, K_P \tan \delta\, D^2 \tag{12.4}$$

where, γ = effective unit weight of soil.

Conditions for Statical Equilibrium

The magnitudes of all the resisting forces and their points of application are as given in Fig. 12.4. At limiting state, all the forces acting on the well should satisfy the following conditions of statical equilibrium. They are:

1. The sum of all the vertical forces must be equal to zero, i.e. $\Sigma V_f = 0$.
2. Sum of all the horizontal forces must be equal to zero, $\Sigma H_f = 0$.
3. Sum of he moments of all the forces about point O on the base (or any other point) must be equal to zero, i.e. $\Sigma M_f = 0$.

We may now write

1. $\Sigma V_f = W - R_b - P_f \tan \delta = 0$ \hfill (12.5)

2. $\Sigma H_f = P_u - P_f + F_b + F_c = 0$ \hfill (12.6)

3. $\Sigma M_f = M_{gu} + P_u D + \dfrac{1}{3}\, P_f D - R_b\, e_b = 0$ \hfill (12.7)

From Eq. (12.5), we have

$$R_b = W - P_f \tan \delta, \qquad (12.8)$$

where, $W = W_T + W_s$.

The grip length can be foundout from Eq. (12.7). Substituting in Eq. (12.7) for P_f and R_b and simplifying, we have an equation for D as

$$D^2 = \frac{12 \left(M_{gu} + P_u D - W e_b \right)}{L m K_P \gamma \left[2D + 3 \tan \delta \left(B - 2 e_b \right) \right]} \qquad (12.9a)$$

Equation (12.9a) may be expressed as

$$D^2 = \frac{WD \left(\alpha_1 \alpha_2 - \alpha_3 \right)}{\gamma \, m K_P A \left(\alpha_4 + \alpha_5 \right)} \qquad (12.9b)$$

where, $\quad \alpha_1 = \dfrac{P_u}{W}, \alpha_2 = \dfrac{H}{D} + 1,$

$\qquad \alpha_3 = \dfrac{e_b}{D}, \alpha_4 = \dfrac{1}{6} \dfrac{D}{B},$

$\qquad \alpha_5 = \dfrac{1}{4} \tan \delta \, (B - 2 e_b),$

$\qquad A = L \times B,$

$\qquad m = $ as per Eq. (12.1),

$\qquad \gamma = $ effective unit weight of soil,

$\qquad \delta = $ angle of wall friction,

$\quad M_{gu} = P_u H.$

The depth D from Eq. (12.9) can be obtained either by trial and error method or graphically.

Eccentricity of Base Reaction

Experimental investigations carried out by Kapur (1971) indicate that the eccentricity of base reaction e_b, expressed as a ratio e_b/B increases linearly with the lateral load P. The maximum value of e_b has been found out to be $B/3$, the maximum toe pressure q_t has been found out to be about 3 times the average base pressure q_a. Figures 12.5 (a) and (b) give the findings of Kapur on eccentricity. However, an eccentricity of $B/6$ is suggested for computing the grip length.

Load Factor

When the load factor is 2, that is when $P_u/P_t = 2$ (where P_u the ultimate lateral load to be considered for computing the grip length D, is equal to twice the maximum lateral load P_t that the well is likely to experience), the eccentricity of the reaction at the base as per Fig. 12.5 (a) is $B/6$. The maximum toe pressure at this eccentricity is about $1.5q_a$ and there is no tension at the heel. If a factor of safety of 3 is used to determine the allowable vertical pressure q_a, the factor of safety of the toe pressure at a load factor of 2 is $3/1.5 = 2$, which is greater than 1.5 that is normally allowed under the maximum lateral load P_u. Load factor is not considered here for the vertical dead and live loads as these loads can be assessed fairly accurately. However, the total vertical load W may be multiplied by a suitable load factor if required.

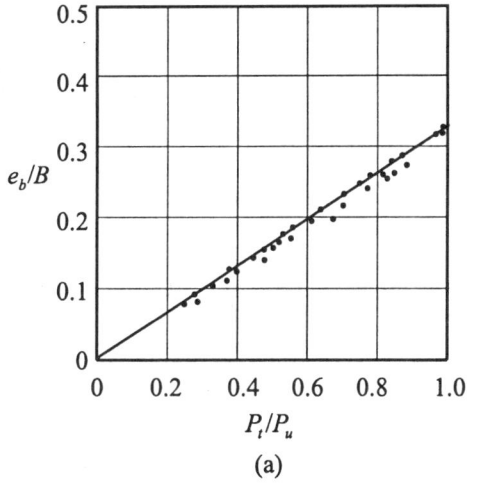

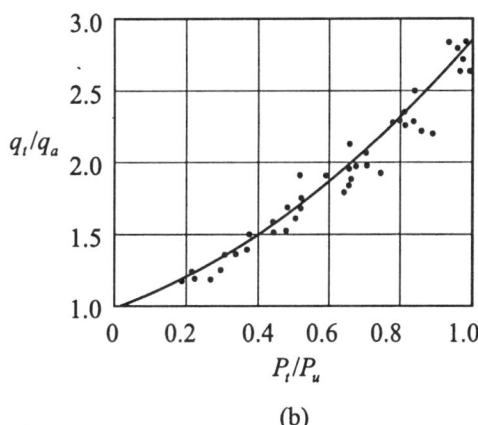

Fig. 12.5 The effect of eccentricity on toe pressure: (a) e_b/B vs. P_t/P_u curves, (b) q_t/q_a vs. P_t/P_u curves (Kapur, 1971)

Shape Factor

The investigation of Kapur (1971) indicates that at equal displacements a square well is about 20 percent more resistant than a circular well of the same cross sectional area. We may therefore write

$$P_{us} = 1.2\, P_{uc} \qquad (12.10)$$

where, P_{us} = ultimate lateral resistance of a square well,

P_{uc} = ultimate lateral resistance of a circular well of the same cross-sectional area.

12.5 GRIP LENGTHS OF WELLS IN COHESIVE SOILS

Figure 12.6 shows the external loads and the resisting forces acting on a well embedded in cohesive soil. Broms (1964) assumes a maximum lateral soil reaction of $9c_u$ per unit area where, c_u is undrained shear strength, which remains constant with depth. He has assumed the soil reaction as zero up to a depth equal to 1.5 times the width of pile which is applicable only to small diameter piles. In the case of well foundations for bridges, the grip length is considered below the maximum scour level and as such there is no necessity to consider zero soil reaction below this depth. The forces that act on a well in cohesive soils are shown in Fig. 12.6.

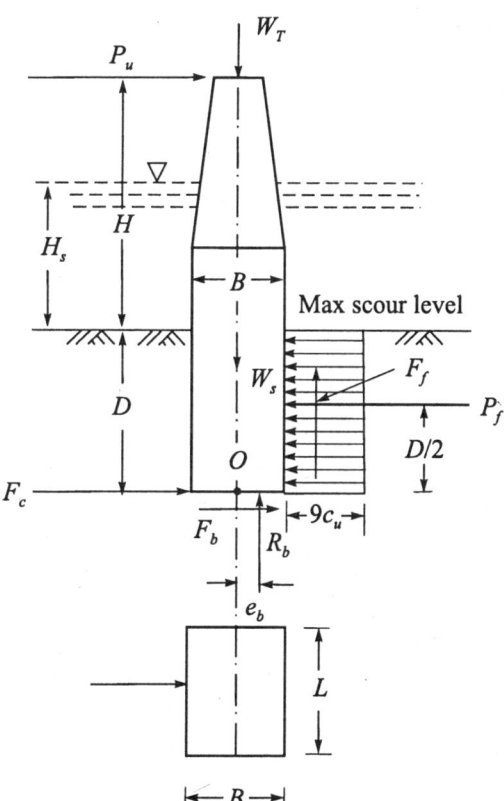

Fig. 12.6 Stability analysis of a well foundation embedded in cohesive soil

We may write the following equations

$$P_f = 9c_u LD \tag{12.11}$$

$$F_f = \alpha c_u \, DL \tag{12.12}$$

$$R_b = W - F_f \tag{12.13}$$

where, c_u = undrained shear strength of soil,

α = adhesions factor.

Taking moments about O we have

$$P_u D + M_{gu} = \frac{1}{2} \times 9c_u LD^2 + \frac{1}{2} \alpha c_u DLB + (W - \alpha c_u DL) e_b \tag{12.14}$$

Simplifying Eq. (12.4), we have

$$D = \frac{2\left(P_u D + M_{gu} - We_b\right)}{c_u L \left[9D + \alpha\left(B - e_b\right)\right]} \tag{12.15}$$

where, $M_{gu} = P_u H$, P_u = the ultimate lateral load.

e_b may be taken as equal to $B/6$ under ultimate lateral load condition. The adhesion factor α may be selected according to consistency of the soil as explained in Chapter 8.

Equation (12.15) can be solved by choosing such value of D which balances both sides of the equation.

12.6 DETERMINATION OF SCOUR DEPTH IN COHESIONLESS SOILS

It has been stated earlier that the grip length D for the well foundation of a bridge pier is the depth below the scour level. Methods or procedures for predicting scour depth vary and one requires sound engineering judgement based on hydraulic and hydrological information, engineering geology and records of performance of adjacent structures for this purpose.

Scour that may occur at a bridge site can be categorised as follows:

1. General scour that would occur in the stream without the bridge.
2. The scour that would occur at the bridge site because of the constriction in waterway caused by the bridge and the approach embankment.
3. The local scour that occurs because of distortion of the flow pattern in the immediate vicinity of the bridge piers and abutments.

General Scour

The depth of scour is normally measured from the high flood level (HFL). No reliable method is available for estimating the depth of scour. The formula that is commonly used for estimating the depth of general scour (which includes the effect of the constriction of water-way also) is the Lacey's formula which is expressed as

$$D_s = 0.473 \left(\frac{Q}{s^2 f}\right)^{1/3} \tag{12.16}$$

where, D_s = Depth of scour measured from *HFL*,

Q = maximum design discharge in cubic metre per sec.

$$s = \frac{4.8\sqrt{Q}}{B_w}$$, a factor which is limited to unity,

B_w = actual width of water spread in metre measured at HFL,

f = Silt factor which depends on grain size.

Local Scour

The local scour occurs at the pier points below the depth of general scour. The formula that is sometimes used for estimating the depth of local scour D_{ls} is

$$D_{ls} = 1.4 \, C_s B_a \tag{12.17}$$

where, B_a = Average width of pier below the *HFL* and above the general scour level,

C_s = Coefficient which depends on the pier shape. A value of 1.0 for cylindrical piers and 1.4 for rectangular pier is normally assumed.

The depth of local scour D_{ls} is sometimes increased by 20 per cent if the discharge in the river exceeds 3000 cubic metre per second.

The total depth of scour D_{ts} be expressed as follows:

1. For discharges upto 3000 cubic metre per second

$$D_{ts} = D_s + D_{ls} \tag{12.18}$$

2. For discharges greater than 3000 cubic metre per second

$$D_{ts} = D_s + 1.2 D_{ls} \tag{12.19}$$

Silt Factor *f*

According to Lacey, the silt factor f depends on the average grain size and density of boundary materials in the river. Assuming an average specific gravity of 2.65 for the river bed material which is normally sand, Lacey gives the following formula for determining the silt factor

$$f = 1.76 \sqrt{d_m} \tag{12.20}$$

where, d_m = average particle size in mm.

Method of Determining Average Particle Size d_m

The average grain size, d_m, of any given sample of soil may be found out by carrying out sieve analysis. A typical example of determining d_m is given in Table 12.1 and 12.2. Table 12.1 gives the percent by weight of the soil particles retained on sieve sizes and Table 12.2 shows the method of determining d_m which is self-explanatory.

Representative samples of the bed materials at different depths up to the normal depth of scour have to be tested and the mean diameter of particle of each sample to be found out. The average of the mean diameters of all the samples tested gives the particle size which has to be used in Eq. (12.20) for computing the silt factor.

Table 12.1 Result of sieve analysis

Sieve designation	Sieve opening in mm	Weight of soil retained gm	Per cent retained
5.60 mm	5.60	0	0
4.00 mm	4.00	0	0
2.80 mm	2.80	16.2	4.05
1.00	1.00	76.5	18.30
425 micron	0.425	79.2	18.85
180 micron	0.180	150.4	35.85
75 micron	0.75	41.0	9.80
Pan	–	55.4	13.2

Table 12.2 Computation of average diameter of particle

Sieve no.	Average sieve size mm	Percentage weight retained	Product of columns 2 and 3
1	*2*	*3*	*4*
4.00 mm to 2.80	3.40	4.05	13.750
2.80 mm to 1.00 mm	1.90	18.30	34.700
1.00 mm to 425 microns	0.712	18.85	13.400
425 microns to 180 microns	0.302	35.85	10.800
180 microns to 75 microns	0.127	9.80	1.250
75 microns and below	0.0375	13.20	0.495
	Total	100.05	74.395

$$\text{Average diameter} = \frac{74.395}{100} = 0.74395 \text{ mm}$$

12.7 THICKNESS OF STEINING OF WELLS

The walls of wells are called as *steining*. The steining may be constructed of cement concrete or brick masonry. The cement concrete steining is normally reinforced to take care of stresses developed during sinking or due to changes in temperature condition. Similarly the brick masonry steining are also reinforced.

The thickness of steining is fixed by taking the following into consideration:

1. It should be possible to sink the well without excessive kentledge.
2. The wells should not get damaged during sinking.
3. If the well develops tilts and shifts during sinking it should be possible to rectify the tilts and shifts without damaging the well.
4. The well should be able to resist safely the earth pressure developed during a sand blow or other conditions like sudden drop that may be experienced during sinking.
5. Stresses at various levels of the steining should be within permissible limits under all load conditions that may be transferred to the well either during sinking or during service.

There is no recognized procedure for computing the thickness of steining. Some of the Codes of Practices on bridges propose empirical methods for computing the thickness. The method that are suggested by the Indian Roads Congress is given below.

Plain Cement Concrete Steining

The thickness, T_s of the steining should not be less than 45 cm nor less than that given by the following equations:

1. For circular or dumbbell shaped wells
   (i) In sandy and silty strata

$$T_s = 1.0 \left(\frac{D_f}{100} + \frac{d_0}{10} \right) \tag{12.21}$$

   (ii) In soft clay strata

$$T_s = 1.1 \left(\frac{D_f}{100} + \frac{d_0}{10} \right) \tag{12.22}$$

   (iii) In hard clay strata

$$T_s = 1.25 \left(\frac{D_f}{100} + \frac{d_0}{10} \right) \tag{12.23}$$

   (iv) In soils where boulders, kankars, shale or laterite or such hard materials are net with

$$T_s = 1.25 \left(\frac{D_f}{100} + \frac{d_0}{8} \right) \tag{12.24}$$

2. For rectangular or double D shaped wells
   (i) In sandy strata

$$T_s = 1.0 \left(\frac{D_f}{100} + \frac{L}{10} \right) \tag{12.25}$$

   (ii) In soft clay strata

$$T_s = 1.1 \left(\frac{D_f}{100} + \frac{L}{10} \right) \tag{12.26}$$

   (iii) In hard clay strata

$$T_s = 1.15 \left(\frac{D_f}{100} + \frac{L}{10} \right) \tag{12.27}$$

   (iv) In soils where boulders, kankars, shale or laterite or such hard materials are met with

$$T_s = 1.20 \left(\frac{D_f}{100} + \frac{L}{10} \right) \tag{12.28}$$

where, D_f = fully designed depth of the well below the bed level of river existing at the time of sinking in cm,

d_0 = outside diameter of the well in cm,

L = length of rectangular well in cm.

The length L depends upon the number of dredging wells provided. For a single dredging well, L is the longest side. If there are more than one dredging wells, the longest side of a dredging well is to be considered.

Brick Masonry Steining

The minimum thickness, T_s of the steining of circular brick masonry wells should not be less than 45 cm and also should not be less than the values given by the following equations.

(*i*) In sandy strata

$$T_s = \left(\frac{D_f}{40} + \frac{d_0}{8}\right)$$ (12.29)

(*ii*) In soft clay strata

$$T_s = 1.1\left(\frac{D_f}{40} + \frac{d_0}{8}\right)$$ (12.30)

(*iii*) In hard clay strata

$$T_s = 1.25\left(\frac{D_f}{40} + \frac{d_0}{8}\right)$$ (12.31)

Distance between Wells

When groups of wells are near each other, special care is needed to ensure that they do not fail in the course of sinking and also do not cause disturbance to wells already sunk. The minimum clearance between the centre to centre of wells should be 11/2 times the external diameter.

12.8 EXAMPLES

Example 12.1

A well 6.4 m external diameter has been sunk for the existing Yamuna Bridge at Agra, India. The well is founded in sandy soil. The following information is available (Kapur, 1971).

Dimensions of well

External diameter of well	= 6.4 m.
Cross sectional area	= 32.2 sq m.
Grip length provided	= 11.3 m.
Height of bearing level above maximum scour level	= 26.6

Design loads

Type of Load	Symbol	Seismic condition	Non-seismic condition
Total vertical load including weight of peir and well (considering buoyancy effect)	W	1930 tonnes	1930 tonnes
Total lateral load at scour level	P_t	85 tonnes	37 tonnes
Total moment at scour level	M_g	2240 tonne-metre	1050 tonne-metre

Soil properties

Relative density $= 60\%$

Submerged unit weight of soil $= 0.962 \ T/m^3$

Angle of internal friction $= 33° \ 42'$

Angle of wall friction $= 27° \ 42'$

Compute the grip length as per the limit state equilibrium method and check up the length provided is adequate or not.

Solution

Compute D for Seismic condition.

Use Eq. (12.9b).

$$D^2 = \frac{WD(\alpha_1\alpha_2 - \alpha_3)}{\gamma \ m \ K_P A(\alpha_4 + \alpha_5)}$$

The well is a circular one. Computation is carried out on a square well of equivalent sectional area. The equivalent width, B, is

$$B = \sqrt{A} = \sqrt{32.2} = 5.67 \ m$$

Computation may be carried out with an assumed eccentricity e_b. This can be taken normally as equal to $B/6$. However, the effect of $e_b = B/3$ on the grip length is also worth studying. The various factors are

$$W = 1930 \ \text{tonnes}$$

With a load factor $F_l = 2$, the ultimate lateral load to be used for determining grip length D for square well is (as per Eq. 12.10)

$$P_u = 1.2 \times 2 \times 85 = 204 \ \text{tonnes}$$

$$\alpha_1 = \frac{P_u}{W} = \frac{204}{1930} = 0.106$$

$$\alpha_2 = \frac{H}{D} + 1 = \frac{26.6}{D} + 1$$

$$\alpha_3 = \frac{e_b}{D}, \ \text{for} \ e_b = \frac{B}{6}, \ \alpha_3 = \frac{5.67}{6} \times \frac{1}{D} = \frac{0.945}{D}$$

$$\text{for} \ e_b = B/3, \ \alpha_3 = \frac{5.67}{3} \times \frac{1}{D} = \frac{1.89}{D}$$

$$\alpha_4 = \frac{1}{6}\frac{D}{B} = \frac{D}{6 \times 5.67} = \frac{D}{34}$$

$$\alpha_5 = \frac{1}{4}\tan\delta\,(B - 2e_b)$$

$$\text{for } e_b = \frac{B}{6}, \alpha_5 = \frac{1}{4}\tan 27.7°\,(5.67 - 2 \times 5.67/6) = 0.44$$

$$\text{for } e_b = \frac{B}{3}, \alpha_5 = 0.248$$

The value of m is assumed as equal to 1 as this happens to be a heavy well. Now substituting all the known factors, we have

$$K_P = \tan^2\,(45° + \phi/2) = \tan^2 61.85° = 3.49$$

$$D^2 = \frac{1930\,D\left[0.106\left(\dfrac{26.6}{D} + 1\right) - C_1\right]}{0.962 \times 1 \times 3.49 \times 32\left(\dfrac{D}{34} + C_2\right)}$$

where,

$$C_1 = \frac{0.945}{D},\; C_2 = 0.44 \text{ for } e_b = \frac{B}{6}$$

$$C_1 = \frac{1.89}{D},\; C_2 = 0.248 \text{ for } e_b = \frac{B}{3}$$

The value of D can be found out by trial and error method for both the cases of e_b. The value of D as obtained are

for $e_b = B/6, D = 8.5$ m

for $e_b = B/3, D = 8$ m.

The increase of eccentricity by 100 percent, decreases the depth D only by 6 percent. Therefore, we can adopt $e_b = B/6$ or $B/3$ without causing significant error in the grip length.

It may also be noted here that an increase in the value of m from 1 to 1.5, decreases the value of D from 8.0 m (for $e_b = B/3$) to 6.5, i.e. a decrease of about 20 percent. If m is increased to 2, the value of D reduces to about 6 m which is a decrease of about 25 percent. This means an increase in the value of m by 100 percent (1 to 2) decreases the value of D by about 25 percent. This helps the designer to choose a proper value for m.

The grip length calculated by assuming $e_b = B/6$ is about 23 percent less than that provided.

Example 12.2

If the well foundation given in Ex. 12.1 is founded in cohesive soil with an undrained cohesive strength $c_u = 5$ tonne/m^2, compute the grip length with all the other details as given in Ex. 12.1 for seismic condition.

Solution

Use Eq. (12.15), assume $e_b = B/6, \alpha = 0.5$.

Now,
$$D = \frac{2\left(P_u D + P_u H - We_b\right)}{c_u B\left[9D + \alpha\left(B - e_b\right)\right]}$$

we have, $P_u = 204$ tonnes,

$\qquad W = 1930$ tonnes.

Substituting all the values known, we have

$$D = \frac{2\left(204D + 204 \times 26.6 - 1930 \times \dfrac{5.67}{6}\right)}{5 \times 5.67\left[9D + 0.6\left(5.67 - \dfrac{5.67}{6}\right)\right]}$$

$$= \frac{2\left(204D + 3602\right)}{28.35\left(9D + 2.835\right)}$$

The solution of this equation gives $D = 6$ m. If the eccentricity. e_b is taken as equal to $B/3$, the grip length $D = 4.5$ m. Possibly it is safer to use $e_b = B/6$.

Foundations on 13 Collapsible and Expansive Soils

13.1 GENERAL CONSIDERATIONS

The structure of soils that experience large loss of strength or great increase in compressibility with comparatively small changes in stress or deformations is said to be *metastable* (Peck *et al*, 1974). Metastable soils include (Peck *et al*, 1974):

1. Extra-sensitive clays such as quick clays.

2. Loose saturated sands susceptible to liquefaction.

3. Unsaturated primarily granular soils in which a loose state is maintained by apparent cohesion, cohesion due to clays at the intergranular contacts or cohesion associated with the accumulation of soluble salts as a binder.

4. Some saprolites either above or below the water table in which a high void ratio has been developed as a result of leaching that has left a network of resistant minerals capable of transmitting stresses around zones in which weaker minerals or voids exist.

Footings on quick clays can be designed by the procedures applicable for clays as explained in Chapter 5. Very loose sands should not be used for support of footings. This chapter deals only with soils under categories 3 and 4 listed above.

There are two types of soils that exhibit volume changes under constant loads with changes in water content. The possibilities are indicated in Fig. 13.1 which represent the result of a pair of tests in a consolidation apparatus on identical undisturbed samples. Curve a represents the e-log p curve for a test started at the natural moisture content and to which no water is permitted access. Curves b and c, on the other hand, correspond to tests on samples to which water is allowed access under all loads until equilibrium is reached. If the resulting e-log p curve, such as curve b, lies entirely below curve a, the soil is said to have collapsed. Under field conditions, at present overburden pressure p_1 and void ratio e_0, the addition of water at the commencement of the tests to sample 1, causes the void ratio to decrease to e_1. The collapsible settlement S_c may be expressed as

$$S_c = \frac{H \Delta e_1}{1 + e_0} \tag{13.1a}$$

where H = the thickness of the stratum in the field.

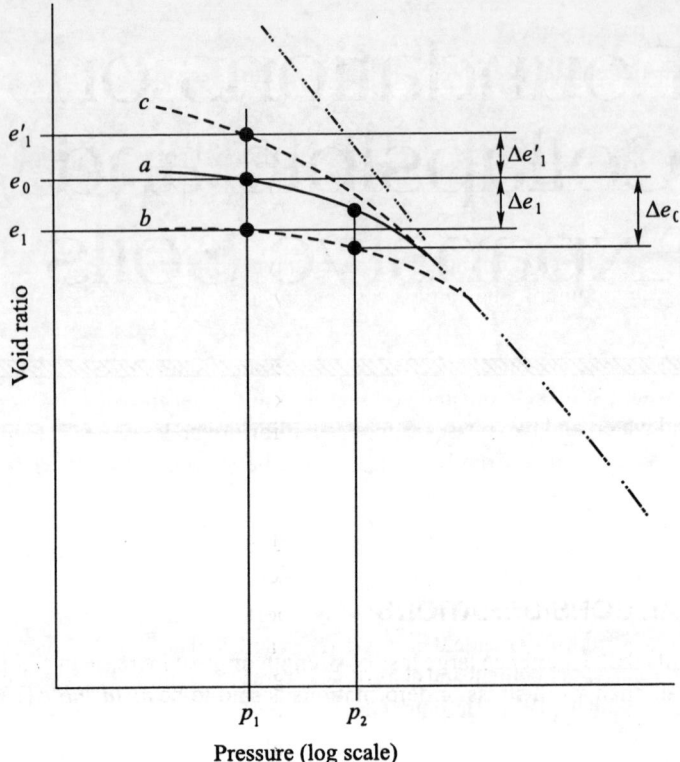

Fig. 13.1 Behaviour of soil in double oedometer or paired confined compression test: (a) Relation between void ratio and total pressure for sample to which no water is added, (b) relation for identical sample to which water is allowed access and which experiences collapse, (c) same as, (b) for sample that exhibits swelling (after Peck *et al*, 1974)

Soils exhibiting this behaviour include true loess, clayey loose sands in which the clay serves merely as a binder, loose sands cemented by soluble salts, and certain residual soils such as those derived from granites under conditions of tropical weathering.

On the other hand, if the addition of water to the second sample leads to curve *c*, located entirely above *a*, the soil is said to have swelled. At a given applied pressure p_1, the void ratio increases to e'_1, and the corresponding rise of the ground is expressed as

$$S_s = \frac{H \Delta e'_1}{1 + e_0} \tag{13.1b}$$

Soils exhibiting this behavior to a marked degree are usually montmorillonitic clays with high plasticity indices.

PART A: COLLAPSIBLE SOILS

13.2 GENERAL OBSERVATIONS

According to Dudley (1970) and Barden *et al* (1973), four factors are needed to produce collapse in a soil structure:

1. An open, partially unstable, unsaturated fabric.
2. A high enough net total stress that will cause the structure to be metastable.
3. A bonding or cementing agent that stabilizes the soil in the unsaturated condition.
4. The addition of water to the soil which causes the bonding or cementing agent to be reduced, and the interaggregate or intergranular contacts to fail in shear, resulting in a reduction in total volume of the soil mass.

Collapsible behaviour of compacted and cohesive soils depends on the percentage of fines, the initial water content, the initial dry density and the energy and the process used in compaction.

Current practice in geotechnical engineering recognizes an unsaturated soil as a four phase material composed of air, water, soil skeleton, and contractile skin. Under the idealization, two phases can flow, that is air and water, and two phases come to equilibrium under imposed loads, that is the soil skeleton and contractile skin. Currently, regarding the behaviour of compacted collapsing soils, geotechnical engineering recognized that

1. Any type of soil compacted at *dry of optimum* conditions and at a low dry density may develop a collapsible fabric or metastable structure (Barden *et al*, 1973).
2. A compacted and metastable soil structure is supported by microforces of shear strength, that is bonds, that are highly dependent upon capillary action. The bonds start losing strength with the increase of the water content and at a critical degree of saturation, the soil structure collapses (Jennings and Knight, 1957; Barden *et al*, 1973).
3. The soil collapse progresses as the degree of saturation increases. There is, however, a critical degree of saturation for a given soil above which negligible collapse will occur regardless of the magnitude of the prewetting overburden pressure (Jennings and Burland, 1962); Houston *et al*, 1989).
4. The collapse of a soil is associated with localized shear failures rather than an overall shear failure of the soil mass.
5. During wetting induced collapse, under a constant vertical load and under K_o-oedometer conditions, a soil specimen undergoes an increase in horizontal stresses.
6. Under a triaxial stress state, the magnitude of volumetric strain resulting from a change in stress state or from wetting, depends on the mean normal total stress and is independent of the principal stress ratio.

The geotechnical engineer needs to be able to identify readily the soils that are likely to collapse and to determine the amount of collapse that may occur. Soils that are likely to collapse are loose fills, altered windblown sands, hillwash of loose consistency, and decomposed granites and acid igneous rocks.

Some soils at their natural water content will support a heavy load but when water is provided they undergo a considerable reduction in volume. The amount of collapse is a function of the relative proportions of each component including degree of saturation, initial void ratio, stress history of the materials, thickness of the collapsible strata and the amount of added load.

Collapsing soils of the loessial type are found in many parts of the world. Loess is found in many parts of the United States, Central Europe, China, Africa, Russia, India, Argentina and elsewhere. Figure 13.2 gives the distribution of collapsible soil in the United States.

Holtz and Hilf (1961) proposed the use of the natural dry density and liquid limit as criteria for predicting collapse. Figure 13.3 shows a plot giving the relationship between liquid limit and dry unit weight of soil, such that soils that plot above the line shown in the figure are susceptible to collapse upon wetting.

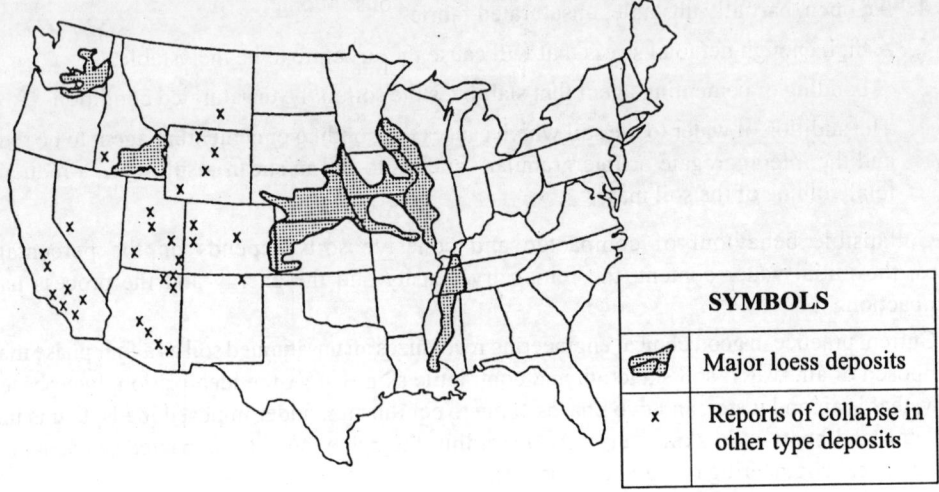

Fig. 13.2 Locations of major loess deposits in the United States along with other sites of reported collapsible soils (After Dudley, 1970)

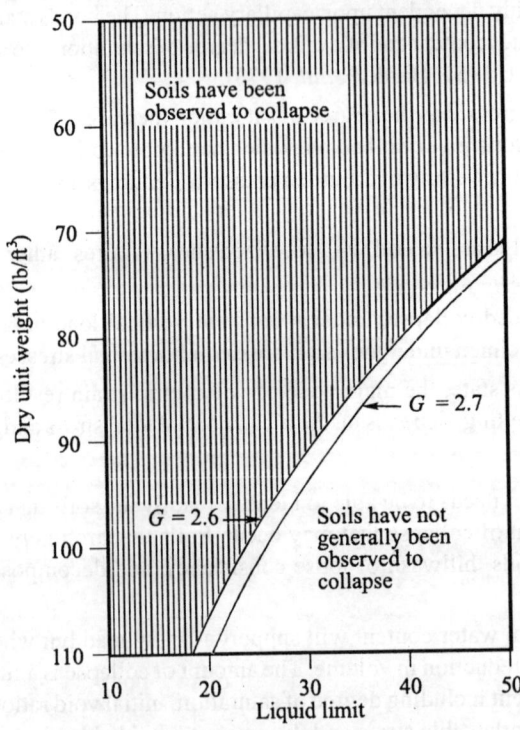

Fig. 13.3 Collapsible and noncollapsible loess (after Holtz and Hilf, 1961)

13.3 COLLAPSE POTENTIAL AND SETTLEMENT

Collapse Potential

A procedure for determining the collapse potential of a soil was suggested by Jennings and Knight (1975). The procedure is as follows:

A sample of an undisturbed soil is cut and fit into a consolidometer ring and loads are applied progressively until about 200 kPa (4 kip/ft^2) is reached. At this pressure the specimen is flooded' with water for saturation and left for 24 hours. The consolidation test is carried on to its maximum loading. The resulting e-log p curve plotted from the data obtained is shown in Fig. 13.4.

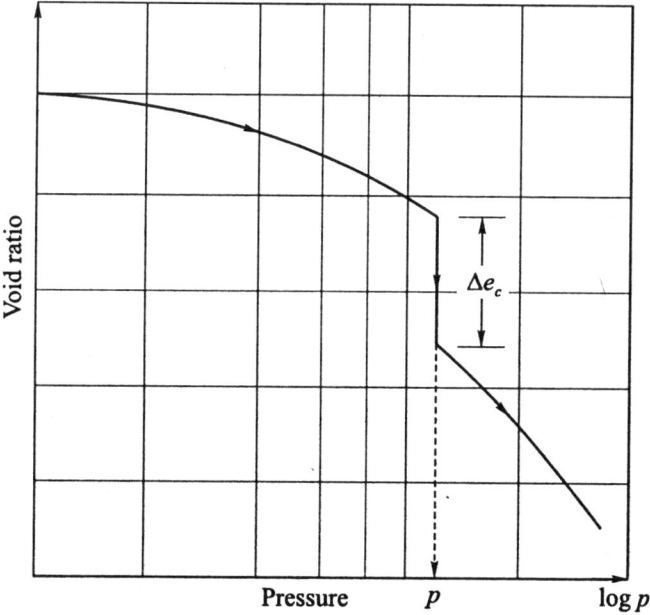

Fig. 13.4 Typical collapse potential test result

The collapse potential C_p is then expressed as

$$C_p = \frac{\Delta e_c}{1 + e_0} \tag{13.2a}$$

in which Δe_c = change in void ratio upon wetting, e_o = natural void ratio.

The collapse potential is also defined as

$$C_p = \frac{\Delta H_c}{H_c} \tag{13.2b}$$

where, ΔH_c = change in the height upon wetting, H_c = initial height.

Jennings and Knight have suggested some value for collapse potential as shown in Table 13.1. These values are only qualitative to indicate the severity of the problem.

13.4 COMPUTATION OF COLLAPSE SETTLEMENT

The double oedometer method was suggested by Jennings and Knight (1975) for determining a quantitative measure of collapse settlement. The method consists of conducting two consolidation tests. Two identical undisturbed soil samples are used in the tests. The procedure is as follows:

1. Insert two identical undisturbed samples into the rings of two oedometers.

Table 13.1 Collapse potential values

$C_p\%$	Severity of problem
0–1	No problem
1–5	Moderate trouble
3–10	Trouble
10–20	Severe trouble
> 20	Very severe trouble

2. Keep both the specimens under a pressure of 1 kN/m² (= 0.15 lb/in²) for a period of 24 hours.
3. After 24 hours, saturate one specimen by flooding and keep the other at its natural moisture content
4. After the completion of 24 hour flooding, continue the consolidation tests for both the samples by doubling the loads. Follow the standard procedure for the consolidation test.
5. Obtain the necessary data from the two tests, and plot e-log p curves for both the samples as shown in Fig. 13.5 for normally consolidated soil.
6. Follow the same procedure for overconsolidated soils and plot the e-log p curves as shown in Fig. 13.6.

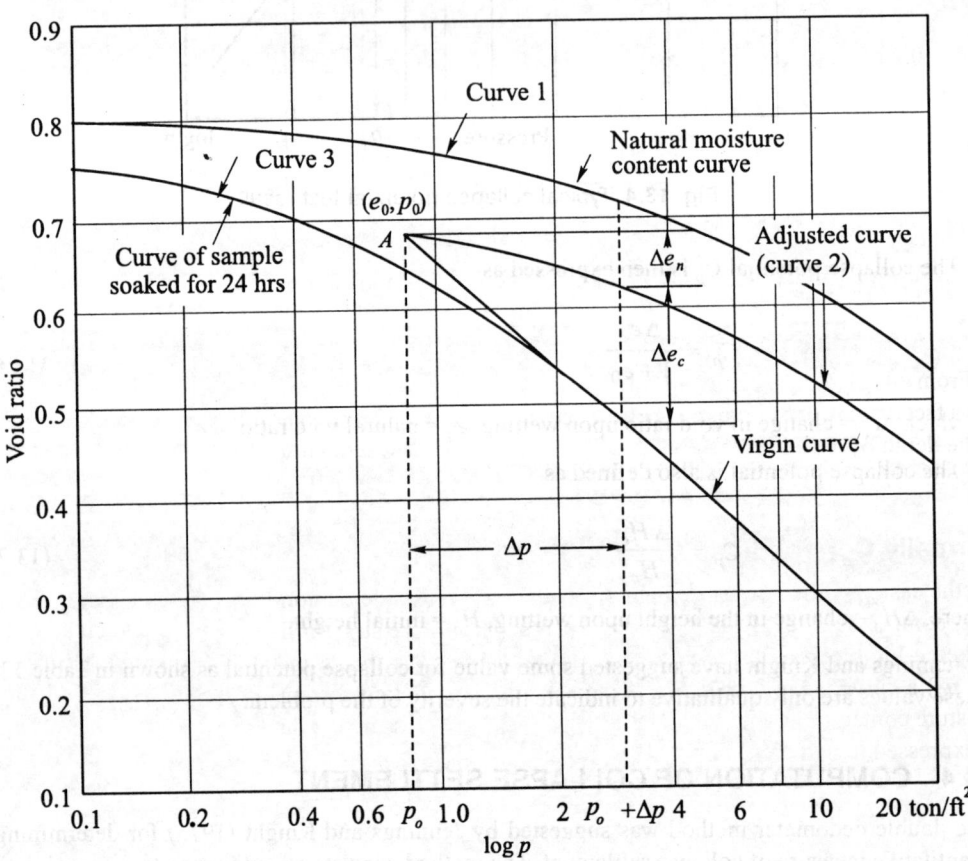

Fig. 13.5 Double consolidation test and adjustments for normally consolidated soil
(Clemence and Finbarr, 1981)

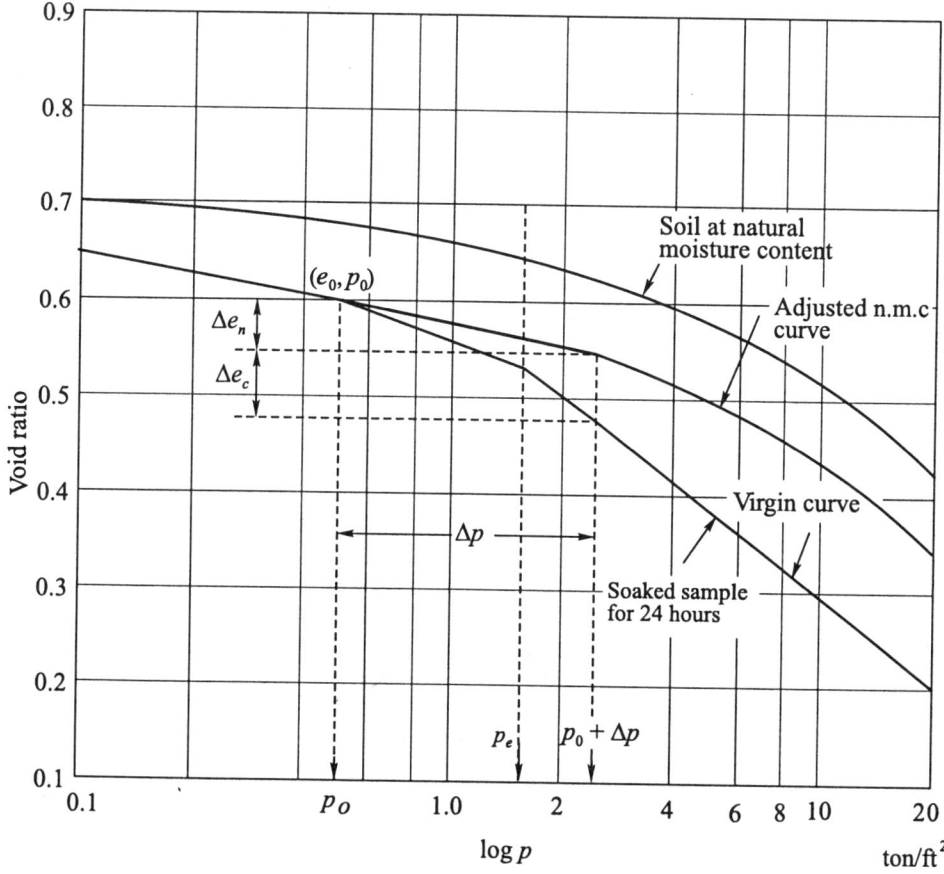

Fig. 13.6 Double consolidation test and adjustments for overconsolidated soil
(Clemence and Finbarr, 1981)

From e-log p plots, obtain the initial void ratios of the two samples after the first 24 hour of loading. It is a fact that the two curves do not have the same initial void ratio. The total overburden pressure p_0 at the depth of the sample is obtained and plotted on the e-log p curves in Figs 13.5 and 13.6. The preconsolidation pressures p_c are found from the soaked curves of Figs 13.5 and 13.6 and plotted.

Normally Consolidated Case

For the case in which p_c/p_0 is about unity, the soil is considered normally consolidated. In such a case, compression takes place along the virgin curve. The straight line which is tangential to the soaked e-log p curve passes through the point (e_0, p_0) as shown in Fig. 13.5. Through the point (e_0, p_0) a curve is drawn parallel to the e-log p curve obtained from the sample tested at natural moisture content. The settlement for any increment in pressure Δp due to the foundation load may be expressed in two parts as

$$S_1 = \frac{\Delta e_n H_c}{1 + e_0} \tag{13.3a}$$

$$S_2 = \frac{\Delta e_c H_c}{1 + e_0} \tag{13.3b}$$

where Δe_n = change in void ratio due to load Δp as per the e-log p curve without change in moisture content,

Δe_c = change in void ratio at the same load Δp with the increase in moisture content (settlement caused due to collapse of the soil structure),

H_c = thickness of soil stratum susceptible to collapse.

From Eqs (13.3a) and (13.3b), the total settlement due to the collapse of the soil structure is

$$S_c = S_1 + S_2 = \frac{H_c}{1 + e_0} (\Delta e_n + \Delta e_c) \tag{13.4}$$

Overconsolidated Case

In the case of an overconsolidated soil the ratio p_c/p_0 is greater than unity. Draw a curve from the point (e_0, p_0) on the soaked soil curve parallel to the curve which represents no change in moisture content during the consolidation stage. For any load $(p_0 + \Delta p) > p_c$, the settlement of the foundation may be determined by making use of the same Eq. (13.4). The changes in void ratios Δe_n and Δe_c are defined in Fig. 13.6.

Example 13.1

A footing of size 10×10 ft is founded at a depth of 5 ft below ground level in collapsible soil of the loessial type. The thickness of the stratum susceptible to collapse is 30 ft. The soil at the site is normally consolidated. In order to determine the collapse settlement, double oedometer tests were conducted on two undisturbed soil samples as per the procedure explained in Section 13.4. The e-log p curves of the two samples are given in Fig. 13.5. The average unit weight of soil $\gamma = 106.6$ lb/ft^3 and the induced stress Δp, at the middle of the stratum due to the foundation pressure, is 4,400 lb/ft^2 ($= 2.20$ t/ft^2). Estimate the collapse settlement of the footing under a soaked condition.

Solution

Double consolidation test results of the soil samples are given in Fig. 13.5. Curve 1 was obtained with natural moisture content. Curve 3 was obtained from the soaked sample after 24 hours. The virgin curve is drawn in the same way as for a normally loaded clay soil.

The effective overburden pressure p_0 at the middle of the collapsible layer is

$$p_0 = 15 \times 106.6 = 1,599 \text{ lb/ft}^2 \text{ or } 0.8 \text{ ton/ft}^2$$

A vertical line is drawn in Fig. 13.5 at $p_0 = 0.8$ ton/ft^2. Point A is the intersection of the vertical line and the virgin curve giving the value of $e_0 = 0.68$. $p_0 + \Delta p = 0.8 + 2.2 = 3.0$ t/ft^2. At $(p_0 + \Delta p)$ $= 3$ ton/ft^2, we have (from Fig. 13.5)

$$\Delta e_n = 0.68 - 0.62 = 0.06$$

$$\Delta e_0 = 0.62 - 0.48 = 0.14$$

From Eq. (13.3)

$$S_1 = \frac{\Delta e_n H_c}{1 + e_0} = \frac{0.06 \times 30 \times 12}{1 + 0.68} = 12.86 \text{ in.}$$

$$S_2 = \frac{\Delta e_c H_c}{1 + e_0} = \frac{0.14 \times 30 \times 12}{1 + 0.68} = 30.00 \text{ in.}$$

Total settlement $S_c = 42.86$ in.

The total settlement would be reduced if the thickness of the collapsible layer is less or the foundation pressure is less.

Example 13.2

Refer to Ex. 13.1. Determine the expected collapse settlement under wetted conditions if the soil stratum below the footing is overconsolidated. Double oedometer test results are given in Fig. 13.6. In this case $p_0 = 0.5$ ton/ft^2, $\Delta p = 2$ ton/ft^2, and $p_c = 1.5$ ton/ft^2.

Solution

The virgin curve for the soaked sample can be determined in the same way as for an overconsolidated clay. Double oedometer test results are given in Fig. 13.6. From this figure:

$$e_0 = 0.6, \Delta e_n = 0.6 - 0.55 = 0.05, \Delta e_c = 0.55 - 0.48 = 0.07$$

As in Ex. 13.1

$$S_1 = \frac{\Delta e_n}{1 + e_0} H_c = \frac{0.05 \times 30 \times 12}{1 + 0.6} = 11.25 \text{ in.}$$

$$S_2 = \frac{\Delta e_c}{1 + e_0} H_c = \frac{0.07 \times 30 \times 12}{1.6} = 15.75 \text{ in.}$$

Total $S_c = 27.00$ in.

13.5 FOUNDATION DESIGN

Foundation design in collapsible soil is a very difficult task. The results from laboratory or field tests can be used to predict the likely settlement that may occur under severe conditions. In many cases, deep foundations, such as piles, piers, etc. may be used to transmit foundation loads to deeper bearing strata below the collapsible soil deposit. In cases where it is feasible to support the structure on shallow foundations in or above the collapsing soils, the use of continuous strip footings may provide a more economical and safer foundation than isolated footings (Clemence and Finbarr, 1981). Differential settlements between columns can be minimized, and a more equitable distribution of stresses may be achieved with the use of strip footing design as shown in Fig. 13.7 (Clemence and Finbarr, 1981).

Load-bearing beams

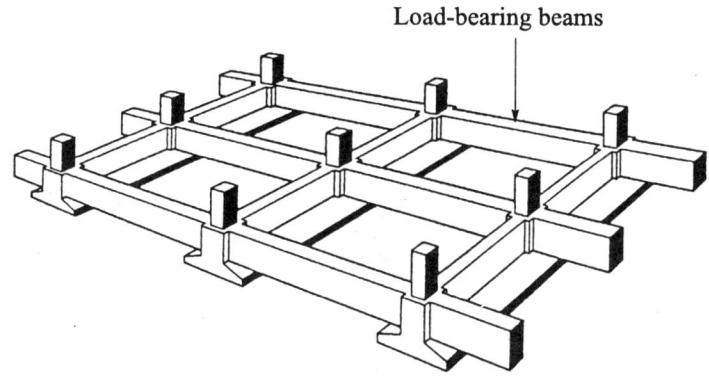

Fig. 13.7 Continuous footing design with load-bearing beams for collapsible soil
(after Clemence and Finbarr, 1981)

13.6   TREATMENT METHODS FOR COLLAPSIBLE SOILS

On some sites, it may be feasible to apply a pretreatment technique either to stabilize the soil or cause collapse of the soil deposit prior to construction of a specific structure. A great variety of treatment methods have been used in the past. Moistening and compaction techniques, with either conventional impact, or vibratory rollers maybe used for shallow depths up to about 1.5 m. For deeper depths, vibroflotation, stone columns, and displacement piles may be tried. Heat treatment to solidify the soil in place has also been used in some countries such as Russia. Chemical stabilization with the use of sodium silicate and injection of carbon dioxide have been suggested (Semkin *et al*, 1986).

Field tests conducted by Rollins *et al*, (1990) indicate that dynamic compaction treatment provides the most effective means of reducing the settlement of collapsible soils to tolerable limits. Prewetting, in combination with dynamic compaction, offers the potential for increasing compaction efficiency and uniformity, while increasing vibration attenuation. Prewetting with a 2 percent solution of sodium silicate provides cementation that reduces the potential for settlement. Prewetting with water was found to be the easiest and least costly treatment, but it proved to be completely ineffective in reducing collapse potential for shallow foundations. Prewetting must be accompanied by preloading, surcharging or excavation to be effective.

PART B: EXPANSIVE SOILS

13.7   DISTRIBUTION OF EXPANSIVE SOILS

The problem of expansive soils is widespread throughout world. The countries that are facing problems with expansive soils are Australia, the United States, Canada, China, Israel, India, and Egypt. The clay mineral that is mostly responsible for expansiveness belongs to the montmorillonite group. Figure 13.8 shows the distribution of the montmorillonite group of minerals in the United States. The major concern with expansive soils exists generally in the western part of the United States. In the

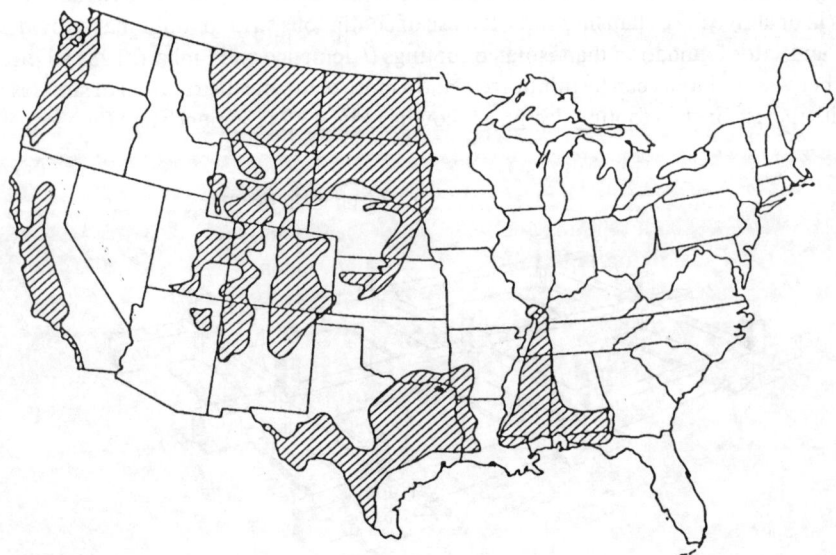

Fig. 13.8 General abundance of montmorillonite in near outcrop bedrock formations in the United States (Chen, 1988)

northern and central United States, the expansive soil problems are primarily related to highly overconsolidated shales. This includes the Dakotas, Montana, Wyoming and Colorado (Chen, 1988). In Minneapolis, the expansive soil problem exists in the Cretaceous deposits along the Mississippi River and a shrinkage/swelling problem exists in the lacustrine deposits in the great Lakes Area. In general, expansive soils are not encountered regularly in the eastern parts of the central United States.

In eastern Oklahoma and Texas, the problems encompass both shrinking and swelling. In the Los Angeles area, the problem is primarily one of desiccated alluvial and colluvial soils. The weathered volcanic material in the Denver formation commonly swells when wetted and is a cause of major engineering problems in the Denver area.

The six major natural hazards are earthquakes, landslides, expansive soils, hurricane, tornado and flood. A study points out that expansive soils tie with hurricane wind/storm surge for second place among America's most destructive natural hazards in terms of dollar losses to buildings. According to the study, it was projected that by the year 2000, losses due to expansive soil would exceed 4.5 billion dollars annually (Chen, 1988).

13.8 GENERAL CHARACTERISTICS OF SWELLING SOILS

Swelling soils, which are clayey soils, are also called expansive soils. When these soils are partially saturated, they increase in volume with the addition of water. They shrink greatly on drying and develop cracks on the surface. These soils possess a high plasticity index. Black cotton soils found in many parts of India belong to this category. Their colour varies from dark grey to black. It is easy to recognize these soils in the field during either dry or wet seasons. Shrinkage cracks are visible on the ground surface during dry seasons. The maximum width of these cracks may be up to 20 mm or more and they travel deep into the ground. A lump of dry black cotton soil requires a hammer to break. During rainy seasons, these soils become very sticky and very difficult to traverse.

Expansive soils are residual soils which are the result of weathering of the parent rock. The depths of these soils in some regions may be up to 6 m or more. Normally the water table is met at great depths in these regions. As such the soils become wet only during rainy seasons and are dry or partially saturated during the dry seasons. In regions which have well-defined, alternately wet and dry seasons, these soils swell and shrink in regular cycles. Since moisture change in the soils bring about severe movements of the mass, any structure built on such soils experiences recurring cracking and progressive damage. If one measures the water content of the expansive soils with respect to depth during dry and wet seasons, the variation is similar to the one shown in Fig. 13.9.

During dry seasons, the natural water content is practically zero on the surface and the volume of the soil reaches the shrinkage limit. The water content increases with depth and reaches a value w_n at a depth D_{us}, beyond which it remains almost constant. During the wet season the water content increases and reaches a maximum at the surface. The water content decreases with depth from a maximum of w_n at the surface to a constant value of w_n at almost the same depth D_{us}. This indicates that the intake of water by the expansive soil into its lattice structure is a maximum at the surface and nil at depth D_{us}. This means that the soil lying within this depth D_{us} is subjected to drying and wetting and hence cause considerable movements in the soil. The movements are considerable close to the ground surface and decrease with depth. The cracks that are developed in the dry seasons close due to lateral movements during the wet seasons.

The zone which lies within the depth D_{us} may be called the *unstable zone* (or active zone) and the one below this the *stable zone*. Structures built within this unstable *zone* are likely to move up and down according to seasons and hence suffer damage if differential movements are considerable.

If a structure is built during the dry season with the foundation lying within the unstable zone, the base of the foundation experiences a *swelling pressure* as the partially saturated soil starts taking in water during the wet season. This swelling pressure is due to constraints offered by the foundation for

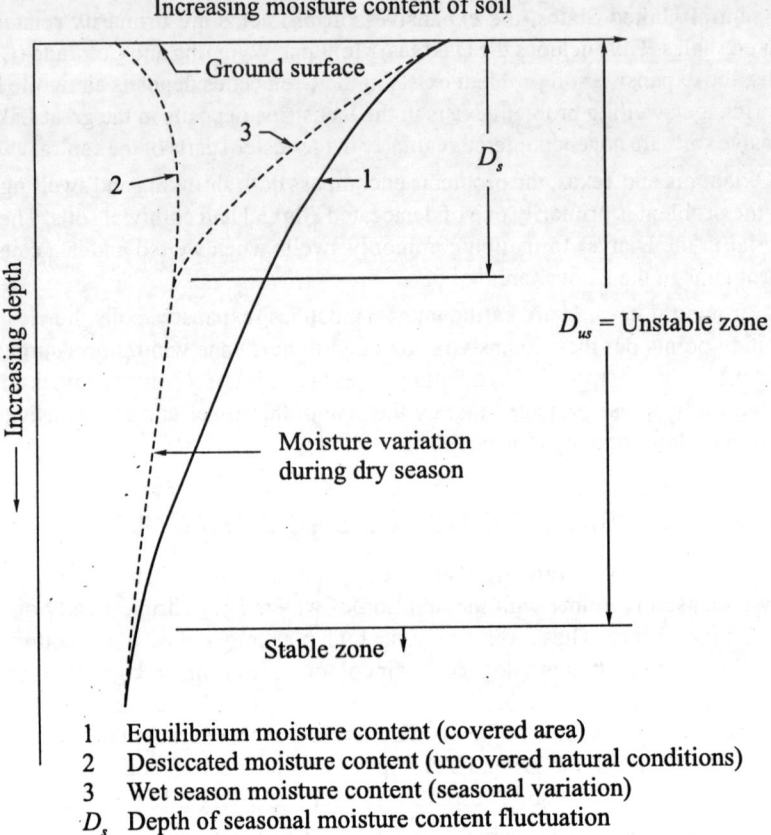

Increasing moisture content of soil

1   Equilibrium moisture content (covered area)
2   Desiccated moisture content (uncovered natural conditions)
3   Wet season moisture content (seasonal variation)
D_s Depth of seasonal moisture content fluctuation
D_{us} Depth of desiccation or unstable zone

Fig. 13.9 Moisture content variation with depth below ground surface (Chen, 1988)

free swelling. The maximum swelling pressure may be as high as 2 MPa (20 tsf). If the imposed bearing pressure on the foundation by the structure is less than the swelling pressure, the structure is likely to be lifted up at least locally which would lead to cracks in the structure. If the imposed bearing pressure is greater than the swelling pressure, there will not be any problem for the structure. If on the other hand, the structure is built during the wet season, it will definitely experience settlement as the dry season approaches, whether the imposed bearing pressure is high or low. However, the imposed bearing pressure during the wet season should be within the allowable bearing pressure of the soil. *The better practice is to construct a structure during the dry season and complete it before the wet season.*

In covered areas below a building there will be very little change in the moisture content except due to lateral migration of water from uncovered areas. The moisture profile is depicted by curve 1 in Fig. 13.9.

13.9   CLAY MINERALOGY AND MECHANISM OF SWELLING

Clays can be divided into three general groups on the basis of their crystalline arrangement. They are:

1. Kaolinite group.
2. Montmorillonite group (also called the smectite group).
3. Illite group.

The kaolinite group of minerals are the most stable of the groups of minerals. The kaolinite mineral is formed by the stacking of the crystalline layers of about 7 Å thick one above the other with the base of the silica sheet bonding to hydroxyls of the gibbsite sheet by hydrogen bonds. Since hydrogen bonds are comparatively strong, the kaolinite crystals consists of many sheet stackings that are difficult to dislodge. The mineral is, therefore, stable and water cannot enter between the sheets to expand the unit cells.

The structural arrangement of the montmorillonite mineral is composed of units made of two silica tetrahedral sheets with a central alumina-octahedral sheet. The silica and gibbsite sheets are combined in such a way that the tips of the tetrahedrons of each silica sheet and one of the hydroxyl layers of the octahedral sheet form a common layer. The atoms common to both the silica and gibbsite layers are oxygen instead of hydroxyls. The thickness of the silica-gibbsite-silica unit is about 10 Å thick structural units. The soils containing a considerable amount of montmorillonite minerals will exhibit high swelling and shrinkage characteristics. The illite group of minerals has the same structural arrangement as the montmorillonite group. The presence of potassium as the bonding materials between units makes the illite minerals swell less.

13.10   DEFINITION OF SOME PARAMETERS

Expansive soils can be classified on the basis of certain inherent characteristics of the soil. It is first necessary to understand certain basis parameters used in the classification.

Swelling Potential

Swelling potential is defined as the percentage of swell of a laterally confined sample in an oedometer test which is soaked under a surcharge load of 7 kPa (1 lb/in^2) after being compacted to maximum dry density at optimum moisture content according to the AASHTO compaction test.

Swelling Pressure

The swelling pressure p_s, is defined as the pressure required for preventing volume expansion in soil in contact with water. It should be noted here that the swelling pressure measured in a laboratory oedometer is different from that in the field. The actual field swelling pressure is always less than the one measured in the laboratory.

Free Swell

Free swell S_f is defined as

$$S_f = \frac{V_f - V_i}{V_i} \times 100 \tag{13.5}$$

where V_i = initial dry volume of poured soil,

V_f = final volume of poured soil.

According to Holtz and Gibbs (1956), 10 cm^3 (V_i) of dry soil passing thorough a No. 40 sieve is poured into a 100 cm^3 graduated cylinder filled with water. The volume of settled soil is measured after 24 hours which gives the value of V_f. Bentonite-clay is supposed to have a free swell value ranging from 1200 to 2000 percent. The free swell value increases with plasticity index. Holtz and Gibbs suggested that soils having a free-swell value as low as 100 percent can cause considerable damage to lightly loaded structures and soils having a free swell value below 50 percent seldom exhibit appreciable volume change even under light loadings.

13.11 EVALUATION OF THE SWELLING POTENTIAL OF EXPANSIVE SOILS BY SINGLE INDEX METHOD (CHEN, 1988)

Simple soil property tests can be used for the evaluation of the swelling potential of expansive soils (Chen, 1988). Such tests are easy to perform and should be used as routine tests in the investigation of building sites in those areas having expansive soil. These tests are

1. Atterberg limits tests
2. Linear shrinkage tests
3. Free swell tests
4. Colloid content tests

Atterberg Limits

Holtz and Gibbs (1956) demonstrated that the plasticity index, I_p, and the liquid limit, w_l, are useful indices for determining the swelling characteristics of most clays. Since the liquid limit and the swelling of clays both depend on the amount of water a clay tries to absorb, it is natural that they are related. The relation between the swelling potential of clays and the plasticity index has been established as given in Table 13.2.

Table 13.2 Relation between swelling potential and plasticity index, I_p

Plasticity index I_p (%)	Swelling potential
0–15	Low
10–35	Medium
20–55	High
55 and above	Very high

Linear Shrinkage

The swell potential is presumed to be related to the opposite property of linear shrinkage measured in a very simple test. Altmeyer (1955) suggested the values given in Table 13.3 as a guide to the determination of potential expansiveness based on shrinkage limits and linear shrinkage.

Table 13.3 Relation between swelling potential, shrinkage limits, and linear shrinkage

Shrinkage limit %	Linear shrinkage %	Degree of expansion
< 10	> 8	Critical
10–12	5–8	Marginal
> 12	0–5	Non-critical

Colloid Content

There is a direct relationship between colloid content and swelling potential as shown in Fig. 13.10 (Chen, 1988). For a given clay type, the amount of swell will increase with the amount of clay present in the soil.

13.12 CLASSIFICATION OF SWELLING SOILS BY INDIRECT MEASUREMENT

By utilising the various parameters as explained in Section 13.11, the swelling potential can be evaluated without resorting to direct measurement (Chen, 1988).

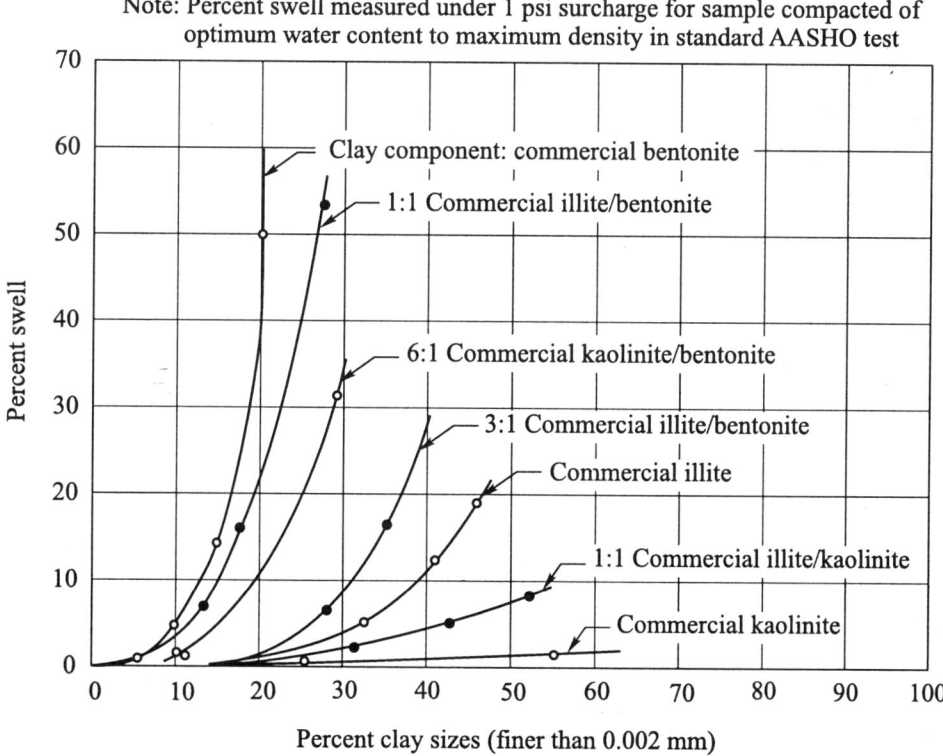

Note: Percent swell measured under 1 psi surcharge for sample compacted of optimum water content to maximum density in standard AASHO test

Fig. 13.10 Relationship between percentage of swell and percentage of clay sizes for experimental soils (after Seed *et al*, 1962)

USBR Method

Holtz and Gibbs (1956) developed this method which is based on the simultaneous consideration of several soil properties. The typical relationships of these properties with swelling potential are shown in Fig. 13.11. Table 13.4 has been prepared based on the curves presented in Fig. 13.11 by Holtz and Gibbs (1956).

The relationship between the swell potential and the plasticity index can be expressed as follows (Chen, 1988)

$$S_p = Be^{A (Ip)} \tag{13.6}$$

where, $A = 0.0838$,
   $B = 0.2558$,
   I_p = plasticity index.

Figure 13.12 shows that with an increase in plasticity index, the increase of swelling potential is much less than predicted by Holtz and Gibbs. The curves given by Chen (1988) are based on thousands of tests performed over a period of 30 years and as such are more realistic.

Active Method

Skempton (1953) defined activity by the following expression

$$A = \frac{I_p}{C} \tag{13.7}$$

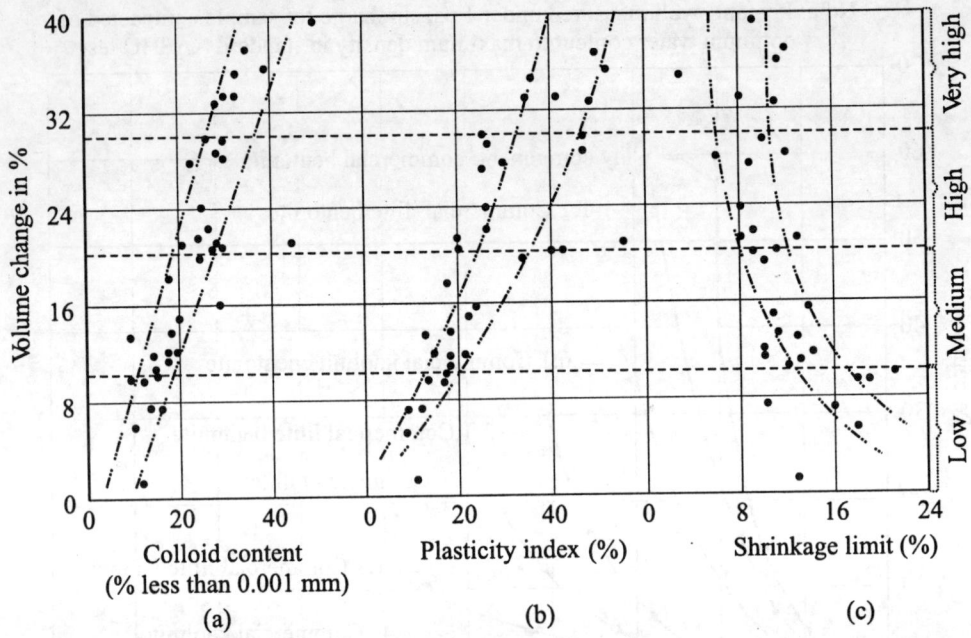

Fig. 13.11 Relation of volume change to (a) colloid content, (b) plasticity index, and (c) shrinkage limit (air-dry to saturated condition under a load of 1 lb per sq in) (Holtz and Gibbs, 1956)

Table 13.4 Data for making estimates of probable volume changes for expansive soils
(*Source:* Chen, 1988)

Data from index tests*			Probable expansion,	
Colloid content, percent finer than 0.001 mm	Plasticity index	Shrinkage limit	percent total vol. change	Degree of expansion
> 28	> 35	< 11	> 30	Very high
20–13	25–41	7–12	20–30	High
13–23	15–28	10–16	10–30	Medium
< 15	< 18	> 15	< 10	Low

*Based on vertical loading of 1.0 psi. (after Holtz and Gibbs, 1956)

where I_p = plasticity index,

C = percentage of clay size finer than 0.002 mm by weight.

The activity method as proposed by Seed, Woodward, and Lundgren, (1962) was based on remolded, artificially prepared soils comprising of mixtures of bentonite, illite, kaolinite and fine sand in different proportions. The activity for the artificially prepared sample was defined as

$$A = \frac{I_p}{C - n}$$

where $n = 5$ for natural soils and, $n = 10$ for artificial mixtures.

The proposed classification chart is shown in Fig. 13.13. This method appears to be an improvement over the USBR method.

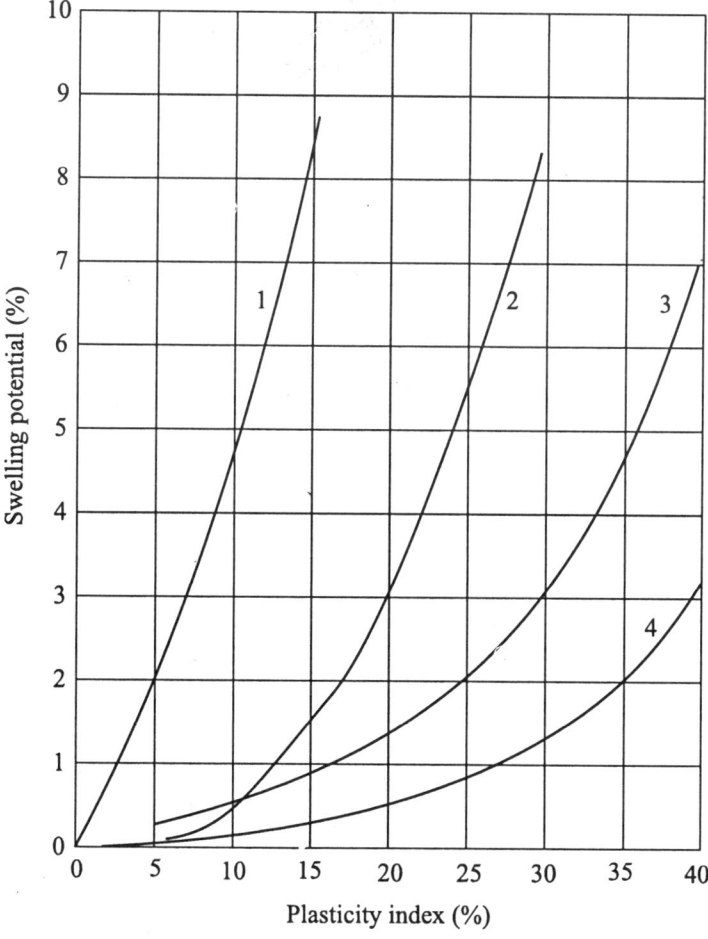

1. Holtz and Gibbs (Surcharge pressure 1 psi)
2. Seed, Woodward and Lundgren (Surcharge pressure 1 psi)
3. Chen (Surcharge pressure 1 psi)
4. Chen (Surcharge pressure 6.94 psi)

Fig. 13.12 Relationships of volume change to plasticity index (*Source:* Chen, 1988)

The Potential Volume Change Method (PVC)

A determination of soil volume change was developed by Lambe under the auspices of the Federal Housing Administration (*Source:* Chen, 1988). Remolded samples were specified. The procedure is as given below.

The sample is first compacted in a fixed ring consolidometer with a compaction effect of 55,000 ft-lb per cu ft. Then an initial pressure of 200 psi is applied, and water added to the sample which is partially restrained by vertical expansion by a proving ring. The proving ring reading is taken at the end of 2 hours. The reading is converted to pressure and is designated as the *swell index*. From Fig. 13.14, the swell index can be converted to potential volume change. Lambe established the categories of PVC rating as shown in Table 13.5.

The PVC method has been widely used by the Federal Housing Administration as well as the Colorado State Highway Department (Chen, 1988).

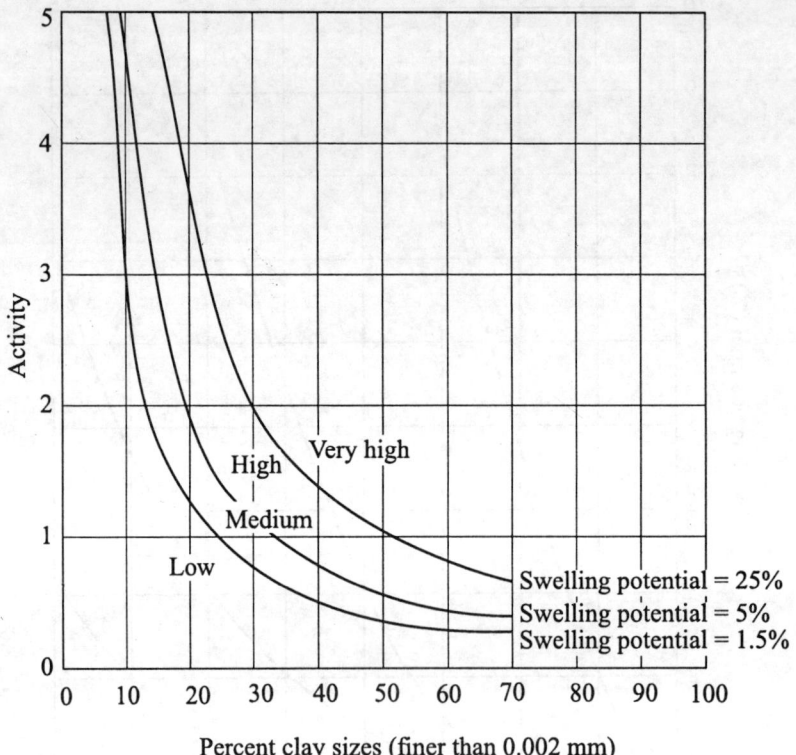

Fig. 13.13 Classification chart for swelling potential (after Seed, Woodward, and Lundgren, 1962)

Figure 13.15 (a) shows a soil volume change metre (ELE International Inc). This metre measures both shrinkage and swelling of soils, ideal for measuring swelling of clay soils, and fast and easy to operate.

Expansion Index (EI)—Chen (1988)

The ASTM Committee on Soil and Rock suggested the use of an *Expansion Index* (EI) as a unified method to measure the characteristics of swelling soils. It is claimed that the EI is a basic index property of soil such as the liquid limit, the plastic limit and the plasticity index of the soil.

The sample is sieved through a No 4 sieve. Water is added so that the degree of saturation is between 49 and 51 percent. The sample is then compacted into a 4 inch diameter mold in two layers to give a total compacted depth of approximates 2 inches. Each layer is compacted by 15 blows of 5.5 lb hammer dropping from a height of 12 inches. The prepared specimen is allowed to consolidate under 1 lb/in^2 pressure for a period of 10 minutes, then inundated with water until the rate of expansion ceases.

The expansion index is expressed as

$$EI = \frac{\Delta h}{h_i} \times 1000 \qquad (13.9)$$

where Δh = change in thickness of sample, in.,

       h_i = initial thickness of sample, in.

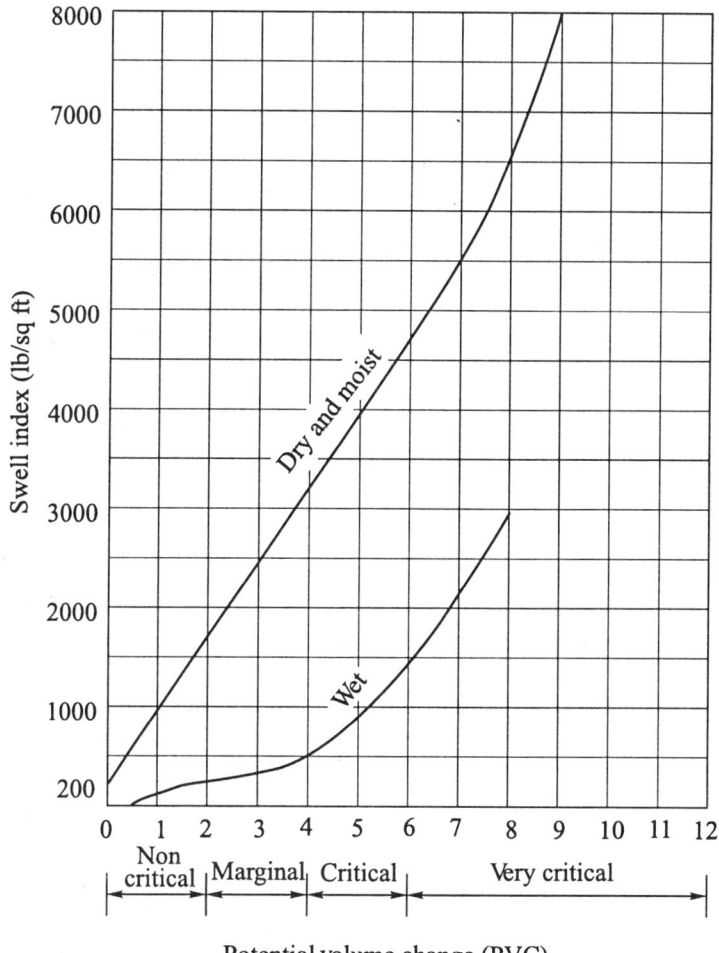

Fig. 13.14 Swell index *vs.* potential volume change (from 'FHA soil PVC meter publication,' Federal Housing Administration Publication no. 701) (*Source:* Chen, 1988)

Table 13.5 Potential volume change rating (PVC)

PVC rating	Category
Less than 2	Non-critical
2 – 4	Marginal
4 – 6	Critical
> 6	Very critical

(*Source:* Chen, 1988)

The classification of a potentially expansive soil is based on Table 13.6.

This method offers a simple testing procedure for comparing expansive soil characteristics.

Figure 13.15 (b) shows an ASTMD-829 expansion index test apparatus (ELE International Inc). This is a completely self-contained apparatus designed for use on determining the expansion index of soils.

(a)   (b)

Fig. 13.15 (a) Soil volume change metre, (b) expansion index test apparatus
(*Courtesy:* Soiltest)

Table 13.6 Classification of potentially expansive soil

Expansion Index, EI	Expansion potential
0–20	Very low
21–50	Low
51–90	Medium
91–130	High
> 130	Very high

(*Source:* Chen, 1988)

Swell Index

Vijayvergiya and Gazzaly (1973) suggested a simple way of identifying the swell potential of clays, based on the concept of the swell index. They defined the swell index, I_s, as follows

$$I_s = \frac{w_n}{w_l} \tag{13.10}$$

where w_n = natural moisture content in percent,

       w_l = liquid limit in percent.

The relationship between I_s and swell potential for a wide range of liquid limit is shown in Fig. 13.16. Swell index is widely used for the design of post-tensioned slabs on expansive soils.

Prediction of Swelling Potential

Plasticity index and shrinkage limit can be used to indicate the swelling characteristics of expansive soils. According to Seed *et al*, (1962), the swelling potential is given as a function of the plasticity index by the formula

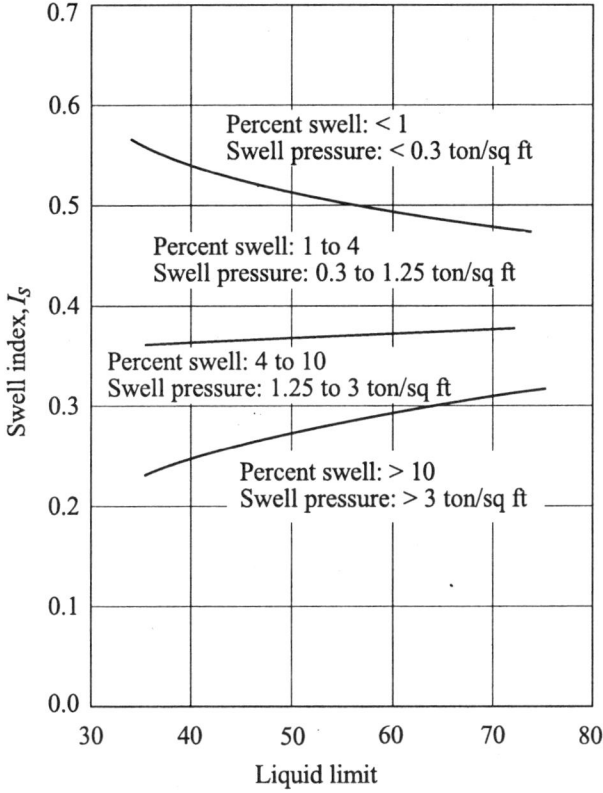

Fig. 13.16 Relationship between swell index and liquid limit for expansive clays
(*Source:* Chen, 1988)

$$S_p = 60k \, I_p^{2.44} \tag{13.11}$$

where S_p = swelling potential in percent,

       I_p = plasticity index in percent,

       $k = 3.6 \times 10^{-5}$, a factor for clay content between 8 and 65 percent.

13.13 SWELLING PRESSURE BY DIRECT MEASUREMENT

ASTM defines swelling pressure which prevents the specimen from swelling or that pressure which is required to return the specimen to its original state (void ratio, height) after swelling. Essentially, the methods of measuring swelling pressure can be either stress controlled or strain controlled (Chen, 1988).

In the stress controlled method, the conventional oedometer is used. The samples are placed in the consolidation ring trimmed to a height of 0.75 to 1 inch. The samples are subjected to a vertical pressure ranging from 500 psf to 2000 psf depending upon the expected field conditions. On the completion of consolidation, water is added to the sample. When the swelling of the sample has ceased the vertical stress is increased in increments until it has been compressed to its original height. The stress required to compress the sample to its original height is commonly termed the *zero volume change swelling pressure*. A typical consolidation curve is shown in Fig. 13.17.

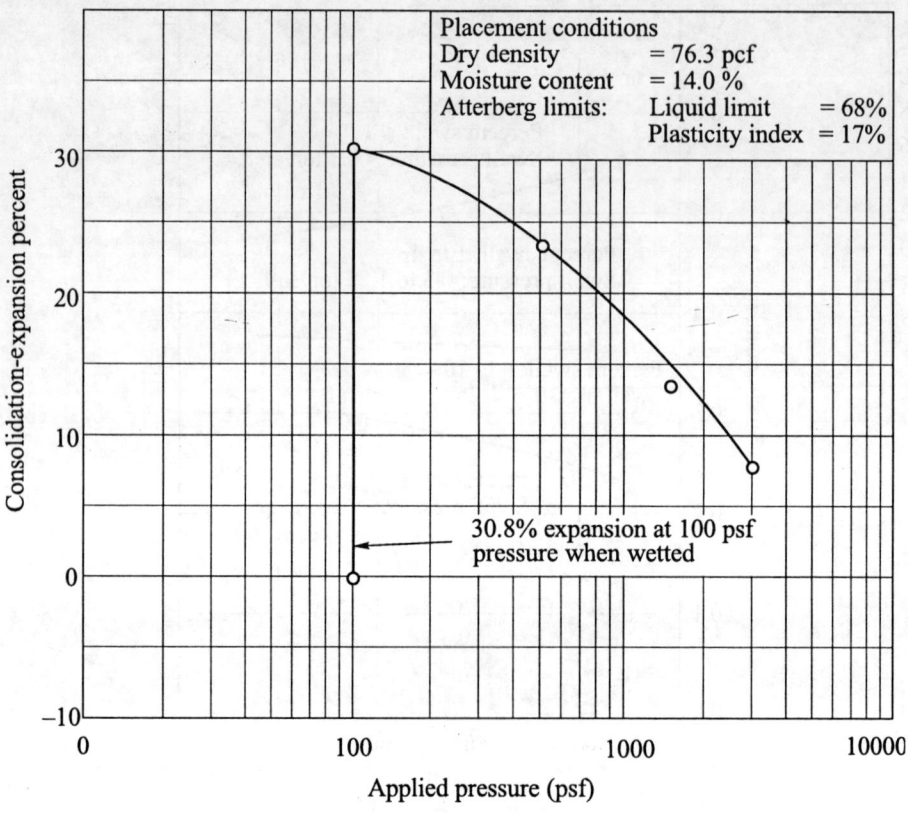

Fig. 13.17 Typical stress controlled swell-consolidation curve

Prediction of Swelling Pressure

Komornik *et al*, (1969) have given an equation for predicting swelling pressure as

$$\log p_s = 2.132 + 0.0208w_l + 0.00065\gamma_d - 0.0269w_n \tag{13.12}$$

where p_s = swelling pressure in kg/cm^2,

w_l = liquid limit (%),

w_n = natural moisture content (%),

γ_d = dry density of soil in kg/cm^3.

13.14 EFFECT OF INITIAL MOISTURE CONTENT AND INITIAL DRY DENSITY ON SWELLING PRESSURE

The capability of swelling decreases with an increase of the initial water content of a given soil because its capacity to absorb water decreases with the increase of its degree of saturation. It was found from swelling tests on black cotton soil samples, that the initial water content has a small effect on swelling pressure until it reaches the shrinkage limit, then its effect increases (Abouleid, 1982). This is depicted in Fig. 13.18 (a).

The effect of initial dry density on the swelling percent and the swelling pressure increases with an increase of the dry density because the dense soil contains more clay particles in a unit volume and consequently greater movement will occur in a dense soil than in a loose soil upon wetting (Abouleid, 1982). The effect of initial dry density on swelling pressure is shown in Fig. 13.18 (b).

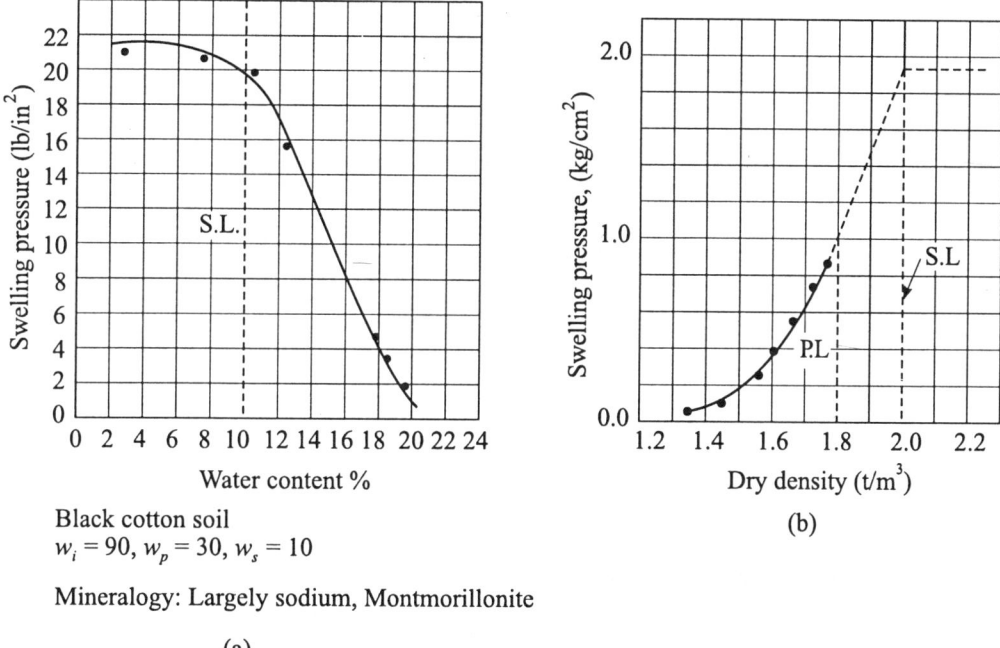

Fig. 13.18 (a) Effect of initial water content on swelling pressure of black cotton soil, (b) effect of initial dry density on swelling pressure of black cotton soil (*Source:* Abouleid, 1982)

13.15 ESTIMATING THE MAGNITUDE OF SWELLING

When footings are built in expansive soil, they experience lifting due to the swelling or heaving of the soil. The amount of total heave and the rate of heave of the expansive soil on which a structure is founded are very complex. The heave estimate depends on many factors which cannot be readily determined. Some of the major factors that contribute to heaving are:

1. Climatic conditions involving precipitation, evaporation, and transpiration affect the moisture in the soil. The depth and degree of desiccation affect the amount of swell in a given soil horizon.

2. The thickness of the expansive soil stratum is another factor. The thickness of the stratum is controlled by the depth to the water table.

3. The depth to the water table is responsible for the change in moisture of the expansive soil lying above the water table. No swelling of soil takes place when it lies below the water table.

4. The predicted amount of heave depends on the nature and degree of desiccation of the soil immediately after construction of a foundation.

5. The single most important element controlling the swelling pressure as well as the swell potential is the in-situ density of the soi. On the completion of excavation, the stress condition in the soil mass undergoes changes, such as the release of stresses due to elastic rebound of the soil. If construction proceeds without delay, the structural load compensates for the stress release.

6. The permeability of the soil determines the rate of ingress of water into the soil either by gravitational flow or diffusion, and this in turn determines the rate of heave.

Various methods have been proposed to predict the amount of total heave under a given structural load. The following methods, however, are described here.

1. The Department of Navy method (1982).
2. The South African method [also known as the Van Der Merwe method (1964)].

The Department of Navy Method
Procedure for estimating total swell under structural load

1. Obtain representative undisturbed samples of soil below the foundation level at intervals of depth. The samples are to be obtained during the dry season when the moisture contents are at their lowest.
2. Load specimens (at natural moisture content) in a consolidometer under a pressure equal to the ultimate value of the overburden plus the weight of the structure. Add water to saturate the specimen. Measure the swell.
3. Compute the final swell in terms of percent of original sample height.
4. Plot swell *vs.* depth.
5. Compute the total swell which is equal to the area under the percent swell *vs.* depth curve.

Procedure for estimating undercut

The procedure for estimating undercut to reduce swell to an allowable value is as follows:

1. From the percent swell *vs.* depth curve, plot the relationship of total swell *vs.* depth at that height. Total swell at any depth equals area under the curve integrated upward from the depth of zero swell.
2. For a given allowable value of swell, read the amount of undercut necessary from the total swell *vs.* depth curve.

Van Der Merwe method (1964)

Probably the nearest practical approach to the problem of estimating swell is that of Van Der Merwe. This method starts by classifying the swell potential of soil into very high to low categories as shown in Fig. 13.19. Then assign potential expansion (PE) expressed in in./ft of thickness based on Table 13.7.

Procedure for estimating swell

1. Assume the thickness of an expansive soil layer or the lowest level of ground water.
2. Divide this thickness (z) into several soil layers with variable swell potential.
3. The total expansion is expressed as

$$\Delta H_e = \sum_{i=1}^{i=n} \Delta_i \tag{13.13}$$

where ΔH_e = total expansion (in.),

$$\Delta_i = (PE)_i \, (\Delta D)_i \, (F)_i, \tag{13.14}$$

$$(F)_i = \log^{-1}\left(-\frac{D_i}{20}\right) = \text{reduction factor for layer } i,$$

z = total thickness of expansive soil layer (ft),

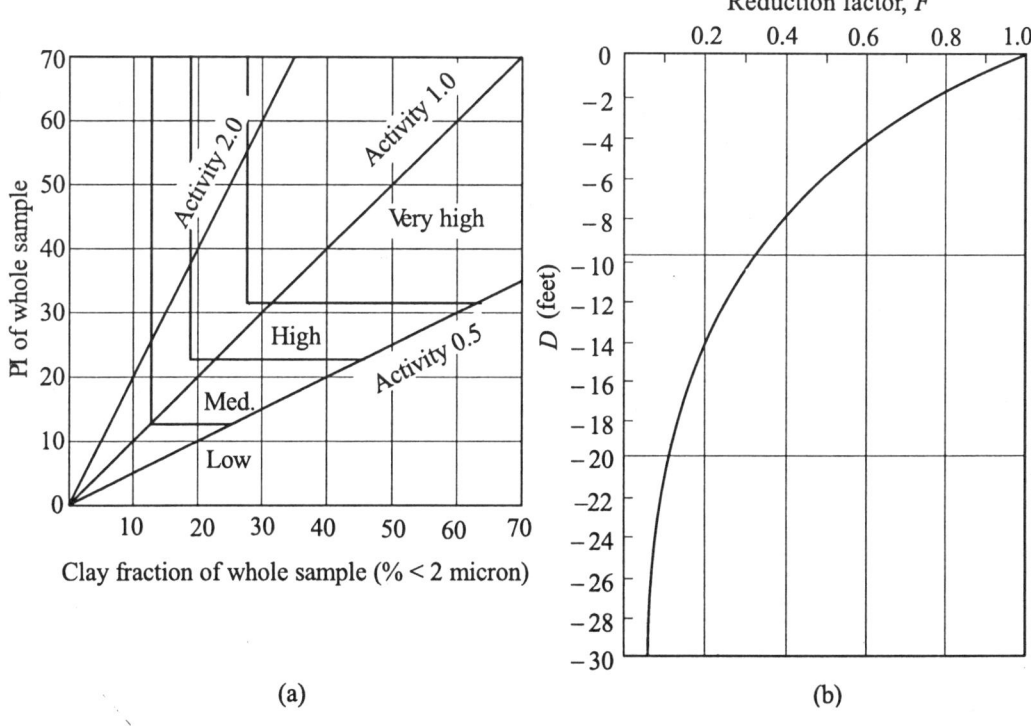

Fig. 13.19 Relationships for using Van Der Merwe's prediction method: (a) Potential expansiveness, (b) reduction factor (Van der Merwe, 1964)

Table 13.7 Potential expansion

Swell potential	Potential expansion (PE) in./ft
Very high	1
High	1/2
Medium	1/4
Low	0

D_i = depth to midpoint of ith layer (ft),

$(\Delta D)_i$ = thickness of ith layer (ft).

Figure 13.19 (b) gives the reduction factor plotted against depth.

13.16 DESIGN OF FOUNDATIONS IN SWELLING SOILS

It is necessary to note that all parts of a building will not equally be affected by the swelling potential of the soil. Beneath the center of a building where the soil is protected from sun and rain the moisture changes are small and the soil movements the least. Beneath outside walls, the movement are greater. Damage to buildings is greatest on the outside walls due to soil movements.

Three general types of foundations can be considered in expansive soils. They are:

1. Structures that can be kept isolated from the swelling effects of the soils.
2. Designing of foundations that will remain undamaged in spite of swelling.

3. Elimination of swelling potential of soil.

All three methods are in use either singly or in combination, but the first is by far the most widespread. Figure 13.20 show a typical type of foundation under an outside wall. The granular fill provided around the shallow foundation mitigates the effects of expansion of the soils.

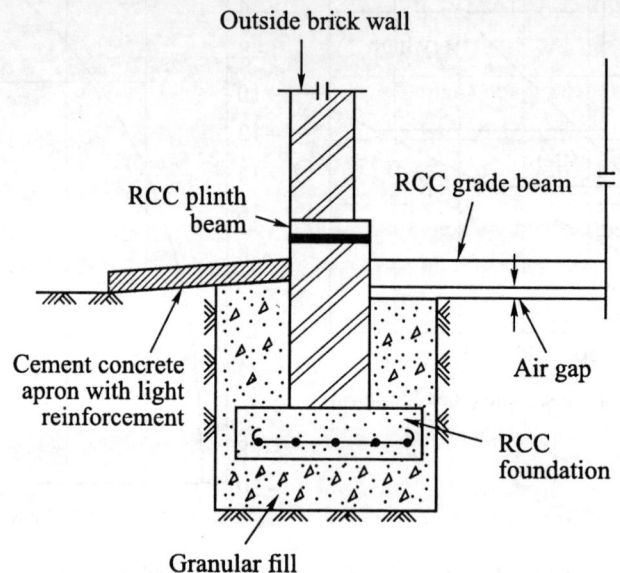

Fig. 13.20 Foundation in expansive soil

13.17 DRILLED PIER FOUNDATIONS IN EXPANSIVE SOILS

Drilled piers are commonly used to resist uplift forces caused by the swelling of soils. Drilled piers, when made with an enlarged base, are called, *belled piers* and when made without an enlarged base are referred to as *straight-shaft piers*.

Woodward, *et al*, (1972) commented on the empirical design of piers: "Many piers, particularly where rock bearing is used, have been designed using strictly empirical considerations which are derived from regional experience". They further stated that "when surface conditions are well established and are relatively uniform, and the performance of past constructions well documented, the design by experience approach is usually found to be satisfactory."

The principle of drilled piers is to provide a relatively inexpensive way of transferring the structural loads down to stable material or to a stable zone where moisture changes are improbable. There should be no direct contact between the soil and the structure with the exception of the soils supporting the piers.

Straight-Shaft Piers in Expansive Soils

Figure 13.21 (a) shows a straight-shaft drilled pier embedded in expansive soil. The following notations are used.

L_1 = length of shaft in the unstable zone (active zone) affected by wetting.

L_2 = length of shaft in the stable zone unaffected by wetting.

d = diameter of shaft.

Q = structural dead load = qA_b.

q = unit dead load pressure.

A_b = base area of pier.

When the soil in the unstable zone takes water during the wet season, the soil tries to expand which is partially or wholly prevented by the rough surface of the pile shaft of length L_1. As a result there will be an upward force developed on the surface of the shaft which tries to pull the pile out of its position. The upward force can be resisted in the following ways.

1. The downward dead load Q acting on the pier top.
2. The resisting force provided by the shaft length L_2 embedded in the stable zone.

Two approaches for solving this problem may be considered. They are:

1. The method suggested by Chen (1988).
2. The O'Neill (1988) method with belled pier.

Two cases may be considered. They are:

1. The stability of the pier when no downward load Q is acting on the top. For this condition a factor of safety of 1.2 is normally found sufficient.
2. The stability of the pier when Q is acting on the top. For this a value of $F_s = 2.0$ is used.

Equations for Uplift Force Q_{up}

Chen (1988) suggested the following equation for estimating the uplift force Q_{up}

$$Q_{up} = \pi \, d \, \alpha_u \, p_s \, L_1 \qquad (13.15)$$

where d = diameter of pier shaft,

α_u = coefficient of uplift between concrete and soil = 0.15,

p_s = swelling pressure,

= 10,000 psf (480 kN/m^2) for soil with high degree of expansion

= 5,000 (240 kN/m^2) for soil with medium degree of expansion

The depth (L_1) of the unstable zone (wetting zone) varies with the environmental conditions. According to Chen (1988) the wettingzone is limited only to the upper 5 feet of the pier. It is possible for the wetting zone to extend beyond 10–15 feet in some countries and limiting the depth of unstable zone to a such a low value of 5 ft may lead to unsafe conditions for the stability of structures. However, it is for designers to decide this depth L_1 according to local conditions. With regards to swelling pressure p_s, it is unrealistic to fix any definite value of 10,000 or 5,000 psf for all types of expansive soils under all conditions of wetting. It is also not definitely known if the results obtained from laboratory tests truly represent the in situ swelling pressure. Possibly one way of overcoming this complex problem is to relate the uplift resistance to undrained cohesive strength of soil just as in the case of friction piers under compressive loading. Eq. (13.15) may be written as

$$Q_{up} = \pi \, d \, \alpha_s \, c_u \, L_1 \qquad (13.16)$$

where α_s = adhesion factor between concrete and soil under a swelling condition,

c_u = unit cohesion under undrained conditions,

It is possible that the value of α_s may be equal to 1.0 or more according to the swelling type and environmental conditions of the soil. Local experience will help to determine the value of α_s This approach is simple and pragmatic.

Resisting Force

The length of pier embedded in the stable zone should be sufficient to keep the pier being pulled out of the ground with a suitable factor of safety. If L_2 is the length of the pier in the stable zone, the resisting force Q_R is the frictional resistance offered by the surface of the pier within the stable zone. We may write

$$Q_R = \pi\, d\, L_2\, \alpha\, c_u \tag{13.17}$$

where α = adhesion factor under compression loading,

c_u = undrained unit cohesion of soil.

The value of α may be obtained from Fig. 11.15.

Two cases of stability may be considered:

1. Without taking into account the dead load Q acting on the pier top, and using $F_s = 1.2$

$$Q_{up} = \frac{Q_R}{1.2} \tag{13.18a}$$

2. By taking into account the dead load Q and using $F_s = 2.0$

$$(Q_{up} - Q) = \frac{Q_R}{2.0} \tag{13.18b}$$

For a given shaft diameter d Eqs (13.18a) and (13.18b) help to determine the length L_2 of the pier in the stable zone. The one that gives the maximum length L_2 should be used.

Belled Piers

Piers with a belled bottom are normally used when large uplift forces have to be resisted. Figure 13.21 (b) shows a belled pier with all the forces acting.

The uplift force for a belled pier is the same as that applicable for a straight shaft. The resisting force equation for the pier in the stable zone may be written as (O'Neill, 1988).

$$Q_{R1} = \pi\, d\, L_2\, \alpha\, c_u \tag{13.19a}$$

$$Q_{R2} = \frac{\pi}{4}\, (d_b^2 - d^2)\, (cN_c + \gamma L_2) \tag{1319b}$$

where d_b = diameter of the underream,

N_c = bearing capacity factor,

c = unit cohesion under undrained condition,

γ = unit weight of soil.

The values of N_c are given in Table 13.8 (O'Neill, 1988).

Two cases of stability may be written as before.

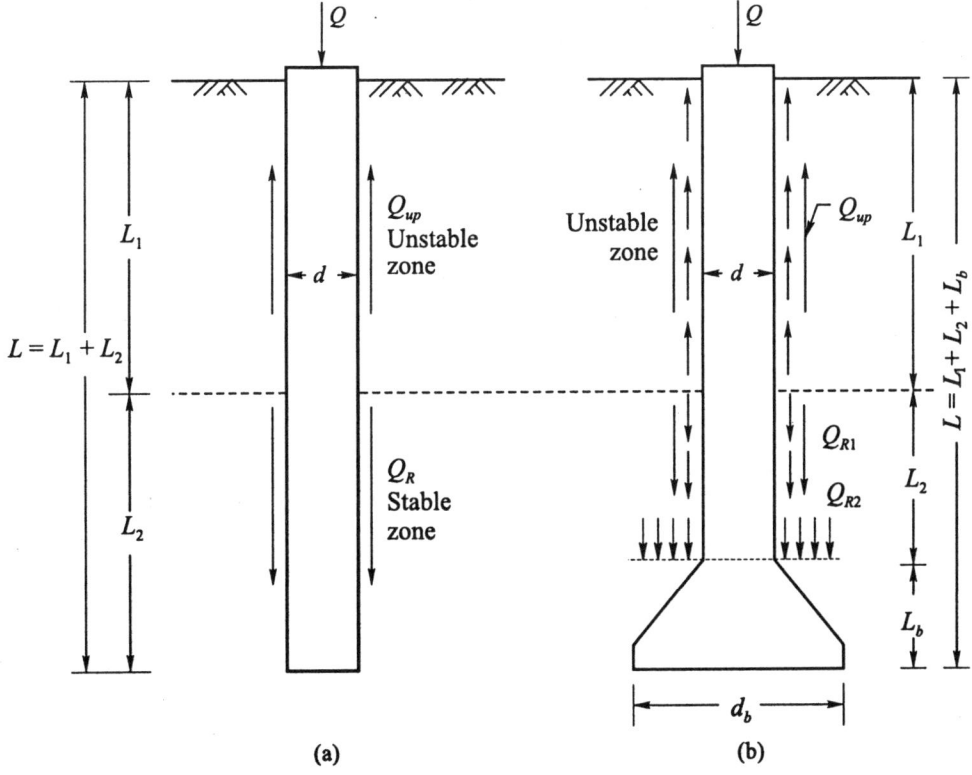

Fig. 13.21 Drilled pier in expansive soil

Table 13.8 Values of N_c

L_2/d_b	N_c
1.7	4
2.5	6
≥ 5.0	9

1. Without taking the dead load Q and using $F_s = 1.2$

$$Q_{up} = \frac{1}{1.2}(Q_{R1} + Q_{R2})$$

2. By taking into account the dead load and $F_s = 2.0$

$$(Q_{up} - Q) = \frac{1}{2.0}(Q_{R1} + Q_{R2})$$

For a given shaft diameter d and base diameter d_b, the above equations help to determine the value of L_2. The one that gives the maximum value for L_2 has to be used in the design.

Figure 13.22 gives a typical foundation design with grade beams and drilled piers (Chen, 1988). The piers should be taken sufficiently below the unstable zone of wetting in order to resist the uplift forces.

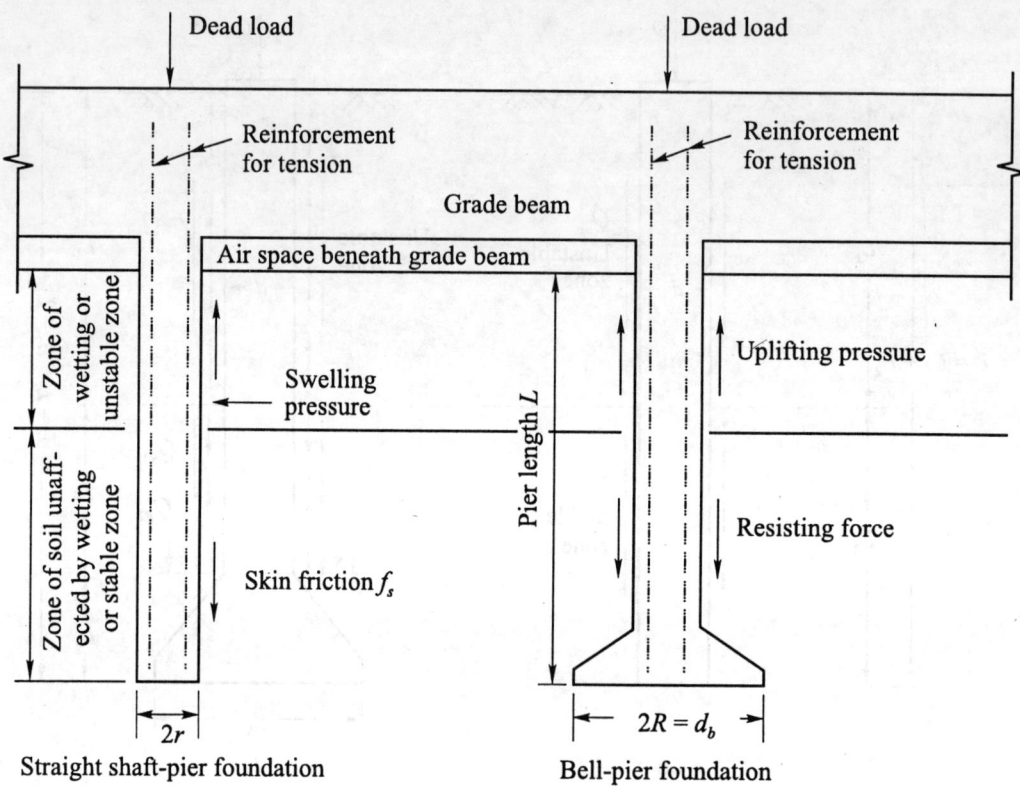

Fig. 13.22 Grade beam and pier system (Chen, 1988)

Example 13.3

A footing founded at a depth of 1 ft below ground level in expansive soil was subjected to loads from the superstructure. Site investigation revealed that the expansive soil extended to a depth of 8 ft below the base of the foundation, and the moisture contents in the soil during the construction period were at their lowest. In order to determine the percent swell, three undisturbed samples at depths of 2, 4 and 6 ft were collected and swell tests were conducted per the procedure described in Section 13.16. Figure Ex. 13.3 (a) shows the results of the swell tests plotted against depth. A line passing through the points is drawn. The line indicates that the swell is zero at 8 ft depth and maximum at a base level equal to 3%. Determine (a) the total swell, and (b) the depth of undercut necessary for an allowable swell of 0.03 ft.

Solution

(a) The total swell is equal to the area under the percent swell *vs.* depth curve in [Fig. Ex. 13.3 (a)].

Total swell = $1/2 \times 8 \times 3 \times 1/100 = 0.12$ ft

(b) Depth of undercut

From the percent swell *vs.* depth relationship given by the curve in Fig. 13.3 (a), total swell at different depths are calculated and plotted against depth in Fig. 13.3 (b). For example the total swell at depth 2 ft below the foundation base is

Total swell = $1/2 \, (8-2) \times 2.25 \times 1/100 = 0.067$ ft plotted against depth 2 ft. Similarly total swell at other depths can be calculated and plotted. Point *B* on curve in Fig. 13.3 (b) gives the allowable swell of 0.03 ft at a depth of 4 ft below foundation base. That is, the undercut necessary in clay is 4 ft which may be replaced by an equivalent thickness of non-swelling compacted fill.

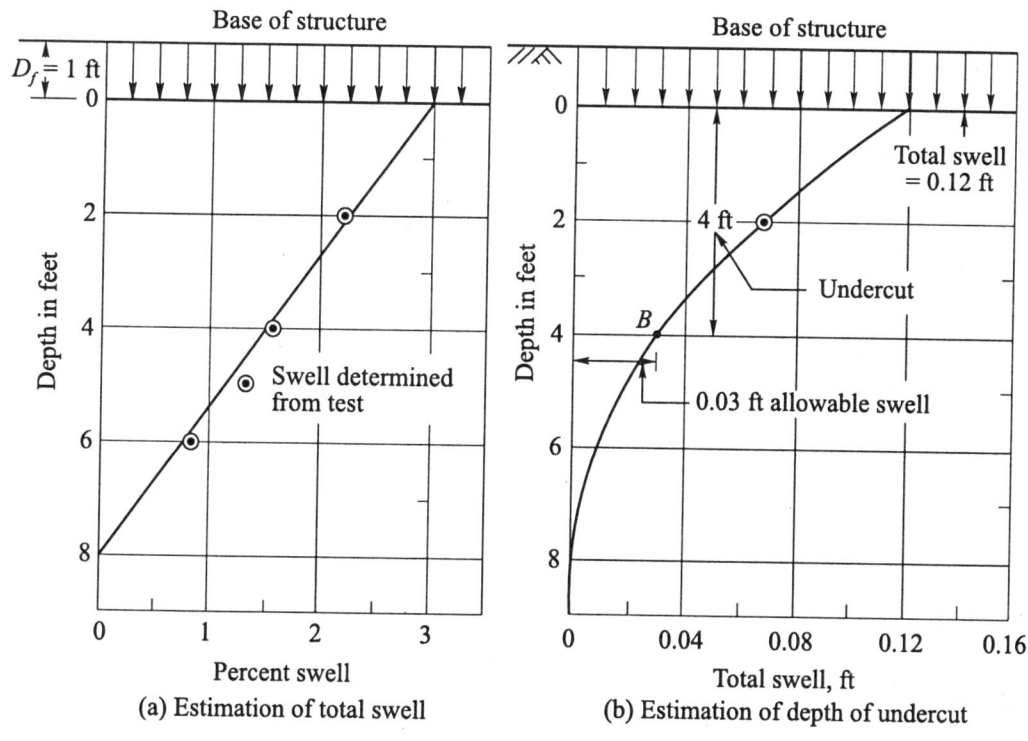

(a) Estimation of total swell

(b) Estimation of depth of undercut

Fig. Ex. 13.3

Example 13.4

Figure Ex. 13.4 shows that the soil to a depth of 20 ft is an expansive type with different degrees of swelling potential. The soil mass to a 20 ft depth is divided into four layers based on the swell potential rating given in Table 13.7. Calculate the total swell per the Van Der Merwe method.

Solution

The procedure for calculating the total swell is explained in Section 13.16. The details of the calculated results are tabulated below.

The details of calculated results

Layer No.	Thickness ΔD (ft)	PE	D ft	F	ΔH_s (in.)
1	5	0	2.5	0.75	0
2	8	1.0	9.0	0.35	2.80
3	2	0.5	14.0	0.20	0.20
4	5	1.0	17.5	0.13	0.65
					Total 3.65

In the table above D = depth from ground level to the mid-depth of the layer considered, F = reduction factor.

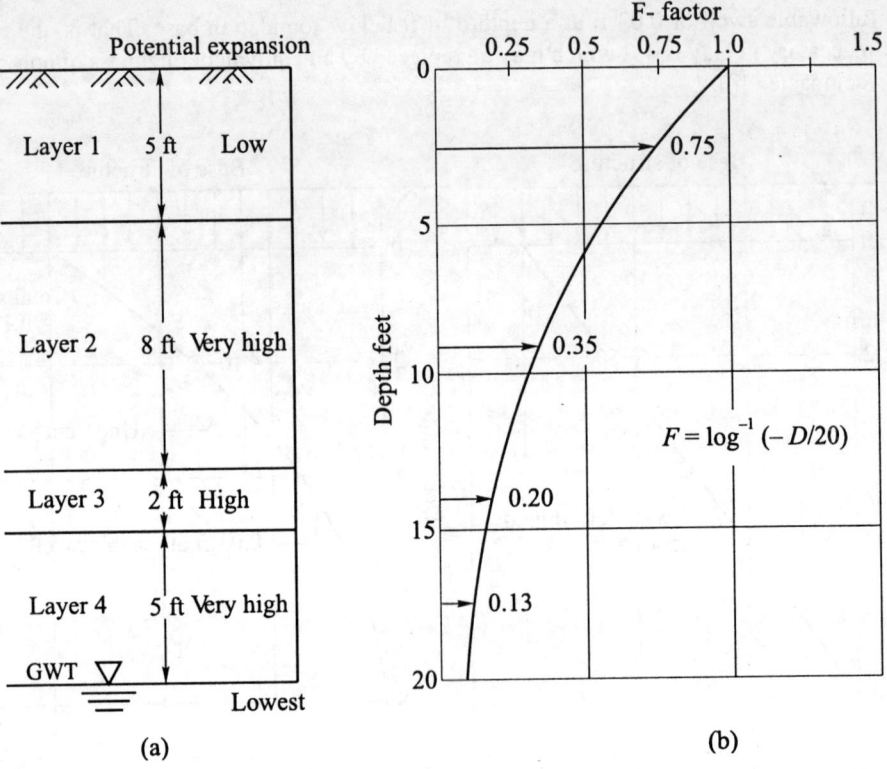

Fig. Ex. 13.4

Example 13.5

A drilled pier [refer to Fig. 13.21 (a)] was constructed in expansive soil. The water table was not encountered. The details of the pier and soil are:

$L = 20$ ft, $d = 12$ in., $L_1 = 5$ ft, $L_2 = 15$ ft, $p_s = 10{,}000$ lb/ft^2, $c_u = 2089$ lb/ft^2, SPT $(N) = 25$ blows per foot,

Required

(a) Total uplift capacity Q_{up}.

(b) Total resisting force due to surface friction.

(c) Factor of safety without taking into account the dead load Q acting on the top of the pier.

(d) Factor of safety with the dead load acting on the top of the pier.

Assume $Q = 10$ kips. Calculate Q_{up} by Chen's method (Eq. 13.15).

Solution

(a) Uplift force Q_{up} from Eq. (13.15)

$$Q_{up} = \pi \, d \, \alpha_u p_s L_1 = \frac{3.14 \times (1) \times 0.15 \times 10{,}000 \times 5}{1000} = 23.55 \text{ kips}$$

(b) Resisting force Q_R

From Eq. (13.17)

$$Q_R = \pi d \, (L - L_1) \, \alpha c_u$$

where $c_u = 2089$ lb/ft$^2 \approx 100$ kN/m^2

$$\frac{c_u}{p_a} = \frac{100}{101} \approx 1.0 \text{ where } p_a$$

$\qquad = $ atmospheric pressure $= 101$ kPa

From Fig. 11.15, $\alpha = 0.55$ for $c_u/p_a \approx 1.0$

Now substituting the known values

$$Q_R = 3.14 \times 1 \, (20 - 5) \times 0.55 \times 2000$$

$$= 51{,}810 \text{ lb} \approx 52 \text{ kips}$$

(c) Factor of safety with $Q = 0$

From Eq. (13.18a)

$$F_s = \frac{Q_R}{Q_{up}} = \frac{52}{23.5}$$

$$= 2.2 > 1.2 \text{ required OK.}$$

(d) Factor of safety with $Q = 10$ kips

From Eq. (13.18b)

$$F_s = \frac{Q_R}{\left(Q_{up} - Q\right)} = \frac{52}{(23.5 - 10)} = \frac{52}{13.5} = 3.9 > 2.0 \text{ required OK.}$$

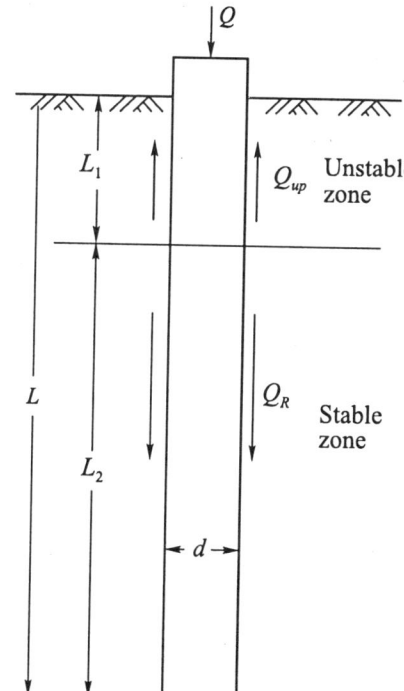

Fig. Ex. 13.5

Example 13.6

Solve Ex. 13.5 with $L_1 = 10$ ft. All the other data remain the same.

Solution

(a) Uplift force Q_{up}

$$Q_{up} = 23.5 \times (10/5) = 47.0 \text{ kips}$$

where $Q_{up} = 23.5$ kips for $L_1 = 5$ ft

(b) Resisting force Q_R

$$Q_R = 52 \times (10/15) \approx 34.7 \text{ kips}$$

where $Q_R = 52$ kips for $L_2 = 15$ ft

(c) Factor of safety for $Q = 0$

$$F_s = \frac{34.7}{47.0} = 0.74 < 1.2 \text{ as required not OK.}$$

(d) Factor of safety for $Q = 10$ kips

$$F_s = \frac{34.7}{(47 - 10)} = \frac{34.7}{37} = 0.94 < 2.0 \text{ as required not OK.}$$

The above calculations indicate that if the wetting zone (unstable zone) is 10 ft thick the structure will not be stable for $L = 20$ ft.

Example 13.7

Determine the length of pier required in the stable zone for $F_s = 1.2$ where $Q = 0$ and $F_s = 2.0$ when $Q = 10$ kips. All the other data given in Ex. 13.6 remain the same.

Solution

(a) Uplifting force Q_{up} for L_1 (10 ft) = 47 kips

(b) Resisting force for length L_2 in the stable zone.

$$Q_R = \pi d\, \alpha\, c_u\, L_2 = 3.14 \times 1 \times 0.55 \times 2000\, L_2 = 3.454\, L_2 \text{ lb/ft}^2$$

(c) $Q = 0$, minimum $F_s = 1.2$

$$\text{or} \quad 1.2 = \frac{Q_R}{Q_{up}} = \frac{3,454\, L_2}{47,000}$$

solving we have $L_2 = 16.3$ ft.

(d) $Q = 10$ kips. Minimum $F_s = 2.0$

$$F_s = 2.0 = \frac{Q_R}{\left(Q_{up} - Q\right)} = \frac{3,454\, L_2}{(47,000 - 10,000)} = \frac{3,454\, L_2}{37,000}$$

Solving we have $L_2 = 21.4$ ft.

The above calculations indicate that the minimum $L_2 = 21.4$ ft or say 22 ft is required for the structure to be stable with $L_1 = 10$ ft. The total length $L = 10 + 22 = 32$ ft.

Example 13.8

Figure Ex. 13.8 shows a drilled pier with a belled bottom constructed in expansive soil. The water table is not encountered. The details of the pier and soil are given below:

$L_1 = 10$ ft, $L_2 = 10$ ft, $L_b = 2.5$ ft, $d = 12$ in., $d_b = 3$ ft, $c_u = 2000$ lb/ft^2, $p_s = 10,000$ lb/ft^2, $\gamma = 110$ lb/ft^3.

Required

(a) Uplift force Q_{up}.

(b) Resisting force Q_R.

(c) Factor of safety for $Q = 0$ at the top of the pier.

(d) Factor of safety for $Q = 20$ kips at the top of the pier.

Solution

(a) Uplift force Q_{up}

As in Ex. 13.6 $Q_{up} = 47$ kips

(b) Resisting force Q_R

$$Q_{R1} = \pi d\, L_2\, \alpha\, c_u$$

$$\alpha = 0.55 \text{ as in Ex. 13.5}$$

Substituting known values

$$Q_{R1} = 3.14 \times 1 \times 10 \times 0.55 \times 2000$$
$$= 34540 \text{ lbs} = 34.54 \text{ kips}$$

$$QR_2 = \frac{\pi}{4} (d_b^2 - d^2)(cN_c + \gamma L_2)$$

where $d_b = 3$ ft, $c = 2000$ lb/ft^2,

$N_c = 7.0$ from Table 13.8 for L^2/d_b

$$= 10/3 = 3.33$$

Substituting known values

$$Q_{R2} = \frac{3.14}{4} [3^2 - (1)^2]$$

$$(2000 \times 7.0 + 110 \times 10)$$
$$= 6.28 (15,100)$$
$$= 94,828 \text{ lbs} = 94.8 \text{ kips}$$

$$Q_R = Q_{R1} + Q_{R2}$$
$$= 34.54 + 94.8 = 129.3 \text{ kips}$$

(c) Factor of safety for $Q = 0$

$$F_s = \frac{Q_R}{Q_{up}}$$

$$= \frac{129.3}{47.0} = 2.75 > 1.2 \text{ OK}$$

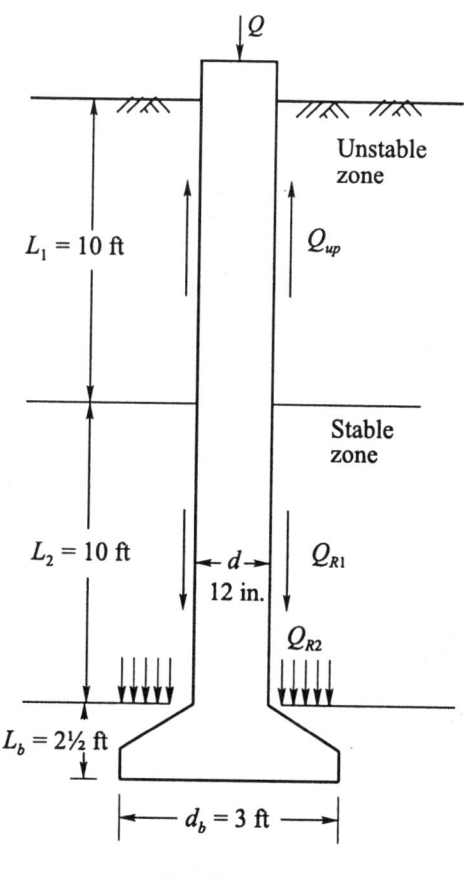

Fig. Ex. 13.8

(d) Factor of safety for $Q = 20$ kips

$$F_s = \frac{Q_R}{(Q_{up} - Q)} = \frac{129.3}{(47 - 20)} = 4.79 > 2.0 \text{ as required OK.}$$

The above calculations indicate that the design is over conservative. The length L_2 can be reduced to provide an acceptable factor of safety.

13.18 ELIMINATION OF SWELLING

The elimination of foundation swelling can be achieved in two ways. They are:

1. Providing a granular bed and cover below and around the foundation (Fig. 13.19).
2. Chemical stabilization of swelling soils.

Figure 13.19 gives a typical example of the first type. In this case, the excavation is carried out up to a depth greater than the width of the foundation by about 20 to 30 cm. Freely draining soil, such as a mixture of sand and gravel, is placed and compacted up to the base level of the foundation. A Reinforced concrete footing is constructed at this level. A mixture of sand and gravel is filled up loosely over the fill. A reinforced concrete apron about 2 m wide is provided around the building to prevent moisture directly entering the foundation. A cushion of granular soils below the foundation absorbs the effect of swelling, and thereby its effect on the foundation will considerably be reduced.

A foundation of this type should be constructed only during the dry season when the soil has shrunk to its lowest level. Arrangements should be made to drain away the water from the granular base during the rainy seasons.

Chemical stabilisation of swelling soils by the addition of lime may be remarkably effective if the lime can be mixed thoroughly with the soil and compacted at about the optimum moisture content. The appropriate percentage usually ranges from about 3 to 8 percent. The lime content is estimated on the basis of pH tests and checked by compacting, curing and testing samples in the laboratory. The limit has the effect of reducing the plasticity of the soil, and hence its swelling potential.

13.19 PROBLEMS

13.1 A building was constructed in a loessial type normally consolidated collapsible soil with the foundation at a depth of 1 m below ground level. The soil to a depth of 6 m below the foundation was found to be collapsible on flooding. The average overburden pressure was 56 kN/m². Double consolidometer tests were conducted on two undisturbed samples taken at a depth of 4 m below ground level, one with its natural moisture content and the other under soaked conditions per the procedure explained in Section 13.4. The following data were available.

Applied pressure, kN/m²	10	20	40	100	200	400	800
Void ratios at natural moisture content	0.80	0.79	0.78	0.75	0.725	0.68	0.61
Void ratios in the soaked condition	0.75	0.71	0.66	0.58	0.51	0.43	0.32

Plot the e-log p curves and determine the collapsible settlement for an increase in pressure $\Delta p = 34$ kN/m² at the middle of the collapsible stratum.

13.2 Soil investigation at a site indicated overconsolidated collapsible loessial soil extending to a great depth. It is required to construct a footing at the site founded at a depth of 1.0 m below ground level. The site is subject to flooding. The average unit weight of the soil is 19.5 kN/m³. Two oedometer tests were conducted on two undisturbed samples taken at a depth of 5 m from ground level. One test was conducted at its natural moisture content and the other on a soaked condition per the procedure explained in Section 13.4.

The following test results are available.

Applied pressure, kN/m²	10	20	40	100	200	400	800	2000
Void ratio under natural moisture condition	0.795	0.79	0.787	0.78	0.77	0.74	0.71	0.64
Void ratio under soaked condition	0.775	0.77	0.757	0.730	0.68	0.63	0.54	0.37

The swell index determined from the rebound curve of the soaked sample is equal to 0.08. Required:

(a) Plots of e-log p curves for both tests.

(b) Determination of the average overburden pressure at the middle of the soil stratum.

(c) Determination of the preconsolidation pressure based on the curve obtained from the soaked sample.

(d) Total collapse settlement for an increase in pressure $\Delta p = 710$ kN/m².

13.3 A footing for a building is founded 0.5 m below ground level in an expansive clay stratum which extends to a great depth. Swell tests were conducted on three undisturbed samples taken at different depths and the details of the tests are given below.

Depth (m) below GL	Swell %
1	2.9
2	1.75
3	0.63

Required:

(a) The total swell under structural loadings.

(b) Depth of undercut for an allowable swell of 1 cm.

13.4 Figure Prob. 13.4 gives the profile of an expansive soil with varying degrees of swelling potential. Calculate the total swell per the Van Der Merwe method.

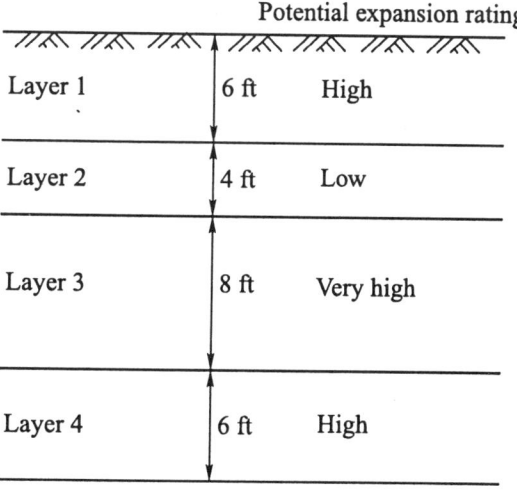

Potential expansion rating

Layer 1	6 ft	High
Layer 2	4 ft	Low
Layer 3	8 ft	Very high
Layer 4	6 ft	High

Fig. Prob. 13.4

13.5 Figure Prob. 13.5 depicts a drilled pier embedded in expansive soil. The details of the pier and soil properties are given in the figure.

Determine:

(a) The total uplift capacity.

(b) Total resisting force.

(c) Factor of safety with no load acting on the top of the pier.

(d) Factor of safety with a dead load of 100 kN on the top of the pier.

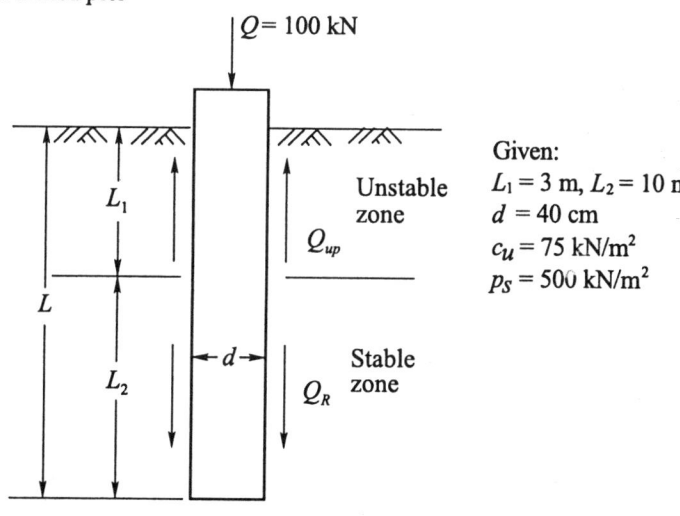

Given:
$L_1 = 3$ m, $L_2 = 10$ m
$d = 40$ cm
$c_u = 75$ kN/m^2
$p_S = 500$ kN/m^2

Fig. Prob. 13.5

Calculated Q_{up} by Chen's method.

13.6 Solve problem 13.5 using Eq. (13.16)

13.7 Figure Prob. 13.7 shows a drilled pier with a belled bottom. All the particulars of the pier and soil are given in the figure.

Required:

(a) The total uplift force.

(b) The total resisting force.

(c) Factor of safety for $Q = 0$

(d) Factor of safety for $Q = 200$ kN.

Use Chen's method for computing Q_{up}.

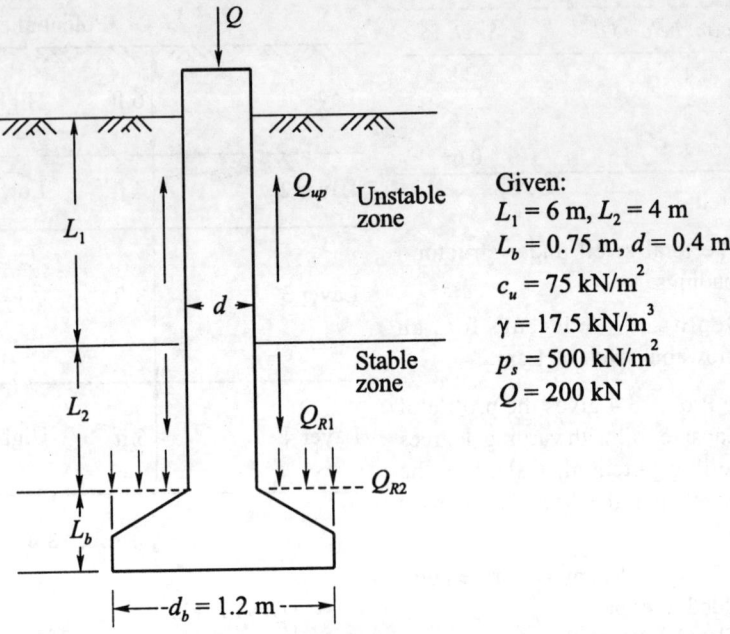

Given:
$L_1 = 6$ m, $L_2 = 4$ m
$L_b = 0.75$ m, $d = 0.4$ m
$c_u = 75$ kN/m^2
$\gamma = 17.5$ kN/m^3
$p_s = 500$ kN/m^2
$Q = 200$ kN

Fig. Prob. 13.7

13.8 Solve Prob. 13.7 by making use of Eq. (13.16) for computing Q_{up}.

13.9 Refer to Fig. Prob. 13.9. The following data are available:

$L_1 = 15$ ft, $L_2 = 13$ ft, $d = 4$ ft, $d_b = 8$ ft and $L_b = 6$ ft.

All the other data are given in the figure.

Required:

(a) The total uplift force

(b) The total resisting force

(c) Factor of safety for $Q = 0$

(d) Factor of safety for $Q = 60$ kips.

Use Chen's method for computing Q_{up}.

13.10 Solve Prob. 13.9 using Eq. (13.16) for computing Q_{up}.

13.11 If the length L_2 is not sufficient in Prob. 13.10, determine the required length to get $F_s = 3.0$.

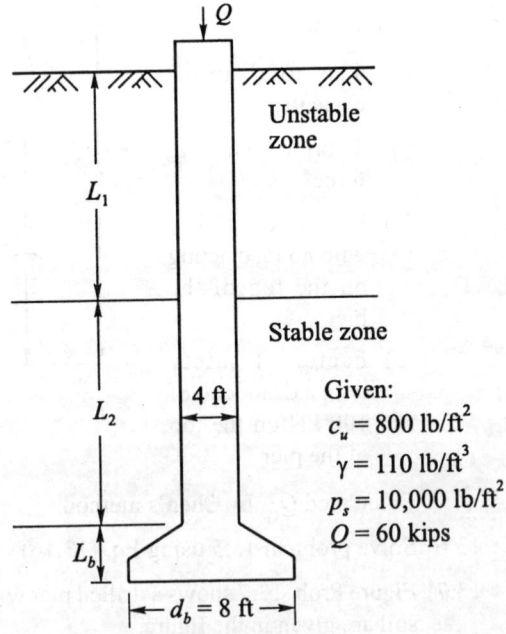

Unstable zone

Stable zone

Given:
$c_u = 800$ lb/ft^2
$\gamma = 110$ lb/ft^3
$p_s = 10,000$ lb/ft^2
$Q = 60$ kips

Fig. Prob. 13.9

Cellular Cofferdams

14.1   INTRODUCTION

A cofferdam is a temporary structure. It is normally constructed to divert water in a river or to keep away water from an enclosed area in order to construct a permanent structure. A bridge pier in a standing water may be constructed by first constructing cofferdam around the site and then pumping out water from the site to enable the construction work to go on in a dry condition. During the entire period of construction a certain amount of pumping is constantly needed to keep away the water that leaks through the dam and foundation. Cofferdams are of many types. The relative merits of the various types are:

1. Cantilever sheet pile cofferdams [Fig. 14.1 (a)]. They are suitable in cases where the height of dam is very small. Such dams are normally subjected to large leakage and flood damage.
2. Braced cofferdams [Fig. 14.1 (c)]. They are economical for small to moderate heights. They are also susceptible to floods damage.
3. Earth embankments. There is no limitation to height but the construction occupies more time and space.
4. Double-wall cofferdams [Fig. 14.1 (g)]. They are suitable for moderate heights.
5. Cellular cofferdams [Fig. 14.1 (e)]. These dams are suitable from moderate to large heights. They can constantly be used in excavation in water. When the excavation is in large area overlain by water, as in a river or lake bottom, cellular cofferdams are generally used to provide a water barrier. This type of structure is widely used to provide a dry work area where dams are constructed in rivers, and for waterfront construction.

The cellular structure shown in Fig. 14.2 is economical since stability is achieved by using a soil cell fill for mass, which is relatively cheap. The internal bracings which would obstruct the work area are avoided. The sheet piling may be pulled out and reused. It is not usually necessary to drive the piling to great depths in the natural soil, thus avoiding damage to the piles during driving. To achieve a cell which is stable against bursting it is necessary that the sheet piling be driven so that continuity of interlocks is maintained. The cellular type is more water tight than the braced cofferdam. The construction problem is relatively simple. The design procedures of double-wall cofferdams and cellular cofferdams are fundamentally the same. The basic principles of design of these dams are discussed in this chapter.

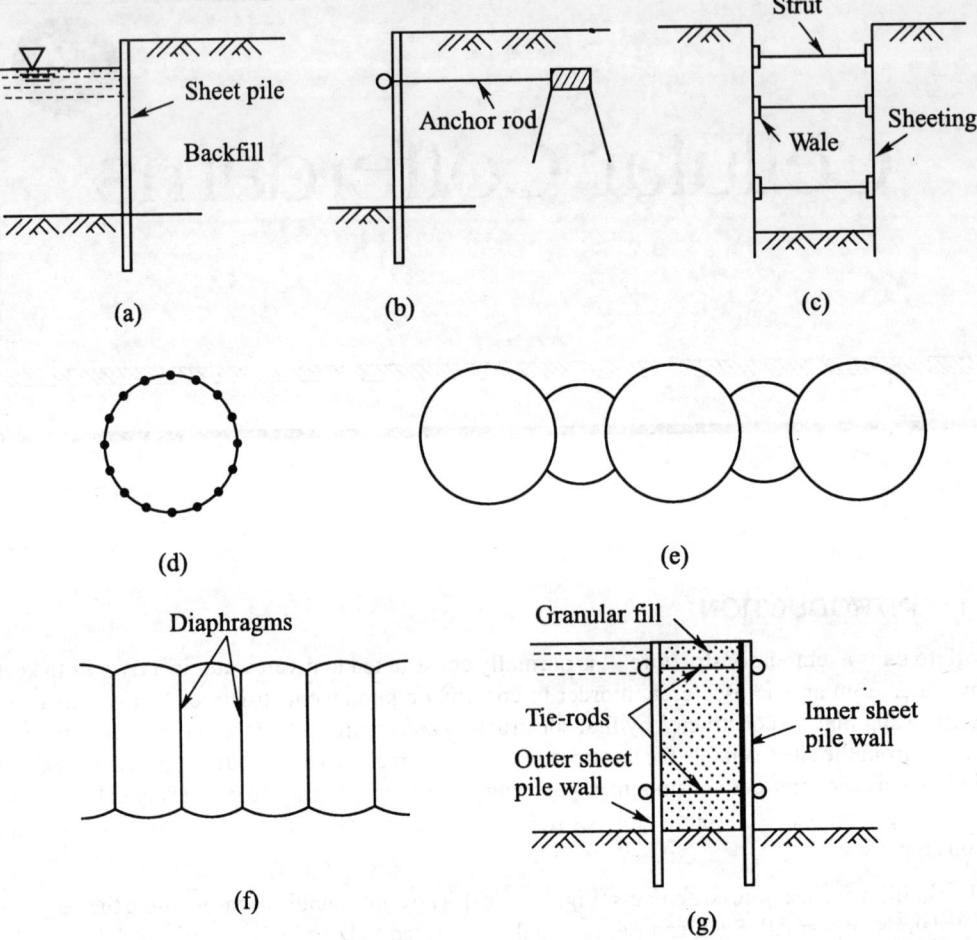

Fig. 14.1 Use of sheet piles: (a) Cantilever sheet piles, (b) anchored bulk head, (c) braced sheeting in cuts, (d) single cell cofferdam, (e) cellular cofferdam, (f) cellular cofferdam, diaphragm type, and (g) double sheet pile walls

14.2 CELLULAR COFFERDAMS

Cellular cofferdams are basically of two types as shown in Fig. 14.2. They are:

1. Circular type.
2. Diaphragm type.

Circular Type

Circular type consists of individual large diameter circles connected together by arcs of smaller diameter. These arcs usually intersect the circles at a point at 30° or 45° with the longitudinal axis of the cofferdam. They are often perpendicular to the circle, but occasionally different angles may be used.

Advantages of the Curcular Type

1. The circular type can be used singularly in a group or at end.
2. It will not collapse in the event of failure of adjoining cells (due to interlock damage, sudden floods, etc.)

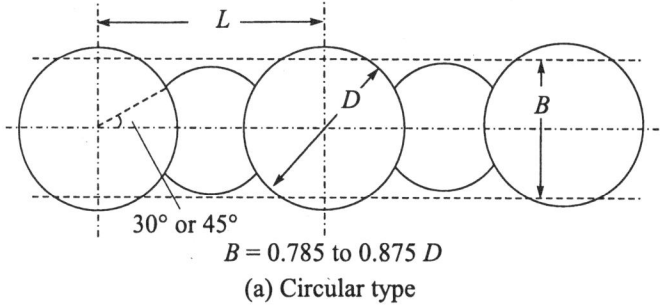

30° or 45°

$B = 0.785$ to $0.875 D$

(a) Circular type

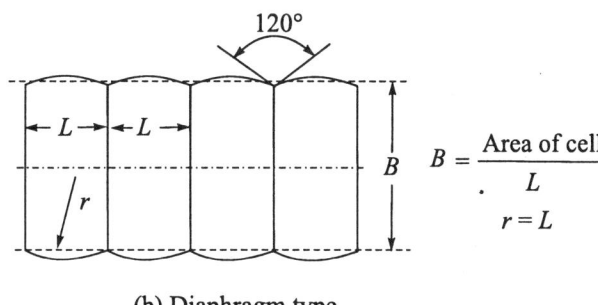

120°

$$B = \frac{\text{Area of cell}}{L}$$

$$r = L$$

(b) Diaphragm type

Fig. 14.2 Cellular cofferdams

3. Each cell can be filled independent of the other without hampering the progress of work.
4. It requires less number of piles per lineal metre of cofferdam as compared with the diaphragm type of an equal design.

Diaphragm Type

It consists of two series of arcs connected together by diaphragms perpendicular to the axis of the cofferdam. Generally the radii of these arcs are made equal to the distances between the diaphragms. At the intersection point the two arcs and the diaphragm make an angle of 120° between each other.

Advantages of the Diaphragm Type

1. It has uniform interlock stress throughout the section at any given level. The stress is smaller than at the point of circular cell of comparable design.
2. It can be widened readily by increasing the length of the diaphragm if it is required for stability. This will not increase the interlock stress which is a function of the radius of the arc.

14.3   COMPONENTS OF CELLULAR COFFERDAMS

A cellular cofferdam is made of:

1. Steel cells.
2. Cell fill.

Steel Cells

Steel cells are fabricated out of stell sheet piles. In large diameter cells any two adjacent piles are almost on a straight line. In smaller cells, however, each sheet pile must deflect at a relatively large angle from the straight line in order to form the desired circle.

Cell Fills

The material used for the cell fill should have the following properties.

1. The fill should be free-draining granular soil with little fine particles.
2. It should possess high angle of friction.
3. The fill should be as dense as possible.
4. It should possess large resistance to scour and leakage. Well graded soils are most suitable.

Normally, natural deposits of mixed sand and gravel possess all of these desirable properties and as such are the best materials for the cell fill.

14.4   DIMENSIONS OF CELLULAR COFFERDAM

The height of a cofferdam is fixed on the basis of the maximum flood level in the river where it is to be constructed. Its diameter is based on the safety requirements. The desigh of the cofferdam begins with a tentative proportion which is subsequently analysed for stability and other safety requirements. The design is usually made on the basis of a section one metre long with a uniform average width B which is used in the design calculations. For the design purposes, a simple procedure may be used whereby the average width is determined such that the area in the rectangular section and that in the actual cofferdam are equal.

Let, B = Average width

   L = distance centre to centre of cell

Then, $B = \dfrac{\text{Area of main cell} + \text{Area of connecting cell}}{L}$

The average width B of cells ranges from $0.785D$ to $0.875D$ where D is the diameter of cell.

14.5   STABILITY OF CELLULAR COFFERDAMS

There are a few methods of design of cellular cofferdams available in technical literature. The method that is described in this book is the one that is largely used by TVA (1957) engineers. The tentative dimensions that are initially assumed are to be checked for their safety requirements. The preliminary dimensions assumed for cofferdams should satisfy the following stability requirements.

Cofferdams on Rock

1. Resistance to sliding

The lateral and vertical forces that act on a cellular cofferdam are shown in Fig. 14.3 (a). The forces are:

P = The lateral pressure due to water and the submerged weight of soil below the river bed. This pressure tends to push the cell away from its position.

W = Effective weight of the fill material which is a sum of the submerged weight below the saturation line (assumed at 2 : 1 slope) and the total weight above the saturation line.

P_p = Passive pressure. If there is no berm on the wall side, the passive pressure which resists the movement of the cell is due to the soil below the river bed level. If a berm is provided to add to stability, the passive pressure due to the berm should also be taken into account.

F_f = The frictional resistance that is developed at the base of the cell. This is equal to $W \tan \phi$ for soil to soil sliding at the base.

A cofferdam should provide adequate resistance to sliding on the base caused by the lateral pressure P (taken per unit length of cell) acting on the cell face on the river side. The factor of safety, F_s, against sliding may be written as

$$F_s = \frac{P_p + F_f}{P} \tag{14.1}$$

The factor of safety should be greater than 1. A maximum value of 1.25 may be used in the analysis.

2. Resistance to Overturning

The cofferdam should be stable against overturning. The stability to overturning may be checked by two methods.

(a) *Check for the resultant weight to fall within the middle-third of the base of cell* [Fig. 14.3 (b)].

The overturning moment M_0 due to the lateral pressure P is

$$M_0 = P\bar{y} \tag{14.2}$$

Where $\bar{y}$ is the point of application of P above the base.

The resisting Moment M_r due to reaction at the base is

$$M_r = We = \gamma HBe \tag{14.3}$$

Where

γ = effective unit weight of the fill material,

H = Height of fill,

B = Average width of cell,

e = Eccentricity.

When $M_0 = M_r$, we have

$$e = \frac{P\bar{y}}{\gamma HB} \tag{14.4}$$

Since soil cannot take tension, the safety requirement is

$$e = \frac{P\bar{y}}{\gamma HB} \le \frac{B}{6} \tag{14.5}$$

Equation (14.5) indicates that the width of the cell should be increased as the height of the cell increases so that the reaction may fall within the middle third.

(b) *Check for frictional resistance of the steel piling on the cell fill material.*

The resistance to overturning may be analysed in a different way as shown in Fig. 14.3 (c). One may argue that as the cell tends to tip over, the soil will pour out at the heel. For this to occur the frictional resistance of the steel piling on the cell fill is to be fully developed. On this side of the cell the water pressure P_w is pushing the piling against the fill so that the friction force per unit length of cell is $P_w \tan \delta$ where δ is the angle of wall friction between soil and the pile. The resisting shear force acts downward on the wall of the pile as shown in the Fig. 14.3 (c). The overturning moment about the toe of the cell (point A in the figure) is

$$M_0 = P_w \bar{y}' \tag{14.6}$$

The resisting moment is

$$M_r = P_w B \tan \delta \tag{14.7}$$

If the cofferdam is to be stable, the resisting moment should be more than the overturning moment or factor of safety F_s may be written as

$$F_s = \frac{P_w B \tan \delta}{P_w \bar{y}'} = \frac{B \tan \delta}{\bar{y}'} \tag{14.8}$$

A factor of safety of 1.1 to 1.25 is generally suggested.

3. Cell Shear

Overturning moments on a cell develops shear stress on a vertical plane through the centre line of the cell as shown in Fig. 14.3 (d). For stability the shearing resistance along this plane which is the sum of soil shear resistance and resistance in the interlocks, must be equal to or greater than the shear due to overturning effects. Referring to Fig. 14.3 (d), and assuming a linear pressure distribution across the base of the cell, the overturning moment due to the overturning shear force developed at the base is

$$M_0 = 2/3 \ BV \tag{14.9}$$

Where $2/3 \ B$ = lever arm of the shearing force V acting as a couple.

Solving for the overturning shear on the plane through the centre line, we have

$$V = 1.5 \ \frac{M_0}{B} \tag{14.10}$$

The total resisting shear, F_{cr} = Resisting shear S_r in the soil, + Resisting shear in the locks, F_c
The resisting shear in the soil. S_r, is given by the expression

$$S_r = \frac{1}{2} \gamma H^2 \bar{K}_A \tan \phi \tag{14.11}$$

where $\bar{K}_A$ = Coefficient of earth pressure which is found to be greater than the active earth pressure coefficient K_A. The expression for determining $\bar{K}_A$ as proposed by TVA (1957) is

$$K_A = \frac{\cos^2 \phi}{2 - \cos^2 \phi} \tag{14.12}$$

where ϕ = angle of the internal friction of the fill material.

The frictional force in the interlock is computed as follows:

Hoop tension per unit depth of wall in the interlock at any depth from the surface $T = p_a r$

where, p_a = Lateral pressure on the wall at any depth z from the surface of fill,

r = Radius of the cell.

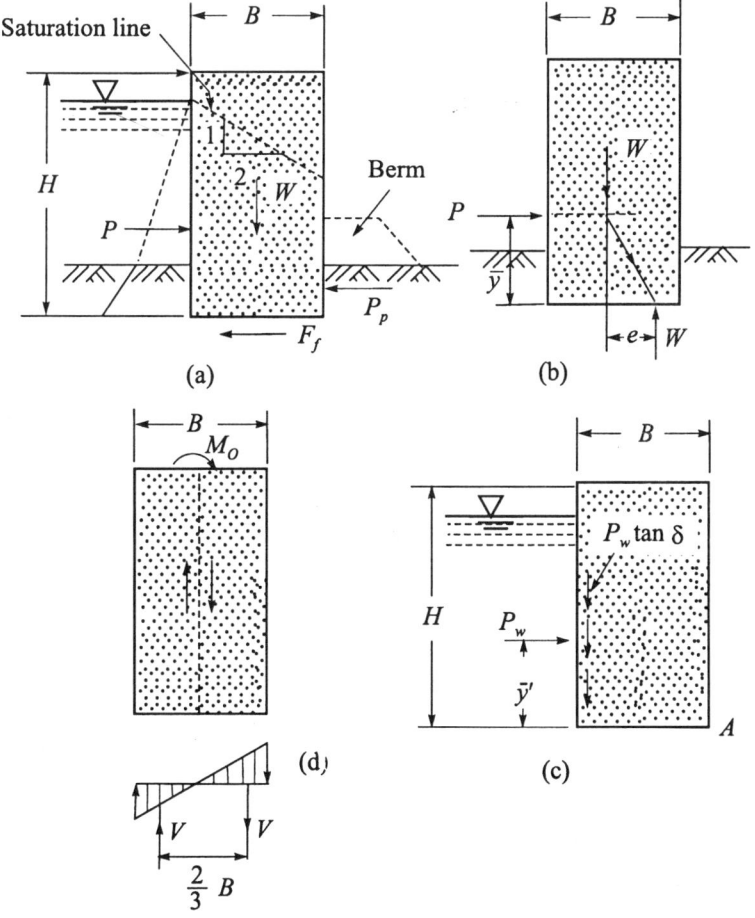

Fig. 14.3 Stability of cofferdams

Since p_a increases with the depth, the total hoop-tension force for cell of total depth H is

$$T = \frac{1}{2}\, \gamma\, H_2\, K_A r \tag{14.13}$$

where, γ = effective unit weight of fill material,

K_A = Coefficient of earth pressure due to Rankine or Coulomb.

If a sheet pile rests on a rock bed or is embedded in soil, a restraining force preventing the lateral movement of the cell is developed at the base. This effect causes the maximum pressure to be developed at a depth $H' \approx 0.75\, H_c$ = height of the sheet piling above the point of fixity, or embedment. The depth of the point of fixity from the top is taken. For all practical purposes as the average of the total height of the cell and the depth of water above $0.75\, H_c$, the river bed.

The triangular distribution of earth pressure p_a for the entire height of sheet piling is shown in Fig. 14.4 (a). When fixity condition prevails, the soil pressure distribution will be as shown in Fig. 14.4 (b). The area of this triangle gives the total lateral earth pressure acting on the cell. If we assume the pressure distribution on the entire height H of piling with the fixity condition prevailing at depth, the total lateral earth pressure is

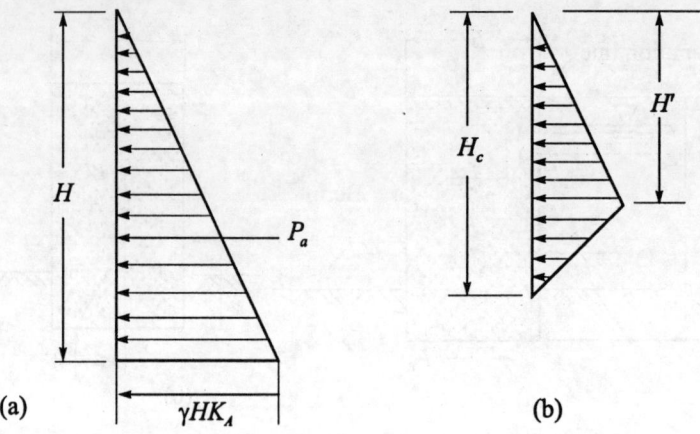

Fig. 14.4 Earth pressure distribution on the sheet-piling wall

$$P_a = \frac{1}{2} \gamma H (0.75 H_c) K_A \tag{14.14}$$

Therefore, the adjusted hoop-tension force is

$$T = P_a r = \frac{1}{2} \gamma H (0.75 H_c) K_A r$$

$$= 3/8 \, \gamma \, HH_c K_A r \tag{14.15}$$

The friction force in the interlock is, therefore

$$\bar{F}_i = Tf = 3/8 \, \gamma \, HH_c K_A rf \tag{14.16}$$

where $f =$ Coefficient of interlock friction which may be taken as 0.3.

Since the analysis is usually performed on a unit length of cell (say 1 metre length), the shear resistance per unit length of cell is

$$F_i = \frac{\bar{F}_i}{L} = 3/8 \, \gamma \, HH_c K_A f \frac{r}{L} \tag{14.17}$$

where $L =$ distance between cross-walls for diaphragm type cellular dams, and $L = r$, the radius, for circular cells.

The hoop tension per unit length of circular cell is, therefore

$$F_i = 3/8 \, \gamma \, HH_c K_A f \tag{14.18}$$

The total cell shear resistance, F_{cr} is, therefore, for a circular cell type cofferdam

$$F_{cr} = \frac{1}{2} \gamma H^2 \bar{K}_A \tan \phi + 3/8 \, \gamma \, HH_c K_A f \tag{14.19}$$

The factor of safety, F_s for the overturning shear force may be written as

$$F_s = \frac{F_{cr}}{V} \tag{14.20a}$$

or
$$F_s = F_{cr} \frac{2B}{3M_0}$$

(14.20b)

The factor of safety recommended varies from 1.1 to 1.25.

4. Resistance to Bursting

The cell should be stable against bursting pressure. The critical locations are in the interlock points and in the T's and Y's used for the connecting area. The bursting pressure is the maximum hoop-tension developed in the interlock at a depth H' from the surface of fill. The maximum pressure p_a at depth $H' \approx 0.75H_c$ as shown in Fig. 14.4 (b) is equal to

$$p_a = \gamma H' K_A$$

The bursting pressure T at depth H', is therefore

$$T = p_a r + \gamma_w H'', \text{ per unit depth of wall}$$

(14.21)

where

$$\gamma_w = \text{Unit weight of water,}$$
$$H'' = \text{Depth of water above the point of maximum pressure } p_a.$$

The allowable value for the interlock tension T, depends on the size and shape of the rolled steel pile sections. The computed maximum interlock tension, should not exceed the maximum stress specified by the manufacturers of steel sheet piles.

According to the Tennesse Valley Authority (TVA), the stress T' in a 90°-Tee, used for the connecting arc, can be obtained as an approximation from

$$T' = \frac{T}{\cos \theta}$$

where T is obtained from Eq. (14.21), and θ is the angle of intersection of the connecting arcs as shown in Fig. 14.2 (a)

Cofferdams on Deep Layers of Sand or Clay

The principle of analysis cofferdams founded on deep layers of sand or clay is the same as that applied to cofferdams on rock. In addition, the following requirements must be satisfied.

1. The sheet piling must be driven to such a depth at which level the bearing capacity of the soil should be at least 1.5 times the maximum vertical pressure transmitted to the soil by the cellfill.

    The maximum pressure transmitted to the soil by the cellfill when the sheet pile is subjected to lateral force may be written as

$$F_v = \frac{1}{2} \gamma H^2 K_A \tan \delta$$

(14.23)

where

$$F_v = \text{Vertical pressure per unit length of sheet of piling,}$$
$$\gamma = \text{Effective unit weight of cell fill,}$$
$$H = \text{Height of cell,}$$
$$K_A = \text{Coefficient of active earth pressure,}$$
$$\delta = \text{Angle of friction between cell fill and piling.}$$

2. Cellular cofferdams to be founded on sand bed should be designed to prevent boiling at the toe due to seepage of water.

Figure 14.5 gives the section of a cellular cofferdam founded on sand with the flow nets drawn. Due to high permeability of the soil, water retained on behind the dam percolates below its base at a relatively large velocity and rise up in front of the toe. If the seepage pressure in front of the toe is more than the buoyant weight of the soil, boiling of sand or quick sand condition develops. The danger of boiling can readily be eliminated by the use of loaded filter. As an alternative, the sheet piling may be driven to a great depth in order to eliminate the boiling condition. The depth of sheet piling below the bed level required for this purpose is approximately equal to $0.67 H$ where H is the height sheet pile above bed.

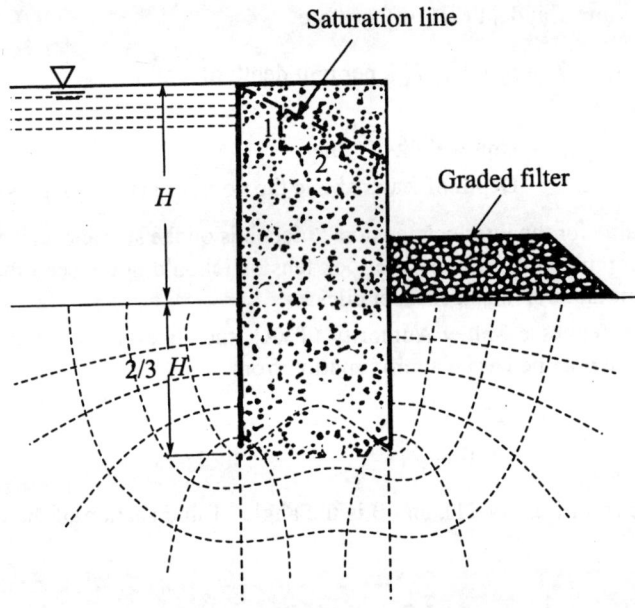

Fig. 14.5 Seepage in cellular cofferdam is sand

3. Cellular Cofferdams founded on clay should be investigated for the bearing capacity of the clay at the toe end of the dam. The maximum height of the Cofferdam above the bed level is a function of the undrained shear strength of clay. The equation for the critical height $\bar{H}_c$ may be written as

$$\bar{H}_c = 5.7 \frac{c_u}{\gamma} \qquad (14.24)$$

where c_u = undrained shear strength of clay,

γ = effective unit weight of cell fill.

If a minimum factor of safety of 1.5 is used, the allowable height of cofferdam above the clay bed is

$$H = 3.8 \frac{c_u}{\gamma} \qquad (14.25)$$

14.6 EXAMPLES

Example 14.1

Design a cellular cofferdam diaphragm type, with the dimensions given in Fig. Ex. 14.1. The other available data are

   (*i*) Unit weight of fill above saturation line,

$$\gamma = 1.7 \ t/m^3$$

  (*ii*) Submerged unit weight of fill and soil outside the cell,

$$\gamma = 1.1 \ t/m^3$$

 (*iii*) Angle of internal friction of fill and soil,

$$\phi = 30°$$

 (*iv*) Frictional co-efficient of fill on rock,

$$f = 0.57$$

  (*v*) Interlock friction,

$$f = 0.3$$

 (*vi*) Interlock tension allowed,

$$T = 1450 \ kg/cm$$

 (*vii*) Frictional coefficient of steel on fill,

$$f = 0.4$$

(*viii*) Allowable steel tensile stress,

$$f_t = 1500 \ kg/cm^2$$

Solution

Let B = actual width of cell

          b = average width of cell = $A + 1.82r$

         A = distance between the centres of curvature

         r = radius of the curved portion of cell.

1. *Sliding stability:* The average width b is used in the stability analysis. Weight of fill above rock level per unit length of cell

$$= W = W_1 + W_2$$

where,

     W_1 = weight of fill above saturation line,

     W_2 = Weight of fill below saturation line,

$$W_1 = \frac{1}{2} \times 1.7 \ (1 + 1 + b/2) \ b \ \text{tonnes},$$

$$W_2 = \frac{1}{2} \times 1.1 \ (15 + 15 - b/2) \ b \ \text{tonnes},$$

Therefore

$$W = 18.2b - 0.025b^2$$

The frictional resistance due to the cell weight at the rock level (neglecting the weight of the steel pile wall) is

$$F_f = FW = 0.57 (18.2b - 0.025b^2)$$

Passive pressure

$$P_p = \frac{1}{2} \gamma_b h^2 K_P = \frac{1}{2} \times 1.1 \times 62 \times \tan^2 (45° + 30/2) = 59 \text{ tonnes}$$

The driving force P_d is

$$P_d = P_w + P_a$$

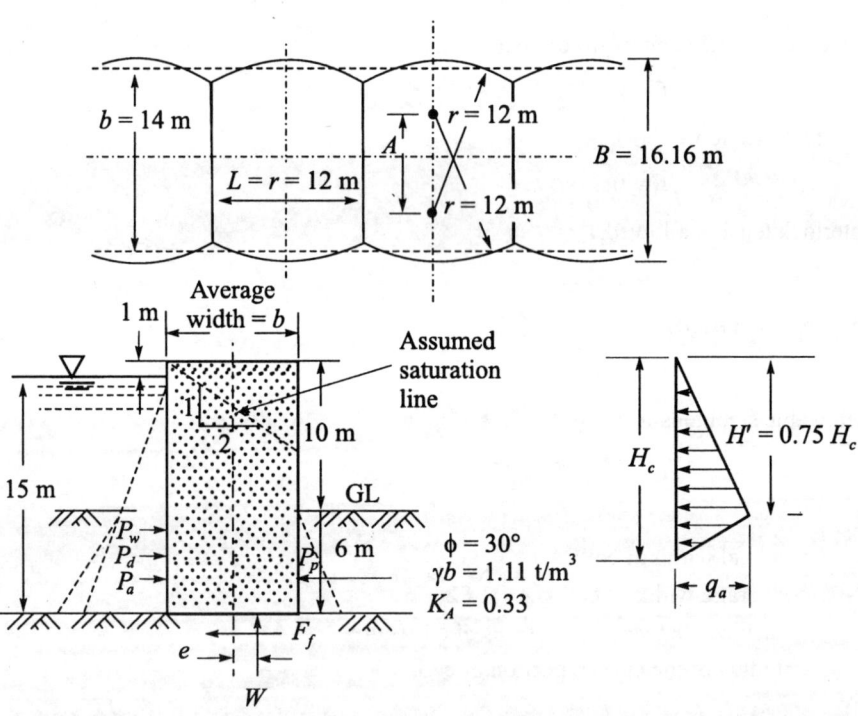

Fig. Ex. 14.1

where,

P_w = Total pressure due to water,

P_a = Active pressure due to soil,

$$P_w = \frac{1}{2} \times 1 \times 15^2 = 112.5 \text{ tonnes,}$$

$$P_a = \frac{1}{2} \times 1.1 \times 6^2 \times 0.33 = 6.5 \text{ tonnes,}$$

$$P_d = 112.5 + 6.5 = 119.0 \text{ tonnes.}$$

Let the factor of safety $F_s = 1.25$. Therefore

$$1.25 = \frac{F_f + P_p}{P_d}$$

$$= \frac{0.57\left(18.2b - 0.025b^2\right) + 59}{119}$$

The quadratic equation in b may be written as

$$0.014b^2 - 10.4b + 90 = 0$$

Solving for b, we have

$$b = 8.8 \text{ m}$$

2. *Width to satisfy overturning with a factor of safety 1.25 overturning moment M_0.*

$$M_0 = P_w \times \frac{15}{3} + P_a \times \frac{6}{3} - P_p \times \frac{6}{3}$$

$$= 112.5 \times 5 + 6.5 \times 2 - 59 \times 2$$

$$= 457.5 \text{ m-tonnes}$$

The maximum allowable eccentricity is

$$e = \frac{b}{6}$$

For stability, we may write the equation

$$W_e = M_0 F_s$$

Therefore,

$$(18.2b - 0.025b^2)\frac{b}{6} = 457.5 \times 1.25 = 572$$

The value of $b = 14$ m by trial and error method. Next check overturning from shear of piling on cell fill.

Summing moments about the toe, we obtain

$$fb P_w = M_0 F_s$$

or

$$b = \frac{M_0 F_s}{f P_w} = \frac{457.5 \times 1.25}{0.4 \times 112.5} = 12.7 \text{ m}$$

The controlling width, $b = 14$ m.

3. *Check shear along centre line of cell and interlock friction. Assume the radius of Cell $r = L$, the distance between diaphragms.*

The total weight of soil in the cell per metre length

$$= 18.2 \times 14 - 0.025 \times 14^2 = 250 \text{ tonnes}$$

The average unit weight of the fill in the cell

$$= \gamma_a = \frac{250}{16 \times 14} = 1.15 \text{ tonnes/m}^3$$

The lateral pressure coefficient for $\phi = 30°$ is

$$\bar{K}_A = \frac{\cos^2 \phi}{2 - \cos^2 \phi} = \frac{0.75}{2 - 0.75} = 0.6$$

where, $\tan \phi = 0.577$

The soil shear resistance S_r along the centre line of the cell is (assuming average unit weight γ_a)

$$S_r = \frac{1}{2} \gamma_a H^2 \bar{K}_A \tan \phi$$

$$= \frac{1}{2} \times 1.15 \times 16^2 \times 0.6 \times 0.577 = 51 \text{ tonnes}$$

For computing interlock shear F_c, assume $H_c = \dfrac{16 + 9}{2}$

$$= 12.5 \text{ m}$$

$$F_c = 3/8 \, \gamma_a \, HH_c K_A f$$

$$= 3/8 \times 1.15 \times 16 \times 12.5 \times 0.33 \times 0.3$$

$$= 8.6 \text{ tonnes.}$$

The shear on the centre line of the cell due to overturning is

$$V = \frac{1.5 \, M_0}{b} = \frac{1.5 \times 457.5}{14} = 49 \text{ tonnes}$$

The safety factor is $F_s = \dfrac{S_r + F_c}{V} = \dfrac{51 + 8.6}{49}$

$$= 1.22 \approx 1.25$$

4. *Check for interlock tension:* The depth at which maximum pressure occurs on the cell wall is

$$H' = 3/4 \, H_c = 3/4 \times 12.5 = 9.37 \text{ m};$$

$$H'_w = 9.0 - (10 - 9.37) = 8.37 \text{ m}$$

$$q_a = \gamma_a H' K_A + \gamma_w H_w'$$

$$= 1.15 \times 9.37 \times 0.33 + 1 \times 8.37$$

$$= 3.56 + 8.37 = 11.93 \text{ tonness}/\text{m}^2$$

Interlock tension, $T = \dfrac{q_a r}{100}$ tonnes/cm, where r is in metres

$$= \frac{11.93 r}{100} = 0.1193 r \text{ tonnes/cm}$$

For T less than or equal to 1.45 tonnes/cm

$$r = \frac{1.45}{0.1193} = 12.1 \text{ m} \approx 12 \text{ m}$$

The interlock tension $T = 0.1193 \times 12 = 1.43 \; T/\text{cm}$

Assuming thickness of web = 1.25 cm, the web stress

$$= \frac{1.43}{1.25} = 1.14 \text{ tonnes/sq. cm} < 1.5 \text{ tonnes/cm}^2$$

The final cell dimensions are (Fig. Ex. 14.1)

The distance between diaphragm $L = r = 12$ m

Actual width of cell,

$$B = 14 = A + 2r = A + 24$$

Average width of cell,

$$b = 14 = A + 1.82r$$

$$= A + 21.84$$

Solving for B, we have $B = 16.16$ m

With $B = 16.16$ m, the distance $A = -7.84$ m

Example 14.2

Design a circular type cofferdam by making use of the data given in Ex. 14.1.

Solution

All computations remain the same as in Ex. 14.1. The diameter D of the cell is obtained from

$$D = \frac{b}{0.875} = \frac{14}{0.875} = 16 \text{ m}$$

Hoope tension $\quad T = \dfrac{q_a r}{100} = \dfrac{11.93 \times 16}{100 \times 2}$

$$= 0.95 \text{ tonnes/cm} < 1.45 \text{ T/cm}$$

The stress in the Tee is

$$T' = \frac{0.95}{\cos 45°} = 1.39 \text{ tonnes/cm}$$

The final dimensions for a circular cell are

$$b = 14 \text{ m}, \qquad D = 16 \text{ m}$$

14.7 QUESTIONS AND PROBLEMS

14.1 What are the different types of cellular cofferdams? Discuss their advantages and disadvantages.

14.2 What are the components of a cellular cofferdam? What are the desirable properties of fill in a cell?

14.3 Find the depth of embedment of a diaphragm type cofferdam of total width $B = 7$ m and height above ground level = 7 m. Provide a factor of safety of 1.2. Use the other data given in Ex. 14.1. The dam retains water on one side up to the top.

Machine Foundations Subjected to Dynamic Loads ⑮

15.1 INTRODUCTION

Static and Dynamic Loads

Foundations are subjected to either static or dynamic loads or a combination of both. The word *static* implies that the loads are imposed slowly and gradually on the foundations in such a way as to avoid any vibration of the foundation-soil system; whereas the load is said to be *dynamic* if the load coming on the foundation leads to vibration of the whole system. In both the cases, stresses and strains are induced into the foundation-soil system. *Soil Mechanics,* as it is normally understood, deals with static loads only. The design of foundations subjected to static loads have been dealt with in the earlier chapters.

The sources of dynamic loads are many. The most of the violent types of dynamic loads are due to earth quakes and bomb blasts. Earthquakes produce random ground motions which would lead to violent vibration of the foundation-soil system if the structures lie close to the epicenters of earthquakes. As a result of such vibrations, footings may settle and structures may collapse. The damage to the structures would be all the more severe if they are founded on saturated sandy strata. Under violent vibrations the sandy strata may liquefy loosing its strength to support structures.

Foundations may also be subjected to vibrations due to driving of piles in the vicinity, due to landing of aircrafts, wind and water action on the structures, etc.

The vibrations due to the above causes are transient and intense. It is not within the scope of this book to analyse these causes and design the foundations.

Dynamic Loads due to Vibrating Machines

The vibrations that are of interest in this chapter are the ones due to the operation of machines such as reciprocating and rotary types and hammers. The vibrations caused to the foundation soil system by these machines can be analysed based on the *principle of harmonic motion.* The behaviour of the system can be predicted by reducing the foundation-soil system to an idealised lumped parameters. The simplest system is the *classical single-degree of freedom system* with *viscous damping* consisting of a *mass, spring and a dashpot.* Experience has indicated that this concept provides a very satisfactory model with which to make a dynamic analysis, even though the real system may not physically resemble the mathematical model. The analysis of the *lumped-parameter system* is based on the assumption that the vibrations induced into the foundation-soil system by the operation of a machinery follow a simple-harmonic motion with one degree of freedom. (The *degree of freedom* of a system depends on the number of independent coordinates necessary to describe the motion of a system).

For some systems a single-degree of freedom model will not represent accurately the dynamic soil response. In these cases a model having two or more degrees-of-freedom may be required. The analysis then becomes considerably more complicated and the use of a digital or analog computer becomes necessary. Fortunately, the major part of the foundation problems arising from vibrating machinery can be analysed by assuming a simple harmonic application of force.

Methods of Analysis

In this chapter the following methods of analysis of machine foundation vibratory problems are discussed.

1. The classical single-degree-of-freedom-system with viscous damping comprising of a mass, spring and a dashpot.
2. The dynamic response of machine foundation resting on elastic half-space.

The principal difficulty which currently exists in vibration analysis consists in determining the necessary soil parameters which serve as inputs into the differential equation describing the vibratory motion. A brief review of the methods of determining soil parameters are also presented in this chapter.

The end product of the design procedure is the determination of a foundation-soil-system which satisfactorily supports equipments or machinery. The supported unit may be the source of dynamic loads applied to the system or it may require isolation from external excitation. In each case, the criteria for satisfactory operation of the unit dictate the design requirements. A brief discussion on the design criteria is presented.

When two or more vibrating machineries are close to each other or machineries lying close to the walls of the building structure, there is every possibility of the vibration of one machinery interfering with the vibration of the adjoining machinery or affecting the foundation of the building structure. A brief discussion on the methods to be followed for screening of one from the other is also presented.

15.2 BASIC THEORIES OF VIBRATION

The lumped parameter analysis of foundation vibratory problems represented by a mass, spring and a dashpot can be better understood by studying the problem under the following headings.

1. Free vibrations of a mass-and-spring system without damping.
2. Free vibrations of a mass-and-spring system with damping.
3. Forced vibrations of mass-and-spring system without damping.
4. Forced vibrations of mass-and-spring system with damping.

Before taking up the study of the basic theories vibrations, it is better to understand first the properties of simple harmonic motion.

15.3 SIMPLE HARMONIC MOTION

Graphical Representation of Equation of Displacement

The simplest form of *periodic* motion is *harmonic motion*. Simple harmonic motion is defined as the motion of a point in a straight line, such that the acceleration of the point is proportional to the distance of the point from some fixed origin. An example is the motion of a weight suspended by a spring [Fig. 15.1 (b)] and set into vertical oscillation by being pulled down beyond the static position [Fig. 15.1 (d)] and released. If the spring is frictionless and weightless, the weight W oscillates about he static position, OO, undamped. The maximum displacement, z_{max}, with respect to OO is termed as the amplitude A. *The total displacement at the extreme positions of the weight W is $2A$* sometimes referred to as double amplitude [Fig. 15.1 (e)], the displacement at any position of the weight may be called as z (The displacements are measured with respect to a point p marked above the weight).

A graphical method of representing the motion of the weight, oscillating with simple harmonic motion, is shown in [Fig. 15.1 (g)]. The actual line of oscillation of the point p in the vertical direction may be taken as the projection on the vertical line of the point a rotating at uniform velocity along a circle having its centre at O. The displacement-time curve is also shown in [Fig. 15.1 (g) (iii)].

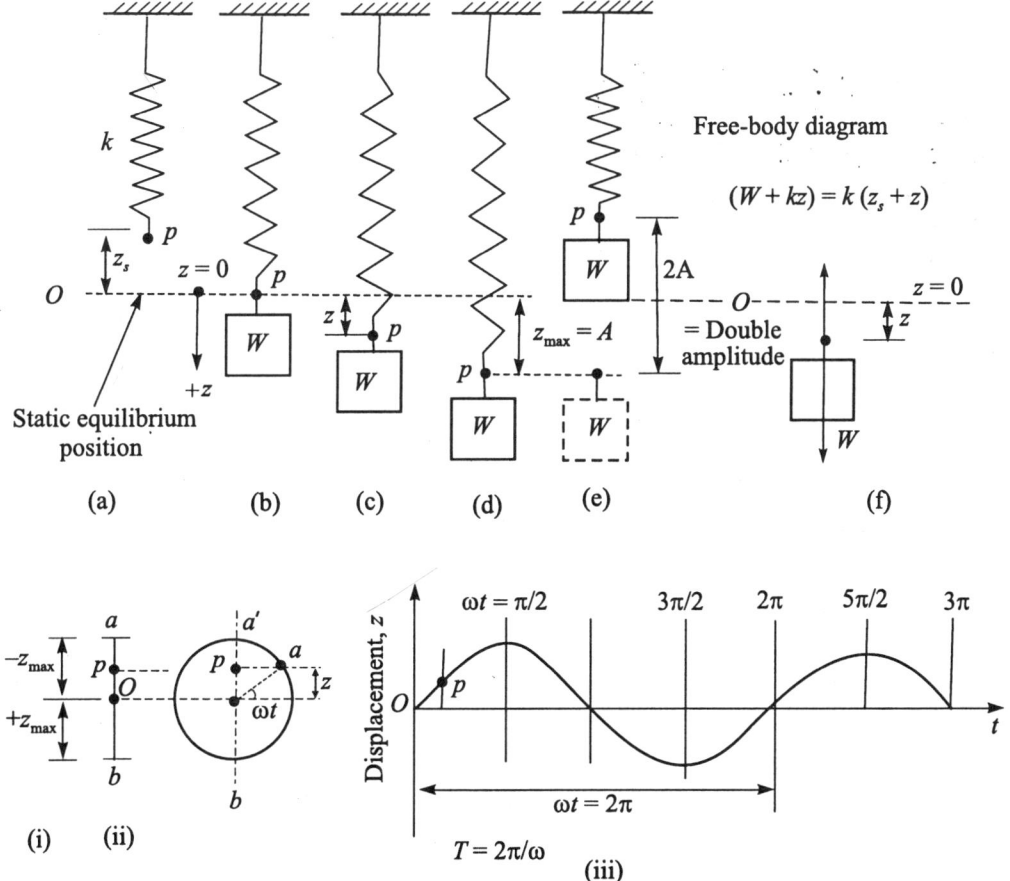

(g) Graphical representation of simple harmonic motion

Fig. 15.1 Principles of free-vibration without damping: (a) Unstretched spring, (b) static position with weight on spring, (c) weight in motion, (d) weight in a maximum downward position, (e) weight in maximum upward position, (f) free-body diagram, (g) representation of harmonic motion, (i) actual path of p moving in simple harmonic motion, (ii) p determined from projection of equivalent circular motion, (iii) time-displacement curve for motion of p

The equation of motion is a sine function represented by the equation

$$z = A \sin \omega t \tag{15.1}$$

where, ω = circular frequency in radians per unit of time. With reference to [Fig. 15.1 (g)], the angular speed of the point a around the circle is represented by ω. Since the function repeats itself after 2π radians, a cycle of motion is completed when

$$\omega t = 2\pi$$

or

$$t = T = \frac{2\pi}{\omega} \tag{15.2}$$

where, T is referred to as the *period* of motion. The number of oscillation in terms of cycles per unit of time is given by

$$f = \frac{1}{T} = \frac{\omega}{2\pi} \tag{15.3}$$

The units of cycles per second are also called as Hertz (Hz).

Vector Representation of Harmonic Motion

The use of rotating vectors to represent the harmonic motion helps to have a physical insight to the mechanism of vibration.

Consider the displacement equation Eq. (15.1)

$$z = A \sin \omega t$$

By successive differentiation of Eq. (15.1), we get the equations of velocity and acceleration as

$$\text{velocity,} \quad \frac{dz}{dt} = \dot{z} = \omega A \cos \omega t = \omega A \sin (\omega t + \pi/2) \tag{15.4a}$$

$$\text{acceleration,} \quad \frac{d^2z}{dt^2} = \ddot{z} = -\omega^2 A \sin \omega t = \omega^2 A \sin (\omega t + \pi) \tag{15.4b}$$

It is clear from the above equations, that velocity leads the displacement by 90° and acceleration leads displacement by 180°.

If a vector of length A is rotated counter clockwise about the origin as shown in Fig. 15.2 (a) its projection on to the vertical axis would be equal to $A \sin \omega t$ which is exactly the expression for displacement given in Eq. (15.1). It follows that the velocity can be represented by the vertical projection of a vector of length ωA positioned 90° ahead of displacement vector. Likewise, acceleration can be represented by a vector of length $\omega^2 A$ located 180° ahead of the displacement vector. A plot of all the three quantities is shown in Fig. 15.2 (b).

15.4 FREE VIBRATION OF A MASS-AND-SPRING SYSTEM WITHOUT DAMPING

Figure 15.1 gives the different positions of the oscillating system under free vibration: (*a*) Position of spring without the weight W being attached; (*b*) represents the position of the spring with weight under static equilibrium condition. With the weight W acting, the spring gets stretched vertically a distance z_s under static condition which can be related to a spring constant k as

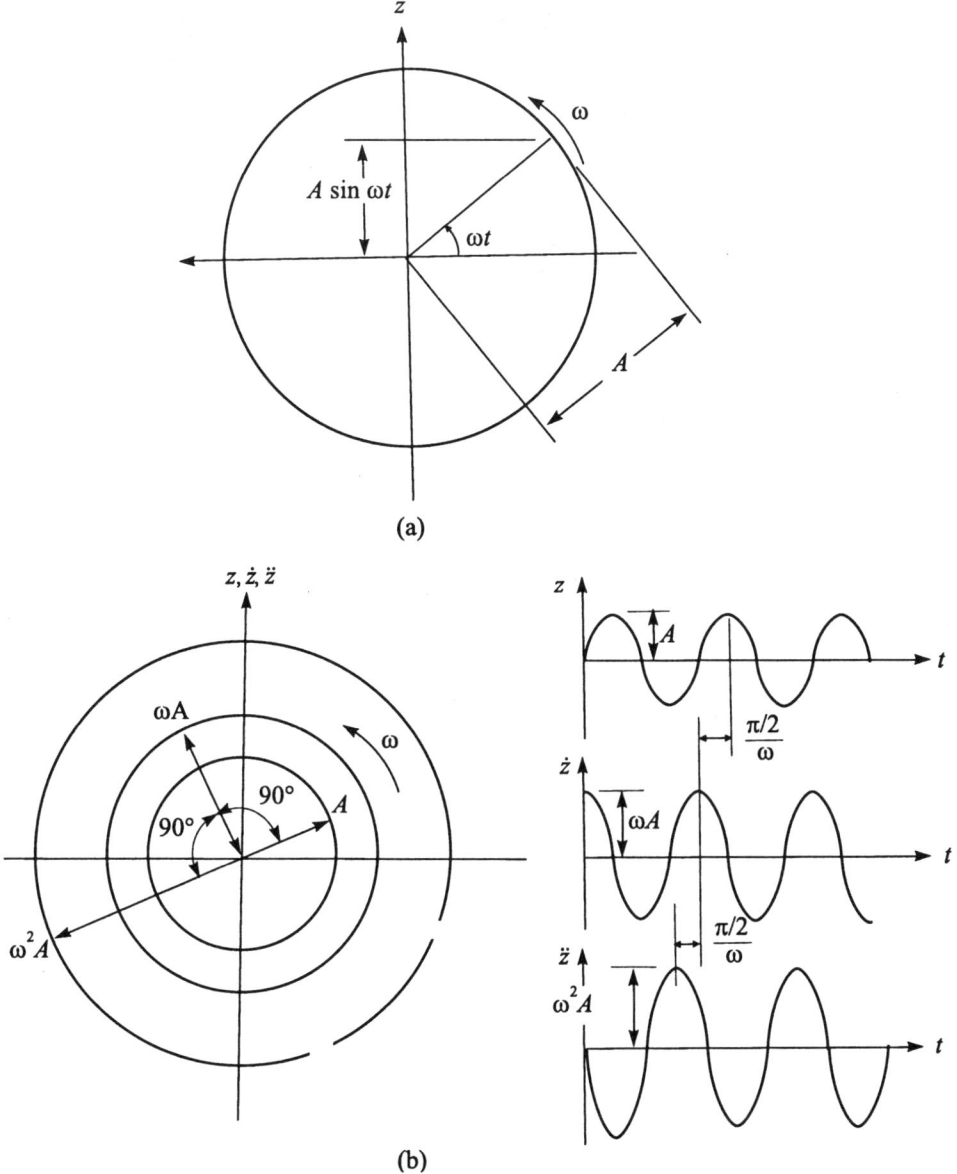

Fig. 15.2 Vector method of representing harmonic motion: (a) Vector representation of harmonic motion, (b) vector representation of harmonic displacement, velocity and acceleration

$$k = \frac{W}{z_s} \tag{15.5}$$

which is expressed as a *static force per unit deflection of the spring.*

The differential equation of motion is obtained from Newton's second saw which states that the net unbalanced force on a constant mass system is equal to the mass of the system multiplied by its acceleration. If the system shown in Fig. 15.1 is displaced a distance z [Fig. 15.1 (c)] from the at rest position ($z = 0$), the force in the spring will be equal to ($W + kz$) [Fig. 15.1 (f)]. Equilibrium of the system requires that

$$\Sigma F = m\ddot{z}$$

where, ΣF = sum of the vertical forces,

m = mass of the system,

$$\ddot{z} = \text{acceleration} = \frac{d^2 z}{dt^2}$$

Therefore,

$$-(W + kz) + W = m \frac{d^2 z}{dt^2}$$

or

$$\frac{d^2 z}{dt^2} + \frac{kz}{m} = 0 \qquad (15.6)$$

The value of z that satisfies Eq. (15.6) must be a function of t whose second derivative with respect to t is equal to the original function multiplied by k/m. The general solution of Eq. (15.6) may be written as

$$z = C_1 \sin \sqrt{\frac{k}{m}} t + C_2 \cos \sqrt{\frac{k}{m}} t \qquad (15.7)$$

where, C_1 and C_2 are arbitrary constants which can be evaluated from the initial conditions of the system. The quantity $\sqrt{k/m}$ corresponds to the undamped natural circular frequency of the system, designated as

$$\omega_n = \sqrt{\frac{k}{m}} \text{ rad/sec} \qquad (15.8)$$

and the undamped natural frequency is

$$f_n = \frac{1}{2\pi} \sqrt{\frac{k}{m}} \text{ cycles/sec} \qquad (15.9)$$

The time required for one complete cycle (one revolution) is called as the *natural period* which is expressed as

$$T_n = \frac{2\pi}{\omega_n} = 2\pi \sqrt{\frac{m}{k}} \qquad (15.10)$$

By substituting W/g for m, and z_s for W/k, we have from Eq. (15.9)

$$f_n = \frac{1}{2\pi} \sqrt{\frac{g}{z_s}} \qquad (15.11)$$

and

$$T_n = \frac{1}{f_n} = 2\pi \sqrt{\frac{z_s}{g}} \qquad (15.12)$$

where, g = acceleration due to gravity ≈ 981 cm/sec^2.

It is clear from the above equations that for free vibration of an undamped single-degree of freedom system, the motion is harmonic and occurs at a natural frequency of f_n. The amplitude of motion is determined from the initial conditions. We may write,

at $t = 0$, $\qquad\qquad\qquad z = z_0$ and $\dot{z} = v_0$ $\qquad\qquad\qquad\qquad\qquad$ (15.13)

Substituting these conditions in Eq. (15.7), and writing $\sqrt{k/m} = w_n$, we have

$$C_1 = \frac{v_0}{\omega_n}, \ C_2 = z_0 \qquad\qquad\qquad\qquad (15.14)$$

Now Eq. (15.7) may be expressed as

$$z = \frac{v_0}{\omega_n} \sin \omega_n t + z_0 \cos \omega_n t \qquad\qquad\qquad\qquad (15.15)$$

Example

A mass supported by a spring gives a static deflection of 1.50 mm. Determine its natural frequency of oscillation.

Solution

From Eq. (15.11), we have

$$f_n = \frac{1}{2\pi} \sqrt{\frac{g}{z_s}} = \frac{1}{2 \times 3.14} \sqrt{\frac{981 \times 10}{1.5}}$$

$$= 12.89 \ \text{cycles/sec} = 12.89 \ \text{Hz}.$$

15.5 FREE VIBRATIONS WITH VISCOUS DAMPING

Figure 15.3 (a) shown a schematic arrangement of mass-spring-dashpot system under free vibration. The simplest mathematical element is the *viscous damper* or otherwise called as *dashpot*. The force in the dashpot under dynamic loading is directly proportional to the velocity of the oscillating mass. (An example of viscous damping is a hydraulic shock absorber). The dashpot exerts a force which acts to oppose the motion of the mass. The free-body diagram of the mass is given in Fig. 15.3 (a). The equation of motion is

$$SF = m\ddot{z}$$

or $\qquad\qquad -(W + kz) - c \dfrac{dz}{dt} + W = m \dfrac{d^2z}{dt^2}$

where, $\quad c$ = the coefficient of viscous damping expressed as force per unit velocity ($FL^{-1}T$).

The differential equation for free vibration with viscous damping is

$$\frac{d^2z}{dt^2} + \frac{c}{m}\frac{dz}{dt} + \frac{kz}{m} = 0 \qquad\qquad\qquad\qquad (15.16)$$

Let the solution to Eq. (15.16) be in the form

$$z = e^{\lambda t} \tag{15.17}$$

which satisfies the Eq. (15.16), where λ is a constant to be determined, and e the base of Naperian logarithm.

Substituting $e^{\lambda t}$ for z in Eq. (15.16), we have

$$\left(\lambda^2 + \frac{c}{m}\lambda + \frac{k}{m} \right) e^{\lambda t} = 0 \tag{15.18}$$

which gives, $\qquad \lambda^2 + \dfrac{c}{m}\lambda + \dfrac{k}{m} = 0 \tag{15.19}$

The solution of λ may now be written as

$$\lambda_1 = -\frac{c}{2m} + \sqrt{\left(\frac{c}{2m}\right)^2 - \frac{k}{m}} \tag{15.20a}$$

$$\lambda_2 = -\frac{c}{2m} - \sqrt{\left(\frac{c}{2m}\right)^2 - \frac{k}{m}} \tag{15.20b}$$

Three possible types of damping arises from the roots of Eq. (15.20). They are:

Case 1: Roots real if $\left(\dfrac{c}{2m}\right)^2 > \dfrac{k}{m}$

Case 2: Roots equal if $\left(\dfrac{c}{2m}\right)^2 = \dfrac{k}{m}$

Case 3: Roots imaginary or complex if $\left(\dfrac{c}{2m}\right)^2 < \dfrac{k}{m}$

Case 1: When $\left(\dfrac{c}{2m}\right)^2 > \dfrac{k}{m}$

For this case, the two roots of Eq. (15.19) are real as well as negative. The general solution to Eq. (15.16) may be expressed as

$$z = C_1 \exp(\lambda_1 t) + C_2 \exp(\lambda_2 t) \tag{15.21}$$

In Case 1, since λ_1 and λ_2 are both negative, z will decrease exponentially without change in sign as shown in Fig. 15.3 (b). In this case no oscillation will occur and the *system is said to be overdamped.*

Case 2: When $\left(\dfrac{c}{2m}\right)^2 = \dfrac{k}{m}$

The roots are equal. This is similar to overdamped case except that it is possible for the sign to change once as shown in Fig. 15.3 (b). The value of c required to satisfy the above condition is called the *critical damping coefficient* c_c. Since $\left(\dfrac{c}{2m}\right)^2 = \dfrac{k}{m}$, we may write

$$c = c_c = 2 \sqrt{km} \tag{15.22a}$$

As per Eq. (15.8), $k = \omega_n^2 m$.

Substituting for k in Eq. (15.22a), we have

$$c_c = 2m \, \omega_n \tag{15.22b}$$

Case 3: When $\left(\dfrac{c}{2m}\right)^2 < \dfrac{k}{m}$

For systems with damping less than the critical damping, the roots of Eq. (15.19) will be complex conjugates. By introducing the relationship for c_c, the roots λ_1 and λ_2 become

$$\lambda_1 = \omega_n \left(-D + i \sqrt{1 - D^2}\right) \tag{15.23a}$$

$$\lambda_2 = \omega_n \left(-D - i \sqrt{1 - D^2}\right) \tag{15.23b}$$

where, $D = \dfrac{c}{c_c}$ called as the damping factor.

Substituting Eq. (15.23) into Eq. (15.21) and simplifying, the general solution then becomes

$$z = \exp\left(-\omega_n D_t\right) \left(C_3 \sin \omega_n t \sqrt{1 - D^2} + C_4 \cos \omega_n t \sqrt{1 - D^2}\right) \tag{15.24}$$

where, C_3 and C_4 are arbitrary constants.

Equation (15.24) indicates that the motion will be oscillating and the decay in amplitude will be proportional to $\exp\left(-\omega_n dt\right)$ as shown by the dashed curve in Fig. 15.3 (c). Examination of Eq. (15.24) indicates that the frequency of free vibrations is less than that undamped natural circular frequency and that as $D \to 1$, the frequency approaches zero. The natural frequency for damped oscillation in terms of undamped natural frequency is given by

$$\omega_{dn} = \omega_n \sqrt{1 - D^2} \tag{15.25}$$

where, ω_{dn} = damped natural frequency.

Figure 15.3 (c) indicates that there is a decrement in the successive peak amplitudes. By making use of Eq. (15.24), it is possible to write an equation for the ratios of the successive peak amplitudes as follows.

Let z_1 and z_2 are the amplitude of two successive peaks at times t_1 and t_2 respectively as shown in Fig. 15.3 (c). The ratios of peak amplitudes is given by

$$\frac{z_1}{z_2} = \exp\left(\frac{2\pi D}{\sqrt{1 - D^2}}\right) \tag{15.26}$$

The logarithmic decrement is defined as

$$\delta = ln\ \frac{z_1}{z_2} = \frac{2\pi D}{\sqrt{1-D^2}} \approx 2\pi D \qquad (15.27)$$

when D is small

It can be seen from Eq. (15.27) that one of the properties of viscous damping is that the decay of vibrations is such that the ratios of amplitudes of any two successive peaks is a constant. Thus, the logarithmic decrement can be obtained from any two peak amplitudes z_1 and z_{1+n} from the relationship.

$$\delta = \frac{1}{n}\ ln\ \frac{z_1}{z_{1+n}} \qquad (15.28)$$

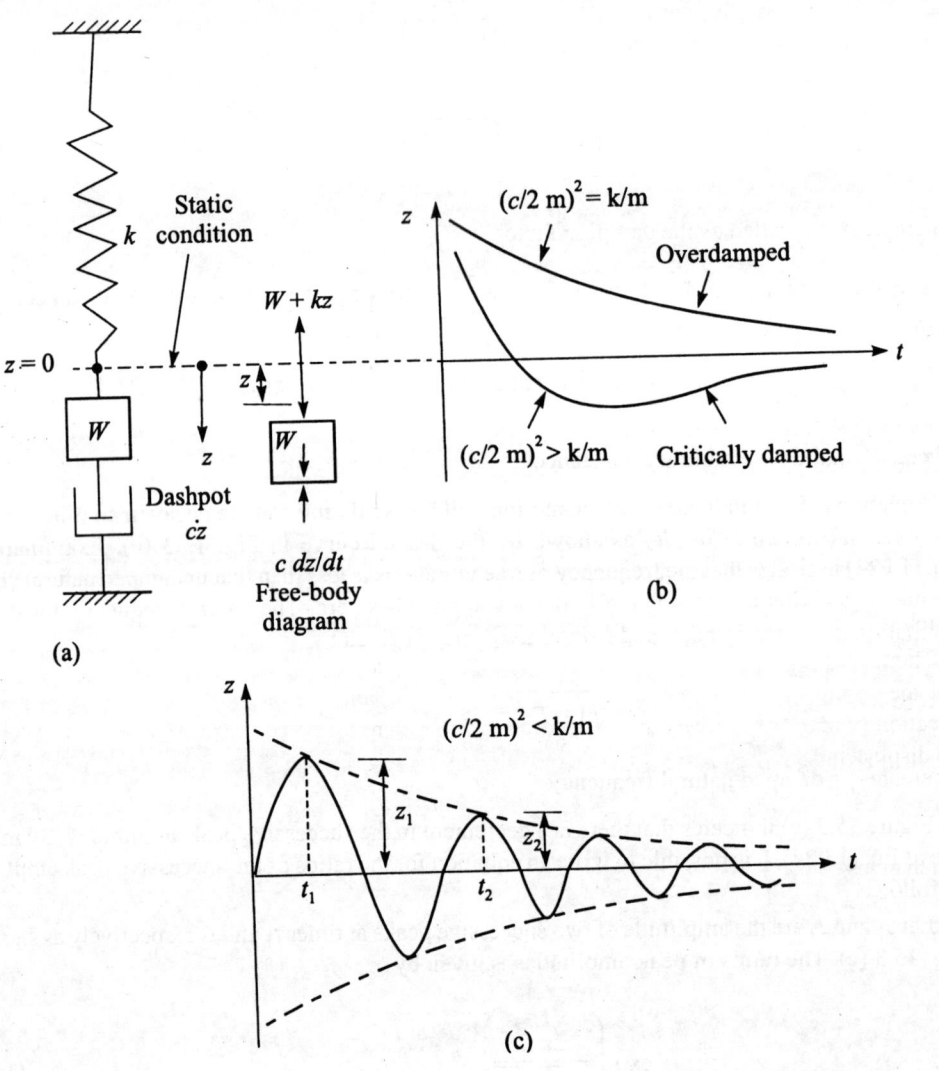

Fig. 15.3 Free vibration system with viscous damping: (a) Forces acting on a system under free vibrations with damping, (b) overdamped and critically damped system, (c) under damped system

15.6 FORCED VIBRATIONS OF MASS-AND-SPRING SYSTEM WITHOUT DAMPING

If a mass which is resting on an elastic medium is subjected to a *periodic oscillating force,* the mass will oscillate, and the period and amplitude of the oscillation will be influenced by the period and the amplitude of the exciting force. Such periodic forces may be applied to a foundation by unbalanced rotating machinery, or by the inertia forces of compressor pistons, or by other means.

Assume that the periodic force is expressed as

$$Q = Q_0 \sin \omega t \tag{15.29}$$

where, ω = the circular frequency of the force of excitation (radians/sec).

The free-body diagram of the mass m is as shown in Fig. 15.4 (a). The equation of motion for this case is $\Sigma F = m\ddot{z}$. Summing up all the vertical forces, we have

$$W - (W + kz) + Q_0 \sin \omega t = m \frac{d^2z}{dt^2}$$

The differential equation of motion is

$$m \frac{d^2z}{dt^2} + kz = Q_0 \sin \omega t \tag{15.30}$$

The solution to Eq. (15.30) includes the solution for free vibrations, Eq. (15.7), along with the solution which satisfies the right hand side of Eq. (15.30). Eq. (15.30) can be solved by parts or by using the concept of rotating vectors. Since the concept of rotating vectors gives a physical feeling for the problem, only this method is explained here. For the solution by parts readers may refer to Converse (1962).

The motion of the system is assumed to be harmonic, since the applied force is harmonic. The motion of the system may be of the form as per Eq. (15.1).

$$z = A \sin \omega t \tag{15.31}$$

which is represented by motion vectors in Fig. 15.4 (b).

The forces acting on the mass are represented by *force vectors* in Fig. 15.4 (c). It may be noted that the spring force acts opposite to the displacement, and the inertia force acts opposite to the direction of acceleration. The *exciting force vector* of amplitude Q_0 is shown acting in phase with the displacement vector. For equilibrium, we have

$$Q_0 + m\omega^2 A - kA = 0 \tag{15.32}$$

which gives

$$A = \frac{Q_0}{k - m\omega^2} = \frac{\dfrac{Q_0}{k}}{1 - \left(\dfrac{\omega}{\omega_n}\right)^2} \tag{15.33}$$

Now, from Eq. (15.7), (15.31) and (15.33), the complete solution for forced vibrations under undamped conditions may be written as

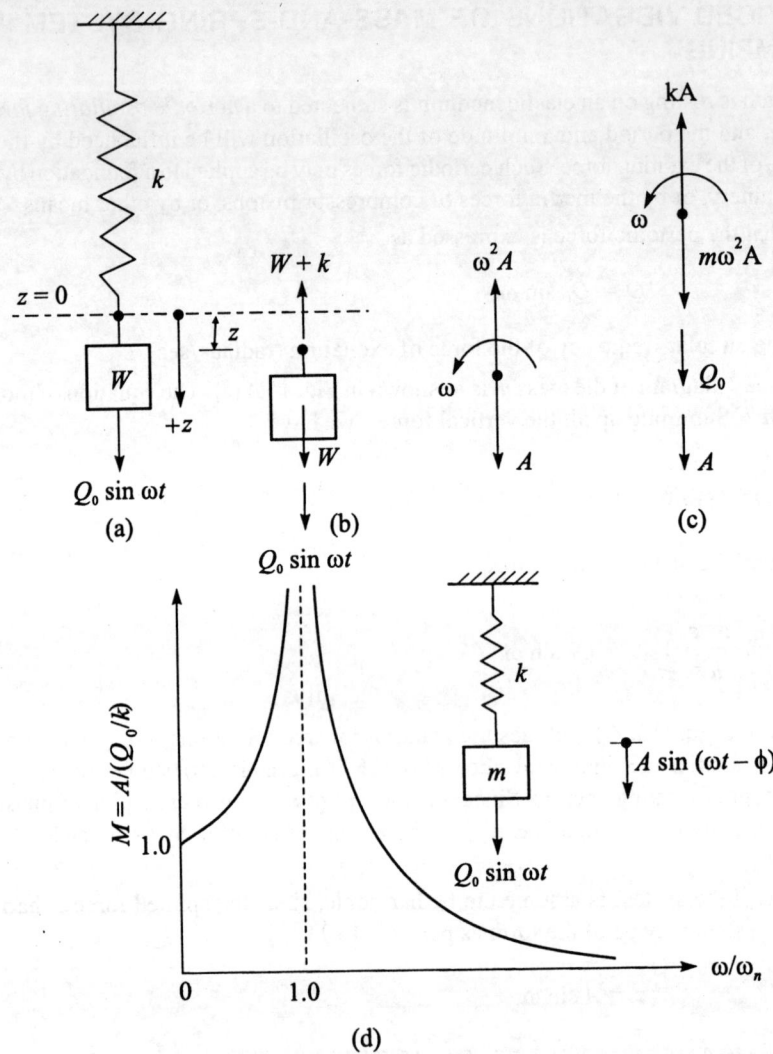

Fig. 15.4 Forced vibrations of mass-and-spring system: (a) Free-body diagram, (b) motion vectors, (c) force vectors, and (d) magnification factor *vs* ω/ω_n

$$z = \frac{\dfrac{Q_0}{k}}{1 - \left(\dfrac{\omega}{\omega_n}\right)^2} \sin \omega t + C_1 \sin \omega_n t + C_2 \cos \omega_n t \qquad (15.34)$$

The last two terms in Eq. (15.34) will eventually vanish because of damping in the real system, and the equation that gives the steady state solution is

$$z = \frac{\dfrac{Q_0}{k}}{1 - \left(\dfrac{\omega}{\omega_n}\right)^2} \sin \omega t \qquad (15.35)$$

From Eq. (15.35), the amplitude A is obtained, as

$$z_{max} = A = \frac{\dfrac{Q_0}{k}}{1 - \left(\dfrac{\omega}{\omega_n}\right)^2} \tag{15.36}$$

or

$$M = \frac{A}{\dfrac{Q_0}{K}} = \frac{1}{1 - \left(\dfrac{\omega}{\omega_n}\right)^2} \tag{15.37}$$

where, M = dynamic magnification factor,

$\dfrac{Q_0}{k}$ = static deflection of the system under the action of Q_0.

If the phase angle ϕ (will be defined in the next section) between the force and displacement is taken into account, then the Eq. (15.31) will be written as

$$z = A \sin (\omega t - \phi) \tag{15.38}$$

Figure 15.4 (d) gives a relationship between the magnification factor M and ω/ω_n. The value of M becomes infinite when $\omega = \omega_n$ because no damping is included in the model.

15.7 FORCED VIBRATIONS OF MASS-AND-SPRING SYSTEM WITH VISCOUS DAMPING

The properties of many foundation soil system can be approximated by introducing viscous damping into the system, since damping is always present in one form or another. Figure 15.5 (a) represents the system to be analysed with the free-body-diagram.

$$\Sigma F = m\ddot{z}$$

or

$$W - (W + kz) - c\,\frac{dz}{dt} + Q_0 \sin \omega t = m\,\frac{d^2z}{dt^2}$$

which reduces to

$$m\,\frac{d^2z}{dt^2} + c\,\frac{dz}{dt} + kz = Q_0 \sin \omega t \tag{15.39}$$

Using the same method used for the undamped case, the solution to Eq. (15.39) may be found out by using the concept of rotation vectors.

Since the exciting force vector Q_0 is placed with a phase angle ϕ ahead of the displacement vector A the equation of displacement may be expressed as

$$z = \sin (\omega t - \phi) \tag{15.40}$$

The relative position of motion vectors are shown in Fig. 15.5 (b). The force vectors are given in Fig. 15.5 (c). These vectors act opposite to that of the corresponding motion vectors, whereas the exciting force vector Q_0 is shown at ϕ degrees ahead of the displacement vector A. Summing up the vectorial forces in the horizontal and vertical directions for equilibrium condition, we have

$$kA - m\omega^2 A - Q_0 \cos\phi = 0 \qquad (15.41a)$$

$$c\omega A - Q_0 \sin\phi = 0 \qquad (15.41b)$$

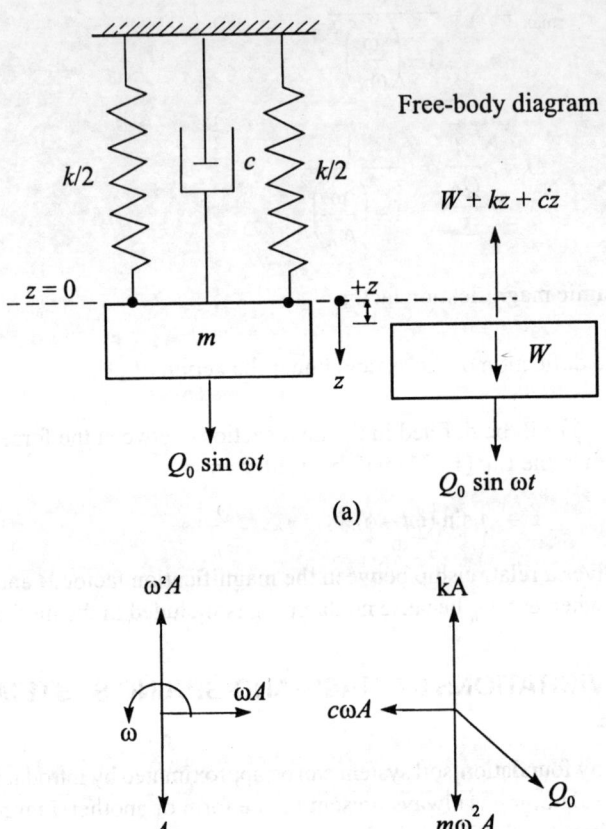

Fig. 15.5 Forced mass-spring-dashpot system: (a) Mass-spring-dashpot system, (b) motion vectors, and (c) force vectors

Solving the two equations for A and ϕ, we have

$$A = \frac{Q_0}{\sqrt{\left(k - m\omega^2\right)^2 + \left(c\omega\right)^2}} \qquad (15.42)$$

$$\tan\phi = \frac{c\omega}{k - m\omega^2} \qquad (15.43)$$

Equation (15.42) may be written in another form as

$$A = \frac{Q_0/k}{\sqrt{\left(1 - \frac{m}{k}\omega^2\right)^2 + \left(\frac{c\omega}{k}\right)^2}}$$

or

$$A = \frac{Q_0 / k}{\sqrt{\left[1 - \left(\dfrac{\omega}{\omega_n}\right)^2\right]^2 + \left(2D\dfrac{\omega}{\omega_n}\right)^2}}$$

(15.44)

since,

$$\left(\frac{c\omega}{k}\right)^2 = \left(\frac{c}{c_c}\frac{c\omega}{m\omega_n^2}\right)^2 = \left(\frac{c}{c_c}\frac{2m\omega_n\omega}{m\omega_n^2}\right)^2 = \left(2D\frac{\omega}{\omega_n}\right)^2$$

and

$$\frac{m}{k} = \frac{1}{\omega_n^2}$$

Since $\dfrac{Q_0}{k} = z_s$, the static deflection of the system under the action of Q_0, the magnification faction M may be expressed as

$$M = \frac{A}{Q_0 / k} = \frac{1}{\sqrt{\left[1 - \left(\dfrac{\omega}{\omega_n}\right)^2\right]^2 + \left(2D\dfrac{\omega}{\omega_n}\right)^2}}$$

(15.45)

It may be seen that Eq. (15.45) reduces to Eq. (15.37) if there is no viscous damping. In the same way, Eq. (15.43) may be written as

$$\tan\phi = \frac{2D\left(\dfrac{\omega}{\omega_n}\right)}{1 - \left(\dfrac{\omega}{\omega_n}\right)^2}$$

(15.46)

The Eqs (15.45) and (15.46) are plotted in Fig. 15.6 (a) for various values of D. These curves are referred to here as *response curves* for *constant-force-amplitude-excitation*.

It may be seen from Fig. 15.6 (a), that maximum amplitude occurs at a frequency slightly less than the undamped natural circular frequency ω_n where $\omega/\omega_n = 1$. The frequency at maximum amplitude will be referred to as *resonant frequency, f_m*, for constant force amplitude Q_0. We may write.

$$f_m = f_n\sqrt{1 - 2D^2} = \frac{1}{2\pi}\sqrt{\frac{k}{m}}\sqrt{1 - 2D^2}$$

(15.47)

Equation (15.47) may be obtained by differentiating Eq. (15.45) with respect to (ω/ω_n) and equating to zero $\left[\text{that is, } \dfrac{\partial(A/Q_0/k)}{\partial(\omega/\omega_n)} = 0\right]$.

The magnification factor M at the frequency of f_m is given by

$$M_{max} = \frac{1}{2D\sqrt{1 - D^2}}$$

(15.48)

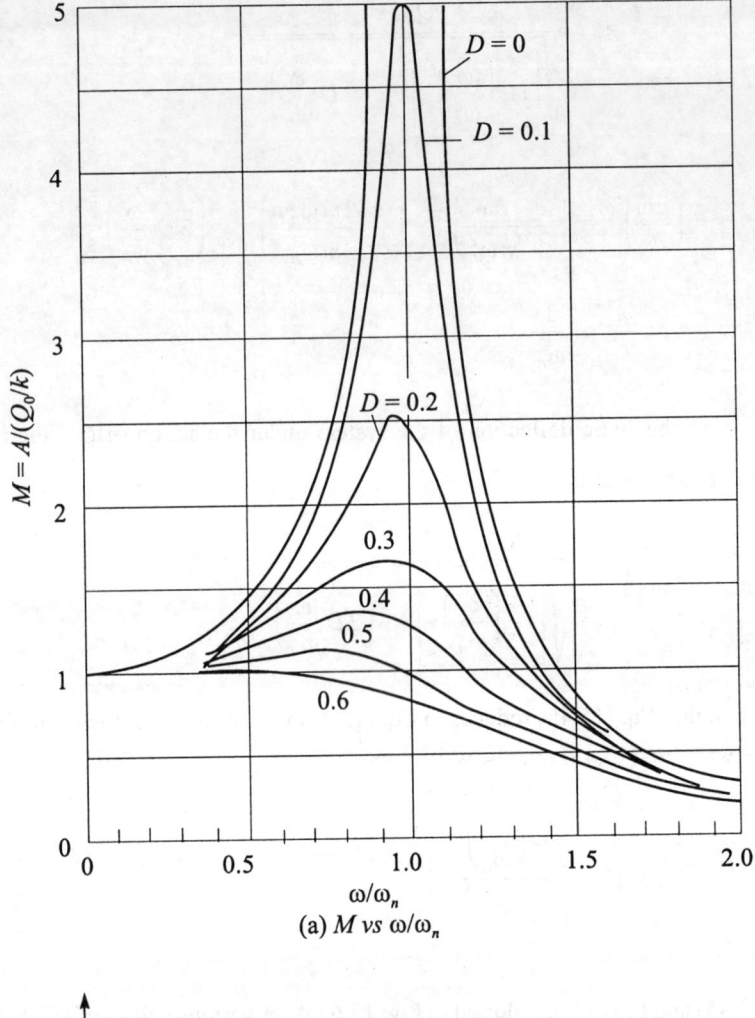

(a) M vs ω/ω_n

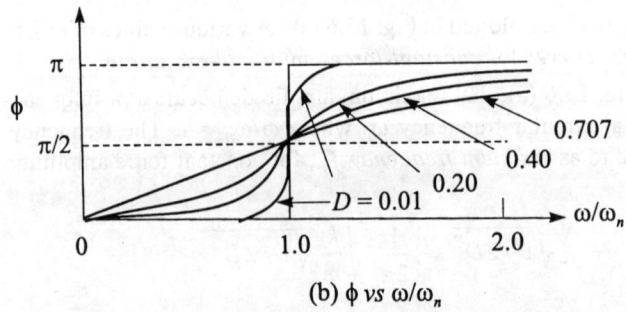

(b) ϕ vs ω/ω_n

Fig. 15.6 Response curves for a viscously damped single-degree of freedom system

when $D = 1/\sqrt{2} = 0.71, f_m = 0$, which means that the maximum response is the static response. The relationship between ϕ and ω/ω_n as per Fig. 15.6 (b) reveals the following characteristics.

for $\qquad \dfrac{\omega}{\omega_n} = 1, \qquad \phi = 90,$

for $\qquad \dfrac{\dot{\omega}}{\omega_n} < 1 \qquad \phi < 90°$

for $\qquad \dfrac{\omega}{\omega_n} > 1 \qquad \phi > 90°$

15.8 FORCED FREQUENCY DEPENDENT EXCITING FORCE WITH VISCOUS DAMPING

In many practical problems, the vibrating systems of machine foundations have unbalanced rotating masses which imparts into the foundation-soil system frequency dependent exciting forces. Fig. 15.7 (a) gives a common type vibrating generator consisting of two counteracting eccentric masses m_1 at an eccentricity e

The phase relationship between the masses is such that they reach their top positions simultaneously. The rotating force of each mass is

$$Q_1 = m_1 e \, \omega^2$$

The total force in the vertical position is therefore

$$Q_v = 2Q_1 = 2m_1 e \, \omega^2 = m_e e \, \omega^2 \qquad (15.49)$$

But when the masses are in a horizontal position the forces cancel each other. The vibrating force at any position may be represented by

$$Q = m_e e \, \omega^2 \sin \omega t = \bar{Q}_0 \sin \omega t \qquad (15.50)$$

where, $\qquad \bar{Q}_0 = m_e e \, \omega^2$

As per Eq. (15.29) the periodic force is expressed as

$$Q = Q_0 \sin \omega t$$

where, Q_0 may be replaced by $\bar{Q}_0$ for a frequency dependent exciting force. We may now write

$$\frac{\bar{Q}_o}{k} = \frac{m_e e \omega^2}{k} = \left(\frac{m_e e}{m}\right)\left(\frac{m}{k}\right) \omega^2 = \frac{m_e e}{k}\left(\frac{\omega}{\omega_n}\right)^2 \qquad (15.51)$$

since $\dfrac{m}{k} = \dfrac{1}{\omega_n^2}$ as per Eq. (15.8).

The equation for amplitude A for frequency dependent exciting force can be written from Eq. (15.44) by substituting $\dfrac{\bar{Q}_o}{k}$ for $\dfrac{Q_0}{k}$ as (after simplifying)

or $\qquad A = \dfrac{{}' \dfrac{m_e e \omega^2}{m}\left(\dfrac{\omega}{\omega_n}\right)^2}{\sqrt{\left[1 - \left(\dfrac{\omega}{\omega_n}\right)^2\right]^2 + \left(2D\,\dfrac{\omega}{\omega_n}\right)^2}} \qquad (15.52)$

Now from Eq. (15.52), we may write

$$\frac{A}{\frac{m_e e}{m}} = \frac{(\omega/\omega_n)^2}{\sqrt{\left[1-\left(\frac{\omega}{\omega_n}\right)^2\right]^2 + \left(2D\frac{\omega}{\omega_n}\right)^2}} = \left(\frac{\omega}{\omega_n}\right)^2 M \tag{15.53}$$

where, M is the *dynamic magnification factor* for the *constant-force-amplitude* case. It may be noted here that the Eq. (15.46) for ϕ remains the same for this case also.

Figure 15.7 (b) gives a plot of Eq. (15.53) for various values of D. In this case, the damped resonant frequency, f_{mr}, occurs above the undamped natural frequency f_n, and the relationship between the two is given by

$$f_{mr} = f_n \frac{1}{\sqrt{1-2D^2}} = \frac{1}{2\pi}\sqrt{\frac{k}{m}}\frac{1}{\sqrt{1-2D^2}} \tag{15.54}$$

The ordinate of f_{mr} is given by

$$\left[\frac{A}{(m_e e/m)}\right]_{\text{max}} = \frac{1}{2D\sqrt{1-D^2}} \tag{15.55}$$

It may be noted here that m is the total vibrating mass which includes the eccentric mass m_e also. Further the amplitude A approaches the value $m_e e/m$ as the ratio ω/ω_n increases beyond the resonant frequency. For this case the vibration amplitude A *reaches a value e* since at this stage $m_e \approx m$. This phenomena is the basis for adding more weight to a system vibrating above its resonant frequency in order to reduce its vibration amplitude.

15.9 PROPERTIES OF RESPONSE CURVES

The shapes of experimental response curves of a single-degree-of-freedom system help to determine the properties of the system. Consider an experimental response curve given in Fig. 15.7 (c) starting from zero amplitude for a rotating-mass-type excitation and plotted against frequency of excitation f. A line op drawn from the origin tangent to the response curve gives a point p which corresponds to the undamped natural frequency f_n of the system. If any other line ob is drawn cutting the curve at two points a and b the natural frequency f_n of the system can be found out from the equation

$$f_n = \sqrt{f_a f_b} \tag{15.56}$$

where, f_a and f_b are the frequencies at points a and b respectively. It is possible, therefore, that from a single experimental curve several independent values for f_n can be obtained and the average of the values may be taken as the *undamped natural frequency f_n* of the system.

Figure 15.7 (d) gives another curve for a single-degree of freedom system acted upon by a constant-force amplitude. The response curve in this case starts from a finite value of A at $f = 0$. A classical method of measuring damping makes use of the relative width of the curve. From this curve the logarithmic decrement δ can be calculated from the equation

$$\delta = \frac{\pi}{2} \cdot \frac{f_b^2 - f_a^2}{f_m^2}\sqrt{\frac{A^2}{A_{\text{max}}^2 - A^2}}\frac{\sqrt{1-2D^2}}{1-D^2} \tag{15.57}$$

The logarithmic decrement δ as per Eq. (15.27) is

$$\delta = \frac{2\pi D}{\sqrt{1 - D^2}} \tag{15.27}$$

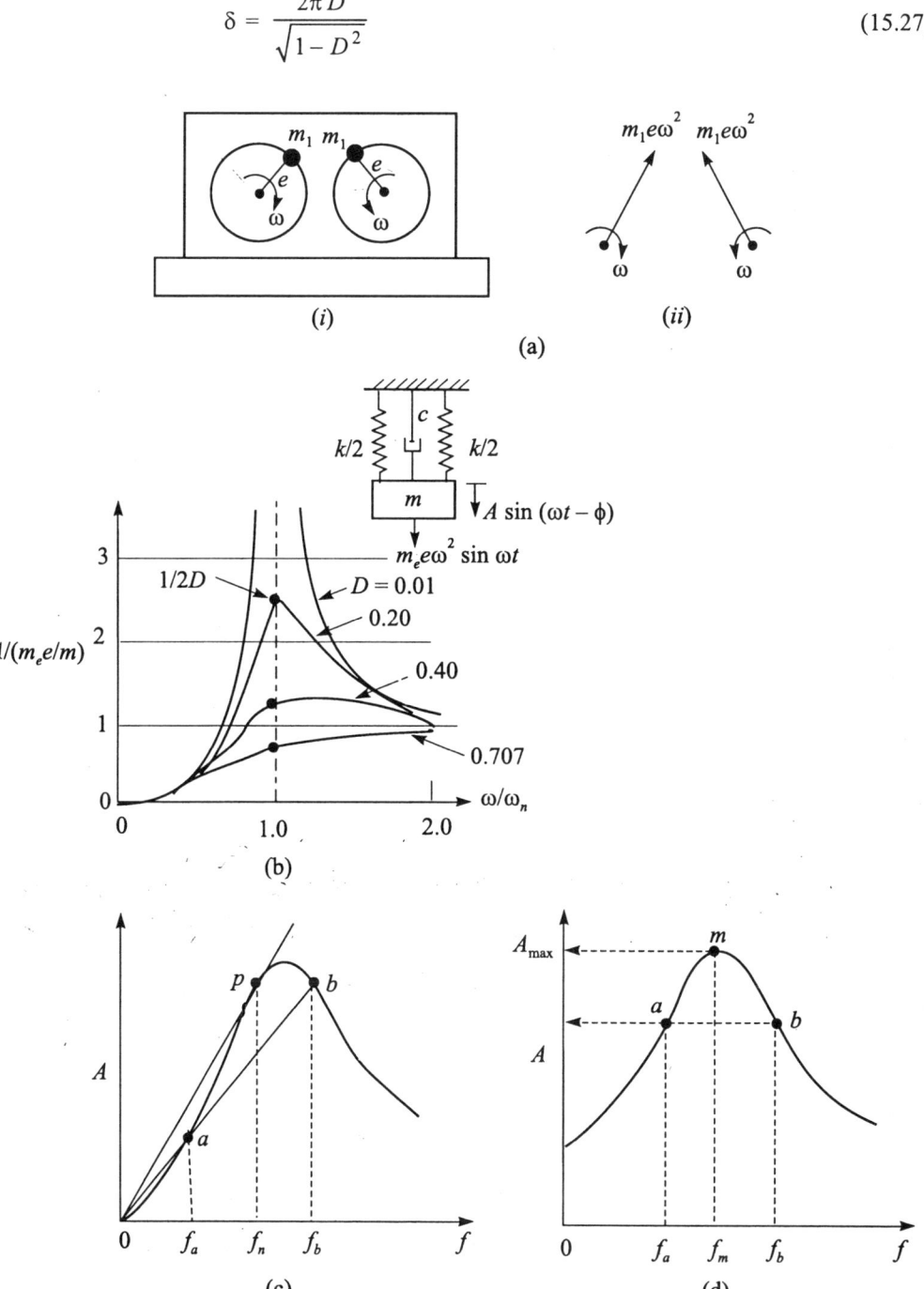

Fig. 15.7 Response curves for rotating-mass-type-excitation and properties of response curves: (a) (i) Counteracting masses, (ii) counteracting forces, (b) response curves for rotating mass type excitation of viscously damped single degree of freedom system, (c) response curve from rotating-mass-type excitation, (d) response curve from constant-force amplitude

The value of D can be determined by trial and error method by equating Eq. (15.57) with Eq. (15.27) since D is found on both sides of the equation. In Eq. (15.57) all values except D is known from the experimental curve, Fig. 15.7 (d). However a simplified approach is:

1. Assume $\dfrac{\sqrt{1-2D^2}}{1-D^2} \approx 1$ for small values of D.

2. Choose a on the experimental curve Fig. 15.7 (d) such that

$$A = 0.707\ A_{max}$$

If the above approximations are considered, Eq. (15.57) reduces to

$$\delta = \frac{(f_b - f_a)\ \pi}{f_m} \tag{15.58}$$

It is also shown in Eq. (15.27), for small values of D,

$$\delta \approx 2\pi D \tag{15.59}$$

Equating Eqs (15.58) and (15.59), we have

$$D = \frac{1}{2}\ \frac{f_b - f_a}{f_m} \tag{15.60}$$

Thus, Eq. (15.60) gives a simple method for determining the *damping factor* D from an experimental response curve obtained from a constant-force-amplitude system.

15.10 MACHINE FOUNDATIONS SUBJECTED TO STEADY STATE VIBRATIONS

Introduction

The basic theories of vibrations have been dealt with in the earlier sections. How far these theories are applicable for the design of machine foundations are required to be understood. Machine foundations are heavy rigid structures embedded at a certain depth below the ground surface. The theories that have been discussed earlier are applicable to surface foundations only. In these theories, the shapes of the foundations are also not considered. Subsequent discussion involve the effect of shape and embedment on the vibrations of foundations.

Machine foundations, when acted upon by dynamic loads, may vibrate in any one of the six degrees of freedom as illustrated in Fig. 15.8 (a). The six modes of vibration comprise of translations along anyone of the X, Y and Z directions, rocking (rotation) about X or Y axes and yawing (torsion) about the Z-axis (vertical axis). Of the six degrees of freedom of vibration, vertical vibration and rotation about the vertical axis (Z-axis) occur independent of the other modes of vibration. A block subjected to dynamic loads may have to be analysed by making use only four modes of motion, namely,

1. vertical vibration,
2. translation along X or Y-axis,
3. rocking about X or Y-axis,
4. rotation about the Z-axis.

When a foundation is subjected to rocking about X or Y axis, it is always accompanied with translation along Y or X axis. Such a motion is called as *coupled motion*. However, when a foundation is subjected to purely vertical vibration through the centroid of the mass, it will not be accompanied with any of the other modes of vibration.

Methods of Analysis

There are currently in use four methods of analysis of machine foundations. They are:

1. Elastic half-space theory.
2. Elastic half-space Analog.
3. Lumped-mass or parameter method.
4. Elastic soil-spring method of Barkan (1962).

Elastic Half-Space Theory

In the case of elastic half-space theory the elastic medium (soil) is considered as homogeneous and isotropic. The theory assumes that the waves generated due to vibration of circular footings resting on elastic-half space move away in radial directions from the footing. As it moves away, it carries with it some of the energy put into the soil. Since this energy is then not available to participate in a resonance phenomenon, a damping effect is introduced. This type of damping is called as *geometrical damping or radiation damping*. The elastic half-space theory, therefore, does not consider the dashpot system of viscous damping as used in the lumped parameter method. The basic soil parameters used for the development of the theory are the *shear modulus G*, the mass density ρ, and the *Poisson's ratio* μ. The theory makes use of certain mathematical simplifications which are not quite realistic. However, the analystical solution serves as a useful guide for a rational means of evaluating the spring and damping constants which are used in the Analog method.

Elastic half-space theory for vertical oscillation was first developed by Reissner (1936), which was later on extended by Quinlan (1953) and Sung (1953). Hsieh (1962) improved the basic equations for *geometrical damping* as presented by Reissner. Reissner (1937), and Reissner and Sagoci (1944) presented analytical solutions for the torsional oscillation of a circular footing resting on elastic half-space. The analytical solutions for the rocking mode were presented by Arnold, Bycroft, and Warburton (1955), and Bycroft (1956). Sliding oscillation of a circular disc was analysed by Arnold, Bycroft and Warburton (1955), and by Bycroft (1956).

The efforts of all these investigators were directed towards evolving analytical solutions for computing resonant frequency and displacement of circular footings under dynamic loads. Since the procedures are quite involved, it is beyond the slope of this book to present them here. Those who are interested may refer to Richart *et al* (1970).

Elastic Half-Space Analog

Lysmer (1965), and Lysmer and Richart (1966) developed a procedure by which vertical vibrations of rigid circular footings on elastic half-space could be represented satisfactorily by a mass-spring dashpot system if the damping constants and spring constants were chosen correctly. They made use of the equations developed analytically by elastic half-space theory for computing these constants. Thus, the method developed by them, called the *analog method,* is a bridge between the elastic half-space theory and the lumped-parameter method of mass-spring-dashpot system.

Based on Lysmer's approach, Hall (1967) presented methods for solving problems connected with rocking and sliding vibration modes of rigid circular footing. In the same way, the analog method based on elastic half-space theory presented by Reissner (1937) and Reissner and Sagoci (1944) for Torsional oscillation of rigid circular footing has been developed. The analog method of predicting the dynamic response of soil has been presented in this Chapter.

Lumped-Parameter Method

The theory of lumped-parameter method for single-degree of freedom has already been discussed in detail in the earlier sections. The accuracy of this method depends upon the accuracy by which the spring and damping constants re evaluated. The lumped-parameter method provides the necessary basic theory for the elastic half-space analog and the Barkan's elastic soil-spring analog. Therefore, lumped-parameter method as such has not been discussed further in this book. However the basic differential equations developed for the different degrees of freedom are given.

Elastic Soil-spring Analog

Barkan (1962) by using the concept of elastic sub-grade reaction has simplified lumped-parameter method of vibration analysis. His method is quite popular in India. This method is briefly discussed in this chapter.

Dynamic Loads

Rotating machinery designed to operate at a constant speed for long periods of time includes turbines, axial compressor, centrifugal pumps, turbogenerator sets and fans. In the case of each it is theoretically possible to balance the moving parts to produce no unbalanced forces during rotation. However, in practice some unbalance always exists, and its magnitude includes factors introduced by design, manufacture, installation and maintenance. These factors may include an axis of rotation which does not pass through the centre of gravity of a rotating component; an axis of rotation which does not pass through the principal axis of inertia of a unit, thereby, introducing longitudinal couples; gravitational deflection of shaft; misalignment during installation etc. The cumulative result of the unbalanced forces must not be great enough to cause vibrations of the machine-foundation system and exceed the design criteria.

In certain types of machines, unbalanced forces are developed on purpose, for example in vibroflots, vibratory rollers for surface compaction etc. In these cases, the exciting-force-amplitude can be evaluated from Eq. (15.49).

$$Q_0 = m_e e \, \omega^2 \tag{15.61}$$

in which m_e is the total unbalanced mass, and e is the eccentric radius to the centre of gravity of the total unbalanced mass.

The different arrangements of the rotating mass is shown in Fig. 15.8 (b). In Fig. 15.8 (b1) the central petal force developed by a single rotating mass, Eq. (15.61), is a vector force Q_0 which acts outward from the centre of rotation. By combining two rotating masses on parallel shafts within the same mechanism, it is possible to produce an oscillating force with a controlled direction Fig. 15.8 (b2). The counteracting masses can be so arranged that the horizontal force components cancel each other, but the vertical components are added. The vertical component produced as per Eq. (15.50) is

$$Q = Q_0 \sin \omega t = 2m_1 e \, \omega^2 \sin \omega t \tag{15.62}$$

Four masses can also be arranged in different ways as shown in Fig. 15.8 (b3) through 15.8 (b5) with one mass at each end of two parallel shafts. A vertical oscillation force will be developed as per Fig. 15.8 (b3), a torsion couple about the vertical axis as per Fig. 15.8 (b4), and a rocking couple about a horizontal axis as per Fig. 15.8 (b5). Note that for the rocking or torsional forces developed from the four-mass exciter, the torque or moment is given by

$$T_m = 4m_1 e \, \frac{\overline{x}}{2} \, \omega^2 \sin \omega t \tag{15.63}$$

in which $\overline{x}$ = the distance between the weights at the ends of each shaft.

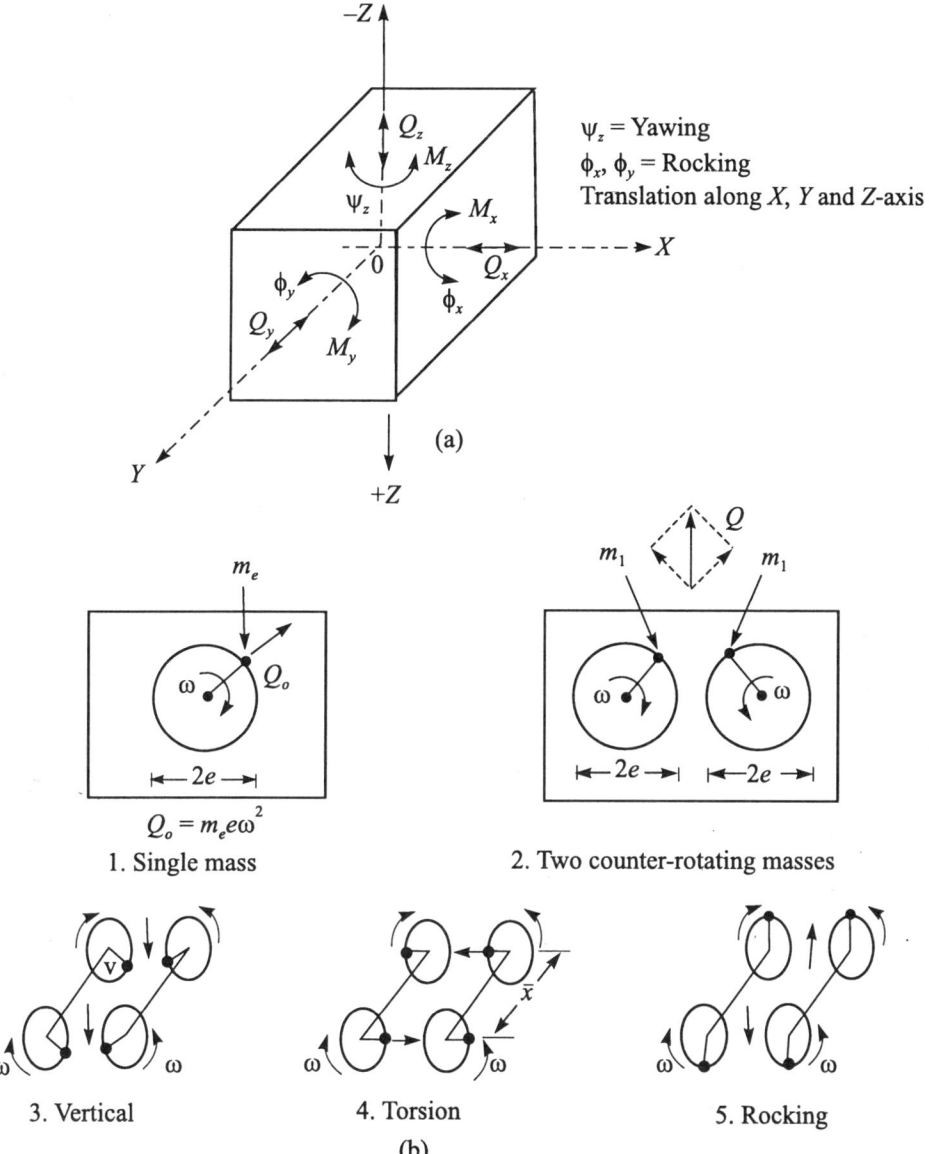

Fig. 15.8 Degrees of Freedom of a block foundation and forces from rotating mass exciters: (a) Degrees of freedom of a block foundation, (b) forces from rotating mass exciters

Differential Equations for the Different Modes of Vibration

The general differential equations of motion for the different modes of forced vibration are of the same form as applicable for vertical vibration (Eq. 15.39). The equations are (with respect to Z, X and Y coordinate system):

For vertical vibration

$$m\ddot{z} + c_z\dot{z} + k_z z = Q_z\,(t) \tag{15.64a}$$

For horizontal sliding

$$m\ddot{x} + c_x\dot{x} + k_xx = Q_x(t) \tag{15.64b}$$

$$m\ddot{y} + c_y\dot{y} + k_yy = Q_y(t) \tag{15.64c}$$

For yawing about Z-axis.

$$I_{\psi0}\ddot{\psi} + c_{\psi z}\dot{\psi} + k_\psi\psi = M_z(t) \tag{15.64d}$$

For rocking

$$I_{\phi0}\ddot{\phi} + c_{\phi x}\dot{\phi} + k_\phi\phi = M_x(t) \tag{15.64c}$$

$$I_{\phi0}\ddot{\psi} + c_{\phi y}\dot{\phi} + k_\phi\phi = M_y(t) \tag{15.64f}$$

where, m = mass of foundation + mass of machine

$c_z, c_x, c_y, c_{\psi z}, c_{\phi x}, c_{\phi y}$ = the damping constants for the corresponding modes of vibration,

ψ and ϕ = angles of rotation,

$k_z, k_x, k_y, k_\psi, k_\phi$ = spring constants for the corresponding modes of vibration,

$I_{\psi o}, I_{\phi o}$ = mass moment of inertia of the machine and foundation about the corresponding axis of rotation.

The general differential equations given form the basis for the elastic half-space theory, analog method, and the elastic soil-spring method.

The method of computing the mass moments of inertia is given in Fig. 15.9.

15.11 VIBRATION ANALYSIS OF RIGID CIRCULAR FOOTINGS BY ELASTIC HALF-SPACE ANALOG METHOD

Vertical Oscillation

With the elastic half-space as the mathematical model, Reissner (1936) developed an analytical solution for the periodic vertical displacement z_0 at the centre of a circular loaded area of the surface. The parameters used to describe the properties of the elastic body (which was assumed as homogeneous and isotropic) were the shear modulus G the Poisson's ratio μ, and the mass density ρ ($= \gamma/g$) where, γ is the unit weight of the material and g the acceleration due to gravity. The vibrating footing was represented by an oscillating mass which produced a periodic vertical pressure distributed uniformly over a *circular area of radius* r_0 on the surface of the half-space. Reissner's solution for z_0 was found out in accordance with field tests. Lysmer (1965) and Lysmer and Richart (1966) showed that the vertical response of rigid circular foundation can be represented quite well by a lumped-mass-spring dash pot system by making use of the theoretical solutions for spring and damping constants as established by the theory of elasticity.

The following parameters were developed for the solution of the problem.

Dimensionless frequency, $\qquad a_0 = \omega r_0\sqrt{\dfrac{\rho}{G}} = \dfrac{\omega r_0}{v_s} \tag{15.65}$

Modified dimensionless mass ratio

$$B_z = \frac{1-\mu}{4}\frac{m}{\rho r_0^3} \tag{15.66}$$

Vertical spring constant, $\qquad k_z = \dfrac{4Gr_0}{1-\mu} \tag{15.67}$

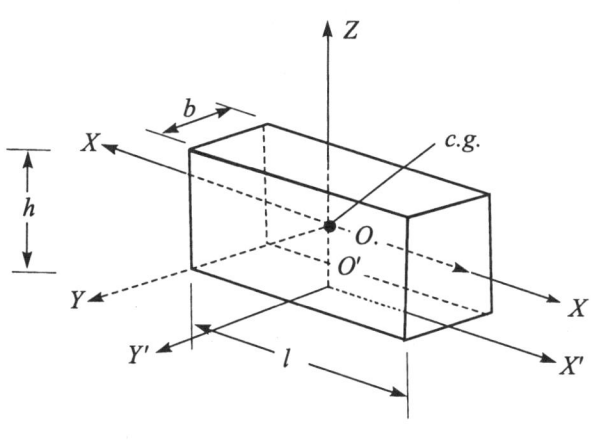

Rectangular prism

Z, X, Y axes pass through the centre of gravity of the mass. X' and Y' axes pass through the centroid of base area. Mass moment of inertia about

Z-axis : $\dfrac{1}{12}$ m $(b^2 + l^2)$

X-axis : $\dfrac{1}{12}$ m $(b^2 + h^2)$

Y-axis : $\dfrac{1}{12}$ m $(l^2 + h^2)$

X'-axis : $\dfrac{1}{12}$ m $(b^2 + h^2) + \dfrac{mh^2}{4}$

Y'-axis : $\dfrac{1}{12}$ m $(l^2 + h^2) + \dfrac{mh^2}{4}$

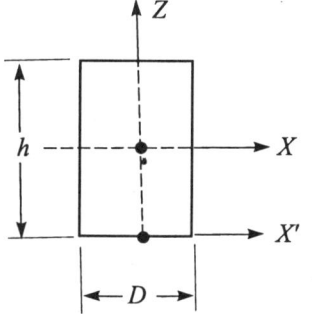

Soild circular cylinder

Mass moment of inertia about

Z-axis : $\dfrac{mD^2}{8}$

X-axis : $\dfrac{1}{12}$ m $(3/4D^2 + h^2)$

X'-axis : $\dfrac{1}{12}$ m $(3/4D^2 + h^2) + \dfrac{mh^2}{4}$

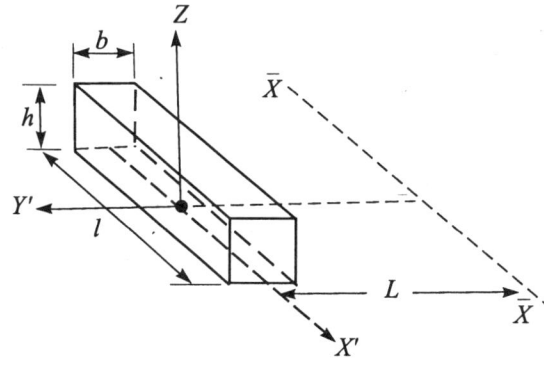

Mass moment of inertia of a Rectangular prism about $\bar{X}\ \bar{X}$ axis $\bar{X}$ -axis and is parallel to $\bar{X}$ - axis lies in the plane of X'-axis

MMI : $\dfrac{1}{12}$ m $(b^2 + h^2) + \dfrac{mh^2}{4} + mL^2$

Fig. 15.9 Mass-moment of inertia

Damping constant, $\qquad\qquad c_z = \dfrac{3.4 r_0^2}{1 - \mu}\ \sqrt{\rho G}$ $\qquad\qquad$ (15.68)

Amplitudes of vibration, for constant force excitation Q_0

$$A_z = \frac{(1-\mu)Q_0}{4Gr_0} M \qquad (15.69)$$

for rotating-mass excitation

$$A_z = \frac{m_e e}{m} M_r \qquad (15.70)$$

where, v_s = velocity of progapation of the shear wave in the elastic body,

M = magnification factor = A_z/z_s,

z_s = static displacement produced by constant force amplitude $Q_o = Q_0/k_z$,

M_r = magnification factor = $A_z/m_e e/m$,

m_z = mass of foundation and the machinery.

Figure 15.10 (a) gives a plot of M vs. a_0 for various values of B_z; similarly Fig. 15.10 (b) gives curves M_r vs. a_0 for various values of B_z.

From Figs 15.10 (a) and (b), the values of M and M_r at the peaks of each response curves and the values of a_0 at the *peak* can be established. These values may then be plotted as B_z vs. a_{0m} Fig. 15.11 (a), or B_z vs. M_m or M_{rm} [Fig. 15.11 (b)]. Figures 15.11 (a) and (b) provide a simple means for evaluating the maximum amplitude of vertical motion of a rigid circular footing and the frequency at which this occurs for both *constant force* and *rotating-mass excitation*.

Since the values of spring constant k_z [Eq. (15.67)] and the damping constant c_z [Eq. (15.68)] are known, the equation of motion for Lysmer's analog for vertical oscillation of a rigid circular footing on the elastic-half space maybe expressed [as per Eq. (15.64a)] as

$$m\ddot{z} + \frac{3.4 r_0^2}{1-\mu} \sqrt{\rho G} \, \dot{z} + \frac{4Gr_0}{1-\mu} z = Q_z(t) \qquad (15.71)$$

The various other parameters connected with the vertical vibration of the system may be expressed as follows.

Critical damping, $\qquad c_c = 2\sqrt{k_z m} = 2\sqrt{\dfrac{4Gr_0 m}{(1-\mu)}} \qquad (15.72)$

Damping ratio, $\qquad D_z = \dfrac{c_z}{c_c} = \dfrac{0.425}{\sqrt{B_z}} \qquad (15.73)$

For excitation by a force of *constant amplitude* Q_0, the resonant frequency is

$$f_m = \frac{1}{2\pi} \sqrt{\frac{k_z}{m}} \sqrt{1 - D_z^2} = \frac{1}{2\pi} \frac{v_s}{r_0} \frac{\sqrt{B^2 - 0.36}}{B_z} \qquad (15.74)$$

For rotating-mass excitation force, $Q_0 = m_e e \omega^2$, the resonant frequency is

$$f_{mr} = \frac{v_s}{2\pi r_0} \sqrt{\frac{0.9}{B_z - 0.45}} \qquad (15.75)$$

Equation (15.75) gives good answers only for $B_z \geq 1$

As per Eq. (15.48), the maximum magnification factor M is

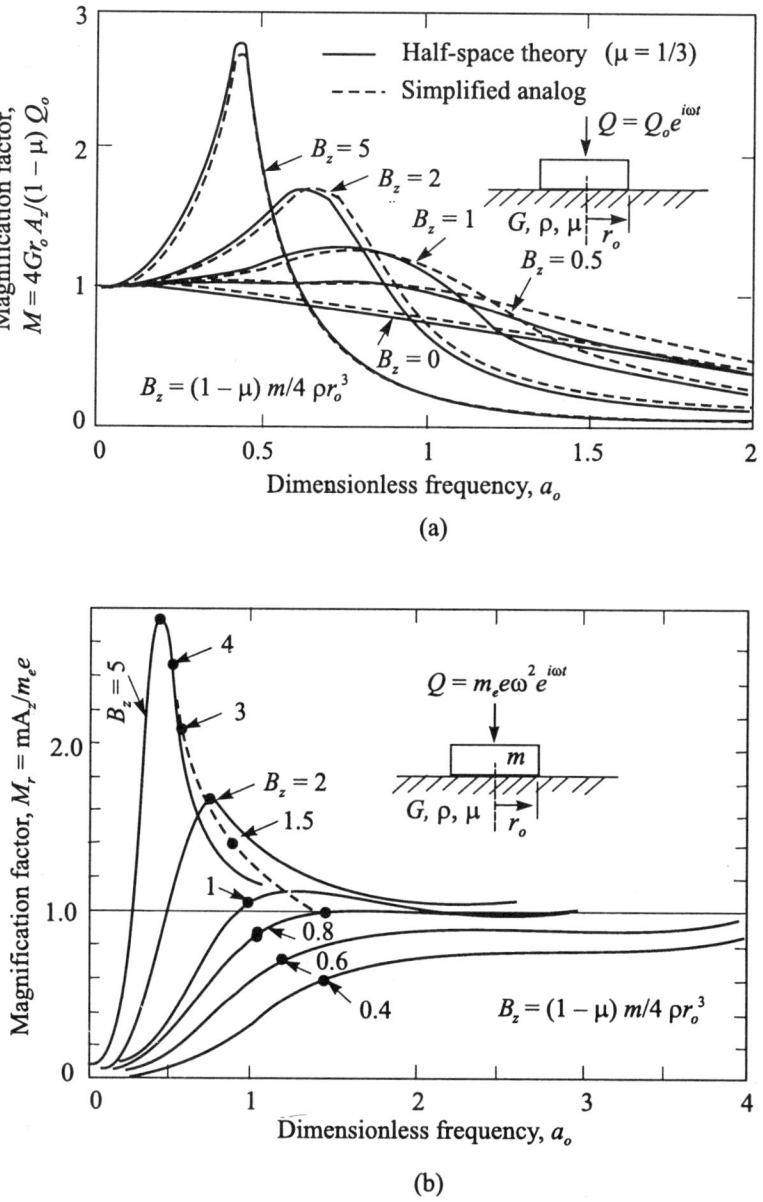

Fig. 15.10 Response curves of rigid circular footings subjected to vertical oscillation by elastic half-space theory and analog: (a) Response of rigid circular footing to vertical force developed by constant force excitation (from Lysmer and Richart, 1966), (b) response of rigid circular footing to vertical force developed by rotating mass exciter ($Q_1 = m_e e\omega^2$) (Lysmer and Richart, 1966)

$$M = \frac{1}{2D\sqrt{1-D^2}} \tag{15.76}$$

Now substituting for D from Eq. (15.73), and the value of M in Eq. (5.69) the maximum amplitude of displacement A_{zm} for constant force amplitude Q_0 is

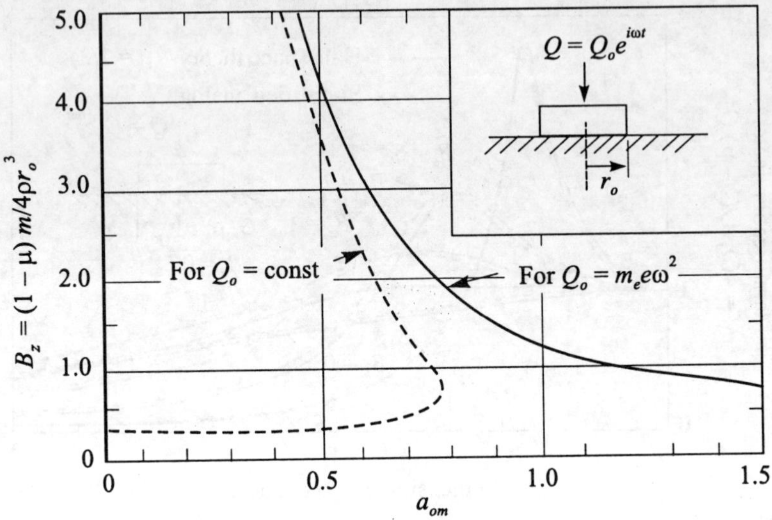

Fig. 15.11 Vertical oscillation of rigid circular footing on elastic half-space: (a) Mass ratio *vs.* dimensionless frequency at resonance, (b) mass ratio *vs.* magnification factor at resonance. (Richart *et al*, 1970)

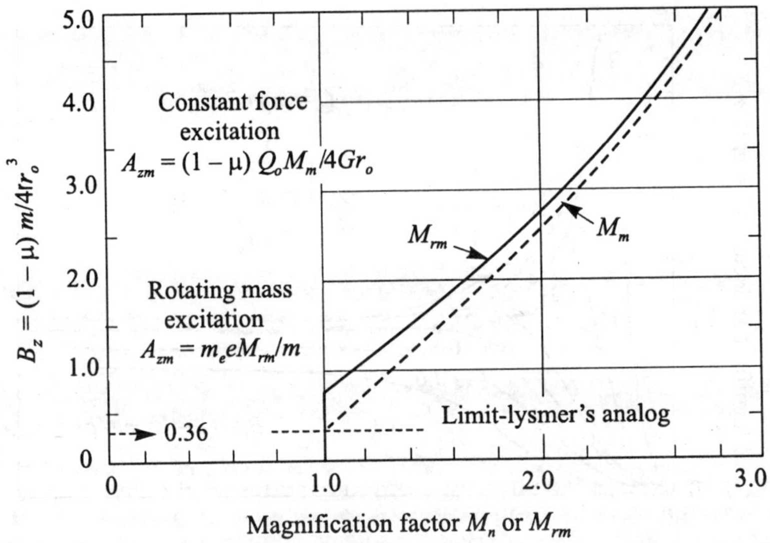

$$A_{zm} = \frac{Q_0 (1-\mu)}{4Gr_0} \frac{B_z}{0.85\sqrt{B_z - 0.18}} \qquad (15.77a)$$

In the same way, A_{zm} for rotating-mass excitation is

$$A_{zm} = \frac{m_e e}{m} \frac{B_z}{0.85\sqrt{B_z - 0.18}} \qquad (15.77b)$$

The phase angle ϕ is determined from

$$\tan \phi = \frac{0.85 a_0}{B_z a_0^2 - 1} \tag{15.78}$$

Sliding Oscillation

The angle method for the vibration analysis of footings under sliding mode was developed by Hall (1967) based on the theoretical half-space theory of Bycroft (1956).

The various dynamic parameters developed by him for the sliding mode are as follows. The footing here is considered to be in sliding oscillation mode in the X-direction.

Spring constant, $\quad k_x = \dfrac{32 \left(1 - \mu\right)}{7 - 8\mu} G r_0 \tag{15.79}$

Damping constant, $\quad c_x = \dfrac{18.4 \left(1 - \mu\right)}{7 - 8\mu} r_0^2 \sqrt{\rho G} \tag{15.80}$

Mass ratio, $\quad B_x = \dfrac{7 - 8\mu}{32 \left(1 - \mu\right)} \dfrac{m}{\rho r_0^3} \tag{15.81}$

Frequency factor as per Eq. (15.65)

$$a_0 = \omega r_0 \sqrt{\frac{\rho}{G}} = \frac{\omega r_0}{v_s} \tag{15.65}$$

Critical damping $\quad c_c = 2 \sqrt{\dfrac{32 \left(1 - \mu\right) m}{\left(7 - 8\mu\right)} G r_0} \tag{15.82}$

Damping factor $\quad D_x = \dfrac{c_x}{c_c} = \dfrac{0.288}{\sqrt{B_x}} \tag{15.83}$

The maximum amplitudes of displacement, A_{xm}, may be obtained from the following expressions. For constant–force amplitude, Q_0

$$A_{xm} = \frac{7 - 8\mu}{32 \left(1 - \mu\right)} \frac{Q_0}{G r_0} M_{xm} \tag{15.84}$$

For rotating-mass type excitation,

$$A_{xm} = \frac{m_e e}{m} M_{xrm} \tag{15.85}$$

where, M_{xm} = maximum value of magnification factor that can be established from curves giving the relationship between a_0 and M_x as a function of B_x for constant force amplitude Q_0,

M_{xrm} = maximum value of magnification factor from the curves showing the relationship between a_0 and M_{xr} as a function of B_x for the rotating-mass type excitation.

The curves in Figs 15.12 have been developed in the same way the curves are developed for vertical oscillation and shown in Fig. 15.11. The curves in Fig. 15.12. are useful for computing the resonant frequencies and the corresponding maximum amplitudes for the sliding mode of vibration of rigid circular footings.

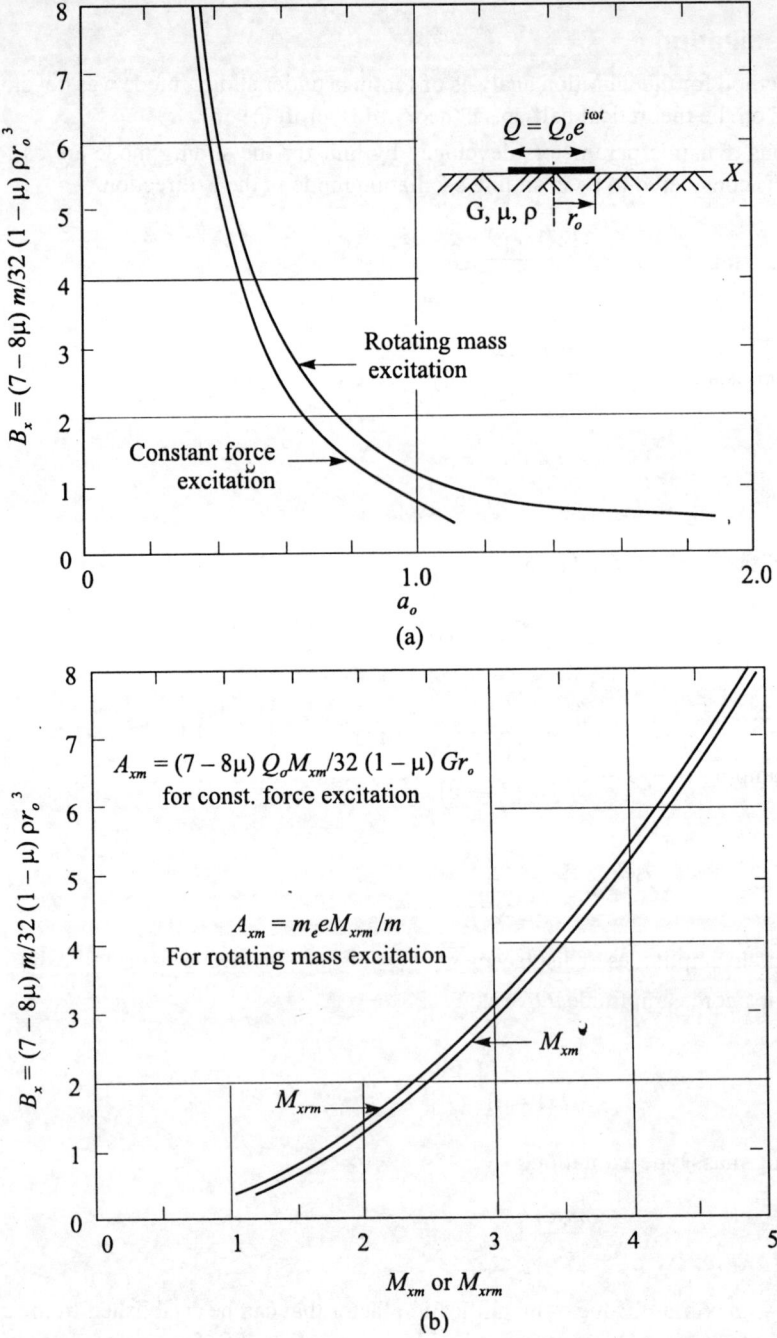

Fig. 15.12 Sliding oscillation of rigid circular disk on elastic half-space: (a) Mass ratio *vs.* dimensionless frequency at resonance, (b) mass ratio *vs.* magnification factor at resonance (Richart *et al*, 1970)

Rocking Mode of Vibrations

Hall (1962) developed the analog method for computing resonant frequencies and the corresponding maximum amplitudes of displacements under the rocking mode based on the elastic half-space theory of Bycroft (1956). The various parameters developed for the solution of the problem are given below.

$$\text{Spring constant,} \quad k_\phi = \frac{8 G r_0^3}{3(1-\mu)} \tag{15.86}$$

$$\text{Damping constant,} \quad c_\phi = \frac{0.8 r_0^4 \sqrt{G\rho}}{(1-\mu)(1+B_\phi)} \tag{15.87}$$

$$\text{Mass ratio,} \quad B_\phi = \frac{3(1-\mu)}{8} \frac{I_\phi}{r_0^5} \tag{15.88}$$

$$\text{The damping ratio,} \quad D_\phi = \frac{c_\phi}{c_c} = \frac{0.15}{(1+B_\phi)\sqrt{B_\phi}} \tag{15.89}$$

Maximum magnification factor

$$M_{\phi m} \approx \frac{1}{2 D_\phi} \tag{15.90}$$

The maximum amplitudes of displacements are expressed as below.
For constant force amplitude, Q_0

$$A_{\phi m} = \frac{3(1-\mu)}{8} \frac{T_\phi}{6 r_0^3} M_{\phi m} \tag{15.91}$$

For rotating mass-type excitation,

$$A_{\phi m} = \frac{m_e e \overline{z}}{I_\phi} M_{\phi rm} \tag{15.92}$$

where, I_ϕ = mass moment of intertie of the footing about the axis of rotation,

T_ϕ = external rocking moment,

$\overline{z}$ = vertical distance above point O [Fig. 15.13 (a)] of a horizontally oscillating force $m_e e \omega^2$.

As in the other modes of vibration, Fig. 15.13 (a) gives relationships between a_{om} and B_ϕ, and Fig. 15.13 (b) B_ϕ vs. $M_{\phi m}$ or $M_{\phi rm}$.

Torsional Oscillation

Reissner (1937) and Reissner and Sagoci (1944) presented analytical solutions for the torsional oscillation of a circular footing resting on the surface of elastic half-space. Based on the theoretical solution, an analog method was developed. The various parameters used in the solution are given below.

$$\text{Spring constant,} \quad k_\psi = \frac{16}{3} G r_0^3 \tag{15.93}$$

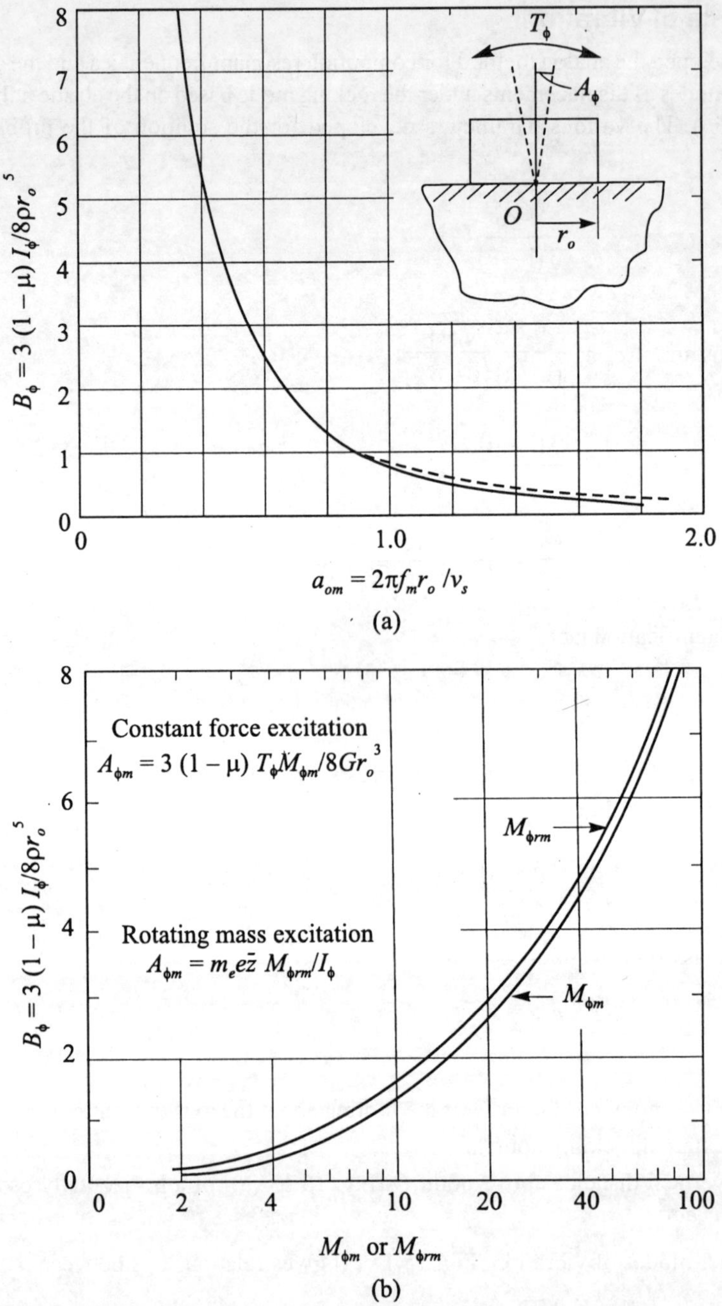

Fig. 15.13 Rocking of rigid circular footings on elastic half-space: (a) Mass ratio *vs.* dimensionless frequency at resonance, (b) mass ratio *vs.* magnification factor at resonance (Richart *et al*, 1970)

Mass ratio, $$B_\psi = \frac{I_\psi}{\rho r_0^5}$$ (15.94)

Damping ratio, $$D_\psi = \frac{0.50}{1 + 2 B_\psi}$$ (15.95)

The amplitudes of vibration are as follows.

For constant force amplitude

$$A_{\psi m} = \frac{3}{16} \frac{T_\psi}{G r_0^3} M_{\psi m} \tag{15.96a}$$

For rotating mass–type excitation

$$A_{\psi m} = \frac{m_e e \bar{x}}{I_\psi} M_{\psi m} \tag{15.96b}$$

$$T_\psi = m_e e \bar{x} \omega^2 \tag{15.96c}$$

where, T_ψ = the exciting torque,

$\bar{x}$ = horizontal moment arm of unbalanced weight from the axis of rotation.

As in the other modes of vibration, Fig. 15.14 (a) and (b) give values of a_{0m}, $M_{\psi m}$ and M_{mr} as functions of B_ψ.

Mass-and-Damping Ratios

The variation of damping ratio D with mass ratio B for the various modes of vibration are shown in Fig. 15.15. The expressions for B_z, B_x, B_ϕ and B_ψ are given by Eqs (15.66), (15.81), (15.88) and (15.94) respectively. It is clear from Fig. 15.15, appreciable damping is associated with translational mode of vibration. On the other hand, damping is quite low for rotational modes of vibration, particularly for values $B_\psi > 2$ in torsional oscillation, and for $B_\psi > 1$ in rocking.

Effect of Foundation Shape

The analog method considered only circular rigid footings for the computation of the various parameters. Solutions for strip foundation were presented by Quinlan (1953) for vertical vibrations, and by Awojobi and Grootenhuis (1965) for vertical and rocking vibrations. Rectangular footings were considered by Elorduy, Nieto and Szekely (1967). The vertical, sliding, rocking and torsional modes of vibration applicable to rectangular foundation were dealt with by Kobori, Minai Suzuki and Kusakabe (1966). However, Whitman and Richart (1967) presented a solution for rectangular foundation which is simple in form and can be used as a first estimate. This can be accompanied by converting the rectangular base into equivalent circular base having a radius r_0. If the dimensions of a rectangular footing is $2b \times 2l$, the expressions for r_0 for the various modes of vibration are:

For translation, $r_0 = \sqrt{\dfrac{4bl}{\pi}}$ (15.97a)

For rocking, $r_0 = \sqrt[4]{\dfrac{16bl^3}{3\pi}}$ (15.97b)

For torsion, $r_0 = \sqrt[4]{\dfrac{16bl\left(b^2 + l^2\right)}{6\pi}}$ (15.97c)

where, $2b$ = width of foundation (along the axis of rotation for the case of rocking) and $2l$ is the length of the foundation (in the plane of rotation for rocking). Model tests (Chae, 1969) have indicated that the use of Eq. (15.97) would lead to conservative evaluation of the amplitudes of vibration.

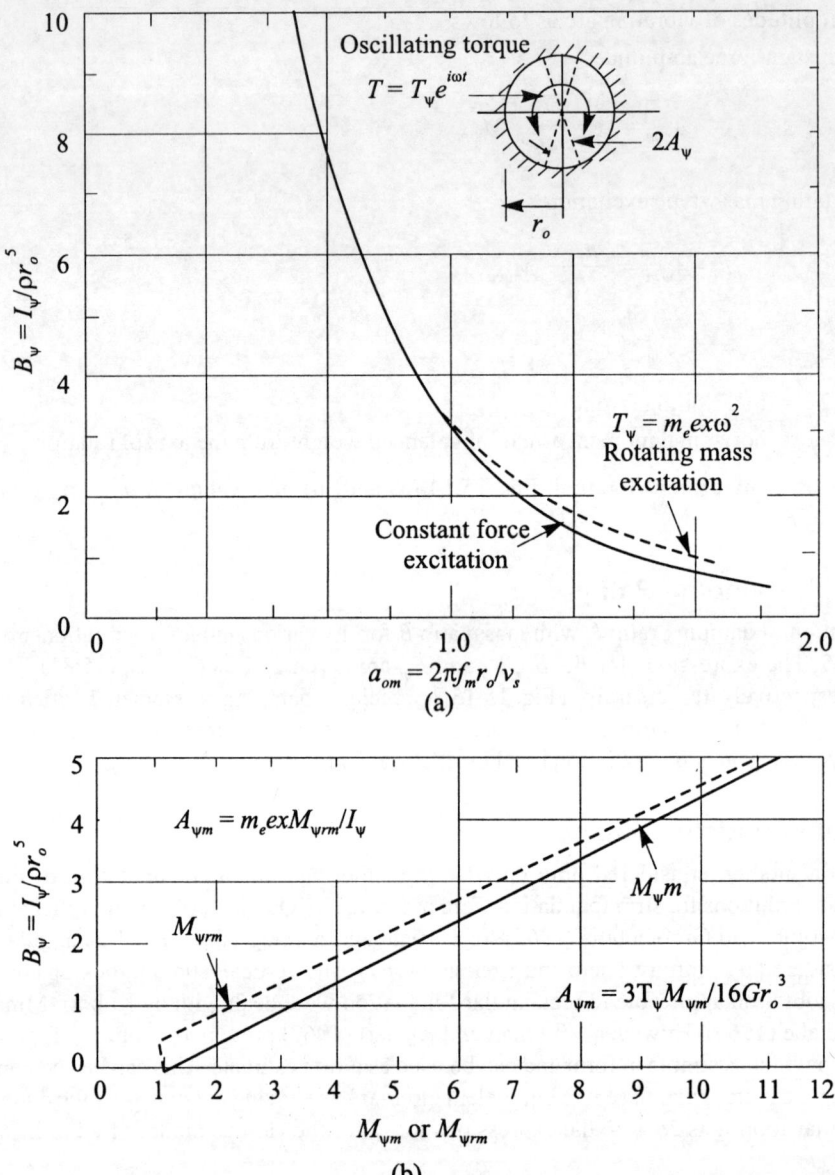

Fig. 15.14 Tosional oscillation of rigid circular footing on elastic half-space: (a) Mass ratio *vs.* dimensionless frequency at resonance, (b) mass ratio *vs.* magnification factor at resonance (Richart *et al*, 1970)

Effect of Foundation Depth

The works of Paw (1952), Barkan (1962), and Fry (1963) clearly established that embedment of foundation blocks increases the resonant frequency and decreases the amplitudes of vibratory motion. A study of the effect of embedment f circular rigid footings on static spring constants was presented by Kaldjian (1969). His results are shown in Fig. 15.16 (a). Curve 1 represents a rigid footing which adheres to the soil along the vertical surface, thereby, developing resistance by pressure on the base. Curve 2 corresponds approximately to the situation of an embedded foundation without the effect of side friction.

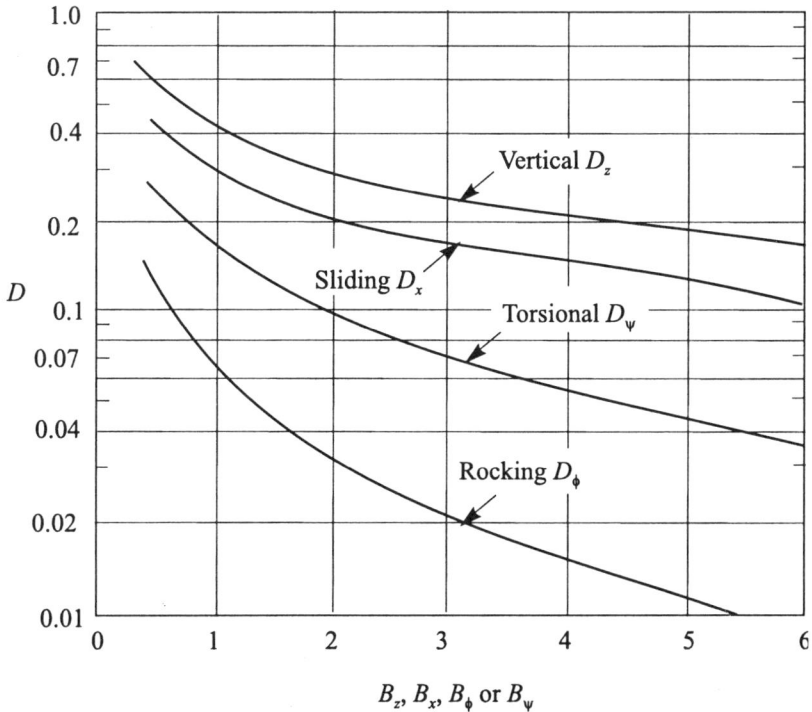

Fig. 15.15 Equivalent damping ratio for oscillation of rigid circular footing on elastic half-space (Richart *et al*, 1970)

Elastic Constants of Soils

The elastic constants considered in the analog study are modulus of elasticity in shear, G, and Poisson's ratio μ. For design purpose the Poisson's ratios recommended are

1. for Cohesive soils $\mu = 0.4$,
2. for Cohesionless soils $\mu = 0.33$.

Values of the shear modulus G may be evaluated in the field or from samples taken to the laboratory. The value of G can be obtained from the following equation from shear wave velocity v_s

$$G = \rho v_s^2 \qquad (15.98)$$

where, ρ = mass density $(F - \sec^2 L^{-4}) = \gamma/g$,

$\quad \gamma$ = unit weight of soil (FL^{-3}),

$\quad g$ = acceleration due to gravity $(L\text{-sec}^{-2})$.

G is also related to E and m as follows

$$G = \frac{E}{2(1 + \mu)} \qquad (15.99a)$$

Sometimes G is evaluated based on the void ratio e and the effective overburden pressure σ_0' as per the following expression for round-grained sands $(e < 0.80)$,

$$G = \frac{6900(2.17 - e)^2}{1 + e}(\sigma_0')^{0.5} \qquad (15.99b)$$

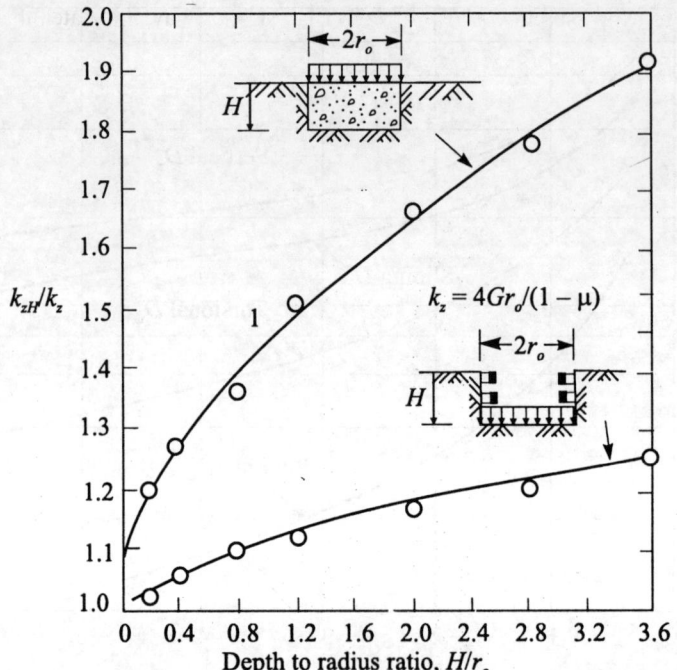

(a) Effect of depth of embedment on the spring constant
for vertically loaded circular footings (*from Kaldjian,* 1969)

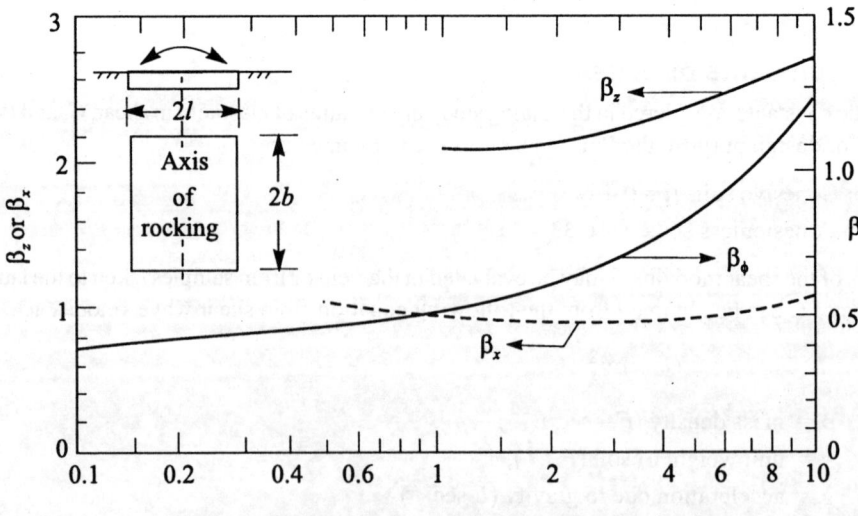

(b) Coefficients β_z, β_x, and β_ϕ for rectangular footings
(*after Whitman and Richart,* 1967)

Fig. 15.16 (a) Effect of foundation depth on k_z (Kaldljian, 1969), (b) coefficients β_z β_x and β_ϕ for
rectangular footings (Whitman and Richart, 1967)

and, for angular-grained materials ($e > 0.6$),

$$G = \frac{3230\,(2.97 - e)^2}{1 + e}\,(\sigma_0')^{0.5} \tag{15.99c}$$

in which G and σ_0' are expressed in kPa, Hardin and Black (1968) have indicated that Eq. (15.99b) is also reasonable approximation for the shear modulus of normally consolidated clays. Typical values of G for preliminary design are given in Table 15.1 (Bowles, 1988)

Table 15.1 Representative values of shear modulus G

Type of soil	G (MPa)
Clean dense quartz sand	$12-20$
Micaceous fine sand	16
Loamy sand	$17-24$
Dense sand gravel	70
Wet soft silty clay	$9-15$
Dry soft silty clay	$17-21$
Dry silty clay	$25-35$
Medium clay	$12-30$
Sandy Clay	$12-30$

Table 15.2 Velocity, v_s of shear waves

Soil	v_s m/sec
Moist clay	150
Loess at natural moisture content	260
Dense sand and gravel	250
Fine grained sand	110
Medium grained sand	160
Medium-sized gravel	180

Formulae for Spring Constants

Theory of elasticity can provide useful formulae for spring constants k for footings of simple shapes. Eqs (15.67), (15.79), (15.86) and (15.93) give formulae for circular footings for vertical, horizontal, rocking and torsion modes of vibration respectively. The spring constants for rectangular footings for vertical horizontal and rocking modes of vibration are also available as given in Table 15.3.

Table 15.3 Spring constants for rigid rectangular footing on elastic half-space

Motion	Spring constant	Reference
Vertical	$k_z = \dfrac{G}{1-\mu}\,\beta_z$	Barkan (1962)
Horizontal	$k_x = 4\,(1+\mu)\,G\beta_x\,\sqrt{bl}$	Barkan (1962)
Rocking	$k_\phi = \dfrac{8G}{1-\mu}\,\beta_\phi bl^2$	

where, $2b$ = width of foundation along the axis of rotation

$2l$ = length of foundation perpendicular to the axis of rotation

β_z, β_x and β_ϕ = functions of l/b [Fig. 15.16 (b)]

15.12 ELASTIC SOIL-SPRING METHOD OF VIBRATION ANALYSIS OF FOUNDATIONS (BARKAN, 1962)

Elastic Soil-Spring Constants

Barkan (1962), by making use of the concept of subgrade reaction, developed the following elastic soil-spring constants for the various modes of vibration of footings.

1. Coefficient of *elastic uniform compression* of soil C_u for vertical mode of vibration, expressed in units of FL^{-3}.
2. Coefficient of *elastic non-uniform compression* of soil, C_ϕ, for rocking mode of vibration, expressed in units of FL^{-3}.
3. Coefficient of *elastic uniform shear* of soil C_τ, for translational mode of vibration, expressed in units of FL^{-3}.
4. Coefficient of *elastic non-uniform shear* of soil, C_ψ, for torsional mode of vibration, expressed in units of FL^{-3}.

Coefficient of elastic uniform compression of soil

The value of C_u can be obtained by conducting either a *cyclic vertical plate load test* or a *block resonance test*. Figure 15.17 (a) represents a typical graph of the results of a plate load test performed by Barkan (1962). The area of the plate used for the test was 1.40 m². The test was conducted on loess soil. The maximum stress transferred to the soil was increased in each loop. The difference between the total settlement and the residual settlement (at zero load) gives the elastic settlement

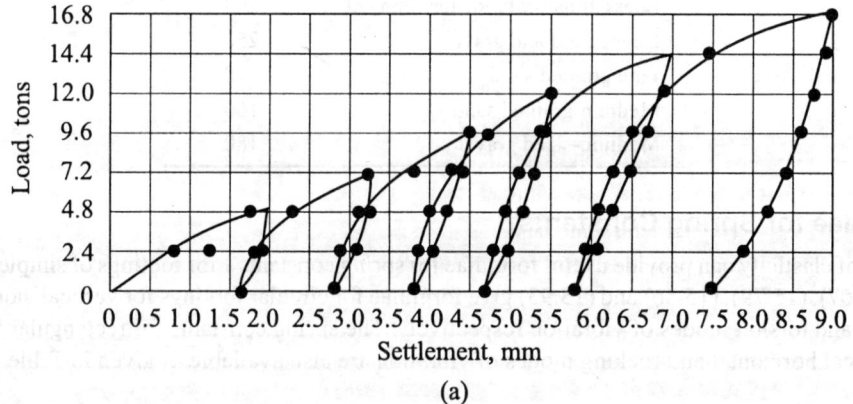

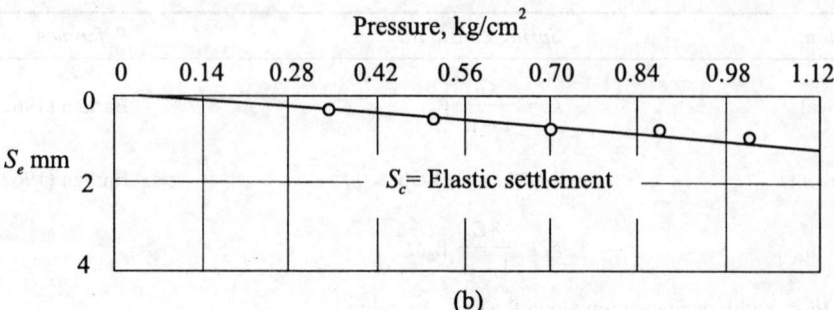

Fig. 15.17 Cyclic plate load test for determining elastic uniform compression of soil: (a) Results of load test of a 1.4 m² plate on loess, (b) pressure *vs.* elastic settlements (Barkan, 1962)

(rebound) for each stress level of the loop. The maximum stress of each loop is plotted against the corresponding elastic settlement in Fig. 15.17 (b) which gives a linear relationship. The vertical reaction pressure p_z may now be expressed as

$$p_z = C_u S_e \tag{15.100a}$$

or
$$C_u = \frac{p_z}{S_e} \tag{15.100b}$$

where, C_u = coefficient of proportionality called as the coefficient of elastic uniform compression of the soil,

S_e = elastic settlement of the bearing plate due to the external pressure.

Further investigations (Barkan, 1962) revealed that C_u is a function of the area of bearing for the same soil. The expression given for C_u is

$$C_u = \frac{1.13 E}{1 - \mu^3} \frac{1}{\sqrt{A}} \tag{15.101}$$

where, E = Young's modulus of soil,

μ = Poisson's ratio,

A = Bearing area of footing.

Table 15.4 gives the probable values of E for various types of soil (Barkan, 1962).

Table 15.4 Values of E for different types of soil

Type of soil	$E, kg/cm^2$
Plastic silty clay with sand and organic silt	310
Brown saturated silty clay with sand	440
Dense silty clay with some sand	2950
Medium moist sand	540
Gray sand with gravel	540
Fine saturated sand	850
Medium sand	830
Loess	1000 – 1300

If C_{u1} and C_{u2} are the coefficients of elastic uniform compressions for foundations of areas A_1 and A_2 respectively, they are related by the formula

$$C_{u2} = C_{u1} \sqrt{\frac{A_1}{A_2}} \tag{15.102}$$

It has been found by investigation that the value of C_u decreases with the increase in the area of the foundation base. The decrease from computation has been found to be greater than the experimental values. Barkan (1962) therefore, recommends that a standard area of 10 m² may be used for computing C_u for areas greater than 10 m².

The vertical spring constant k_z may be expressed by the relationship

$$k_z = C_u A = \frac{P_z}{S_e} \tag{15.103}$$

Now the relationship between C_u and the natural frequency f_{nz} for vertical mode of vibration may be expressed as (From Eq. 15.9)

$$f_{nz}^2 = \frac{1}{4\pi^2} \frac{k_z}{m} = \frac{1}{4\pi^2} \frac{C_u A}{m}$$

or

$$C_u = \frac{4\pi^2 f_{nz}^2 m}{A} \tag{15.104}$$

where, m = mass of footing and the machinery.

If a block resonance test is conducted, C_u can be determined from the known experimental resonant frequency of the system f_{nz}. As an alternative C_u can be found out from a cyclic plate load tests. The resonant frequency of the system can be determined from Eq. (15.104) from the known value C_u.

Table 15.5 gives probable values of C_u (Barkan, 1962) for the different types of soil for a preliminary design of foundation vibrating system.

Table 15.5 Recommended design values of C_u (Barkan, 1962) for $A = 10$ m^2

Category	Soil type	permissible static load kg/cm^2	C_u kg/cm^3
I	Weak soils (clays and silty clays with sand, in a plastic state; clayey and silty sands; also soils of category II and III with laminae of organic silt peat	up to 1.5	up to 3.0
II	Soils of medium strength (clays and silty clays with sand close to the plastic limit; sand)	1.5–3.5	3.0–5.0
III	Strong soils (clays and silty clays with sand of hard consistency; gravels and gravelly sands; loess and loessial soils)	3.5–5.0	5.0–10.0
IV	Rocks	> 5.0	> 10.0

Coefficient of elastic nonuniform compression of soil, C_ϕ

Rocking of a footing about X or Y-axis produces on the base nonuniform base pressure. The rotation angle is designated as ϕ. The corresponding static spring constant k_ϕ is expressed as

$$k_\phi = C_\phi I_b \tag{15.105}$$

where, I_b = moment of inertia of the base contact area about the horizontal axis (X or Y) normal to the plane of rocking and passing through the centroid of the contact area,

C_ϕ = coefficient of elastic nonuniform compression of soil. The relationship between C_u and C_ϕ is expressed as (Barkan, 1962)

$$C_\phi = 2C_u \tag{15.106}$$

Coefficient of elastic uniform shear of soil, C_τ

When a footing is in a translational mode parallel to either X or Y-axis, shearing resistance will be developed at the base of contact area. The elastic spring constant for the sliding mode may be expressed as

$$k_x \text{ (or } k_y) = C_\tau A \qquad (15.107)$$

where, C_τ = coefficient of elastic uniform shear of soil,

A = base contact area.

The relationship between C_u and C_τ is expressed as

$$C_\tau = 0.5 C_u \qquad (15.108)$$

Coefficient of nonuniform shear of soil, C_ψ

If a foundation is acted upon by a moment with respect to the vertical axis, it will rotate about this axis. Tests show that the angle of rotation, ψ of a foundation is proportional to the external moment and the resistance offered by the soil at the base. We may therefore, write

$$M_z = C_\psi I_z \psi \qquad (15.109)$$

where, M_z = external moment,

I_z = polar moment of inertia of contact base area of foundation,

ψ = angle of rotation in radians.

The rotational (or torsional) spring constant k_ψ may now be expressed as

$$k_\psi = C_\psi I_z \qquad (15.110)$$

The relationship between C_ψ and C_u is

$$C_\psi = 1.5 C_\tau = 0.75 C_u \qquad (15.111)$$

Basic Assumptions in the Theory of Vibration

Barkan (1962) found solution to the problem of vibrating machine foundation purely based on the concept of subgrade reaction. He assumed that the foundation vibrations as a problem of a solid body resting on weightless springs, the latter serving as a model for the soil. As a result of this concept, it is assumed that there is a linear relationship between the soil reacting on a vibrating foundation and the displacement of the foundation. Thus the relationship between the displacements and the reactions will be determined in terms of the coefficients of elastic uniform and nonuniform compressions, as well as elastic shear. The analysis also assumed that the soil underlying the foundation does not have inertial properties as described by the coefficients.

The foundation under machines with a steady regime of work are usually designed in such a way that there is a significant difference between the frequency ω of the machine and the resonant frequency ω_n of the foundation soil system. For low frequency machines, the ratio $\omega/\omega_n \leq 0.5$ and for high frequency machines $\omega/\omega_n \geq 1.5$. The damping effect is negligible for machines which fall within this regime and as such can be ignored. Neglecting of damping in the analysis of vibration of foundations would lead to conservative results.

Analysis of Foundations Subjected to Vertical Mode of Vibration

The analysis of foundation soil system in the vertical mode of vibration is based on the principle of mass-and-spring analogy without the effect of damping. Figure 15.18 (a) shows a foundation block subjected to vertical vibration under constant amplitude force as

$$Q = Q_0 \sin \omega t \tag{15.112}$$

acting through the centre of gravity of the block. The differential equation of motion is

$$m\ddot{z} + k_z z = Q_0 \sin \omega t \tag{15.113}$$

The maximum amplitude of motion as per Eq. (15.33) is

$$A_z = \frac{Q_0}{k_z - m\omega^2} = \frac{Q_0}{C_u A - m\omega^2} = \frac{Q_0}{m\left(\omega_{nz}^2 - \omega^2\right)} \tag{15.114a}$$

where,
$$k_z = C_u A \tag{15.114b}$$

$$\omega_{nz}^2 = \frac{C_u A}{m} \text{ or } \omega_{nz} = \sqrt{\frac{C_u A}{m}} \tag{15.114c}$$

Footings Subjected to Purely Translational Mode of Vibration

Figure 15.18 (b) shows a foundation subjected horizontal exciting force. The force is assumed to act parallel to any one of the axes X or Y in the plane of the axis passing through the centroid of the foundation. The equations of force and forced vibrations will be analogous to the equations of vertical vibrations of a foundation, in which C_τ is used instead of C_u, thus the equation of forced horizontal vibration will be (acting in the direction of X-axis)

$$m\ddot{x} + k_x x = Q_x \sin \omega t \tag{15.115}$$

where, x = horizontal displacement of the centre of gravity of the foundation.

From Eq. (15.115), we have

$$\omega_{nx} = \sqrt{\frac{k_x}{m}} = \sqrt{\frac{C_\tau A}{m}} \tag{15.116a}$$

or
$$f_{nx} = \frac{1}{2\pi} \sqrt{\frac{C_\tau A}{m}} \tag{15.116b}$$

The amplitude A_x is

$$A_x = \frac{Q_x}{m\left(\omega_{nx}^2 - \omega_n^2\right)} \tag{15.117}$$

where, f_{nx} = natural frequency of the foundation soil system for pure sliding.

Foundation Subjected to Pure Rocking Mode of Vibration (Rotational mode)

Figure 15.18 (c) shows a foundation block subjected to pure rocking by external periodic moment $M_y(t)$ about Y-axis in the X-Z plane, where

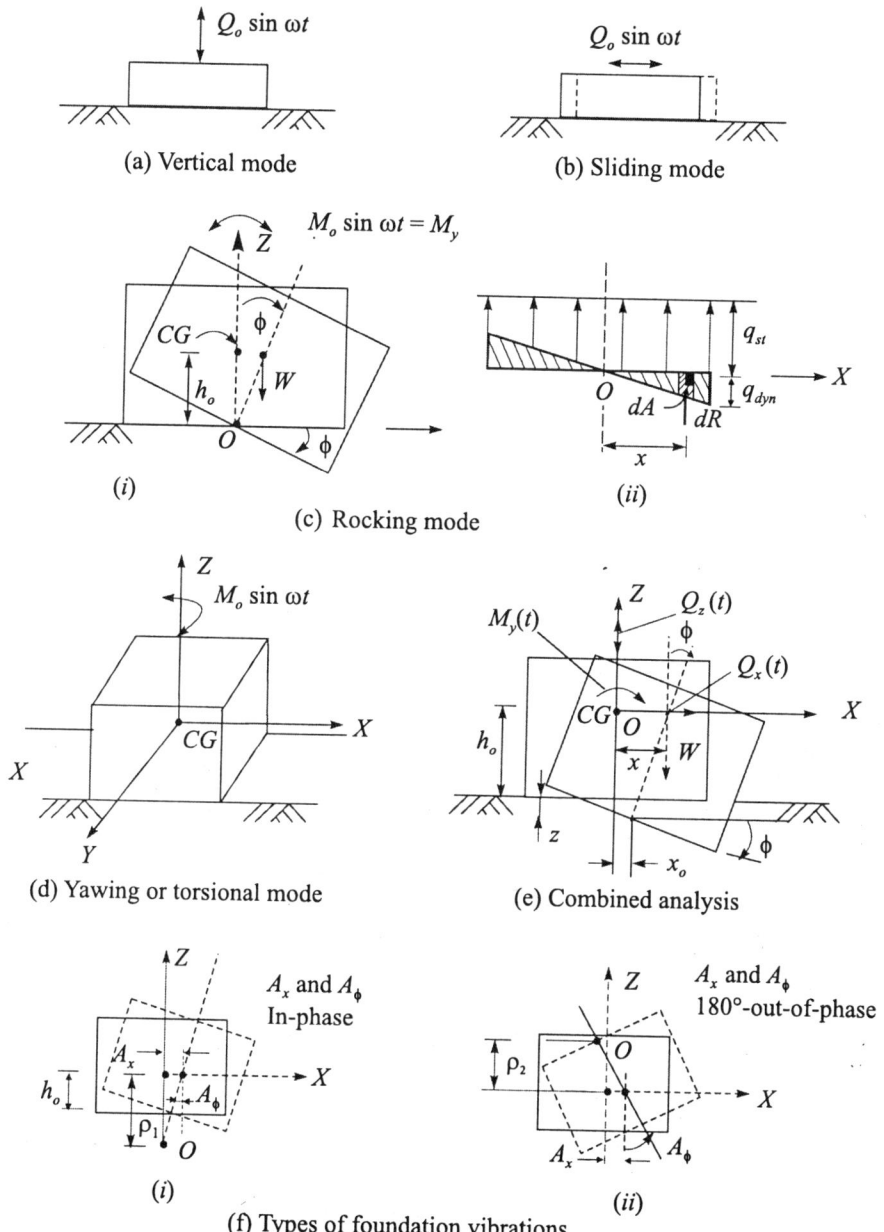

$Q_o \sin \omega t$

(a) Vertical mode

$Q_o \sin \omega t$

(b) Sliding mode

$M_o \sin \omega t = M_y$

(i)

(ii)

(c) Rocking mode

$M_o \sin \omega t$

(d) Yawing or torsional mode

$M_y(t)$ $Q_z(t)$ $Q_x(t)$

(e) Combined analysis

A_x and A_ϕ In-phase

(i)

A_x and A_ϕ 180°-out-of-phase

(ii)

(f) Types of foundation vibrations

Fig. 15.18 Analysis of uncoupled and coupled vibrations by elastic soil-spring method

$$M_y(t) = M_0 \sin \omega_t \qquad (15.118)$$

The position of foundation is determined by one independent variable ϕ, the angle of rotation of the foundation about the axis rotation. Figure 15.18 (c). At any instant when the angle of rotation is ϕ, the equation of motion will be

$$I_{\phi 0} \ddot{\phi} = \Sigma M$$

where, $I_{\phi 0}$ = mass moment of inertia of the foundation about the axis of rotation O,

ΣM = sum of all external moments with respect to the same axis,

$\ddot{\phi}$ = angular acceleration of the block.

Let W = weight of foundation acting through the centre of gravity of the block,

h_o = height of centre of gravity of the block above the axis of rotation O at the base.

Moment due to weight of block

Now, the external moment due to weight W with respect to the axis of rotation at any instant of time t and angle of rotation ϕ is

$$M_w = Wh_o\,\phi$$

Moment due to soil reaction

Let dA = an element of foundation area in contact with the soil at a distance x from the axis of rotation O [Fig. 15.17 (c)],

dR = soil reaction over this area $= x\phi C_\phi dA$,

where, C_ϕ = the coefficient of elastic nonuniform compression of soil,

ϕx = displacement of the centre of area dA.

The moment of dR about O is

$$dM_r = -xdR = -C_\phi x^2\,\phi dA$$

Assuming that there is no tension at the base, the total reactive moment M_r may be expressed as

$$M_r = -C_\phi\,\phi \int_A x^2 dA = -C_\phi I_b\,\phi \qquad (15.119)$$

where, I_b = moment of inertia of the base in contact with the soil with respect to the axis of rotation of the foundation.

Now, summing up the moments we have,

$$I_{\phi o}\ddot{\phi} = Wh_o\phi - C_\phi I_b\phi + M_o \sin \omega t$$

or $I_{\phi o}\ddot{\phi} + (C_\phi I_b - Wh_o)\,\phi = M_o \sin \omega t \qquad (15.120)$

The equation for free-rocking vibration with respect to Y-axis may be obtained by equating M_o to zero.

$$I_{\phi o}\ddot{\phi} + (C_\phi I_b - Wh_o)\,\phi = 0 \qquad (15.121)$$

The solution for natural circular frequency $\omega_{n\phi}$ is given by

$$\omega_{n\phi} = \sqrt{\dfrac{C_\phi I_b - Wh_o}{I_{\phi o}}} \qquad (15.122a)$$

In practice Wh_o is quite small and may be neglected. The reduced $\omega_{n\phi}$ is

$$\omega_{n\phi} = \sqrt{\dfrac{C_\phi I_b}{I_{\phi o}}} \qquad (15.122b)$$

or
$$f_{n\phi} = \frac{1}{2\pi} \sqrt{\frac{C_\phi I_b}{I_{\phi o}}} \qquad (15.122c)$$

Since
$$I_b = \frac{BL^3}{12}, \text{ we have}$$

$$f_{n\phi} = \frac{1}{2\pi} \sqrt{\frac{C_\phi}{I_{\phi o}} \frac{BL^3}{12}} \qquad (15.122d)$$

where, B = width and L = length of side of foundation (The width is parallel to the axis of rotation).

The expression for maximum angular amplitude is

$$A_\phi = \frac{M_o}{I_b \left(\omega_{n\phi}^2 - \omega^2\right)} \qquad (15.123)$$

It is clear from Eq. (15.22d) that the length of the side of the foundation base in contact with the soil and perpendicular to the axis of rotation has considerable effect on the natural frequency of rocking vibrations of the foundation. The natural frequency and the amplitude of motion change with the change in length whereas the width parallel to the axis of rotation has little effect on these values.

If L is the length of the foundation perpendicular to the axis of rotation, then the maximum vertical amplitude A of vibration along the edge B of the foundation is

$$A = \frac{1}{2} L A_\phi = \frac{Q_0 L}{2 I_b \left(\omega_{n\phi}^2 - \omega^2\right)} \qquad (15.124)$$

Rocking vibrations occur mostly where unbalanced horizontal components of exciting forces and moments occur due to the machine being installed on the tops of high foundation blocks.

Foundations Subjected to Torsional Vibrations

In addition to the types of vibrations discussed above, vibrations in shear may have a form of rotational vibration with respect to the vertical axis passing through the centre of gravity of the foundation and the centroid of the area of its base [Fig. 15.18 (d)]. The equation of motion for a foundation subjected to periodic torsional moments in the horizontal plane normal to the vertical axis may be expressed as

$$I_{\psi o} \ddot{\psi} + C_\psi I_z \psi = M_z \sin \omega t \qquad (15.125)$$

where, $I_{\psi o}$ = mass moment of inertia of foundation block and machine about the vertical axis of rotation,

I_z = polar moment of inertia of the base area,

ψ = angle of torsion of the foundation,

C_ψ = coefficient of elastic non-uniform shear of motion,

$M_z \sin \psi t$ = exciting moment acting in the horizontal plane.

The expression for natural frequency is

$$f_{n\psi} = \frac{1}{2\pi} \sqrt{\frac{C_\psi I_z}{I_{\psi o}}}$$

(15.126)

where, the torsional spring constant k_ψ is

$$k_\psi = C_\psi I_z$$

(15.127)

The maximum angular displacement may be expressed as

$$A_\psi = \frac{M_z}{I_\psi \left(\omega_{n\psi}^2 - \omega^2\right)}$$

(15.128)

15.13 VIBRATION ANALYSIS OF FOUNDATIONS SUBJECTED TO SIMULTANEOUS VERTICAL, SLIDING AND ROCKING OSCILLATIONS BY ELASTIC SOIL-SPRING METHOD (BARKAN, 1962)

The foregoing sections were concerned with the uncoupled modes of vibrations of foundations. Normally a rocking vibration about Y or X-axis is associated with translation mode of vibration along the X or Y-axis respectively. The vibration in the vertical direction (Z-axis) is independent of the other two types of vibrations. If a foundation is acted by exciting loads having no vertical components, then no vertical vibration of the foundation develops. In this case the foundation will undergo rotation about the Y-axis (or X-axis) and horizontal displacement in the direction of X-axis (or Y-axis). Such vibrations are called as *coupled vibrations*.

The fact that vertical vibrations of foundations are independent of vibrations in the directions of X (or Y) leads to the conclusion that we have to investigate the effect of the coupled vibration together on the response of foundation-soil-system. The effect of vertical vibration on the system can be analysed independently as dealt with in Section 15.12.

Let the foundation block in Fig. 15.18 (e) be acted upon by exciting forces $Q_x(t)$ and $M_y(t)$. Under the action of these forces, the foundation will undergo a two dimensional motion determined by the values of two independent parameters, that is, the projection of x displacements of foundation centre of gravity on the X-axis and the angle ϕ of rotation with respect to Y-axis which passes through the centre of gravity of the foundation and machine, perpendicular to the plane of vibrations.

By projecting all the force sat any time t acting on the foundation on the X-axis and adding on the same axis the intertial forces, the equations of motion may be written as

$$m\ddot{x} = \Sigma X_i$$

(15.129)

$$I_{\phi g}\ddot{\phi} = \Sigma M_i$$

where, m = mass of foundation with machinery,

     $I_{\phi g}$ = moment of inertia of the mass with respect to the Y-axis passing through the centre of gravity of the mass,

     ϕ = angle of rotation of the vertical axis passing through the centre of gravity of the mass,

    ΣX_i = the sum of the x-components of all the external forces,

    ΣM_i = the sum of the moments of all forces about the Y-axis,

Let W = weight of the foundation,

    x_0 = displacement of the centroid of the base contact area at time t,

x = the corresponding displacement of the centre of gravity of the mass in the X-direction,

A = contact area of foundation,

C_τ = coefficient of elastic uniform shear of soil,

I_b = moment of inertia of foundation base with respect to the axis passing through the centroid of the area and perpendicular to the plane of vibration,

$I_{\phi o}$ = moment of inertia of the mass about an axis parallel to the Y-axis and passing through the centroid of the base area, and perpendicular to the plane of vibration,

h_o = height of centre of gravity of mass from the base.

1. *Horizontal reaction of elastic resistance of soil*

   Due to displacement x_o of foundation in the positive direction of X-axis. Figure 15.18 (e), there will be a reaction X_1, at the base in the opposite direction which is expressed as

   $$X_1 = -C_\tau A x_o \tag{15.130a}$$

   From Fig. 15.18 (e), $x_o = x - h_o\phi$, therefore,

   $$X_1 = -C_\tau A (x - h_o\phi) \tag{15.130b}$$

   The moment of this force about the Y-axis is

   $$M_1 = C_\tau A h_o (x - h_o\phi) \tag{15.131a}$$

   which acts in the clockwise direction and hence positive.

2. *Reactive resistance of soil induced by rotation of foundation base area*

   As per Eq. (15.119), the resisting moment M_2 is

   $$M_2 = -C_\phi \phi I_b \tag{15.131b}$$

3. *Moment due to the weight W of foundation*

   $$M_3 = W h_o \phi \tag{15.131c}$$

where, $x = h_o\phi$, eccentricity of foundation weight W. Now substituting the respective forces in Eq. (15.129), we have

$$m\ddot{x} + C_\tau A x - C_\tau A h_o \phi = Q_x \sin \omega t$$

$$I_{\phi o}\ddot{\phi} - C_\tau A h_o x + (C_\phi I_b - W h_o + C_\tau A h_o^2) \phi = M_y \sin \omega t \tag{15.132}$$

The equations of motion Eq. (15.132) give rise to two natural frequencies ω_{n1} and ω_{n2}. These natural frequencies can be obtained by solving Eq. (15.132) after putting the RHS = 0. Since the details of solution are quite involved, only the final equation for natural frequencies is given here as

$$\omega_n^4 = \frac{\omega_{n\phi}^2 + \omega_{nx}^2}{\eta}\, \omega_n^2 + \frac{\omega_{n\phi}^2 + \omega_{nx}^2}{\eta} = 0 \tag{15.133a}$$

The roots of Eq. 15.133 (a) may be written as

$$\omega_{n1,2}^2 = \frac{1}{2\eta}\left[\omega_{n\phi}^2 + \omega_{nx}^2 \pm \sqrt{\left(\omega_{n\phi}^2 + \omega_{nx}^2\right)^2 - 4\eta\omega_{n\phi}^2\omega_{nx}^2}\right] \tag{15.133b}$$

where,

$$\eta = \frac{I_{\phi g}}{I_{\phi o}}$$

$$\omega_{n\phi}^2 = \frac{C_\phi I_b - W h_o}{I_{\phi o}} \quad \text{from Eq. (15.122a),}$$

$$\omega_{nx}^2 = \frac{C_\tau A}{m} \quad \text{from Eq. (15.116a).}$$

It can be shown that the smaller of the two natural frequencies ω_{n1} and ω_{n2} (for example ω_{n2}) is smaller than the smallest of the two limiting frequencies $\omega_{n\phi}$ and ω_{nx} and the larger natural frequency ω_{n1} is always larger than $\omega_{n\phi}$ and ω_{nx}.

Amplitudes of motion

The amplitudes of motion from Eq. (15.132) be found out in two steps

1. Consider only $Q_x \sin \omega t$ is acting and $M_y \sin \omega t = 0$.
2. Consider $Q_x \sin \omega t = 0$ and only $M_y \sin \omega t$ is acting.

The procedure is quite involved and as such not given here. The final solution for the amplitudes of motion are as given below (when both Q_x and M_y acting)

$$A_x = \frac{\left(C_\tau A h_o^2 + C_\phi I_b - W h_o - I_{\phi g} \omega^2\right) Q_x + \left(C_\tau A h_o\right) M_y}{\Delta\left(\omega^2\right)} \qquad (15.134)$$

$$A_\phi = \frac{\left(C_\tau A h_o\right) Q_x + \left(C_\tau A - m\omega^2\right) M_y}{\Delta\left(\omega^2\right)} \qquad (15.135)$$

where, $\Delta\left(\omega^2\right) = mI_{\phi g}\left(\omega_{n1}^2 - \omega^2\right)\left(\omega_{n2}^2 - \omega^2\right)$.

The amplitudes of motion when only Q_x or My (t) is acting may be expressed as follows:

1. When $Q_x(t) = 0$ and only $M_y(t)$ acting

$$A_x = \frac{\left(C_\tau A h_o\right) M_y}{\Delta\left(\omega^2\right)} \qquad (15.136)$$

$$A_y = \frac{\left(C_\tau A - m\omega^2\right) M_y}{\Delta\left(\omega^2\right)} \qquad (15.137)$$

2. When $M_y(t) = 0$ and only $Q_x(t)$ acting

$$A_x = \frac{\left(C_\tau A h_o^2 + C_\phi I_b - W h_o - I_{\phi g} \omega^2\right) Q_x}{\Delta\left(\omega^2\right)} \qquad (15.138)$$

$$A_\phi = \frac{\left(C_\tau A h_o\right) Q_x}{\Delta\left(\omega^2\right)} \qquad (15.139)$$

There are four possible types of foundation vibrations which can be found out as follows. The ratios of the amplitudes of motion for the case $Q_x(t) = 0$, Eq. (15.136) and (15.137), may be expressed as

$$\rho = \frac{A_x}{A_\phi} = \frac{C_\tau A h_o}{C_\tau A - m\omega^2} = \frac{\omega_{nx}^2 h_o}{\omega_{nx}^2 - \omega^2} \tag{15.140}$$

1. If the frequency of excitation ω is smaller than ω_{nx}, then $\rho \approx h_o$, that is the axis of rotation lies along the vertical axis passing through the centroid of the base area. The foundation undergoes only rocking, and sliding is absent.

2. The value of ρ [Eq. (15.140)] increases with the decrease in the value of denominator of Eq. (15.140). This happens when ω increases. In such cases the foundation undergoes simultaneous rocking and sliding vibrations Fig. 15.18 (f1). The two motions are in phase with each other

3. The value of ρ tends to infinity if the operating frequency ω is closer to ω_{nx}. In such a case, the foundation will undergo only sliding and rocking vibrations will be absent.

4. If the operating frequency ω is greater than ω_{nx}, ρ will be negative. Then the amplitudes of motion A_x and A_ϕ will be out of phase by 180° Fig. 15.18 (f2).

If the vibration of footing is as per Case 2 mentioned above, the centre of rotation will lie at a depth ρ_1 below the centre of gravity of the mass. When the amplitudes of motion A_x and A_ϕ are out of phase by 180°, the foundation vibrates around a point which lies higher than the centre of gravity and at a distance ρ_2 determined from Eq. (15.140).

15.14 MACHINE FOUNDATIONS SUBJECTED TO IMPACT LOADS (BARKAN, 1962)

Introduction

The forge hammer foundations are subjected to repeated impact loads. The various components of a forge hammer foundation with the machinery is shown in Fig. 15.19 (a). It consists of a *die* kept on an *anvil* which will be forged to the desired shapes by repeated blows of a falling hammer. The side frame of the hammer is normally connected to the *steel anvil*. Oak-timber or plywood serve as elastic pads for the anvil. The reinforcement below the pad consists 2 to 4 horizontal grillages formed by 8 to 12 mm bars and spaced 10 to 20 cm apart. Near the foundation surface in contact with soil, the reinforcement consists of 1 or 2 horizontal grillages formed by 12 to 20 mm bars and spaced 15 to 30 cm apart. Distances between the grillages are 10 to 15 cm in the part of the foundation under the anvil and 15 to 30 cm near the foundation contact surface. Spring pads are placed between the anvil and the foundation to absorb shocks of impact.

Hammer

The weight of the *hammer* includes the weight of ram, side frames which guide the ram, ram cylinder with anchor plate etc., but excluding the anvil weight. The *anvil weight* may be taken as equal to 15 to 20 times the weight of the ram. The total weight of the hammer and anvil is taken to be 25 to 30 times the weight of the *ram*. The ratio between he weight of the foundation and the hammer is normally very high, as an example, for each 10 kN of dropping weight, the foundation weight may go up to 400 kN (Barkan, 1962).

The hammers used for forging might be drop hammers or pneumatic hammers. The modern forging practice mostly employs double-acting hammers. In these hammers, steam or compressed air acts on the ram not only while it is being lifted but also during its drop. Therefore, the velocity and kinetic energy are considerably larger at the moment of impact of the ram against the workpiece. In single-acting hammers, the ram which is rigidly tied to the piston by means of a rod, is lifted by the pressure of steam released through a valve located under the piston and opened when the latter is in its extreme low position. After the piston is raised to the height desired, the access of the steam to the cylinder under the piston is stopped, the valve opens, and the steam or compressed air escapes. The piston together with the ram drops at increasing speed. After the access of steam is discontinued and the exhaust valve opens, the steam cannot escape from the space in the cylinder under the piston. Therefore, a counterpressure against the ram drop is created, resulting in a loss both in the ram velocity and the kinetic energy of its drop.

Velocity of Hammer Fall before and after Impact

Velocity of hammer before impact

Single acting hammers and the like are called as hammers with an *unrestricted drop*. The pneumatic double-acting hammers are called as hammers with *restricted ram drop*.

The velocity v of the ram drop under the action of unrestricted motion equals

$$v = \alpha \sqrt{2gh} \qquad (15.141)$$

where, g = acceleration of gravity,

h = height of ram drop,

α = coefficient which takes into account counter pressure and friction forces.

The value of α has been found to be about 0.90, but for all practical purposes $\alpha = 1$ may be used (Barkan, 1962).

The velocity of forced motion of ram under the action of its own weight and the steady steam or air pressure p may be expressed as

$$v = \alpha \sqrt{\frac{2g\left(W_d + ap\right)h}{W_d}} \qquad (15.142)$$

where, a = area of piston,

p = pressure on piston,

W_d = total weight of dropping parts,

α = correction coefficient.

The correction coefficient a has been found to lie within a range of 0.45 to 0.80, but an average value of 0.65 has been recommended (Barkan, 1962).

Velocity of Hammer and Anvil after Impact

In the dynamic response of foundations subjected to repeated impact dynamic loads, it is necessary to know the velocity of *anvil* after impact.

Let, v_1 = velocity of ram before impact,

v_2 = rebound velocity of ram after impact,

v_a = velocity of anvil after impact.

It may be noted here that the foundation is *motionless* before the impact and as such its velocity before the impact is zero.

The principle of impact between ram and the anvil is that the momentum of colliding bodies before and after impact is constant. Therefore, we may write

$$m_d v_1 = m_d v_2 + m_a v_a \qquad (15.143)$$

where, m_d = mass of the dropping weight,

m_a = mass of anvil including the weight of frame (if mounted on it).

In Eq. (15.143) there are two unknowns v_2 and v_a. We require therefore another equation for solving for v_2 and v_a.

In order to derive a second equation, Newton's hypothesis concerning the _restitution of impact_ is used. According to this hypothesis, _if there occurs an impact between two bodies, moving in relation to each other, the relative velocity after the impact is proportional to the relative velocity before the impact_. The ratio between these two depends only on the material of the bodies which undergo the impact. The foundation was motionless before the impact as such the relative velocity of the ram is v_1. After impact, the rebound velocity of ram v_2 and that of the anvil with the frame (if attached) is v_a. The relative velocity after impact is equal to $(v_a - v_2)$. The _coefficient of elastic restitution e_ may now be written as

$$e = \frac{v_a - v_2}{v_1} \qquad (15.144)$$

or $$v_2 = (v_a - ev_1)$$

Substituting for v_2 in Eq. (15.144) and simplifying, we have

$$v_a = \frac{v_1 (1 + e)}{(1 + n)} \qquad (15.145)$$

where, $n = \dfrac{m_a}{m_d}$

For perfectly elastic bodies $e = 1$, and $e = 0$ for the impact of a rigid body (ram) against a plastic one (when the _die_ is at a very high temperature). For real bodies the value of e lies within the range of $0 < e < 1$.

As per Barkan (1962), $e \approx 0.1$ when the workpiece is at a high temperature and just before subjected to shaping. As the number of blows on the _workpiece_ increases, the temperature of the workpiece decreases, the impact rigidity increases, and consequently the value of e increases. The ultimate value of $e \approx 0.5$. Since higher values of e leads to greater amplitudes of motion, a value $e = 0.5$ is recommended for the design of hammer foundations.

When once the value of e is known (or assumed), the velocity of the anvil v_a can be found out from Eq. (15.145).

Thickness of Timber Pad Under the Anvil

Timber sleepers are laid in layers in the form of grillages under the anvil. The thickness of the pad varies with the weight of the striking part of the hammer and the type of hammer, from about 20 cm for 10 kN hammer to a maximum of about 100 cm for hammers of over 30 kN.

Thickness of the pad should be so selected that the dynamic stresses induced into the pad by the impact loads do not exceed permissible values. For example,

Oak	300 to 350 kg/cm²,
Pine	200 to 250 kg/cm²,
Larch	150 to 200 kg/cm².

Coefficient of Elastic Uniform Compression of Soil

The Coefficient of uniform compression C_u as used for rotary type foundations applies for shallow foundations. The depths of foundations of forge hammers are normally greater than that for rotary type. In deeper foundations the effect of side friction and the overburden effect increases the value of C_u. Barkan (1962) recommends for hammer foundation a value C_u as equal to

$$\bar{C}_u = 3C_u \qquad (15.146)$$

where, C_u = Coefficient of elastic uniform compression for rotary type foundations at shallow depths.

Permissible Amplitudes

The permissible amplitudes for the foundations of forge hammers are normally greater than that allowed for other types of foundations.

Table 15.6 gives the permissible amplitudes for the various weights of hammers.

Table 15.6 Permissible amplitudes for Hammer foundations (Barkan, 1962)

Weight of Hammer kN	*Permissible amplitude (mm)*
10	1.0
20	2.0
> 30	3 to 4

Dynamic Analysis

The hammer-anvil-pad foundation soil system is assumed to have two-degrees of freedom. The pad between the anvil and the foundation is assumed to be an elastic body with a spring constant k_2 and the soil below the foundation another elastic body with a spring constant k_1.

The computation set-up is therefore, reduced to a system of three bodies, namely, the ram which is the striking body, the anvil which is separated by the foundation by an elastic connection (spring constant k_2) and the foundation on an elastic base (with a spring constant k_1) as shown schematically in Fig. 15.19 (b). The free-body diagram is also shown by the side. There are two equations of motion for the whole system. Consider the free-body diagram shown in Fig. 15.19 (b). The line AA and BB represent the static equilibrium positions for the *anvil* and the foundation respectively before the impact. After the impact at any instant of time t, the anvil undergoes a displacement z_2 with reference to AA, and the foundation undergoes a displacement z_1 with reference to the line BB. Due to these displacements, the spring k_1 gets compressed by an amount equal to z_1 but the spring k_2 gets compressed by an amount equal to $(z_2 - z_1)$, (where $z_2 > z_1$). As a result of these displacements, the following differential equation can be written.

$$m_1 \ddot{z}_1 + k_1 z_1 - k_2 (z_2 - z_1) = 0 \qquad (15.147a)$$

$$m_2 \ddot{z}_2 + k_2 (z_2 - z_1) = 0 \qquad (15.147b)$$

where, m_1 = mass of foundation,

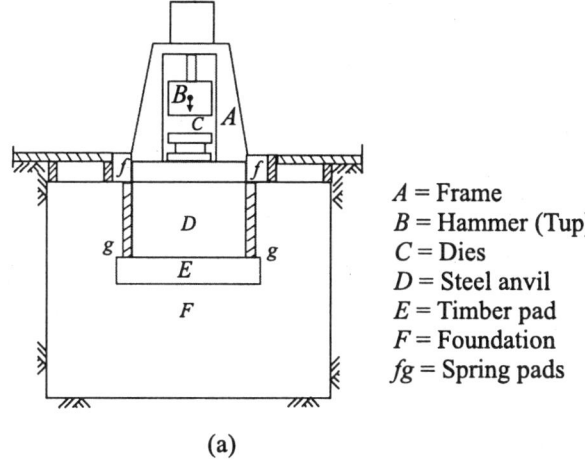

(a)

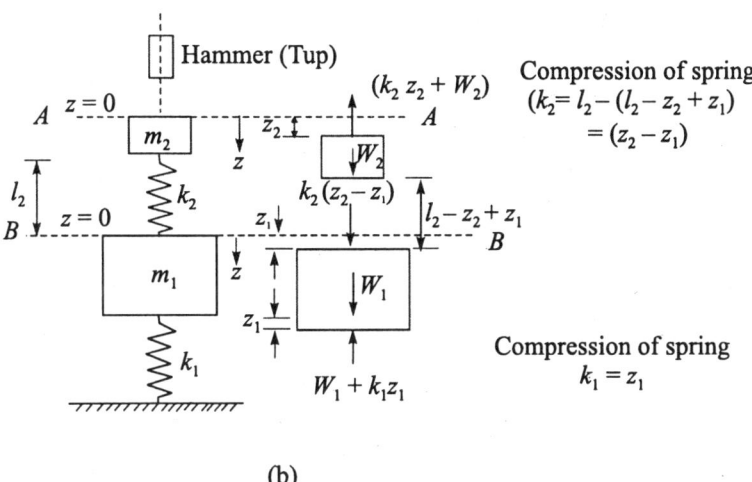

(b)

Fig. 15.19 (a) Forge Hammer foundation, (b) two-spring system and free-body of analysis

m_2 = mass of anvil,

k_1 = soil spring constant = $\bar{C}_u A_1$,

A_1 = contact area of foundation,

k_2 = coefficient of rigidity (equivalent spring constant) of the pad between the anvil and the foundation = $(E/t) A_2$,

E = Young's modulus of material of pad,

t = thickness of pad,

A_2 = base area of pad,

z_1, z_2 = displacement of foundation and anvil respectively from equilibrium position.

The development of solutions to Eq. (15.147) is quite involved and as such not given here.

From the differential equations Eq. (15.147), it is possible to develop a frequency equation of the fourth degree as given below.

$$\omega_n^4 - (1 + N)(\omega_{na}^2 + \omega_{nl}^2)\,\omega_n^2 + (1 + N)\,\omega_{na}^2\,\omega_{nl}^2 = 0 \qquad (15.148)$$

where, $N = \dfrac{m_2}{m_1}$

ω_{na} = the limiting natural frequency for the anvil resting on the elastic pad expressed as

$$\omega_{na} = \sqrt{\frac{k_2}{m_2}} \qquad (15.149b)$$

ω_{nl} = limiting natural frequency for the whole foundation soil system expressed as

$$\omega_{nl} = \sqrt{\frac{k_1}{m_1 + m_2}} \qquad (15.149c)$$

Here ω_{na} is the natural frequency of vibration of anvil on a motionless foundation. Whereas, w_{nl} is the natural frequency of the entire installation by assuming the pad below the anvil is infinitely rigid.

The natural frequencies ω_{n1} and ω_{n2} are determined as roots of Eq. (15.148). The roots may be expressed as

$$\omega_{n1,2} = \frac{1}{2}\left[(1+N)\left(\omega_{na}^2 + \omega_{nl}^2\right)\right] \pm$$

$$\sqrt{\left[(1+N)\left(\omega_{na}^2 + \omega_{nl}^2\right)\right]^2 - 4\left(1+N\right)\left(\omega_{na}^2\,\omega_{nl}^2\right)} \qquad (15.150)$$

The amplitudes z_1 and z_2 are obtained by solving Eq. (15.147). The details of solutions are not given here. The final solutions may be expressed as

$$z_1 = \frac{\left(\omega_{na}^2 - \omega_{n2}^2\right)\left(\omega_{na}^2 - \omega_{n1}^2\right)}{\omega_{na}^2\left(\omega_{n1}^2 - \omega_{n2}^2\right)}\,v_a\left[\frac{\sin \omega_{n1} t}{\omega_{n1}} - \frac{\sin \omega_{n2} t}{\omega_{n2}}\right] \qquad (15.151a)$$

$$z_2 = \frac{v_a}{\omega_{n1}^2 - \omega_{n2}^2}\left[\frac{\omega_{na}^2 - \omega_{n2}^2}{\omega_{n1}}\sin \omega_{n1} t - \frac{\omega_{na}^2 - \omega_{n1}^2}{\omega_{n2}}\sin \omega_{n2} t\right]$$

Barkan (1962) showed from field observations that vibrations occurred at lower frequency only and as such, it may be assumed that the amplitude of vibrations for $\sin \omega_{n1} t$ (where $\omega_{n1} > \omega_{n2}$) equals zero. Then the approximate expressions for dynamic displacement of the foundation and anvil will be as follows.

$$z_1 = -\frac{\left(\omega_{na}^2 - \omega_{n1}^2\right)\left(\omega_{na}^2 - \omega_{n2}^2\right)}{\omega_{na}^2\left(\omega_{n1}^2 - \omega_{n2}^2\right)\omega_{n2}}\,v_a \sin \omega_{n2} t \qquad (15.152a)$$

$$z_2 = -\frac{\omega_{na}^2 - \omega_{n1}^2}{\left(\omega_{n1}^2 - \omega_{n2}^2\right)\omega_{n2}}\,v_a \sin \omega_{n2} t \qquad (15.152b)$$

The maximum amplitudes of motion occurs when $\sin \omega n_2 t = 1$. The expressions for maximum amplitudes are, for foundation soil system

$$z_1 = A_z = -\frac{\left(\omega_{na}^2 - \omega_{n2}^2\right)\left(\omega_{na}^2 - \omega_{n1}^2\right)}{\omega_{na}^2 \left(\omega_{n1}^2 - \omega_{n2}^2\right)\omega_{n2}} v_a \tag{15.153a}$$

for amplitude of anvil

$$z_2 = A_a = -\frac{\omega_{na}^2 - \omega_{n1}^2}{\left(\omega_{n1}^2 - \omega_{n2}^2\right)\omega_{n2}} v_a \tag{15.153b}$$

The maximum stress σ in the pad may be obtained from the expression

$$\sigma = \frac{k_2 \left(z_1 + z_2\right)}{A_2} \tag{15.154}$$

Tentative Determination of Foundation Weight and Base Area

For preliminary calculations, the foundation weight per unit of dropping weight may be calculated by (Barkan, 1962) as

$$n_f = 8.0 \, (1 + e) \, v - n_a \tag{15.155}$$

where, n_f = foundation weight per unit of dropping weight = W_f / W_d,

W_f = weight of foundation with backfill,

W_d = weight of dropping parts,

e = the Coefficient of restitution,

v = velocity of dropping weight just before impact,

$$n_a = \frac{\text{weight of anvil and frame}}{\text{actual weight of dropping parts}} = \frac{W_a}{W_d}$$

W_a = weight of anvil and frame.

Some numerical values of hammer coefficients are given Table 15.7.

The total weight of foundation (including the backfill) is

$$W_f = n_f W_d \tag{15.156}$$

Table 15.7 Values of some Hammer coefficients (Barkan 1962)

Type of Hammer	v, m/sec	e	n_a	n_f
Stamping Hammers:				
Double-acting				
(stamping of steel pieces)	6.5	0.5	30	48
Unrestricted Hammers:				
stamping of steel pieces	4.5	0.5	20	34
Stamping of nonferrous metals	4.5	0.0	–	16
Forge hammers proper:				
Double acting	6.5	0.25	30	35
Unrestricted	4.5	0.25	20	25

Tentative determination of base area (Barkan, 1962)

The equation proposed is

$$a_f = \frac{20\,(1+e)}{q_a} \qquad\qquad (15.157)$$

where, a_f = base area per unit of dropping weight,

q_a = allowable bearing pressure of soil.

The total base area for a dropping weight W_d may be found from

$$A_f = a_f\,W_d \qquad\qquad (15.158)$$

15.15 DESIGN CRITERIA FOR MACHINE FOUNDATIONS

Stability of a Foundation

A foundation in general should satisfy two basic requirements. They are:

1. It should be stable against shear failure.
2. The settlement must be within permissible limits.

Based on the two criteria mentioned above, the safe bearing pressure of a foundation has to be worked out for a satisfactory functioning of the foundation. These criteria should be satisfied first for static loads, and then for dynamic loads. Under dynamic loads there should not be further compaction of the soil beyond permissible elastic limits.

Types of Machine Foundations and Machines

The foundations for a machine should be designed to suit its particular requirements keeping in view the stability of the foundation and the cost structure. The foundations can be of box type or block type, or any other type according to the type of machinery to be installed.

Foundations are required for the following types of machines.

1. Rotary type engines or generators with imperfectly balanced rotating parts.
2. Reciprocating engines.
3. Impact Machines.

Steam turbines, motor generators and centrifugal pumps come under the first category. Air or gas compressors and reciprocating pumps represent the second type. Forge hammers and the like which impart repeated blows come under the last category of machines.

The rotary machines are classified on the basis of their speed as follows.

Low frequency	up to 1500 rpm
Medium frequency	1500 to 3000 rpm
High frequency	Greater than 3000 rpm.

Reciprocating machines run at operating speeds less than 600 rpm; Rotary machines like turbogenerators and compressors may have speeds more than 3000 rpm and even go up to 10,000 rpm. There are motor generators which operate at much lower speeds than turbogenerators. Their speeds may range from 300 to 400 rpm (Barkan, 1962).

Permissible Amplitudes for Rotary Type Machines (Barkan, 1962)

Type of machine	Amplitude (mm)
(a) For medium frequency machines	
Vertical vibration	0.06 to 0.04
Horizontal vibration	0.09 to 0.07
(b) For High frequency machines	
Vertical vibration	0.03 to 0.02
Horizontal vibration	0.05 to 0.04
(c) For low frequency machines	0.02
(less than 500 rpm)	

The permissible amplitudes for impact type machines are given in Section 15.14.

Weight of Machine Foundation

The use of heavy foundations to eliminate excessive vibrations have been in use from early times. Manufacturers of machines invariably recommend the weight and size of the foundation suitable for their equipments. As a thumb rule, the weight of foundation for rotary type machines may range from 2 to 2.5 times the weight of the engine.

Design Criteria

The end product of the design procedure is the determination of a foundation soil system which satisfactorily supports equipment or machinery. The design criteria should satisfy the following.

1. The settlement and bearing pressure of foundation must be within permissible limits under static loads.
2. Under the action of dynamic loads the following criteria should be satisfied.
   (a) The operating frequency of machine should not be close to the natural frequency of the foundation-soil system to avoid resonance.
   (b) For low speed machines and high speed machines the ratio of the operating frequency and natural frequency of the system (ω/ω_n) should comply the following.
   Low speed $\omega/\omega_n \le 0.5$.
   High speed $\omega/\omega_n \ge 1.5$.
   (c) The amplitudes of motion at operating frequencies should not exceed the permissible limits.
   (d) The vibrations generated by a machine should not cause adverse effects on adjoining machines or structures.
   (e) The vibrations should not cause annoyance to persons standing close by.

The design criteria most often encountered relate to the dynamic response of the foundation-soil system. These are expressed in terms of the limiting amplitudes of vibration at a particular frequency or a limiting value of a peak velocity or acceleration. Figure 15.20 indicates the order of magnitudes which may be involved in the criteria for dynamic response. Five curves limit the zones of different sensitivities of response by persons ranging from *not noticeable* to *severe*. The envelope described by the shaded line as limit for machines and machine foundations indicates the *limit for safety* and *not a limit for satisfactory operation of machines.* Below 2000 cycles/min, this limit represents a peak velocity of 1.0 inch per second, and above 2000 cycles per minute, it corresponds to acceleration of 0.5 g.

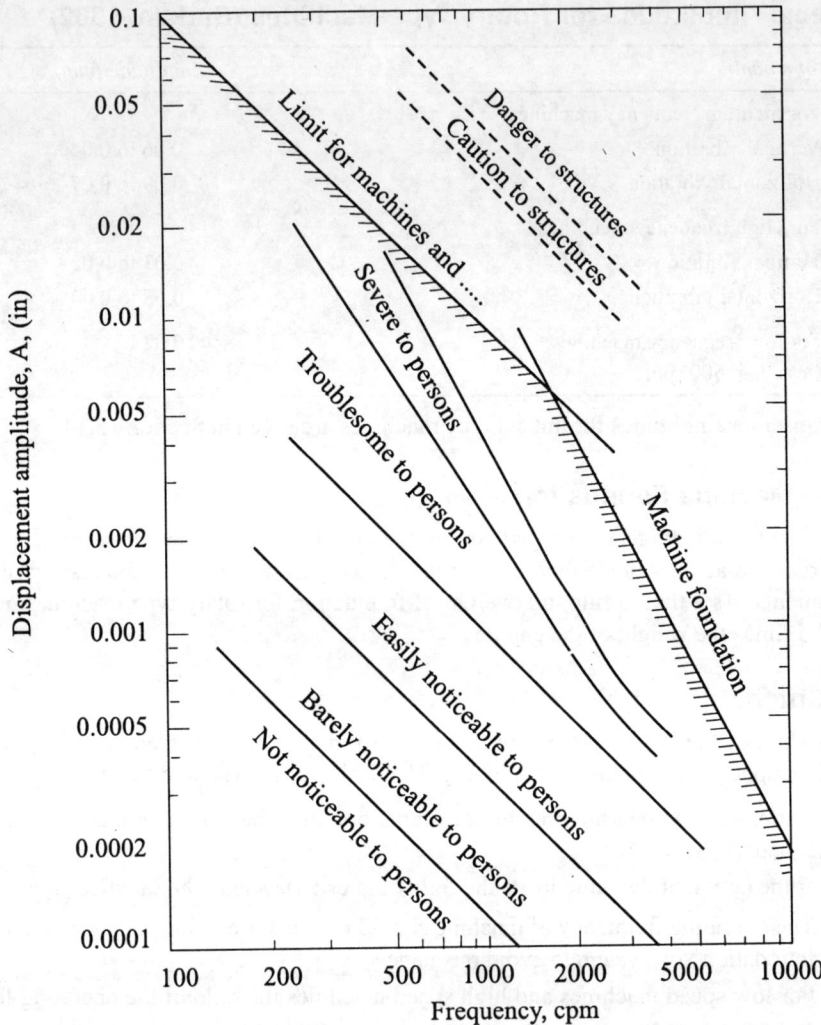

Fig. 15.20 General limits of displacement amplitude for a particular frequency of vibration (Richart, 1962)

The design criteria for impact machine foundations has been discussed in Section 15.14.

15.16 SCREENING VIBRATIONS

A machine foundation which is a source of vibration may transit vibrations to adjoining foundations of machines or structures. If the vibration at the source is excessive, that is if the amplitude of motion is beyond the permissible limits, then the transmitted vibrations to adjoining foundations may adversely affect its satisfactory performance. In such cases there are two remedial measures. They are:

1. Reduce the vibration at the source by providing spring pads within the system itself.

2. Isolate the source of vibration from the adjoining structures by providing around the source screening trenches.

The built in isolators in the first-case may be rubber or composite pads, springs or spring-damper systems, and pneumatic springs. Cork is also used as a satisfactory spring material so long it does not get wet and start rotting. Asbestos fibres have also been found to be an effective isolator.

In the second case trenches of suitable width and depth are dug around the source of vibration and filled with bentonite slurry. This arrangement has been found to be effective in certain cases. This type of isolation at the source is called as *active isolation* as shown in schematically in Fig. 15.21 (a) this type of arrangement reduces the amount of energy radiated away from the source.

In another case the source of vibration might be due to passing of traffic adjoining a sensitive instrument factory which needs to be protected from the incoming vibrations. In these cases also trenches may be dug around the machinery to be protected from vibrations and filled with bentonite slurry. Such types of isolation (screening away from the source) is known as *passive isolation* shown schematically in Fig. 15.21 (b).

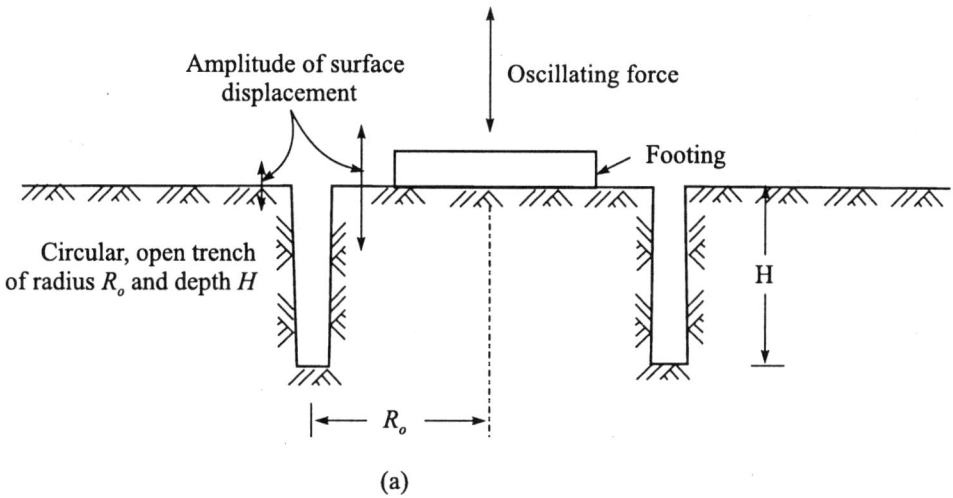

(a)

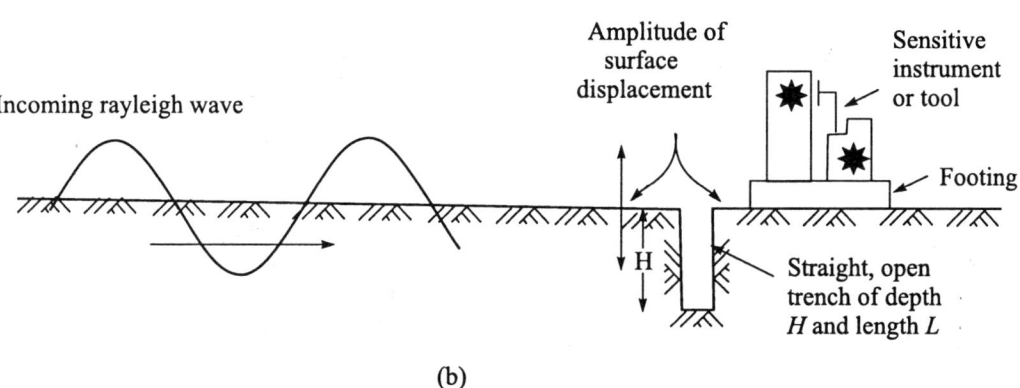

(b)

Fig. 15.21 Screening of vibrations (from Woods, 1968): (a) Schematic of vibration isolation using a circular trench surrounding the source of vibrations—active isolation, (b) schematic of vibration isolation using a straight trench to create a quiescent zone—passive isolation

15.17 EXAMPLES

Example 15.1

Example of footing subjected to steady state vertical oscillation by Elastic-Analog method. (Richart *et al*, 1970).

Given

Soil at the site	=	Silty Clay (CL).
Unit weight of soil	=	117 lb/ft^3.
Velocity of shear wave, v_s	=	460 ft/sec.
Shear modulus, G	=	5340 lb/in^2.
Poisson's Ratio μ	=	0.355.
Diameter of footing	=	62 inches.
Weight of eccentric mass	=	1356 lb
Eccentric setting	=	0.105 in.
Total weight of footing + 4-mass oscillator	=	30970 lb.

Required

(a) Resonant frequency f_{mr}.

(b) The amplitude of motion A_{zm}.

Solution

From Eq. (15.66), $B_z = \dfrac{1-\mu}{4} \dfrac{m_z}{\rho r_o^3} = \dfrac{1-\mu}{4} \dfrac{W}{\gamma r_o^3}$

$$= \frac{1-0.355}{4} \frac{30970}{117\,(2.583)^3} = 2.48$$

where, $r = 31.0$ ins $= 2.583$ ft.

Now from Fig. 15.11 (a), for $B_z = 2.48$, $a_{om} = 0.67$.

The resonant natural frequency f_{mr} may be expressed as

$$f_{mr} = \frac{\omega_{mr}}{2\pi}$$

From Eq. (15.65), $\omega_{mr} = \dfrac{a_{om} v_s}{r_o}$

Substituting, $f_{mr} = \dfrac{a_{om} v_s}{2\pi r_o} = \dfrac{0.67 \times 460}{2 \times 3.14 \times 2.583} = 19$ cycle/sec $= 19$ Hz

The value of f_{mr} can also be obtained from Eq. (15.75)

$$f_{mr} = \frac{v_s}{2\pi r_o} \sqrt{\frac{0.9}{B_z - 0.45}} = \frac{460}{2 \times 3.14 \times 2.583} \sqrt{\frac{0.9}{2.48 - 0.45}}$$

$$= 18.9 \text{ Hz}$$

The amplitude of vertical oscillation depends on B_z and the magnitude of the exciting force. From Fig. 15.10 (b), the magnification factor M_{rm} is equal to 1.86 for $B_z = 2.48$.

From Eq. (15.70),

$$A_{zm} = \frac{m_e e}{m} M_{rm} = \frac{W_e e}{W} M_{rm}$$

Substituting,

$$A_{zm} = \frac{1356 \times 0.105}{30,970} \times 1.86 = 0.0086 \text{ in.}$$

Alternately the value of A_{zm} can also be calculated from Eq. (15.77b)

$$A_{zm} = \frac{m_e e B_z}{m(0.85)\sqrt{B_z - 0.18}}$$

or

$$A_{zm} = \frac{W_e e B_z}{W(0.85)\sqrt{B_z - 0.18}}$$

Substituting,

$$A_{zm} = \frac{1356 \times 0.105 \times 2.48}{30970 \times 0.85 \sqrt{2.48 - 0.18}}$$

$$= 0.0088 \text{ in.}$$

Example 15.2

Design of Foundation size for a Vertical Single-Cylinder Compressor by Elastic-half space analog method.

Given

Operating frequency of compressor	= 450 rpm.
Constant amplitude force of excitation Q_o	= 10, 900 lb.
Soil	= Silty clay.
Shear wave Velocity v_s	= 806 ft/sec.
Shear modulus, G	= 14000 lb/in².
Poisson's ratio, μ	= 0.33.
Unit weight of soil	= 100 lb/ft³.
Permissible amplitude	= 0.0020 ins.

Required

(a) The size of foundation block to keep the amplitude within the permissible limits.

(b) The natural frequency of the system.

Solution

The first approximation for the foundation plan dimensions may be obtained from the base area required to limit the static displacement caused by $Q_o = 11,400$ lb to a value of 0.002 in. The equivalent rigid circular footing will be used in both the static and dynamic analysis, although a rectangular foundation plan is required. The static deflection, z_s, can be calculated from Eq. (15.69) by putting the magnitude factor $M = 1$ for static condition and equating to $z_s = 0.002$ ins

$$A_z = z_s = \frac{(1-\mu) Q_o}{4 G r_o} = 0.002 \text{ ins.}$$

or

$$r_o = \frac{(1-0.33) \times 11,400}{4 \times 14,000 \times 0.02} = 67.9 \text{ inches or } 5.66 \text{ ft.}$$

Assume $r_o = 6$ ft for which the base area $A = 113$ ft^2. For the compressor, use a rectangular foundation block of 17×7 ft which gives $A = 119$ ft^2. Let the height of block $= 3$ ft. Assuming unit weight of concrete $\gamma_c = 150$ lb/ft^3,

The weight of the foundation block W_f is

$$W_f = 16 \times 7 \times 3 \times 150 = 50,400 \text{ lb}$$

- Total weight $\quad W = 50,400 + 10,900 = 61,300$ lb

Then for the equivalent circular footing of $r_o = 6$ ft.

$$B_z = \frac{1-\mu}{4} \frac{W}{\gamma r_o^3} = \frac{(1-0.33) \, 61,300}{4 \times 100 \times 6^3} = 0.473$$

From Fig. 15.15 the damping ratio $D_z = 0.6$ for $B_z = 0.473$. The natural frequency f_n for the system depends upon the oscillating mass and spring constant. From Eq. (15.67), we have

$$k_z = \frac{4 G r_o}{1-\mu} = \frac{4 \times 14,000 \times 72}{(1-0.33)} = 6.048 \times 10^6 \text{ lb/in.}$$

Then from Eq. (15.9)

$$f_n = \frac{1}{2\pi} \sqrt{\frac{k_z}{m}} = \frac{1}{2 \times 3.14} \sqrt{\frac{6.048 \times 10^6 \times 386}{61,300}}$$

$$= 31.1 \text{ cycles/sec}$$

or $\quad f_n = 1864$ cycles per minute.

The damping ratio $\quad D_z = \dfrac{0.425}{\sqrt{B_z}}$ from Eq. (15.73)

$$= \frac{0.425}{\sqrt{0.473}} \approx 0.6$$

The frequency ratio $\quad = \dfrac{f}{f_n} = \dfrac{450}{1864} = 0.24$

For $\dfrac{f}{f_n} = 0.24$, $D_z = 0.6$, the magnification factor $M = 1.02$ from Fig. 15.5 (d) or $M = \dfrac{A_z}{z_s} = 1.02$

or $A_z = 1.02 \times z_s = 1.02 \times .002 \approx 0.002$ ins.

This value of motion satisfies the design criteria.

Example 15.3

Analysis of a Rocking Machine Foundation by Elastic half-space Analog.

Given

Figure Ex. 15.3 gives the plan and elevation of a proposed foundation to support rotating machinery. The upper slab is at the first floor level, and the lower slab rests directly on the soil at a depth 1.5 ft below the top of the basement slab.

$G = 12,300$ lb/in^2, $v_s = 720$ ft/sec, $\mu = 0.25$ and $\gamma = 110$ lb/ft^3.

Total weight of the foundation including the machinery $W = 272,100$ lb.

Height of centre of gravity above Point $O = 11.2$ ft Fig. Ex. 15.3.

Required

The dynamic-response of the foundation soil system for the horizontal and vertical forces generated by rotating machinery.

Solution

The foundation is rectangular

The area $\qquad A = 34 \times 8 = 272$ sq ft

and $\qquad\qquad r_o = \sqrt{\dfrac{272}{\pi}} = \sqrt{\dfrac{272}{3.14}} = 9.30$ ft

Vertical oscillation

The mass ratio from Eq. (15.66) is

$$B_z = \frac{1-\mu}{4} \frac{W}{\gamma r_o^3} = \frac{0.75 \times 272,100}{4 \times 100 \times (9.3)^3} = 0.58.$$

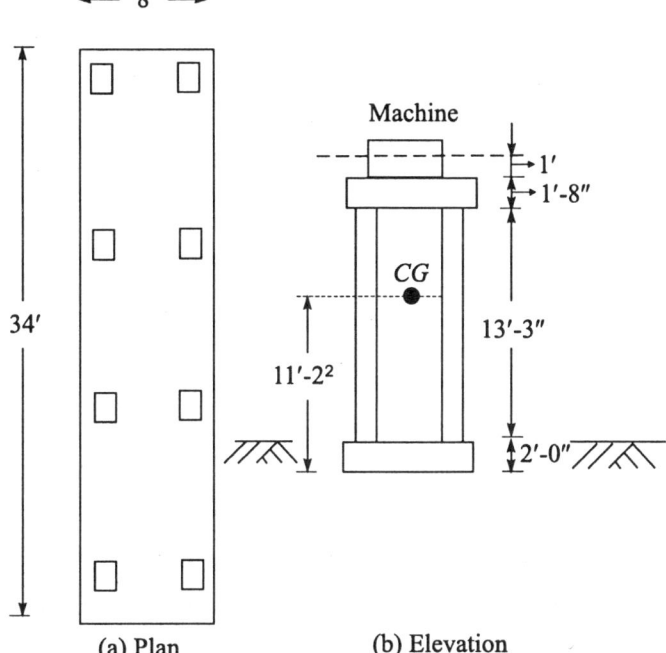

(a) Plan $\qquad\qquad\qquad$ (b) Elevation

Fig. Ex. 15.3

From Fig. 15.11 (b) the magnification M_m for $B_z = 0.58$ is 1.1 and from Fig. 15.15 the damping ratio $D = 0.56$. This indicates that the vertical motion is highly damped and from Fig. 15.5 (d), the vertical motion will be only slightly greater than the static displacement produced by the input force, and as such the vertical vibration need not be considered in the analysis.

Rocking Vibration

For rocking vibration excited by the horizontal component of the machine forces, the equivalent radius of an equivalent base area is required to be calculated by using Eq. (15.97b)

$$r_o = \sqrt[4]{\frac{16bl^3}{3\pi}} = \sqrt[4]{\frac{2l(2b)^3}{3\pi}} = \left(\frac{34 \times 8^3}{3 \times 3.14}\right)^{1/4} = 6.5 \text{ ft}$$

where, $2l = 34$ ft, $2b = 8$ ft.

Then from Eq. (15.88), the mass ratio B_ϕ is

$$B_\phi = \frac{3(1-\mu)}{8} \frac{I_\phi}{\rho r_o^5}$$

where I_ϕ = mass moment of inertia in rocking about point O at the centroid of the base of the footing.

From Fig. Ex. 15.3

$$I_\phi = \frac{W}{12g}(l^2 + h^2) + \frac{W}{g} h_o^2$$

$$= \frac{272100}{12g}(8^2 + 18^2.92) + \frac{272100}{g} \times 11^2.2$$

$$= \frac{4.47 \times 10^7}{g} \text{ ft lb sec}^2$$

Therefore, $$B_\phi = \frac{2.25}{8} \times \frac{4.47 \times 10^7}{110(6.55)^5} = 9.5$$

Figure 15.13 (b) indicates that dynamic magnification factor $M_{\phi m}$ for $B_\phi = 9.50$ is greater than 100. The actual value can be calculated from Eq. (15.90) and (15.89) as follows:

$$D_\phi = \frac{0.15}{(1 + B_\phi)\sqrt{B_\phi}} = 0.0047$$

From Eq. (15.90),

$$M_{\phi m} \approx \frac{1}{2D_\phi} = \frac{1}{2 \times .0047} = 106$$

With this low value of damping ratio, or high magnification factor, the peak of the amplitude frequency response curve will occur at a frequency almost identical with the natural frequency.

Resonant frequency

From Eq. (15.65), the dimensionless frequency at maximum amplitude may be written as

$$a_{om} = \frac{\omega_m \, r_o}{v_s} = \frac{2\pi \, f_m r_o}{v_s} \quad \text{or} \, f_m = \frac{a_{om} v_s}{2\pi \, r_o}$$

From Fig. 15.13 (a), for $B_\phi = 9.5$, the value of $a_{om} \approx 0.30$.

Therefore $\qquad f_m = \dfrac{0.30 \times 720}{2 \times 3.14 \times 6.55} = 5.25 \text{ cycles/sec} = 315 \text{ cycle per minute.}$

Alternate method

As a check, the resonant frequency for the lumped-mass system can be evaluated through Eq. (15.9). If we consider the foundation as a rectangular foundation, the formula for k_ϕ from Table 13.3 is

$$k_\phi = \frac{8G}{1-\mu} \beta_\phi b l^2 = \frac{8 \times 12300 \times 0.4 \times 17 \times 4^2}{0.75}$$

$$= 2.055 \times 10^9 \text{ ft lb/radian,}$$

where, $\qquad b = \dfrac{34}{2} = 17 \text{ ft}, \, l = \dfrac{8}{2} = 4 \text{ ft},$

$\qquad \beta = 0.4$ from Fig. 15.16 (b) for $\dfrac{l}{b} = \dfrac{4}{17} = 0.24,$

$\qquad 2l = $ length perpendicular to the axis of rotation and,

$\qquad 2b = $ width parallel to the axis of rotation.

Then from Eq. (15.9)

$$f_n = \frac{1}{2\pi} \sqrt{\frac{k_\phi}{I_\phi}} = \frac{1}{2 \times 3.14} \sqrt{\frac{2.055 \times 10^9 \times 32.2}{4.47 \times 10^7}}$$

$$\approx = 5.9 \text{ cps} \approx 354 \text{ cpm.}$$

The resonant frequencies obtained by both the method compare very well. It is evident that the foundation would experience a severe rocking oscillation at a frequency in the range 320–350 rpm. The foundation requires re-design to resist rocking. One way of doing this is to reduce the height and increase the width of the block.

Example 15.4

Design of Foundation for a Two-Cylinder Vertical Compressor by the use of Elastic Soil-Spring Constants.

Given

Design Data

The design diagram of foundation is given in Fig. Ex. 15.4. The other details are given below. The foundation is to be founded on silty clay soil having the spring constants.

$\qquad C_u = 5 \times 10^3 \text{ T/m}^3, \, C_\phi = 10 \times 10^3 \text{ T/m}^3 \text{ and } C_\tau = 2.5 \times 10^3 \text{ T/m}^3.$

Weight of compressor	= 12 T.
Weight of motor	= 4 T.
Operating speed	= 480 rp.

Dynamic amplitude of each compressor	= 3 T.
Phase difference	= $\pi/2$.
Distance c/c of cylinders	= 1.3 m.
Distance between compressor and motor	= 2.3 m.
Working load level of motor and compressor above foundation top surface	= 0.8 m.

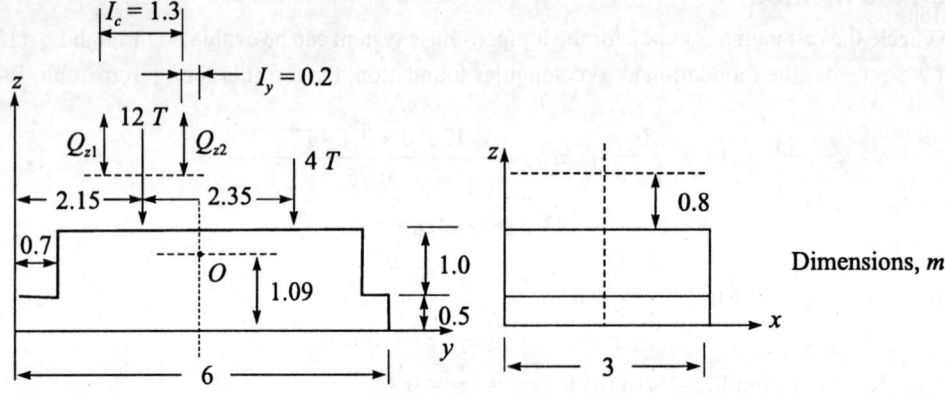

Fig. Ex. 15.4

Required

The amplitudes of motion for the simultaneous vertical, horizontal and rotational modes of vibrations.

Solution

(a) Centre of gravity of the system

Assume the coordinate axes as shown in Fig. Ex. 15.4. Let x_o, y_o and z_o are the coordinates O of the centre of gravity of the system expressed as

$$x_o = \frac{\Sigma m_i x_i}{m}, y_o = \frac{\Sigma m_i y_i}{m}, z_o = \frac{\Sigma m_i z_i}{m}$$

where, m_i = masses of single elements of a system,

x_i, y_i and z_i = Coordinates of the centres of gravity of single elements with respect to axes,

m = mass of the system,

Table Ex. 15.4 gives the details of the results of computation of static moments of single elements of the system, where, A = Dimensions of elements, metre, (B) Mass of element, T-sec^2/m, C = Coordinates, D = Static moment T-sec^2

Using the data in Table Ex. 15.4, we have

$$x_o = \frac{10.35}{6.91} = 1.5 \text{ m}, y_o = \frac{20.49}{6.91} = 2.97 \text{ m},$$

$$z_o = \frac{7.52}{6.91} = 1.09 \text{ m}.$$

The coordinates of O are shown in Fig. Ex. 15.4.

Table Ex. 15.4 Computation for determining the centre of gravity of the system

Elements of system	A			B	C			D		
	a_x	a_y	a_z	$T.sec^2/m$	x_i	y_i	z_i	$m_i x_i$	$m_i y_i$	$m_i z_i$
Foundation slab	3	6	0.5	2.02	1.5	3.0	0.25	3.03	6.06	0.55
Upper part of foundation	3	4.8	1.0	3.25	1.5	3.0	1.0	4.85	9.75	3.25
Compressor	–	–	–	1.23	1.5	2.15	2.3	1.85	2.64	2.82
Motor	–	–	–	0.41	1.5	4.5	2.3	0.62	1.84	0.90
Total				6.91				10.35	20:49	7.52

From the figure, we can calculate the eccentricities in the x and y directions as follows.

$$e_x = 0, \quad e_y = \frac{3.0 - 2.97}{3.0} \times 100 \approx 0.7\%$$

Since the values are quite small, it can be ignored in the computation.

Now we have the total weight W of the system,

$$W = mg = 6.91 \times 9.81 = 67.5 \text{ tonnes.}$$

Contact area $\qquad A = 6 \times 3 = 18 \text{ m}^2.$

Static pressure, $\qquad p_{st} = \dfrac{W}{A} = \dfrac{67.5}{18} = 3.8 \text{ T/m}^2 = 0.38 \text{ kg/cm}^2.$

(*b*) *Possible forms of Foundation Vibrations and design values of exciting loads*

Let Q_{z1} and Q_{z2} represent the periodic forces of each of the compressors in the vertical direction expressed as

$$Q_{z1} = 3.0 \cos \omega t, \quad Q_{z2} = -3.0 \sin \omega t$$

The resultant vertical component of the disturbing forces equals

$$Q_z = Q_{z1} \cos \omega t - Q_{z2} \sin \omega t$$
$$= 3 (\cos \omega t - \sin \omega t) = 4.2 \cos (\omega t + \pi/4)$$

The design value of the vertical component of the exciting loads will be

$$Q_z = 4.2 \text{ T}$$

This load will induce vertical forced vibrations of the foundation.

Let $\quad l_c =$ distance between cylinder axes $= 1.3$ m,

$\qquad l_y =$ distance between the second cylinder and the centre of gravity of complete system $= 0.2$ m.

Due to the asymmetric position of the compressor, the foundation will be subjected to the action of the disturbing moment M_x with respect to the x-axis. The magnitude of this moment is

$$M_x = Q_{z1} (l_c + l_y) + Q_{z2} l_y$$

Substituting $\qquad M_x = 3.0 (1.3 + 0.2) \cos \omega t - 3.0 \times 0.2 \sin \omega t$

$$= 4.6 \cos (\omega t + \pi/4)$$

The design value of the disturbing moment should equal to its greatest magnitude, that is,

$$M_x \, (\text{max}) = 4.6 \, \text{Tm}$$

Vibration takes place in the plane parallel to yz under the action of this moment. They will be accompanied by a simultaneous sliding of the foundation in the direction of the y-axis and a rotation of the foundation with respect to an axis parallel to the x-axis and passing through the centre of gravity of the system.

(c) *Vertical amplitude of motion and natural frequency of the system*

From Eq. (15.114d)

$$f_{nz} = \frac{1}{2\pi} \sqrt{\frac{C_u A}{m}}$$

where,　$C_u = 5 \times 10^3 \, \text{T/m}^2$, $A = 18 \, \text{m}^2$, $m = 6.91 \, \text{T.sec}^2/\text{m}$.

Substituting,　$f_{nz} = \dfrac{1}{2 \times 3.14} \times \sqrt{\dfrac{5 \times 10^3 \times 18}{6.91}} = 18.17 \, \text{cps}$

or　$\omega_{n2} = 114 \, \text{rad/sec}$.

From Eq. (15.114a)

$$A_z = \frac{Q_o}{m\left(\omega_{nz}^2 - \omega^2\right)}$$

$$= \frac{4.2 \times 10^3}{6.91\left[(114)^2 - (16\pi)^2\right]} = 0.058 \, \text{mm}$$

where, ω = operating frequency $= \dfrac{480 \times 2\pi}{60} = 16\pi \, \text{radians/sec}$.

(d) *Base and mass Moments of Inertia*

1. Base moment of inertia parallel to the X-axis passing through the centroid of the base

$$I_b = \frac{3 \times 6^3}{12} = 54 \, \text{m}^4$$

2. Mass moments of inertia with respect to the axis as in (1) above.

(i) Compressor

$$I_{\phi c} = m_1 (0.85^2 + 2.3^2) = 1.23 \times 6.01 = 7.4 \, \text{Tm sec}^2.$$

(ii) Motor

$$I_{\phi m} = m_2 (1.5^2 + 2.3^2) = 0.41 \times 7.55 = 3.1 \, \text{Tm sec}^2.$$

(iii) Foundation slab (1)

$$I_{\phi 1} = \frac{m_1}{12}(a_{y1}^2 + a_{z1}^2) + m_1 h_1^2 = \frac{2.02}{12}(6^2 + 0.5^2) + 2.02 \times 0.25^2$$

$$= 6.1 \, \text{Tm sec}^2$$

where, h_1 = vertical distance between the centre of gravity of mat 1 and the foundation contact area.

Upper part of foundation slab

$$I_{\phi 2} = \frac{3.25}{12} (4.8^2 + 1.0^2) + 3.25 \times 1^2.0 = 9.8 \text{ Tm sec}^2.$$

Total moment of inertia of the mass of the whole system with respect to the axis at the base is

$$I_{\phi o} = 7.4 + 3.1 + 6.1 + 9.8 = 26.4 \text{ Tm sec}^2.$$

The moment of Inertia of the whole system with respect to the axis passing through the centre of gravity of the whole system and perpendicular to the plane of vibration is

$$I_{\phi g} = I_{\phi o} - mh_o^2 = 26.4 - 6.91 \times 1^2.09 = 18.3 \text{ Tm sec}^2$$

$$\eta = \frac{I_{\phi g}}{I_{\phi o}} \quad \frac{18.3}{26.4} = 0.69$$

(e) *Amplitudes of motion and Natural frequency.*

Natural frequencies ω_{n1} *and* ω_{n2}

The limiting natural frequencies are:

For rocking mode, From Eq. (15.122a)

$$\omega_{n\phi} = \sqrt{\frac{C_\phi I_b - Wh_o}{I_{\phi o}}} = \sqrt{\frac{10 \times 10^3 \times 54 - 67.5 \times 1.09}{26.4}}$$

$$= 143.28 \text{ rad/sec}$$

or $\qquad \omega_{n\phi}^2 = 20.5 \times 10^3 \text{ sec}^{-2}.$

The limiting frequency of vibrations in shear from Eq. (15.116a) is

$$\omega_{ny}^2 = \frac{C_\tau A}{m} = \frac{2.5 \times 10^3 \times 18}{6.91} = 6.5 \times 10^3 \text{ sec}^{-2}.$$

The frequency equation for the foundation Eq. (15.133a) is

$$\phi_n^4 = \frac{\omega_{n\phi}^2 + \omega_{ny}^2}{\eta} \omega_n^2 + \frac{\omega_{n\phi}^2 \omega_{ny}^2}{\eta} = 0$$

Substituting, we have

$$\omega_n^4 - \frac{(20.5 + 6.5) \times 10^3}{0.69} \omega_n^2 + \frac{20.5 \times 6.5 \times 10^6}{0.69} = 0$$

or $\qquad \omega_n^4 - 39.2 \times 10^3 \omega_n^2 + 193 \times 10^6 = 0$

By solving this equation, we have

$$\omega_{n1}^2 = 33.4 \times 10^3 \text{ sec}^{-2} \text{ or } \omega_{n1} = 183 \text{ rad/sec}$$

$$\omega_{n2}^2 = 5.8 \times 10^3 \text{ sec}^{-2} \text{ or } \omega_{n2} = 76 \text{ rad/sec}$$

Amplitudes of motion A_y *and* A_ϕ

From Eq. (15.136) and (15.137)

$$A_y = \frac{\left(C_\tau A h_o\right) M_x}{\Delta\left(\omega^2\right)}, A_\phi = \frac{\left(C_\tau - m\omega^2\right) M_x}{\Delta\left(\omega^2\right)}$$

$$\Delta\left(\omega^2\right) = mI_{\phi g}\left(\omega_{n1}^2 - \omega^2\right)\left(\omega_{n2}^2 - \omega^2\right)$$

$$= 6.91 \times 18.2 \,(33.4 - 2.5)\,(5.8 - 2.5) \times 10^6$$

$$= 13.8 \times 10^9$$

$$A_y = \frac{2.5 \times 10^3 \times 18 \times 1.09}{13.8 \times 10^9} = 0.016 \times 10^{-3} \text{ m} = 0.016 \text{ mm}$$

$$A_\phi = \frac{\left(2.5 \times 10^3 \times 18 - 6.91 \times 2.5 \times 10^3\right) 4.6}{13.8 \times 10^9}$$

$$= 0.009 \times 10^{-3} \text{ radians.}$$

The maximum horizontal displacement of the foundation top surface in the plane yz is

$$A = A_y + h_t A_\phi = (0.016 + 0.41 \times .009)\, 10^{-3} \text{ m}$$

$$= 0.020 \times 10^{-3} \text{ m} \approx 0.02 \text{ mm}$$

Example 15.5
Design of Hammer Foundation by making use of elastic soil-spring constants.

Given
A foundation is required to be designed for a double-acting stamping hammer with the following specifications:

Soil at the site: Clay soil mixed with some sand and silt.

The static allowable bearing pressure $q_a = 20$ T/m^2.

Elastic uniform compression of soil with the correction applied

$$\bar{C}_u = 3C_u = 12 \times 10^3 \text{ T/m}^3$$

Weight of dropping parts $W_d = 3.5$ T.

Height of drop, $h = 100$ cm.

Piston area $a = 0.15$ m^2.

Steam pressure $p = 80$ T/m^2.

Base area of anvil, $A_2 = 4.75$ m^2.

Thickness of pad under anvil $t = 60$ cm.

Modulus of elastic of pad $= 50 \times 10^3$ T/m^2.

Required
(a) Amplitude of vibration of the foundation and anvil
(b) Amplitude of vibration of the anvil together with the frame
(c) The dynamic stress in the pad under the anvil.

Solution
1. Velocity of dropping weight

From Eq. (15.142),

$$v = \alpha \sqrt{\frac{2g\left(W_d + ap\right)h}{W_d}}$$

Assuming $\alpha = 0.65$, we have

$$v = 0.65 \sqrt{\frac{2 \times 9.81\left(3.5 + 80 \times 0.15\right) \times 1}{3.5}}$$

$$= 6.1 \text{ m/sec.}$$

2. Foundation weight W_f for preliminary computation

   Assume coefficient of restitution $e = 0.5$ From Eq. (15.153a)

   $$n_f = 8.0\,(1 + e)\,v - n_a$$

where,
$$n_a = \frac{W_2}{W_d} = \frac{90}{3.5} = 25.7$$

Now,
$$n_f = 8.0\,(1 + 0.5)\,6.1 - 25.7 = 47.3$$

The required foundation weight W_f (together with backfill) is

$$W_f = n_f W_d = 47.3 \times 3.5 = 166 \text{ T}$$

Contact area required for the foundation, from Eq. (15.154a), and (15.154b)

$$A_f = W_d a_f$$

$$a_f = \frac{20\,(1 + e)\,v}{q_a} = \frac{20\,(1 + 0.5)\,6.1}{20}$$

$$= 9.2 \text{ m}^2$$

$$A_f = 3.5 \times 9.2 = 32.2 \text{ m}^2$$

3. Preliminary dimensions of foundation as per the data given in (2) above.

   The preliminary dimensions of foundation are given in Fig. Ex. 15.5

   As per this design.

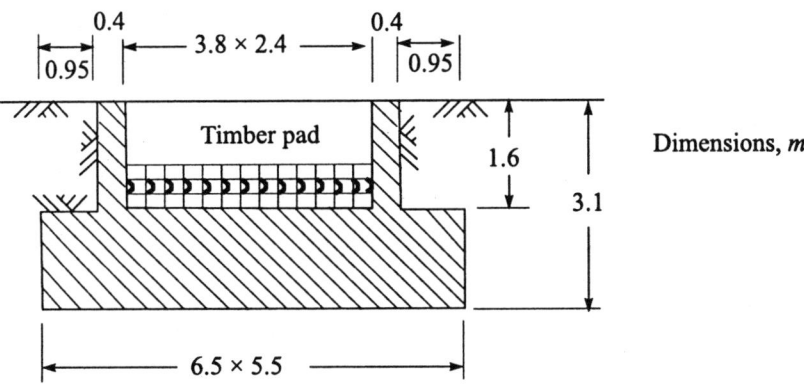

0.4   3.8 × 2.4   0.4

0.95   0.95

Timber pad

1.6   Dimensions, *m*

3.1

6.5 × 5.5

Fig. Ex. 15.5

Total weight of foundation with backfill = 161.4 T

Contact area $A_f = 6.5 \times 5.5 = 35.7$ m^2

4. Amplitudes of foundation vibration

Pad under the anvil

$$E_2 = 50 \times 10^3 \text{ T/m}^2$$

$$t = 0.60 \text{ m}; A_2 = 4.75 \text{ m}^2$$

Therefore

$$k_2 = \frac{E}{t} A_2 = \frac{50 \times 10^3 \times 4.75}{0.60}$$

$$= 39.5 \times 10^4 \text{ T/m}$$

Mass of hammer

$$= m_2 = \frac{90}{9.81} = 9.18 \text{ T sec}^2/\text{m.}$$

From Eq. (15.149b), the limiting frequency of natural vibrations of the anvil on the timber pad is

$$\omega_{na}^2 = \frac{k_2}{m_2} = \frac{39.5 \times 10^4}{9.18} = 43 \times 10^3 \text{ sec}^{-2}$$

The spring constant k_1 of soil below the base is

$$k_1 = \bar{C}_u A = 12 \times 10^3 \times 35.7 = 43.8 \times 10^4 \text{ T/m}$$

The mass of foundation W_f is

$$m_1 = \frac{W_f}{g} = \frac{161.4}{9.81} = 16.5 \text{ T sec}^2/\text{m}$$

The limiting frequency of natural vibrations of the whole system, from Eq. (15.149c), is

$$\omega_l^2 = \frac{k_1}{m_1 + m_2} = \frac{42.8 \times 10^4}{16.5 + 9.18} = 16.7 \times 10^3 \text{ sec}^{-2}$$

From Eq. (15.149a), $N = \dfrac{m_2}{m_1} = \dfrac{9.18}{16.5} = 0.557$

Now as per Eq. (15.148), we have,

$$\omega_n^4 - (1 + 0.557)(43 \times 10^3 + 16.7 \times 10^3)\omega_n^2 + (1 + 0.557) \times 43 \times 10^3 \times 16.7 \times 10^3 = 0$$

or $\quad \omega_n^4 - 92.5 \times 10^3 \omega_n^2 + 1115 \times 10^6 = 0$

Solving this equation we obtain,

$$\omega_{n1}^2 = 78.5 \times 10^3 \text{ sec}^{-2} \text{ or } \omega_{n1} = 280 \text{ sec}^{-1}$$

$$\omega_{n2}^2 = 14.1 \times 10^3 \text{ sec}^{-2} \text{ or } \omega_{n2} = 119 \text{ sec}^{-1}$$

Amplitudes of vibration of foundation

Velocity of motion of the anvil with the frame from Eq. (15.145) is

$$v_a = \frac{v_1(1 + e)}{1 + n}, \text{ where } v_1 = 6.1 \text{ m/sec}, n = \frac{90}{3.5} = 25.71$$

Therefore, $\qquad v_a = \dfrac{6.1\left(1+0.5\right)}{1+25.71} = 0.342 \text{ m/sec}$

From Eq. (15.151a)

$$z_1 = A_z = \frac{\left(\omega_{na}^2 - \omega_{n2}^2\right)\left(\omega_{na}^2 - \omega_{n1}^2\right)}{\omega_{na}^2\left(\omega_{n1}^2 - \omega_{n2}^2\right)} \, v_f$$

$$= \frac{\left(43.0 \times 10^3 - 14.1 \times 10^3\right)\left(43.0 \times 10^3 - 78.5 \times 10^3\right)}{43 \times 10^3\left(78.5 \times 10^3 - 14.1 \times 10^3\right)} \times 0.342$$

$$= 1.07 \text{ mm}$$

Amplitude of vibration of Anvil

From Eq. (15.151b)

$$z_2 = A_a = \frac{\left(\omega_{na}^2 - \omega_{n1}^2\right) v_a}{\left(\omega_{n1}^2 - \omega_{n2}^2\right)\omega_{n2}}$$

$$= \frac{\left(43.0 \times 10^3 - 78.5 \times 10^3\right) 0.342}{\left(78.5 \times 10^3 - 14.1 \times 10^3\right) 119} = 1.6 \times 10^{-3} \text{ m} = 1.6 \text{ mm}$$

The computed amplitude of vibration is within the permissible limit as per Table 15.6

Dynamic stress in the pad

From Eq. (15.152), we have

$$\sigma = \frac{k_2\left(z_1 + z_2\right)}{A_2}$$

$$= \frac{39.5 \times 10^4\left(1.07 \times 10^{-3} + 1.6 \times 10^{-3}\right)}{4.75}$$

$$= 222 \text{ T/m}^2$$

The stress in the pad is also within the permissible limit.

Geotextiles Reinforced Earth and Ground Anchors ●16

16.1 GEOTEXTILES

Introduction

Geotextiles are porous fabric manufactured from synthetic materials such as polypropylene, polyester, polyethylene, nylon, polyvinyl chloride and various mixtures of these. They are available in thicknesses ranging from 10 to 300 mils (1 mil = 1/1000 inch) in widths up to 30 ft, in roll lengths up to 2000 ft. The permeabilities of geotextiles sheets are comparable in range from coarse gravel to fine sand. They are either woven from continuous monofilament fibres or non-woven made by the use of thermal or chemical bonding of continuous fibres and pressed through rollers into a relatively thin fabric. These fabrics are sufficiently strong and durable even in hostile soil environment. They possess a *pH* resistance of 3 to 11.

The use of geotextiles in geotechnical and construction engineering has been growing in popularly for the last many years. Geotextiles can be used in so many ways. They are used as *soil separators*, used in filtration and drainage, used as a reinforcement material to increase the stability of earth mass, used for the control of erosion, etc. Some of the uses of geotextiles are described below.

Geotextiles as Separators

A properly graded filter prevents the erosion of soil in contact with it due to seepage forces. To prevent the movement of erodible soils into or through filters, the pore spaces between the filter particles should be small enough to hold some of the protected materials in place. If the filter material is not properly designed, smaller particles from the protected area move into the pores of the filter material and may prevent proper functioning of drainage.

As an alternative, geotextile can be used as a filter material in place of filter soil as shown for an earth dam in Fig. 16.1 (a). The other uses of geotextiles as a separator are:

1. Separation of natural soil subgrade from the stone aggregates used for the pavement of roads, etc.
2. As a water proofing agent to prevent cracks in existing asphalt pavements.

Geotextiles as Reinforcement

Geotextiles with good tensile strength can contribute to the load carrying capacity of soil which is poor in tension and good in compression.

Geotextiles placed between a natural subgrade below and stone aggregates above in unpaved roads, serve not only as separators but also increase the bearing capacity of the subgrade to take heavier traffic loads. Here, geotextiles function as reinforcers as shown in Fig. 16.1 (b).

Another major way in which geotextiles can be used as reinforcement is in the construction of fabric-reinforced retaining walls and embankments. This technology is borrowed from the technology for reinforced earth walls described in Section 16.2. Geotextiles have been used to form such walls which can provide both the facing element and stability simultaneously. The process of construction of the wall with granular backfill is shown in Fig. 16.1 (c). The procedure is to as follows.

1. Level the working surface.

2. Lay geotextile sheet 1 of proper width on the surface with 1.5 to 2 m at the wall face draped over temporary wooden form as shown in (C1).

3. Backfill over this sheet with granular soil compact it by using a roller of suitable weight.

4. After compaction, fold the geotextile sheet as shown in (C2) Lay down second sheet and continue the process as before. The completed wall is shown in (C).

The front face of the wall can be protected by the use of shotcrete or gunite. Shotcrete is a low water content sand and cement mixture, often with additives, which is sprayed on to the surface at high pressures in a manner similar to gunite. The design of geotextile reinforced walls is similar in principle to that of reinforced earth walls.

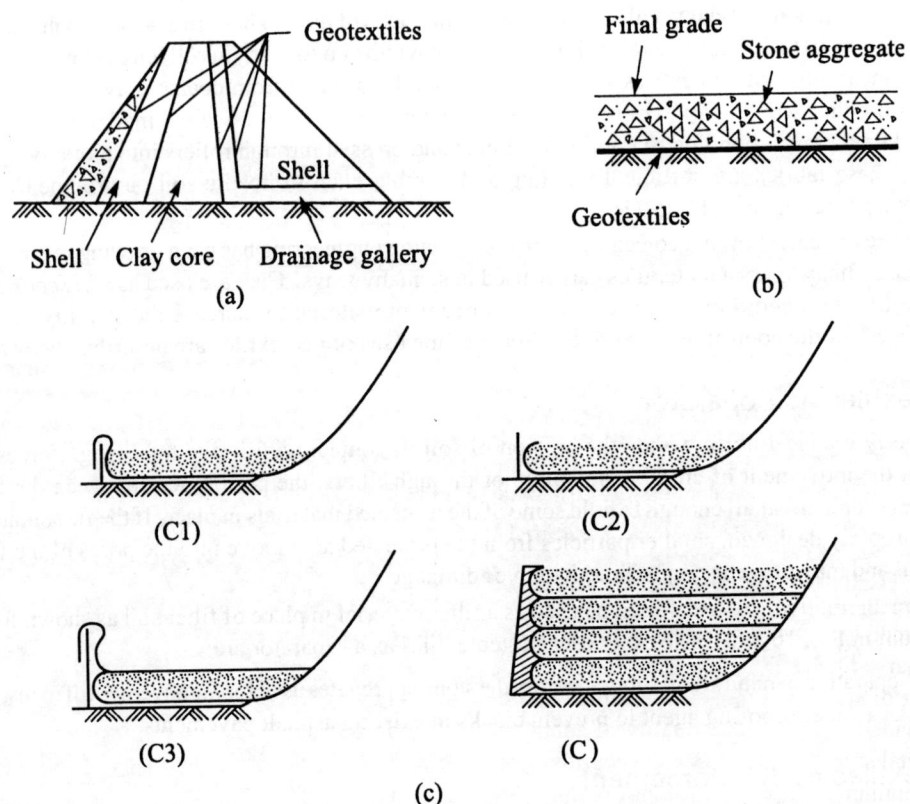

Fig. 16.1 Some uses of geotextiles: (a) Geotextiles to replace filter soils in earth dams, (b) geotextiles as separator and reinforcer on road subgrade, and (c) geotextile fabric construction

Geotextiles in Filtration and Drainage

Geotextile sheets have been successfully used to control erosion of land surfaces. Erosions of exposed and surfaces may occur due to the falling rain water or due to flowing water in rivers, etc. Figure 16.2 (b) shows a schematic sketch for the protection of the banks of flowing water.

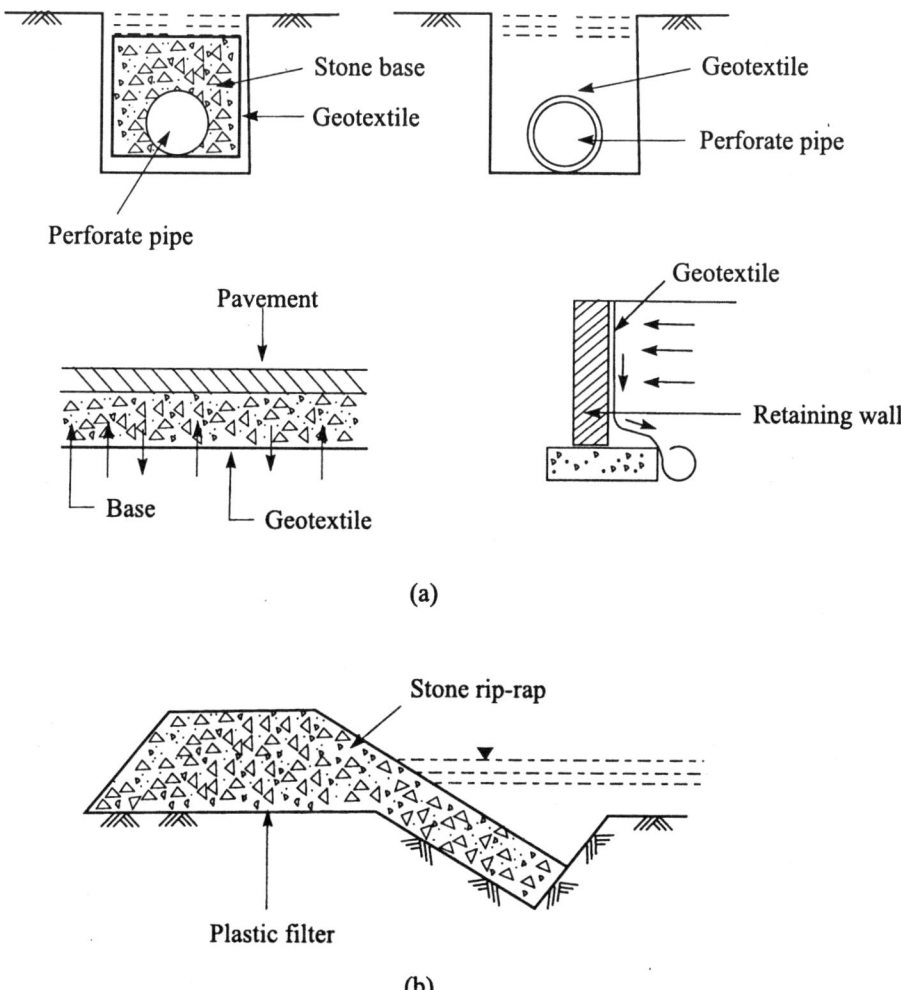

Fig. 16.2 Geotextiles for: (a) Filtration and drainage, and (b) erosion control

16.2 REINFORCED EARTH AND GENERAL CONSIDERATIONS

Reinforced earth is a construction material composed of soil fill strengthened by the inclusion of rods, bars, fibers or nets which interact with the soil by means of frictional resistance. The concept of strengthening soil with rods or fibers is not new. Throughout the ages attempts have been made to improve the quality of adobe brick by adding straw. The present practice is to use thin metal strips, geotextiles, and geogrids as reinforcing materials for the construction of reinforced earth retaining walls. A new era of retaining walls with reinforced earth was introduced by Vidal (1969). Metal strips were used as reinforcing material as shown in Fig. 16.3 (a). here the metal strips extend from the panel back into the soil to serve the dual role of anchoring the facing units and being restrained through the frictional stresses mobilised between the strips and the backfill soil. The backfill soil creates the lateral pressure and interacts with the strips to resist it. The walls are relatively flexible compared to massive gravity structures. These flexible walls offer many advantages including significant lower cost per square metre of exposed surface. The variations in the types of facing units, subsequent to Vidal's introduction of the reinforced earth walls, are many. A few of the types that are currently in use are (Koerner, 1999).

1. Facing panels with metal strip reinforcement.

2. Facing panels with wire mesh reinforcement.

3. Solid panels with tie back anchors.

4. Anchored gabion walls.

5. Anchored crib walls.

6. Geotextile reinforce walls.

7. Geogrid reinforced walls.

In all cases, the soil behind the wall facing is said to be *mechanically stabilized earth* (MSE) and the wall system is generally called an MSE wall.

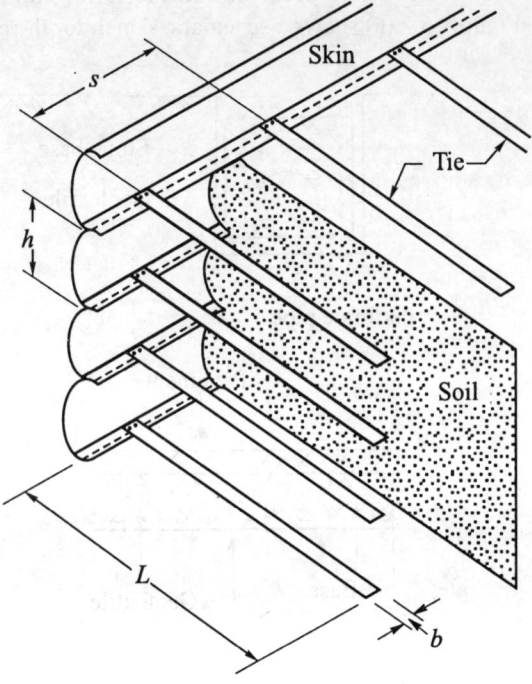

Fig. 16.3 (a) Component parts and key dimensions of reinforced earth wall (Vidal, 1969)

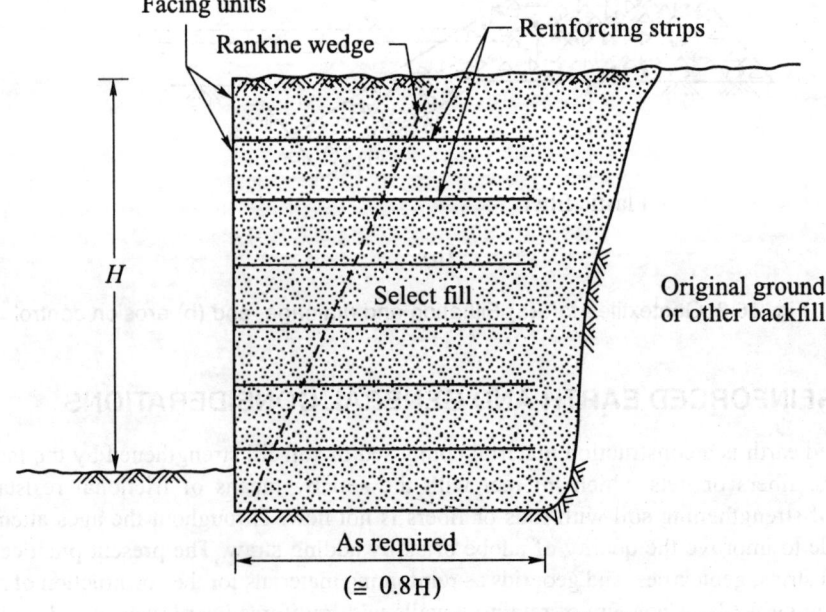

Line details of a reinforced earth wall in place

Fig. 16.3 (b) Reinforced earth walls (Bowles, 1996)

The three components of a MSE wall are the facing unit, the backfill and the reinforcing material. Figure 16.3 (b) shows a side view of a wall with metal strip reinforcement and Fig. 16.3 (c) the front face of a wall under construction (Bowles, 1996).

Modular concrete blocks, currently called segmental retaining walls [SRWS, Fig. 16.4 (a)] are most common as facing units. Some of the facing units are shown in Fig. 16.4. Most interesting in regard to SRWS are the emerging block systems with openings, pouches, or planting areas within them. These openings are soil-filled and planted with vegetation that is indigenous to the area [Fig. 16.4 (b)]. Further possibilities in the area of reinforced wall systems could be in the use of polymer rope, straps, or anchor ties to the facing in units or to geosynthetic layers, and extending them into the retained earth zone as shown in Fig. 16.4 (c).

Front face of a reinforced earth wall under construction far a bridge approach fill using patented precast concrete wall face units

Fig. 16.3 (c) Reinforced earth walls (Bowles, 1996)

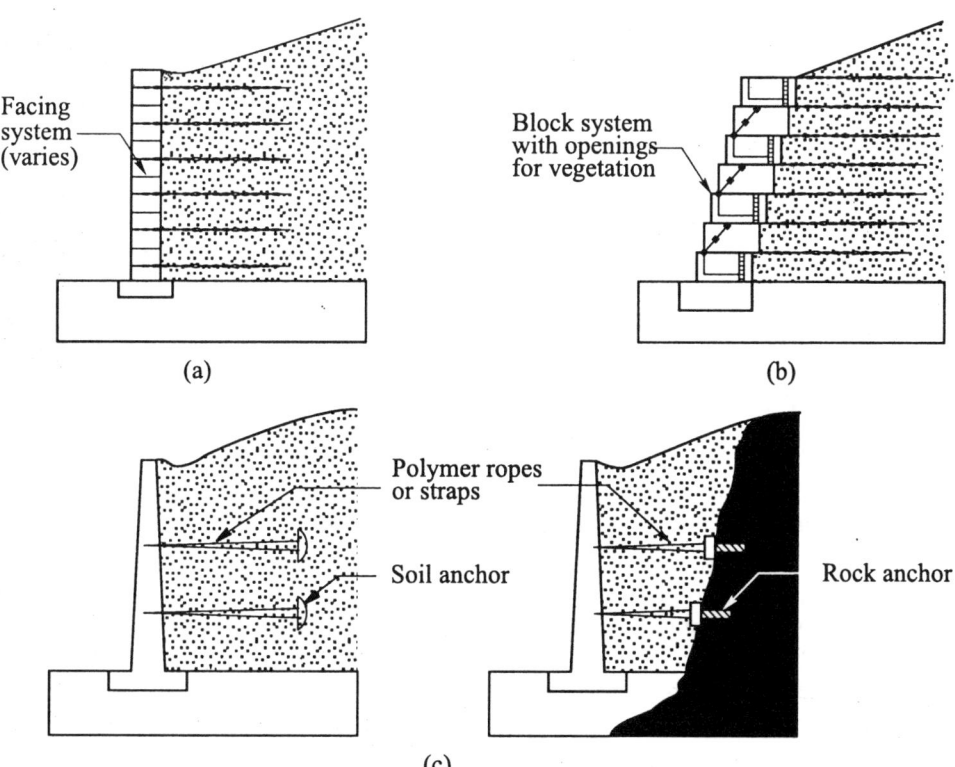

Fig. 16.4 Geosynthetic use for reinforced walls and bulkheads (Koerner, 2000): (a) Geosynthetic reinforced wall, (b) geosynthetic reinforced *live wall*, (c) future types of geosynthetic anchorage

A recent study (Koerner 2000) has indicated that geosynthetic reinforced walls are the least expensive of any wall type and for all wall height categories (Fig. 16.5).

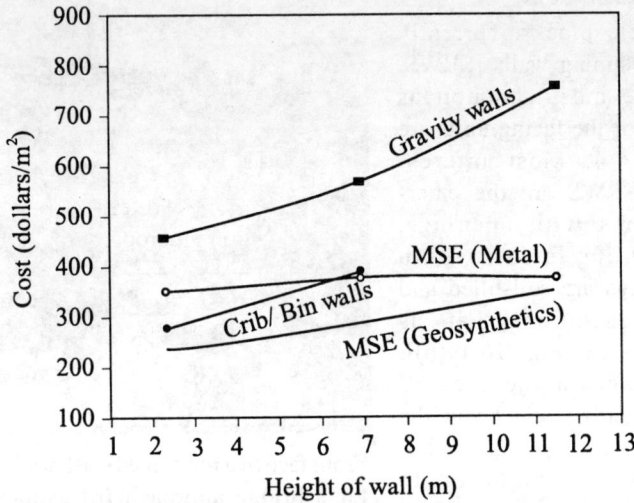

Fig. 16.5 Mean values of various categories of retaining wall costs (Koerner, 2000)

16.3 BACKFILL AND REINFORCING MATERIALS

Backfill

The backfill, is limited to cohesionless, free draining material (such as sand), and thus the key properties are the density and the angle of internal friction.

Reinforcing Material

The reinforcements may be strips or rods of metal or sheets of geotextile, wire grids or geogrids (grids made from plastic).

Geotextile is a permeable geosynthetic comprised solely of textiles. Geotextiles are used with foundation soil, rock, earth or any other geotechnical engineering-related material as an integral part of a human made project, structure, or system (Koerner, 1999). AASHTO (M288–96) provides (Table 16.1) geotextile strength requirements (Koerner, 1999). The tensile strength of geotextile varies with the geotextile designation as per the design requirements. For example, a woven slit-film polypropylene (weighing 240 g/m^2) has a range of 30 to 50 kN/m. The friction angle between soil and geotextiles varies with the type of geotextile and the soil. Table 16.2 gives values of geotextile friction angles (Koerner, 1999).

The test properties represent an idealized condition and therefore result in the maximum possible numerical values when used directly in design. Most laboratory test values cannot generally be used directly and must be suitably modified for in-situ conditions. For problems dealing with geotextiles the ultimate strength T_u should be reduced by applying certain reduction factors to obtain the allowable strength T_a as follows (Koerner, 1999).

$$T_a = T_u \left(\frac{1}{RF_{ID} \times RF_{CR} \times RF_{CD} \times RF_{BD}} \right)$$

(16.1)

where T_a = allowable tensile strength,

        T_u = ultimate tensile strength,

    RF_{ID} = reduction factor for installation damage,

    RF_{CR} = reduction factor for creep,

    RF_{BD} = reduction factor for biological degradation, and

    RF_{CD} = reduction factor for chemical degradation.

Typical values for reduction factors are given in Table 16.3.

Table 16.1 AASHTO M288–96 geotextile strength property requirements

	Test methods	Units	*Geotextile classification* † ‡*					
			Case 1		*Case 2*		*Case 3*	
			Elongation < 50%	*Elongation ≥ 50%*	*Elongation < 50%*	*Elongation ≥ 50%*	*Elongation < 50%*	*Elongation ≥ 50%*
Grab strength	ASTM D4632	N	1400	900	1100	700	800	500
Sewn seam Strength ‡	ASTM D4632	N	1200	810	990	630	720	450
Tear strength	ASTM D4533	N	500	350	400	250	300	180
Puncture strength	ASTM D4833	N	500	350	400	2505	300	180
Burst strength	ASTM D3786	kPa	3500	1700	2700	1300	2100	950

* As measured in accordance with ASTM D4632. Woven geotextiles fail at elongations (strains) < 50%, while nonwovens fail at elongation (strains) > 50%.

* When sewnseams are required. Overlap seam requirements are application specific.

* The required MARY tear strength for woven monofilament geotextiles is 250 N.

Table 16.2 Peak soil-to-geotextile friction angles and efficiencies in selected cohesionless soils*

Geotextile type	Concrete sand (f = 30°)	Rounded sand (f = 28°)	Silty sand (f = 26°)
Woven, monofilament	26° (84%)	–	–
Woven, slit-film	24° (77%)	24° (84%)	23° (87%)
Nonwoven, heat-bonded	26° (84%)	–	–
Nonwoven, needle-punched	30° (100%)	26° (92%)	25° (96%)

* Numbers in parentheses are the efficiencies. Values such as these should not be used in final design. Site specific geotextiles and soils must be individually tested and evaluated in accordance with the particular project conditions : saturation, type of liquid, normal stress, consolidation time, shear rate, displacement amount, and so on. (Koerner, 1999).

Table 16.3 Recommended reduction factor values for use in Eq. (16.1)

Application Area	Installation Damage	Creep*	Chemical Degradation	Biological Degradation
		Range of Reduction Factors		
Separation	1.1 to 2.5	1.5 to 2.5	1.0 to 1.5	1.0 to 1.2
Cushioning	1.1 to 2.0	1.2 to 1.5	1.0 to 2.0	1.0 to 1.2
Unpaved roads	1.1 to 2.0	1.5 to 2.5	1.0 to 1.5	1.0 to 1.2
Walls	1.1 to 2.0	2.0 to 4.0	1.0 to 1.5	1.0 to 1.3
Embankments	1.1 to 2.0	2.0 to 3.5	1.0 to 1.5	1.0 to 1.3
Bearing capacity	1.1 to 2.0	2.0 to 4.0	1.0 to 1.5	1.0 to 1.3
Slope stabilization	1.1 to 1.5	2.0 to 3.0	1.0 to 1.5	1.0 to 1.3
Pavement overlays	1.1 to 1.5	1.0 to 2.0	1.0 to 1.5	1.0 to 1.1
Railroads (filter/sep.)	1.5 to 3.0	1.0 to 1.5	1.5 to 2.0	1.0 to 1.2
Flexible forms	1.1 to 1.5	1.5 to 3.0	1.0 to 1.5	1.0 to 1.1
Silt fences	1.1 to 1.5	1.5 to 2.5	1.0 to 1.5	1.0 to 1.1

* The low end of the range refers to applications which have relatively short service lifetimes and / or situations where creep deformations are not critical to the overall system performance (Koerner, 1999).

Geogrid

A geogrid is defined as a geosynthetic material consisting of connected parallel sets of tensile ribs with apertures of sufficient size to allow strike-through of surrounding soil, stone, or other geotechnical material (Koerner, 1999).

Geogrids are matrix like materials with large open spaces called apertures, which are typically 10 to 100 mm between the ribs, called *longitudinal* and *transverse* respectively. The primary function of geogrids is clearly reinforcement. The mass of geogrids ranges from 200 to 1000 g/m^2 and the open area varies from 40 to 95%. It is not practicable to give specific values for the tensile strength of geogrids because of its wide variation in density. In such cases one has to consult manufacturer's literature for the strength characteristics of their products. The allowable tensile strength, T_a, maybe determined by applying certain reduction factors to the ultimate strength T_u as in the case of geotextiles. The equation is

$$T_a = T_u \left(\frac{1}{RF_{ID} \times RF_{CR} \times RF_{BD} \times RF_{CD}} \right) \qquad (16.2)$$

The definition of the various terms in Eq. (16.2) is the same as in Eq. (19.9). However, the reduction factors are different. These values are given in Table 16.4 (Koerner, 1999).

Table 16.4 Recommended reduction factor values for use in Eq. (16.2) for determining allowable tensile strength geogrids

Application Area	RF_{ID}	RF_{CR}	RF_{CD}	RF_{BD}
Unpaved roads	1.1 to 1.6	1.5 to 2.5	1.0 to 1.5	1.0 to 1.1
Paved roads	1.2 to 1.5	1.5 to 2.5	1.1 to 1.6	1.0 to 1.1
Embankments	1.1 to 1.4	2.0 to 3.0	1.1 to 1.4	1.0 to 1.2
Slopes	1.1 to 1.4	2.0 to 3.0	1.1 to 1.4	1.0 to 1.2
Walls	1.1 to 1.4	2.0 to 3.0	1.1 to 1.4	1.0 to 1.2
Bearing capacity	1.2 to 1.5	2.0 to 3.0	1.1 to 1.6	1.0 to 1.2

Metal Strips

Metal reinforcement strips are available in widths ranging from 75 to 100 mm and thickness on the order of 3 to 5 mm, with 1 mm on each face excluded for corrosion (Bowles, 1996). The yield strength of steel may be taken as equal to about 35000 lb/in^2 (240 MPa) or as per any code of practice.

16.4 CONSTRUCTION DETAILS

The method of construction of MSE walls depends upon the type of facing unit and reinforcing material used in the system. The facing unit which is also called the *skin* can be either flexible or stiff, but must be strong enough to retain the backfill and allow fastenings for the reinforcement to be attached. The facing units require only a small foundation from which they can be built, generally consisting of a trench filled with mass concrete giving a footing similar to those used in domestic housing. The segmental retaining wall sections of dry-laid masonry blocks, are shown in Fig. 16.4 (a). The block system with openings for vegetation is shown in Fig. 16.4 (b).

The construction procedure with the use of geotextiles is explained in Fig. 16.6A. Here, the geotextile serve both as a reinforcement and also as a facing unit. The procedure is described below .(Koerner, 1985) with reference to Fig. 16.6A.

1. Start with an adequate working surface and staging area (Fig. 16.6A).
2. Lay a geotextile sheet of proper width on the ground surface with 4 to 7 ft at the wall face draped over a temporary wooden form (b).

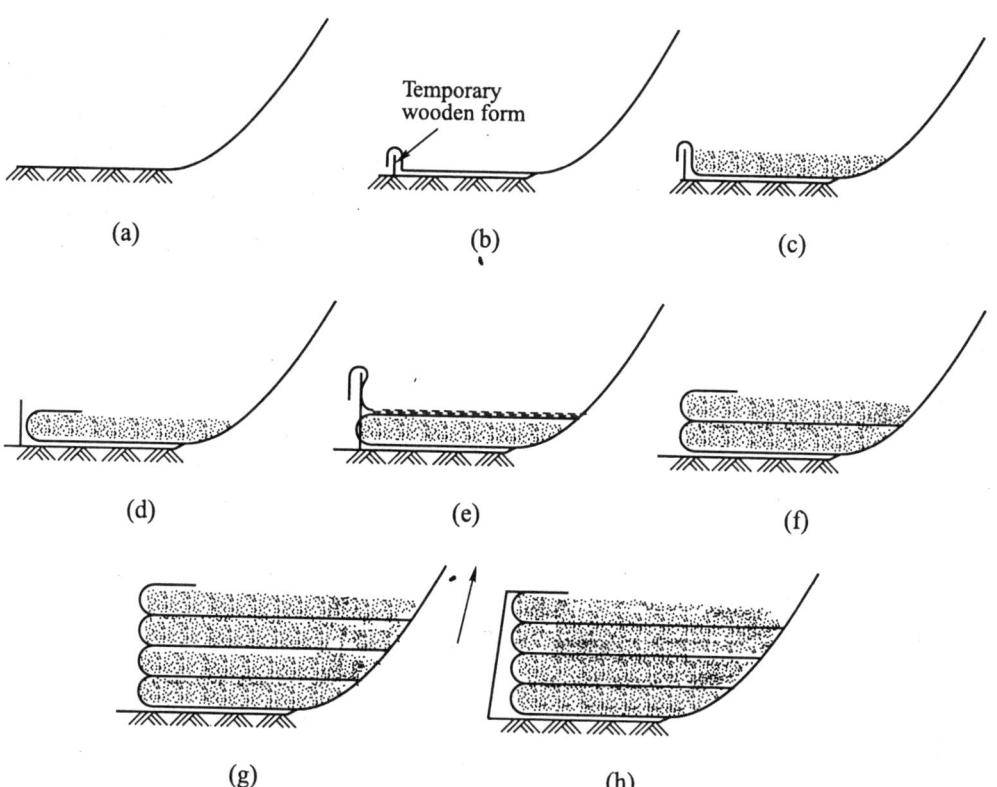

Fig. 16.6 A General construction procedures for using geotextiles in fabric wall construction (Koerner, 1985)

3. Backfill over this sheet with soil. Granular soils or soils containing a maximum 30 percent silt and/or 5 percent clay are customary (c).

4. Construction equipment must work from the soil backfill and be kept off the unprotected geotextile. The spreading equipment should be a wide-tracked bulldozer that exerts little pressure against the ground on which it rests. Rolling equipment likewise should be of relatively light weight.

5. When the first layer has been folded over the process should be repeated for the second layer with the temporary facing form being extended from the original ground surface or the wall being stepped back about 6 inches so that the form can be supported from the first layer. In the latter case, the support stakes must penetrate the fabric.

6. This process is continued until the wall reaches its intended height.

7. For protection against ultraviolet light and safety against vandalism the faces of such walls must be protected. Both shotcrete and gunite have been used for this purpose.

Figure 16.6B shows complete geotextile walls (Koerner, 1999).

Fig. 16.6 B Geotextile walls (Koerner, 1999)

16.5 DESIGN CONSIDERATIONS FOR A REINFORCED EARTH WALL (OR MSE)

The design of a MSE (Mechanically Stabilized Earth) wall involves the following steps:

1. Check for internal stability, addressing reinforcement spacing and length.

2. Check for external stability of the wall against overturning, sliding, and foundation failure.

The general considerations for the design are:

1. Selection of backfill material : granular, freely draining material is normally specified. However, with the advent of geogrids, the use of cohesive soil is gaining ground.

2. Backfill should be compacted with care in order to avoid damage to the reinforcing material.

3. Rankine's theory for the active state is assumed to be valid.

4. The wall should be sufficiently flexible for the development of active conditions.
5. Tension stresses are considered for the reinforcement outside the assumed failure zone.
6. Wall failure will occur in one of three ways.
   (*a*) tension in reinforcements
   (*b*) bearing capacity failure
   (*c*) sliding of the whole wall soil system.
7. Surcharges are allowed on the backfill. The surcharges may be permanent (such as a roadway) or temporary.
   (*a*) Temporary surcharges within the reinforcement zone will increase the lateral pressure on the facing unit which in turn increases the tension in the reinforcements, but does not contribute to reinforcement stability.
   (*b*) Permanent surcharges within the reinforcement zone will increase the lateral pressure and tension in the reinforcement and will contribute additional vertical pressure for the reinforcement friction.
   (*c*) Temporary or permanent surcharges outside the reinforcement zone contribute lateral pressure which tends to overturn the wall.
8. The total length L of the reinforcement goes beyond the failure plane AC by a length L_e. Only length L_e (effective length) is considered for computing frictional resistance. The length L_R lying within the failure zone will not contribute for frictional resistance [Fig. 16.7 (a)].
9. For the propose of design the total length L remains the same for the entire height of wall H. Designers, however, may use their discretion to curtail the length at lower levels. Typical ranges in reinforcement spacing are given in Fig. 16.8.

16.6 DESIGN METHOD

The following forces are considered:

1. Lateral pressure on the wall due to backfill.
2. Lateral pressure due to surcharge if present on the backfill surface.
3. The vertical pressure at any depth z on the strip due to
   (*a*) overburden pressure p_o only.
   (*b*) overburden pressure p_o and pressure due to surcharge.

Lateral Pressure
Pressure due to overburden

Lateral earth pressure due to overburden

At depth z $\qquad p_a = p_{oz} K_A = \gamma z K_A$ $\hfill$ (16.3a)

At depth H $\qquad p_a = p_{oH} K_A = \gamma H K_A$ $\hfill$ (16.3b)

Total active earth pressure

$$P_a = \frac{1}{2} \gamma H^2 K_A \qquad\qquad (16.4)$$

Pressure due to surcharge (a) of limited width, and (b) uniformly distributed

(*a*) From Eq. (2.33)

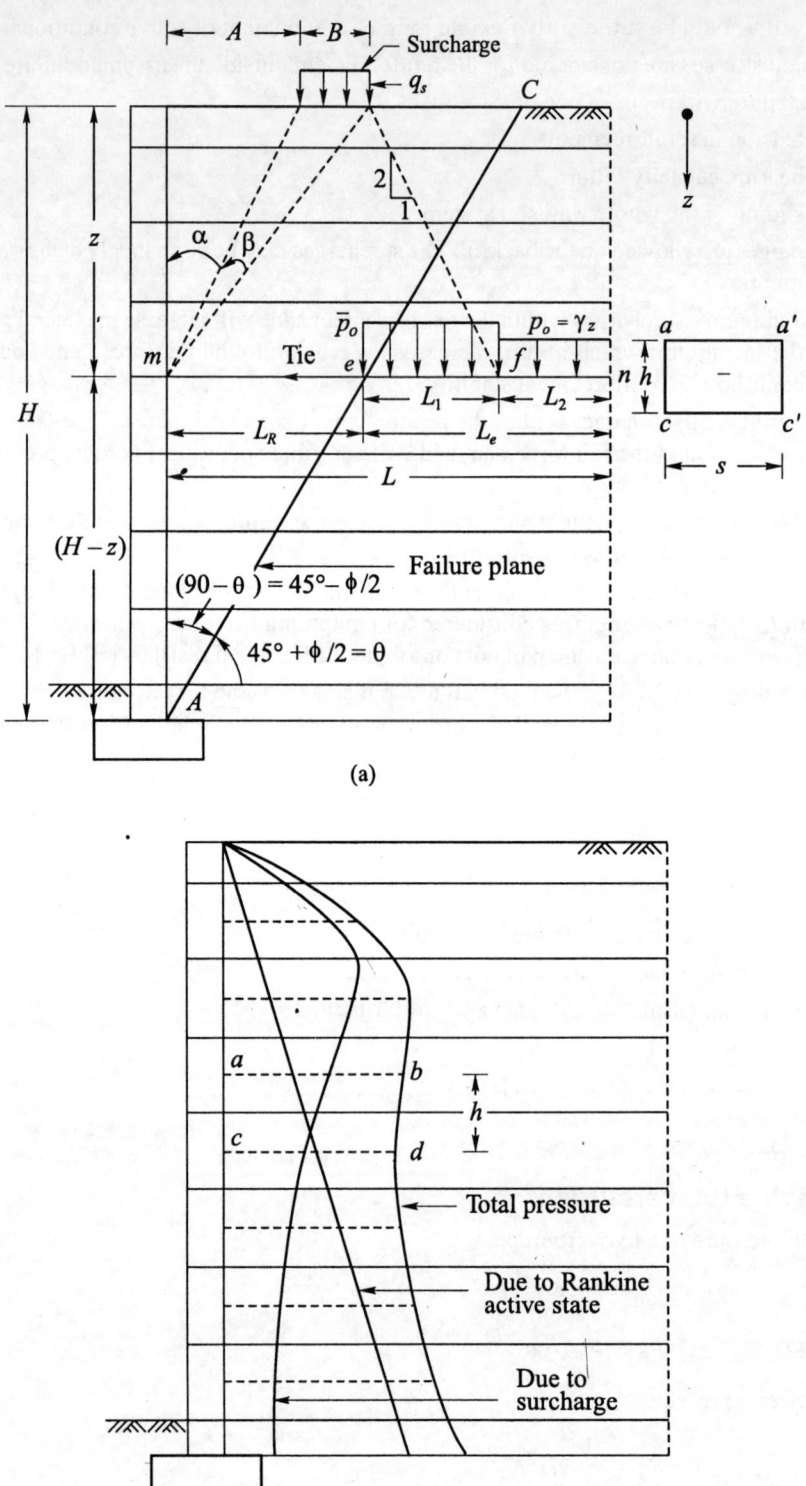

(a)

(b)

Fig. 16.7 Principles of MSE wall design: (a) Reinforced earth-wall profile with surcharge load, (b) lateral pressure distribution diagrams

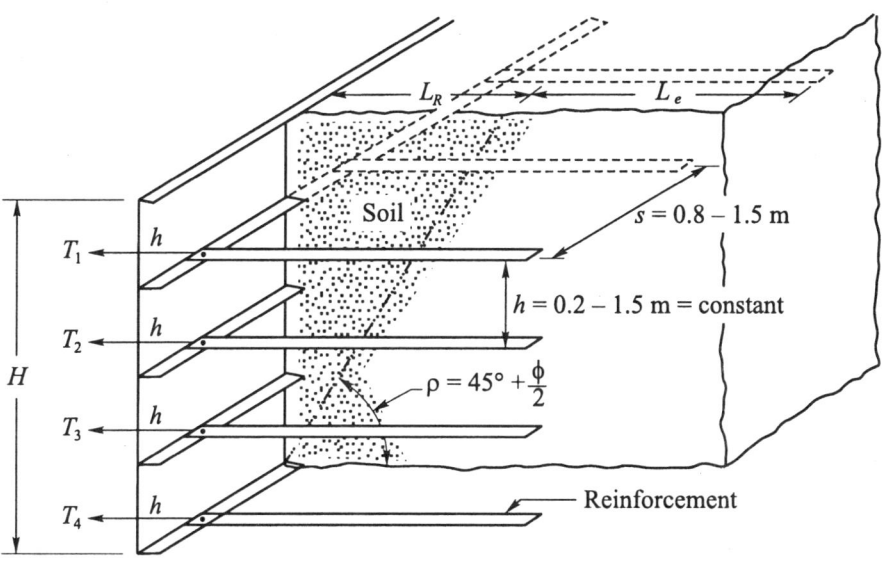

Fig. 16.8 Typical range in strip reinforcement spacing for reinforced earth walls (Bowles, 1996)

$$q_h = \frac{2q_s}{\pi} (\beta - \sin \beta \cos 2\alpha) \tag{16.5a}$$

(b) $q_h = q_s K_A$ (16.5b)

Total lateral pressure due to overburden and surcharge at any depth z

$$p_h = p_a + q_h = (\gamma z K_A + q_h) \tag{16.6}$$

Vertical pressure

Vertical pressure at any depth z due to overburden only

$$p_o = \gamma z \tag{16.7a}$$

due to surcharge (limited width)

$$\Delta q = \frac{q_s B}{B + z} \tag{16.7b}$$

where the 2 : 1 (2 vertical : 1 horizontal) method is used for determining Δq at any depth z.

Total vertical pressure due to overburden and surcharge at any depth z.

$$\bar{p}_o = p_o + \Delta q \tag{16.7c}$$

Reinforcement and distribution

Three types of reinforcements are normally used. They are:

1. Metal strips
2. Geotextiles
3. Geogrids.

Galvanized steel strips of widths varying from 5 to 100 mm and thickness from 3 to 5 mm are generally used. Allowance for corrosion is normally made while deciding the thickness at the rate of 0.001 in. per year and the life span is taken as equal to 50 years. The vertical spacing may range from 20 to 150 cm (8 to 60 in.) and can vary with depth. The horizontal lateral spacing may be on the order of 80 to 150 cm (30 to 60 in.). The ultimate tensile strength may be taken as equal to 240 MPa (35,000 lb/in.^{2}). A factor of safety in the range of 1.5 to 1.67 is normally used to determine the allowable steel strength f_a.

Figure 16.8 depicts a typical arrangement of metal reinforcement. The properties of geotextiles and geogrids have been discussed in Section 16.3. However, with regards to spacing, only the vertical spacing is to be considered. Manufacturers provide geotextiles (or geogrids) in rolls of various lengths and widths. The tensile force per unit width must be determined.

Length of reinforcement

From Fig. (16.7a)

$$L = L_R + L_e = L_R + L_1 + L_2 \tag{16.8}$$

where, $L_R = (H - z) \tan (45° - \phi/2)$,

L_e = effective length of reinforcement outside the failure zone,

L_1 = length subjected to pressure $(p_o + \Delta q) = \bar{p_o}$;

L_2 = length subjected to p_o only.

Strip tensile force at any depth z

The equation for computing T is

$$T = p_h \times h \times s/\text{strip} = (\gamma z K_A + q_h) \, h \times s \tag{16.9a}$$

The maximum tie force will be

$$T(\text{max}) = (\gamma H K_A + q_{hH}) \, h \times s \tag{16.9b}$$

where, $p_h = \gamma z K_A + q_h$,

q_h = lateral pressure at depth z due to surcharge,

$q_{hH} = q_h$ at depth H,

h = vertical spacing,

s = horizontal spacing.

$$T = P_a + P_q \tag{16.10}$$

where, $P_a = 1/2\gamma \, H^2 K_A$—Rankine's lateral force

P_q = lateral force due to surcharge

Frictional resistance

In the case of strips of width b both sides offer frictional resistance. The frictional resistance F_R offered by a strip at any depth z must be greater than the pullout force T by a suitable factor of safety. We may write

$$F_R = 2b\left[(p_o + \Delta q) L_1 + p_o L_2\right] \tan \delta \leq TF_s \tag{16.11}$$

or
$$F_R = 2b\left[\bar{p}_o L_1 + p_o L_2\right] \tan \delta \leq TF_s \tag{16.12}$$

where F_s may be taken as equal to 1.5.

The friction angle δ between the strip and the soil may be taken as equal to ϕ for a rough strip surface and for a smooth surface δ may lie between 10 to 25°.

Sectional area of metal strips

Normally, the width b of the strip is assumed in the design. The thickness t has to be determined based on T (max) and the allowable stress f_a in the steel. If f_y is the yield stress of steel, then

$$f_a = \frac{f_y}{F_s \,(\text{steel})} \tag{16.13}$$

Normally, F_s (steel) ranges from 1.5 to 1.67. The thickness t may be obtained from

$$t = \frac{T\,(\text{max})}{bf_a} \tag{16.14}$$

The thickness of t is to be increased to take care of the corrosion effect. The rate of corrosion is normally taken as equal to 0.001 in/yr for a life span of 50 years.

Spacing of geotextile layers

The tensile force T per unit width of geotextile layer at any depth z may be obtained from

$$T = p_h h = (\gamma z K_A + q_h)\, h \tag{16.15}$$

where q_h = lateral pressure either due to a stripload or due to uniformly distributed surcharge.

The maximum value of the computed T should be limited to the allowable value T_a as per Eq. (16.1). As such we may write Eq. (16.15) as

$$T_a = TF_s = (\gamma z K_A + q_h)\, hF_s \tag{16.16}$$

or
$$h = \frac{T_a}{(\gamma z K_A + q_h)\, F_s} = \frac{T_a}{p_h F_s} \tag{16.17}$$

where F_s = factor of safety (1.3 to 1.5) when using T_a.

Equation (16.17) is used for determining the vertical spacing of geotextile layers.

Frictional resistance

The frictional resistance offered by a geotextile layer for the pullout force T_a may be expressed as

$$F_R = 2\left[(\gamma z + \Delta q)\, L_1 + \gamma z L_z\right] \tan \delta \geq T_a F_s \tag{16.18}$$

Equation (16.18) expresses frictional resistance per unit width and both sides of the sheets are considered.

Design with geogrid layers

A tremendous number of geogrid reinforced walls have been constructed in the past 10 years (Koerner, 1999). The types of permanent geogrid reinforced wall facings are as follows (Koerner, 1999):

1. *Articulated precast panels* are discrete precast concrete panels with inserts for attaching the geogrid.

2. *Full height precast panels* are concrete panels temporarily supported until backfill is complete.

3. *Cast-in-place concrete panels* are often wrap-around walls that are allowed to settle and, after 1/2 to 2 years, are covered with a cast-in-place facing panel.

4. *Masonry block facing walls* are an exploding segment of the industry with many different types currently available, all of which have the geogrid embedded between the blocks and held by pins, nubs, and/or friction.

5. *Gabion facings* are polymer or steel-wire baskets filled with stone, having a geogrid held between the baskets and fixed with rings and/or friction.

The frictional resistance offered by a geogrid against pullout may be expressed as (Koerner, 1999)

$$F_R = 2C_i C_r L_e p_o \tan \phi \geq TF_s \tag{16.19}$$

where C_i = interaction coefficient = 0.75 (may vary),
  C_r = coverage ratio = 0.8 (may vary).

All the other notations are already defined. The spacing of geogrid layers may be obtained from

$$h = \frac{T_a C_r}{p_h} \tag{16.20}$$

where, p_h = lateral pressure per unit length of wall.

16.7 EXTERNAL STABILITY

The MSE wall system consists of three zones. They are:

1. The reinforced earth zone.
2. The backfill zone.
3. The foundation soil zone.

The reinforced earth zone is considered as the wall for checking the internal stability whereas all three zones are considered for checking the external stability. The soils of the first two zones are placed in layers and compacted whereas the foundation soil is a normal one. The properties of the soil in each of the zones may be the same or different. However, the soil in the first two zones is normally a free draining material such as sand.

It is necessary to check the reinforced earth wall (width = B) for external stability which includes overturning, sliding and bearing capacity failure. These are illustrated in Fig. 16.9. Active earth pressure of the backfill acting on the internal face AB of the wall is taken in the stability analysis. The resultant earth thrust P_a is assumed to act horizontally at a height $H/3$ above the base of the wall. The methods of analysis are the same as for concrete retaining walls.

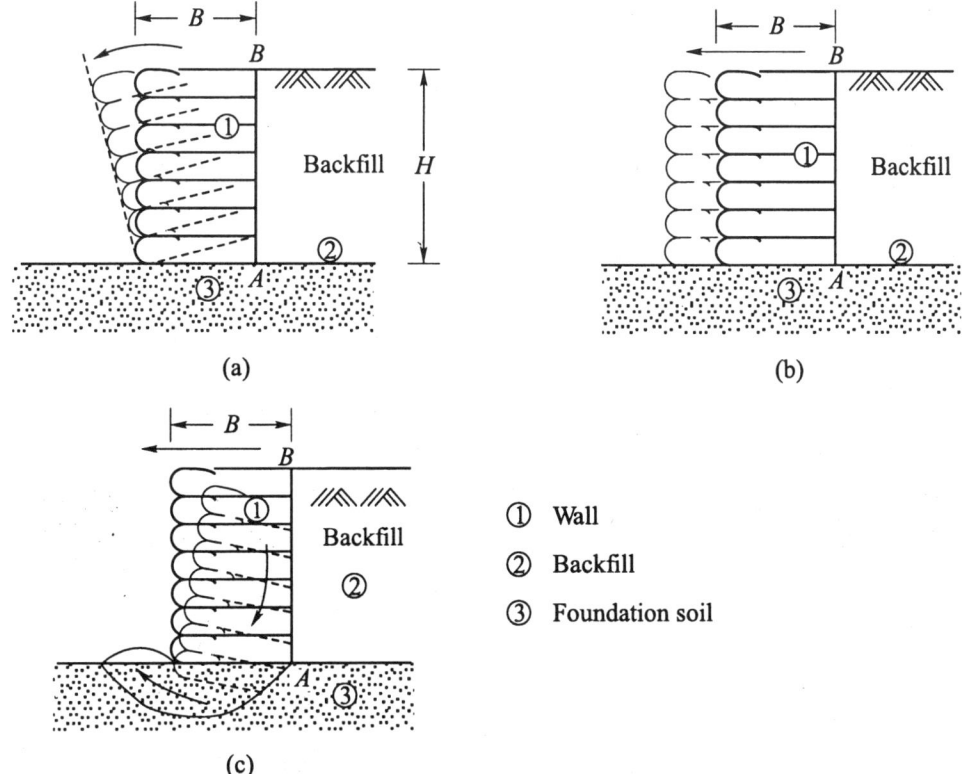

Fig. 16.9 External stability considerations for reinforced earth walls: (a) Overturning considerations, (b) sliding considerations, and (c) foundation considerations

---

Example 16.1

A typical section of a retaining wall with the backfill reinforced with metal strips is shown in Fig. Ex. 16.1. The following data are available:

Height $H = 9$ m; $b = 100$ mm; $t = 5$ mm; $f_y = 240$ MPa; F_s for steel = 1.67; F_s on soil friction = 1.5; $\phi = 36°$; $\gamma = 17.5$ kN/m³; $\delta = 25°$; $h \times s = 1 \times 1$ m.

Required

(a) Lengths L and L_e at varying depths.

(b) The largest tension T in the strip.

(c) The allowable tension in the strip.

(d) Check for external stability.

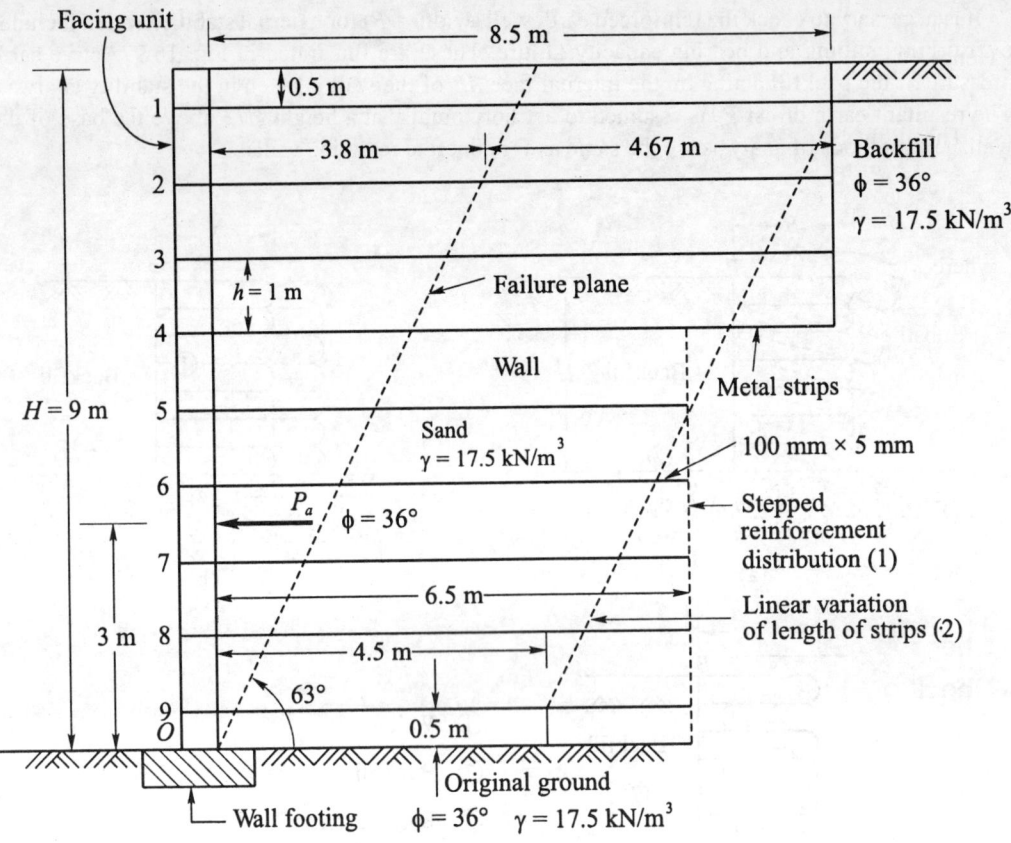

Fig. Ex. 16.1

Solution

From Eq. (16.9a), the tension in a strip at depth z is

$$T = \gamma z \, K_A \, sh \text{ for } q_h = 0$$

where $\gamma = 17.5 \text{ kN/m}^3$, $K_A = \tan^2 (45° - 36/2) = 0.26$, $s = 1$ m; $h = 1$ m.

Substituting

$$T = 17.5 \times 0.26 \,(1)\,[1]\, z = 4.55z \text{ kN/strip}$$

$$L_e = \frac{F_s T}{2\gamma \, zb \tan \delta} = \frac{1.5 \times 4.55 z}{2 \times 17.5 \times 0.1 \times 0.47 \times z} = 4.14 \text{ m}$$

This shows that the length $L_e = 4.14$ m is a constant with depth. Fig. Ex. 16.1 shows the positions of L_e for strip numbers 1, 2 ⋯⋯ 9. The first strip is located 0.5 m below the backfill surface and the 9th at 8.5 below with spacings at 1 m apart. Tension in each of the strips may be obtained by using the equation $T = 4.55 \, z$. The total tension ΣT as computed is

$$\Sigma T = 184.29 \text{ kN/m since } s = 1 \text{ m}.$$

As a check the total active earth pressure is

$$P_a = \frac{1}{2}\gamma \, H_2 \, KA = \frac{1}{2}\, 17.5 \times 92 \times 0.26 = 184.28 \text{ kN/m} = \Sigma T$$

The maximum tension is in the 9th strip, that is, at a depth of 8.5 below the backfill surface. Hence

$$T = \gamma z\, K_A\, sh = 17.5 \times 8.5 \times 0.26 \times 1 \times 1 = 38.68 \text{ kN/strip}$$

The allowable tension is

$$T_a = f_a tb$$

where

$$f_a = \frac{240 \times 10^3}{1.67} = 143.7 \times 10^3 \text{ kN/m}^2$$

Substituting

$$T_a = 143.7 \times 10^3 \times 0.005 \times 0.1 \approx 72 \text{ kN} > T \text{ OK.}$$

The total length of strip L at any depth z is

$$L = L_R + L_e = (H - z) \tan (45 - \phi/2) + 4.14 = 0.51\,(9 - z) + 4.14 \text{ m}$$

where

$$H = 9 \text{ m}$$

The lengths as calculated have been shown in Fig. Ex. 16.1. It is sometimes convenient to use the same length L with depth or stepped in two or more blocks or use a linear variation as shown in the figure.

Check for External Stability

Check of bearing capacity

It is necessary to check the base of the wall with the backfill for the bearing capacity per unit length of the wall. The width of the wall may be taken as equal to 4.5 m (Fig. Ex. 16.1). The procedure as explained under shallow Foundation may be followed. For all practical purposes, the shape, depth, and inclination factors may be taken as equal to 1.

Check for sliding resistance

$$F_s = \frac{\text{Sliding resistance } F_R}{\text{Driving force } P_a}$$

where

$$F_R = W \tan \delta = \frac{4.5 + 8.5}{2} \times 17.5 \times 9 \tan 36°$$

$$= 1024 \times 0.73 = 744 \text{ kN}$$

where $\delta = \phi = 36°$ for the foundation soil, and W = weight of the reinforced wall

$$P_a = 184.28 \text{ kN}$$

$$F_s = \frac{744}{184.28} = 4 > 1.5 \text{ OK}$$

Check for overturning

$$F_s = \frac{M_R}{M_o}$$

From Fig. Ex. 16.1 taking moments of all forces about O, we have

$$M_R = 4.5 \times 9 \times 17.5 \times \frac{4.5}{2} + \frac{1}{2} \times 9 \times (8.5 - 4.5)\,(4.5 + \frac{4}{3}) \times 17.5$$

$$= 1595 + 1837 = 3432 \text{ kNm}$$

$$M_o = P_a \times \frac{H}{3} = 184.28 \times \frac{9}{3} = 553 \text{ kNm}$$

$$F_s = \frac{3432}{553} = 6.2 > 2 \text{ OK.}$$

Example 16.2

A section of a retaining wall with a reinforced backfill is shown in Fig. Ex. 16.2. The backfill surface is subjected to a surcharge of 30 kN/m².

Required

(a) The reinforcement distribution.

(b) The maximum tension in the strip.

(c) Check for external stability.

Given: $b = 100$ mm, $t = 5$ mm, $f_a = 143.7$ MPa, $c = 0$, $\phi = 36°$, $\delta = 25°$, $\gamma = 17.5$ kN/m³, $s = 0.5$ m, and $h = 0.5$.

Solution

From Eq. (16.9a)

$$T = (\gamma z K_A + q_h) \, h \times s = (p_o + q_h) \, A_c$$

where $\quad \gamma = 17.5 \text{ kN/m}^3, K_A = 0.26, A_c = h \times s = (0.5 \times 0.5) \text{ m}^2$

From equation

$$q_h = \frac{2 q_s}{\pi} [\beta - \sin \beta \cos 2\alpha]$$

Refer to Fig. Ex. 16.2 for the definition of α and β.

$q_s = 30 \text{ kN/m}^2$

The procedure for calculating length L of the strip for one depth $z = 1.75$ m (strip number 4) is explained below. The same method is valid for the other strips.

Strip No. 4 depth z = 1.75 m

$$p_a = \gamma z K_A = 17.5 \times 1.75 \times 0.26 = 7.96 \text{ KN/m}^2$$

From Fig. Ex. 16.2 $\quad \beta = 19.07° = 0.3327$ radians

$$\alpha = 29.74°$$

$$q_s = 30 \text{ kN/m}^2$$

$$q_h = \frac{2 \times 30}{3.14} [0.3327 - \sin 19.07° \cos 59.5°] = 3.19 \text{ kN/m}^2$$

Figure Ex. 16.2 shows the surcharge distribution at a 2 (vertical) to 1 (horizontal) slope. Per the figure at depth $z = 1.75$ m, $L_1 = 1.475$ m from the failure line and $L_R = (H - z) \tan (45° - \phi/2)$ = 2.75 tan (45° − 36°/2) = 1.4 m from the wall to the failure line. It is now necessary to determine L_2 [Refer to Fig. 16.7 (a)].

Now $\quad\quad\quad T = (7.96 + 3.19) \times 0.5 \times 0.5 = 2.79 \text{ kN/strip}$

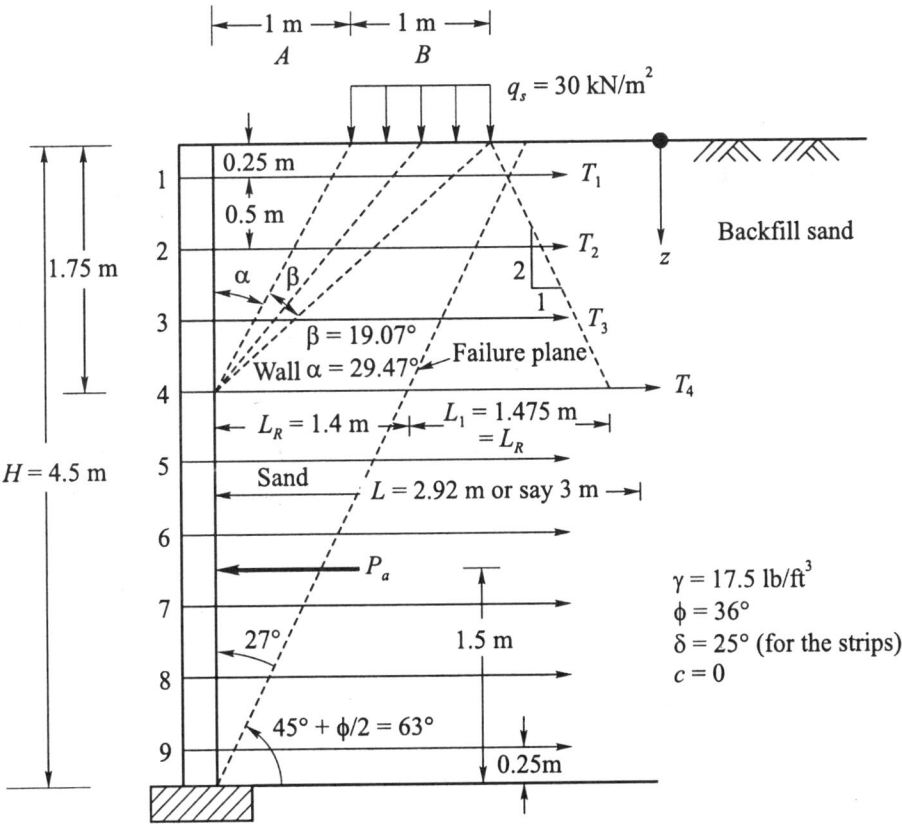

Fig. Ex. 16.2

The equation for the frictional resistance per strip is

$$F_R = 2b\,(\gamma z + \Delta q)\,L_1 \tan \delta + (\gamma z\,L_2 \tan \gamma)\,2b$$

From the 2 : 1 distribution Δq at $z = 1.75$ m is

$$\Delta q = \frac{Q}{B+z} = \frac{30 \times 1}{1+1.75} = 10.9 \text{ kN/m}^2$$

$$p_o = 17.5 \times 1.75 = 30.63 \text{ kN/m}^2$$

Hence $\quad \bar{p}_o = 10.9 + 30.63 = 41.53 \text{ kN/m}^2$

Now equating frictional resistance F_R to tension in the strip with $F_s = 1.5$, we have

$$F_R = 1.5\,T. \text{ Given } b = 100 \text{ mm. Now from Eq. (16.12)}$$

$$F_R = 2b \tan \delta\,(\bar{p}_o L_1 + p_o L_2) = 1.5\,T$$

Substituting and taking $\delta = 25°$, we have

$$2 \times 0.1 \times 0.47\,(41.53 \times 1.475 + 30.63\,L_2) = 1.5 \times 2.79$$

Simplifying

$$L_2 = -0.546 \text{ m} \approx 0$$

Hence $\quad L_e = L_1 + 0 = 1.475 \text{ m}$

$$L = L_R + L_e = 1.4 + 1.475 = 2.875 \text{ m}$$

L can be calculated in the same way at other depths.

Maximum tension T

The maximum tension is in strip number 9 at depth $z = 4.25$ m

Allowable $\quad\quad T_a = f_a bt = 143.7 \times 10^3 \times 0.1 \times 0.005 = 71.85$ kN

$$T = (\gamma z K_A + q_h) sh$$

where $\quad\quad \gamma z K_A = 17.50 \times 4.25 \times 0.26 = 19.34$ kN/m^2

$\quad\quad\quad\quad q_h = 0.89$ kN/m^2 from equation for q_h at depth $z = 4.25$ m

Hence $\quad\quad T = (19.34 + 0.89) \times 1/2 \times 1/2 = 5.05$ kN/strip < 71.85 kN OK.

Example 16.3 (Koerner, 1999)

Figure Ex. 16.3 shows a section of a retaining wall with geotextile reinforcement. The wall is backfilled with a granular soil having $\gamma = 18$ kN/m^3 and $\phi = 34°$.

A woven slit-film geotextile with warp (machine) direction ultimate wide-width strength of 50 kN/m and having $\delta = 24°$ (Table 16.2) is intended to be used in its construction.

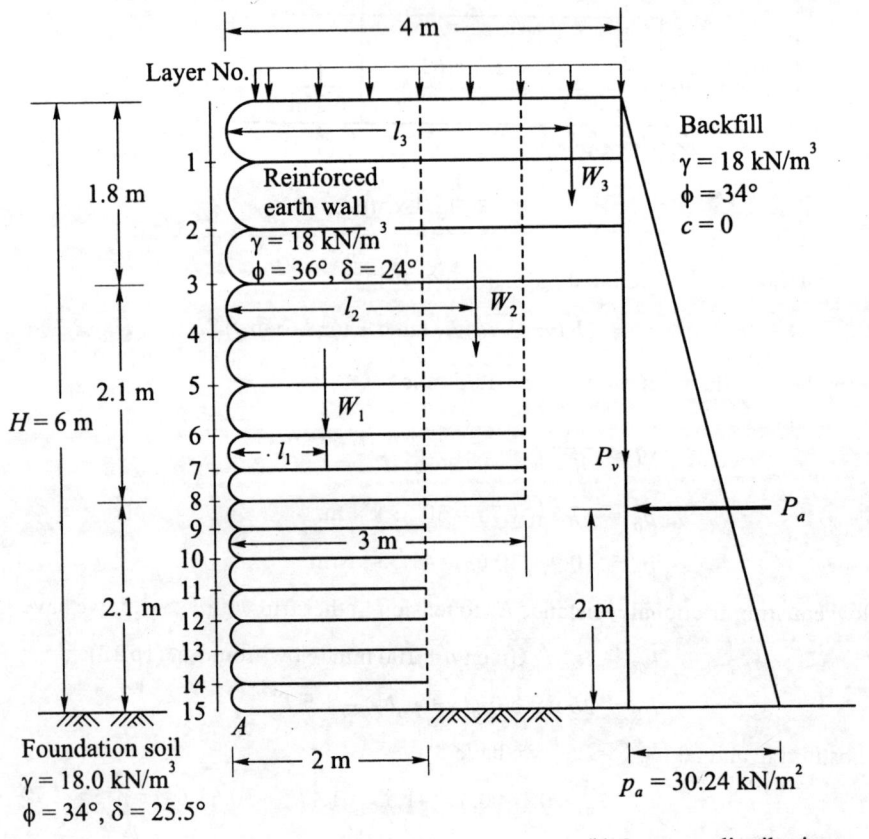

(a) Geotextile layers   (b) Pressure distribution

Fig. Ex. 16.3

The orientation of the geotextile is perpendicular to the wall face and the edges are to be overlapped to handle the weft direction. A factor of safety of 1.4 is to be used along with site-specific reduction factors (Table 16.3).

Required:

(a) Spacing of the individual layers of geotextile.
(b) Determination of the length of the fabric layers.
(c) Check the overlap.
(d) Check for external stability.

The backfill surface carries a uniform surcharge dead load of 10 kN/m^2.

Solution

(a) The lateral pressure p_h at any depth z is expressed as

$$p_h = p_a + q_h$$

where $\qquad p_a = \gamma z\, K_A, q_h = q K_A, K_A = \tan^2(45° - 36/2) = 0.26$

Substituting

$$p_h = 18 \times 0.26z + 0.26 \times 10 = 4.68z + 2.60$$

From Eq. (16.1), the allowable geotextile strength is

$$T_a = T_u \left(\frac{1}{RF_{ID} \times RF_{CR} \times RF_{CD} \times RF_{BD}} \right)$$

$$= 50 \left(\frac{1}{1.2 \times 2.5 \times 1.15 \times 1.1} \right) = 13.2 \text{ kN/m}$$

From Eq. (16.9a), the expression for allowable stress in the geotextile at any depth z may be expressed as

$$T = T_a = p_h\, h\, F_s$$

$$h = \frac{T_a}{p_h F_s}$$

where $\quad h$ = vertical spacing (lift thickness),
$\quad\quad\ T_a$ = allowable stress in the geotextile,
$\quad\quad\ p_h$ = lateral earth pressure at depth z,
$\quad\quad\ F_s$ = factor of safety = 1.4.

Now substituting

$$h = \frac{13.2}{\left[4.68(z) + 2.60\right]1.4} = \frac{13.2}{6.55(z) + 3.64}$$

At $z = 6$ m, $\qquad h = \dfrac{13.2}{6.55 \times 6 + 3.64} = 0.307$ m or say 0.30 m

At $z = 3.3$ m, $\qquad h = \dfrac{13.2}{6.55 \times 3.3 + 3.64} = 0.52$ m or say 0.50 m

At $z = 1.3$ m, $\quad h = \dfrac{13.2}{6.55 \times 1.3 + 3.64} = 1.08$ m, but use 0.65 m for a suitable distribution.

The depth 3.3 m or 1.3 m are used just as a trial and error process to determine suitable spacings. Fig. Ex. 16.3 shows the calculated spacings of the geotextiles.

(b) *Length of the fabric layers*

From Eq. (16.18) we may write

$$L_e = \frac{T \, F_s}{2\gamma \, z \, \tan \delta} = \frac{p_h \, h \, F_s}{2\gamma \, z \, \tan \delta} = \frac{h \left(4.68 \, z + 2.60\right) 1.4}{2 \times 18 \, z \, \tan 24°}$$

$$= L_e = \frac{h \left(6.55 z + 3.64\right)}{16 \, z}$$

From Fig. (16.7) the expression for L_R is

$$L_R = (H - z) \tan (45° - \phi/2) = (H - z) \tan (45° - 36/2) = (6.0 - z)(0.509)$$

The total length L is

$$L = L_R + L_e$$

The computed L and suggested L are given in a tabular form below.

Layer No.	Depth z (m)	Spacing h (m)	L_e (m)	L_e (min) (m)	L_R (m)	L (cal) (m)	L (suggested) (m)
1	0.65	0.65	0.49	1.0	2.72	3.72	4.0
2	1.30	0.65	0.38	1.0	2.39	3.39	–
3	1.80	0.50	0.27	1.0	2.14	3.14	–
4	2.30	0.50	0.26	1.0	1.88	2.88	3.0
5	2.80	0.50	0.25	1.0	1.63	2.63	–
6	3.30	0.50	0.24	1.0	1.37	2.37	–
7	3.60	0.30	0.14	1.0	1.22	2.22	–
8	3.90	0.30	0.14	1.0	1.07	2.07	–
9	4.20	0.30	0.14	1.0	0.92	1.92	2.0
10	4.50	0.30	0.14	1.0	0.76	1.76	–
11	4.80	0.30	0.14	1.0	0.61	1.61	–
12	5.10	0.30	0.14	1.0	0.46	1.46	–
13	5.40	0.30	0.14	1.0	0.31	1.31	–
14	5.70	0.30	0.14	1.0	0.15	1.15	–
15	6.00	0.30	0.13	1.0	0.00	1.00	–

It may be noted here that the calculated values of L_e are very small and a minimum value of 1.0 m should be used.

(c) *Check for the overlap*

When the fabric layers are laid perpendicular to the wall, the adjacent fabric should overlap a length L_o. The minimum value of L_o is 1.0 m. The equation for L_o may be expressed as

$$L_o = \frac{h \, p_h F_s}{2 \times 2\gamma \, z \, \tan \delta} = \frac{h \left[4.68 \left(z\right) + 2.60\right] 1.4}{4 \times 18 \left(z\right) \tan 24°}$$

The maximum value of L_o is at the upper layer at $z = 0.65$. Substituting for z, we have

$$L_o = \frac{0.65\left[4.68\left(0.65\right) + 2.60\right]1.4}{4 \times 18\left(0.65\right)\tan 24°} = 0.25 \text{ m}$$

Since this value of L_o calculated is quite low, use $L_o = 1.0$ m for all the layers.

(*d*) *Check for external stability*

The total active earth pressure P_a is

$$P_a = \frac{1}{2}\gamma H^2 KA = \frac{1}{2} \times 18 \times 6^2 \times 0.28 = 90.7 \text{ kN/m}$$

$$F_s = \frac{\text{Resisting moment } M_R}{\text{Driving moment } M_o} = \frac{W_1 l_1 + W_2 l_2 + W_3 l_3 + P_v l_4}{P_a\left(H/3\right)}$$

where $W_1 = 6 \times 2 \times 18 = 216$ kN and $l_1 = 2/2 = 1$ m,

$W_2 = (6 - 2.1) \times (3 - 2)(18) = 70.2$ kN, and $l_2 = 2.5$ m,

$W_3 = (6 - 4.2)(4 - 3)(18) = 32.4$ kN and $l_3 = 3.5$ m,

$$F_s = \frac{213 \times \left(1\right) + 70.2 \times \left(2.5\right) + 32.4\left(3.5\right)}{90.7 \times \left(2\right)} = 2.78 > 2 \text{ OK.}$$

Check for sliding

$$F_s = \frac{\text{Total resisting force } F_R}{\text{Total driving force } F_d}$$

$$F_R = (W_1 + W_2 + W_3)\tan \delta$$

$$= (216 + 70.2 + 32.4)\tan 25.5°$$

$$= 318.6 \times 0.477 = 152 \text{ kN}$$

$$F_d = P_a = 90.7 \text{ kN}$$

Hence, $F_s = \dfrac{152}{90.7} = 1.68 > 1.5 - \text{OK.}$

Check for a foundation failure

Consider the wall as a surface foundation with $D_f = 0$. Since the foundation soil is cohesionless, we may write

$$q_u = \frac{1}{2}\gamma B N_\gamma$$

Use Terzaghi's theory. For $\phi = 34°$, $N_\gamma = 38$, and $B = 2$ m

$$q_u = \frac{1}{2} \times 18 \times 2 \times 38 = 684 \text{ kN/m}^2$$

The actual load intensity on the base of the backfill

$$q \text{ (actual)} = 18 \times 6 + 10 = 118 \text{ kN/m}^2$$

$$F_s = \frac{684}{118} = 5.8 > 3 \text{ which is acceptable}$$

Example 16.4 (Koerner, 1999)

Design a 7 m high geogrid-reinforced wall when the reinforcement vertical maximum spacing must be 1.0 m. The coverage ratio is 0.80 (Refer to Fig. Ex. 16.4). Given : $T_u = 156$ kN/m, $C_r = 0.80$, $C_i = 0.75$. The other details are given in the figure.

Solution

Internal stability

From Eq. (16.6)

$$p_h = (\gamma z K_A + q_h) = \gamma z K_A + q_s K_A$$

$$K_A = \tan^2 (45° - \phi/2) = \tan^2 (45° - 32/2) = 0.31$$

$$p_h = (18 \times z \times 0.31) + (15 \times 0.31) = 5.58z + 4.65$$

1. For geogrid vertical spacing.

   Given $\quad\quad T_u = 156$ kN/m

   From Eq. (16.2) and Table 16.4, we have

$$T_a = T_u \left(\frac{1}{RF_{ID} \times RF_{CR} \times RF_{BD} \times RF_{CD}} \right) = 40 \text{ kN/m}$$

$$T_a = 156 \left(\frac{1}{1.2 \times 2.5 \times 1.3 \times 1.0} \right)$$

But use $T_{design} = 28.6$ kN/m with $F_s = 1.4$ on T_a

From Eq. (16.20)

$$T_{design} = \frac{h \, p_h}{C_r}$$

$$28.6 = h \, \frac{5.58 z + 4.65}{0.8}$$

or $\quad\quad\quad h = \frac{22.9}{5.58 z + 4.65}$

Maximum depth for $h = 1$ m is

$$1.0 = \frac{22.9}{5.58 z + 4.65} \quad \text{or } z = 3.27 \text{ m}$$

Maximum depth for $h = 0.5$ m

$$0.5 = \frac{22.9}{5.58 z + 4.65} \quad \text{or } z = 7.37 \text{ m}$$

The distribution of geogrid layers is shown in Fig. Ex. 16.4.

2. Embedment length of geogrid layers.

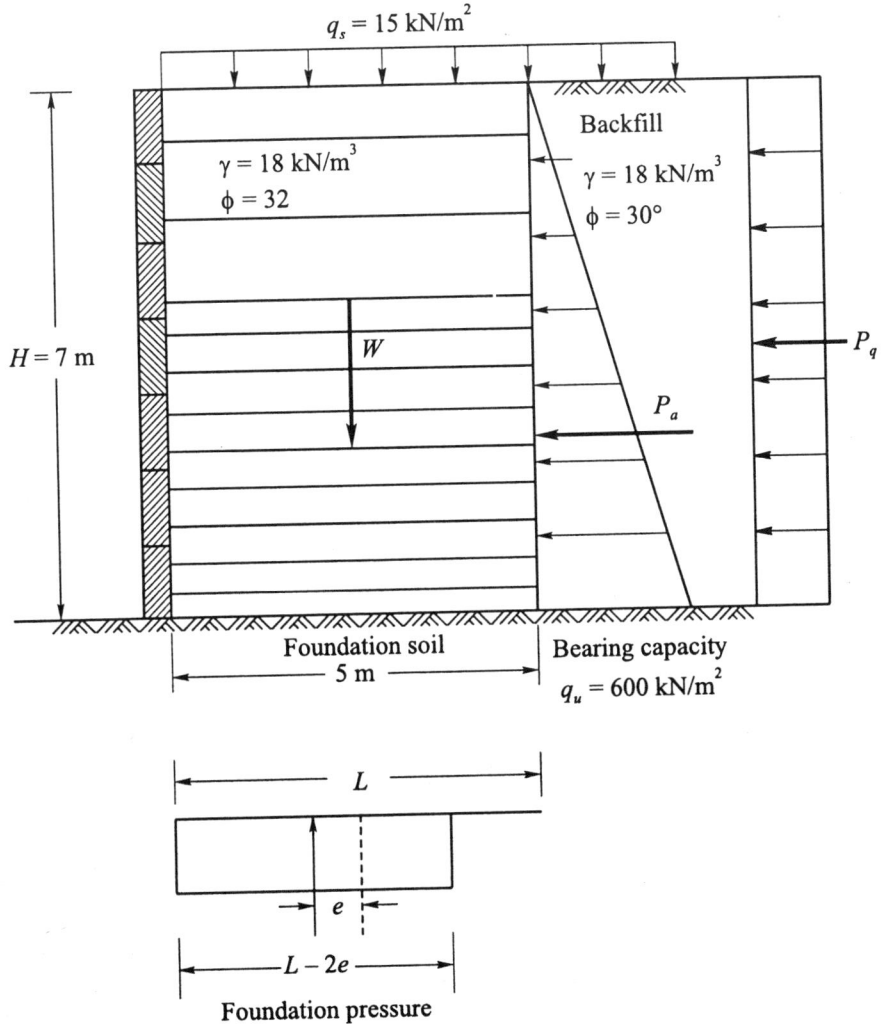

$q_s = 15 \text{ kN/m}^2$

Backfill

$\gamma = 18 \text{ kN/m}^3$
$\phi = 32$

$\gamma = 18 \text{ kN/m}^3$
$\phi = 30°$

$H = 7 \text{ m}$

W

P_q

P_a

Foundation soil
5 m

Bearing capacity
$q_u = 600 \text{ kN/m}^2$

L

e

$L - 2e$

Foundation pressure

Fig. Ex. 16.4

From Eqs (16.19) and (16.16)

$$2 \, C_1 \, C_r \, L_e \, p_o \tan \phi = T_H F_s = p_h \, h \, F_s$$

Substituting known values

$$2 \times 0.75 \times 0.8 \times (L_e) \times 18 \times (z) \tan 32° = h \, (5.58z + 4.65) \, 1.5$$

Simplifying $L_e = \dfrac{(0.62 \, z \; 0.516) \, h}{z}$

The equation for L_R is

$$L_R = (H - z) \tan (45° - \phi/2) = (7 - z) \tan (45° - 32/2)$$

$$= 3.88 - 0.554 \, (z)$$

From the above relationships the spacing of geogrid layers and their lengths are given below.

Layer no.	Depth (m)	Spacing h (m)	L_e (m)	L_e (min) (m)	L_R (m)	L (cal) (m)	L (required) (m)
1	0.75	0.75	0.98	1.0	3.46	4.46	5.0
2	1.75	1.00	0.92	1.0	2.91	3.91	5.0
3	2.75	1.00	0.81	1.0	2.36	3.36	5.0
4	3.25	0.50	0.39	1.0	2.08	3.08	5.0
5	3.75	0.50	0.38	1.0	1.80	2.80	5.0
6	4.25	0.50	0.37	1.0	1.52	2.52	5.0
7	4.75	0.50	0.36	1.0	1.25	2.25	5.0
8	5.25	0.50	0.36	1.0	0.97	1.97	5.0
9	5.75	0.50	0.36	1.0	0.69	1.69	5.0
10	6.25	0.50	0.35	1.0	0.42	1.42	5.0
11	6.75	0.50	0.35	1.0	0.14	1.14	5.0

External stability

Pressure distribution

$$P_a = \frac{1}{2}\gamma H^2 K_A = \frac{1}{2} \times 17 \times 7^2 \tan^2 (45° - 30/2) = 138.8 \text{ kN/m}$$

$$P_q = q_s K_A H = 15 \times 0.33 \times 7 = 34.7 \text{ kN/m}$$

Total ≈ 173.5 kN/m

1. *Check for sliding (neglecting effect of surcharge)*

$$F_R = W \text{ and } \delta = \gamma \times H \times L \tan 25° = 18 \times 7 \times 5.0 \times 0.47 = 293.8 \text{ kN/m}$$

$$P = P_a + P_q = 173.5 \text{ kN/m}$$

$$F_s = \frac{293.8}{173.5} = 1.69 > 1.5 \text{ OK}$$

2. *Check for overturning*

Resisting moment $M_R = W \times \dfrac{L}{2} = 18 \times 7 \times 5 \times \dfrac{5}{2} = 1575$ kN-m

Overturning moment

$$M_O = P_a \times \frac{H}{3} + P_q \times \frac{H}{2}$$

or $\qquad M_O = 138.8 \times \dfrac{7}{3} + 34.7 \times \dfrac{7}{2} = 445.3$ kN-m

$$F_s = \frac{1575}{445.3} = 3.54 > 2.0 \text{ OK}$$

3. *Check for bearing capacity*

Eccentricity $\qquad e = \dfrac{M_O}{W + q_s L} = \dfrac{445.3}{18 \times 7 \times 5 + 15 \times 5} = 0.63$

$$e = 0.63 < \frac{L}{6} = \frac{5}{6} = 0.83 \text{ OK}$$

Effective length = $L - 2e = 5 - 2 \times 0.63 = 3.74$ m

Bearing pressure = $(18 \times 7 + 15)\left(\dfrac{5}{3.74}\right) = 189$ kN/m^2

$$F_s = \frac{600}{189} = 3.17 > 3.0 \text{ OK}$$

16.8 EXAMPLES OF MEASURED LATERAL EARTH PRESSURES

Backfill Reinforced with Metal Strips

Laboratory tests were conducted on retaining walls with backfills reinforced with metal strips (Lee *et al*, 1973). The walls were built within a 30 in. × 48 in. × 2 in. wooden box. Skin elements were made from 0.012 in aluminum sheet. The strips (ties) used for the tests were 0.155 in wide and 0.0005 in thick aluminum foil. The backfill consisted of dry Ottawa No. 90 sand. The small walls built of these materials in the laboratory were constructed in much the same way as the larger walls in the field. Two different sand densities were used : loose, corresponding to a relative density, $D_r = 20\%$, and medium dense, corresponding to $D_r = 63\%$, and the corresponding angles of internal friction were 31° and 44° respectively. SR-4 strain gages were used on the ties to determine tensile stresses in the ties during the tests.

Examples of the type of earth pressure data obtained from two typical tests are shown in Fig. 16.10. Data in Fig. 16.10 (a) refer to a typical test in loose sand whereas data in 16.10 (b) refer to test in dense sand. The ties lengths were different for the two tests. For comparison, Rankine lateral earth pressure variation with depth is also shown. It may be seen from the curve that the measured values of the earth pressure follow closely the theoretical earth pressure variation up to two thirds of the wall height but fall off comparatively to lower values in the lower portion.

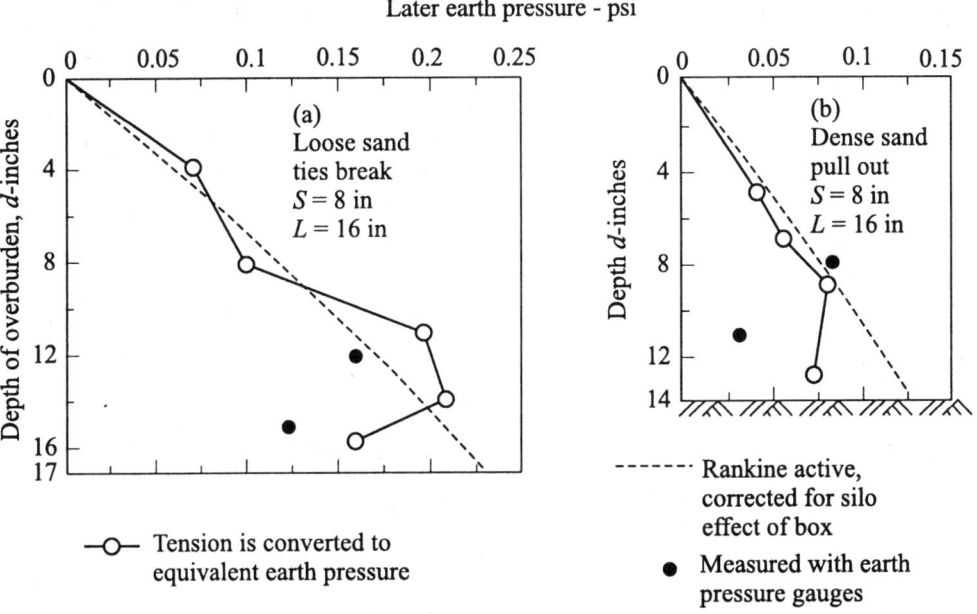

Fig. 16.10 Typical examples of measured lateral earth pressures just prior to wall failure
(1 in. = 25.4 mm; 1 psi = 6.89 kN/m^2) (Lee *et al*, 1973)

Field Study of Retaining Walls with Geogrid Reinforcement

Field studies of the behaviour of geotextile or geotextile or geogrid reinforced permanent wall studies are fewer in number. Berg *et al*, (1986) reported the field behaviour of two walls with geogrid reinforcement. One wall in Tucson, Arizona, 4.6 m high, used a cumulative reduction factor of 2.6 on ultimate strength for allowable strength T_a and a value of 1.5 as a global factor of safety. The second wall was in Lithonia, Georgia, and was 6 m high. It used the same factors and design method. Figure 16.11 presents the results for both the walls shortly after construction was complete. It may be noted that the horizontal pressures at various wall heights are overpredicted for each wall that is, the wall designs that were used appear to be quite conservative.

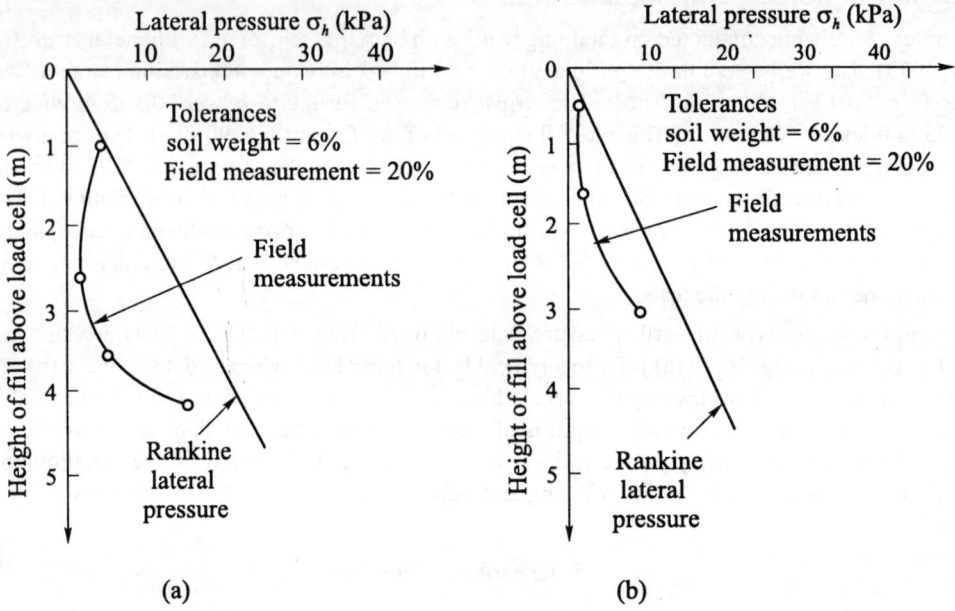

Fig. 16.11 Comparison of measured stresses to design stresses for two geogrid reinforced walls: (a) Results of Tucson, Arizona, wall, (b) results of Lithonia, Georgia, wall (Berg *et al*, 1986)

16.9 GROUND ANCHORS

Introduction

An anchor, whether it is placed in soil or rock is essentially a sub-structural member which transmits a tensile force from the main structure to the surrounding ground. The shear strength of the surrounding ground is used to resist this tensile force and in general, attempts are made to fasten the anchor to firm ground well away from the structure. An anchor may comprise a pile in tension, a dead-man, a gravity block, a rock bolt, or a specially installed tension unit. However, the most common anchor consists of a high strength steel tendon installed at the required inclination and to the required depth to resist the applied load in an efficient manner so that the tendon material is stressed to economic levels and the ground in which it is embedded is also realistically stressed. The details of an anchor in vertical and inclined directions are shown in Fig. 16.12 (a). The tensile force in the anchor is that force which is necessary for equilibrium between the anchor, the structure to which it is attached and the ground in which the anchor is embedded so that the movements of the structure and surrounding ground are kept to acceptable levels. The tendon is usually a high strength steel member (bar, wire

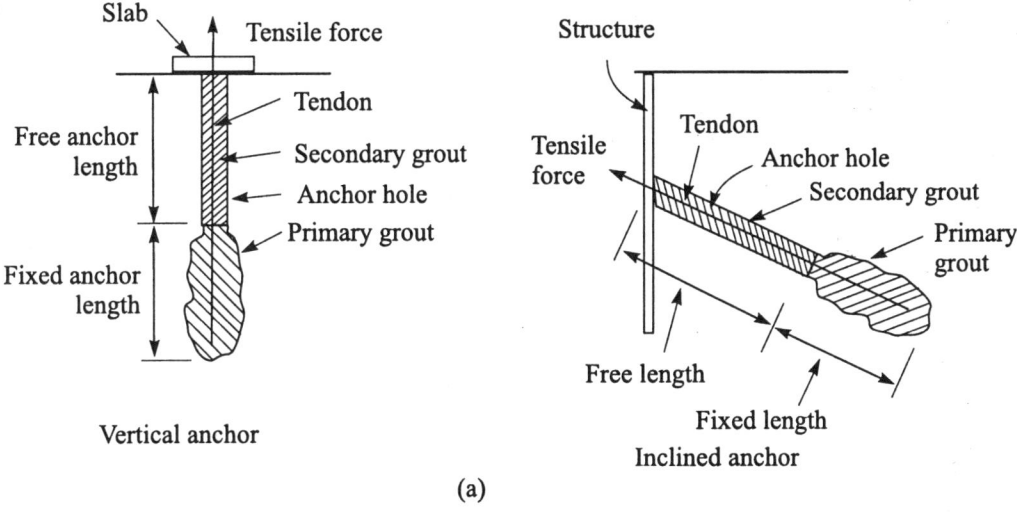

Vertical anchor

Inclined anchor

(a)

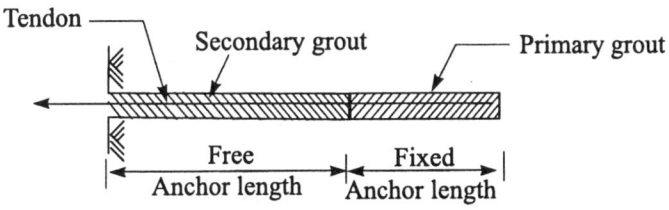

Type 1: Anchor cylinder filled with grout

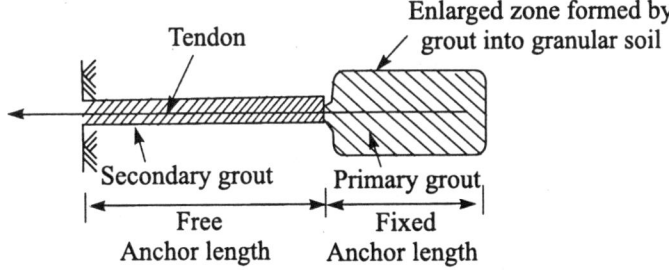

Type 2: Anchor cylinder enlarged by grout injected under high but controlled pressure

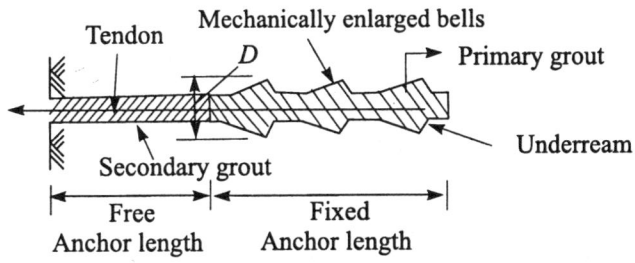

Type 3: Anchor cylinder mechanically enlarged bells

(b)

Fig. 16.12 Types of ground anchors: (a) Details of ground anchor used in vertical and inclined directions, (b) types of ground anchors

or strand) surrounded by cement grout or other fixing agent. The tendon has to be protected against corrosive agents because of the hostile environment in which it is located. The fixed anchor length is that part of the anchor tendon farthest away from the structure over which the tensile force is transmitted to the surrounding ground. The free anchor length is that part of tendon between the top of the fixed anchor length and the structure over which no tension force is transmitted to the surrounding ground. This is achieved by placing frictionless sleeves around the tendon. These sleeves also act as corrosion protection in the free anchor length.

The Various Situations for the Use of Anchors

The various situations where anchoring of the structures are required are given in Fig. 16.13. The anchors may be rock anchors or soil anchors. Figure 16.13 (a) is an example of a dry dock or hydraulic structure where there is uplight pressure on the base of the slab. To prevent flotation this uplift has to be balanced by the self weight of the structure or by the use of tension piles or anchors. The substrata below the base of the slab may be rock or soil, and accordingly *rock* or *soil anchors* are to be used.

Soil anchors are eminently suited for earth retaining walls where deep excavations are involved for the foundations of buildings as shown in Fig. 16.13 (b). The walls may be temporary or permanent. Anchors eliminate the incredibly complex bracing system and the excavation will be free of any obstruction.

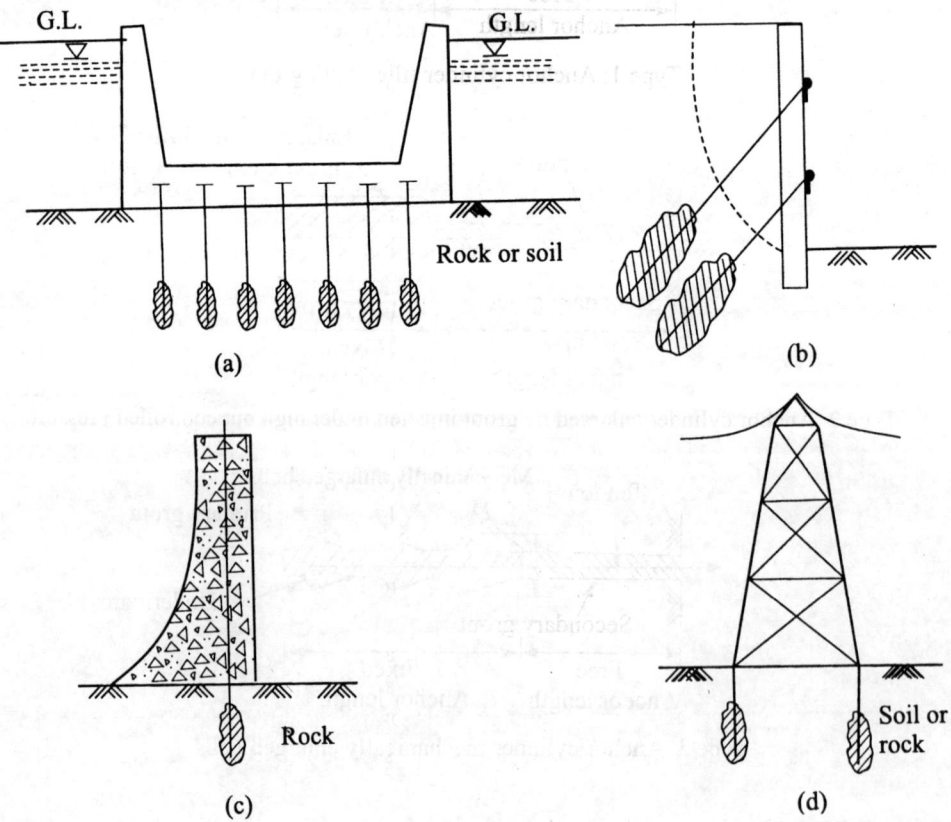

Fig. 16.13 Uses of anchors: (a) Dry rock, (b) deep excavation in sand, (c) concrete dam, and (d) transmission line tower

Rock anchors are the common features for the tying up of a concrete dam to the base to prevent overturning of the structure Fig. 16.13 (c). In the same way rock or soil anchors can be used on the foundations of tall structures such as chimneys, transmission line towers etc. [Figure 16.13 (d)].

Anchor Types

There are three basic types of ground anchor systems. These are shown schematically in Fig. 16.12 (b). The first type comprises a cylindrical hole filled with grout or other fixing agent, depending on the load to be mobilised. The second type is a cylinder which is enlarged by grout injected into the sides of the bore hole under high but controlled pressures in order to form an elongated bulb which acts as the anchorage. The third type is a cylinder with underreams one or more. These underreams or otherwise called a enlarged bells may be formed mechanically by making use of a special cutting device called under reamer. The diameter of the bulb may go up to 4 times the diameter of the bore-hole. The longitudinal spacing of the underreams may go upto three times the diameter of the bulb.

The soil anchors with underreams are normally used in clay soils.

The Various Parts of an Anchor System and the Method of Installation

The anchor system comprises of the following:

1. A drilled circular hole of 100 to 150 mm shaft diameter with or without underreams.
2. A central anchor rod called as tendon.
3. Protective sheath for the tendon in the free anchor length.
4. A primary grout and a secondary grout injected under pressure.
5. An anchor head.

The method of installation of an anchor is as follows:

1. Holes are drilled, usually by means of a drill rig or other reasonably stable piece of coring equipment of the desired diameter and the desired depth in rocky or soil strata. Suitable drilling equipments should be used for drilling in granular or cohesive soils. Holes are drilled either vertically or at an inclination as required with or without casing pipes.
2. The anchor is inserted centrally into the hole. Numerous types of anchors have been used such as structural steel shapes, smooth and deformed bars, steel cables etc.
3. A grout, usually cement, is pumped into the space between the hole and the tendon at a predetermined grouting pressure. The pressures vary for both the primary and secondary grout. Normally a sheath is provided around the tendon in the secondary grout area to protect it from corrosion. If a casing pipe is used, it will be withdrawn during grouting. The grout is allowed to cure.
4. A reaction and locking system is installed on the anchor at the ground surface or on a structural surface which is called as *anchor head.*
5. A tensile stress is applied to the anchor by a hydraulic jack to a value somewhat greater than the anchor's design load. This value generally varies from 1.25 to 2 times the design load. The anchors are also prestressed sometimes to control the longitudinal movement.
6. With the load applied to the anchor, the lockoff system is engaged and the applied tension from the jack is released.

Design of Rock and Soil Anchors

An anchor may fail in any one or more of the following modes:

1. By failure of the grout/tendon bond.
2. By failure of the ground/grout bond.
3. By failure within the ground mass.
4. By failure of the tendon steel or a component.

Rock Anchor

A rock anchor system is designed mostly empirically. Different grouts, grout pressures and rock systems result in different bond resistances, which when multiplied by the surface area of the anchor in question, give the theoretical anchor loads. Examples of approximate resistance values for different kinds of rock are given in Table 16.5. Obviously, the calculated load for an anchor cannot exceed the yield stress of the anchor steel itself, and often this is the limiting factor in the design. Field tests should always be conducted on selected anchors in different rock strata as well as for different construction methods.

Table 16.5 Rock-grout bond stresses for rock anchors
(after Little John *et al*, 1975)

Rock type	Bond Stresses between grout and rock lb/in^2
Sand Stone	120–250
Soft Shales	30–120
Slates and hard shales	120–200
Soft limestone	150–220
Hard limestone	300–400
Granite and basalt	250–800

Soil Anchors

The capacity of a soil anchor depends upon the type of anchor system used, the soil it is anchored into, the type and pressure of the grout etc. all which means the design of soil anchors must proceed empirically. However, some theoretical relationships have been established (Hanna 1982). For belled soil anchors of the type shown in Fig. 16.12 (b) we may write for cohesive soils as

$$P = \frac{\pi}{4} (D^2 - d^2) N_c c_u + \pi D l c_u \qquad (16.21)$$

where, P = ultimate soil anchor capacity,
  D = under-ream diameter,
  d = shaft diameter,
  l = the distance between adjacent under-ream,
  c_u = undrained shear strength of clay,
  N_c = bearing capacity factor = 9.

If the soil anchor contains more than two under reams, only the top two under reams are considered for shear failure along the periphery of the bulb.

In cohesive soil with a straight bore without any under ream, the ultimate anchor capacity may be expressed as

$$P = \pi d L S \qquad (16.22)$$

where, d = diameter of the primary grout bulb,

L = fixed anchor length,

S = soil to grout bond stress.

Approximate values of soil to grout bond stress for soil anchors are given in Table 16.6.

Table 16.6 Soil-to-grout bond stress for grouted soil anchors in cohesive soils
(after Weatherby, 1982)

Soil type	SPT value N	Bond stress between grout and soil $kips/ft^2$
Soft Clay	2–4	0.50–0.75
Silty Clay	3–6	0.50–1.00
Sandy Clay	3–6	0.75–1.00
Medium Clay	4–8	0.75–1.00
Firm Clay	6–12 `	1.00–1.50
Stiff Clay	8–15	1.00–2.00
Very Stiff Clay	15–30	1.50–2.50
Hard Clay	Over 30	1.50–4.0

For grouted soil anchors in granular soils, the theoretical equation for ultimate anchor capacity may be written as

$$P = \pi dL \, (c_a + \sigma \tan \delta) \tag{16.23}$$

where, d = anchor shaft diameter,

L = fixed anchor length,

c_a = adhesion of grout to soil,

δ = friction angle between grout and soil,

σ = normal stress in shear plane (related to, but somewhat less than, the grout pressure).

However, the ultimate anchor capacity as per Eq. (16.23) should be limited to the structural strength of the tendon.

16.10 PROBLEMS

16.1 A typical section of a wall with granular backfill reinforced with metal strips is given in Fig. Prob. 16.1. The following data are available.

$H = 6$ m, $b = 75$ mm, $t = 5$ mm, $f_y = 240$ MPa, F_s for steel = 1.75, F_s on soil friction = 1.5. The other data are given in the figure. Spacing : $h = 0.6$ m, and $s = 1$ m.

Required:

(a) Lengths of tie at varying depths

(b) Check for external stability

16.2 Solve the Prob. 16.1 with a uniform surcharge acting on the backfill surface. The intensity of surcharge is 20 kN/m^2.

16.3 Figure Prob. 16.3 shows a section of a MSE wall with geotextile reinforcement.

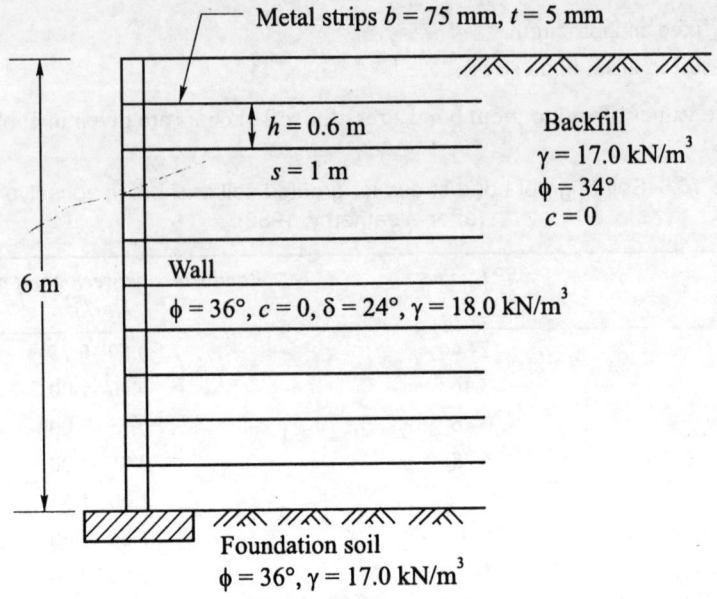

Fig. Prob. 16.1

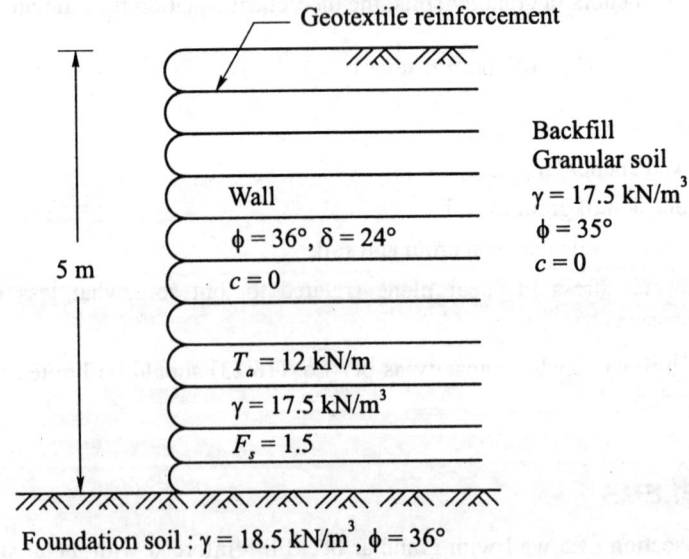

Fig. Prob. 16.3

Required:

(a) Spacing of the individual layers of geotextile.

(b) Length of geotextile in each layer.

(c) Check for external stability.

16.4 Design a 6 m high geogrid-reinforced wall (Fig. Prob. 16.4), where the reinforcement maximum spacing must be at 1.0 m. The coverage ratio $C_r = 0.8$ and the interaction coefficient $C_i = 0.75$, and $T_a = 26$ kN/m. (T_{design}).

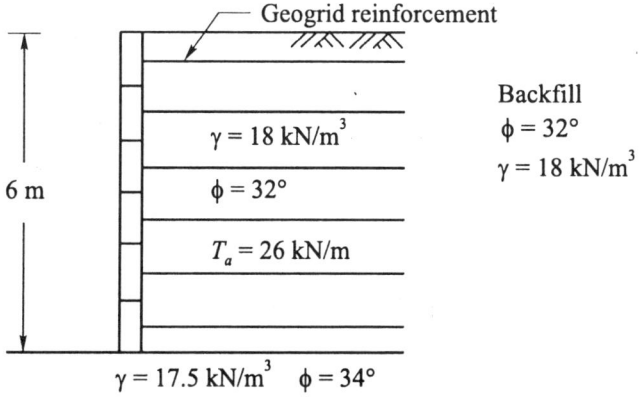

Fig. Prob. 16.4

Given: Reinforced soil properties : $\gamma = 18$ kN/m^3 $\phi = 32°$
        Foundation soil   : $\gamma = 17.5$ kN/m^3 $\phi = 34°$

Soil Improvement

17.1 INTRODUCTION

General practice is to use shallow foundations for the foundations of buildings and other such structures, if the soil close to the ground surface possesses sufficient bearing capacity. However, where the top soil is either loose or soft, the load from the superstructure has to be transferred to deeper firm strata. In such cases, pile or pier foundations are the obvious choice.

There is also a third method which may in some cases prove more economical than deep foundations or where the alternate method may become inevitable due to certain site and other environmental conditions. This third method comes under the heading *foundation soil improvement*. In the case of earth dams, there is no other alternative than compacting the remolded soil in layers to the required density and moisture content. The soil for the dam will be excavated at the adjoining areas and transported to the site. There are many methods by which the soil at the site can be improved. Soil improvement is frequently termed *soil stabilisation,* which in its broadest sense is alteration of any property of a soil to improve its engineering performance. Soil improvement

1. Increases shear strength
2. Reduces permeability
3. Reduces compressibility

The methods of soil improvement considered in this chapter are:

1. Mechanical compaction
2. Dynamic compaction
3. Vibroflotation
4. Preloading
5. Sand and stone columns
6. Use of admixtures
7. Injection of suitable grouts
8. Use of geotextiles

17.2 MECHANICAL COMPACTION

Mechanical compaction is the least expensive of the methods and is applicable in both cohesionless and cohesive soils. The procedure is to remove first the weak soil up to the depth required, and refill or replace the same in layers with compaction. If the soil excavated is cohesionless or a sand-silt clay mixture, the same can be replaced suitably in layers and compacted. If the soil excavated is a fine sand, silt or soft clay, it is not advisable to refill the same as these materials, even under compaction, may not give sufficient bearing capacity for the foundations. Sometimes it might be necessary to transport good soil to the site from a long distance. The cost of such a project has to be studied carefully before undertaking the same.

The compaction equipment to be used on a project depends upon the size of the project and the availability of the compacting equipment. In projects where excavation and replacement are confined to a narrow site, only tampers or surface vibrators may be used. On the other hand, if the whole area of the project is to be excavated and replaced in layers with compaction, suitable roller types of heavy equipment can be used. Cohesionless soils can be compacted by using vibratory rollers and cohesive soils by sheepsfoot rollers.

The control of field compaction is very important in order to obtain the desired soil properties. Compaction of a soil is measured in terms of the dry unit weight of the soil. The dry unit weight, γ_d, may be expressed as

$$\gamma_d = \frac{\gamma_t}{1+w} \tag{17.1}$$

where, γ_t = total unit weight,

   w = moisture content.

Factors Affecting Compaction

The factors affecting compaction are:

1. The moisture content
2. The compactive effort

The compactive effort is defined as the amount of energy imparted to the soil. With a soil of given moisture content, increasing the amount of compaction results in closer pacing of soil particles and increased dry unit weight. For a particular compactive effort, there is only one moisture content which gives the maximum dry unit weight. The moisture content that gives the maximum dry unit weight is called the *optimum moisture content*. If the compactive effort is increased, the maximum dry unit weight also increases, but the optimum moisture content decreases. If all the desired qualities of the material are to be achieved in the field, suitable procedures should be adopted to compact the earthfill. The compactive effort to the soil is imparted by mechanical rollers or any other compacting device. Whether the soil in the field has attained the required maximum dry unit weight can be determined by carrying out appropriate laboratory tests on the soil. The following tests are normally carried out in a laboratory.

1. Standard Proctor test (ASTM Designation D-698), and
2. Modified Proctor test (ASTM Designation D-1557)

17.3 LABORATORY TESTS ON COMPACTION

Standard Proctor Compaction Test

Proctor (1933) developed this test in connection with the construction of earth fill dams in California. The standard size of the apparatus used for the test is given in Fig. 17.1. Table 17.1 gives the

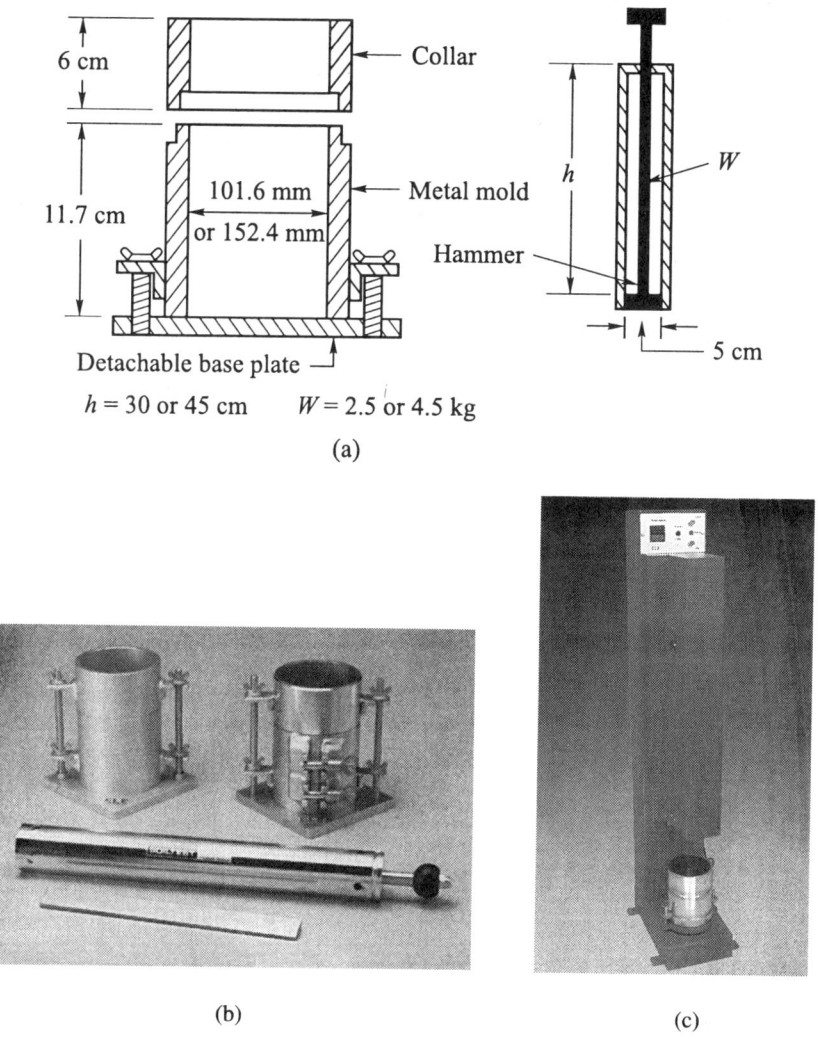

Fig. 17.1 Proctor compaction apparatus: (a) Diagrammatic sketch, (b) photograph of mold, and (c) automatic soil compactor (*Courtesy:* Soiltest)

standard specifications for conducting the test (ASTM designation D-698). Three alternative procedures are provided based the soil material used for the test.

Test Procedure

A soil at a selected water content is placed in layers into a mould of given dimensions (Table 17.1) and (Fig. 17.1), with each layer compacted by 25 or 56 blows of a 5.5 lb (2.5 kg) hammer dropped from a height of 12 in (305 mm), subjecting the soil to a total compactive effort of about 12,375 fl-lb/ft^3 (600 kNm/m^3). The resulting dry unit weight is determined. The procedure is repeated for a sufficient number of water contents to establish a relationship between the dry unit weight and the water content of the soil. This data, when, plotted, represents a curvilinear relationship known as the compaction curve or moisture-density curve. The values of water content and standard maximum dry unit weight are determined from the compaction curve as shown in Fig. 17.2.

Table 17.1 Specification for standard Proctor compaction test

Item	Procedure A	B	C
1. Diameter of mould	4 in. (101.6 mm)	4 in. (101.6 mm)	6 in. (152.4 mm)
2. Height of mould	4.584 in. (116.43 mm)	4.584 in. (116.43 mm)	4.584 in. (116.43 mm)
3. Volume of mould	0.0333 ft^3 (944 cm^3)	0.0333 ft^3 (944 cm^3)	0.075 ft^3 (2124 cm^3)
4. Weight of hammer	5.5 lb (2.5 kg)	5.5 lb (2.5 kg)	5.5 lb (2.5 kg)
5. Height of drop	12.0 in. (304.8 mm)	12.0 in. (304.8 mm)	12.0 in. (304.8 mm)
6. No. of layers	3	3	3
7. Blows per layer	25	25	56
8. Energy of compaction	12,375 ft-lb/ft^3 (600 kN-m/m^3)	12,375 ft-lb/ft^3 (600 kN-m/m^3)	12,375 ft-lb/ft^3 (600 kN-m/m^3)
9. Soil material	Passing No. 4 sieve (4.75 mm). May be used if 20% or less retained on No. 4 sieve	Passing No. 4 sieve (4.75 mm). Shall be used if 20% or more retained on No. 4 sieve and 20% or less retained on 3/8 in (9.5 mm) sieve	Passing No. 4 sieve (4.75 mm). Shall be used if 20% or more retained on 3/8 in. (9.5 mm) sieve and less than 30% retained on 3/4 in. (19 mm) sieve

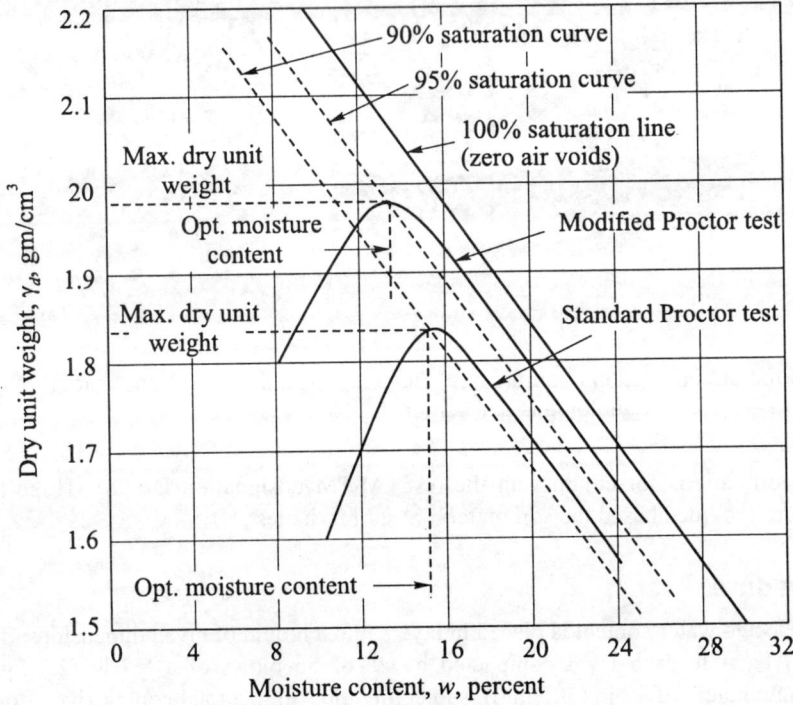

Fig. 17.2 Moisture-dry unit weight relationship

Modified Proctor Compaction Test (ASTM Designation: D 1557)

This test method covers laboratory compaction procedures used to determine the relationship between water content and dry unit weight of soils (compaction curve) compacted in a 4 in. or 6 in. diameter mould with a 10 lb (5 kg) hammer dropped from a height of 18 in. (457 mm) producing a compactive effort of 56,250 ft-lb/ft³ (2,700 kN-m/m³). As in the case of the standard test, the code provides three alternative procedures based on the soil material tested. The details of the procedures are given in Table 17.2.

Table 17.2 Specification for modified proctor compaction test

Item	Procedure		
	A	*B*	*C*
1. Mould diameter	4 in. (101.6 mm)	4 in. (101.6 mm)	6 in. (152.4 mm)
2. Volume of mould	0.0333 ft³ (944 cm³)	0.0333 ft³ (944 cm³)	0.075 ft³ (2124 cm³)
3. Weight of hammer	10 lb (4.54 kg)	10 lb (4.54 kg)	10 lb (4.54 kg)
4. Height of drop	18 in. (457.2 mm)	18 in. (457.2 mm)	18 in. (457.2 mm)
5. No. of layers	5	5	5
6. Blows/layers	25	25	56
7. Energy of compaction	56,250 ft-lb/ft³ (2700 kN-m/m³)	56,250 ft-lb/ft³ (2700 kN-m/m³)	56,250 ft-lb/ft³ (2700 kN-m/m³)
8. Soil material	May be used if 20% or less retained on No. 4 sieve.	Shall be used if 20% or more retained on No. 4 sieve and 20% or less retained on the 1/8 in. sieve	Shall be used if more-than 20% retained on 3/8 in. sieve and less-than 30% retained on the 3/4 in. sieve (19 mm)

Test Procedure

A soil at a selected water content is placed in five layers into a mold of given dimensions, with each layer compacted by 25 or 56 blows of a 10 lb (4.54 kg) hammer dropped from a height of 18 in. (457 mm) subjecting the soil to a total compactive effort of about 56,250 ft-lb/ft³ (2700 kN-m/m³). The resulting dry unit weight is determined. The procedure is repeated for a sufficient number of water contents to establish a relationship between the dry unit weight and the water content for the soil. This data, when plotted, represents a curvilinear relationship known as the compaction curve or moisture-dry unit weight curve. The value of the optimum water content and maximum dry unit weight are determined from the compaction curve as shown in Fig. 17.2.

Determination of Zero Air Voids Line

Referring to Fig. 17.3, we have

Degree of saturation, $\qquad S = \dfrac{V_w}{V_v}$

Water content, $\qquad w = \dfrac{W_w}{W_s}$

Dry weight of solids, $\qquad W_s = V_s G \gamma_w = G \gamma_w$ since $V_s = 1$

$$V_w = \frac{W_w}{\gamma_w} = \frac{wG\gamma_w}{\gamma_w} = wG$$

Therefore $\qquad S = \dfrac{wG}{V_v}$ $\hfill$ (17.2)

or
$$V_v = \frac{wG}{S}$$

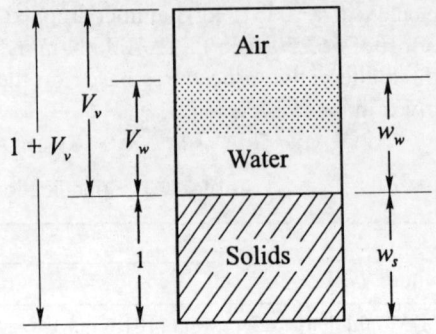

Dry unit weight
$$\gamma_d = \frac{W_s}{1 + V_v} = \frac{G\gamma_w}{1 + \frac{wG}{S}} \quad (17.3)$$

In Eq. (17.3), since G and γ_w, remain constant for a particular soil, the dry unit weight is a function of water content for any assumed degree of saturation. If $S = 1$, the soil is fully saturated (zero air voids). A curve giving the relationship between γ_d and w may be drawn by making use of Eq. (17.3) for $S = 1$. Curves may be drawn for different degrees of saturation such as 95, 90, 80 etc. percents. Figure 17.2 gives typical curves for different degrees of saturation along with moisture-dry unit weight curves obtained by different compactive efforts.

Fig. 17.3 Block diagram for determining zero air voids line

Example 17.1

A proctor compaction test was conducted on a soil sample, and the following observations were mad:

Water content, percent	7.7	11.5	14.6	17.5	19.7	21.2
Mass of wet soil, g	1739	1919	2081	2033	1986	1948

If the volume of the mold used was 950 cm³ and the specific gravity of soils grains was 2.65, make necessary calculations and draw, (*i*) compaction curve and (*ii*) 80% and 100% saturation lines.

Solution

From the known mass of the wet soil sample and volume of the mould, wet density or wet unit weight is obtained by the equations,

$$\rho_t \, (\text{g/cm}^3) = \frac{M}{V} = \frac{\text{Mass of wet sample in gm}}{950 \text{ cm}^3} \quad \text{or } \gamma_t = (\text{KN/m}^3) \approx 9.81 \times \rho_t \, (\text{g/cm}^3)$$

Then from the wet density and corresponding moisture content, the dry density or dry unit weight is obtained from,

$$\rho_d = \frac{\rho_t}{1 + w} \text{ or } \gamma_d = \frac{\gamma_t}{1 + w}$$

Thus for each observation, the wet density and then the dry density are calculated and tabulated as follows:

Water content, percent	7.7	11.5	14.6	17.5	19.7	21.2
Mass of wet sample, g	1739	1919	2081	2033	1986	1984
Wet density, g/cm³	1.83	2.02	2.19	2.14	2.09	2.05
Dry density, g/cm³	1.70	1.81	1.91	1.82	1.75	1.69
Dry unit weight kN/m³	16.7	17.8	18.7	17.9	17.2	16.6

Hence, the compaction curve, which is a plot between the dry unit weight and moisture content can be plotted as shown in the Fig. Ex. 17.1. The curve gives,

Maximum dry unit weight, $MDD = 18.7$ kN/m^3

Optimum moisture content, $OMC = 14.7$ percent

For drawing saturation lines, make use of Eq. (17.3), viz.

$$\gamma_d = \frac{G\gamma_w}{1 + \frac{wG}{S}}$$

where, $G = 2.65$, given, S = degree of saturation 80% and 100%, w = water content, may be assumed as 8%, 12%, 16%, 20% and 24%.

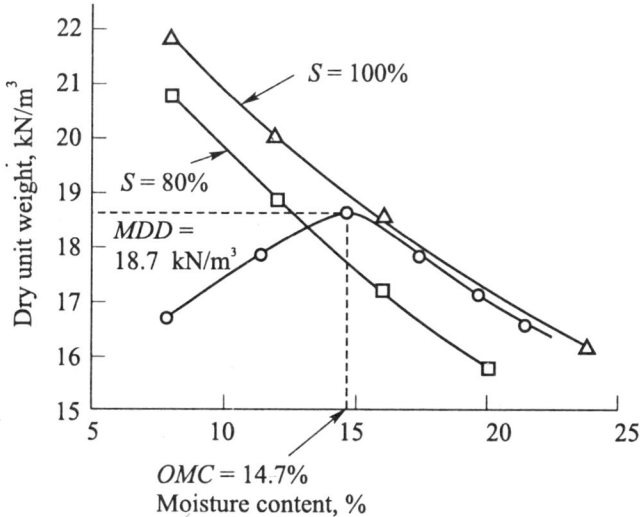

Fig. Ex. 17.1

Hence for each value of saturation and water content, find γ_d, and tabulate:

Water content, percentage	8	12	16	20	24
γ_d kN/m^3 for $S = 100\%$	21.45	19.73	18.26	17.0	15.69
γ_d kN/m^3 for $S = 80\%$	20.55	18.61	17.00	15.64	14.49

With these calculations, saturation lines for 100% and 80% are plotted, as shown in the Fig. Ex. 17.1.

Also the saturation, corresponding to $MDD = 18.7$ kN/m^3 and $OMC = 14.7\%$ can be calculated as,

$$18.7 = \gamma_d = \frac{G\gamma_w}{1 + \frac{wG}{S}} = \frac{2.65 \times 9.81}{1 + \frac{0.147 \times 2.65}{S}}$$

which gives $S = 99.7\%$

Example 17.2

A small cylinder having volume of 600 cm^3 is pressed into a recently compacted fill of embankment filling the cylinder. The mass of the soil in the cylinder is 1100 g. The dry mass of the soil is 910 g.

Determine the void ratio and the saturation of the soil. Take the specific gravity of the soil grains as 2.7.

Solution

Wet density of soil $\qquad \rho_t = \dfrac{1100}{600} = 1.83 \text{ g/cm}^3 \text{ or } \gamma_t = 17.99 \text{ kN/m}^3$

Water content, $\qquad w = \dfrac{1100 - 910}{910} = \dfrac{190}{910} = 0.209 = 20.9\%$

Dry unit weight, $\qquad \gamma_d = \dfrac{\gamma_t}{1 + w} = \dfrac{17.99}{1 + 0.209} = 14.88 \text{ kN/m}^3$

From, $\qquad \gamma_d = \dfrac{G\gamma_w}{1 + e} \text{ we have } e = \dfrac{G\gamma_w}{\gamma_d} - 1$

Substituting and simplifying

$$e = \dfrac{2.7 \times 9.81}{14.88} - 1 = 0.78$$

From, $\qquad S_e = wG \text{ or } S = \dfrac{wG}{e} = \dfrac{0.209 \times 2.7 \times 100}{0.78} = 72.35\%$

17.4 EFFECT OF COMPACTION ON ENGINEERING BEHAVIOUR

Effect of Moisture Content on Dry Density

The moisture content affects the behaviour of the soil. When the moisture content is low, the soil is stiff and difficult to compress. Thus, low unit weight and high air contents are obtained (Fig. 17.2). As the moisture content increases, the water acts as a lubricant, causing the soil to soften and become more workable. This results in a denser mass, higher unit weights and lower air contents under compaction. The water and air combination tend to keep the particles apart with further compaction, and prevent any appreciable decrease in the air content of the total voids, however, continue to increase with moisture content and hence the dry unit weight of the soil falls.

To the right of the peak of the dry unit weight-moisture content curve (Fig. 17.2), lies the saturation line. The theoretical curve relating dry density with moisture content with no air voids is approached but never reached since it is not possible to expel by compaction all the air entrapped in the voids of the soil.

Effect of Compactive Effort on Dry Unit Weight

For all types of soil with all methods of compaction, increasing the amounts of compaction, that is, the energy applied per unit weight of soil, results in an increase in the maximum dry unit weight and a corresponding decrease in the optimum moisture content as can be seen in Fig. 17.4.

The details of compaction are given in the following table.

No.	Layers	Blows per Layer	Hammer weight (lb)	Hammer drop (in.)
1	5	55	10	18 (mod. AASHTO)
2	5	26	10	18
3	5	12	10	18 (std. AASHTO)
4	3	25	5½	12

Note: 6 in. diameter mould used for all tests

Shear Strength of Compacted Soil

The shear strength of a soil increases with the amount of compaction applied. The more the soil is compacted, the greater is the value of cohesion and the angle of shearing resistance. Comparing the shearing strength with the moisture content for a given degree of compaction, it is found that the greatest shear strength is attained at a moisture content lower than the optimum moisture content for maximum dry unit weight. Figure 17.5 shows the relationship between shear strength and moisture-dry unit weight curves for a sandy clay soil. It might be inferred from this that it would be an advantage to carry out compaction at the lower value of the moisture content. Experiments, however, have indicated that soils compacted in this way tend to take up moisture and become saturated with a consequent loss of strength.

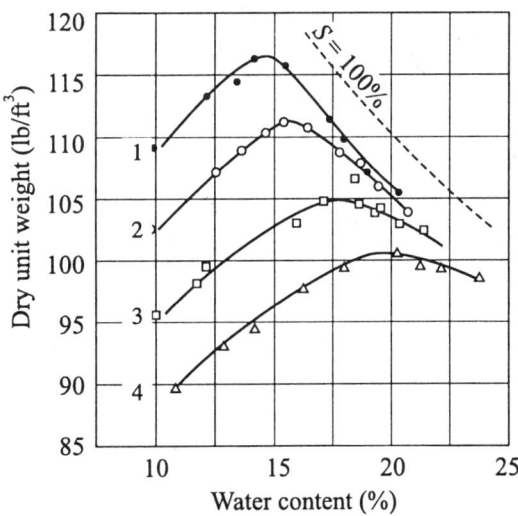

Fig. 17.4 Dynamic compaction curves for a silty clay (from Turnbull, 1950)

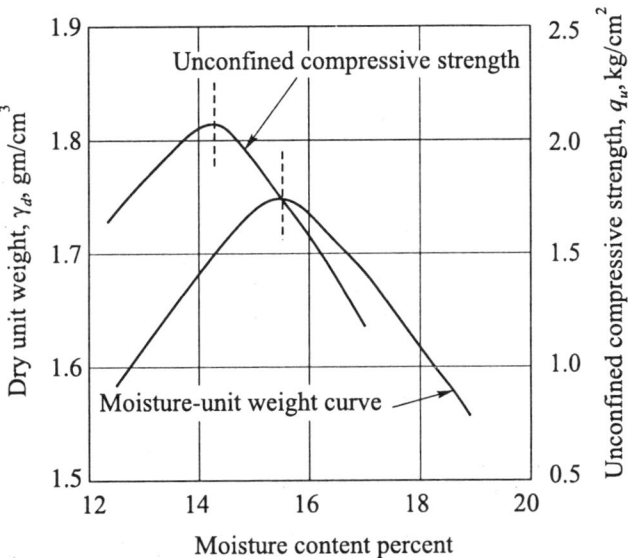

Fig. 17.5 Relationship between compaction and shear strength curves

Effect of Compaction on Structure

Figure 17.6 illustrates the effects of compaction on clay structure (Lambe, 198a). Structure (or fabric) is the term used to describe the arrangement of soil particles and the electric forces between adjacent particles.

The effects of compaction conditions on soil Structure, and thus on the engineering behaviour of the soil, vary considerably with soil type and the actual conditions under which the behaviour is determined.

At low water content, w_A in Fig. 17.6, the repulsive forces between particles are smaller than the attractive forces, and as such the particles flocculate in a disorderly array. As the water content increases beyond w_A, the repulsion between particles increases, permitting the particles to disperse, making particles arrange themselves in an orderly way. Beyond w_B the degree of particle parallelism increases, but the density decreases. Increasing the compactive effort at any given water content increases the orientation of particles and therefore gives a higher density as indicated in Fig. 17.6.

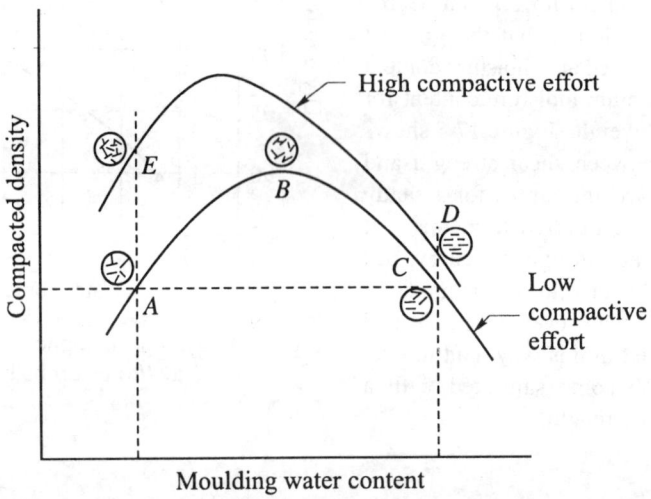

Fig. 17.6 Effects of compaction on structure (from Lambe, 1958a)

Effect of Compaction on Permeability

Figure 17.7 depicts the effect of compaction on the permeability of a soil. The figure shows the typical marked decrease in permeability that accompanies an increase in moulding water content on the dry side of the optimum water content. A minimum permeability occurs at water contents slightly above optimum moisture content (Lambe, 1958a), after which a slight increase in permeability occurs. Increasing the compactive effort decreases the permeability of the soil.

Effect of Compaction on Compressibility

Figure 17.8 illustrates the difference in compaction characteristics between two saturated clay samples at the same density, one compacted on the dry side of optimum and one compacted on the wet side (Lambe, 1958b). At low stresses the sample compacted on the wet side is more compressible than the one compacted on the dry side. However, at high applied stresses the sample compacted on the dryside is more compressible than the sample compacted on the wet side.

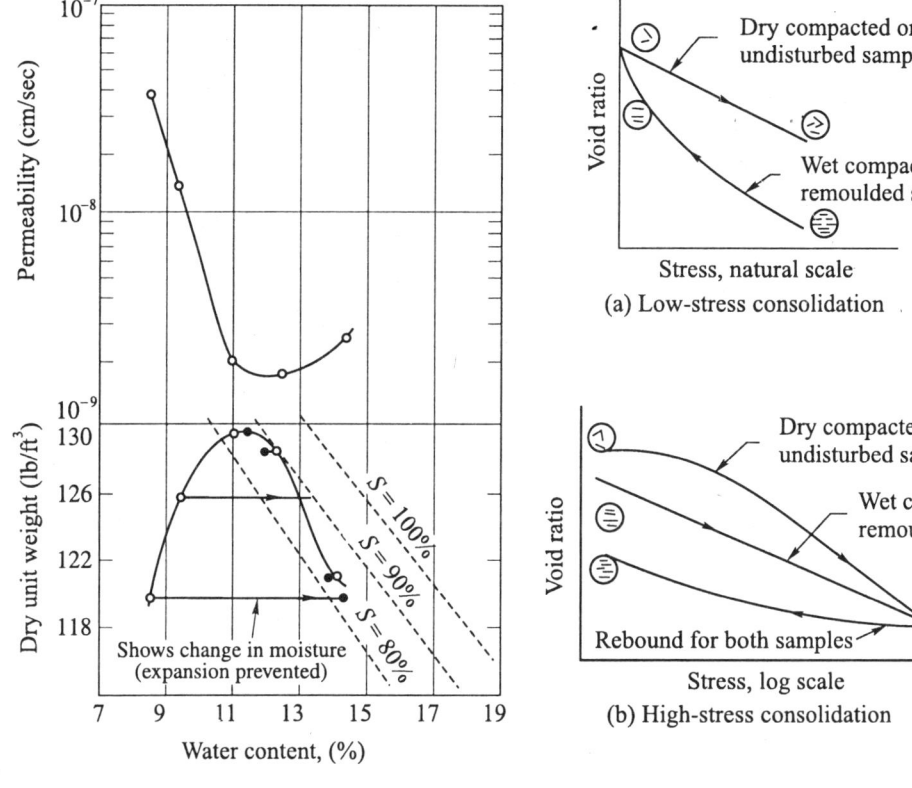

Fig. 17.7 Compaction-permeability tests on Siburua clay (from Lambe, 1962)

Fig. 17.8 Effect of one-dimensional compression on structure (Lambe, 1958b)

17.5 FIELD COMPACTION AND CONTROL

The necessary compaction of subgrades of roads, earth fills, and embankments may be obtained by mechanical means. The equipment that are normally used for compaction consists of

1. Smooth wheel rollers
2. Rubber tired rollers
3. Sheepsfoot rollers
4. Vibratory rollers

Laboratory tests on the soil to be used for construction in the field indicate the maximum dry density that can be reached and the corresponding optimum moisture content under specified methods of compaction. The field compaction method should be so adjusted as to translate laboratory condition into practice as far as possible. The two important factors that are necessary to achieve the objectives in the field are

1. The adjustment of the natural moisture content in the soil to the value at which the field compaction is most effective.
2. The provision of compacting equipment suitable for the work at the site.

The equipment used for compaction are briefly described below:

Smooth Wheel Roller

There are two types of smooth wheel rollers. One type has two large wheels, one in the rear and a similar single drum in the front. This type is generally used for compacting base courses. The equipment weighs from 50 to 125 kN (Fig. 17.9). The other type is the tandem roller normally used for compacting paving mixtures. This roller has large single drums in the front and rear and the weights of the rollers range from 10 to 200 kN.

Fig. 17.9 Smooth wheel roller (*Courtesy:* Caterpillar, USA

Rubber Tired Roller

The maximum weight of this roller may reach 2000 kN. The smaller rollers usually have 9 to 11 tires on two axles with the tires spaced so that a complete coverage is obtained with each pass. The tire loads of the smaller roller are in the range of 7.5 kN and the tire pressures in the order of 200 kN/m^2. The larger rollers have tire loads ranging from 100 to 500 kN per tire, and tire pressures range from 400 to 1000 kN/m^2.

Sheepsfoot Roller

Sheepsfoot rollers are available in drum widths ranging from 120 to 180 cm and in drum diameters ranging from 90 to 180 cm. projections like a sheepsfoot are fixed on the drums. The lengths of these projections range from 17.5 cm to 23 cm. The contact area of the tamping foot ranges from 35 to 56 sq. cm. The loaded weight per drum ranges from about 30 kN for the smaller sizes to 130 kN for the larger sizes (Fig. 17.10).

Fig. 17.10 Sheepsfoot roller (*Courtesy:* Vibromax America Inc.)

Vibratory Roller

The weights of vibratory rollers range from 120 to 300 kN. In some units vibration is produced by weights placed eccentrically on a rotating shaft in such a manner that the forces produced by the rotating weights are essentially in a vertical direction. Vibratory rollers are effective for compacting granular soils (Fig. 17.11).

Selection of Equipment for Compaction in the Field

The choice of a roller for a given job depends on the type of soil to be compacted and percentage of compaction to be obtained. The types of rollers that are recommended for the soils normally met are:

Type of soil	Type of roller recommended
Cohesive soil	Sheepsfoot roller, or Rubber tired roller
Cohesionless soil	Rubber-tired roller or Vibratory roller.

Fig. 17.11 Vibratory drum on smooth wheel roller (*Courtesy:* Caterpillar, USA)

Method of Compaction

The first approach to the problem of compaction is to select suitable equipment. If the compaction is required for an earth dam, the number of passes of the roller required to compact the given soil to the required density at the optimum moisture content has to be determined by conducting a field trial test as follows:

The soil is well mixed with water which would give the optimum water content as determined in the laboratory. It is then spread out in a layer. The thickness of the layer normally varies from 15 to 22.5 cm. The number of passes required to obtain the specified density has to be found by determining the density of the compacted material after every definite number of passes. The density may be checked for different thickness in the layer. The suitable thickness of the layer and the number of passes required to obtain the required density will have to be determined.

In cohesive soils, densities of the order of 95 percent of standard Proctor can be obtained with practically any of the rollers and tampers; however, vibrators are not effective in cohesive soils. Where high densities are required in cohesive soils in the order of 95 percent of modified Proctor, rubber tired rollers with tire load in the order of 100 kN and tire pressure in the order of 600 kN/m^2 are effective.

In cohesionless sands and gravels, vibrating type equipment is effective in producing densities up to 100 percent of modified Proctor. Where densities are needed in excess of 100 percent of modified Proctor such as for base courses for heavy duty air fields and highways, rubber tired rollers with tire loads of 130 kN and above and tire pressure of 1000 kN/m^2 can be used to produce densities up to 103 to 104 percent of modified Proctor.

Field Control of Compaction
Methods of control of density

The compaction of soil in the field must be such as to obtain the desired unit weight at the optimum moisture content. The field engineer has therefore to make periodic checks to see whether the compaction is giving desired results. The procedure of checking involves:

1. Measurement of the dry unit weight, and
2. Measurement of the moisture content.

There are many methods for determining the dry unit weight and/or moisture content of the soil in-situ. The important methods are:

1. Sand cone method,

2. Rubber balloon method,
3. Nuclear method, and
4. Proctor needle method.

Sand Cone Method (ASTM Designation D-1556)

The sand for the sand cone method consists of a sand pouring jar shown in Fig. 17.12. The jar contains uniformly graded clean and dry sand. A hole about 10 cm in diameter is made in the soil to be tested up to the depth required. The weight of soil removed from the holde is determined and its water content is also determined. Sand is run into the hole from the jar by opening the valve above the cone until the hole and the cone below the valve is completely filled. The valve is closed. The jar is calibrated to give the weight of the sand that just fills the hole, that is, the difference in weight of the jar before and after filling the hole after allowing for the weight of sand contained in the cone is the weight of sand poured into the hole.

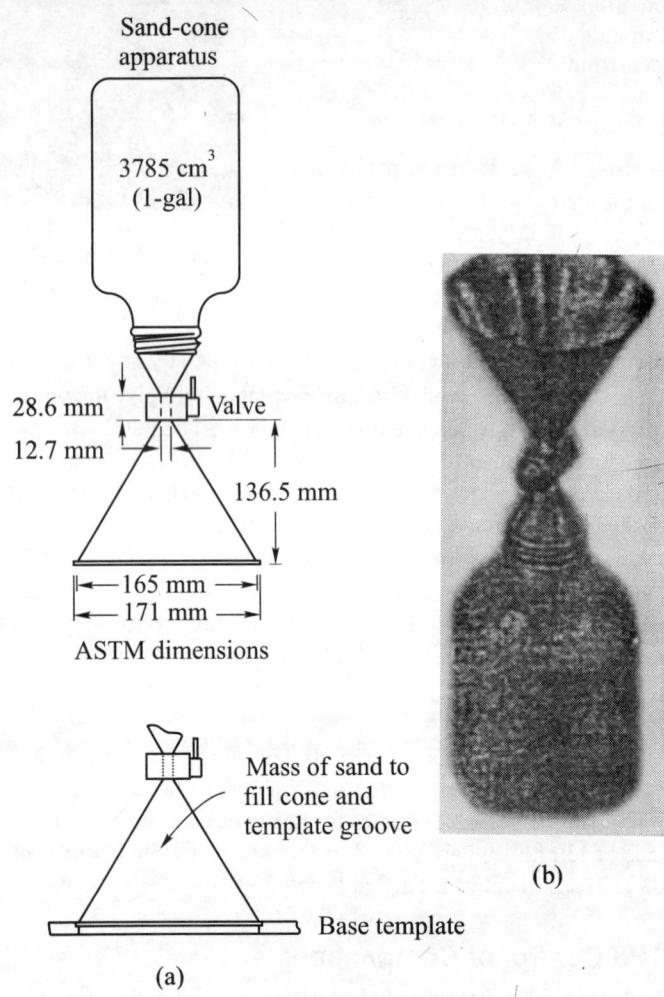

Let W_s = weight of dry sand poured into the hole

G = specific gravity of sand particles

W = weight of soil taken out of the hole

w = water content of the soil

Volume of sand in the hole = volume of soil taken out of the hole

Fig. 17.12 Sand-cone apparatus: (a) Schematic diagram, and (b) photograph

That is,

$$V = \frac{W_s}{G\gamma_w} \qquad (17.4a)$$

The bulk unit weight of soil,

$$\gamma_t = \frac{W}{V} = \frac{WG\gamma_w}{W_s} \qquad (17.4b)$$

The dry unit weight of soil,

$$\gamma_d = \frac{\gamma_t}{1 + w}$$

Rubber Balloon Method (ASTM Designation: D 2167)

The volume of an excavated hole in a given soil is determined using a liquid-filled calibrated cylinder for filling a thin rubber membrane. This membrane is displaced to fill the hole. The in-place unit weight is determined by dividing the wet mass of the soil removed by the volume of the hole. The water (moisture) content and the in-place unit weight are used to calculate the in-place dry unit weight. The volume is read directly on the graduated cylinder. Figure 17.13 shows the equipment.

Nuclear Method

The modern instrument for rapid and precise field measurement of moisture content and unit weight is the Nuclear density/Moisture metre. The measurements made by the metre are non-destructive

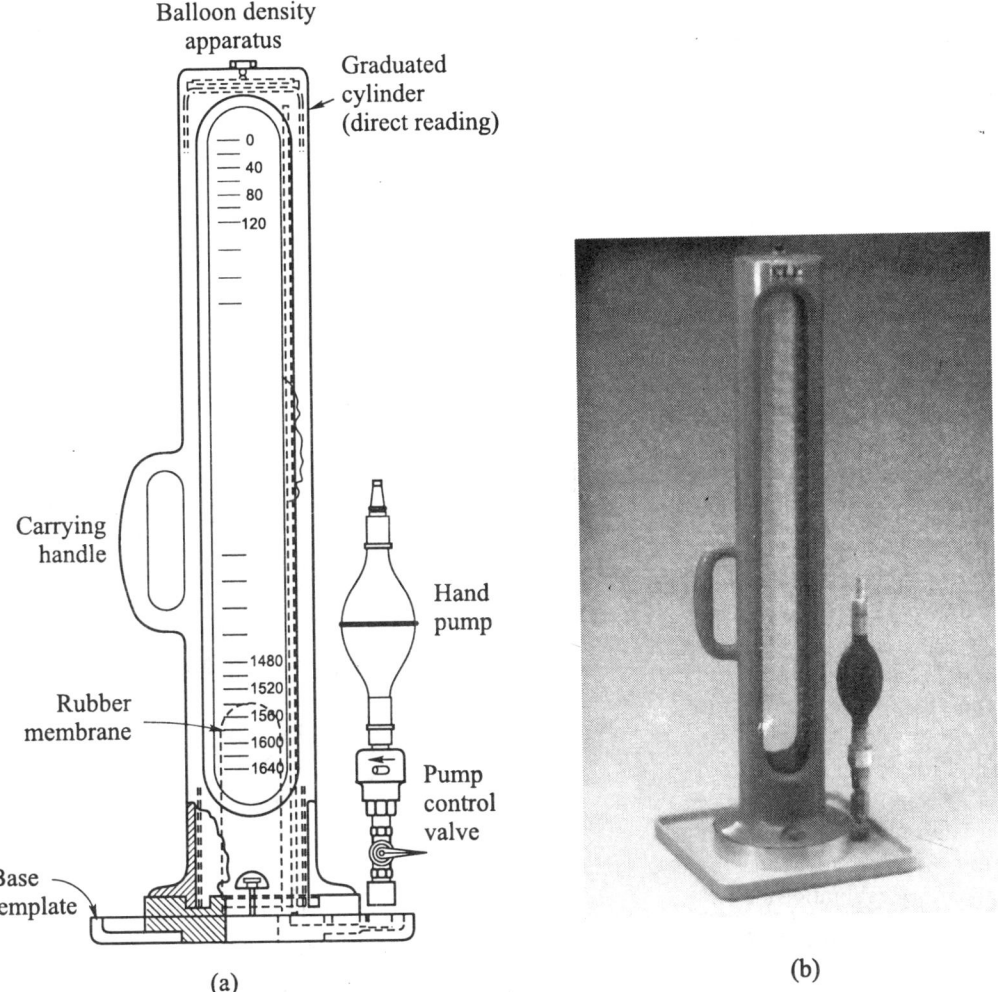

(a)   (b)

Fig. 17.13 Rubber balloon density apparatus: (a) Diagrammatic sketch, and (b) a photograph

and require no physical or chemical processing of the material being tested. The instrument may be used either in drilled holes or on the surface of the ground. The main advantage of this equipment is that a single operator can obtain an immediate and accurate determination of the *in-situ* dry density and moisture content.

Proctor Needle Method

The proctor needle method is one of the methods developed for rapid determination of moisture contents of soils *in-situ*. It consists of a needle attached to a spring loaded plunger, the stem of which is calibrated to read the penetration resis-tance of the needle in lbs/in^2 or kg/cm^2. The needle is supplied with a series of bearing points so that a wide range of penetration resistances can be measured. The bearing areas that are normally provided are 0.05, 0.1, 0.25, 0.50 and 1.0 sq. in. The apparatus is shown in Fig. 17.14. A proctor penetrometer set is shown in Fig. 17.15 (ASTMD-1558).

Laboratory penetration resistance curve

A suitable needle point is selected for a soil to be compacted. If the soil is cohesive, a needle with a larger bearing area is selected. For cohesionless soils, a needle with a smaller bearing area will be sufficient. The soil sample is compacted in the mould.

The penetrometer with a known bearing area of the tip is forced with a gradual uniform push at a rate of about 1.25 cm per sec to a depth of 7.5 cm into the soil. The penetration resistance in kg/cm^2 is read off the calibrated shaft of the penetrometer. The water content of the soil and the corresponding dry density are also determined. The procedure is repeated for the same soil compacted at different moisture contents. Curve giving the moisture-density and penetration resistance-moisture content relationship are plotted as shown in Fig. 17.16.

To determine the moisture content in the field, a sample of the wet soil is compacted into the mold under the same conditions as used in the laboratory for obtaining the penetration resistance curve. The Proctor needle is forced into the soil and its resistance is determined. The moisture content is read from the laboratory calibration curve.

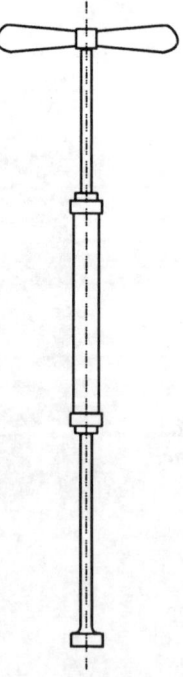

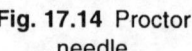

Fig. 17.14 Proctor needle

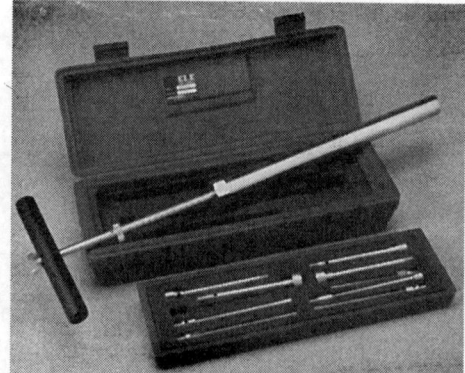

Fig. 17.15 Proctor penetrometer set (*Courtesy:* Soiltest)

This method is quite rapid, and is sufficiently accurate for fine-grained cohesive soils. However, the presence of gravel or small stones in the soil makes the reading on the Proctor needle less reliable. It is not very accurate in cohesionless sands.

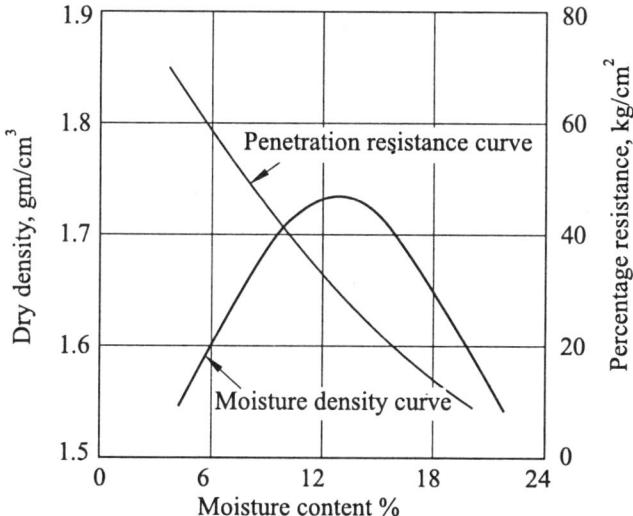

Fig. 17.16 Field method of determining water content by proctor needle method

Example 17.3

The following observations were recorded when a sand cone test was conducted for finding the unit weight of a natural soil:

Total density of sand used in the test = 1.4 g/cm^2.

Mass of the soil excavated from hole = 950 g.

Mass of the sand filling the hole = 700 g.

Water content of the natural soil = 15 percent.

Specific gravity of the soil grains = 2.7.

Calculate: (*i*) the wet unit weight, (*ii*) the dry unit weight, (*iii*) the void ratio, and (*iv*) the degree of saturation.

Solution

Volume of the hole
$$V_p = \frac{700}{1.4} = 500 \text{ cm}^3$$

Wet density of natural soil, $\rho_t = \dfrac{950}{500} = 1.9 \text{ g/cm}^3$ or $\gamma_t = 18.64 \text{ kN/m}^3$

Dry density
$$\rho_d = \frac{\rho_t}{1+w} = \frac{1.9}{1+0.15} = 1.65 \text{ g/cm}^3$$

$$\rho_d = \frac{G}{1+e}\ \rho_w = \frac{2.7}{1+e} \times 1 \text{ or } 1.65 + 1.65e = 2.7$$

Therefore
$$e = \frac{2.7 - 1.65}{1.65} = 0.64$$

and
$$S = \frac{wG}{e} = \frac{0.15 \times 2.7 \times 100}{0.64} = 63\%$$

Example 17.4

Old records of a soil compacted in the past gave compaction water content of 15% and saturation 85%. What might be the dry density of the soil?

Solution

The specific gravity of the soil grains is not known, but as it varies in a small range of 2.6 to 2.7, it can suitably be assumed. An average value of 2.65 is considered here.

Hence
$$e = \frac{wG}{S} = \frac{0.15 \times 2.65}{0.85} = 0.47$$

and dry density $\rho_d = \frac{G}{1+e} \rho_w = \frac{2.65}{1+0.47} \times 1 = 1.8 \text{ g/cm}^3$ or dry unit weight = 17.66 kN/m^3.

Example 17.5

The following data are available in connection with the construction of an embankment:

(a) soil from borrow pit: Natural density = 1.75 Mg/m^3, Natural water content = 12%.
(b) soil after compaction: density = 2 Mg/m^3, water content = 18%.

For every 100 m^3 of compacted soil of the embankment, estimate:

(i) the quantity of soil to be excavated from the borrow pit, and
(ii) the amount of water to be added.

Note: 1 g/cm^3 = 1000 kg/m^3 = $10^3 \times 10^3$ g/m^3 = 1 Mg/m^3 where Mg stands for Megagram = 10^6 g.

Solution

The soil is compacted in the embankment with density of 2 Mg/m^3 and with 18% water content.

Hence, for 100 m^3 of soil.

Mass of compacted wet soil = $100 \times 2.0 = 200$ Mg = 200×10^3 kg

Mass of compacted dry soil = $\frac{200}{1+w} = \frac{200}{1+0.18} = 169.5$ Mg = 169.5×10^3 kg

Mass of wet soil to be excavated = $169.5 (1+w) = 169.5 (1+0.12) = 189.84$ Mg

Volume of the wet soil to be excavated = $\frac{189.84}{1.75} = 108.48$ m^3

Now, in the natural state, the moisture present in 169.5×10^3 kg of dry soil would be $169.5 \times 10^3 \times 0.12 = 20.34 \times 10^3$ kg and the moisture which the soil will possess during compaction is $169.5 \times 10^3 \times 0.18 = 30.51 \times 10^3$ kg.

Hence, mass of water to be added for every 100 m^3 of compacted soil is $(30.51 - 20.34) 10^3 = 10.17 \times 10^3$ kg.

Example 17.6

A sample of sol compacted according to the standard Proctor test has a density of 2.06 g/cm^3 at 100% compaction and at an optimum water content of 14%. What is the dry unit weight? What is the dry unit weight at zero air-voids? If the voids become filled with water what would be the saturated unit weight? Assume $G = 2.67$.

Solution

Refer to Fig. Ex. 17.6. Assume V = total volume = 1 cm^3. Since water content is 14% we may write,

$$\frac{M_w}{M_s} = 0.14 \text{ or } M_w = 0.14\, M_s$$

and since, $M_w + M_s = 2.06$ g

$$0.14\, M_s + M_s = 1.14\, M_s = 2.06$$

or

$$M_s = \frac{2.06}{1.14} = 1.807 \text{ g}$$

$$M_w = 0.14 \times 1.807 = 0.253 \text{ g}$$

By definition, $\rho_d = \dfrac{M_s}{V} = \dfrac{1.807}{1} = 1.807$ g/cm^3 or $\gamma_d = 1.807 \times 9.81 = 17.73$ kN/m^3

The volume of solids (Fig. Ex. 17.6) is

$$V_s = \frac{1.807}{2.67} = 0.68 \text{ g/cm}^3$$

The volume of voids = $1 - 0.68 = 0.32$ cm^3

The volume of water = 0.253 cm^3

The volume of air = $0.320 - 0.253 = 0.067$ cm^3

If all the air is squeezed out of the samples the dry density at zero air voids would be, by definition,

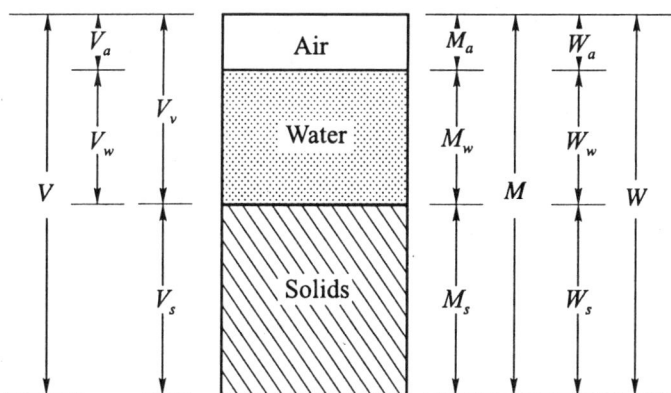

Fig. Ex. 17.6

$\rho_d = \dfrac{1.807}{0.68 + 0.253} = 1.94$ g/cm^3 or $\gamma_d = 1.94 \times 9.81 = 19.03$ kN/m^3 on the other hand, if the air

voids also were filled with water,

The mass of water would be = $0.32 \times 1 = 0.32$ g

The saturated density is $\rho_{\text{sat}} = \dfrac{1.807 + 0.32}{1} = 2.13$ g/cm3 or $\gamma_{\text{sat}} = 2.13 \times 9.81 = 20.90$ kN/m^3

17.6 COMPACTION FOR DEEPER LAYERS OF SOIL

Three types of dynamic compaction for deeper layers of soil are discussed here. They are:

1. Vibroflotation.
2. Dropping of a heavy weight.
3. Blasting.

Vibroflotation

The vibroflotation technique is used for compacting granular soil only. The vibroflot is a cylindrical tube containing water jets at top and bottom and equipped with a rotating eccentric weight, which develops a horizontal vibratory motion as shown in Fig. 17.17. The vibroflot is sunk into the soil using the lower jets and is then raised in successive small increments, during which the surrounding material is compacted by the vibration process. The enlarged hole around the vibroflot is backfilled with suitable granular material. This method is very effective for increasing the density of a sand deposit for depths up to 30 m. Probe spacings of compaction holes should be on a grid pattern of about 2 m to produce relative densities greater than 70 percent over the entire area. If the sand is coarse, the spacings may be somewhat larger.

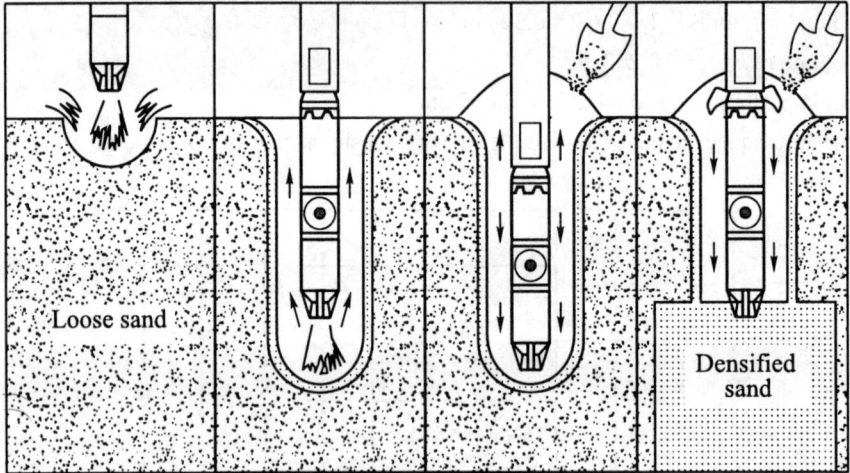

Fig. 17.17 Compaction by using vibroflot (Brown, 1977)

In soft cohesive soil and organic soils the vibroflotation technique has been used with gravel as the backfill material. The resulting densified stone column effectively reinforces softer soils and acts as a bearing pile for foundations.

Dropping of a Heavy Weight

The repeated dropping of a heavy weight on to the ground surface is one of the simplest of the methods of compacting loose soil.

The method, known as deep dynamic compaction or deep dynamic consolidation may be used to compact cohesionless or cohesive soils. The method uses a crane to lift a concrete or steel block, weighing up to 500 kN and up to heights of 40 to 50 m, from which height it is allowed to fall freely on to the ground surface. The weight leaves a deep pit at the surface. The process is then repeated either at the same location or sequentially over other parts of the area to be compacted. When the required number of repetitions is completed over the entire area, the compaction at depth is completed. The soils near the surface, however, are in a greatly disturbed condition. The top soil may then be levelled and compacted, using normal compacting equipment. The principal claims of this method are:

1. Depth of recompaction can reach up to 10 to 12 m.
2. All soils can be compacted.
3. The method produces equal settlements more quickly than do static (surcharge type) loads.

The depth of recompaction, D, in metres is approximately given by Leonards, *et al*, (1980) as

$$D \approx \frac{1}{2} (Wh)^{1/2} \tag{17.5}$$

where W = weight of falling mass in metric tons,

          h = height of drop in metres.

Blasting

Blasting, through the use of buried, time-delayed explosive charges, has been used to densify loose, granular soils. The sands and gravels must be essentially cohesionless with a maximum of 15 percent of their particles passing the No. 200 sieve size and 3 percent passing 0.005 mm size. The moisture condition of the soil is also important for surface tension forces in the partially saturated state limit the effectiveness of the technique. Thus the soil, as well as being granular, must be dry or saturated, which requires sometimes prewetting the site via construction of a dike and reservoir system.

The technique requires careful planning and is used at a remote site. Theoretically, an individual charge densifies the surrounding adjacent soil and soil beneath the blast. It should not lift the soil situated above the blast, however, since the upper soil should provide a surcharge load. The charge should not create a crater in the soil. Charge delays should be timed to explode from the bottom of the layer being densified upward in a uniform manner. The uppermost part of the stratum is always loosened, but this can be surface-compacted by vibratory rollers. Experience indicates that repeated blasts of small charges are more effective than a single large charge for achieving the desired results.

17.7   PRELOADING

Preloading is a technique that can successfully be used to densify soft to very soft cohesive soils. Large-scale construction sites composed of weak silts and clays or organic materials (particularly marine deposits), sanitary land fills, and other compressible soils may often be stabilised effectively and economically by preloading. Preloading compresses the soil. Compression takes place when the water in the pores of the soil is removed which amounts to artificial consolidation of soil in the field. In order to remove the water squeezed out of the pores and hasten the period of consolidation, horizontal and vertical drains are required to be provided in the mass. The preload is generally in the form of an imposed earth fill which must be left in place long enough to induce consolidation. The process of consolidation can be checked by providing suitable settlement plates and piezometers. The greater the surcharge load, shorter the time for consolidation. This is a case of three-dimension consolidation.

Two types of vertical drains considered are:

1. Cylindrical sand drains
2. Wick (prefabricated vertical) drains

Sand Drains

Vertical and horizontal and drains are normally used for consolidating very soft clay, silt and other compressible materials. The arrangement of sand drains shown in Fig. 17.18 is explained below:

1. It consists of a series of vertical sand drains or piles. Normally medium to coarse sand is used.
2. The diameter of the drains are generally not less than 30 cm and the drains are placed in a square grid pattern at distances of 2 to 3 metres apart. Economy requires a careful study of the effect of spacing the sand drains on the rate of consolidation.
3. Depth of the vertical drains should extend up to the thickness of the compressible stratum.
4. A horizontal blanket of free draining sand should be placed on the top of the stratum and the thickness of this may be up to a metre.
5. Soil surcharge in the form of an embankment is constructed on top of the sand blanket in stages.

The height of surcharge should be so controlled as to keep the development of pore water pressure in the compressible strata at a low level. Rapid loading may induce high pore water pressures resulting in the failure of the stratum by rupture. The lateral displacement of the soil may shear off the sand drains and block the drainage path.

The application of surcharge squeezes out water in radial directions to the nearest sand drain and also in the vertical direction to the sand blanket. The dashed lines shown in Fig. 17.18 (b) are drawn midway between the drains. The planes passing through these lines may be considered as impermeable

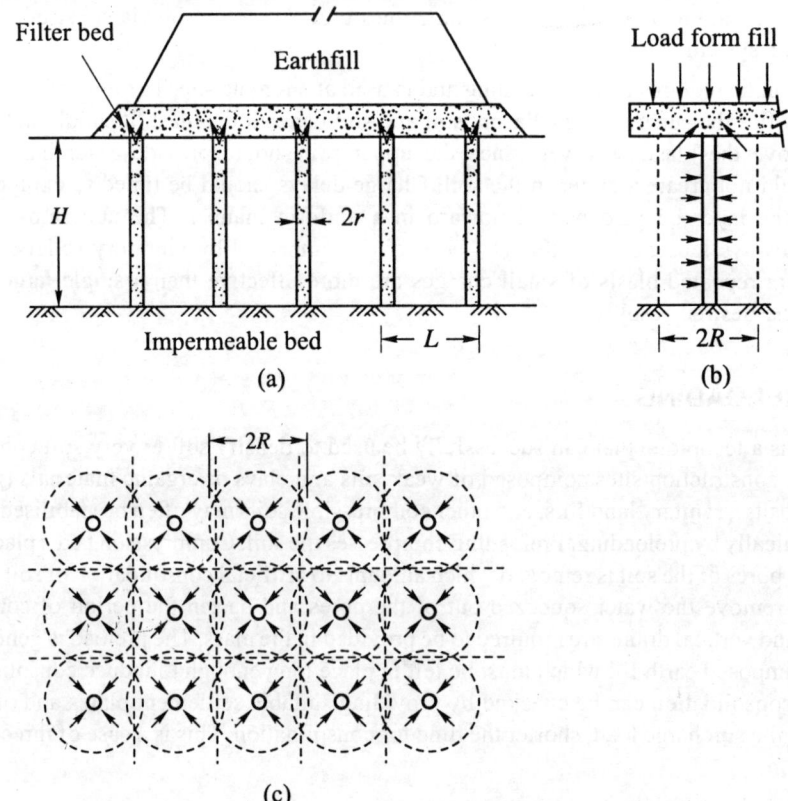

Fig. 17.18 Consolidation of soil using sand drains: (a) Vertical section, (b) section of single drain, and (c) plan of sand drains

membranes and all the water within a block has to flow to the drain at the centre. The problem of computing the rate of radial drainage can be simplified without appreciable error by assuming that each block can be replaced by a cylinder of radius R such that

$$\pi R^2 = L^2$$

where L is the side length of the prismatic block.

The relation between the time t and degree of consolidation $U_z\%$ is determined by the equation

$$U_z\% = 100 f(T)$$

wherein,

$$T = \frac{c_v t}{H^2} \tag{17.6}$$

If the bottom of the compressible layer is impermeable, then H is the full thickness of the layer.

For radial drainage, Rendulic (1935) has shown that the relation between the time t and the degree of consolidation $U_r\%$ can be expressed as

$$U_r\% = 100 f(T) \tag{17.7}$$

wherein,

$$T_r = \frac{c_v r}{4R^2} t \tag{17.8}$$

is the time factor. The relation between the degree of consolidation $U_r\%$ and the time factor T_r depends on the value of the ratio R/r. The relation between T_r and $U_r\%$ for ratios of R/r equal to 1, 10 and 100 in Fig. 17.19 are expressed by curves C_1, C_{10} and C_{100} respectively.

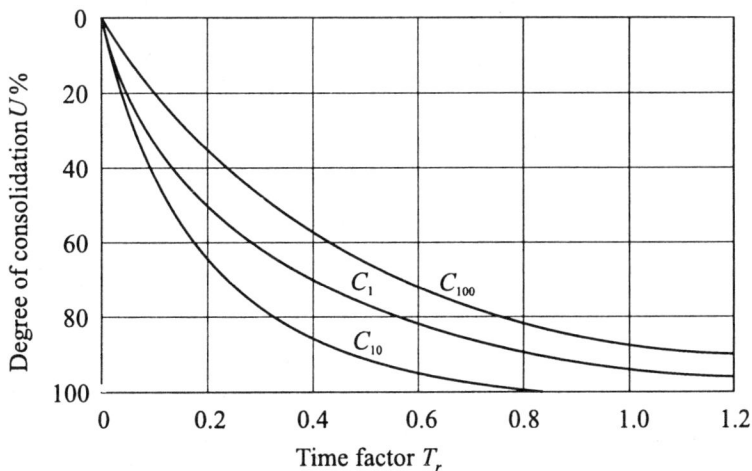

Fig. 17.19 T_r vs. U_r

Installation of Vertical Sand Drains

The sand drains are installed as follows:

1. A casing pipe of the required diameter with the bottom closed with a loose-fit-cone is driven up to the required depth.

2. The cone is slightly separated from the casing by driving a mandrel into the casing.

3. The sand of the required gradation is poured into the pipe for a short depth and at the same time the pipe is pulled up in steps. As the pipe is pulled up, the sand is forced out of the pipe by applying pressure on to the surface of the sand. The procedure is repeated till the holes is completely filled with sand.

The sand drains may also be installed by jetting a hole in the soil or by driving an open casing into the soil, washing the soil out of the casing, and filling the hole with sand afterwards.

Sand drains have been used extensively in many parts of the world for stabilizing soils for port development works and for foundations of structures in reclaimed areas on the sea coasts. It is possible that sand drains may not function satisfactorily if the soil surrounding the well gets remolded. This condition is referred to as *smear.* Though theories have been developed by considering different thickness of smear and different permeability, it is doubtful whether such theories are of any practical use since it would be very difficult to evaluate the quality of the smear in field.

Wick (Prefabricated Vertical) Drains

Geocomposites used as drainage media have completely taken over certain geotechnical application areas. Wick drains, usually consisting of plastic fluted or nubbed cores that are surrounded by a geotextile filter, have considerable tensile strength. Wick drains do not require any sand to transmit flow. Most synthetic drains are of a strip shape. The strip drains are generally 100 mm wide and 2 to 6 mm thick, Fig. 17.20 shows typical core shapes of strip drains (Hausmann, 1990).

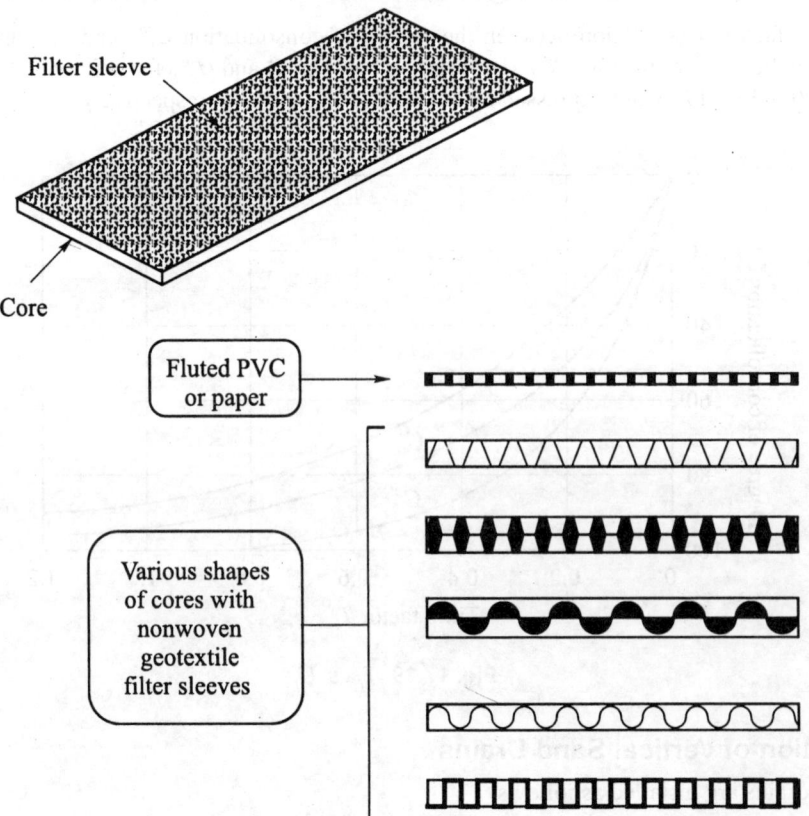

Fig. 17.20 Typical core shapes of strip drains (Hausmann, 1990)

Wick drains are installed by using a hollow lance. The wick drain is threaded into a hollow lance, which is pushed (or driven) through the soil layer, which collapses around it. At the ground surface the ends of the wick drains (typically at 1 to 2 m spacing) are interconnected by a granular soil drainage layer or geocomposite sheet drain layer. There are a number of chimerically available wick drain manufacturers and installation contractors who provide information on the current products, styles, properties, and estimated costs (Koerner, 1999).

With regards to determining wick drain spacings, the initial focal point is on the time for the consolidation of the subsoil to occur. Generally the time for 90% consolidation (t_{90}) is desired. In order to estimate the time t, it is first necessary to estimate an equivalent sand drain diameter for the wick drain used. The equations suggested by Koerner (1999) are

$$d_{sd} = \sqrt{\frac{d_v^2}{n_s}} \qquad (17.9a)$$

$$d_v = \sqrt{\frac{4 b t n_d}{\pi}} \qquad (17.9b)$$

where d_{sd} = equivalent sand drain diameter,
d_v = equivalent void circle diameter,
b, t = width and thickness of the wick drain,
n_s = porosity of sand drain,

$$n_d = \frac{\text{Void area of wick drain}}{\text{total cross-sectional area of strip}} = \frac{\text{Void area of wick drain}}{b \times t}$$

It may be noted here that equivalent sand drain diameters for various commercially available wick drains vary from 30 to 50 mm (Korner, 1999).

The equation for estimating the time t for consolidation is (Koerner, 1999)

$$t = \frac{D^2}{8 c_h} \left(\ln \frac{D}{d} - 0.75 \right) \ln \frac{1}{1 - U} \qquad (17.20)$$

where t = time for consolidation,
c_h = coefficient of consolidation of soil for horizontal flow,

d = equivalent diameter of strip drain = $\dfrac{\text{circumference}}{\pi}$,

D = sphere of influence of the strip drain,
  (*a*) for a triangular pattern, $D = 1.05 \times$ spacing D_t
  (*b*) for a square pattern, $D = 1.13 \times$ spacing D_s
D_t = distance between drains in triangular spacing,
D_s = distance for square pattern, and
U = average degree of consolidation.

Advantages of Using Wick Drains (Koerner, 1999)

1. The analytic procedure is available and straightforward in its use.
2. Tensile strength is definitely afforded to the soft soil by the installation of the wick drains.

3. There is only nominal resistance to the flow of water if it enters the wick drain.
4. Construction equipment is generally small.
5. Installation is simple, straightforward and economic.

Example 17.7

What is the equivalent sand drain diameter of a wick drain measuring 96 mm wide and 2.9 mm thick that is 92% void in its cross section? Use an estimated sand porosity of 0.3 for typical sand in a sand drain.

Solution

The area of wick drain $= b \times t = 96 \times 2.9 = 279 \text{ mm}^2$

Void area of wick drain $= n_d \times b \times t = 0.92 \times 279 = 257 \text{ mm}^2$

The equivalent circle diameter (Eq. 17.9b) is

$$d_v = \sqrt{\frac{4\,btn_d}{\pi}} = \sqrt{\frac{4 \times 257}{3.14}} = 18.1 \text{ mm}$$

The equivalent sand drain diameter [Eq. (17.19a)] is

$$d_{sd} = \sqrt{\frac{d_v^2}{n_s}} = \sqrt{\frac{18.1^2}{0.3}} = 33 \text{ mm}$$

Example 17.8

Calculate the times required for 50, 70 and 90% consolidation of a saturated clayey silt soil using wick drains at various triangular spacings. The wick drains measure 100×4 mm and the soil has a $c_h = 6.5 \ 10^{-6} \text{ m}^2/\text{min}$.

Solution

In the simplified formula the equivalent diameter d of a strip drain is

$$d = \frac{\text{circumference}}{\pi} = \frac{100 + 100 + 4 + 4}{3.14} = 66.2 \text{ mm}$$

Using Eq. (17.10)

$$t = \frac{D^2}{8c_h}\left(\ln\frac{D}{d} - 0.75\right)\ln\left(\frac{1}{1-U}\right)$$

substituting the known values

$$t = \frac{D^2}{8\left(6.5 \times 10^{-6}\right)}\left(\ln\frac{D}{0.0062} - 0.75\right)\ln\left(\frac{1}{1-U}\right)$$

The times required for the various degrees of consolidation are tabulated below for assumed theoretical spacings of wick drains.

Wick drain spacings D (m)	Time in days for various degrees of consolidation (U)		
	50%	70%	90%
2.1	110	192	367
1.8	77	133	254
1.5	49	86	164
1.2	29	50	95
0.9	14	24	46
0.6	4.8	8.4	16
0.3	0.6	1.1	2.1

For the triangular pattern, the spacing D_t is

$$D_t = \frac{D}{1.05}$$

17.8   SAND COMPACTION PILES AND STONE COLUMNS

Sand Compaction Piles

Sand compaction piles consists of driving a hollow steel pipe with the bottom closed with a collapsible plate down to the required depth; filling it with sand, and withdrawing the pipe while air pressure is directed against the sand inside it. The bottom plate opens during withdrawal and the sand backfills the voids created earlier during the driving of the pipe. The in-situ soil is densified while the pipe is being withdrawn, and sand backfill prevents the soil surrounding the compaction pipe from collapsing as the pipe is withdrawn. The maximum limits on the amount of fines that can be present are 15 percent passing the No. 200 sieve (0.075 mm) and 3 percent passing 0.005 mm. The distance between the piles may have to be planned according to the site conditions.

Stone columns

The method described for installing sand compaction piles or the vibroflot described earlier can be used to construct stone columns. The size of the stones used for this purpose range from about 6 to 40 mm. Stone columns have particular application in soft inorganic, cohesive soils and are generally inserted on a volume displacement basis.

The diameter of the pipe used either for the construction of sand drains or sand compaction piles can be increased according to the requirements. Stones are placed in the pipe instead of sand, and the technique of constructing stone columns remains the same as that for sand piles.

Stone columns are placed 1 to 3 m apart over the whole area. There is no theoretical procedure for predicting the combined improvement obtained, so it is usual to assume the foundation loads are carried only by the several stone columns with no contribution from the intermediate ground (Bowles, 1996).

Bowles (1996) gives an approximate formula for the allowable bearing capacity of stone columns as

$$q_a = \frac{K_P}{F_s} (4c + \sigma_r') \tag{17.11}$$

where $K_P = \tan^2 (45° + \phi'/2)$,

ϕ' = drained angle of friction of stone,

c = either drained cohesion (suggested for large areas) or the undrained shear strength c_u,

σ'_r = effective radial stress as measured by a pressuremeter (but may use $2c$ if pressuremeter data are not available),

F_s = factor of safety, 1.5 to 2.0.

The total allowable load on a stone column of average cross-section area A_c is

$$Q_a = q_a A_c \qquad (17.12)$$

Stone columns should extend through soft clay to firm strata to control settlements. There is no end bearing in Eq. (17.12) because the principal load carrying mechanism is local perimeter shear.

Settlement is usually the principal concern with stone columns since bearing capacity is usually quite adequate (Bowles, 1996). There is no method currently available to compute settlement on a theoretical basis.

Stone columns are not applicable to thick deposits of peat or highly organic silts or clays (Bowles, 1996). Stone columns can be used in loose sand deposits to increase the density.

17.9 SOIL STABILISATION BY THE USE OF ADMIXTURES

The physical properties of soils can often economically be improved by the use of admixtures. Some of the more widely used admixtures include lime, portland cement and asphalt. The process of soil stabilisation first involves mixing with the soil a suitable additive which changes its property and then compacting the admixture suitably. This method is applicable only for soils in shallow foundations or the base courses of roads, airfield pavements, etc.

Soil–Lime Stabilisation

Lime stabilisation improves the strength, stiffness and durability of fine grained materials. In addition, lime is sometimes used to improve the properties of the fine grained fraction of granular soils. Lime has been used as a stabiliser for soils in the base courses of pavement systems, under concrete foundations, on embankment slopes and canal linings.

Adding lime to soils produces a maximum density under a higher optimum moisture content than in the untreated soil. Morever, lime produces a decrease in plasticity index.

Lime stabilisation has been extensively used to decrease swelling potential and swelling pressures in clays. Ordinarily the strength of wet clay is improved when a proper amount of lime is added. The improvement in strength is partly due to the decrease in plastic properties of the clay and partly to the pozzolanic reaction of lime with soil, which produces a cemented material that increases in strength with time. Lime-treated soils, in general, have greater strength and a higher modulus of elasticity than untreated soils.

Recommended percentages of lime for soil stabilisation vary from 2 to 10 percent. For coarse soils such as clayey gravels, sandy soils with less than 50 percent silt-clay fraction, the percent of lime varies from 2 to 5, whereas for soils with more than 50 percent silt-clay fraction, the percent of lime lies between 5 and 10. Lime is also used with fly ash. The fly ash may vary from 10 to 20 percent, and the percent of lime may lie between 3 and 7.

Soil–Cement Stabilisation

Soil–cement is the reaction product of an intimate mixture of pulverized soil and measured amounts of portland cement and water, compacted to high density. As the cement hydrates, the mixture becomes a hard, durable structural material. Hardened soil-cement has the capacity to bridge over local weak points in a subgrade. When properly made, it does not soften when exposed to wetting and drying, or freezing and thawing cycles.

Portland cement and soil mixed at the proper moisture content has been used increasingly in recent years to stabilise soils in special situations. Probably the main use has been to build stabilised bases under concrete pavements for highways and airfields. Soil cement mixtures are also used to provide wave protection on earth dams. There are three categories of soil-cement (Mitchell and Freitag, 1959). They are:

1. Normal soil-cement usually contains 5 to 14 percent cement by weight and is used generally for stabilizing low plasticity soils and sandy soils.
2. Plastic soil-cement has enough water to produce a wet consistency similar to mortar. This material is suitable for use as water proof canal linings and for erosion protection on steep slopes where road building equipment may not be used.
3. Cement-modified soil is a mix that generally contains less than 5 percent cement by volume. This forms a less rigid system than either of the other types, but improves the engineering properties of the soil and reduces the ability of the soil to expand by drawing in water.

The cement requirement depends on the gradation of the soil. A well graded soil containing gravel, coarse sand and fine sand with or without small amounts of silt or clay will require 5 percent or less cement by weight. Poorly graded sands with minimal amount of silt will require about 9 percent by weight. The remaining sandy soils will generally require 7 percent. Non-plastic or moderately plastic silty soils generally require about 10 percent, and plastic clay soils require 13 percent or more.

Bituminous Soil Stabilisation

Bituminous materials such as asphalts, tars, and pitches are used in various consistencies to improve the engineering properties of soils. Mixed with cohesive soils, bituminous materials improve the bearing capacity and soil strength at low moisture content. The purpose of incorporating bitumen into such soils is to water proof them as a means to maintain a low moisture content. Bituminous materials added to sand act as a cementing agent and produces a stronger, more coherent mass. The amount of bitumen added varies from 4 to 7 percent for cohesive materials and 4 to 10 percent for sandy materials. The primary use of bituminous materials is in road construction where it may be the primary ingredient for the surface course or be used in the subsurface and base courses for stabilizing soils.

17.10 SOIL STABILISATION BY INJECTION OF SUITABLE GROUTS

Grouting is a process whereby fluid like materials, either in suspension, or solution form, are injected into the subsurface soil or rock.

The purpose of injecting a grout may be any one or more of the following:

1. To decrease permeability.
2. To increase shear strength.
3. To decrease compressibility.

Suspension-type grouts include soil, cement, lime, asphalt emulsion, etc. while the solution type grouts include a wide variety of chemicals. Grouting proves especially effective in the following cases:

1. When the foundation has to be constructed below the ground water table. The deeper the foundation, the longer the time needed for construction, and therefore, the more benefit gained from grouting as compared with dewatering.
2. When there is difficult access to the foundation level. This is very often the case in city work, in tunnel shafts, sewers, and subway construction.

3. When the geometric dimensions of the foundation are complicated and involves many boundaries and contact zones.

4. When the adjacent structures require that the soil of the foundation strata should not be excavated (extension of existing foundations into deeper layers).

Grouting has been extensively used primarily to control ground water flow under earth and masonry dams, where rock grouting is used. Since the process fills soil voids with some type of stabilizing material grouting is also used to increase soil strength and prevent excessive settlement.

Many different materials have been injected into soils to produce changes in the engineering properties of the soil. In one method a casing is driven and injection is made under pressure to the soil at the bottom of the hole as the casing is withdrawn. In another method, a grouting hole is drilled and at each level in which injection is desired, the drill is withdrawn and a collar is placed at the top of the area to be grouted and grout is forceed into the soil under pressure. Another method is to perforate the casing in the area to be grouted and leave the casing permanently in the soil.

Penetration grouting may involve portland cement or fine grained soils such as bentonite or other materials of a particulate nature. These materials penetrate only a short distance through most soils and are primarily useful in very coarse sands or gravels. Viscous fluids, such as a solution of sodium silicate, may be used to penetrate fine grained soils. Some of these solutions from gels that restrict permeability and improve compressibility and strength properties.

Displacement grouting usually consists of using a grout like portland cement and sand mixture which when forced into the soil displaces and compacts the surrounding material about a central core of grout. Injection of lime is sometimes used to produce lenses in the soil that will block the flow of water and reduce compressibility and expansion properties of the soil. The lenses are produced by hydraulic fracturing of the soil.

The injection and grouting methods are generally expensive compared with other stabilisation techniques and are primarily used under special situations as mentioned earlier. For a detailed study on injections, readers may refer to Caron *et al*, (1975).

17.11 SOIL STABILISATION BY ELECTRICAL AND THERMAL METHODS

Electrical Method

The electrical method is used to densify the in situ cohesive soils. The method is called as electro-osmosis. It consists of placing in the soil to be stabilised a number of electrodes and then passing a direct current between them. The electric current induces a flow of water from the anode to the cathode (This is because of the attraction of cations, and of the unbalanced, negatively charged clay particles themselves, to the anode). The cathode is generally a perforated metal pipe which is used as a well point for removing the water. The anode can be any type of metal rod. Typical electro-osmotic stabilisation configuration are shown in Fig. 17.21. In general, both the cathodes and anodes should be placed about 2 m apart beneath the lowest elevation to be stabilised. Typical spacings of the cathodes (the well points) are 6 to 9 m apart with the anodes being placed midway between them.

The flow rate to a cathode well point can be estimated using a modification of Darav's law as follows:

$$q = k_e i_e A \tag{17.13}$$

where, q = flow rate, m³/s,

k_e = electroosmotic coefficient of permeability based on voltage,
1×10^{-9} to 7×10^{-9} m/s per V/m,

i_e = electrical potential gradient, V/m,

A = cross-sectional area, m².

The consumption of power varies from 1 to 10 kW per m³ of stabilised soil.

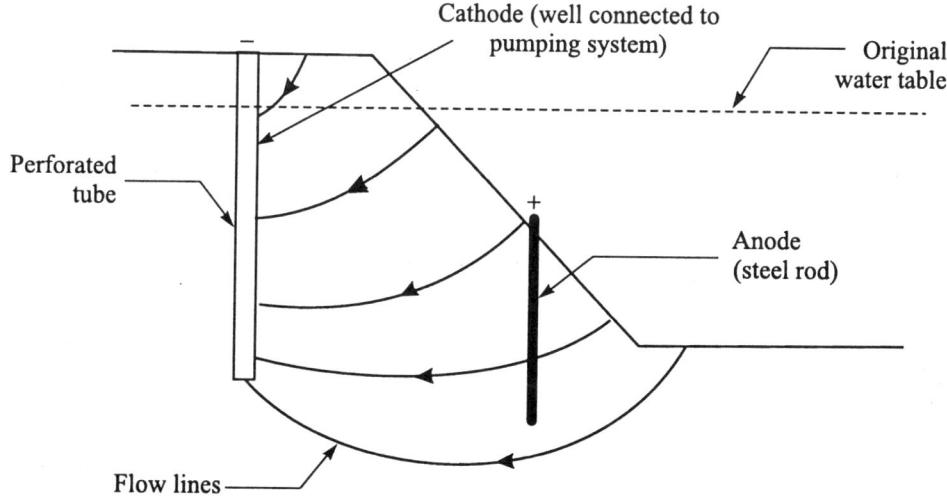

Fig. 17.21 Electroosmotic stabilisation of soils

Thermal Methods

Heat is very rarely used to stabilise soils. However it is technically possible to stabilise saturated clays by heat. Russians have stabilised deep deposits of partly saturated loess by burning a mixture of liquid fuel and air injected into the ground through a network of pipes. A temperature of 100° C causes drying and increase in the strength of clays. A permanent change in the structure of clays is possible at a temperature of 500° C, and at 1000° C there will be fusion of clay particles transforming lay into a solid substance much like brick. However, the economics of using heat precludes its use in construction projects.

Ground freezing appears to be gaining popularity in some cases. It is accomplished by bringing a refrigerant into the proximity of soil porewater that is stationary. The porewater around the refrigerant pipe begins to freeze, and with continued exposure the ice layers expand in all directions. A series refrigerant pipes layed close to each other will help to form a continuous wall of ice. The freezed soil possesses high strength and low permeability. It can stabilise a wide range of soil types. Freezing technique has been successfully used in sinking tunnel shafts, advancing tunnels in running ground, providing lateral restraint for excavations, etc.

17.12 PROBLEMS

17.1 Differentiate: (*i*) Compaction and consolidation, and (*ii*) Standard proctor and modified proctor tests.

17.2 Draw an ideal 'compaction curve' and discuss the effect of moisture on the dry unit weight of soil.

17.3 Explain: (*i*) the unit, in which the compaction is measured, (*ii*) 95 percent of proctor density, (*iii*) zero air-voids line, and (*iv*) effect of compaction on the shear strength of soil.

17.4 What are the types of rollers used for compacting different types of soils in the field? How do you decide the compactive effort required for compacting the soil to a desired density in the field?

17.5 What are the methods adopted for measuring the density of the compacted soil? Briefly describe the one which will suit all types of soils.

17.6 A soil having a specific gravity of solids $G = 2.75$, is subjected to proctor compaction test in a mould of volume $V = 945$ cm^3. The observations recorded are as follows:

Observation number	1	2	3	4	5
Mass of wet sample, g	1389	1767	1824	1784	1701
Water content, percentage	7.5	12.1	17.5	21.0	25.1

What are the values of maximum dry unit weight and the optimum moisture content ? Draw 100% saturation line.

17.7 A field density test was conducted by sand cone method. The observation data are given below:

(a) Mass of jar with cone and sand (before use) = 4950 g, (b) mass of jar with cone and sand (after use) = 2280 g, (c) mass of soil from the hole = 2925 g, (d) dry density of sand = 1.48 g/cm^3, (e) water content of the wet soil = 12%. Determine the dry unit weight of compacted soil.

17.8 If a clayey sample is saturated at a water content of 30%, what is its density ? Assume a value for specific gravity of solids.

17.9 A soil in a borrow pit is at a dry density of 1.7 Mg/m^3 with a water content of 12%. If a soil mass of 2000 cubic metre volume is excavated from the pit and compacted in an embankment with a porosity of 0.32, calculated the volume of the embankment which can be constructed out of this material. Assume $G = 2.70$.

17.10 In a Proctor compaction test, for one observation, the mass of the wet sample is missing. The oven dry mass of this sample was 1800 g. The volume of the mould used was 950 cm^3. If the saturation of this sample was 80 percent, determine (i) the moisture content, and (ii) the total unit weight of the sample. Assume $G = 2.70$.

17.11 A field-compacted sample of a sandy loam was found to have a wet density of 2.176 Mg/m^3 at a water content of 10%. The maximum dry density of the soil obtained in a standard Proctor test was 2.0 Mg/m^3. Assume $G = 2.65$. Compute ρ_d, S, n and the percent of compaction of the field sample.

17.12 A proposed earth embankment is required to be compacted to 95% of standard proctor dry density. Tests on the material to be used for the embankment give $\rho_{max} = 1.984$ Mg/m^3 at an optimum water content of 12%. The borrow pit material in its natural condition has a void ratio of 0.60. If $G = 2.65$, what is the minimum volume of the borrow required to make 1 cu.m of acceptable compacted fill?

17.13 The following data were obtained from a field density test on a compacted fill of sandy clay. Laboratory moisture density tests on the fill material indicated a maximum dry density of 1.92 Mg/m^3 at an optimum water content of 11%. What was the percent compaction of the fill? Was the fill water content above or below optimum.

Mass of the moist removed from the test hole	= 1038 g
Mass of the soil after oven drying	= 914 g
Volume of the test hole	= 478.55 cm^3

17.14 A field density test performed by sand-cone method gave the following data.

Mass of the soil removed + pan	= 1590 g
Mass of the pan	= 125 g
Volume of the test hole	= 750 cm^3
Water content information	
Mass of the wet soil + pan	= 404.9 g
Mass of the dry soil + pan	= 365.9 g
Mass of the pan	= 122.0 g

Compute: ρ_d, γ_d, and the water content of the soil. Assume $G = 2.67$.

Braced-Cuts and Drainage ⑱

PART A: BRACED-CUTS

18.1 GENERAL CONSIDERATIONS

Shallow excavations can be made without supporting the surrounding material if there is adequate space to establish slopes at which the material can stand. The steepest slopes that can be used in a given locality are best determined by experience. Many building sites extend to the edges of the property lines. Under these circumstances, the sides of the excavation have to be made vertical and must usually be supported by bracings.

Common methods of bracing the sides when the depth of excavation does not exceed about 3 m are shown in Figs 18.1 (a) and (b). The practice is to drive vertical timber planks known as sheeting along the sides of the excavation. The sheeting is held in place by means of horizontal beams called *wales* that in turn are commonly supported by horizontal *struts* extending from side to side of the excavation. The struts are usually of timber for widths not exceeding about 2 m. For greater widths metal pipes called *trench braces* are commonly used.

When the excavation depth exceeds about 5 to 6 m, the use of vertical timber sheeting will become uneconomical. According to one procedure, *steel sheet piles* are used around the boundary of the excavation. As the soil is removed from the enclosure, wales and struts are inserted. The wales are commonly of steel and the struts may be of steel or wood. The process continues until the excavation is complete. In most types of soil, it may be possible to eliminate sheet piles and to replace them with a series of *H* piles spaced 1.5 to 2.5 m apart. The *H* piles, known as *soldier piles* or *soldier beams,* are driven with their flanges parallel to the sides of the excavation as shown in Fig. 18.1 (b). As the soil next to the piles is removed horizontal boards known as *lagging* are introduced as shown in the figure and are wedged against the soil outside the cut. As the general depth of excavation advances from one level to another, wales and struts are inserted in the same manner as for steel sheeting.

If the width of a deep excavation is too great to permit economical use of struts across the entire excavation, *tiebacks* are often used as an alternative to cross-bracings as shown in Fig. 18.1 (c). Inclined holes are drilled into the soil outside the sheeting or *H* piles. Tensile reinforcement is then inserted and concreted into the hole. Each tieback is usually prestressed before the depth of excavation is increased.

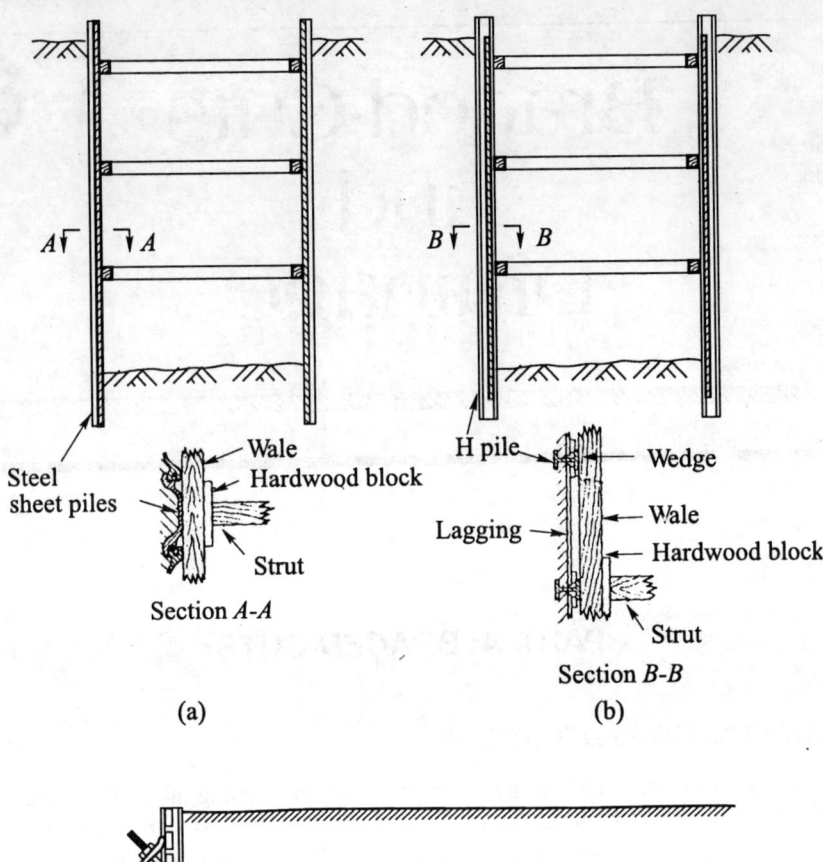

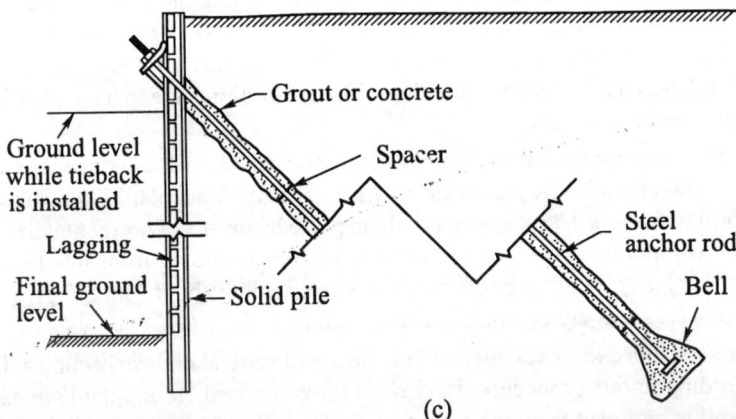

Fig. 18.1 Cross-sections, through typical bracing in deep excavation: (a) Sides retained by steel sheet piles, (b) sides retained by H piles and lagging, (c) one of several tieback systems for supporting vertical sides of open cut. Several sets of anchors may be used, at different elevations (Peck, 1969)

18.2 LATERAL EARTH PRESSURE DISTRIBUTION ON BRACED-CUTS

Since most open cuts are excavated in stages within the boundaries of sheet pile walls or walls consisting of soldier piles and lagging, and since struts are inserted progressively as the excavation precedes, the walls are likely to deform as shown in Fig. 18.2. Little inward movement can occur at the top of the cut after the first strut is inserted. The pattern of deformation differs so greatly from that required for Rankine's state that the distribution of earth pressure associated with retaining

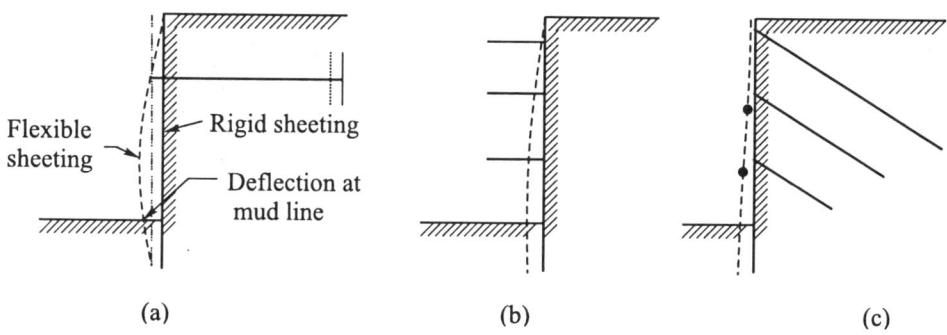

Fig. 18.2 Typical pattern of deformation of vertical walls: (a) Anchored bulkhead, (b) braced-cut, and (c) tieback cut (Peck *et al*, 1974)

walls is not a satisfactory basis for design (Peck *et al*, 1974). The pressures against the upper portion of the walls are substantially greater than those indicated by the equation.

$$p_a = \frac{1 - \sin \phi}{1 + \sin \phi} \, p_v \tag{18.1}$$

for Rankine's condition

where p_v = vertical pressure,

ϕ = friction angle.

Apparent Pressure Diagrams

Peck (1969) presented pressure distribution diagrams on braced-cuts. These diagrams are based on a wealth of information collected by actual measurements in the field. Peck called these pressure diagrams *apparent pressure envelopes* which represent fictitious pressure distributions for estimating strut loads in a system of loading. Figure 18.3 gives the apparent pressure distribution diagrams as proposed by Peck.

Deep Cuts in Sand

The apparent pressure diagram for sand given in Fig. 18.3 was developed by Peck (1969) after a great deal of study of actual pressure measurements on braced-cuts used for subways.

The pressure diagram given in Fig. 18.3 (b) is applicable to both loose and dense sands. The struts are to be designed based on this apparent pressure distribution. The most probable value of any individual strut load is about 25 percent lower than the maximum (Peck, 1969). It may be noted here that this apparent pressure distribution diagram is based on the assumption that the water table is below the bottom of the cut.

The pressure p_a is uniform with respect to depth. The expression for p_a is

$$p_a = 0.65 \, \gamma H K_A \tag{18.2}$$

where, $K_A = \tan^2 (45° - \phi/2)$,

γ = unit weight of sand.

Cuts in Saturated Clay

Peck (1969) developed two apparent pressure diagrams, one for soft to medium clay and the other for stiff fissured clay. He classified these clays on the basis of non-dimensional factors (stability number N_s) as follows.

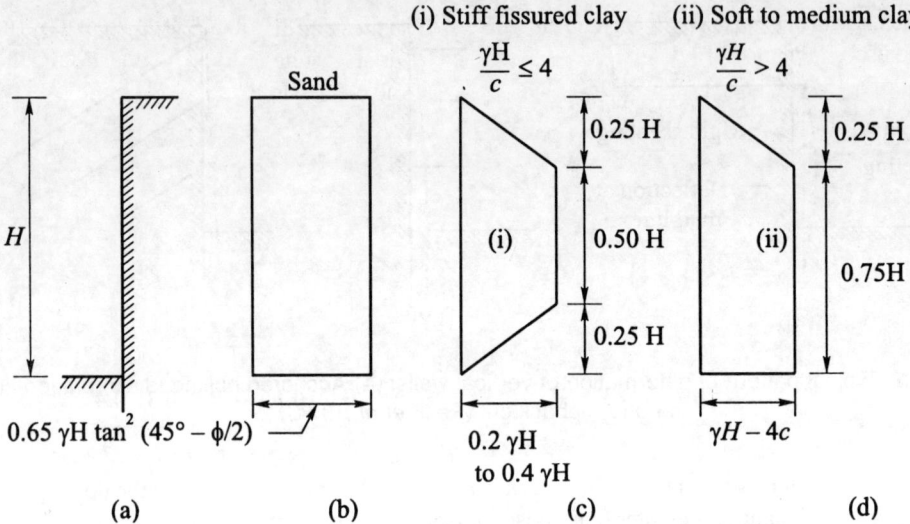

Fig. 18.3 Apparent pressure diagram for calculating loads in struts of braced cuts: (a) Sketch of wall of cut, (b) diagram for cuts in dry or moist sand, (c) diagram for clays if $\gamma H/c$ is less than 4, and (d) diagram for clays if $\gamma H/c$ is greater than 4 where c is the average undrained shearing strength of the soil (Peck, 1969)

Stiff fissured clay

$$N_s = \frac{\gamma H}{c} \leq 4 \qquad\qquad (18.3)$$

Soft to medium clay

$$N_s = \frac{\gamma H}{c} > 4 \qquad\qquad (18.4)$$

where γ = unit weight of clay, c = undrained cohesion ($\phi = 0$)

The pressure diagrams for these two types of clays are given in Fig. 18.3 (c) and (d) respectively. The apparent pressure diagram for soft to medium clay [Fig. 18.3 (d)] has been found to be conservative for estimating loads for design supports. [Figure 18.3 (c)] shows the apparent pressure diagram for stiff-fissured clays. Most stiff clays are weak and contain fissures. Lower pressures should be used only when the results of observations on similar cuts in the vicinity so indicate. Otherwise a lower limit for $p_a = 0.3\ \gamma H$ should be taken. Figure 18.4 gives a comparison of measured and computed pressures distribution for cuts in London, Oslo and Houston clays.

Cuts in Stratified Soils

It is very rare to find uniform deposits of sand or clay to a great depth. Many times layers of sand and clays overlying one another the other are found in nature. Even the simplest of these conditions does not lend itself to vigorous calculations of lateral earth pressures by any of the methods available. Based on field experience, empirical or semi-empirical procedures for estimating apparent pressure diagrams may be justified. Peck (1969) proposed the following unit pressure for excavations in layered soils (sand and clay) with sand overlying as shown in Fig. 18.5.

When layers of sand and soft clay are encountered, the pressure distribution shown in Fig. 18.3 (d) may be used if the unconfined compressive strength q_u is substituted by the average $\bar{q}_u$ and the unit weight of soil γ by the average value $\bar{\gamma}$ (Peck, 1969). The expressions for $\bar{q}_u$ and $\bar{\gamma}$ are

$$\bar{q}_u = \frac{1}{H} (\gamma_1 K_s h_1^2 \tan \phi + h_2 n q_u) \tag{18.5}$$

$$\bar{\gamma} = \frac{1}{H} (\gamma_1 h_1 + \gamma_2 h_2) \tag{18.6}$$

Values of γH

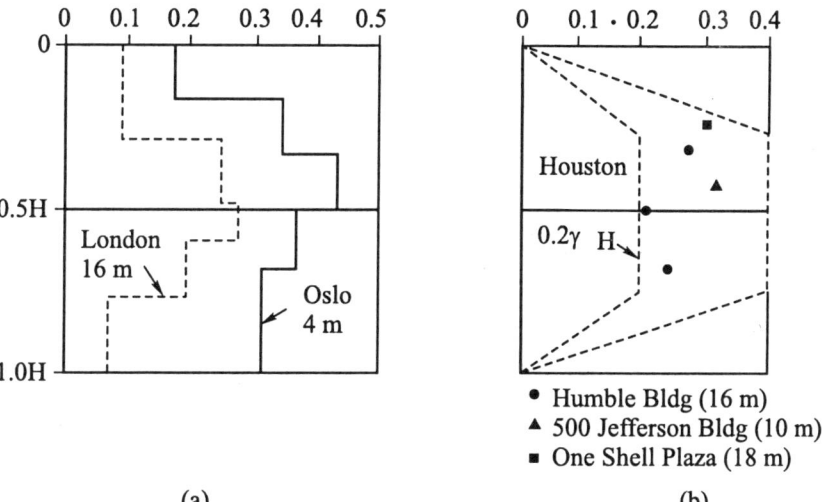

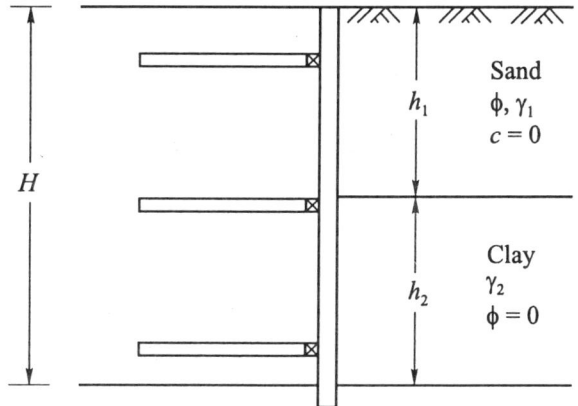

(a) (b)

Fig. 18.4 Maximum apparent pressures for cuts in stiff clays: (a) Fissured clays in London and Oslo, (b) stiff slickensided clays in Houston (Peck, 1969)

where H = total depth of excavation,

γ_1, γ_2 = unit weights of sand and clay respectively,

h_1, h_2 = thickness of sand and clay layers respectively,

K_s = hydrostatic pressure ratio for the sand layer, may be taken as equal to 1.0 for design purposes,

ϕ = angle of friction of sand,

n = coefficient of progressive failure varies from 0.5 to 1.0 which depends upon the creep characteristics of clay. For Chicago clay n varies from 0.75 to 1.0,

q_u = unconfined compression strength of clay.

Fig. 18.5 Cuts in stratified soils

18.3 STABILITY OF BRACED-CUTS IN SATURATED CLAY

A braced-cut may fail as a unit due to unbalanced external forces or heaving of the bottom of the excavation. If the external forces acting on opposite sides of the braced cut are unequal, the stability of the entire system has to be analysed. If soil on one side of a braced cut is removed due to some unnatural forces the stability of the system will be impaired. However, we are concerned here about the stability of the bottom of the cut. Two cases may arise. They are:

1. Heaving in clay soil.
2. Heaving in cohesionless soil.

Heaving in Clay Soil

The danger of heaving is greater if the bottom of the cut is soft clay. Even in a soft clay bottom, two types of failure are possible. They are:

Case 1: When the clay below the cut is homogeneous at least up to a depth equal $0.7\,B$ where B is the width of the cut.

Case 2: When a hard stratum is met within a depth equal to $0.7\,B$.

In the first case a full plastic failure zone will be formed and in the second case this is restricted as shown in Fig. 18.6. A factor of safety of 1.5 is recommended for determining the resistance here. Sheet piling is to be driven deeper to increase the factor of safety. The stability analysis of the bottom of the cut as developed by Terzaghi (1943) is as follows.

Case 1: Formation of Full Plastic Failure Zone below the Bottom of Cut

Figure 18.6 (a) is a vertical section through a long cut of width B and depth H in saturated cohesive soil ($\phi = 0$). The soil below the bottom of the cut is uniform up to a considerable depth for the formation of a full plastic failure zone. The undrained cohesive strength of sol is c. The weight of the blocks of clay on either side of the cut tends to displace the underlying clay toward the excavation. If the underlying clay experiences a bearing capacity failure, the bottom of the excavation heaves and the earth pressure against the bracing increases considerably.

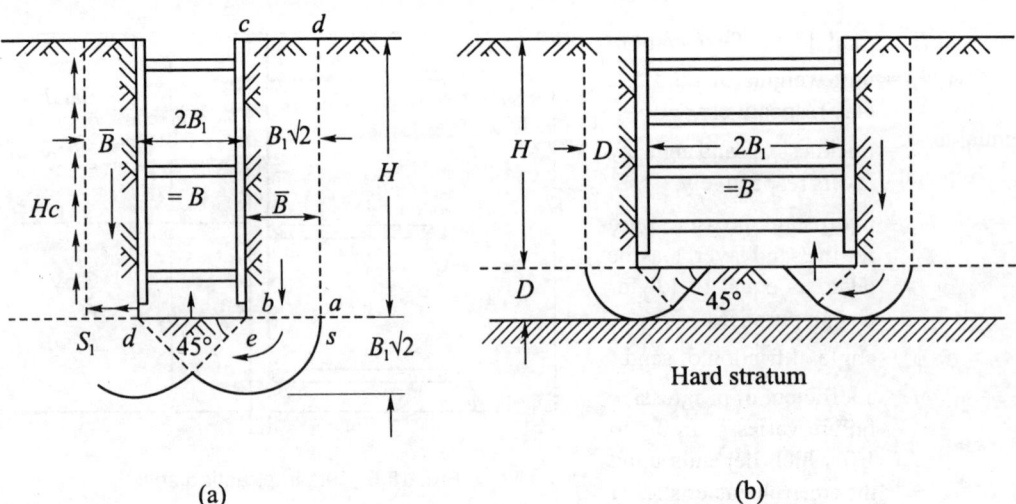

Fig. 18.6 Stability of braced cut: (a) Heave of bottom of timbered cut in soft clay if no hard stratum interferes with flow of clay, (b) as before, if clay rests at shallow depth below bottom of cut on hard stratum (after Terzaghi, 1943)

The anchorage load block of soil $a\,b\,c\,d$ in Fig. 18.6 (a) of width $\bar{B}$ (assumed) at the level of the bottom of the cut per unit length may be expressed as

$$Q = \gamma H \bar{B} - cH = \bar{B}H\left(\gamma - \frac{c}{\bar{B}}\right) \tag{18.7}$$

The vertical pressure q per unit length of a horizontal, ba is

$$q = \frac{Q}{\bar{B}} = H\left(\gamma - \frac{c}{\bar{B}}\right) \tag{18.8}$$

The bearing capacity q_u per unit area at level ab is

$$q_u = N_c c = 5.7c \tag{18.9}$$

where $N_c = 5.7$

The factor of safety against heaving is

$$F_s = \frac{q_u}{q} = \frac{5.7\,c}{H\left(\gamma - \dfrac{c}{\bar{B}}\right)} \tag{18.10}$$

Because of the geometrical condition, it has been found that the width $\bar{B}$ cannot exceed $0.70B$. Substituting this value for $\bar{B}$,

$$F_s = \frac{5.7c}{H\left(\gamma - \dfrac{c}{0.7B}\right)} \tag{18.11}$$

This indicates that the width of the failure slip is equal to $\bar{B}\sqrt{2} = 0.7B$.

Case 2: When the Formation of Full Plastic Zone is Restricted by the Presence of a Hard Layer

If a hard layer is located at a depth D below the bottom of the cut (which is less than $0.7B$), the failure of the bottom occurs as shown in Fig. 18.6 (b). The width of the strip which can sink is also equal to D.

Replacing $0.7B$ by D in Eq. 18.11, the factor of safety is represented by

$$F_s = \frac{5.7\,c_u}{H\left(\gamma - \dfrac{c}{D}\right)} \tag{18.12}$$

For a cut in soft clay with a constant value of c_u below the bottom of the cut, D in Eq. (18.12) becomes large, and F_s approaches the value

$$F_s = \frac{5.7\,c_u}{\gamma\,H} = \frac{5.7}{N_s} \tag{18.13}$$

where

$$N_s = \frac{\gamma\,H}{c_u} \tag{18.14}$$

is termed the *stability number.* The stability number is a useful indicator of potential soil movements. The soil movement is smaller for smaller values of N_s.

The analysis discussed so far is for long cuts. For short cuts, square, circular or rectangular, the factor of safety against heave can be found in the same way as for footings.

18.4 BJERRUM AND EIDE (1956) METHOD OF ANALYSIS

The method of analysis discussed earlier gives reliable results provided the width of the braced cut is larger than the depth of the excavation and that the braced cut is very long. In the cases where the braced-cuts are rectangular, square or circular in plan or the depth of excavation exceeds the width of the cut, the following analysis should be used.

In this analysis the braced cut is visualized as a deep footing whose depth and horizontal dimensions are identical to those at the bottom of the braced cut. This deep footing would fail in an identical manner to the bottom braced cut failed by heave. The theory of Skempton for computing N_c (bearing capacity factor) for different shapes of footing is made use of. Fig. 18.7 gives values of N_c as a function of H/B for long, circular or square footings. For rectangular footings, the value of N_c may be computed by the expression

$$N_c \text{ (rect)} = (0.84 + 0.16 \, B/L) \, N_c \text{ (sq)} \tag{18.15}$$

where L = length of excavation

     B = width of excavation

The factor of safety for bottom heave may be expressed as

$$F_s = \frac{cN_c}{\gamma H + q} \geq 1.5$$

where γ = effective unit weight of the soil above the bottom of the excavation

     q = uniform surcharge load (Fig. 18.7)

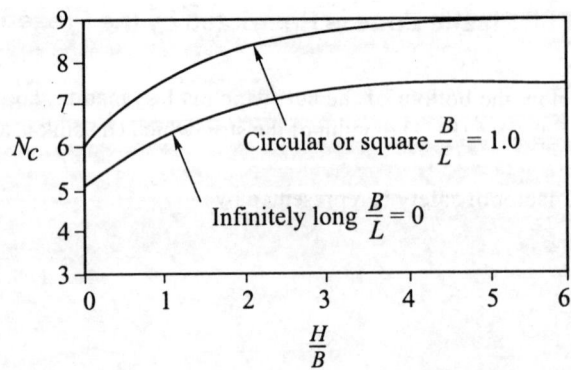
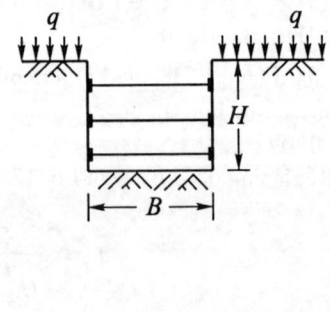

Fig. 18.7 Stability of bottom excavation (after Bjerrum and Eide, 1956)

Example 18.1

A long trench is excavated in medium dense sand for the foundation of a multistorey building. The sides of the trench are supported with sheet pile walls fixed in place by struts and wales as shown in Fig. Ex. 18.1. The soil properties are:

$$\gamma = 18.5 \text{ kN/m}^3, c = 0 \text{ and } \phi = 38°$$

Determine

(a) The pressure distribution on the walls with respect to depth.

(b) Strut loads. The struts are placed horizontally at distances $L = 4$ m centre to centre.

(c) The maximum bending moment for determining the pile wall section.

(d) The maximum bending moments for determining the section of the wales.

Solution

(a) For a braced cut in sand use the apparent pressure envelope given in Fig. 18.3 (b). The equation for p_a is

$$p_a = 0.65 \, \gamma H \, K_A = 0.65 \times 18.5 \times 8 \tan^2 (45 - 38/2) = 23 \text{ kN/m}^2$$

Figure Ex. 18.1 (b) shows the pressure envelope.

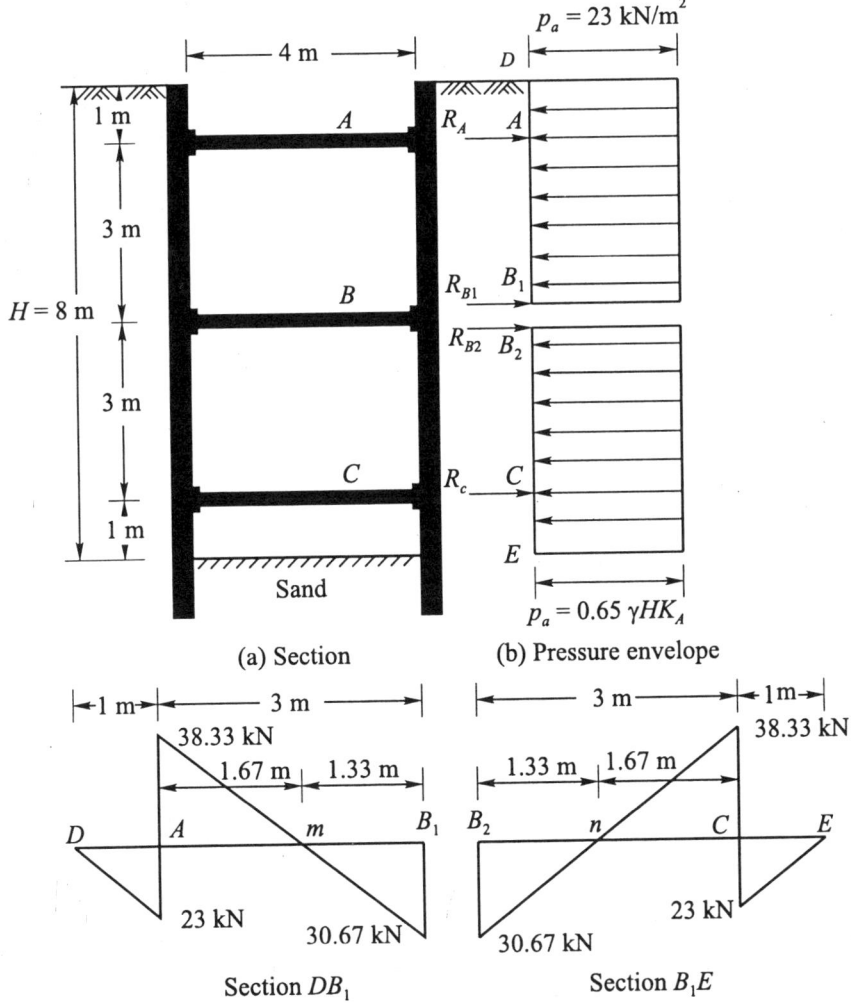

(a) Section

(b) Pressure envelope

Section DB_1

Section B_1E

(c) Shear force distribution

Fig. Ex. 18.1

(b) Strut loads

The reactions at the ends of struts A, B and C are represented by R_A, R_B and R_c respectively. For reaction R_A, take moments about B

$$R_A \times 3 = 4 \times 23 \times \frac{4}{2} \text{ or } R_A = \frac{184}{3} = 61.33 \text{ kN}$$

$$R_{B1} = 23 \times 4 - 61.33 = 30.67 \text{ kN}$$

Due to the symmetry of the load distribution,

$$R_{B1} = R_{B2} = 30.67 \text{ kN, and } R_A = R_C = 61.33 \text{ kN.}$$

Now the strut loads are (for $L = 4$ m)

Strut A, $\qquad P_A = 61.33 \times 4 \approx 245$ kN

Strut B, $\qquad P_B = (R_{B1} + R_{B2}) \times 4 = 61.34 \times 4 \approx 245$ kN

Strut C, $\qquad P_C = 245$ kN

(c) Moment of the pile wall section

To determine moments at different points it is necessary to draw a diagram showing the shear force distribution.

Consider sections DB_1 and B_2E of the wall in Fig. Ex. 18.1 (b). The distribution of the shear forces are shown in Fig. 18.1 (c) along with the points of zero shear.

The moments at different points may be determined as follows

$$M_A = \frac{1}{2} \times 1 \times 23 = 11.5 \text{ kN-m}$$

$$M_C = \frac{1}{2} \times 1 \times 23 = 11.5 \text{ kN-m}$$

$$M_m = \frac{1}{2} \times 1.33 \times 30.67 = 20.4 \text{ kN-m}$$

$$M_n = \frac{1}{2} \times 1.33 \times 30.67 = 20.4 \text{ kN-m}$$

The maximum moment $M_{max} = 20.4$ kN-m. A suitable section of sheet pile can be determined as per standard practice.

(d) Maximum moment for wales

The bending moment equation for wales is

$$M_{max} = \frac{RL^2}{8}$$

where R = maximum strut load = 245 kN

L = spacing of struts = 4 m

$$M_{max} = \frac{245 \times 4^2}{8} = 490 \text{ kN-m}$$

A suitable section for the wales can be determined as per standard practice.

Example 18.2

Figure Ex. 18.2 (a) gives the section of a long braced cut. The sides are supported by steel sheet pile walls with struts and wales. The soil excavated at the site is stiff clay with the following properties

$$c = 800 \text{ lb/ft}^2, \phi = 0, \gamma = 115 \text{ lb/ft}^3$$

Determine

(a) The earth pressure distribution envelope.

(b) Strut loads.

(c) The maximum moment of the sheet pile section.

The struts are placed 12 ft apart centre to centre horizontally.

Solution

(a) The stability number N_s from Eq. (18.3a) is

$$N_s = \frac{\gamma H}{c} = \frac{115 \times 25}{800} = 3.6 < 4$$

The soil is stiff fissured clay. As such the pressure envelope shown in Fig. 18.3 (c) is applicable. Assume $p_a = 0.3 \, \gamma H$

$$p_a = 0.3 \times 115 \times 25 = 863 \text{ lb/ft}^2$$

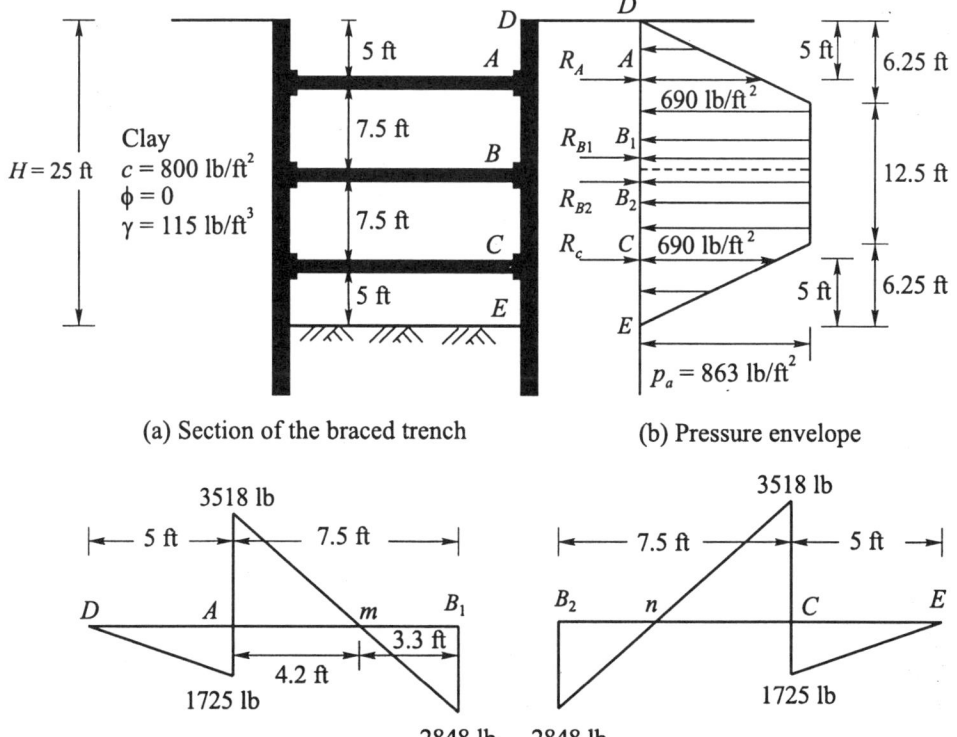

(a) Section of the braced trench

(b) Pressure envelope

(c) Shear force diagram

Fig. Ex. 18.2

The pressure envelope is drawn as shown in Fig. Ex. 18.2 (b).

(b) Strut loads

Taking moments about the strut head B_1 (B)

$$R_A \times 7.5 = \frac{1}{2} \, 863 \times 6.25 \left(\frac{6.25}{3} + 6.25 \right) + 863 \times \frac{(6.25)^2}{2}$$

$$= 22.47 \times 10^3 + 16.85 \times 10^3 = 39.32 \times 10^3$$

$$R_A = 5243 \text{ lb/ft}$$

$$R_{B1} = \frac{1}{2} \times 863 \times 6.25 + 863 \times 6.25 - 5243 = 2848 \text{ lb/ft}$$

Due to symmetry

$$R_A = R_C = 5243 \text{ lb/ft}$$

$$R_{B2} = R_{B1} = 2848 \text{ lb/ft}$$

Strut loads are:

$$P_A = 5243 \times 12 = 62{,}916 \text{ lb} = 62.92 \text{ kips}$$

$$P_B = 2 \times 2848 \times 12 = 68{,}352 \text{ lb} = 68.35 \text{ kips}$$

$$P_C = 62.92 \text{ kips}$$

(c) Moments

The shear force diagram is shown in Fig. Ex. 18.2 (c) for sections DB_1 and B_2E

Moment at $\quad A = \dfrac{1}{2} \times 5 \times 690 \times \dfrac{5}{3} = 2{,}875 \text{ lb-ft/ft of wall}$

Moment at $\quad m = 2848 \times 3.3 - 863 \times 3.3 \times \dfrac{3.3}{2} = 4699 \text{ lb-ft/ft}$

Because of symmetrical loading

Moment at A = Moment at C = 2875 lb-ft/ft of wall

Moment at m = Moment at n = 4699 lb-ft/ft of wall

Hence, the maximum moment = 4699 lb-ft/ft of wall.

The section modulus and the required sheet pile section can be determined in the usual way.

18.5 PIPING FAILURES IN SAND CUTS

Sheet piling is used for cuts in sand and the excavation must be dewatered by pumping from the bottom of the excavation. Sufficient penetration below the bottom of the cut must be provided to reduce the amount of seepage and to avoid the danger of piping.

Piping is a phenomenon of water rushing up through pipe-shaped channels due to large upward seepage pressure. When piping takes place, the weight of the soil is counteracted by the upward hydraulic pressure and as such there is no contact pressure between the grains at the bottom of the excavation. Therefore, it offers no lateral support to the sheet piling and as a result the sheet piling may collapse. Further the soil will become very loose and may not have any bearing power. It is therefore, essential to avoid piping. Piping can be reduced by increasing the depth of penetration of sheet piles below the bottom of the cut.

18.6 PROBLEMS

18.1 Figure Prob. 18.1 shows a braced cut in medium dense sand. Given $\gamma = 18.5 \text{ kN/m}^3$, $c = 0$ and $\phi = 38°$. (*a*) Draw the pressure envelope, (*b*) determine the strut loads, and (*c*) determine the maximum moment of the sheet pile section.

The struts are placed laterally at 4 m centre to centre.

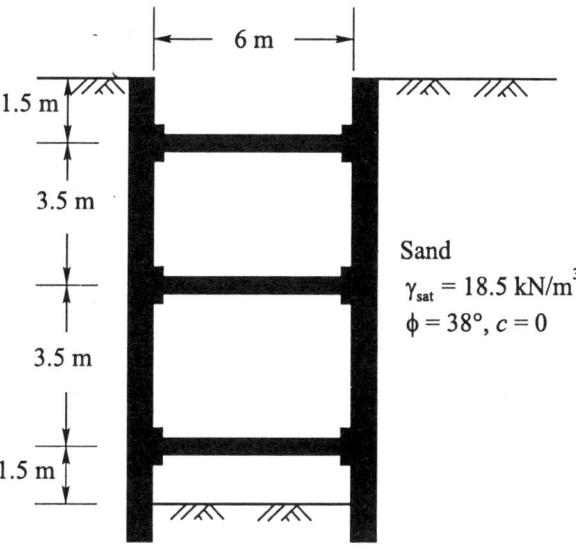

Fig. Prob. 18.1

18.2 Figure Prob. 18.2 shows the section of a braced cut in clay. Given: $c = 650 \text{ lb/ft}^2$, $\gamma = 115 \text{ lb/ft}^3$. (*a*) Draw the earth pressure envelope, (*b*) determine the strut loads, and (*c*) determine the maximum moment of the sheet pile section.

Assume that the struts are placed laterally at 12 ft centre to centre.

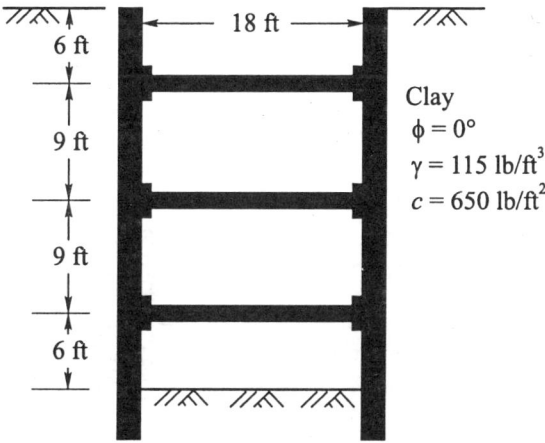

Fig. Prob. 18.2

PART B: DRAINAGE

18.7 INTRODUCTION

When the depth of excavation is greater than the distance to the free water surface in a pervious soil having a coefficient of permeability greater than about 10^{-3} cm/sec, the soil must be drained to permit construction of foundations in the dry. If the coefficient of permeability of the soil is within the range 10^{-3} to 10^{-5} cm/sec, the quantity of water that seeps into the excavation may be inconsequential but drainage may still be required to maintain the stability of the sides and bottom of the excavation. If the coefficient of permeability is smaller than about 10^{-7} cm/sec, the soil is likely to possess sufficient cohesion to overcome the influence of the seepage forces and drainage may not be required even if the excavation extends for a considerable depth below the water table.

After completion of structures with basements, it is often necessary to maintain the water level in a lowered position. This requires the installation of permanent drains.

18.8 DITCHES AND SUMPS

Where space permits, ditches may be used to lower the water table in sand or in other materials made pervious by cracks or joints. In silty or fine sands, the side slopes must ordinarily be relatively flat on account of the seepage pressures exerted by the entering water.

The relatively flat slopes required for open ditches in sand generally preclude the use of ditches for lowering the water table more than a very few feet. However, open ditches are commonly used in the bottom of an excavation to collect the water that seeps into the hole. Such ditches lead to sumps from which the water is pumped.

A *sump* is a pit with its bottom below the level of the ditches that enter it. Considerable care must often be exercised to prevent sand and silt beside and beneath the sump from washing in and being pumped out with the water. To reduce the loss of sand by pumpage and to prevent the consequent instability, it is often desirable to line the sides of the sump and to cover the bottom with coarse-grained material that acts as a filter. Such a sump for an open cut in sand is shown in Fig. 18.8. A large diameter pipe, set vertically, with filter material in the lower part, is often satisfactory.

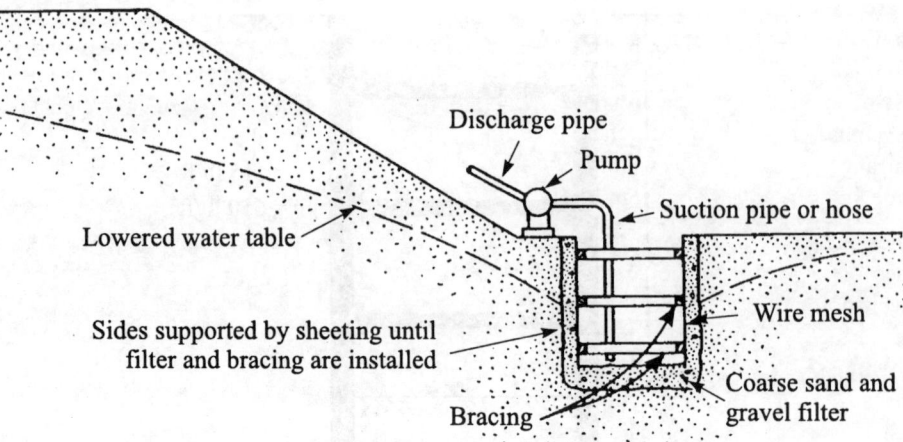

Fig. 18.8 Filter-protected sump for open cut in sand

Drainage for either temporary or permanent construction may also be accomplished by excavating trenches rather than ditches, placing drain tiles or perforated pipes in the trenches, and filling over the drains with pervious material. To prevent washing the fine material out of the backfill or surrounding soil, it may be necessary to surround the drains with granular material satisfying the requirements for a filter. The width of the openings in the drain pipe should preferably be equal to about the 60 per cent size (D_{60}) of the surrounding material.

18.9 WELL POINTS

The water table in granular materials may be lowered by means of well points. A *well point* is a perforated pipe about 3 ft long and 1½ in. diameter covered by a cylindricalscreen to prevent the enhance of fine particles. It is attached to the bottom of a 1½ or 2 in. riser pipe inserted vertically in the ground. The point can usually be jetted into the ground without driving, although stiff strata sometimes require the use of a punching device or of an auger. On the job, a line of well points spaced at 2 to 5 ft is connected to a 6, 8, or 10 in. header pipe laid on the surface of the ground. The header, in turn, is connected to a suction pump. The various parts of the assembly are shown in Fig. 18.9.

If the depth of excavation below the water table is greater than about 15 ft, several stages of well points are required. The first excavation is made to a depth of about 15 ft, whereupon additional well points are jetted into the ground before the next 15 ft are excavated. The points are generally arranged in such a manner that the sides of the excavation consist of slopes connected by flat berms containing drainage ditches. This arrangement, known as a *multiplestage setup,* is shown in Fig. 18.10.

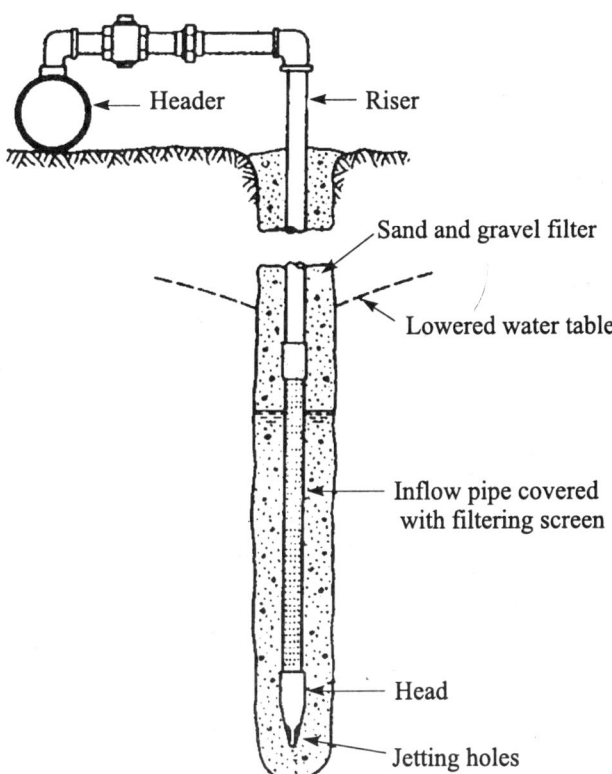

Fig. 18.9 Details of well-point assembly

When the quantity to be pumped per well point is small, a *jet-eductor system* may be used in place of a multiple-stage setup. Each well point is installed in the bottom of a cased hole. The well point is attached to the under side of a jet-eductor pump, which in turn is connected to the surface by two pipes, one for incoming high-pressure water that operates the pump, and the other for the return water including that furnished by the well point. The efficiency is low because most of the water taken from the system was previously injected to operate the pumps. However, a single-stage system can lower the water table as much as 100 ft. When limitations of space preclude the use of a multiple-stage system, jet eductors may prove economical.

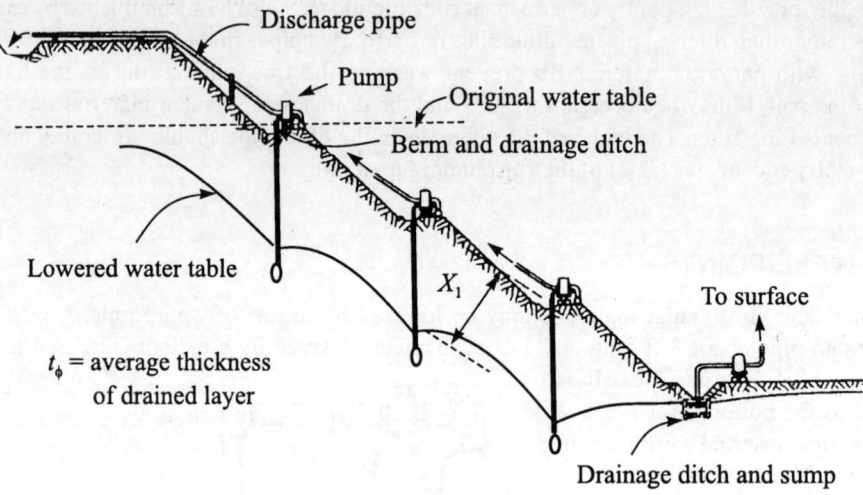

Fig. 18.10 Multiple-stage well-point set-up

If the permeability is less than about 10^{-4} cm/sec, drainage cannot be accomplished simply by pumping from well points because the capillary forces prevent the flow of water from the pores of the soil. However, drainage can be accomplished by consolidation. This may be done by means of the vacuum method of well-point operation (Fig. 18.11). In this method, the well points are set in holes about 8 in. in diameter made by either auguring or jetting. A filter of medium to coarse sand is then placed around the point and pipe to within about 2 or 3 ft of the surface. Above the filter an impervious material such as clay is tamped to form a seal. Special techniques may be required in holes that fail to stand open.

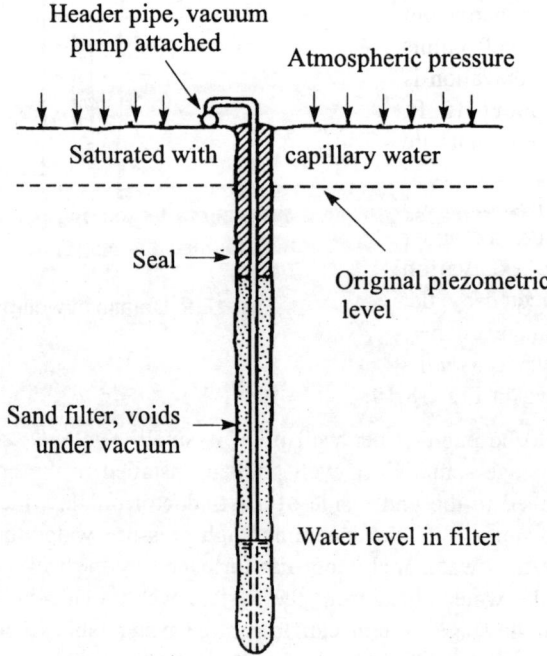

Fig. 18.11 Vacuum well-point installation

The pumps for such an installation must be capable of maintaining a vacuum in the well points and the surrounding filter. The pressure around the well points is thereby reduced to a small fraction of atmospheric pressure whereas the surface of the ground is acted upon by the weight of the atmosphere. Thus, the soil becomes consolidated under a pressure of about 1 ton/sq ft.

The vacuum process is highly effective in silts and organic silts, but the time required to achieve consolidation and stability is likely to be several weeks.

18.10 DEEP-WELL PUMPS

For very deep excavations the multiple-stage well-point setup has the disadvantage that the water level is pulled down rather abruptly at the edges of the excavation. As a consequence, the hydraulic gradient near the excavation is quite large and the resulting seepage pressures may lead to instability of the side slopes. Under these circumstances, it is safer and sometimes more economical to install large-diameter drain-age wells equipped with deep-well pumps. A typical arrangement for such wells is shown in Fig. 18.12. The spacing, which commonly varies from 20 to 200 ft, depends on several factors, including the permeability of the soil and the depth of the permeable stratum.

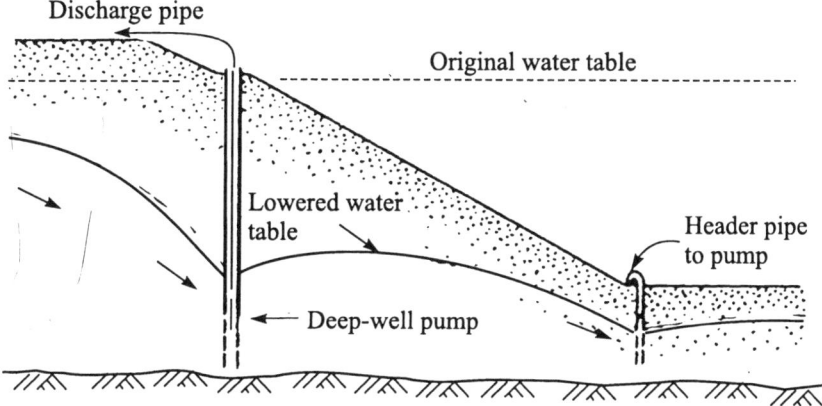

Fig. 18.12 Drainage by means of deep-well pumps

The drainage wells for such an installation consist of cased holes with diameters commonly ranging from 6 to 24 in. The casing is perforated in the pervious zones. The pumping unit consists of a submersible multistage turbine pump and motor mounted on a common vertical shaft. A 10-in. pump of this type is capable of discharging about 1000 gpm against a head of 80 ft and requires about a 30 hp motor.

18.11 SAND DRAINS

In many instances, it is necessary to build a structure or to construct an embankment on fine-grained soils with low shearing resistance. The initial strength of the soils may be too low to support the weight of the structure without failure. However, if the weak soils can be drained rapidly enough to let consolidation occur at nearly the same rate as the load is applied, the strength of the material may increase sufficiently to permit safe construction.

For the purpose of accelerating drainage in relatively impervious deposits, vertical drains may be installed. These drains commonly consist of columns of sand about 2 ft in diameter arranged in a pattern of squares or triangles at a spacing of 10 to 15 ft. The ground surface at the top of the drains is covered by a pervious blanket, and the structure or embankment is constructed on top of the blanket (Fig. 18.13). As the weight increases, the water is squeezed out of the subsoil into the drains from which it escapes through the drainage blanket to ditches. The rate at which consolidation will occur can be controlled by varying the spacing and diameter of the drains.

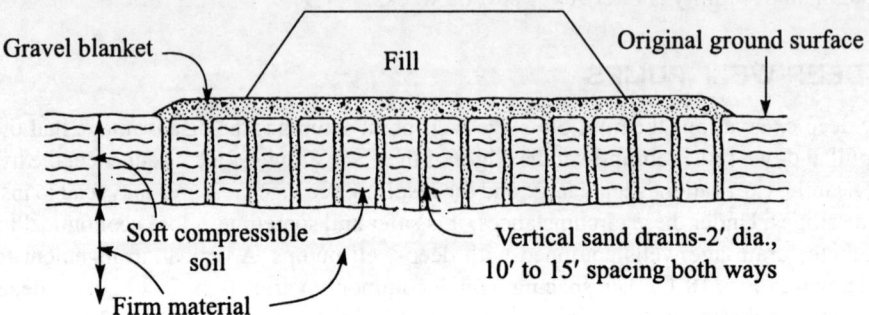

Fig. 18.13 Sand-drain installation

The act of installing sand drains may considerably disturb the structure of the soil, whereupon the permeability and strength may be decreased and the compressibility increased. The disturbance is especially great if the drains are formed with the aid of a mandrel that displaces the soil. Failure to minimize or to take account of these unfavorable effects has led to many unsatis-factory installations.

SI Units in Geotechnical Engineering

Introduction

There has always been some confusion with regards to the system of units to be used in engineering practices and other commercial transactions. FPS (Foot-pound-second) and MKS (Metre-Kilogram-second) systems are still in use in many parts of the world. Sometimes a mixture of two or more systems are in vogue making the confusion all the greater. Though the SI (Le System International d'Unites or the International System of Units) units was first conceived and adopted in the year 1960 at the Eleventh General Conference of Weights and Measures held in Paris, the adoption of this coherent and systematically constituted system is still slow because of the past association with the FPS system. The conditions are now gradually changing and possibly in the near future the SI system will be the only system of use in all academic institutions in the world over. It is therefore essential to understand the basic philosophy of the SI units.

The Basics of the SI System

The SI system is a fully coherent and rationalized system. It consists of six basic units and two supplementary units, and several derived units (Table 1).

Table 1 Basic units of interest in geotechnical engineering

S.no.	Quantity	Unit	SI symbol
1.	Length	Metre	m
2.	Mass	Kilogram	kg
3.	Time	Second	S
4.	Electric current	Ampere	A
5.	Thermodynamic temperature	Kelvin	K

Supplementary Units

The supplementary units include the *radian* and *steradian*, the units of plane and solid angles, respectively.

Derived Units

The derived units used by geotechnical engineers are tabulated in Table 2.

Table 2 Derived units

Quantity	Unit	SI symbol	Formula
Acceleration	metre per second squared	–	m/sec^2
Area	square metre	–	m^2
Density	kilogram per cubic metre	–	kg/m^3
Force	newton	N	$kg\text{-}m/s^2$
Pressure	pascal	Pa	N/m^2
Stress	pascal	Pa	N/m^2
Moment or torque	newton-metre	N-m	$kg\text{-}m^2/s^2$
Unit weight	newton per cubic metre	N/m^3	$kg/s^2\,m^2$
Frequency	hertz	Hz	cycle/sec
Volume	cubic metre	m^3	–
Volume	litre	L	$10^{-3}\,m^3$
Work (energy)	joule	J	N-m

Prefixes are used to indicate *multiples* and *submultiples* of the basic and derived units as given below.

Factor	Prefix	Symbol
10^6	mega	M
10^3	kilo	k
10^{-3}	milli	m
10^{-6}	micro	μ

Mass

Mass is a measure of the amount of matter an object contains. The mass remains the same even if the object's temperature and its location change. Kilogram, kg, is the unit used to measure the quantity of mass contained in an object. Sometimes *Mg* (*megagram*) and gram (*g*) are also used as a measure of mass in an object.

Time

Although the second (s) is the basic SI time unit, minutes (min), hours (h), days (d) etc. may be used as and where required.

Force

As per Newton's second law of motion, force, F, is expressed as $F = Ma$, where, M = mass expressed in kg, and a is acceleration in units of m/sec^2. If the acceleration is g, the standard value of which is $9.80665\ m/sec^2 \approx 9.81\ m/s^2$, the force F will be replaced by W, the weight of the body. Now the above equation may be written as $W = Mg$.

The correct unit to express the weight W, of an object is the *newton* since the weight is the gravitational force that causes a downward acceleration of the object.

Newton, N, is defined as the force that causes a 1 kg mass to accelerate 1 m/s².

or
$$1 \text{ N} = 1 \frac{\text{kg-m}}{s^2}$$

Since, a *newton*, is too small a unit for engineering usage, multiples of newtons expressed as *kilonewton, kN,* and *meganewton, MN,* are used. Some of the useful relationships are

$$1 \text{ kilonewton, kN} = 10^3 \text{ newton} = 1000 \text{ N}$$

$$\text{meganewton, MN} = 10^6 \text{ newton} = 10^3 \text{ kN} = 1000 \text{ kN}$$

Stress and Pressure

The unit of *stress* and *pressure* in SI units is the *pascal (Pa)* which is equal to 1 newton per square metre (N/m^2). Since a *pascal* is too small a unit, multiples of pascals are used as *prefixes* to express the unit of stress and pressure. In engineering practice kilopascals or megapascals are normally used. For example,

$$1 \text{ kilopascal} = 1 \text{ kPa} = 1 \text{ kN/m}^2 = 1000 \text{ N/m}^2$$

$$1 \text{ megapascal} = 1 \text{ MPa} = 1 \text{ MN/m}^2 = 1000 \text{ kN/m}^2$$

Density

Density is defined as mass per unit volume. In the SI system of units, mass is expressed in kg/m^3. In many cases, it may be more convenient to express density in megagrams per cubic metre or in gm per cubic centimetre. The relationships may be expressed as

$$1 \text{ g/cm}^3 = 1000 \text{ kg/m}^3 = 10^6 \text{ g/m}^3 = 1 \text{ Mg/m}^3$$

It may be noted here that the density of water, ρ_w is exactly 1.00 g/cm³ at 4 °C, and the variation is relatively small over the range of temperatures in ordinary engineering practice. It is sufficiently accurate to write

$$\rho_w = 1.00 \text{ g/cm}^3 = 10^3 \text{ kg/m}^3 = 1 \text{ Mg/m}^3$$

Unit Weight

Unit weight is still the common measurement in geotechnical engineering practice. The relationship between unit weight, γ, and density ρ, may be expressed as $\gamma = \rho g$.

For example, if the density of water, $\rho_w = 1000 \text{ kg/m}^3$, then

$$\gamma_w = \rho_w g = 1000 \frac{\text{kg}}{\text{m}^3} \times 9.81 \frac{\text{m}}{s^2} = 9810 \frac{\text{kg}}{\text{m}^3} \cdot \frac{\text{m}}{s^2}$$

Since,
$$1 \text{ N} = 1 \frac{\text{Kg m}}{s^2}, \quad \gamma_w = 9810 \frac{\text{N}}{\text{m}^3} = 9.81 \text{ kN/m}^3$$

Table 3 Conversion factors

To convert	SI to FPS			FPS to SI		
	From	To	Multiply by	From	To	Multiply by
Length	m	ft	3.281	ft	m	0.3048
	m	in	39.37	in	m	0.0254
	cm	in	0.3937	in	cm	2.54
	mm	in	0.03937	in	mm	25.4
Area	m^2	ft^2	10.764	ft^2	m^2	929.03×10^{-4}
	m^2	in^2	1550	in^2	m^2	6.452×10^{-4}
	cm^2	in^2	0.155	in^2	cm^2	6.452
	mm^2	in^2	0.155×10^{-2}	in^2	mm^2	645.16
Volume	m^3	ft^3	35.32	ft^3	m^3	28.317×10^{-3}
	m^3	in^3	61,023.4	in^3	m^3	16.387×10^{-6}
	cm^3	in^3	0.06102	in^3	cm^3	16.387
Force	N	lb	0.2248	lb	N	4.448
	kN	lb	224.8	lb	kN	4.448×10^{-3}
	kN	kip	0.2248	kip	kN	4.448
	kN	US ton	0.1124	US ton	kN	8.896
Stress	N/m^2	lb/ft^2	20.885×10^{-3}	lb/ft^2	N/m^2	47.88
	kN/m^2	lb/ft^2	20.885	lb/ft^2	kN/m^2	0.04788
	kN/m^2	$US ton/ft^2$	0.01044	$US ton/ft^2$	kN/m^2	95.76
	kN/m^2	kip/ft^2	20.885×10^{-3}	kip/ft^2	kN/m^2	47.88
	kN/m^2	lb/in^2	0.145	lb/in^2	kN/m^2	6.895
Unit weight	kN/m^3	lb/ft^3	6.361	lb/ft^3	kN/m^3	0.1572
	kN/m^3	lb/in^3	0.003682	lb/in^3	kN/m^3	271.43
Moment	N-m	lb-ft	0.7375	lb-ft	N-m	1.3558
	N-m	lb-in	8.851	lb-in	N-m	0.11298
Moment in inertia	mm^4	in^4	2.402×10^{-6}	in^4	mm^4	0.4162×10^6
	m^4	in^4	2.402×10^6	in^4	m^4	0.4162×10^{-6}
Section modulus	mm^3	in^3	6.102×10^{-5}	in^3	mm^3	0.16387×10^5
	m^3	in^3	6.102×10^4	in^3	m^3	0.16387×10^{-4}
	m/min	ft/min	3.281	ft/min	m/min	0.3048
Hydraulic	cm/min	ft/min	0.03281	ft/min	cm/min	30.48
conductivity	m/sec	ft/sec	3.281	ft/sec	m/sec	0.3048
	cm/sec	in/sec	0.3937	in/sec	cm/sec	2.54
Coefficient	cm^2/sec	in^2/sec	0.155	in^2/sec	cm^2/sec	6.452
of consolidation	$m^2/year$	in^2/sec	4.915×10^{-5}	in^2/sec	$m^2/year$	20.346×10^3
	cm^2/sec	ft^2/sec	1.0764×10^{-3}	ft^2/sec	cm^2/sec	929.03

Table 4 Conversion factors—general

To convert from	To	Multiply by
Angstrom	inches	3.9370079×10^{-9}
	feet	3.28084×10^{-10}
	units millimetres	1×10^{-7}
	centimetres	1×10^{-8}
	metres	1×10^{-10}
Microns	inches	3.9370079×10^{-5}
US gallon (gal)	cm^3	3785
	m^3	3.785×10^{-3}
	ft^3	0.133680
	litres	3.785
Pounds	dynes	4.44822×10^5
	grams	453.59243
	kilograms	0.45359243
Tons (short or US tons)	kilograms	907.1874
	pounds	2000
	kips	2
Tons (metric)	grams	1×10^6
	kilograms	1000
	pounds	2204.6223
	kips	2.2046223
	tons (short or US tons)	1.1023112
$kips/ft^2$	lbs/in^2	6.94445
	lbs/ft^2	1000
	US tons/ft^2	0.5000
	kg/cm^2	0.488244
	metric ton/ft^2	4.88244
Pounds/in^3	gms/cm^3	27.6799
	kg/m^3	27679.905
	lbs/ft^3	1728
Poise	kN-sec/m^2	10^{-4}
	poise	10^{-3}
millipoise	kN-sec/m^2	10^{-7}
	gm-sec/cm^2	10^{-6}
ft/min	ft/day	1440
	ft/year	5256×10^2
ft/year	ft/min	1.9025×10^{-6}
cm/sec	m/min	0.600
	ft/min	1.9685
	ft/year	1034643.6

References

1. Abduljauwad, SN, Al-Sulaimani, GJ, Basunbul, A, and Al-Buraim, I (1998). "Laboratory and Field Studies of Response of Structures to Heave of Expansive Clay", *Geotechnique*, 48 No. 1.

2. Abouleid, AF (1982). "Measurement of Swelling and Collapsible Soil Properties", *Foundation Engineering*, Edited by George Pilot. Presses de le Cole Nationale des Ponts et Chaussees, Paris, France.

3. Alizadeh, M, and Davisson, MT (1970), "Lateral Load Test on Piles-Arkansas River Project", J of SMFD, SM 5, Vol. 96.

4. Alpan, I. (1967). "The Empirical Evaluation of the Coefficient K_o and K_{or} Soil and Foundation", *J Soc. Soil Mechanics Foundation Engineering.*

5. Altmeyer, WT (1955). "Discussion of Engineering Properties of Expansive Clays", *Proc. ASCE*, Vol. 81, No. SM 2.

6. Amar, S, Jézéquel, JF (1972). "Essais en Place et en Laboratoire sus Sols Coberents Comparaisons des Resultats", *Bulletin de Liaison des Laboratoires des Ponts et Chaussees*, No. 58.

7. American Association of State Highway and Transportation Officials (1982). AASHTO Materials, Part I, Specifications, Washington, DC.

8. American Society for Testing and Materials (1994). Annual Book of ASTM Standards, Vol. 04.08, Philadelphia, Pa.

9. American Society of Civil Engineers, New York (1972). "Performance of Earth and Earth-supported Structures", Vol. 1, Part 2, *Proc. of the Specialty Conf.* Purdue University, Lafayette, Indiana.

10. Anderson, P (1956). "Substructure Analysis and Design", Second Ed., The Rotand Press Co., New York.

11. Applied Technology Council, (1978). "Tentative Provisions for the Developing of Seismic Regulations for Buildings", Publication ATC-3-06.

12. Arnold, RN, GN Bycroft, and GB Warburton, (1955). "Forced Vibrations of a Body on an In Infinite Elastic Solid", J Appl. Mech., Trans. ASME, Vol. 77.

13. Atterberg, A. (1911). "Uber die Physikalisehe Bodenuntersuchung Und Uber die Plastizitat der Tone", *Int. Mitt. Für Bodenkunde*, Vol. 1, Berlin.

14. Awad, A and Petrasovits, G (1968). "Consideration on the Bearing Capacity of Vertical and Batter Piles Subjected to Forces Acting in Different Directions", *Proc. 3rd Budapest Conf. SM and FE*, Budapest.

15. Azzouz, AS, Kriezek, RJ, and Corotis, RB (1976). "Regression Analysis of Soil Compressibility", *Soils Foundation*, Tokyo, Vol. 16, No. 2.

16. Baguelin, F, Jezequel, JF and Shields, DH (1978). "The Pressuremeter and Foundation Engineering", Trans Tech. Publications, Clausthal, Germany.

17. Balaam, NP, Poulos, HG, and Booker, JR (1975). "Finite Element Analysis of the Effects of Installation on Pile Load-settlement Behaviour", *Geotechnical Engineering*, Vol. 6, No. 1.

18. Balla, A, (1962). "Bearing Capacity of Foundations", JSMFD, Proc. ASCE.

19. Barden, L, McGown, A, and Collins, K (1973). "The Collapse Mechanism in Partly Saturated Soil", *Engineering Geology*, Vol. 7.

20. Barkan, DD, (1962). "Dynamics of Bases and Foundations", McGraw-Hill Book Co., New York.

21. Barden, L, McGown, A, and Collins, K (1973). "The Collapse Mechanism in Partly Saturated Soil", *Engineering Geology*, Vol. 7.

22. Baver, LD (1940). "Soil Physics", John Wiley and Sons, New York.

23. Bazara, AR, (1967). "Use of Standard Penetration Test for Estimating Settlements of Shallow Foundations on Sand", PhD Thesis, University of Illinois, USA.

24. Bell, AL, (1915). "The Lateral Pressure and Resistance of Clay and Supporting Power of Clay Foundations", *Century of Soil Mechanics*, ICF, London.

25. Berezantsev, VG, Khristoforov, V, and Golubkov, V (1961). "Load Bearing Capacity and Deformation of Piled Foundations", *Proc. 5th Int. Conf. SM and FE*, Vol. 2.

26. Berg, RR, Bonaparte, R, Anderson, RP, and Chouery, VE (1986). "Design Construction and Performance of two Tensar Geogrid Reinforced Walls", *Proc. 3rd Intl. Conf. Geotextiles*, Vienna: Austrian Society of Engineers.

27. Bergado, DT, Anderson, LR, Miura, N, Balasubramaniam, AS (1996). "Soft Ground Improvement", Published by ASCE Press, ASCE, New York.

28. Biot, MA, and Clingan, FM (1941). "General Theory of Three-Dimensional Consolidation", *J Applied Physics*, Vol. 12.

29. Bishop, AW (1954). "The use of Pore Pressure Coefficients in Practice", *Geotechnique*, Vol. 4, No. 4, London.

30. Bishop, AW (1955). "The use of Slip Circle in the Stability Analysis of Earth Slopes", *Geotechnique*, Vol. 5, No. I.

31. Bishop, AW, (1960-61). "The Measurement of Pore Pressure in the Triaxial Tests", *Proc. Conf. Pore Pressure and Suction in Soils*, Butterworths, London.

32. Bishop, AW, and Bjerrum, L (1960). "The Relevance of the Triaxial Test to the Solution of Stability Problems", *Proc. Res. Conf. Shear Strength of Soils*, ASCE.

33. Bishop, AW, and Henkel, DJ (1962). "The Measurement of Soil Properties in the Triaxial Test", Second Ed., Edward Arnold, London.

34. Bishop, AW, and Morgenstern, N (1960). "Stability Coefficients for Earth Slopes", *Geotechnique*, Vol. 10, No 4, London.

35. Bjerrum, L (1972). "Embankments on Soft Ground", *ASCE Conf. on Performance of Earth and Earth-supported Structures*, Purdue University.

36. Bjerrum, L (1973). "Problems of Soil Mechanics and Construction on Soft Days", *Proc. 8th Int. Conf. on Soil Mechanics and Foundation Engineering*, Moscow, 3.

37. Bjerrum, L, and Eide, O (1956). "Stability of Strutted Excavations in Clay", *Geotechnique*, Vol. 6, No. 1.

38. Bjerrum, L, Casagrande, A, Peck, RB, and Skempton, AW, eds (1960). "From Theory to Practice in Soil Mechanics", Selections from the Writings of Karl Terzaghi, John Wiley and Sons.

39. Bjerrum, L, and Simons, NE (1960). "Comparison of Shear Strength Characteristics of Normally Consolidated Clay", *Proc. Res. Conf. Shear Strength Cohesive Soils*, ASCE.

40. Boussinesq, JV (1885). "Application des Potentiels á L 'Etude de L' Equilibre et Due Mouvement Des Solides Elastiques", Gauthier-Villard, Paris.

41. Bowles, JE, (1988). "Foundation Analysis and Design", McGraw-Hill Book Company, New York.

42. Bowles JE (1992). "Engineering Properties of Soils and their Measurement", McGraw Hill, New York.

43. Bowles, JE (1996). "Foundation Analysis and Design", McGraw-Hill, New York.

44. Briaud, JL, Smith, TD, and Tucker, LM (1985). "A Pressuremeter Method for Laterally Loaded Piles", Int. Conf. of SM and FE, San Francisco.

45. Broms, BB [1964 (a)]. "Lateral Resistance of Piles in Cohesive Soils", JSMFD, ASCE, Vol. 90, SM 2.

46. Broms, BB [1964 (b)]. "Lateral Resistance of Piles in Cohesionless Soils", JSMFD, ASCE, Vol. 90, SM 3.

47. Broms, BB (1966). "Methods of Calculating the Ultimate Bearing Capacity of Piles—A Summary", *Soils-Soils*, No. 18-19: 21–32.

48. Broms, BB (1987). "Soil Improvement Methods in Southeast Asia for Soft Soil", *Proc. 8 ARC on SM and FE*, Kyoto, Vol. 2.

49. Broms, BB (1995). "Fabric Reinforced Soil", *Developments in Deep Foundations and Ground Improvement Schemes*, Balasubramanium, et al., (eds), Balkema Rotterdam, ISBN 9054105933.

50. Brown, JD, and Meyerhof, GG (1969). "Experimental Study of Bearing Capacity in Layered Clays", 7th ICSFME, Vol. 2.

51. Brown, RE (1977). "Drill Rod Influence on Standard Penetration Test," JGED, ASCE, Vol. 103, SM 3.

52. Brown, RE (1977). "Vibroflotation Compaction of Cohesioinless Soils", *J of Geotechnical Engineering Div.*, ASCE, Vol. 103, No. GT 12.

53. Burland, JB (1973). "Shaft Friction of Piles in Clay—A Simple Fundamental Approach", *Ground Engineering*, Vol. 6, No. 3.

54. Burland, JB, and Burbidge, MC (1985). "Settlement of Foundations on Sand and Gravel", Proc. ICE, Vol. 78.

55. Burland, JB, and Wroth, CP (1974). "Allowable and Differential Settlement of Structures Including Damage and Soil Structure Interaction," Proceedings, Conf. on Settlement of Structures Cambridge University, England.

56. Burmister, DM (1943). "The Theory of Stresses and Displacements in Layer Systems and Application to Design of Airport Runways", *Proc. Highway Res. Board*, Vol. 23.

57. Button, SJ (1953). "The Bearing Capacity of Footings on a Two-layer Cohesive Sub-soil", *3rd ICEMFE*, Vol. 1.

58. Bycroft, GN, (1956). "Forced Vibrations of a Rigid Circular Plate on a Semi-Infinite Elastic Space and on an Elastic Structure", Philiosophical Trans. Royal Society, London, Vol. 248.

59. Capazzoli, L (1968). "Test Pile Program at St. Gabriel, Louisiana, Louis", *J Capazzoli and Associates.*

60. Caquot, A, and Kerisel, J (1984). "Tables for the Calculation of passive Pressure, Active Pressure, and Bearing Capacity of Foundations", (Translated by MA Bec, London), Gauthier-Villars Paris.

61. Caquot, A, and Kerisel, J (1956). "Traité de Méchanique des Sols", 2nd Ed., Gauthier Villars, Paris.

62. Caron, PC, and Thomas, FB (1975). "Injection", *Foundation Engineering Handbook*, Edited by Winterkorn and Fang, Van Nastrand Reinhold Company, New York.

63. Carrillo, N (1942). "Simple Two and Three-Dimensional Cases in the Theory of Consolidation of Soils", *J Math. Physics* Vol. 21.

64. Carter, JP, and Kulhawy, FH (1988). "Analysis and Design of Drilled Shaft Foundations Socketed into Rock", Final Report Project 1493-4, EPRI EL-5918, Geotechnical Group, Cornell University, Ithaca.

65. Casagrande, A (1931). "The Hydrometer Method for Mechanical Analysis of Soils and other Granular Materials", Cambridge, Massachusetts.

66. Casagrande, A (1932). "Research of Atterberg Limits of Soils", *Public Roads*, Vol. 13, No. 8.

67. Casagrande, A (1936). "The Determination of the Preconsolidation Load and its Practical Significance", *Proc. First Int. Conf. Soil Mechanics and Foundation Engineering*, Vol. 3.

68. Casagrande, A (1937). "Sepage through Dams", *JN Engl.* Water Works Assoc. L1 (2).

69. Casagrande, A (1958-59). "Review of Past and Current Works on Electro-osmotic Stabilisation of Soils", *Harvard Soil Mechanics*, Series No. 45. Harvard University, Cambridge Mass.

70. Casagrande, A (1967). "Classification and Identification of Soils", *Proc. ASCE*, Vol. 73, No. 6, Part 1, New York.

71. Casagrande, A, and Fadum, RE (1940). "Notes on Soil Testing for Engineering Purposes", *Soil Mechanics Series*, Graduate School of Engineering, Harvard University, Cambridge MA, Vol. 8.

72. Casagrande, L (1932). "Naeherungsmethoden zur Bestimmurg von Art' und Menge der Sickerung Durch Geschuettete Daemme", Thesis, Techniche Hochschule, Vienna.

73. Chellis, RD (1961). "Pile Foundations", McGraw Hill, New York.

74. Chen, FH (1988). "Foundations on Expansive Soils". Elsevier Science Publishing Company Inc., New York.

75. Chen, YJ, and Kulhawy, FH (1994). "Case History Evaluation of the Behaviour of Drilled Shaft Under Axial and Lateral Loading", Final Report, Project 1493-04, EPRI TR-104601, Geotechnical Group, Cornell University, Ithacka.

76. Christian, JT, and Carrier, WD (1978). "Janbu, Bjerrum, and Kjaernsli's Chart Reinterpreted", *Canadian Geotechnical Journal*, Vol. 15.

77. Clemence, SP, and Finbarr, AO (1981). "Design Considerations for Collapsible Soils", *J of the Geotechnical Engineering Division*, ASCE, Vol. 107, No. GT 3.

78. Coulomb, CA (1776). "Essai Sur Une Application des Regles des Maximis et Minimum a Queiques", *Problemes de Statique Relatifs a L'Architecture*, Mem. Acad. Roy Pres. Divers Savants, Paris, Vol. 3, Paris.

79. Coyle, HM, and Sulaiman, IH (1967). "Skin Friction for Steel Piles in Sand", *JSMFD*, ASCE, Vol. 93, SM 6.

80. Coyle, HM, and Castello, RR (1981). "New Design Correlations for Piles in Sand", *J of the Geotechnical Engineering Division*, ASCE, Vol. 107, No. GT 17.

81. Culmann, C (1875). *Die Graphische Statik*, Zurich, Meyer and Zeller.

82. D' Appolonia, E, Ellison, RD, and D' Appolonia, DJ (1975). "Drilled Piers", *Foundation Engineering Handbook*, Edited: HF Winterkorn and HY Fang, Van Nostrand Reinhold Company, New York.

83. Darcy, H (1856). "Les Fontaines Publiques de la Ville de Dijon", Paris.

84. Das, BM, and Seeley, GR (1975). "Load-Displacement Relationships for Vertical Anchor Plates", *J of the Geotechnical Engineering Div*, ASCe, Vol. 101, GT 7.

85. Davis, LH (1977). "Tubular Steel Foundation", Test Report RD-1517, Florida Power and Light company, Miami, Florida.

86. Davis, MCR, and Schlosser, F (1997). "Ground Improvement Geosystems", *Proc. Third Int. Conf. on Ground Improvement Geosystems*, London.

87. Davisson, MT (1960). "Behaviour of Flexible Vertical Piles Subjected to Moments, Shear and Axial Load", Ph.D Thesis, University of Illinois, Urbana, USA.

88. Davisson, MT, and Salley, JR (1969). "Lateral Load Tests on Drilled Piers", Performance of Deep Foundations, ASTM STP 444, ASTM.

89. De Beer, EE (1965). "Bearing Capacity and Settlement of Shallow Foundations on Sand", *Proc.* of Symp Held at Duke University.

90. De Ruiter, J, and Beringen, FL (1979). "Pile Foundations for Large North Sea Structures", *Marine Geotechnology*, Vol. 3, No. 3.

91. Dennis, ND, and Olson, RE [1983 (a)]. "Axial Capacity of Steel Pipe Piles in Sand", *Proc. ASCE Conf. on Geotechnical Practice in Offshore Engineering*, Austin, Texas.

92. Dennis, ND, and Olson, RE, [1983 (b)]. "Axial Capacity of Steel Pipe Piles in Clay", *Proc. ASCE Conf. on Geotechnical Practice in Offshore Engineering*, Austin, Texas.

93. Department of the Navy, Naval Facilities Engineering Command (1982). "Design Manual", NAVFAC DM-71.

94. Desai, CS (1974). "Numerical Design-Analysis for Piles in Sands", *J Geotechnical Engineering Division*, ASCE, Vol. 100, No. GT 6.

95. Douglas, BJ (1984). "The Electric Cone Penetrometer Test: A User's Guide to Contracting for Services, Quality Assurance, Data Analysis", The Earth Technology Corporation, Long Beach, California, USA.

96. Douglas, BJ, and Olsen, RS (1981). "Soil Classification Using the Electric Cone Penetrometer, Cone Penetration Testing and Experience", *ASCE FAll Convention.*

97. Downs, DI, and Chieurzzi, R (1966). "Transmission Tower Foundations", *J Power Div.*, ASCE, Vol. 92, PO 2.

98. Dudley, JH (1970). "Review of Collapsible Soils", *J of SM and F Div.*, ASCE, Vol. 96, No. SM 3.

99. Duncan, JM, Evans, LT, Jr, and Ooi PSK (1994). "Lateral Load Analysis of Single Piles and Drilled Shafts", *J of Geotechnical Engineering* ASCE, Vol. 120, No. 6.

100. Dunn, IS, Anderson, IR, and Kiefer, FW (1980). "Fundamentals of Geotechnical Analysis", John Wiley and Sons.

101. Dupuit, J (1863). "Etudes Theoriques et practiques sur le Mouvement des eaux dans les Canaus Decouverts et a travers les Terrains permeables", Dunod, Paris.

102. Eggestad, A (1963). "Deformation Measurement Below a Model Footing on the Surface of Dry Sand", *Proc. European Conf. Soil Mechanics*, Wiesbaden, 1.

103. Elorduy, J, JA Nieto, and EM Szekely, (1967). "Dynamic Response of Bases of Arbitrary Shapes Subjected to Periodic Vertical Loading", Proc. Int. Symp. on Wave Propagation and Dynamic Properties of Earth Materials Albuquerque.

104. Ellison, RD, D' Appolonia, E, and Thiers, GR (1971). "Load-Deformation Mechanism for Bored Piles" *JSMFD*, ASCE, Vol. 97, SM 4.

105. Fadum, RE (1948). "Influence Values for Estimating Stresses in Elastic Foundations," *Proc. Second Int. Conf. SM and FE*, Vol. 3.

106. Fellenius, W (1927). *Erdstatische Berechnungen*, (revised edition, 1939), W Ernstu, Sons, Berlin.

107. Fellenius, W (1947). *Erdstatische Berechnungen Mit Reibung and Kohasion and Unter Annahme Keissy Undrischer Gleitf Lachen*, Wilhelm Ernt and Sohn, 11 Ed., Berlin.

108. Forchheimer, P (1930). *General Hydraulics*, Third Ed., Leipzig.

109. Foster, CR, and Ashlvin, RG (1954). "Stresses and deflections induced by a uniform circular load", *Proceedings*, HRB.

110. Fox EN (1948). "The Mean Elastic Settlements of Uniformly Loaded Area at Depth Below the Ground Surface", *2nd ICSMEF*, Vol. 1.

111. Frohlick, OK (1955). "General Theory of Stability of Slopes", *Geotechnique*, Vol. 5.

112. Ghaly, AM (1997). "Load-Displacement Prediction for Horizontally Loaded Vertical Plates", *J of Geotechnical and Enviornmental Engineering*, ASCE, Vol. 123, No. 1.

113. Gibbs, HJ, and Holtz, WG (1957). "Research on Determining The Density of Sands by Spoon Penetration Testing", *4th ICSMEF*, Vol. 1.

114. Gibson, RE (1953). "Experimental Determination of the True Cohesion and True Angle of Internal Friction in Clays," *Proc. Third Int. Conf. Soil Mechanics Foundation Engineering*, Zurich, Vol. 1.

115. Gibson, RE and Lo, KY (1961). "A Theory of Consolidation for Soils Exhibiting Secondary Compression", Norwegian Geotechnical Institute, Publication No. 41.

116. Gilboy, G (1934). "Mechanics of Hydraulic Fill Dams" *Journal Boston Society of Civil Engineers*, Boston.

117. Glesser, SM (1953). "Lateral Load Tests on Vertical Fixed-Head and Free-Head Pile", ASTM, STP 154.

118. Goodman, RE (1980). "Introduction to Rock Mechanics", Wiley, New York.

119. Grim, RE (1959). "Physico-Chemical Properties of Soils", *ASCE J Soil Mechanics*, 85, SM 2, 1–17.

120. Grim, RE (1962). "Applied Clay Mineralogy", McGraw Hill.

121. Hagerty, DJ, and Nofal, MM (1992). "Design Aids: Anchored Bulkheads in Sand", *Canadian Geotechnical Journal*, Vol. 29, No. 5.

122. Hall, JR, Jr., (1962). "Effect of Amptitude on Damping and Wave Propagationin Granular Materials", PhD Dissertation, University Florida, 1962.

123. Hall, JR, Jr., (1967). "Coupled Rocking and Sliding Oscillations of Rigid Circular Footings", Proc. Int. Symp on Wave Propagation and Dynamic Properties of Earth Materials, Albuquerque, NM, 1967.

124. Hansen, JB (1970). "A Revised and Extended Formula for Bearing Capacity", *Danish Geotechnical Institute Bul.* No. 28, Copenhagen.

125. Hanzawa, H, and Kishida, T (1982). "Determination of *in situ* Undrained Strength of Soft Clay Deposits", *Soils and Foundations*, 22, No. 2.

126. Harr, ME (1962). "Ground Water and Seepage", McGraw Hill, New York.

127. Harr, ME (1966). "Foundations of Theoretical Soil Mechanics", McGraw Hill, New York.

128. Hausmann, MR (1990). "Engineering Principles of Ground Modification", McGraw Hill, New York.

129. Hazen, A (1893). "Some Physical Properties of Sands and Gravels with Special Reference to their use in Filtration", 24th Annual Report, Massachusetts, State Board of Health.

130. Hazen, A (1911). "Discussion of Dams and Sand Foundations", by AC Koenig, Trans ASCE, Vol. 73.

131. Hazen, A (1930). "Hydraulics", Third Ed., Leipzig.

132. Henkel, DJ (1960). "The Shear Strength of Saturated Remoulded Clays", *Proc. Re. Conf. Shear Strength Cohesive Soils*, ASCE.

133. Hetenyi, M (1946). "Beams on Elastic Foundations", The University of Michigan Press, Ann Arbor, MI.

134. Highter, WH, and Anders, JC (1985). "Dimensioning Footing Subjected to Eccentric Loads", *J of Geotechnical Engineering*, ASCE, Vol. III, No. GT 5.

135. Holl, DL (1940). "Stress Transmission on Earths", *Proc. Highway Res. Board*, Vol. 20.

136. Holl, DL (1941). "Plane Strain Distribution of Stress in Elastic Media", Iowa Engineering Expts Station, Iowa.

137. Holtz, RD, and Kovacs, WD (1981). "An Introduction to Geotechnical Engineering", Prentice Hall Inc., New Jersey.

138. Holtz, WG and Gibbs, HJ (1956). "Engineering Properties of Expansive Clays". Trans ASCE 121.

139. Holtz, WG and Hilf, JW (1961). "Settlement of Soil Foundations Due to Saturation", *Proc. 5th Internation on SMFE*, Paris, Vol. 1.

140. Hough, BK (1957). "Basic Soil Engineering", Ronald Press, New York.

141. Housel, WS (1929). "A Practical Method for the Seletion of Foundations Based on Fundamental Research in Soil Mechanics", *Research Bulletin*, No. 13, University of Michigan, AnnArbor.

142. Houston, WN, and Houston, SL (1989). "State-of-the-Practice Mitigation Measures for Collapsible Soil Sites", *Proceedings, Foundation Engineering: Current Principles and Practices*, ASCE, Vol. 1.

143. Hrennikoff, A (1949). "Analysis of Pile Foundations with Batter Piles", *Proc. ASCE*, Vol. 75.

144. Hsiesh, TK, (1962). "Foundation Vibrations", Proc. Institution of Civil Engineers, Vol. 22.

145. Huntington, WC (1957). "Earth Pressures and Retaining Walls", John Wiley and Sons, New York.

146. Hvorslev, MJ (1949). "Surface Exploration and Sampling of Soils for Civil Engineering Purposes". *Water Ways Experimental Station, Engineering Foundations*, New York.

147. Hvorslev, MJ (1960). "Physical Components of the Shear Strength of Saturated Clays", *Proc. Res. Conf. on Shear Strength of Cohesive Soils*, ASCE, Boulder, Colorado.

148. Ireland, HO (1957). "Pulling Tests on Piles in Sand", *Proc. 4th Int. Conf. SM and FE*, Vol. 2.

149. Ismael, NF and Klym, TW (1977). "Behavior of Rigid Piers in Layered Cohesive Soils". *J Geotechnical Engineering Division*, ASCE Vol. 104, No. GT 8.

150. Jakosky, JJ (1950). "Exploration Geophysics", Trija Publishing Company, California USA.

151. Jaky, J (1944). "The Coefficient of Earth Pressure at Rest", *J Soc. Hungarian Arch. Engineering*.

152. Janbu, N (1976). "Static Bearing Capacity of Friction Piles", *Proc. 6th European Conf. on Soil Mechanics and F Engineering* Vol. 1-2.

153. Janbu, N, Bjerrum, L, and Kjaernsli, B (1956). "Veiledning Ved Losning av Fundamentering Soppgaver", Pub. No. 16, Norwegian Geotechnical Institute.

154. Jarquio, R (1981). "Total Lateral Surcharge Pressure Due to Strip Load", *J Geotechnical Engineering Div.*, ASCE, 107, 10.

155. Jennings, JE, and Knight, K (1957). "An Additional Settlement of Foundation Due to a Collapse Structure of Sandy Subsoils on Wetting", *Proc. 4th Int. Conf on SM and Foundation Engineering*, Paris, Vol. 1.

156. Jennings, JE, and Knight, K (1975). "A Guide to Construction on or with Materials Exhibiting Additional Settlements Due to 'Collapse' of Grain Structure", *Proc. Sixth Regional Conference for Africa on SM and F Engineering*, Johannesburg.

157. Jumikis, AR (1962). "Soil Mechanics", D Van Nostrand Co. Inc., New York.

158. Kaldjian, MJ, (1969). "Discussion of Design Procedures for Dynamically Loaded Foundations", by RV Whitman and FE Richart, Jr., JSMFD, Proc. ASCE, Vol. 95, SM 1.

159. Kamon, M, and Bergado, DT (1991). "Ground Improvement Techniques", *Proc. 9 ARC on SMFE*, Bangkok, Vol. 2.

160. Kapur, R, (1971). Lateral Stability Analysis of Well Foundations PhD Thesis, University Roorkee.

161. Karlsson, R, Viberg, L (1967). "Ratio c_u/p in Relation to Liquid Limit and Plasticity Index with Special Reference to Swedish Clays", *Proc. Geotechnical Conf.*, Olso, Norway, Vol. 1.

162. Kenney, TC, Lau, D, and Ofoegbu, GI (1984). "Permeability of Compacted Granular Materials", *Canadian Geotechnical J*, 21 No. 4.

163. Kerisel, J (1961). "Foundations Profondes en Milieu Sableux", *Proc. 5th Int. Conf. SM and FE*, Vol. 2.

164. Kezdi, A (1957). "The Bearing Capacity of Piles and Pile Groups", *Proc. 4th Int. Conf. SM and FE*, Vol. 2.

165. Kezdi, A (1965). "Foundations Profondes en Milies Sableux", *Proc. 5th Int. Conf. SM and FE*, Vol. 2.

166. Kezdi, A (1965). "General Report on Deep Foundations", *Proc. 6th Int. Conf. on SM and FE*, Montreal.

167. Kezdi, A (1975). "Pile Foundations", *Foundations Engineering Handbook*, Edited by HF Winterkorn and HY Fang, Van Nostrand, New York.

168. Kishida, H (1967). "Ultimate Bearing Capacity of Piles Driven into Loose Sand", *Soil and Foundations*, Vol. 7, No. 3.

169. Koerner, RM (1985). "Construction and Geotechnical Methods in Foundation Engineering", McGraw Hill, New York.

170. Koerner, RM (1999). "Design with Geosynthetics". Fourth Edition, Prentice Hall, New Jersey.

171. Koerner, RM (2000). "Emerging and Future Developments of Selected Geosynthetic Applications" Thirty-Second Terzaghi Lecture, Drexel University, Philadelphia.

172. Kogler, F, and Scheidig, A (1962). "Soil Mechanics", by AR Jumikis, D Van Nostrand.

173. Komornik, A, and David, D (1969). "Prediction of Swelling Pressure of Clays", *JSMFD*, ASCE, Vol. 95, SM 1.

174. Kovacs, WD, and Salomone (1982). "SPT Hammer Energy Measurement", JGED, ASCE, GT 4.

175. Kozeny, JS (1931). "Ground Wasserbewegund bei Freim Spiegel, Fluss and Kanalversicheringi", *Wasserkraft and Wasserwitrschaft*, No. 3.

176. Kubo, J (1965). "Experimental Study of the Behaviour of Laterally Loaded Piles", *Proc. 6th Int. Conf. SM and FE*, Vol. 2.

177. Ladd, CC, and Foott, R (1974). "New Design Procedure for Stability of Soft Clays", J Geotechnical Engineering Div., ASCE, Vol. 100, No. GT 7.

178. Ladd, CC, Foott, R, Ishihara, K, Schlosser, F, and Poulos, HG (1977). "Stress Deformation and Strength Characteristics", *Proceedings 9th Int. Conf. SM and FE*, Tokyo, Vol. 2.

179. Lambe TW [1958 (a)]. "The Structure of Compacted Clay", *J of SM and F Div.*, ASCE, Vol. 84, No. SM 2.

180. Lambe,TW [1958 (b)]. "The Engineering Behaviour of Compacted Clay", *J Soil Mechanics, Foundation, Division*, ASCE, Vol. 84, No. SM 2.

181. Lambe TW (1962). "Soil Stabilization", Chapter 4 of Foundation Engineering, GA Leonards (ed.), McGraw Hill, New York.

182. Lambe, TW, and Whitman, RV (1969). "Soil Mechanics", John Wiley and Sons Inc.

183. Leonards, GA (1962). "Foundation Engineering", McGraw Hill.

184. Leonards, GA Cutter, WA, and Holtz, RD (1980). "Dynamic Compaction of Granular Soils", *Jou. of Geotechnical Engineering Div.*, ASCE, Vol. 106, No. GT 1.

185. Lee, KL, and Singh, A (1971). "Relative Density and Relative Compaction", *J Soil Mechanics Foundation, Div.*, ASCE Vol. 97, SM 7.

186. Lee, KL, Adams, BD, and Vagneron, JJ (1973). "Reinforced Earth Retaining Walls", Jou. of SMFD, ASCE, Vol. 99, No. SM 10.

187. Liao, SS, and Whitman, RV (1986). "Overburden Correction Factors for SPT in Sand", JGED, Vol. 112, No. 3.

188. Littlejohn, GS, and Bruce, DA (1975). "Rock Anchors: State of the Art, Part I–Design", *Ground Engineering*, Vol. 8, No. 3.

189. Louden, AG (1952). "The Computation of Permeability from Soil Tests", *Geotechnical*, 3, No. 4.

190. Lowe, J (1974). "New Concepts in Consolidation and Settlement Analysis", *J Geotechnical Engineering Div.*, ASCE, Vol. 100.

191. Loos, W and Breth, H (1949). "Modellversuche Uber Biege Beanspruch-ungen Von Pfahlen and Spunwenden", Der Bauingeniur, Vol 28.

192. Lunne, T, and Christoffersen, HP, (1985). "Interpretation of Cone Penetrometer Data for Offshore Sands", Norwegian Geotechnical Institute, Pub. No. 156.

193. Lunne, T, and Kelven, A (1981). "Role of CPT in North Sea Foundation Engineering", *Symposium on Cone Penetration Testing and Experience*, Geotechnical Engineering Devision ASCE, St. Louis.

194. Lysmer, J, (1965). "Vertical Motionof Rigid Footings", Department of Civil Engineering, University of Michigan, 1965.

195. Lysmer, J, and FE Richart, (1966). "Dynamic Response of Footings to Vertical Loading", JSMFD, Proc. ASCE, Vol. 92, SM 1.

196. Malter, H (1958). "Numerical Solutions for Beams on Elastic Foundations", *J Soil Mechanics and Foundation Div.*, LXXXIV, No. ST 2 Part 1, ASCE.

197. Marchetti, S (1980). "*In situ* Test by Flat Dilatometer", *J of Geotechnical Engineering Division*, ASCE, Vol. 106, No. GT 3.

198. Marsland, A (1971). "Large *in situ* Tests to Measure the Porperties of Stiff Fissure Clay", *Proc. 1st Australian Conf. and Geomech*, Melbourne, 1.

199. Matlock, H (1970). "Correlations for Design of Laterally Loaded Piles in Soft Clay", *Proc. 2nd Offshore Tech. Conf.*, Houston, Vol. 1.

200. Matlock, H, and Reese, LC (1960). "Generalized Solutions for Laterally Loaded Piles", *JSMFD*, ASCE, Vol. 86, N SM 5, Part 1.

201. Matlock, H, and Reese, LC (1961). "Foundation Analysis of Offshore Supported Structures", *Proc. 5th Int. Conf. SM and FE*, Vol. 2.

202. Matsuo, H (1939). "Tests on the Lateral Resistance of Piles (in Japanese)", Research Institute of Civ. Bugg, Min, of Home Affairs, Report No. 46.

203. McDonald, DH, and Skempton, AW (eds) (1955). "A Survey of Comparison betwen Calculated and Observed Settlements of Structures on Clay", *Conf. on Correlation of Calculated and Observed Stresses and Displacements*, Institution of Civil Engineers, London.

204. Means, RE, Parcher, JV (1965). "Physical Properties of Soils". Prentice Hall of India, New Delhi.

205. Menard, L (1957). "An Apparatus for Measuring the Strength of Soils in Place", M.Sc. Thesis, University Illinois, Urbana, USA.

206. Menard, L (1976). "Interpretation and Application of Pressuremeter Test Results to Foundation Design, Soils", *Soils*, No. 26.

207. Mesri, G (1973). "Coefficient of Secondary Compression", *J Soil Mechanics Foundation Div.*, ASCE, Vol. 99, SMI.

208. Mesri, G, and Godlewski, PM (1977). "Time and Stress-compressibility Interrelationship", *J. Geotechnical Engineering*, ASCE, 103, No. 5.

209. Mesri, G, and Choi, YK (1985). "Settlement Analysis of Embankments on Soft Clays", *J Geotechnical Engineering*, ASCE, III, No. 4.

210. Mesri G, and Choi, YK [1985 (b)]. "The Uniqueness of the End-of-Primary (EOP) Void Ratio-effective Stress Relationship", *Proc. 11th Int. Conf. on Soil Mechanics and Foundation Engineering*, San Francisco, No. 2.

211. Mesri, G, and Feng, TW (1986). "Stress-strain–strain Rate Relation for the Compressibility of Sensitive Natural Clays", Discussion, *Geotechnical*, 36, No. 2.

212. Mesri, G, Feng, TW, Ali, S, and Hayat, TM (1994). "Permeability Characteristics of Soft Clays", *Proc. 13th Int. Conf. On Soil Mech and Foundation Engineering*, New Delhi.

213. Meyerhof, GG (1951). "The Ultimate Bearing Capacity of Foundation", *Geotechnique*, Vol. 2, No. 4.

214. Meyerhof, GG (1953). "The Bearing Capacity of Foundations Under Eccentric and Inclined Loads", *3rd ICSMFE*, Vol. 1.

215. Meyerhof, GG (1956). "Penetration Tests and Bearing Capacity of Cohesionless Soils", JSMFD, ASCE, Vol. 82, SM 1.

216. Meyerhof, GG (1957). "Discussions on Sand Density by Spoon Penetration", *4th ICSMFE*, Vol. 3.

217. Meyerhof, GG (1957). "The Ultimate Bearing Capacity of Foundations on Slopes", *4th ICSMFC*, Vol. 1, London.

218. Meyerhof, GG (1959). "Compaction of Sands and Bearing Capacity of Piles", *JSMFD*, ASCE, Vol. 85, SM 6.

219. Meyerhof, GG (1963). "Some Recent Research on Bearing Capacity of Foundation", CGJ Ottawa, Vol. 1.

220. Meyerhof, GG (1965). "Shallow Foundations", *JSMFD*, ASCE, Vol. 91, SM 2.

221. Meyerhof, GG and Adams JI (1968). "The Ultimate Uplift Capacity of Foundations", *Canadian Geotechnical J*, Vol. 5, No. 4.

222. Meyerhof GG (1974). "Ultimate Bearing Capacity of Footings on Sand Layer Overlying Clay", *Canadian Geotechnical J*, Vol. II, No. 2.

223. Meyerhof, GG (1975). "Penetration Testing Outside Europe", General Report, *Proc. of the European Symp. on Penetration Testing*, Vol. 2, Stockholm.

224. Meyerhof, GG (1976). "Bearing Capacity and Settlement of Pile Foundations", *JGED*, ASCE, Vol. 102, GT 3.

225. Meyerhof, GG and Adams JI (1968). "The Ultimate Uplift Capacity of Foundations", *Canadian Geotechnical J*, Vol. 5, No. 4.

226. Meyerhof, GG, and Hanna, AM (1978). "Ultimate Bearing Capacity of Foundations on Layered Soil Under Inclined Load", *Canadian Geotechnical J*, Vol. 15, No. 4.

227. Mindlin, RD (1936). "Forces at a Point in the Interior of a Semi-Infinite Solid", Physics, Vol. 7.

228. Mindlin, RD (1939). "Stress Distribution Around a Tunnel", Trans ASCE. *Am. Soc. Civil Engineers*. Vol. 104.

229. Mitchell, JK, and Freitag, DR (1959). "A Review and Evaluation of Soil-Cement Pavements", *J of SM and FD*, ASCE, Vol. 85, No. SM 6.

230. Mohr, O (1900). Die Elastizitatsgrenze Und Bruseh eines Materials (The Elastic Limit and the Failure of a Material), Zeitschrift Veneins Deuesche Ingenieure, Vol. 44.

231. Mononobe, N (1929). "On the Determination of Earth Pressure During Earthquakes", *Proceedings World Engineering Conference*, Vol. 9.

232. Morgenstern, NR (1963). "Stability Charts for Earth Slopes During Rapid Drawdown", *Geotechnique*, Vol. 13, No. 2.

233. Morrison, CS, and Reese LC (1986). "A Lateral Load Test of a Full Scale Pile Group in Sand". Geotechnical Engineering Center, Bureau of Engineering Research, University of Texas, Austin.

234. Morrison, EE Jr., and Ebeling, RM (1995). "Limit Equilibrium Computation of Dynamic Passive Earth Pressure", *Canadian Geotechnical J*, Vol. 32, No. 3.

235. Murphy, VA (1960). "The Effect of Ground Water Characteristics on the Sismic Design of Structures", *Proce. 2nd World Conf. on Earthquake Engineering*, Tokyo, Japan.

236. Murthy, VNS, and Subba Rao, KS (1995). "Prediction of Nonlinear Behaviour of Laterally Loaded Long Piles" *Foundation Engineer*, Vol. 1, No. 2, New Delhi.

237. Murthy, VNS (1965). "Behaviour of Batter Piles Subjected to Lateral Loads", Ph.D. Thesis, Indian Institute of Technology, Kharagpur, India.

238. Murthy, VNS (1982). "Report on Soil Investigation for the Construction of Cooling Water System", Part II of Farakha Super thermal Power Project, National Thermal Power Corporation, New Delhi.

239. Muskat, M (1946). "The Flow of Homogeneous Fluids through Porous Media", McGraw Hill Co.

240. Nagaraj, TS, and Murthy, BRS (1985). "Prediction of the Preconsolidation Pressure and Recompression Index of Soils", *Geotechnical Testing Journal*, Vol. 8. No. 4.

241. Nelson, JD, Miller, DJ (1992). "Expansive Soils: Problems and Practice in Foundation and Pavement Engineering", John Wiley and Sons, Inc., New York.

242. Newmark, NM (1942). "Influence Charts for Computation of Stresses in *Elastic Soils*", University of Illinois Expt. Stn., *Bulletin* No. 338.

243. Nordlund, RL (1963). "Bearing Capacity of Piles in Cohesionless Soils", JSMFD, ASCE, Vol. 89, SM 3.

244. Norris, GM, and Holtz, RD (1981). "Cone Penetration Testing and Experience", *Proc. ASCE National Convention*, St. Louis, Missouri, Published by ASCE, NY.

245. O'Connor, MJ, and Mitchell, RJ (1977). "An Extension of the Bishop and Morgenstern Slope Stability Charts", *Canadian Geotechnical J*. Vol. 14.

246. O' Neill, MW, and Reese, LC (1999). "Drilled Shafts: Construction Procedures and Design Methods", Report No. FHWA-IF-99-025, Federal Highway Administration Office of Infrastructure/Office of Bridge Technology, HIBT, Washington DC.

247. O' Neill, MW, Townsend, FC, Hassan, KH, Buttler A, and Chan, PS (1996). "Load Transfer for Intermediate Geomaterials", Publication No. FHWA-RD-95-171, Federal Highway Administration, Office of Engineering and Highway Operations R and D, McLean, VA.

248. O' Neill, MW (1988). "Special Topics in Foundations", *Proc. Geotechnical Engineering Division*, ASCE National Convention, Nashville.

249. Okabe, S (1926). "General Theory of Earth Pressure", *J of the Jap. Soc. of Civil Engineers*, Tokyo, Vol. 12, No. 1.

250. Osterberg, JO (1952). "New Piston Type Soil Sampler", *Eng. News. Rec.*, 148.

251. Osterberg, JO (1957). "Influence Values for Vertical Stresses in Semi-Infinite Mass due to Embankment Loading", *Proc. Fourth Int. Conf. Conf. SM and FE*, Vol. 1.

252. Ostwald, W (1919). "A Handbook of Colloid Chemistry", *Trans.* by MH Fisher, Philadelphia, P Balkiston's Son and Co.

253. Palmer, LA, and Thompson, JB (1948). "The Earth Pressure and Deflection Along Embedded Lengths of Piles Subjected to Lateral Thrusts", *Proc. 2nd Int. Conf. SM and FE*, Rotterdam, Vol. 5.

254. Pauker, HE (1889). "An Explanatory Report on the Project of a Sea-battery (in Russian)", *Journal of the Ministry of Ways and Communications*, St. Petersburg,

255. Pauw, A, (1952). "A Rational Design Procedure for Machine Foundations", PhD Thesis, California Institute of Technology.

256. Peck, RB (1969). "Deep Excavations and Tunneling in Soft Ground", *Proc., 7th Int. Conf. on SMFE*, Mexico City.

257. Peck, RB, and Byrant, FG (1953). "The Bearing Capacity Failure of the Transcona Elevator", *Geotechnique*, Vol. 3, No. 5.

258. Peck RB, Hanson, WE, and Thornburn, TH (1974). "Foundation Engineering", John Wiley and Sons Inc., New York.

259. Perloff, WH, and Baron, W (1976). "Soil Mechanics, Principles and Applications", John Wiley and Sons, New York.

260. Petterson, KE (1955). "The Early History of Circular Sliding Surfaces", *Geotechnique*. The Institution of Engineers, London, Vol. 5.

261. Poncelet, JV (1840). "Mémoire sur la Stabilite des Rerétments et de Leurs Foundation", Note Additionelle sur les Relations Analytiques qui Lient Entre Elles la Poussée et la Butee de la Terre. Memorial de 1' officier due genie, Paris, Vol. 13,

262. Poulos, HG (1974). "Elastic Solutions for Soil and Rock Mechanics", John Wiley and Sons, New York.

263. Poulos, HG, and Davis, EH (1980). "Pile Foundation Anslysis and Design", John Wiley and Sons, New York.

264. Prakash, S, and Saran, S (1966). "Static and Dynamic Earth Pressure Behind Retaining Walls", *Proc. 3rd Symp. on Earthquake Engineering*, Roorke, India, Vol. 1.

265. Prakash, S, (1981). "Soil Dynamics", McGraw-Hill Book Company, New York.

266. Prandtl, L (1921). "Uber die Eindringungsflstigkeit Plastischer Baustoffe und die Festigkeit von Schneiden", Zeit. angew. Math., 1, No. 1.

267. Proctor, RR (1933). "Four Articles on the Design and Construction of Rolled Earth-Dams", *Eng. News Record*, Vol. 3.

268. Quinlan, PM, (1953). "The Elastic Theory of Soil Dynamics", Symp. on Dynamic Testing of Soils, ASTM, STP No. 156.

269. Rankine, WJM (1857). "On the Stability of Loose Earth Dams", Phil. Trans. Royal Soc. Vol. 147, London.

270. Reese, LC (1975). "Laterally Loaded Piles", *Proc. of the Seminar Series*, Design, Construction and Performance of Deep Foundations: Geotechnical Group and Continuing Education Committee, San Francisco Section, ASCE, Berkeley.

271. Reese, LC (1985). "Behaviour of Piles and Pile Groups Under Lateral Load", Technical Report No. FHWA/RD-85/106., Federal Highway Administration. Office of Engineering and Highway Operations, Research and Development, Washington DC.

272. Reese, LC, and Matlock, H (1956). "Non-dimensional Solutions for Laterally Loaded Piles with Soil Modulus Assumed Proportional to Depth", *Proc. 8th Texas Conf. SM and FE* Spec. Pub 29, Bureau of Eng. Res., University of Texas, Austin.

273. Reese, LC, Cox, WR, and Koop, FD (1974). "Analysis of Laterally Loaded Piles in Sand", *Proc. 6th Offshore Tech. Conf.* Houston, Texas.

274. Reese, LC and Welch, RC (1975). "Lateral Loadings of Deep Foundations, in Stiff Clay", JGED, ASCE, Vol. 101, GT 7.

275. Reese LC, and O'Neill, MW (1988). "Field Load Tests of Drilled Shafts", *Proc. International Seminar on Deep Foundations and Auger Piles*, Van Impe (ed.), Balkema, Rotterdam.

276. Reissner, E, (1936). "Stationare, Axial Symmetrische Durch Eine Schutlenlnde Masse erregte Schwingungen Eines Homogenen Elastischen Halbraumes Ingenieur", Archiv, Vol. 7. Part 6.

277. Reissner, E, (1937). "Freie and erwungene, Torsionschwingungen des elastischem Halbraumes", Ingenieur-Archiv., Vol. 8, No. 4.

278. Reissner, E and HF Sagoci, (1944). "Forced Torsional Oscillations of an Elastic Half Space", J of Appl Physics, Vol. 15.

279. Rendulic, L (1935). "Der Hydrodynamische Spannungsaugleich in Zentral Entwasserten Tonzylindern", Wasserwirtsch. u. Technik, Vol. 2.

280. Reynolds, O (1883). "An Experimental Investigation of the Circumstances which Determine Whether the Motion of Water shall be Direct or Sinuous and the Law of Resistance in Parallel Channel", Trans. Royal Soc. Vol. 174, London.

281. Richards, R, and Elms, DG (1979). "Seismic Behaviour of Gravity Retaining Walls", *Jou. of the Geotechnical Engineering Div.*, ASCE, Vol. 105, No. GT4.

282. Richart, FE, RD Woods, and JR Hall (1970). "Vibration of Soils and Foundations", Prentice-Hall, Inc., Englewood Cliffs, New Jersey.

283. Robertson, PK, and Campanella, RG [1983 (a)]. "Interpretation of Cone Penetration Tests", Part I— Sand, *CBJ*, Ottawa, Vol. 20, No. 4.

284. Robertson, PK, and Companella, RG [1983 (b)]. "SPT-CPT Correlations", *JGED*, ASCE, Vol. 109.

285. Rowe, PW (1952). "Anchored Sheet Pile Walls", *Proc. Inst. of Civ. Engineers*, Vol. 1, Part 1.

286. Sanglerat, G (1972). "The Penetrometer and Soil Exploration", Elsevier Publishing Co., Amsterdam.

287. Saran, S, and Prakash, S (1968). "Dimensionless Parameter for Static and Dynamic Earth Pressure for Retaining Walls", Indian Geotechnical Journal, Vol. 7, No. 3.

288. Scheidig, A (1934). "Der Loss (Loess)", Dresden.

289. Schmertmann, JH (1955), "The Undisturbed Consolidation Behaviour of Clay", Trans. ASCE, No. 120.

290. Schmertmann, JH (1970). "Static Cone to Compute Static Settlement Over Sand", *JSMFD*, ASCE, Vol. 96, SM 3.

291. Schmertmann, JH (1978). "Guidelines for Cone Penetration Test: Performance and Design". US Department of Transportation, Washington, DC.

292. Schmertmann, JH (1986). "Suggested Method for Performing the Flat Dilatometer Test", *Geotechnical Testing Journal*, ASTM Vol. 9, No 2.

293. Schmertmann, JM (1975). "Measurement of *in situ* Shear Strength", *Proc. ASCE Specially Conf. on In Situ Measurement of Soil Properties*, Raleigh, 2.

294. Scott, RF (1963). "Principles of Soil Mechanics", Addison-Wesely Publishing Co., Inc. London.

295. Seed, HB, Woodward, RJ, and Lundgren, R (1962). "Prediction of Swelling Potential for Compacted Clays", *Journal ASCE*, SMFD, Vol. 88, No. SM 3.

296. Seed, HB, and Whitmann, RV (1970). "Design of Earth Retaining Structures for Dynamic Loads", ASCE, Spec. Conf. Lateral Stresses in the Ground and Design of Earth Retaining Structures.

297. Semkin, VV, Ermoshin, VM, and Okishev, ND (1986). "Chemical Stabilization of Loess Soils in Uzbekistan", Soil Mechanics and Foundation Engineering (trans. from Russian), Vol. 23, No. 5.

298. Sichardt, W (1930). "Grundwasserabsenkung bei Fundierarbeiten", Berlin, Julius Springer.

299. Simons, NE (1960). "Comprehensive Investigation of the Shear Strength of an Undisturbed Drammen Clay", *Proc. Res. Conf. Shear Strength Cohesive Soils*, ASCE.

300. Simons, NE, and Som, NN (1970). "Settlement of Structures on Clay with Particular Emphasis on London Clay Industry Research Institute", Assoc. Report 22.

301. Skempton, AW (1944). "Notes on the Compressibilities of Clays", *QJ Geological Society*, London, C (C: Parts 1 and 2).

302. Skempton, AW (1951), "The Bearing Capacity of Clays", *Proc. Building Research Congress*, Vol. 1, London.

303. Skempton, AW (1953). "The Colloidal Activity of Clays", *Proceedings, 3rd Int. Conf. SM and Fe*, London, Vol. 1.

304. Skempton, AW (1954), "The Pore Pressure Coefficients A and B", *Geotechnique*, Vol. 4.

305. Skempton, AW (1957), "The Planning and Design of the New Hongkong Airport Discussion", *Proc. Int. of Civil Engineers* (London), Vol. 7.

306. Skempton, AW (1959), "*Cast-in-situ* Bored Piles in London Clay", *Geotechnical*, Vol. 9.

307. Skempton, AW, and Northey, RD (1952). "The Sensitivity of Clays", *Geotechnique*, Vol. III.

308. Skempton, AW, and Northey, RD (1954). Sensitivity of Clays", *Geotechnique*, Vol. 3, London.

309. Skempton, AW, and Bjerrum, L (1957). "A Contribution to the Settlement Analysis of Foundations on Clay", *Geotechnique* 7, London.

310. Skempton, AW, Yassin, AA, and Gibson, RE (1953). "Theorie de la Force Portaute des Pieuse Dans le Sable", Annales de 1, Institute Technique, Batiment 6.

311. Skempton, AW, and L Bjerrum, (1957). "A Contribution to the Settlement Analysis of Foundations on Clay", Geotechnique, 7.

312. Smith, RE, and Wahls, HE (1969). "Consolidation Under Constant Rate of Strain" *J Soil Mechanics Foundation Div.*, ASCE, Vol. 95, SM 2.

313. Sowa, VA (1970). "Pulling Capacity of Concrete *cast-in-situ* Bored Pile", *Can. Geotechnical J*, Vol. 7.

314. Spangler MG (1938). "Horizontal Pressures on Retaining Walls due to Concentrated Surface Loads", IOWA State University Engineering Experiment Station, *Bulletin*, No. 140.

315. Spencer, E (1967). "A Method of Analysis of the Stability of Embankments, Assuming Parallel Inter-slice Forces", *Geotechnique*, Vol. 17, No. 1.

316. Stas, CV, and Kulhawy, FH (1984). "Critical Evaluation of Design Methods for Foundations Under Axial Uplift and Compression Loading", EPRI Report EL-3771, Palo Alto, California.

317. Steinbrenner W (1934). "Tafeln zur Setzungsberechnung", Vol. 1, No. 4, Schriftenreihe der Strasse 1, Strasse.

318. Stokes, GG (1856). "On the Effect of the Internal Friction of Fludis on the Motion of Pendulium". Trans. Cambridge Philosophical Society, Vol. 9, Part 2.

319. Sung, TY, *Vibration in Semi-Infinite Solids due to Periodic Surface Loadings*, Symp. on Dynamic Testing of Soils, ASTM, STP No. 156, 1953.

320. Szechy, C (1961). "Foundation Failures", Concrete Publications Limited, London.

321. Tavenas, F and Leroueil, S (1987). "Laboratory and Stress-strain-time Behaviour of Soft Clays", *Proc. Int. Symp on Geotechnical Engineering of Soft Soils*, Mexico City, 2.

322. Taylor, DW (1937). "Stability of Earth Slopes", *J of Boston Soc. of Civil Engineers*, Vol. 24.

323. Taylor, DW (1948). "Fundamentals of Soil Mechanics", John Wiley and Sons, New York.

324. Teng, WC (1969). "Foundation Design", Prentice Hall, Englewood Cliffs, NJ.

325. Terzaghi, K (1925). "Erdbaumechanik". Franz, Deuticke, Vienna.

326. Terzaghi, K (1943). "Theoretical Soil Mechanics", Wiley and Sons, New York.

327. Terzaghi, K (1951). "The Influence of Modern Soil Studies on the Design and Construction of Foundations" Bldg. Research Congr., London, 1951, Div. 1, Part III.

328. Terzaghi, K (1955). "Evaluation of the Co-efficient of Sub-grade Reacton", *Geotechnique*, Institute of Engineers, Vol. 5, No. 4, London.

329. Terzaghi, K, and Frohlich, OK (1936). "Theorie der Setzung von Tonschichten (Theory of Settlement of Clay Layers)", Leipzig, Deuticke.

330. Terzaghi, K, and Peck, RB (1948). "Soil Mechanics in Engineering Practice", Wiley, New York.

331. Terzaghi, K, and Peck, RB (1967). "Soil Mechanics in Engineering Practice", John Wiley and Sons, NY.

332. Terzaghi, K, Peck, RB and Mesri, G (1996). "Soil Mechanics in Engineering Practice", John Wiley and Sons, Inc., Third Ed., New York.

333. Timoshenko, SP, and Goodier, JN (1970). "Theory of Elasticity", Third Ed., McGraw Hill, New York.

334. Tomlinson, MJ (1977). "Pile Design and Construction Practice", Viewpoint Publications, London.

335. Tomlinson, MJ (1986). "Foundation Design and Construction", 5th Ed., New York, John Wiley and Sons.

336. Tschebotarioff, GP (1953). "The Resistance of Lateral Loading of Single Piles and of Pile Group". ASTM Special Publication No. 154, Vol. 38.

337. Tschebotarioff, GP (1958). "Soil Mechanics Foundation and Earth Structures", McGraw Hill, New York.

338. Tsytovich, N (1986). "Soil Mechanics", Mir Publishers, Moscow.

339. Turnbull, WJ (1950). "Compaction and Strength Tests on Soil", Presented at Annual Meeting, ASCE (January).

340. US Army Engineer Water Ways Experiments Station (1957). The Unified Soil Classification System Technical Memo No. 3–357, "Corps of Engineers", Vicksburg, USA.

341. US Corps of Engineers (1953). Technical Memo 3–360, "Filter Experiments and Design Criteria", US Waterways Experimental Station, Vicksburg, Miss.

342. US Department of the Navy (1971). "Design Manual–Soil Mechanics, Foundations, and Earth Structures", NAVF AC DM-7, Washington, DC.

343. Van Der Merwe, DH (1964). "The Prediction of Heave from the Plasticity Index and Percentage Clay Fraction of Soils", *Civil Engineer in South Africa*, Vol. 6, No. 6.

344. Van Wheele, AF (1957). "A Method of Separating the Bearing Capacity of a Test Pile into Skin Friction and Point Resistance" Proc. 4th Int. Conf. SM and FE, Vol. 2.

345. Vander Veen, C, and Boersma, L (1957). "The Bearing Capacity of a Pile Predetermined by a Cone Penetration Test", *Proc. 4th Int. Conf. SM and FE*, Vol. 2.

346. Vandepitte, D, *La Charge, Portante Des Foundations Sur Pieux*, Extrait des Annales Des Travaux Publics De Belgique, 1957.

347. Verbrugge, JC (1981). "Evaluation du Tassement des Pieure a Partir de 1" Essai de Penetration Statique", *Revue Francaise de Geotechnique*, No. 15.

348. Vesic, AS (1956). "Contribution a L'Extude des Foundations Sur Pieux Verticaux et Inclines", Annales des Travaux Publics de Belgique, No. 6.

349. Vesic, AS (1961). "Bending of Beam Resting on Isotropic Elastic Solid", *Jou. Eng. Mechs. Div.*, ASCE, Vol. 87, No. EM 2.

350. Vesic, AS (1963). "Bearing Capacity of Deep Foundation in Sand", Highway Research Record, No. 39, Highway Research Board, Washington DC.

351. Vesic, AS (1963). "Discussion-Session III", *Proc. of the 1st. Int. Conf. on Structural Design of Asphalt Pavements*, University of Michigau.

352. Vesic, AS (1964). "Investigation of Bearing Capacity of Piles in Sand", *Proc. No. Amer. Conf. on Deep Foundations*, Mexico city, Vol. 1.

353. Vesic, AS (1965). "Ultimate Loads and Settlements of Deep Foundations", *Proc. of a Symp* Held at Duke University.

354. Vesic, AS (1967). "A Study of Bearing Capacity of Deep Foundations", Final Report School of Civil Engg., Georgia Inst. Tech., Atlanta, USA.

355. Vesic, AS (1969). "Effect of Scale and Compressibility on Bearing Capacity of Surface Foundations", Discussions, *Proceedings, 7th Int. Conf. SM and FE*. Mexico City, Vol. III.

356. Vesic, AS (1970). "Tests on Instrumented Piles — Ogeechee River Site", *JSMFD*, ASCE, Vol. 96, No. SM 2.

357. Vesic, AS (1972). "Expansion of Cavities in Infinite Soil Mass", *JSMFD*, ASCE, Vol. 98.

358. Vesic, AS (1973). "Analysis of Ultimate Loads of Shallow Foundations", *JSMFD*, ASCE, Vol. 99, SM 1.

359. Vesic, AS (1974). *Bearing Capacity of Shallow Foundations*, Foundation Engineering Handbook, Van Nostrand Reinhold Book Co., NY.

360. Vesic, AS (1975). "Bearing Capacity of Shallow Foundations", *Foundation Engineering Handbook*, Van Nostrand Reinhold Book Co., NY.

361. Vesic, AS (1977). "Design of Pile Foundations", Synthesis of Highway Practice 42, Res. Bd., Washington DC.

362. Vetter, CP, *Design of Pile Foundations*, Trans ASCE, Vol. 104, 1939.

363. Vidal, H (1969). "The Principal of Reinforced Earth", HRR No. 282.

364. Vijayvergiya, VN, and Focht, JA Jr, (1972). "A New Way to Predict the Capacity of Piles in Clay", *4th Annual Offshore Tech. Conf.*, Houston, Vol. 2.

365. Vijayvergiya, VN, and Ghazzaly, OI (1973). "Prediction of Swelling Potential of Natural Clays", *Third International Research and Engineering Conference on Exapansive Clays*.

366. Walsh, KD, Houston, SL, and Houston, WN (1955). "Development of *t-z* Curves for Comented Fine-Grained Soil Deposits", *Journal of Geotechnical Engineering*, ASCE, Vol. 121.

367. Wartanburton, GB, *Forced Vibration of a Body Upon an Elastic Stratum*, J Appl. Mech. Trans. ASME, Vol. 24, 1957.

368. Westergaard, HM (1917). "The Resistance of a Group of Piles", *J Western Society of Engineers*, Vol. 22.

369. Westergaard, HM (1926). Stresses in Concrete Pavement Computed by Theoretical Analysis Public Road, Vol. 7, No. 12, Washington.

370. Westergaard, HM (1938). "A Problem of Elasticity Suggested by Problem in Soil Mechanics, Soft Material Reinforced by Numerous Strong Horizontal Sheets", in *Contribution to the Mechanics of Solids*, Stephen Timoshenko 60th Anniversary Vol., Macmillan, New York.

371. Winterkorn, HF, and Hsai-Yang Fang. (1975). "Foundation Engineering Handbook", Van Nostrand Reinhold Company, New York.

372. Whitman RV, and FE Richart, Jr, *Design Procedures for Dynamically Loaded Foundations*, JSMFD, Proc. ASCE, Vol. 93, SM 6, 1967.

373. Woodward, Jr, RJ, Gardner, WS, and Greer, DM (1972). "Drilled Pier Foundations", McGraw Hill, New York.

374. Yoshimi, Y (1964). "Piles in Cohesionless Soils Subjected to Oblique Pull", *JSMFD*, ASCE, Vol. 90, SM 6.

Index

B

Braced-cuts and drainage
    lateral earth pressure, 754
    stability, 758
    piping failures, 764
    drainage, 766
    ditches and sumps, 766
    well points, 767, 768
    deep-well pumps, 769
    sand drains, 769

C

Cellular cofferdams, 593
    types, 594
    uses, 594
    components, 595
    stability, 596
    seepage, 602
Collapsible and expansive soils, 555
    collapsible soils, 556
    collapsible potential, 558
    collapse settlement, 559
    foundation on collapsible soil, 563
    expansive soil, 564
    swelling potential, 568
    expansion index, 572
    swelling index, 574
    swelling pressure, 575
    foundation on swelling soils, 579
    elimination of swelling, 589

D

Deep Foundation 1
    vertical bearing capacity of single vertical pile, 251
    classification, 251
    types of piles, 252
    cast-in-situ piles, 253, 267
    driven and cast-in-situ piles, 254
    selection of piles, 256
    installation of piles, 256–259
    load transfer mechanism, 259
    factor of safety, 262
    general theory for ultimate bearing capacity, 263
    critical depth, 266
    driven piles, 267
    base bearing capacity, 269, 270, 273, 277, 280, 295, 306, 309
    skin resistance, 274, 275, 278-279, 299
    uplift resistance, 314, 315
Deep Foundation 2
    lateral load on single vertical pile, 319
    differential equation, 321–325
    solution for laterally loaded single pile, 325, 329–339, 351, 367
    modulus of sub-grade reaction, 326
    coefficient of soil modules variation, 328
    non-dimensional solution, 339
    relative stiffness factor, 341
    case studies, 374

p-y curves for the solution, 383
pressuremeter method for the solution, 409
Poulos method for the solution, 419
Lateral load on single batter pile, 423
  mechanism of failure, 423
  model tests on batter piles, 426
  variation of soil modulus, 427
  non-dimensional solution, 428
  relative stiffness factor, 431
  ultimate lateral learning capacity, 432
  lateral resistance, 434
  coefficient of passive pressure, 435
  behaviour of laterally loaded batter pile, 440
  case studies, 441
Deep Foundation 3
  pile groups subjected to vertical and lateral loads, 457
  spacing of piles, 457
  pile group efficiency, 459, 460
  vertical bearing capacity, 461, 462
  settlement of piles group, 463–465
  allowable load on groups of piles, 466
  negative friction on piles, 466-467
  pile groups subjected to vertical and lateral loads, 459
  pile groups subjected to eccentric vertical loads, 479
  anchor piles, 481
  uplift capacity of a pile group, 482
Deep Foundation 4
  drilled pier foundations, 491
  types of drilled piers, 491
  methods of construction, 493
  load transfer mechanism, 500
  general bearing capacity equation, 505
  bearing capacity of base, 505-506, 508
  ultimate skin resistance, 510-511, 513
  settlement of drilled piers, 514-518
  uplift capacity of drilled piers, 526
  lateral bearing capacity of drilled piers, 527, 531
  case study, 534
Deep Foundation 5
  caisson foundations, 539
  types of caissons, 539
  stability analysis, 541
  grip lengths, 542, 545
  scour depth, 546
  thickness of steining, 548

G

Geotechnical properties of soil, 3
  index properties, 4
  grains size distribution, 5
  relative density, 6
  consistency, 6-7
  degree of shrinkage, 6
  activity, 7
  classification of soil, 8
  stress distribution, 10
  pressure isobars, 10
  consolidation settlement, 12, 13
  compression index, 14
  shear strength, 15
  stress paths, 16
  lateral pressure due to surcharge loads, 24

M

Machine foundation, 609
  dynamic loads, 609
  methods of analysis, 610
  basic theories, 610
  simple harmonic motion, 610
  free vibration, 612, 615
  forced vibration, 619, 621
  forced frequency dependent, 625
  magnification factor, 626
  steady state vibration, 628
  elastic resistants, 643
  vibration analysis, 632, 646, 654, 657
  design criteria, 654
  screening vibrations, 666

R

Reinforced earth, 683
  geotextiles, 683
  uses of geotextiles, 684
  reinforcing materials, 688
  construction details, 691
  design method, 693
  external stability, 698
  ground anchors, 712

S

Shallow Foundation 1, 103
  types, 103, 105

requirements for a stable foundation, 104
location and depth, 106
frost action, 108
selection of type, 109
Shallow Foundation 2, 111
bearing capacity defined, 111
types of failure, 113
Terzaghi's bearing capacity theory, 116
Skempton's bearing capacity factor, 122
effect of water table on bearing capacity, 123
general bearing capacities equation, 132
becoming capacity factors, 118, 133-134
effect of soil compressibility on bearing capacities, 138
effect of eccentrically on bearing capacity, 144
SPT and CPT methods for determining bearing capacity, 147
bearing capacity on stratified deposits, 150
bearing capacity of footings on slopes, 157-158, 160
bearing capacity by pressuremeter method, 163
foundation on rock, 172
case history of Transcona grain elevator, 174
Shallow Foundation 3, 185
safe bearing pressure and settlement calculation, 185
effect of settlement on the structure, 186
field plate load tests, 186
maximum and differential settlements, 187-188
effect of size on settlements, 193
design charts, 194, 196
empirical equations based on SPT and CPT, 198-199
foundation settlement, 200
seat of settlement, 201
modulus of elasticity, 202
settlement by elastic methods, 204–207
settlement by using CPT values, 207–210
settlement by pressuremeter method, 214
settlement by using oedometer data, 219-220
settlement by stress path method, 224
Shallow Foundation 4, 233
combined footing and mat foundation, 233

types, 233
safe bearing pressure and mat foundations, 234-235
eccentric loading, 235
coefficient sub-grade reaction, 236
cantilever footing, 238
design of combined footings, 239, 241
design of mat foundation, 241, 242
floating foundation, 243
Soil exploration, 29
boring of holes, 30
rotary drilling, 32
coring bits, 33
calyx drilling, 33
percussion drilling, 34
sampling in soil, 36
rock core sampling, 41
standard penetration test, 42
dilatancy, 46
SPT values, 46, 51
static cone penetrometer, 53
piezocone, 57
soil classification, 59
pressuremeter, 67–69
pressuremeter modulus, 74
flat dilatometer, 80
ground water conditions, 83, 85
artesian flow, 85
geophysical exploration, 88
refraction method, 88
electric resistants method, 91
Soil improvement, 721
mechanical compaction, 722
standard proctor, 722
modified proctor, 724
effect of compaction, 728
field compaction, 731
proctor needle, 736
vibroflotation, 740
blasting, 741
preloading, 741
sand drains, 742
wick drains, 744
sand compaction piles, 747
stone columns, 747
soil stabilization, 748, 750

Reader's Notes

Reader's Notes

Reader's Notes

Reader's Notes

Reader's Notes